Factoring Formulas:

$$A^2 - B^2 = (A + B)(A - B);$$
$$A^2 + 2AB + B^2 = (A + B)^2; \qquad A^2 - 2AB + B^2 = (A - B)^2;$$
$$A^3 + B^3 = (A + B)(A^2 - AB + B^2); \qquad A^3 - B^3 = (A - B)(A^2 + AB + B^2)$$

To factor a polynomial:

A. Always factor out the largest common factor that exists.

B. Look at the number of terms.

Two terms: Determine whether you have a difference of squares first. Next, try to factor as a sum or a difference of cubes. Do *not* try to factor a sum of squares.

Three terms: Determine whether the trinomial is a square. If so, you know how to factor. If not, try trial and error, using the FOIL method or the grouping method.

Four terms: Try factoring by grouping and factoring out a common binomial factor. Next, try grouping into a difference of squares, one of which is a trinomial.

C. Always *factor completely.* If a factor with more than one term can still be factored, you should factor it.

The Principle of Zero Products:

For any real numbers a and b:
If $a \cdot b = 0$, then $a = 0$ or $b = 0$. If $a = 0$ or $b = 0$, then $a \cdot b = 0$.

CHAPTER 4

Simplifying Complex Expressions:

I: By first adding or subtracting

1. Add or subtract, as necessary, to get a single rational expression in the numerator.
2. Add or subtract, as necessary, to get a single rational expression in the denominator.
3. Invert and multiply.
4. Simplify, if possible.

II: By using the LCM

1. Find the LCM of all the denominators *within* the complex rational expression.
2. Multiply the complex rational expression by 1, using the LCM to form the expression for 1.
3. Distribute and simplify, if possible.
4. Factor and simplify, if possible.

To solve an equation involving rational expressions:

1. Multiply on both sides by the LCM of all the denominators.
2. Solve the resulting equation.
3. Check the answer in the original problem.

CHAPTER 5

For any real numbers $\sqrt[k]{a}$ and $\sqrt[k]{b}$,

$$\sqrt[k]{a} \cdot \sqrt[k]{b} = \sqrt[k]{a \cdot b} \quad \text{and} \quad \sqrt[k]{\frac{a}{b}} = \frac{\sqrt[k]{a}}{\sqrt[k]{b}} \quad (b \neq 0).$$

(continued on page E-3)

Algebra for College Students

Marvin L. Bittinger

Indiana University – Purdue University at Indianapolis

David J. Ellenbogen

St. Michael's College

Addison-Wesley Publishing Company

Reading, Massachusetts • Menlo Park, California • New York
Don Mills, Ontario • Wokingham, England • Amsterdam
Bonn • Sydney • Singapore • Tokyo • Madrid • San Juan • Milan • Paris

Sponsoring Editor	Bill Poole
Production Supervisor	Jack Casteel
Production Services	The Book Department, Inc.
Design	Jack Casteel / Marshall Henrichs
Editorial Services	Joyce Grandy
Illustrator	Graphic Typesetting Service
Art Consultants	Loretta Bailey / Connie Hulse
Manufacturing Supervisor	Roy Logan
Cover Design	Peter Blaiwas

PHOTO CREDITS

1, 35, 45, David Ellenbogen **53,** Rick Haston **61,** Joseph Szabo/Photo Researchers, Inc. **70,** Steve Potter/Stock, Boston **71,** AP/Wide World Photos **85,** Mike McGovern/The Picture Cube **88,** Ron MacNeil/Ron MacNeil Photography **89,** David Ellenbogen **92,** AP Laser Photo **121,** Donald Dietz/Stock, Boston **125,** The Photo Works/Photo Researchers, Inc. **139,** Peter Menzel/Stock, Boston **168,** Kevin Horan/Stock, Boston **170,** Gale Zucker/Stock, Boston **174,** (top) John Vogl, (bottom) Christopher Morrow/Stock, Boston **181,** Patricia Hollander Gross/Stock, Boston **189,** David R. Frazier/Photo Researchers, Inc. **194,** Harold Edgerton, MIT **209,** Patricia Hollander Gross/Stock, Boston **229,** Peter Menzel/Stock, Boston **241, 254,** Rick Haston **262,** Paul Nurnberg/The Picture Cube **269, 272,** David Ellenbogen **284,** Rick Haston **288, 299, 330,** David Ellenbogen **337,** Roy Bishop/Stock, Boston **391,** NASA **436, 443,** AP/Wide World Photos **444,** Rick Haston **453,** William Clark/U.S. Department of the Interior **460,** Rick Haston **485,** W.B. Finch/Stock, Boston **503,** U.S. Department of Agriculture **513,** Rick Haston **551,** AP/Wide World Photos **556,** Peter Menzel/Stock, Boston **574,** PSSC Physics, 2nd edition, 1965; D. C. Heath and Company with Educational Development Center, Inc. Newton, MA

Library of Congress Cataloging–in–Publication Data

Bittinger, Marvin L.
Algebra for college students / Marvin L. Bittinger, David J. Ellenbogen. — 1st ed.
p. cm.
Includes index.
ISBN 0-201-19657-3
1. Algebra. I. Ellenbogen, David J. II. Title.
QA152.2.B579 1992
512.9--dc20 91-34030
CIP

1 2 3 4 5 6 7 8 9 10-DO-95949392

For Peggy, Monroe, and Zachary

Preface

This text has been designed to provide students with the background necessary for continued work in a variety of courses including precalculus, statistics, trigonometry, finite mathematics, or a non-trigonometry based calculus course. Because our book emphasizes problem-solving and the *why* behind the topics studied, it will also function well for a one quarter or one semester course in college algebra. ALGEBRA FOR COLLEGE STUDENTS is accompanied by an extensive supplements package that includes videotapes, tutorial software, a student's solutions manual, an instructor's solution manual, an instructor's resource guide, a printed test bank, and computerized testing.

Approach

Our aim in preparing this text was to develop a book tailored specifically to the needs of students who need a review of algebra while making the transition from "skills-oriented" basic algebra courses to the more "concept-oriented" presentation found in higher level college mathematics courses. In order to accomplish this goal, we have streamlined the review of basic algebra as much as possible and introduced a problem-solving approach that introduces critical thinking skills as early as possible.

Following are some aspects of our approach that we believe will help students and faculty meet the challenges of a course that typically covers a wide range of concepts in a limited amount of time.

Problem Solving

The distinguished feature of our approach is our treatment of and emphasis on problem solving. We use problem solving and applications to motivate the material wherever possible, and we include real-life applications and problem-solving techniques throughout the text. We feel that problem solving encourages students to think about how mathematics can be used. It also challenges students and stimulates them to reason at a higher conceptual level than that promoted by the routine of "skill and drill". This "applied problem-solving" approach better prepares students for the challenges they will encounter in later courses or outside the classroom setting.

- After a brief review of preliminaries in Chapter 1, we introduce a five-step process for solving problems: "Familiarize", "Translate", "Carry out", "Check", and "State the answer". We use these steps throughout the text whenever we encounter a problem-solving situation. The repeated use of the same algorithm gives students a sense that they have a starting point for any type of problem they encounter, and frees them to focus on the mathematics necessary to successfully translate the problem situation. (See pp. 45-48.)

- The first step of our problem-solving process is "Familiarize". This critical step involves reading and re-reading the problem, looking up additional information if necessary, making a drawing or a chart, assigning variables, and making educated guesses about the correct answer. It is designed to encourage students to both read and think about the problem carefully. In the early chapters of the book, we concentrate heavily on familiarization, because this leads to successful translation of the problem situation. It also starts students thinking about the different ways in which mathematics can be verbalized. (See pp. 45–48 and 168–169.)
- The second step of our problem-solving process is "Translate". We introduce this step with key words bracketed over the corresponding mathematics along with frequent references to any guesses made in the "Familiarize" step. In this manner, students are better able to see that making and checking guesses can help in translating words to symbols. (See pp. 45–47, 62, 123, and 168.)
- In the third step, "carry out", we normally utilize algebraic techniques that have been recently studied. At times, because our problems deal with real world applications, the mathematics employed in the "carry-out" step may have first appeared in earlier chapters. By being exposed to a wide variety of applications students can see that in the "carry-out" step they are forced not only to integrate concepts covered earlier but to solve equations of their own making. This makes for a deeper understanding of how important equation-solving skills are.
- We emphasize the fourth step, "Check", because we feel that it is critical for students to examine the reasonableness of their answer. (See pp. 45–48, 124, 170, and 263.)
- The fifth step, "State the Answer", requires the student to use words when stating the answer to a problem situation and guarantees that the student completely understands the meaning of his or her answer. (See pp. 69, 263, and 328.)

Content

One of the common pitfalls in the organization of a book of this type is that too many chapters are devoted to review while newer, more challenging topics are relegated to the end of the text where they often appear in an illogical order. This leads initially to student complacency and then confusion. Our carefully planned ordering of topics avoids this difficulty.

Functions

To retain skills and apply them at a more conceptual level in later courses, students must have an intuitive understanding of the material studied. A visual interpretation can help provide this type of understanding, particularly to students who have a visual rather than symbolic orientation. Furthermore, familiarity with functions and graphing techniques makes students more comfortable when these important tools appear in later courses.

Functions and graphing are introduced in Chapter 6, and are integrated throughout the remaining chapters. Because quadratic functions arise in so many applications, and because we have chosen to discuss functions prior to our development of the quadratic formula, we give graphing a special emphasis in Chapter 7, Quadratic Equations and Functions.

Theory of Equations

The study of higher degree polynomial equations really begins with the study of quadratic equations in Chapter 7, and since the study of polynomial equations forms the cornerstone of many courses, we present polynomial functions and equations in Chapter 8. This makes for a natural flow of topics. Additionally in order to prevent mathematics from becoming a study of the mysterious, we have included justification for the theorems that assist us in finding the roots of a polynomial equation. Thus, unlike many authors, we include proofs of the Upper Bound and Lower Bound theorems. Since both our proof of the Lower Bound theorem and our graphing of polynomial functions utilize symmetry, we have reserved our discussion of symmetry for Section 8.2.

Systems of Nonlinear Equations

In order to present a visual, as well as algebraic perspective to the solution of systems of nonlinear equations, we have chosen to first discuss the conic sections in a condensed manner. This ordering of topics better enables students to develop an intuitive feel for how many real solutions a system might possess.

The Binomial Theorem

We provide justification for the Binomial Theorem that does not rely on Pascal's triangle, but instead makes use of the combinatorics that appear in the preceding section. This is consistent with our desire to provide a rigorous, but user friendly approach that gets the most mileage possible out of previously covered topics.

Pedagogy

Verbalization. Wherever appropriate throughout the text, we have discussed how mathematical terms are used in language. In addition, the Summary and Review sections emphasize "Terms to Know", as well as key properties and definitions (see pp. 79 and 80). The "Thinking it Through" exercises, found immediately after the Summary and Review sections, encourage students to use their language skills to write or think out answers to questions that are often open ended in nature (see pp. 82 and 178).

Skill Maintenance Exercises and Cumulative Reviews. Retention of skills is critical to the future success of our students. In nearly all exercise sets, we include Skill Maintenance Exercises that review skills and concepts from earlier sections of the text. These exercises have been carefully selected to review all important concepts and to focus on skills necessary for succeeding material. Additionally, each chapter test includes Skill Maintenance Exercises selected from four text sections that are identified in the appropriate chapter introduction. After every three chapters we have also included a Cumulative Review. These exercises review skills and concepts from all preceding chapters of the text. (See pp. 132 and 136.)

Synthesis Exercises. At the end of every exercise set is a set of synthesis exercises that require students to synthesize skills or concepts from several previously studied sections, or provide additional insight into the present material. These are usually more difficult or more subtle than the exercises found in the main body of the exercise set and make the book flexible enough to use with even the most talented students.

Thinking it Through. Exercises at the end of each chapter encourage students to both think and write about key mathematical ideas that they have encountered in the chapter.

Answers. The answers to the odd exercises in each exercise set are found in the back of the text. In addition, all answers to the Chapter Reviews, the Chapter Tests, and the Cumulative Reviews are provided, accompanied by a parenthetical section number for ease of reference.

Supplements

This text is accompanied by a comprehensive supplements package. Below is a brief list of these supplements, followed by a detailed description of each one.

For the Instructor
Instructor's Solutions Manual
Instructor's Manual and Printed Test Bank
Answer Book
Computerized Testing

For the Student
Student's Solutions Manual
Videotapes
Comprehensive Tutorial Software
Drill and Practice Software

Supplements for the Instructor

All supplements for the instructor are free upon adoption of this text.

Instructor's Solutions Manual

This manual by Judith A. Penna contains worked-out solutions to all even-numbered exercises as well as discussions of the "Thinking it Through" sections.

Instructor's Manual and Printed Test Bank

This is an extensive collection of alternate chapter test forms, including the following:

- 3 alternate test forms for each chapter with questions in the same topic order as the objectives presented in the chapter,
- 3 alternate test forms for each chapter with the questions in a different order,
- 6 alternate forms of the final examination, 3 with questions organized by chapter, 3 with questions scrambled.

Also included are:

- Extra practice problems, and
- Indexes to the videotapes and the software that accompany the text.

Answer Book
The Answer Book contains answers to all the exercises in the text for you to make available to your students, if you wish.

Computerized Testing
OmniTest II (IBM PC).
This algorithm-driven testing system for the IBM allows you to create up to 99 variations of a customized test with just a few keystrokes, choosing from over 300 open-ended and multiple-choice test items. Instructors can also select and print tests by chapter, level of difficulty, type of problem, or according to their own designated coding.

The IBM testing program, OmniTest II also allows users to enter their own test items and edit existing items in an easy-to-use What You See Is What You Get format with variable spacing.

Supplements for the Student

Student's Solutions Manual
This manual by Judith A. Penna contains completely worked out solutions with step-by-step annotations for all the odd-numbered exercises in the text. It is free to adopting instructors and may be purchased by your students from Addison-Wesley Publishing Company.

Videotapes
Using the chalkboard and manipulative aids, Donna DeSpain and John Baumgart lecture in detail, work out exercises, and solve problems from most sections in the text. These tapes are ideal for students who have missed a lecture or who need extra help. A complete set of videotapes is available to qualifying adopters.

Tutorial Software
A variety of tutorial software packages is available to accompany this text. Please contact your Addison-Wesley representative for a software sampler that contains demonstration disks for these packages and a summary of our distribution policy.

Instructional Software for Algebra (Apple II series)
This software covers selected algebra topics. It also gives students brief explanations and examples, followed by practice exercises with interactive feedback for student error.

Drill and Practice Packages

The Math Lab by Chris Avery and Chris Barker, DeAnza College (Apple II series, IBM PC, or Macintosh)
Students choose the topic, level of difficulty, and number of exercises. If they get a wrong answer, *The Math Lab* will prompt them with the first step of the solution. This software also keeps detailed records of student scores.

Professor Weissman's Software by Martin Weissman, Essex County College (IBM PC or compatible)
Professor Weissman's Software generates exercises based on the student's selection of topic and level of difficulty. If they get a wrong answer, the software gives them a step-by-step solution. The level of difficulty increases if students are successful.

The Algebra Problem Solver by Michael Hoban and Kathirgama Nathan, LaGuardia Community College (IBM PC)
After selecting the topic and exercise type, students can enter their own exercises or request an exercise from the computer. In each case, *The Algebra Problem Solver* will give the student detailed, annotated, step-by-step solutions.

Acknowledgments

We wish to express our appreciation to the many people who helped with the development of this book. Barbara Johnson and Laurie A. Hurley merit special thanks for their precise proofreading of the manuscript and Tim Kasten deserves thanks for his many fine suggestions. Judy Penna went beyond the call of duty, providing numerous helpful suggestions, while working on the *Student's Solution Manual,* the *Instructor's Solution Manual,* and the indices. For Judy's outstanding work we are particularly grateful.

A special thank you must be sent to our wives, Elaine Bittinger and Peggy Carey. Were it not for their encouragement and understanding during all the ups and downs of this project, the writing of this book would have been impossible.

In addition, we thank the following professors for their thorough reviewing:

Sabah Al-hadad, California Polytechnic State Univ.
Mary Alter, Ph.D., University of Maryland at College Park
Wayne Andrepont, University of Southwestern Louisiana
Margaret R. Berkes, University of Vermont
Ben Bockstege, Broward Community College–Central Campus
Art Dull, Diablo Valley College
William Dunn, Las Positas Community College
Robert Horvath, El Camino College
John R. Knott, University of Evansville
Jan Vandever, Ph.D., South Dakota State University
Frances Ventola, Brookdale Community College

M.L.B.
D.J.E.

Index of Applications

In addition to the applications highlighted below, there are other applied problems and examples of problem solving in the text. An extensive list of their locations can be found under the heading "Applied problems" in the index.

Geometric Applications

Physical Science

Contents

CHAPTER 1

Important Preliminaries

FEATURE PROBLEM

A triangular sail has an 8-meter-long base and a height of 6.4 meters. How many square meters of cloth does the sail use?

THE MATHEMATICS

The formula for the area of a triangle is $\frac{1}{2} \cdot b \cdot h$. We *evaluate* this expression for $b = 8$ and $h = 6.4$.

One of the most important reasons for studying mathematics is to be able to solve problems. Before we can do so, we must be familiar with certain algebraic manipulations. These manipulations, such as simplifying expressions, utilize properties, symbolism, and vocabulary that will be developed in this chapter.

1.1 The Beginnings of Algebra

This section is intended to introduce some of the basic concepts of algebra. We will study the use of algebraic expressions in problem solving and some of the types of numbers needed for problem solving.

Algebraic Expressions and Their Use

In arithmetic, you worked with expressions such as

$$42 + 58, \qquad 9 \times 12, \qquad 17 - 5, \qquad \text{and} \qquad \frac{5}{7}.$$

In algebra, we will work with expressions such as

$$42 + x, \qquad l \cdot w, \qquad 17 - t, \qquad \text{and} \qquad \frac{d}{y}.$$

Sometimes a letter can stand for various numbers. In that case, we call the letter a **variable.** Sometimes a letter can stand for just one number. In that case, we call the letter a **constant.** Let b represent your date of birth. Then b is a constant. Let a represent your age. Then a is a variable since a changes from year to year.

An **algebraic expression** consists of variables, numbers, and operation signs. Thus all of the preceding expressions are examples of algebraic expressions.

Algebraic expressions frequently arise in problem-solving situations. For example, consider the following chart.

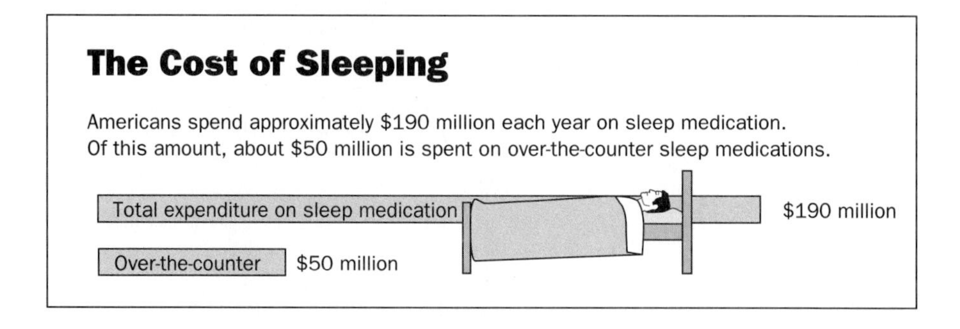

Suppose we wanted to determine how much money was spent on prescription drugs for sleep. We might use algebra to translate the problem into an equation, with x representing the amount spent on prescription sleep medication.

Money spent on over-the-counter sleep medication	plus	Money spent on prescription sleep medication	is	Total expenditure on sleep medication
\$50,000,000	$+$	x	$=$	\$190,000,000

Note that we have an algebraic expression on the left. To find the number x, we can subtract 50,000,000 on both sides of the equation:

$$x = 190{,}000{,}000 - 50{,}000{,}000.$$

Then we carry out the subtraction and obtain the answer: \$140,000,000.

In arithmetic, you probably would do this subtraction right away without considering an equation. In algebra, you will find most problems difficult without first solving an equation.

Translating to Algebraic Expressions

In algebra, we translate problems to equations. The parts of equations are translations of phrases to algebraic expressions. To do this, we need to know what words translate to certain operation symbols:

KEY WORDS

Addition	Subtraction	Multiplication	Division
add	subtract	multiply	divide
sum	difference	product	divided by
plus	minus	times	quotient
increased by	less than	twice	
more than	decreased by	of	
	take from		

Phrase	Algebraic Expression
Five more than some number	$n + 5$ or $5 + n$
Half of a number	$\frac{1}{2}t$ or $\frac{t}{2}$
Five more than three times some number	$5 + 3p$ or $3p + 5$
The difference of two numbers	$x - y$
Six less than the product of two numbers	$mn - 6$
Seventy-six percent of some number	$76\%z$ or $0.76z$

Note that expressions such as mn represent products and may also be written as $m \cdot n$, $m \times n$, or $(m)(n)$.

EXAMPLE 1 Translate to an algebraic expression.

Five less than forty-five percent of the quotient of two numbers

Solution We let m and n represent the two numbers.

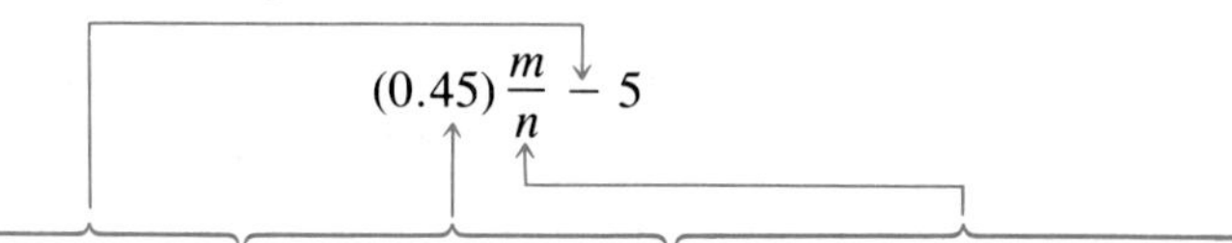

■

Evaluating Algebraic Expressions

When we replace a variable by a number, we say that we are **substituting** for the variable. This process is called **evaluating the expression.**

EXAMPLE 2 Evaluate the expression $3xy + z$ if $x = 2$, $y = 5$, and $z = 7$.

Solution We substitute and carry out the multiplication and addition:

$$\begin{aligned} 3xy + z &= 3 \cdot 2 \cdot 5 + 7 \\ &= 30 + 7 \\ &= 37. \end{aligned}$$

■

EXAMPLE 3 The area of a triangular sail with a base of length b and a height of length h is $\frac{1}{2} \cdot b \cdot h$. Find the area when b is 8 m and h is 6.4 m.

Solution We substitute 8 for b and 6.4 for h and carry out the multiplication:

$$\frac{1}{2} \cdot b \cdot h = \frac{1}{2} \cdot 8 \cdot 6.4 = 25.6 \text{ square meters (sq m).}$$

■

Solutions to Equations

The use of the symbol = ("equals") in the work above indicates that the symbols on either side of the equals sign represent, or name, the same number. An **equation** is a number sentence with the verb = . At the beginning of this section, we saw that $50{,}000{,}000 + x = 190{,}000{,}000$ is a true equation if x is replaced by 140,000,000.

> A replacement or substitution that makes an equation true is called a *solution*. Some equations may have more than one solution, and some may have no solution. When we have found all the solutions, we say that we have *solved* the equation.

EXAMPLE 4 Determine whether 5.6 is a solution of the equation $3x = 16.8$.

Solution We substitute 5.6 for the variable and see whether we get a true sentence.

$$\begin{array}{c|c} 3x & = 16.8 \\ \hline 3(5.6) & 16.8 \\ 16.8 & \end{array}$$

Substituting 5.6 for x and then simplifying the left side

TRUE

The number 5.6 is a solution. In fact, it is the *only* solution. ■

Sets of Numbers

When solving equations, or evaluating algebraic expressions, we often need to concern ourselves with the *type* of numbers used. For example, if we are solving for the optimum number of seats for a lecture hall, a fractional solution must be rounded up or down, since of course it does not make sense to discuss a fractional part of a seat. Three frequently used sets of numbers follow.

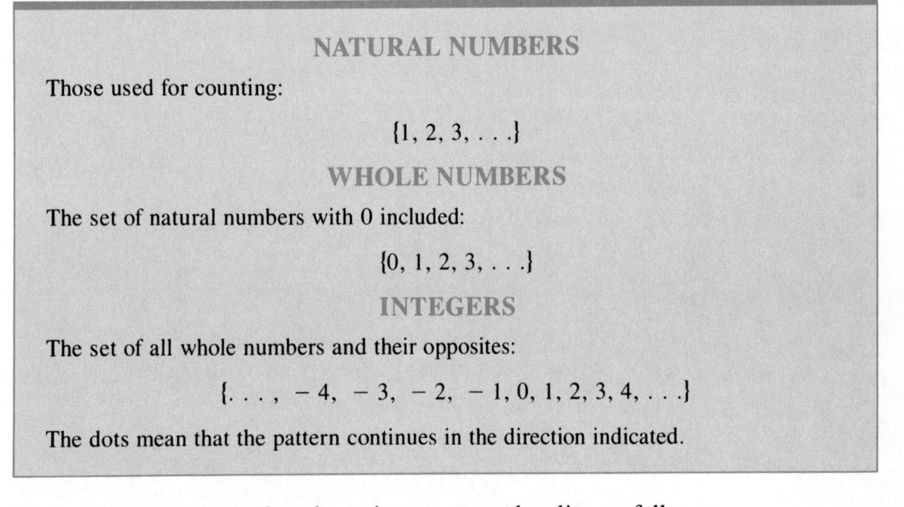

NATURAL NUMBERS

Those used for counting:

$$\{1, 2, 3, \ldots\}$$

WHOLE NUMBERS

The set of natural numbers with 0 included:

$$\{0, 1, 2, 3, \ldots\}$$

INTEGERS

The set of all whole numbers and their opposites:

$$\{\ldots, -4, -3, -2, -1, 0, 1, 2, 3, 4, \ldots\}$$

The dots mean that the pattern continues in the direction indicated.

The integers correspond to the points on a number line as follows:

–4 –3 –2 –1 0 1 2 3 4

To fill in the rest of the points on our number line, we will need to describe two more sets of numbers. To do so, we must first discuss set notation.

Set Notation

The set containing the numbers -2, 1, and 3 can be written $\{-2, 1, 3\}$. This method of writing a set is known as **roster notation.** We used the roster notation for the sets listed above.

We can also name a set by specifying conditions under which a number is in the set. The following symbolism is read as indicated, and the notation is known as **set-builder notation:**

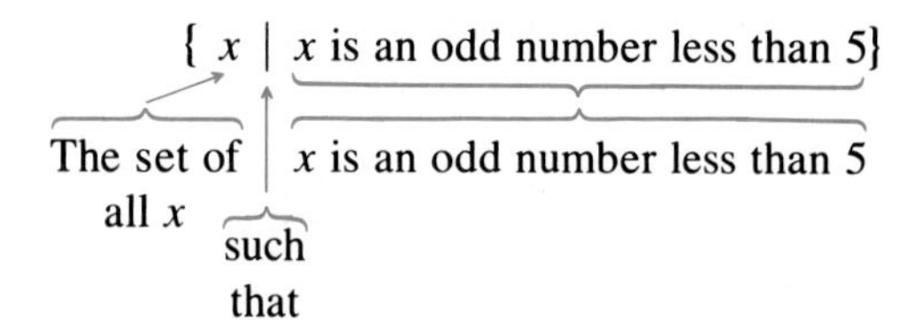

EXAMPLE 5 Name the set consisting of the first four even natural numbers, using both roster notation and set-builder notation.

Solution Using roster notation: $\{2, 4, 6, 8\}$

Using set-builder notation: $\{n|n \text{ is an even number between 1 and 9}\}$ ■

The Greek letter epsilon, $\in$, is used to indicate that an element belongs to a set. Thus if $A = \{2, 4, 6, 8\}$, we might write $4 \in A$ to indicate that 4 *is an element of A.* We might also write $5 \notin A$ to indicate that 5 *is not an element of A.*

EXAMPLE 6 Classify the statement $8 \in \{x|x \text{ is an even number}\}$ as true or false.

Solution Since 8 *is* an element of the set of all x such that x is an even number, the statement is true. In other words, because 8 is even, it belongs to the set. ■

Using set-builder notation, we can now describe the set of all *rational numbers.*

> **RATIONAL NUMBERS**
>
> Numbers that can be expressed as a quotient of an integer and a nonzero integer are called *rational numbers*:
>
> $$\left\{\frac{p}{q} \mid p \text{ is an integer, } q \text{ is an integer, and } q \neq 0\right\}.$$

Rational numbers are generally expressed using fractional or decimal notation.

Notation for Rational Numbers

The rational numbers can be named as quotients of integers.

Fractional Notation. *Fractional notation* consists of symbolism such as the following:

$$\frac{2}{3}, \quad \frac{12}{-7}, \quad \frac{-17}{15}, \quad -\frac{9}{7}.$$

Decimal Notation. In *decimal notation,* rational numbers either *terminate* or *repeat.*

EXAMPLE 7 Written in decimal form, does the number $\frac{5}{8}$ terminate or repeat?

Solution Since $\frac{5}{8}$ means $5 \div 8$, we perform long division to find that $\frac{5}{8} = 0.625$. Thus $\frac{5}{8}$ can be written as a terminating decimal. ■

EXAMPLE 8 Written in decimal form, does the number $\frac{6}{11}$ terminate or repeat?

Solution Using long division, we find that $6 \div 11 = 0.5454\ldots$, so $\frac{6}{11}$ can be written as a repeating decimal. Repeating decimal notation can be abbreviated by writing a bar over the repeating part—in this case $0.\overline{54}$. ■

The set of all rational numbers does not fill up the number line. Numbers such as π, $\sqrt{2}$, and $\sqrt{15}$, which will be used later in the text, can be only approximated by rational numbers. As decimals, these numbers are nonterminating and nonrepeating. Numbers such as π, $\sqrt{2}$, and $\sqrt{15}$ are said to be **irrational.**

The set of all irrational numbers, taken together with the set of all rational numbers, gives us the set of all **real numbers.**

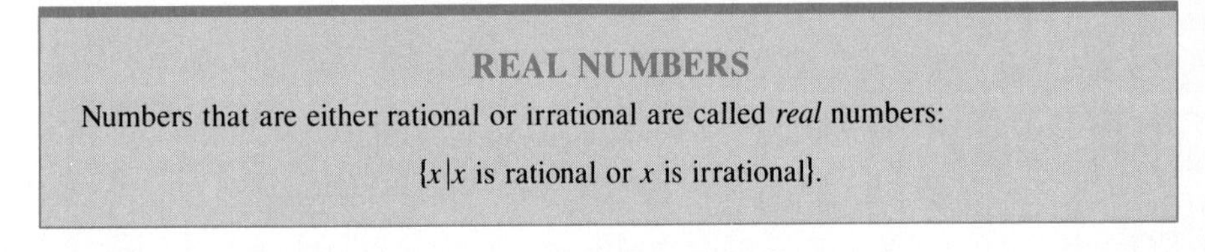

REAL NUMBERS

Numbers that are either rational or irrational are called *real* numbers:

$$\{x \mid x \text{ is rational or } x \text{ is irrational}\}.$$

The set of all real numbers *does* fill up the number line.

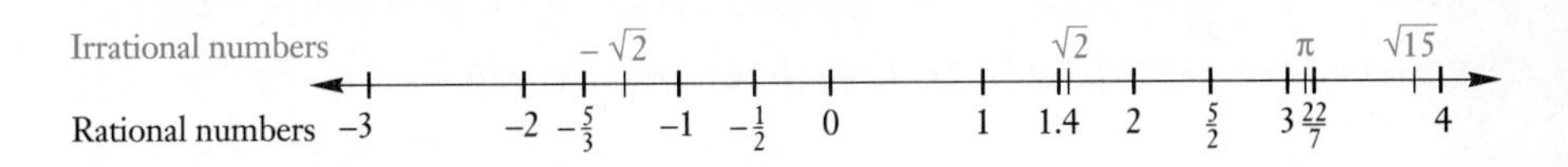

The following figure shows the relationships among various kinds of numbers.

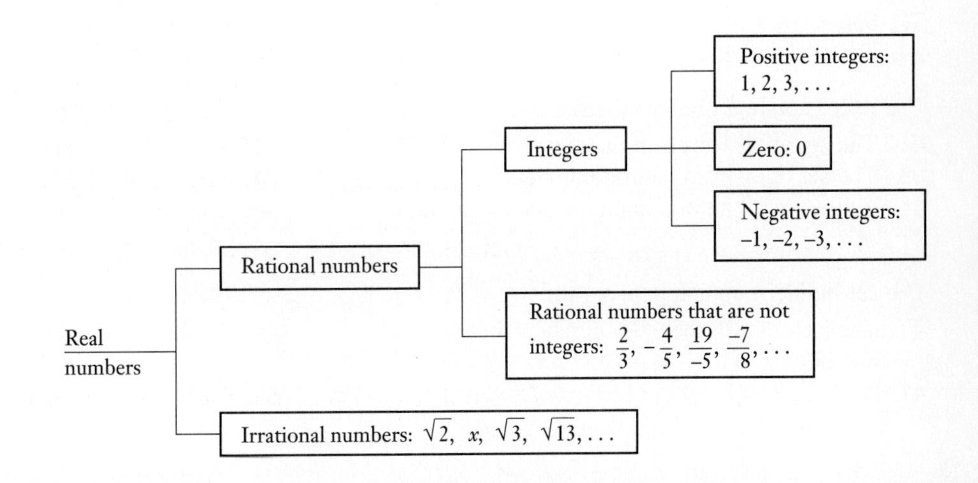

When *all* the members of a set are found in a second set, we say that the first set is a **subset** of the second set. Thus if $A = \{1, 2, 3\}$ and $B = \{1, 2, 3, 4, 5\}$, we can write $A \subseteq B$ to indicate that *A is a subset of B*. We can see from the preceding diagram that if $\mathbb{Z}$ represents the set of all integers and $\mathbb{R}$ represents the set of all real numbers, then $\mathbb{Z} \subseteq \mathbb{R}$. Similar statements may be made using other sets in the diagram.

EXERCISE SET 1.1

Translate the phrase to mathematical language.

1. Four less than some number
2. Six more than some number
3. Twice a number
4. Five times a number
5. Thirty-two percent of some number
6. Forty-seven percent of some number
7. Seven more than half of a number
8. Eight less than twice a number
9. Four less than nineteen percent of some number
10. Three more than eighty-two percent of some number
11. Five more than the difference of two numbers
12. Six less than the sum of two numbers
13. Four less than the product of two numbers
14. Ten more than the product of two numbers
15. One more than thirty-five percent of some number
16. Two less than eight percent of some number

Evaluate the given expression using the values provided.

17. $2x + y$ if $x = 3$ and $y = 7$
18. $3a - b$ if $a = 7$ and $b = 5$
19. $7abc$ if $a = 2$, $b = 1$, and $c = 3$
20. $2xyz$ if $x = 5$, $y = 2$, and $z = 1$
21. $8mn - p$ if $m = 1$, $n = 2$, and $p = 9$
22. $7rs + q$ if $r = 3$, $s = 2$, and $q = 5$
23. $5ab \div c$ if $a = 4$, $b = 2$, and $c = 10$
24. $4xy \div z$ if $x = 6$, $y = 3$, and $z = 12$
25. $2ab - a$ if $a = 5$ and $b = 3$
26. $3xy + y$ if $x = 2$ and $y = 3$
27. $pqr \div q$ if $p = 2$, $q = 3$, and $r = 2$
28. $abc \div a$ if $a = 5$, $b = 3$, and $c = 1$

Determine whether the given value is a solution of the equation.

29. $3;\ x - 2 = 1$
30. $4;\ x + 3 = 7$
31. $7;\ 4 + a = 13$
32. $6;\ 9 - y = 2$
33. $6;\ 13 - y = 7$
34. $7;\ 8 + a = 13$
35. $5;\ 2x + 3 = 13$
36. $4;\ 2n - 3 = 5$
37. $0.4;\ 8n - 1.7 = 2.5$
38. $0.5;\ 3x + 2.9 = 4.4$
39. $\frac{17}{3};\ 3x - 4 = 13$
40. $\frac{9}{4};\ 4x + 2 = 11$

Use roster notation to name the set.

41. The set of all vowels in the alphabet
42. The set of all the days of the week
43. The set of all even natural numbers
44. The set of all odd natural numbers
45. The set of all natural numbers that are multiples of 5
46. The set of all natural numbers that are multiples of 3

Use set-builder notation to name the set.

47. The set of all the odd numbers between 10 and 30
48. The set of all multiples of 4 between 22 and 45
49. $\{0, 1, 2, 3, 4\}$
50. $\{-3, -2, -1, 0, 1, 2\}$

51. The set of all multiples of 5 between 7 and 79

52. The set of all even numbers between 9 and 99

Classify each statement as true or false. The following sets are used:

$\mathbb{N}$ = the set of natural numbers
$\mathbb{W}$ = the set of whole numbers
$\mathbb{Z}$ = the set of integers
$\mathbb{Q}$ = the set of rational numbers
$\mathbb{R}$ = the set of real numbers

53. $7 \in \mathbb{N}$
54. $3.4 \in \mathbb{N}$
55. $-5 \in \mathbb{Z}$
56. $9.\overline{32} \in \mathbb{R}$
57. $7.\overline{4} \in \mathbb{Q}$
58. $3.9 \in \mathbb{Q}$
59. $\sqrt{7} \in \mathbb{R}$
60. $\sqrt{8} \in \mathbb{R}$
61. $7.1 \notin \mathbb{W}$
62. $-3 \notin \mathbb{W}$
63. $\mathbb{N} \subseteq \mathbb{Z}$
64. $\mathbb{N} \subseteq \mathbb{R}$
65. $\mathbb{N} \subseteq \mathbb{W}$
66. $\mathbb{Z} \subseteq \mathbb{W}$
67. $\mathbb{W} \subseteq \mathbb{N}$
68. $\mathbb{Z} \subseteq \mathbb{Q}$
69. $\mathbb{Q} \subseteq \mathbb{R}$
70. $\mathbb{N} \subseteq \mathbb{Q}$

To the student and the instructor: The synthesis exercises found at the end of most exercise sets will challenge you to combine objectives or skills studied in that section or in preceding parts of the text.

SYNTHESIS

Translate to mathematical language.

71. Three times the sum of two numbers

72. Half of the difference of two numbers

73. The quotient of the difference of two numbers and their sum

74. The product of the sum of two numbers and their difference

Use roster notation to name the set.

75. The set of all integers that are whole numbers but not natural numbers

76. The set of all integers that are not whole numbers

Use set-builder notation to name the set.

77. The set of all solutions of the equation $2x + 6 = 2(x + 3)$

78. The set of all solutions of the equation $3x - 6 = 3(x - 2)$

79. To help see that $\sqrt{2}$ is a very "real" point on the number line, you may recall from geometry that when a right triangle has two legs of length 1 unit, the remaining side has length $\sqrt{2}$ units. In this manner, we can "measure" a value for $\sqrt{2}$.

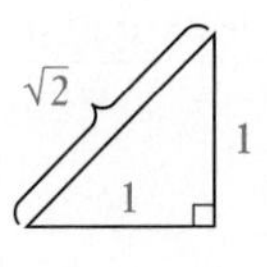

Sketch a triangle that could be used to "measure" $\sqrt{13}$.

1.2 Operations and Properties of Real Numbers

In this section, we review how real numbers (also called *reals,* for short) are added, subtracted, multiplied, and divided. We also study some important rules for the manipulation of algebraic expressions.

The following discussions of absolute value and inequalities will help us when we begin to add real numbers.

Absolute Value

We will find it convenient to have a notation that represents a number's distance from zero on the number line.

> We write $|a|$, read "the absolute value of a," to represent the number of units that a is from zero.

EXAMPLES Find the absolute value.

1. $|-3| = 3$ -3 is 3 units from 0.
2. $|7.2| = 7.2$ 7.2 is 7.2 units from 0.
3. $|0| = 0$ 0 is 0 units from itself. ■

Note that whereas the absolute value of a nonnegative number is the number itself, the absolute value of a negative number is positive.

Inequalities

We need a notation to indicate how any two real numbers compare with each other. For any two numbers on the number line, the one to the left is said to be less than, or smaller than, the one to the right. The symbol $<$ means "is less than" and the symbol $>$ means "is greater than." The symbol $\leq$ means "is less than or equal to" and the symbol $\geq$ means "is greater than or equal to." These symbols are used to form **inequalities.**

In the following figure, note that although $|-3| > |-1|$, we have $-3 < -1$ since -3 is to the left of -1.

−7 −6 −5 −4 −3 −2 −1 0 1 2 3 4 5 6 7

EXAMPLES Write out the meaning of the inequality and determine whether it is a true statement.

Inequality	*Meaning*
4. $-7 < -2$	-7 is less than -2, a true statement since -7 is left of -2.
5. $4 > -1$	4 is greater than -1, a true statement.
6. $-3 \geq -2$	-3 is greater than or equal to -2, a false statement since -3 is left of -2.
7. $5 \leq 6$	5 is less than or equal to 6. Since $5 < 6$ is true, $5 \leq 6$ is true.
8. $6 \leq 6$	6 is less than or equal to 6. Since $6 = 6$ is true, $6 \leq 6$ is true.

■

Addition

We are now ready to review the method for adding any two real numbers.

Rules for Addition of Real Numbers

1. *Positive numbers:* Add the numbers. The result is positive.
2. *Negative numbers:* Add absolute values. Make the answer negative.
3. *A negative and a positive number:* Subtract the smaller absolute value from the larger one. Then:
   - **a)** If the positive number has the larger absolute value, make the answer positive.
   - **b)** If the negative number has the larger absolute value, make the answer negative.
   - **c)** If the numbers have the same absolute value the answer is 0.
4. *One number is zero:* The sum is the other number.

EXAMPLES

9. $-9 + (-5)$ Two negatives. *Think:* Add the absolute values, getting 14. The answer is *negative,* -14.

10. $-3.2 + 9.7$ The absolute values are 3.2 and 9.7. Subtract 3.2 from 9.7 to get 6.5. The larger absolute value came from the positive number, so the answer is *positive,* 6.5.

11. $-\frac{3}{4} + \frac{1}{3} = -\frac{9}{12} + \frac{4}{12}$ The absolute values are $\frac{9}{12}$ and $\frac{4}{12}$. Subtract to get $\frac{5}{12}$. The larger absolute value came from the negative number, so the answer is *negative,* $-\frac{5}{12}$. ■

Some students may wish to now read the review of fractions presented at the end of this section.

Subtraction; Additive Inverses and Opposites

When numbers like 7 and -7 are added, the result is 0. Such numbers are called **additive inverses,** or **opposites,** of one another.

THE LAW OF ADDITIVE INVERSES

For any two numbers a and $-a$,

$$a + (-a) = 0.$$

(When additive inverses, or opposites, are added, their sum is 0.)

EXAMPLES Find the additive inverse, or opposite, of each number.

12. -17.5 The additive inverse of -17.5 is 17.5 because $-17.5 + 17.5 = 0$.

13. $\frac{4}{5}$ The additive inverse of $\frac{4}{5}$ is $-\frac{4}{5}$ because $\frac{4}{5} + \left(-\frac{4}{5}\right) = 0$.

14. 0 The additive inverse of 0 is 0 because $0 + 0 = 0$. ■

To name the additive inverse, we use the symbol $-$ and read the symbolism $-a$ as "the opposite of a" or "the additive inverse of a."

Note that $-a$ does not necessarily denote a negative number. In fact, when a is *negative,* $-a$ is *positive.*

EXAMPLES Find $-x$ for each of the following.

15. If $x = -2$, then $-x = -(-2) = 2$. The opposite of -2 is 2.

16. If $x = \frac{3}{4}$, then $-x = -\frac{3}{4}$. The opposite of $\frac{3}{4}$ is $-\frac{3}{4}$. ■

We can now give a more formal definition of absolute value.

$$|x| = \begin{cases} x & \text{if } x \geq 0, \\ -x & \text{if } x < 0 \end{cases}$$

(The absolute value of x is x if x is nonnegative. The absolute value of x is the opposite of x if x is negative.)

A negative number is said to have a negative "sign" and a positive number a positive "sign." To subtract we can add an opposite. Thus we sometimes say that we "change the sign of the number being subtracted and then add."

EXAMPLES Subtract.

17. $5 - 9 = 5 + (-9)$ **Change the sign and add.**

$= -4$

18. $-1.2 - (-3.7) = -1.2 + 3.7$

$= 2.5$

19. $-\frac{4}{5} - \frac{2}{3} = -\frac{4}{5} + \left(-\frac{2}{3}\right)$

$= -\frac{12}{15} + \left(-\frac{10}{15}\right)$ **Finding a common denominator**

$= -\frac{22}{15}$ ■

Multiplication

Multiplication of real numbers may be regarded as repeated addition or as repeated subtraction. For example,

$$3 \cdot (-2) = (-2) + (-2) + (-2) = -6$$

and

$$(-2)(-5) = -(-5) - (-5) = 5 + 5 = 10.$$

1. To multiply two numbers with *unlike signs,* multiply their absolute values. The answer is *negative.*
2. To multiply two numbers having the *same sign,* multiply their absolute values. The answer is *positive.*

EXAMPLES Multiply.

20. $(-5)(-8) = 40$
21. $(-3)9 = -27$
22. $\left(-\frac{2}{3}\right)\left(\frac{3}{8}\right) = -\frac{1}{4}$ ■

Division

To divide, we use the definition of division. The quotient $a \div b$ (also denoted a/b) is that number c, if it exists, such that $c \cdot b = a$. For example, $10 \div (-2) = -5$ since $(-5)(-2) = 10$; $(-12) \div 3 = -4$ since $(-4)3 = -12$; and $-18 \div (-6) = 3$ since $3(-6) = -18$. We obtain the following rules for division, which are just like the rules for multiplication.

1. To divide two numbers with *unlike signs,* divide their absolute values. The answer is *negative.*
2. To divide two numbers having the *same sign,* divide their absolute values. The answer is *positive.*

EXAMPLES Divide.

23. $\frac{-8}{2} = -4$

24. $\frac{-45}{-15} = 3$

25. $\frac{8}{-2} = -4$

Note that $\frac{-8}{2} = \frac{8}{-2} = -4 = -\frac{8}{2}$. ■

This can be generalized in the following statement.

For any number a and any nonzero number b,

$$\frac{-a}{b} = \frac{a}{-b} = -\frac{a}{b}.$$

Recall that

$$\frac{a}{b} = \frac{a}{1} \cdot \frac{1}{b} = a \cdot \frac{1}{b}.$$

That is, if we prefer, we can multiply by $1/b$ rather than divide by b. Provided that b is not 0, the number $1/b$ is called the **reciprocal,** or **multiplicative inverse,** of b. The fact that b cannot be 0 will be discussed shortly.

THE LAW OF MULTIPLICATIVE INVERSES

For any two numbers a and $1/a$ $(a \neq 0)$,

$$a \cdot \frac{1}{a} = 1.$$

(When multiplicative inverses, or reciprocals, are multiplied, their product is 1.)

EXAMPLES Find the reciprocal, or multiplicative inverse, of the number.

26. $\frac{7}{8}$ The reciprocal of $\frac{7}{8}$ is $\frac{8}{7}$ because $\frac{7}{8} \cdot \frac{8}{7} = 1$.

27. $-\frac{3}{4}$ The reciprocal of $-\frac{3}{4}$ is $-\frac{4}{3}$.

28. -8 The reciprocal of -8 is $\frac{1}{-8}$ or $-\frac{1}{8}$. ■

To divide, we can multiply by a reciprocal. We sometimes say that we "invert and multiply."

EXAMPLES Divide.

29. $-\frac{1}{4} \div \frac{3}{5} = -\frac{1}{4} \cdot \frac{5}{3}$ "Inverting" the $\frac{3}{5}$ and changing the division to a multiplication

$= -\frac{5}{12}$

30. $\frac{2}{3} \div \left(-\frac{4}{9}\right) = \frac{2}{3} \cdot \left(-\frac{9}{4}\right)$

$= -\frac{18}{12}$, or $-\frac{3}{2}$ ■

Note that in Examples 26–30 we have consistently avoided dividing by 0 or, what is equivalent, having a denominator of 0. There is a reason for this. Suppose we were to divide 5 by 0. Then our answer would have to be some number such that when we multiplied it by 0, we got 5. But any number times 0 is 0! Thus we cannot divide 5 or any other nonzero number by 0.

What if we divide 0 by 0? In this case, our solution would need to be some number such that when we multiplied it by 0, we got 0. But then *any* number would work as a solution to $0 \div 0$. This could lead to great confusion so we agree to exclude division of zero by zero also.

> We never divide by 0. If asked to divide a nonzero number by 0, we say the answer is *undefined*. If asked to divide 0 by 0, we say the answer is *indeterminate*.

Multiplication and Division by 1

Whereas division by 0 is problematic, division or multiplication by 1 is easy to perform. The result is always the original number. We can use this fact with fractional notation to help find **equivalent expressions.**

When evaluating algebraic expressions, any replacement that makes us divide by zero is not a meaningful replacement.

> Expressions that have the same value for all meaningful replacements are called *equivalent expressions.*

EXAMPLE 31 Use multiplication by 1 to find an expression equivalent to $x/5$ with a denominator of 15.

Solution We multiply by 1, using $\frac{3}{3}$ as a name for 1:

$$\frac{x}{5} \cdot \frac{3}{3} = \frac{x \cdot 3}{15} = \frac{3x}{15}.$$

For any replacement of x, say $x = 1$, $x/5$ and $3x/15$ have the same value. ■

When we use the word "simplify," we form equivalent expressions.

EXAMPLE 32 Use multiplication by 1 to simplify $-20x/12x$.

Solution

$$-\frac{20x}{12x} = -\frac{5 \cdot 4x}{3 \cdot 4x}$$ We factored the numerator and the denominator, after identifying (mentally) a common factor of $4x$.

$$= -\frac{5}{3} \cdot \frac{4x}{4x}$$ Next we factored the fractional expression, with one factor equal to 1.

$$= -\frac{5}{3}$$ Finally, we leave off the factor of 1.

Note that for all meaningful replacements (that is, we exclude 0 because of the denominator $12x$) $-20x/12x = -5/3$. ■

Fractional Notation: A Brief Review

The following is a review of adding, subtracting, multiplying, and dividing numbers using fractional notation.

Addition.

I. Adding numbers using fractional notation with *like,* or *common, denominators* is done by adding numerators. The denominator remains unchanged.

EXAMPLE 33 $\frac{2}{7}+\frac{3}{7}=\frac{2+3}{7}=\frac{5}{7}.$ ■

II. With unlike denominators we first multiply by a suitable factor of 1. After a common denominator is formed, the numerators are added, as in Example 33.

EXAMPLES Add.

34. $\frac{3}{8}+\frac{1}{4}=\frac{3}{8}+\frac{1}{4}\cdot\frac{2}{2}$ **Multiplying by a factor of 1**

$=\frac{3}{8}+\frac{2}{8}=\frac{5}{8}$

35. $\frac{4}{9}+\frac{5}{6}=\frac{4}{9}\cdot\frac{2}{2}+\frac{5}{6}\cdot\frac{3}{3}$ **Multiplying by factors of 1**

$=\frac{8}{18}+\frac{15}{18}=\frac{23}{18}$ ■

Subtraction. Common denominators are also needed for subtraction.

EXAMPLES Subtract.

36. $\frac{7}{8}-\frac{2}{8}=\frac{7-2}{8}=\frac{5}{8}$

37. $\frac{2}{3}-\frac{4}{5}=\frac{2}{3}\cdot\frac{5}{5}-\frac{4}{5}\cdot\frac{3}{3}$ **Multiplying by factors of 1**

$=\frac{10}{15}-\frac{12}{15}=\frac{10+(-12)}{15}$ **Change the sign and add.**

$=\frac{-2}{15},$ or $-\frac{2}{15}$

38. $-\frac{3}{5}-\frac{1}{10}=\left(-\frac{3}{5}\right)\cdot\frac{2}{2}-\frac{1}{10}$

$=-\frac{6}{10}-\frac{1}{10}=\frac{-6+(-1)}{10}$ **Change the sign and add.**

$=\frac{-7}{10},$ or $-\frac{7}{10}$ ■

Multiplication. To multiply using fractional notation, no common denominator is needed. Simply multiply across—numerator times numerator and denominator times denominator. When possible, simplify by "removing" a factor of 1, as in Example 32.

EXAMPLES Multiply.

39. $\frac{3}{7} \cdot \frac{2}{5} = \frac{3 \cdot 2}{7 \cdot 5} = \frac{6}{35}$

40. $\left(-\frac{2}{9}\right)\left(\frac{4}{7}\right) = -\frac{8}{63}$ **The product of two numbers with unlike signs is negative.**

41. $\left(-\frac{5}{4}\right)\left(-\frac{3}{10}\right) = \frac{15}{40}$ **The product of two negatives is positive.**

$= \frac{3}{8} \cdot \frac{5}{5} = \frac{3}{8}$ **"Removing" a factor of 1** ■

Division. To divide using fractional notation, we find the reciprocal, or multiplicative inverse, of the divisor and then multiply as above.

EXAMPLE 42 $\frac{2}{7} \div \frac{3}{5} = \frac{2}{7} \cdot \frac{5}{3}$ **Find the reciprocal and multiply.**

$= \frac{10}{21}$ ■

EXERCISE SET 1.2

Find the absolute value.

1. $|-7|$ **2.** $|-9|$ **3.** $|9|$ **4.** $|12|$

5. $|-6.2|$ **6.** $|-7.9|$ **7.** $|0|$ **8.** $|3\frac{3}{4}|$

9. $|1\frac{7}{8}|$ **10.** $|0.91|$ **11.** $|-4.21|$ **12.** $|-5.309|$

Write the meaning of the inequality, and determine whether it is a true statement.

13. $-9 \leq -1$ **14.** $-1 \leq -5$ **15.** $-7 > 1$ **16.** $7 \geq -2$

17. $3 \geq -5$ **18.** $9 \leq 9$ **19.** $-9 < -4$ **20.** $7 \geq -8$

21. $-4 \geq -4$ **22.** $2 < 2$ **23.** $-5 < -5$ **24.** $-2 > -12$

Add.

25. $5 + 12$ **26.** $9 + 7$ **27.** $-4 + (-7)$ **28.** $-8 + (-3)$

29. $-5.9 + 2.7$ **30.** $-1.9 + 7.3$ **31.** $\frac{2}{7} + \left(-\frac{3}{5}\right)$ **32.** $\frac{3}{8} + \left(-\frac{2}{5}\right)$

33. $-4.9 + (-3.6)$ **34.** $-2.1 + (-7.5)$ **35.** $-\frac{1}{9} + \frac{2}{3}$ **36.** $-\frac{1}{2} + \frac{4}{5}$

37. $0 + (-4.5)$ **38.** $-3.19 + 0$ **39.** $-7.24 + 7.24$ **40.** $-9.46 + 9.46$

41. $15.9 + (-22.3)$ **42.** $21.7 + (-28.3)$

Find the additive inverse, or opposite.

43. 7.29 **44.** 5.43 **45.** -4.8 **46.** -8.1

47. 0 **48.** $-2\frac{3}{4}$ **49.** $-6\frac{1}{3}$ **50.** $4\frac{1}{5}$

Find $-x$ for each of the following.

51. $x = 7$ **52.** $x = 3$ **53.** $x = -2.7$ **54.** $x = -1.9$ **55.** $x = 1.79$

56. $x = 3.14$ **57.** $x = 0$ **58.** $x = -1$ **59.** $x = -0.03$ **60.** $x = -1.09$

Subtract.

61. $9 - 7$
62. $8 - 3$
63. $4 - 9$
64. $3 - 10$
65. $-6 - (-10)$
66. $-3 - (-9)$
67. $-4 - 13$
68. $-7 - 8$
69. $2.7 - 5.8$
70. $3.7 - 4.2$
71. $-\frac{3}{5} - \frac{1}{2}$
72. $-\frac{2}{3} - \frac{1}{5}$
73. $-3.9 - (-6.8)$
74. $-5.4 - (-4.3)$
75. $0 - (-7.9)$
76. $0 - 5.3$

Multiply.

77. $(-4)7$
78. $(-5)9$
79. $(-3)(-8)$
80. $(-7)(-8)$
81. $(4.2)(-5)$
82. $(3.5)(-8)$
83. $(-7.2)(1)$
84. $1 \cdot 5.9$
85. $(-17.45) \cdot 0$
86. 15.2×0
87. $(-3.2) \times (-1.7)$
88. $(1.9) \cdot (4.3)$

Divide.

89. $\frac{-10}{-2}$
90. $\frac{-15}{-3}$
91. $\frac{-100}{20}$
92. $\frac{-50}{5}$
93. $\frac{73}{-1}$
94. $\frac{-62}{1}$
95. $\frac{0}{-7}$
96. $\frac{0}{-11}$
97. $\frac{-42}{-6}$
98. $\frac{-48}{-6}$

Find the reciprocal, or multiplicative inverse.

99. 5
100. 3
101. -9
102. -7
103. $\frac{2}{3}$
104. $\frac{4}{7}$
105. $-\frac{3}{11}$
106. $-\frac{7}{3}$

Divide.

107. $\frac{2}{3} \div \frac{4}{5}$
108. $\frac{2}{7} \div \frac{6}{5}$
109. $-\frac{3}{5} \div \frac{1}{2}$
110. $\left(-\frac{4}{7}\right) \div \frac{1}{3}$
111. $\left(-\frac{2}{9}\right) \div \left(-\frac{3}{4}\right)$
112. $\left(-\frac{2}{11}\right) \div \frac{4}{7}$
113. $\left(-\frac{3}{8}\right) \div 1$
114. $\left(-\frac{2}{7}\right) \div (-1)$
115. $\frac{7}{3} \div (-1)$
116. $-\frac{5}{4} \div 1$

Use multiplication by 1 to find an expression satisfying the given conditions.

117. Equivalent to $3x/7$ with a denominator of 35
118. Equivalent to $2a/5$ with a denominator of 40
119. Equivalent to $-3a/8$ with a denominator of 32
120. Equivalent to $-4x/3$ with a denominator of 33

Simplify.

121. $\frac{50x}{5}$
122. $\frac{75x}{3}$
123. $\frac{56a}{7}$
124. $\frac{72a}{9}$

SKILL MAINTENANCE

To the student and the instructor: Exercises included for Skill Maintenance review skills previously studied in the text. You can expect such exercises in almost every exercise set.

125. Translate to an algebraic expression:

Five less than seventy percent of a number.

126. Translate to an algebraic expression:

Two more than half a number.

127. Evaluate $7xy - z$ given $x = 2$, $y = 1$, and $z = 20$.

128. Evaluate $5ab + c$ given $a = 2$, $b = 3$, and $c = -7$.

SYNTHESIS

Simplify.

129. $-\{-[-(-9)]\}$
130. $-\{-[-(a-b)]\}$
131. $-|-[-4]|$
132. $-\{-|-(-3)|\}$

1.3 Exponential Notation and Order of Operations

Algebraic expressions can involve exponential notation. In this section, we learn how to write exponential notation and to evaluate algebraic expressions involving exponential notation. We also learn rules for using order of operations in making certain calculations.

Exponential Notation

A product in which the factors are the same is called a **power.** For

$$\underbrace{10 \cdot 10 \cdot 10}_{3 \text{ factors}} \quad \text{we write } 10^3.$$

This is read "ten to the third power." We call the number 3 an **exponent** and say that 10 is the **base.** An exponent of 2 or greater tells how many times the base is used as a factor. For example,

$$a \cdot a \cdot a \cdot a = a^4.$$

Here the exponent is 4 and the base is a. An expression such as a^4 is said to be written in **exponential notation.**

a^n ← This is the exponent.

This is the base. (arrow to a)

EXAMPLE 1 Write exponential notation for $10 \cdot 10 \cdot 10 \cdot 10 \cdot 10$.

Solution

$$10 \cdot 10 \cdot 10 \cdot 10 \cdot 10 = 10^5$$

■

EXAMPLE 2 What is the meaning of 3^5, n^4, $(-2n)^3$, and $50x^2$?

Solution

3^5 means $3 \cdot 3 \cdot 3 \cdot 3 \cdot 3$, $\quad$ n^4 means $n \cdot n \cdot n \cdot n$,

$(-2n)^3$ means $(-2n)(-2n)(-2n)$, $\quad$ $50x^2$ means $50 \cdot x \cdot x$. ■

Look for a pattern in the following:

$$8 \cdot 8 \cdot 8 \cdot 8 = 8^4$$
$$8 \cdot 8 \cdot 8 = 8^3$$
$$8 \cdot 8 = 8^2$$
$$8 = 8^?.$$

We divide by 8 each time.

The exponents decrease by 1 each time. To continue the pattern, we would say that

$$8 = 8^1.$$

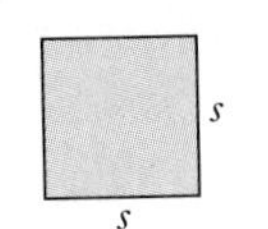

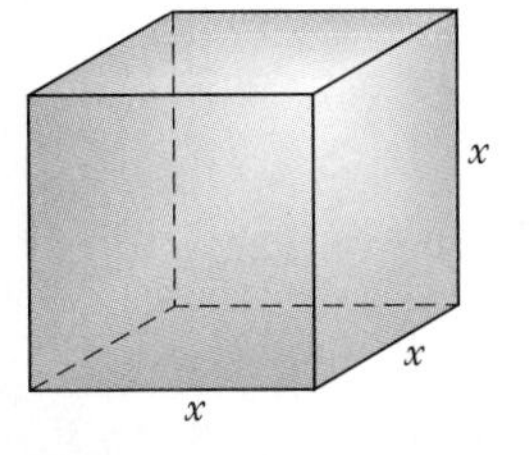

We read exponential notation as follows:

b^n is read *"the nth power of b,"* or simply *"b to the nth,"* or *"b to the n."*

We often read s^2 as "*s-squared.*" The terminology comes from the fact that the area of a square of side s is $s \cdot s$, or s^2.

We often read x^3 as "x-cubed." The terminology comes from the fact that the volume of a cube with length, width, and height x is $x \cdot x \cdot x$, or x^3.

We now summarize our definition of exponential notation.

EXPONENTIAL NOTATION

$b^1 = b$, for any number b.

For any natural number n greater than or equal to 2,

$$b^n \text{ means } \overbrace{b \cdot b \cdot b \cdot b \cdots b}^{n \text{ factors}}.$$

Order of Operations

What does $1 + 2 \cdot 5^2$ mean? If we add 1 and 2 and multiply by 5^2, or 25, we get 75. If we multiply 2 times 25 and add 1, we get 51. Clearly, both results cannot be correct. To help us determine which procedure to use, mathematicians have adopted the following conventions.

Rules for Order of Operations

1. Do all calculations within grouping symbols before operations outside.
2. Evaluate all exponential expressions.
3. Do all multiplications and divisions in order from left to right.
4. Do all additions and subtractions in order from left to right.

EXAMPLE 3 Calculate: $4 + 2(1 - 3)^2$.

Solution

$$\begin{aligned} 4 + 2(1-3)^2 &= 4 + 2(-2)^2 && \textbf{Working within parentheses first} \\ &= 4 + 2(4) && \textbf{Simplifying } (-2)^2 \\ &= 4 + 8 && \textbf{Multiplying} \\ &= 12 && \textbf{Adding} \end{aligned}$$

■

It is important to remember that division should be performed before multiplication if, after calculations within grouping symbols are made and exponential expressions are simplified, the division is encountered first when reading from left to right. Similarly, if subtraction appears before addition we should subtract first.

EXAMPLE 4 Calculate: $7 - (-1 + 3) + 6 \div 2 \cdot (-3)^2$.

Solution

$$\begin{aligned} 7 - (-1 + 3) + 6 \div 2 \cdot (-3)^2 &= 7 - 2 + 6 \div 2 \cdot 9 && \text{Working within parentheses and simplifying } (-3)^2 \\ &= 7 - 2 + 3 \cdot 9 && \text{Dividing} \\ &= 7 - 2 + 27 && \text{Multiplying} \\ &= 5 + 27 && \text{Subtracting} \\ &= 32 && \text{Adding} \end{aligned}$$

■

In addition to the usual grouping symbols—parentheses, brackets, and braces—a fraction bar or absolute value symbols may indicate groupings.

EXAMPLE 5 Calculate:

$$\frac{12|7 - 9| + 4 \cdot 5}{(-3)^4 + 2^3}.$$

Solution We do the calculations in the numerator and in the denominator and divide the results.

$$\begin{aligned} \frac{12|7 - 9| + 4 \cdot 5}{(-3)^4 + 2^3} &= \frac{12|-2| + 4 \cdot 5}{81 + 8} \\ &= \frac{12(2) + 20}{89} \\ &= \frac{44}{89} && \text{Multiplying and adding} \end{aligned}$$

■

The rules for the order of operations are used when evaluating algebraic expressions.

EXAMPLE 6 Evaluate $x^3 - xy + 2$ for $x = -2$ and $y = 5$.

Solution

$$\begin{aligned} x^3 - xy + 2 &= (-2)^3 - (-2)5 + 2 && \text{Substituting} \\ &= -8 - (-10) + 2 && \text{Simplifying the power and the product} \\ &= -8 + 10 + 2 && \text{Adding the opposite} \\ &= 4 && \text{Adding from left to right} \end{aligned}$$

■

Calculators. Most of today's hand-held calculators make use of the order of operations that we have adopted. These calculators are said to possess *algebraic logic.* Because many of these calculations do not have parentheses, we must be especially careful when calculating products and quotients and when raising an expression to a power.

EXAMPLES Write the sequence of keys that should be pressed to calculate each of the following.

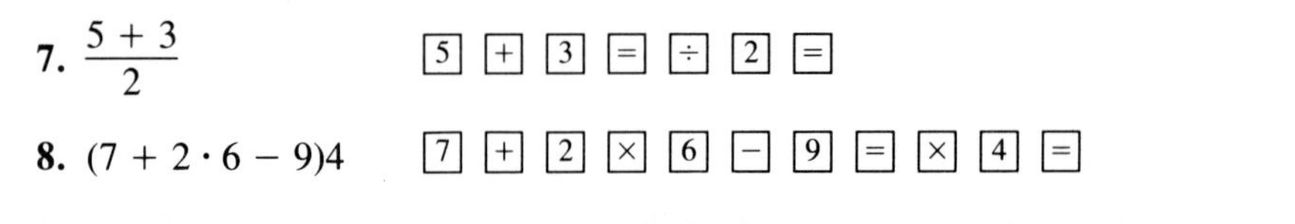

Note that had we not pressed the first equals sign in Example 7, only the 3 would have been divided by 2. Similarly, had we not pressed the first equals sign in Example 8, only the 9 would have been multiplied by 4. A similar use of the equals sign is made when raising a bracketed expression to a power.

EXAMPLE 9 Write the sequence of keys that should be pressed to calculate $(5 + 2 \cdot 3)^4$.

EXERCISE SET 1.3

What is the meaning of each expression?

1. 2^4 **2.** 5^3 **3.** $(1.4)^5$ **4.** $(2.5)^1$
5. $(-3)^2$ **6.** $(-7)^3$ **7.** $(-3x)^1$ **8.** $(9c)^3$
9. $(5ab)^3$ **10.** $(-6xy)^2$ **11.** $(10pq)^1$ **12.** $(-7d)^1$

Write exponential notation for the following.

13. $10 \times 10 \times 10 \times 10 \times 10 \times 10$ **14.** $6 \times 6 \times 6 \times 6$ **15.** $x \cdot x \cdot x \cdot x \cdot x \cdot x \cdot x$
16. $y \cdot y \cdot y$ **17.** $3y \cdot 3y \cdot 3y$ **18.** $5m \cdot 5m \cdot 5m \cdot 5m \cdot 5m$
19. $(-2a)(-2a)(-2a)$ **20.** $(-7x)(-7x)(-7x)(-7x)(-7x)$

Calculate using the rules for order of operations.

21. $5 + 2 \cdot 3^2$ **22.** $9 - 3 \cdot 2^2$ **23.** $12 - (9 - 3 \cdot 2^3)$
24. $19 - (4 + 2 \cdot 3^2)$ **25.** $\frac{5 \cdot 2 - 4^2}{27 - 2^4}$ **26.** $\frac{7 \cdot 3 - 5^2}{9 + 4 \cdot 2}$
27. $\frac{3^4 - (5-3)^4}{1 - 2^3}$ **28.** $\frac{4^3 - (7-4)^2}{3^2 - 7}$ **29.** $5^3 - [2(4^2 - 3^2 - 6)]^3$
30. $7^2 - [3(5^2 - 4^2 - 7)]^2$ **31.** $|2^2 - 7|^3 + 1$ **32.** $|-2 - 3| \cdot 4^2 - 1$
33. $30 - (-5)^2 + 15 \div (-3) \cdot 2$ **34.** $55 - (-9 + 2)^2 + 18 \div 6 \cdot (-2)$
35. $12 - (7 - 5) + 4 \div 3 \cdot 2^3$ **36.** $15 - (3 - 8) + 5 \div 10 \cdot 3^2$

Evaluate.

37. $3 \cdot (a + 10)$ for $a = 12$

38. $b \cdot (7 + b)$ for $b = 5$

39. $(t + 3)^3$ for $t = 4$

40. $(12 - w)^3$ for $w = 7$

41. $(x + 5) \cdot (12 - x)$ for $x = 7$

42. $(y - 4) \cdot (y + 6)$ for $y = 10$

43. $(5y)^3 - 75$ for $y = -2$

44. $(7x)^2 + 59$ for $x = -1$

45. $\dfrac{y + 3}{2y}$ for $y = 5$

46. $\dfrac{4x + 2}{2x}$ for $x = 5$

47. $\dfrac{w^2 + 4}{5w}$ for $w = -4$

48. $\dfrac{b^2 + b}{2b}$ for $b = -5$

49. $(x - 4) \cdot (8 + y)$ for $x = 12$ and $y = -3$

50. $(y + 6) \cdot (9 - x)$ for $x = 7$ and $y = -10$

Evaluate $(4n)^3$ and $4n^3$ for each of the following.

51. When $n = 2$

52. When $n = 3$

53. When $n = -1$

54. When $n = -5$

List the sequence of keys that should be pressed on a calculator to compute each of the following. Assume algebraic logic.

55. $\dfrac{9 + 3 \cdot 4 + 1}{2}$

56. $\dfrac{7 - 2 \cdot 3 + 5}{4}$

57. $8 - 2 \cdot 3 + \dfrac{4}{15}$

58. $12 + 3 \cdot 2 - \dfrac{7}{12}$

59. $\dfrac{9 - 2^8 + 4}{3}$

60. $\dfrac{15 - 3^7 + 8}{14}$

61. $\left(7 - \dfrac{2}{3}\right)^4 - \dfrac{5}{8}$

62. $\left(9 - \dfrac{4}{7}\right)^5 + \dfrac{3}{11}$

SKILL MAINTENANCE

63. Translate to an algebraic expression: The sum of p and q.

64. Use roster notation and then set-builder notation to write the set of all natural numbers that are less than 15.

SYNTHESIS

Find a value of the variable that shows that the two expressions are *not* equivalent.

65. $3x^2$; $(3x)^2$

66. $(a + 2)^3$; $a^3 + 2^3$

67. $\dfrac{x + 2}{2}$; x

68. $\dfrac{y^6}{y^3}$; y^2

Write an algebraic expression for each of the following.

69. A number squared plus 7

70. A number plus the square of 7

71. The square of the sum of 7 and some number

72. A number squared plus 7 squared

73. The numerator is 3 more than some number and the denominator is the square of the numerator.

74. Two numbers are multiplied. One of them is 5 more than the other.

75. Carole is twice as old as Victor was a year ago. Victor's age is now x. Write an expression for Carole's age.

76.
a) The square of the sum of two numbers
b) The sum of the squares of two numbers

1.4 The Commutative, Associative, and Distributive Laws

The Commutative and Associative Laws

Several laws of real numbers are available to assist us in simplifying algebraic expressions. These laws are used often, so it will be helpful to remember their names.

Note that $5 + 8 = 13$ and $8 + 5 = 13$, so that $5 + 8 = 8 + 5$. This concept, that the result of addition is independent of the order in which the numbers are added, is known as the *commutative law of addition*. A similar law holds for multiplication.

THE COMMUTATIVE LAWS

Addition. For any real numbers a and b,

$$a + b = b + a.$$

Multiplication. For any real numbers a and b,

$$ab = ba.$$

The commutative laws are often used to formulate equivalent expressions. Another pair of tools that can produce equivalent expressions, the *associative laws*, have to do with the fact that $2 + (4 + 3) = (2 + 4) + 3$ and, for multiplication, $2 \cdot (4 \cdot 3) = (2 \cdot 4) \cdot 3$.

THE ASSOCIATIVE LAWS

Addition. For any real numbers a, b, and c,

$$a + (b + c) = (a + b) + c.$$

Multiplication. For any real numbers a, b, and c,

$$a \cdot (b \cdot c) = (a \cdot b) \cdot c.$$

Thus, if just addition or just multiplication is involved, we can group numbers as we please.

EXAMPLE 1 Write an expression equivalent to $3x + 4y$, using the commutative law of addition.

Solution

$$3x + 4y = 4y + 3x \quad \text{Using the commutative law to change order}$$

The expressions are equivalent. They will name the same number for all replacements of x and y by real numbers. ■

EXAMPLE 2 Write an expression equivalent to $(3x + 7y) + 9z$, using the associative law of addition.

Solution

$$(3x + 7y) + 9z = 3x + (7y + 9z)$$

The expressions are equivalent. They will name the same number for all replacements of x, y, and z by real numbers. ■

EXAMPLE 3 Write an expression equivalent to

$$\frac{5}{x} \cdot (2yz),$$

using the associative and commutative laws of multiplication.

Solution

$$\frac{5}{x} \cdot (2yz) = \left(\frac{5}{x} \cdot 2y\right) \cdot z \qquad \text{Using the associative law}$$

$$= \left(2y \cdot \frac{5}{x}\right) \cdot z \qquad \text{Using the commutative law in the parentheses}$$ ■

Although you probably don't think about it, you use the associative law of multiplication when you multiply a term such as $8x$ by a number such as 5, to get $40x$:

$$5(8x) = (5 \cdot 8)x = 40x.$$

The Distributive Law

Let's look at two other ways of formulating equivalent expressions. We say that multiplication is distributive over addition, meaning that when we multiply a number by a sum, we can either add first or multiply first. Multiplication is also distributive over subtraction. If we regard subtraction as the addition of additive inverses, the following law holds for both addition and subtraction.

THE DISTRIBUTIVE LAW

For any numbers a, b, and c,

$$a(b + c) = ab + ac \quad \text{and} \quad a(b - c) = ab - ac.$$

EXAMPLE 4 Obtain an expression equivalent to $5x(y + 4)$ by multiplying.

Solution We use the distributive law to get

$$\begin{aligned} 5x(y+4) &= 5xy + 5x\cdot 4 && \text{Using the distributive law} \\ &= 5xy + 5\cdot 4\cdot x && \text{Using the commutative law of multiplication} \\ &= 5xy + 20x. && \text{Simplifying} \end{aligned}$$

The expressions $5x(y + 4)$ and $5xy + 20x$ are equivalent. They will name the same number for all replacements of x and y by real numbers. ■

When we do the opposite of what we did in Example 4, we say that we are **factoring** an expression.

EXAMPLE 5 Obtain an expression equivalent to $3xy - 6x$ by factoring.

Solution We use the distributive law to get

$$3xy - 6x = 3x(y) - 3x(2) = 3x(y - 2).$$

The expressions $3xy - 6x$ and $3x(y - 2)$ are equivalent. They will name the same number for any replacements of x and y by real numbers. ■

Collecting Like Terms

In the expression $5xy + 20x$, the parts separated by the plus sign are called **terms.** Thus $5xy$ and $20x$ are terms in $5xy + 20x$. When terms have variable factors that are exactly the same, we refer to the terms as **like** or **similar** terms. Thus $4x^2$ and $5x^2$ are like terms, but $2a^3$ and $6a^2$ are not. We often simplify expressions by using the distributive law to **collect** or **combine like terms.**

EXAMPLE 6 Collect like terms: $3x + 4x$.

Solution

$$\begin{aligned} 3x + 4x &= (3+4)x && \text{Using the distributive law (in reverse), or factoring} \\ &= 7x \end{aligned}$$

■

EXAMPLE 7 Collect like terms: $2x + 3y^2 + 5x + 8y^2$.

Solution

$$\begin{aligned} 2x + 3y^2 + 5x + 8y^2 &= 2x + 5x + 3y^2 + 8y^2 && \text{Using the commutative law of addition} \\ &= (2+5)x + (3+8)y^2 && \text{Factoring} \\ &= 7x + 11y^2 \end{aligned}$$

■

EXAMPLE 8 Collect like terms: $xy + 3xy - 8xy$.

Solution

$$xy + 3xy - 8xy = (1 + 3 - 8)xy \qquad \text{Using the distributive law and noting that } xy = 1xy.$$
$$= -4xy$$

With practice we can leave out some steps, collecting like terms mentally. The numbers like 4 and 7 in the expression $4 + x + 7$ are constants and are also considered to be like terms.

EXAMPLES Collect like terms. Do as much as possible mentally.

9. $5y + 2y + 4y = 11y$

10. $3x + 7x + 2y = 10x + 2y$

11. $3x^2 + 7x^2 + 2y = 10x^2 + 2y$

12. $8p + q + p + 0.3q = 9p + 1.3q$

13. $4 + x + 7 = x + 11$

14. $3x + 25 + 7y + 8x + 11 = 11x + 7y + 36$

The distributive law enables us to "remove" parentheses, brackets, or braces. Often, after doing so, we can simplify by collecting like terms.

EXAMPLE 15 Simplify the expression $3 + 2[4 + 5(x + 2y)]$.

Solution

$$\begin{aligned} 3 + 2[4 + 5(x + 2y)] &= 3 + 2[4 + 5x + 10y] && \text{Using the distributive law} \\ &= 3 + 8 + 10x + 20y && \text{Using the distributive law} \\ &= 11 + 10x + 20y && \text{Collecting like terms} \end{aligned}$$

EXAMPLE 16 Simplify the expression $-2\{3[x + 2(y - z)] - 4(x + y)\}$.

Solution

$$\begin{aligned} &-2\{3[x + 2(y - z)] - 4(x + y)\} \\ &\quad = -2\{3[x + 2y - 2z] - 4(x + y)\} && \text{Using the distributive law} \\ &\quad = -2\{3x + 6y - 6z - 4(x + y)\} && \text{Using the distributive law} \\ &\quad = -2\{3x + 6y - 6z - 4x - 4y\} && \text{Using the distributive law} \\ &\quad = -2\{-x + 2y - 6z\} && \text{Collecting like terms} \\ &\quad = 2x - 4y + 12z && \text{Using the distributive law} \end{aligned}$$

Multiplying by -1

When we multiply a number by -1, we get its additive inverse, or opposite. For example,

$$-1 \cdot 8 = -8 \qquad \text{(the additive inverse of 8)}$$

and

$$-1(-5) = 5 \qquad \text{(the additive inverse of } -5\text{).}$$

> **THE MULTIPLICATIVE PROPERTY OF -1**
>
> For any real number a,
>
> $$-1 \cdot a = -a.$$
>
> (Negative 1 times a is the additive inverse, or opposite, of a.) Multiplying by -1 changes the sign of any nonzero number.

Using this fact and the distributive law, we can obtain equivalent expressions.

EXAMPLE 17 Obtain an expression equivalent to $-(a - b)$, using the above property of -1.

Solution

$$\begin{aligned} -(a - b) &= -1 \cdot (a - b) && \text{Replacing } - \text{ by } -1 \\ &= -1 \cdot a - (-1) \cdot b && \text{Using the distributive law} \\ &= -a - (-b) && \text{Replacing } -1 \cdot a \text{ by } -a \text{ and } (-1) \cdot b \text{ by } -b \\ &= -a + b, \quad \text{or} \quad b - a \end{aligned}$$

The expressions $-(a - b)$ and $b - a$ are equivalent. They will name the same number for all replacements of a and b by real numbers. ■

Example 17 illustrates something worth remembering because it gives us a useful shortcut—that is,

the opposite or additive inverse of $a - b$ is $b - a$.

EXERCISE SET 1.4

Write an expression equivalent to each of the following. Use a commutative law.

1. $y + 8$ **2.** $x + 3$ **3.** mn **4.** ab

5. $9 + xy$ **6.** $11 + ab$ **7.** $ab + c$ **8.** $rs + t$

9. $x + y^2$ **10.** p^2q^3 **11.** $xt^2 + t$ **12.** $(a + b)^2$

Write an equivalent expression using an associative law.

13. $2 \cdot (8x)$ **14.** $x \cdot (3y)$ **15.** $x + (2y + 5)$ **16.** $(3y + 4) + 10$

17. $\frac{1}{2} + (3a + b)$ **18.** $(7 + 2x) + y$ **19.** $\left(\frac{3}{4}x\right)y$ **20.** $2(xy)$

Obtain an equivalent expression by multiplying.

21. $3(a + 1)$ **22.** $8(x + 1)$ **23.** $4(x - y)$ **24.** $9(a - b)$

25. $-5(2a + 3b)$ **26.** $-2(3c + 5d)$ **27.** $2a(b - c + d)$ **28.** $5x(y - z + w)$

29. $2\pi r(h + 1)$ **30.** $P(1 + rt)$ **31.** $\frac{1}{2}h(a + b)$ **32.** $\pi r(1 + s)$

Obtain an equivalent expression by factoring.

33. $8x + 8y$
34. $7a + 7b$
35. $9p - 9$
36. $12x - 12$
37. $7x - 21$
38. $6y - 36$
39. $xy + x$
40. $ab + a$
41. $2x - 2y + 2z$
42. $3x + 3y - 3z$
43. $3x + 6y - 3$
44. $4a + 8b - 4$

Obtain an equivalent expression by collecting like terms. Use the distributive law.

45. $4a + 5a$
46. $9x + 3x$
47. $8b - 11b$
48. $9c - 12c$
49. $14y + y$
50. $13x + x$
51. $12a - a$
52. $15x - x$
53. $t - 9t$
54. $x - 6x$
55. $5x - 3x + 8x$
56. $3x - 11x + 2x$
57. $5x - 8y + 3x$
58. $9a - 10b + 4a$
59. $7c + 8d - 5c + 2d$
60. $12a + 3b - 5a + 6b$
61. $4x - 7 + 18x + 25$
62. $13p + 5 - 4p + 7$
63. $13x + 14y - 11x - 47y$
64. $17a + 17b - 12a - 38b$

Obtain an equivalent but simpler expression, by first removing parentheses.

65. $a - (2a + 5)$
66. $x - (5x + 9)$
67. $4m - (3m - 1)$
68. $5a - (4a - 3)$
69. $3d - 7 - (5 - 2d)$
70. $8x - 9 - (7 - 5x)$

Simplify.

71. $-2(x + 3) - 5(x - 4)$
72. $-9(y + 7) - 6(y - 3)$
73. $5x - 7(2x - 3)$
74. $8y - 4(5y - 6)$
75. $9a - [7 - 5(7a - 3)]$
76. $12b - [9 - 7(5b - 6)]$
77. $5\{-2 + 3[4 - 2(3 + 5)]\}$
78. $7\{-7 + 8[5 - 3(4 + 6)]\}$
79. $2y + \{7[3(2y - 5) - (8y + 7)] + 9\}$
80. $7b - \{6[4(3b - 7) - (9b + 10)] + 11\}$

SKILL MAINTENANCE

81. Add and simplify: $\frac{11}{12} + \frac{15}{16}$.

82. Subtract and simplify: $\frac{7}{8} - \frac{2}{3}$.

83. Evaluate $(2x)^3$ when $x = 2$.

84. Calculate: $64 - 32 \div 16 - 4$.

SYNTHESIS

85. Evaluate $a \div b$ and $b \div a$ when $a = 64$ and $b = 8$. Is there a commutative law for division of whole numbers?

86. Evaluate $a - (b - c)$ and $(a - b) - c$ for $a = 25$, $b = 9$, and $c = 4$. Is there an associative law for subtraction?

87. Evaluate $a \div (b \div c)$ and $(a \div b) \div c$ for $a = 32$, $b = 8$, and $c = 4$. Is there an associative law for division?

88. When you put money in the bank and draw simple interest, the amount in your account later on is given by the expression $P + Prt$, where P is the principal, r is the rate of interest, and t is the time. Factor the expression.

89. Solve.

a) Factor $17x + 34$. Then evaluate both expressions when $x = 10$.

b) Will you get the same answer for both expressions no matter what the value of x? Why or why not?

90. Find a simpler expression that always has the same value as

$$\frac{3a + 6}{2a + 4}.$$

CHAPTER SUMMARY AND REVIEW 1

TERMS TO KNOW

Variable
Constant
Algebraic expression
Substituting
Evaluating the expression
Equation
Solution
Natural numbers
Whole numbers
Integers
Roster notation
Set-builder notation
Rational numbers
Irrational numbers
Real numbers
Subset
Absolute value
Inequality
Additive inverse
Opposite
Reciprocal
Multiplicative inverse
Undefined
Equivalent expressions
Common denominator
Power
Exponent
Base
Exponential notation
Order of operations
Algebraic logic
Factoring
Term
Collect like terms

IMPORTANT PROPERTIES AND FORMULAS

The law of additive inverses: $a + (-a) = 0$

The law of multiplicative inverses: $a \cdot \frac{1}{a} = 1, \quad a \neq 0$

Commutative laws: $a + b = b + a,\ ab = ba$

Associative laws: $a + (b + c) = (a + b) + c,\ a(bc) = (ab)c$

Distributive law: $a(b + c) = ab + ac$

Absolute value: $|x| = \begin{cases} x & \text{if } x \geq 0, \\ -x & \text{if } x < 0 \end{cases}$

RULES FOR ADDITION

1. *Positive numbers*: Add the numbers. The result is positive.
2. *Negative numbers*: Add absolute values. Make the answer negative.
3. *A negative and a positive number*: Subtract the smaller absolute value from the larger one. Then:
   a) If the positive number had the larger absolute value, make the answer positive.
   b) If the negative number had the larger absolute value, make the answer negative.
   c) If the numbers have the same absolute value, the answer is 0.
4. *One number is zero*: The sum is the other number.

RULES FOR MULTIPLICATION

1. To multiply two numbers with *unlike signs,* multiply their absolute values. The answer is *negative.*
2. To multiply two numbers with the *same sign,* multiply their absolute values. The answer is *positive.*

RULES FOR DIVISION

1. To divide two numbers with *unlike signs,* divide their absolute values. The answer is *negative.*
2. To divide two numbers with the *same sign,* divide their absolute values. The answer is *positive.*

For any number a and any nonzero number b,

$$\frac{-a}{b} = \frac{a}{-b} = -\frac{a}{b}.$$

EXPONENTIAL NOTATION

$b^1 = b$, for any number b.

For any natural number n greater than or equal to 2,

$$b^n \text{ means } \overbrace{b \cdot b \cdot b \cdot b \cdots b}^{n \text{ factors}}.$$

RULES FOR ORDER OF OPERATIONS

1. Do all calculations within grouping symbols before operations outside.
2. Evaluate all exponential expressions.
3. Do all multiplications and divisions in order from left to right.
4. Do all additions and subtractions in order from left to right.

REVIEW EXERCISES

1. Translate the phrase to mathematical language: Five less than thirty-six percent of some number.
2. Evaluate the given expression using the values provided: $5xy - z$ if $x = 2$, $y = 3$, and $z = 12$.

Determine whether the given value is a solution of the equation.

3. 0.6; $4x + 1.5 = 3.9$
4. $\frac{11}{5}$; $6 - 5y = -4$
5. Use roster notation to name the set of all natural numbers that are multiples of 4.
6. Use set-builder notation to name the set of all multiples of 3 between 17 and 49.

Classify each statement as true or false. The following sets are used.

$\mathbb{N}$ = the set of natural numbers
$\mathbb{W}$ = the set of whole numbers
$\mathbb{Z}$ = the set of integers
$\mathbb{Q}$ = the set of rational numbers
$\mathbb{R}$ = the set of real numbers

7. $6.\overline{45} \in \mathbb{Q}$

8. $0 \notin \mathbb{Z}$

9. $\mathbb{W} \subseteq \mathbb{Q}$

10. $\mathbb{R} \subseteq \mathbb{N}$

11. Find the absolute value: $|-3.7|$.

12. Write the meaning of the inequality and determine whether it is a true statement: $-5 \leq -3$.

13. Add: $2.8 + (-4.3)$.

14. Find the additive inverse, or opposite: $3\frac{2}{3}$.

15. Find $-x$ if $x = -2.1$.

Perform the indicated operation and simplify.

16. $-7 - (-3)$

17. $(-3.2)(-5)$

18. $\dfrac{48}{-6}$

19. $\dfrac{3}{8} \div \dfrac{6}{7}$

20. Find the reciprocal, or multiplicative inverse: $\frac{9}{4}$.

21. Use multiplication by 1 to find an expression that is equivalent to $-5x/3$ with a denominator of 18.

22. Simplify: $\dfrac{54a}{6}$.

23. What is the meaning of the expression $(-3ab)^2$?

24. Write exponential notation:

$$4b \cdot 4b \cdot 4b \cdot 4b \cdot 4b.$$

Calculate using the rules for order of operations.

25. $\dfrac{2 \cdot 5 - (5-2)^2}{4^2 - 8}$

26. $|7 - 3^2| \cdot 5 - 4$

Evaluate.

27. $(-3x)^3 + 86$ for $x = 2$

28. $\dfrac{y^2 - 6}{4y}$ for $y = -3$

29. $(2m)^4$ and $2m^4$ for $m = 3$

30. List the sequence of keys that should be pressed on a calculator to compute the following. Assume algebraic logic.

$$\frac{5 + 3^5 - 12}{6}$$

31. Write an equivalent expression using a commutative law: $a + bc^2$.

32. Write an equivalent expression using an associative law: $5 + (3x + 9)$.

33. Obtain an equivalent expression by multiplying: $7x(2 - x)$.

34. Obtain an equivalent expression by factoring: $5x - 15y + 5$.

35. Obtain an equivalent expression by collecting like terms. Use the distributive law.

$$11a - 6b + 5a - 2b$$

36. Obtain an equivalent but simpler expression, by first removing parentheses: $5x + 3 - (2x - 7)$.

37. Simplify: $7b - [-5 - 3(4b + 6)]$.

SYNTHESIS

38. Use roster notation to name the set of all whole numbers that are not positive real numbers.

39. Write an algebraic expression for the sum of the squares of two consecutive integers.

40. Evaluate the expressions $(a \div b) \div c$ and $a \div (b \div c)$ for $a = 20$, $b = 5$, and $c = 1$. Do the results mean that division is associative? Explain.

THINKING IT THROUGH

1. Give five examples of rational numbers that are not integers.
2. Explain the relationships between natural numbers, whole numbers, integers, rational numbers, and real numbers in terms of subsets.
3. Explain and compare the commutative, associative, and distributive laws.
4. Give two expressions that are equivalent. Give two that are not equivalent.

CHAPTER TEST 1

1. Translate the phrase to mathematical language: Seven more than the product of two numbers.
2. Evaluate the given expression using the values provided: $m - 3np$ if $m = 15$, $n = -2$, and $p = 4$.
3. Determine whether the given value is a solution of the equation: $\frac{7}{3}$; $5 - 3m = -2$.
4. Use set-builder notation and roster notation to name the set of odd numbers between 8 and 16.

Classify each statement as true or false. The following sets are used.

$\mathbb{W}$ = the set of whole numbers $\quad \mathbb{Z}$ = the set of integers $\quad \mathbb{Q}$ = the set of rational numbers

5. $\sqrt{5} \in \mathbb{Q}$
6. $\mathbb{W} \subseteq \mathbb{Z}$
7. Find the absolute value: $|-4.9|$.
8. Find the additive inverse, or opposite, of $-2\frac{1}{3}$.

Perform the indicated operation and simplify.

9. $\frac{1}{2} + \left(-\frac{2}{7}\right)$
10. $-4.2 - (-2.8)$
11. $(-4.3)(-3)$
12. $\frac{3}{7} \div \left(-\frac{6}{11}\right)$
13. Simplify: $\dfrac{48x}{-8}$.
14. What is the meaning of the expression $(-4x)^3$?

Calculate using the rules for order of operations.

15. $\dfrac{4 \cdot 7 - 6^2}{11 + 3 \cdot 5}$
16. $7 - |9 - 4^2|^2$
17. Evaluate:

$$\frac{3 + 6y}{3y} \quad \text{for } y = -4.$$

18. List the sequence of keys that should be pressed on a calculator to compute $\left(5 - \frac{3}{8}\right)^3 - \frac{2}{7}$. Assume algebraic logic.
19. Obtain an equivalent expression by multiplying: $-2x(3 - y)$.
20. Obtain an equivalent expression by factoring: $3ab + 6a$.
21. Obtain an equivalent expression by collecting like terms. Use the distributive law:

$$7 - 5x + 13 - 3x.$$

22. Simplify: $-3(4 - b) - 5(2b + 6)$.

SYNTHESIS

23. Simplify: $-(-(-(-|-7|)))$.
24. Use the distributive law to factor the following:

$$(5 + x)(x^2 - 1) + (5 + x)(2x - 3) = (5 + x)(\underline{\qquad\qquad})$$

CHAPTER 2

Equations and Inequalities

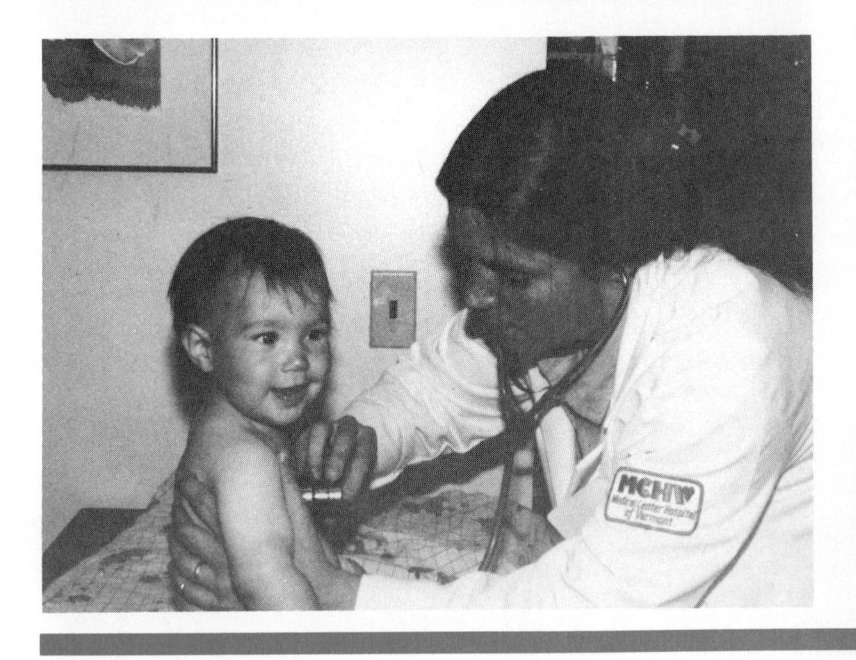

FEATURE PROBLEM

An insurance company offers two types of medical coverage. With plan A, you pay the first \$100 of your medical bills and the insurance company pays 80% of the rest. With plan B, you pay the first \$250 of your bills and the insuror pays 90% of the rest. For what total amounts of medical bills will plan B save you money?

THE MATHEMATICS

Let a be the amount of medical bills. Then the problem translates to the *inequality*

$$100 + 0.2a > 250 + 0.1a.$$

↑ Here your share is 20% (under $0.2a$)

↑ Here your share is 10% (under $0.1a$)

Solving equations and inequalities is a recurring theme in much of mathematics. In this chapter, we will study and use principles that will enable us to solve equations and inequalities in which the variable in question is raised to the first power. Equations and inequalities of this type will then be used to solve applied problems.

2.1 Solving Equations

Solving equations is essential for problem solving in algebra. In this section, we will review and practice the solving of simple equations.

Equivalent Equations

We have already seen in Section 1.1 that the equation $4 + x = 9$ has 5 as a solution. Although this problem may seem very simple to you, it is important for us to understand how such an equation is solved using the principles of algebra.

EQUIVALENT EQUATIONS

Two equations are said to be *equivalent* if they have the same solution(s).

EXAMPLE 1 Determine whether $2x = 6$ and $10x = 30$ are equivalent equations.

Solution When x is replaced by 3, both equations are true, and for any other replacement, both equations are false. Thus the equations are equivalent. ■

EXAMPLE 2 Determine whether $3x = 12$ and $6x/2 = 12$ are equivalent equations.

Solution When x is replaced by 4, both equations are true. For other replacements, both equations are false. Thus the equations are equivalent. ■

EXAMPLE 3 Determine whether $4 + x = 9$ and $x = 5$ are equivalent equations.

Solution When x is replaced by 5, both equations are true. For other replacements, both equations are false. Thus the equations are equivalent. ■

EXAMPLE 4 Determine whether $3x = 4x$ and $3/x = 4/x$ are equivalent equations.

Solution When x is replaced by 0, neither $3/x$ nor $4/x$ is defined, so 0 is *not* a solution to $3/x = 4/x$. Since 0 *is* a solution to $3x = 4x$, we conclude that the equations are not equivalent. ■

Note that in Example 2 we replaced $3x$ by an equivalent *expression*, which produced an equivalent *equation*. Example 3 illustrates the type of equivalent equations that we will work with most—those in which the variable is eventually isolated on one side of an equation, with the solution to the resulting equation easy to find.

The Addition and Multiplication Principles

Consider the equation $a = b$. It says that a and b represent the *same number.* Suppose that $a = b$ is true and then add a number c to a. We will get the same answer if we add c to b, because a and b are the same number. The same is true if we multiply by c.

THE ADDITION AND MULTIPLICATION PRINCIPLES FOR EQUATIONS

If $a = b$ is true, then:

a) $a + c = b + c$ is true for any number c;
b) $a \cdot c = b \cdot c$ is true for any number c.

EXAMPLE 5 Solve: $y - 4.7 = 13.9$.

Solution

$$y - 4.7 = 13.9$$

$$y - 4.7 + 4.7 = 13.9 + 4.7 \quad \text{Using the addition principle; adding 4.7}$$

$$y + 0 = 13.9 + 4.7 \quad \text{The law of additive inverses}$$

$$y = 18.6$$

Check:

$y - 4.7 = 13.9$		
$18.6 - 4.7$	13.9	Substituting 18.6 for y and calculating
13.9		TRUE

The solution is 18.6. ■

In Example 5, why did we add 4.7 on both sides? Because we wanted the variable y alone on one side of the equation. When we added 4.7, we got $y + 0$, which left y alone on the left. Then the other side showed us what calculations to make to find the equation $y = 18.6$ from which the solution, 18.6, could be easily read.

EXAMPLE 6 Solve: $4x = 9$.

Solution

$$4x = 9$$

$$\tfrac{1}{4} \cdot 4x = \tfrac{1}{4} \cdot 9 \quad \text{Using the multiplication principle, we multiply by } \tfrac{1}{4}\text{, the reciprocal of 4.}$$

$$1 \cdot x = \tfrac{9}{4} \quad \text{The law of multiplicative inverses}$$

$$x = \tfrac{9}{4}$$

Check:

$4x = 9$	
$4 \cdot \frac{9}{4}$	9
9	

TRUE

The solution is $\frac{9}{4}$. ■

In Example 6, why did we choose to multiply by $\frac{1}{4}$? Because we wanted x alone on one side of the equation. We multiplied by the reciprocal of 4. Then we got $1 \cdot x$, which simplified to x. This eliminated the 4 on the left.

EXAMPLE 7 Solve: $\frac{2}{3}x + 3 = 11$.

Solution

$$\frac{2}{3}x + 3 = 11$$

$$\frac{2}{3}x + 3 - 3 = 11 - 3 \quad \text{Using the addition principle; adding } -3 \text{ or subtracting } 3$$

$$\frac{2}{3}x = 8 \quad \text{Simplifying; the law of additive inverses}$$

$$\frac{3}{2} \cdot \frac{2}{3}x = \frac{3}{2} \cdot 8 \quad \text{Using the multiplication principle}$$

$$x = 12 \quad \text{Simplifying; the law of multiplicative inverses}$$

We leave the check for the student. The solution is 12. ■

Equations with Grouping Symbols

In Section 1.4 we saw how the distributive law can be used to "remove" grouping symbols. We then simplified by combining like terms. Equations containing grouping symbols are solved in a similar manner—grouping symbols are "removed," and then the addition and multiplication principles are used.

EXAMPLE 8 Solve: $8x - 3(x + 4) = 4[x + 2(3 - x)]$.

Solution

$$8x - 3(x + 4) = 4[x + 2(3 - x)]$$

$$8x - 3x - 12 = 4[x + 6 - 2x] \quad \text{Using the distributive law}$$

$$5x - 12 = 4[-x + 6] \quad \text{Collecting like terms}$$

$$5x - 12 = -4x + 24 \quad \text{Using the distributive law}$$

$$4x + 5x - 12 = 4x - 4x + 24 \quad \text{Using the addition principle; adding } 4x$$

$$9x - 12 = 24 \quad \text{Collecting like terms}$$

$$9x - 12 + 12 = 24 + 12 \quad \text{Using the addition principle}$$

$$9x = 36 \quad \text{Simplifying; the law of additive inverses}$$

$$\frac{1}{9} \cdot 9x = \frac{1}{9} \cdot 36 \quad \text{Using the mulitiplication principle}$$

$$x = 4 \quad \text{Simplifying; the law of multiplicative inverses}$$

Check:

$8x - 3(x+4)$	$= 4[x + 2(3 - x)]$
$8 \cdot 4 - 3(4+4)$	$4[4 + 2(3-4)]$
$32 - 3(8)$	$4[4 + 2(-1)]$
$32 - 24$	$4[4 - 2]$
8	$4 \cdot 2$
	8 TRUE

The solution is 4. ■

You may have wondered whether all of the preceding steps must be performed in a specific order. Often there is more than one "valid" sequence of steps. For instance, in Example 8 we could have "distributed" the 4 earlier.

Don't rush to do steps in your head. As long as you are careful in your use of the principles we've studied, the number of steps in your solution should be of little concern. What *is* important is that each step produces a simpler, yet equivalent, equation. Our goal is always to get the variable alone on one side.

Special Cases

Sometimes we encounter equations that have *infinitely many solutions,* like $x + 3 = 3 + x$; other times we encounter equations that have *no solutions,* like $3 + x = 4 + x$. The rules of algebra can help us to recognize either one of these situations.

EXAMPLE 9 Solve: $x + 3 = 3 + x$.

Solution

$$x + 3 = 3 + x$$

$$x + 3 + (-x) = 3 + x + (-x) \quad \text{Using the addition principle}$$

$$3 = 3 \quad \text{Simplifying}$$

Since the equation $3 = 3$ is true regardless of our choice of x, and because $x + 3 = 3 + x$ is equivalent to $3 = 3$, we see that $x + 3 = 3 + x$ is true for any choice of x. All real numbers are solutions. If it troubles you that the solution set to $3 = 3$ is all real numbers, think of $3 = 3$ as $3 + 0 \cdot x = 3 + 0 \cdot x$. All real numbers are solutions. ■

EXAMPLE 10 Solve: $3x - 5 = 3(x - 2) + 4$.

Solution

$$3x - 5 = 3(x - 2) + 4$$

$$3x - 5 = 3x - 6 + 4 \quad \text{Using the distributive law}$$

$$3x - 5 = 3x - 2$$

$$-3x + 3x - 5 = -3x + 3x - 2 \quad \text{Using the addition principle}$$

$$-5 = -2$$

Since our original equation is equivalent to the equation $-5 = -2$, there is no solution to this problem. There is no choice of x that will solve the original problem. ■

Solution Sets

We will sometimes refer to the set of solutions, or **solution set,** of a particular problem. Thus the solution set for Example 8 is $\{4\}$. The solution set for Example 10 is the set containing *no* elements, denoted by $\{\ \}$ or $\emptyset$, and referred to as the **empty set.** The solution set in Example 9 may be written simply as $\mathbb{R}$, the set of all real numbers.

EXERCISE SET 2.1

Determine whether the given pair of equations are equivalent.

1. $3x = 12$ and $2x = 8$

2. $5x = 20$ and $15x = 60$

3. $2x - 1 = -7$ and $x = -3$

4. $x + 2 = -5$ and $x = -7$

5. $x + 5 = 11$ and $3x = 18$

6. $x - 3 = 7$ and $3x = 24$

7. $13 - x = 4$ and $2x = 20$

8. $3x - 4 = 8$ and $3x = 12$

9. $5x = 2x$ and $\dfrac{4}{x} = 3$

10. $6 = 2x$ and $5 = \dfrac{2}{3 - x}$

Solve. Don't forget to check.

11. $x - 5.2 = 9.4$

12. $y + 4.3 = 11.2$

13. $9y = 72$

14. $7x = 63$

15. $4x - 12 = 60$

16. $4x - 6 = 70$

17. $5y + 3 = 28$

18. $7t + 11 = 74$

19. $2y - 11 = 37$

20. $3x - 13 = 29$

21. $-4x - 7 = -35$

22. $-9y + 8 = -91$

23. $5x + 2x = 56$

24. $3x + 7x = 120$

25. $9y - 7y = 42$

26. $8t - 3t = 65$

27. $-6y - 10y = -32$

28. $-9y - 5y = 28$

29. $7y - 1 = 23 - 5y$

30. $15x + 20 = 8x - 22$

31. $5 - 4a = a - 13$

32. $8 - 5x = x - 16$

33. $3m - 7 = -7 - 4m - m$

34. $5x - 8 = -8 + 3x - x$

35. $5r - 2 + 3r = 2r + 6 - 4r$

36. $5m - 17 - 2m = 6m - 1 - m$

37. $\frac{1}{4} + \frac{3}{8}y = \frac{3}{4}$

38. $\frac{1}{5} + \frac{3}{10}x = \frac{4}{5}$

39. $-\frac{5}{2}x + \frac{1}{2} = -18$

40. $0.9y - 0.7 = 4.2$

41. $0.8t - 0.3t = 6.5$

42. $1.4x + 5.02 = 0.4x$

43. $2(x + 6) = 8x$

44. $3(y + 5) = 8y$

45. $80 = 10(3t + 2)$

46. $27 = 9(5y - 2)$

47. $180(n - 2) = 900$

48. $210(x - 3) = 840$

49. $5y - (2y - 10) = 25$

50. $8x - (3x - 5) = 40$

51. $0.7(3x + 6) = 1.1 - (x + 2)$

52. $0.9(2x + 8) = 20 - (x + 5)$

53. $\frac{1}{8}(16y + 8) - 17 = -\frac{1}{4}(8y - 16)$

54. $\frac{1}{6}(12t + 48) - 20 = -\frac{1}{8}(24t - 144)$

55. $a + (a - 3) = (a + 2) - (a + 1)$

56. $0.8 - 4(b - 1) = 0.2 + 3(4 - b)$

57. $5[2 + 3(x - 1)] = 4$

58. $3[t - 4(t + 7)] = -3$

59. $5 + 2(x - 3) = 2[5 - 4(x + 2)]$

60. $3[2 - 4(x - 1)] = 3 - 4(x + 2)$

61. $2\{9 - 3[2x - 4(x + 1)]\} = 5(2x + 8)$

62. $3\{7 - 2[3x + 4(x - 1)]\} = 7(6 - 3x)$

Find the solution set.

63. $4x - 2x - 2 = 2x$

64. $2x + 4 + x = 4 + 3x$

65. $2 + 9x = 3(3x + 1) - 1$

66. $4 + 7x = 7(x + 1)$

67. $-8x + 5 = 14 - 8x$

68. $-8x + 5 = 5 - 8x$

69. $2\{9 - 3[-2x - 4]\} = 12x + 42$

70. $3\{7 - 2[7x - 4]\} = -42x + 45$

SKILL MAINTENANCE

71. Write the set consisting of the positive integers less than 10, using both roster notation and set-builder notation.

72. Write the set consisting of the negative integers greater than -9, using both roster notation and set-builder notation.

Perform the indicated operation.

73. $-9.4 + 7.2$

74. $-7.9 + 4.5$

75. $-7 - (-5.3)$

76. $-9 - (-3.7)$

77. $(-9)(-6)$

78. $(-4)(-9)$

79. $(-12) \div (-3)$

80. $(-15) \div (-5)$

SYNTHESIS

Solve. Remember to check.

81. 🖩 $0.0008x = 0.00000564$

82. 🖩 $43.008z = 1.201135$

83. 🖩 $4.23x - 17.898 = -1.65x - 42.454$

84. 🖩 $-0.00458y + 1.7787 = 13.002y - 1.005$

85. $x - \{3x - [2x - (5x - (7x - 1))]\} = x + 7$

86. $3x - \{5x - [7x - (4x - (3x + 1))]\} = 3x + 5$

87. $7x - 2\{3x + 4[x - 3(x - 2(x + 1))]\} = 14$

88. $5x - 3\{2x + 4[2x - 5(x - 3(x + 1))]\} = 8$

89. $17 - 3\{5 + 2[x - 2]\} + 4\{x - 3(x + 7)\} = 9\{x + 3[2 + 3(4 - x)]\}$

90. $23 - 2\{4 + 3[x - 1]\} + 5\{x - 2(x + 3)\} = 7\{x - 2[5 - (2x + 3)]\}$

2.2 Problem Solving

We now begin to study and practice the "art" of problem solving. Although we are interested mainly in the use of algebra to solve problems, much of what we say here applies to solving all kinds of problems.

What do we mean by a *problem*? Perhaps you've already used algebra to solve some "real-world" problems. What procedure did you use? Was there anything in your approach that could be used to solve problems of a more general nature? These are some questions that we will attempt to answer in this section.

In this text, we do not restrict the use of the word "problem" to computational situations involving arithmetic or algebra, such as $589 + 437$ or $3x + 5x$. We mean instead some question to which we need or want to find an answer. Perhaps this can best be illustrated with some sample problems:

1. How can I schedule my classes so that all of them are in the morning?
2. What is the fastest way to get from New York City to Washington, D.C.?
3. Can I eat 3000 calories a day and still lose weight?
4. An airplane traveling at a speed of 210 mph in still air encounters a head wind of 46 mph. How long will it take for the plane to travel 369 mi into the head wind?

These problems are all different, but there are some similarities. We cannot give rules for problem solving, but there is a general *strategy* that can be used. Some problems can be solved using algebra and some cannot, but the overall strategy can be used in any case.

The Five-Step Strategy

Since you have already studied some algebra, you have had some experience with problem solving. The following steps make up a strategy that you may already have used in algebra. They constitute a good strategy for problem solving in general.

FIVE STEPS FOR PROBLEM SOLVING IN ALGEBRA

1. *Familiarize* yourself with the problem situation.
2. *Translate* to mathematical language.
3. *Carry out* some mathematical manipulation.
4. *Check* your possible answer in the original problem.
5. *State* the answer clearly.

The First Step: Familiarization

Of the five steps, probably the most important is the first: becoming familiar with the problem situation. Here are some hints for familiarization.

The First Step in Problem Solving with Algebra

Familiarize yourself with the problem situation.

1. If a problem is given in words, read it carefully.
2. Reread the problem, perhaps aloud. Try to verbalize the problem to yourself.
3. List the information given and the question to be answered. Choose a variable or variables to represent any unknown(s) and clearly state what each variable represents. Be descriptive! For example, let l = length, d = distance, and so on.
4. Find further information. Look up a formula at the back of the book or in a reference book. Talk to a reference librarian or an expert in the field.
5. Make a table of the information given and the information you have collected. Look for patterns that may help in the translation to an equation.
6. Make a drawing and label it with known information. Also indicate unknown information, using specific units if given.
7. Guess or estimate an answer.

EXAMPLE 1 How might you familiarize yourself with the situation given in the first sample problem: "How can I schedule my classes so that all of them are in the morning?"

Solution Clearly you will need to find further information in order to solve this problem. You might:

a) List all courses that you are *required* to take.
b) Get a schedule of course offerings and study it.
c) Talk to counselors.

When the information is known, it might be wise to make a table or chart to assist with your schedule selection. ■

EXAMPLE 2 How might you familiarize yourself with the situation given in the fourth sample problem: "How long will it take for the plane to travel 369 mi into the head wind?"

Solution First read the question *very* carefully. This may even involve speaking aloud. You may need to reread the problem one or more times to fully understand what information you are given and what information is desired. A sketch is often helpful.

Plane
210 mph
46-mph wind

In this case, a table can be constructed to clearly list the relevant information.

Speed of Plane in Still Air	210 mph
Speed of Head Wind	46 mph
Speed of Plane in Head Wind	?
Distance to Be Traveled	369 mi
Time Required	?

As a next step in the familiarization process, we should determine, possibly with the aid of outside references, what relationships exist between the various quantities in the problem. With some effort it can be learned that a head wind's speed should be subtracted from the plane's speed in still air if we are to determine the speed of the plane in the head wind. We might consult a physics book to note that

$$\textbf{Distance} = \textbf{Speed} \times \textbf{Time.}$$

We rewrite part of the table, letting t represent the time, in hours, required for the plane to fly 369 mi into the head wind.

Speed of Plane in Head Wind	$210 - 46 = 164$ mph
Distance to Be Traveled	369 mi
Time Required	t

At this point, we could attempt to guess an answer. Suppose that the plane flew into the head wind for 2 hr. Then the plane will have traveled

$$\begin{array}{ccccc} \text{Speed} & \times & \text{Time} & = & \text{Distance.} \\ 164 & \times & 2 & = & 328 \text{ mi} \end{array}$$

Since $328 \neq 369$, we conclude that our guess was wrong. However, an examination of how we checked our guess gives added insight into the problem. We notice that a better guess, when multiplied by 164, would yield a number closer to 369. ■

The Other Steps for Problem Solving

The second step in problem solving is to translate the situation to mathematical language. In algebra, this usually consists of forming an equation.

The Second Step in Problem Solving with Algebra

Translate the situation of the problem to mathematical language. In some cases, translation can be done by writing an algebraic expression, but most problems in this text can be solved by translating to an equation.

In the third step of our process, we work with the results of the first two steps. Often this will require us to use the algebra that we have studied.

The Third Step in Problem Solving with Algebra

Carry out some mathematical manipulation. If you have translated to an equation, this means to solve the equation.

To properly complete the problem-solving process, we should always **check** our solution and then **state** the solution in a clear and precise manner. Normally our check consists of returning to the original problem and determining whether all its conditions have been satisfied. If our answer checks, we write a complete English sentence to state what the solution is. Our five steps are listed again. Try to apply them regularly in your work in mathematics.

FIVE STEPS FOR PROBLEM SOLVING IN ALGEBRA

1. *Familiarize* yourself with the problem situation.
2. *Translate* to mathematical language.
3. *Carry out* some mathematical manipulation.
4. *Check* your possible answer in the original problem.
5. *State* the answer clearly.

Problem Solving

At this point our study of algebra is still in a beginning stage. Consequently, we have a small number of algebraic tools with which to work problems. As the number of tools in our algebraic "tool box" increases, so will the level of difficulty of the problems to be solved. For now our problems may seem simple; however, to gain practice with the problem-solving process, you should attempt to proceed through all five steps. Later some steps may be skipped or shortened.

EXAMPLE 3 A student paid \$1187.20 for a computer. If the price paid included a 6% sales tax, what was the price of the computer itself?

Solution

1. **Familiarize** yourself with the problem. We note that the tax is calculated from, and then added to, the computer's price. We let

$$C = \text{the computer's price.}$$

Let's guess that the computer's price was \$1000. To check the guess, we calculate the amount of tax, $(0.06) \times 1000 = \$60$, and add it to \$1000:

$$\begin{aligned}(0.06)(\$1000) + \$1000 &= \$60 + \$1000 \\ &= \$1060.\end{aligned}$$

Our guess was wrong, but it helped us familiarize ourselves with the problem. The manner in which we manipulated our guess will guide us in the next step.

2. **Translate** the problem to mathematical language. Our guess leads us to the following translation:

$$\underbrace{\text{6\% of the computer's price}}_{(0.06)C} \quad \underbrace{\text{plus}}_{+} \quad \underbrace{\text{the computer's price}}_{C} \quad \underbrace{\text{is}}_{=} \quad \underbrace{\text{the price with sales tax.}}_{\$1187.20}$$

3. **Carry out** the algebraic manipulation:

$$\begin{aligned}0.06C + 1C &= 1187.20 \\ 1.06C &= 1187.20 \quad \text{Collecting like terms} \\ \frac{1}{1.06} \cdot 1.06C &= \frac{1}{1.06} \cdot 1187.20 \quad \text{Using the multiplication principle} \\ C &= 1120.\end{aligned}$$

4. **Check** the answer in the original problem. To do this, we calculate 6% of 1120,

$$(0.06)1120 = 67.20,$$

and add it to 1120,

$$67.20 + 1120 = 1187.20.$$

We see that \$1120 checks in the original problem.

5. **State** the answer clearly. The computer itself cost \$1120.00. ■

EXAMPLE 4 A piece of wood molding 100 in. long is to be cut into two pieces, and those pieces are each to be cut into the shape of a square frame. The length of a side of one square is to be $1\frac{1}{2}$ times the length of a side of the other. How should the wood be cut?

Solution

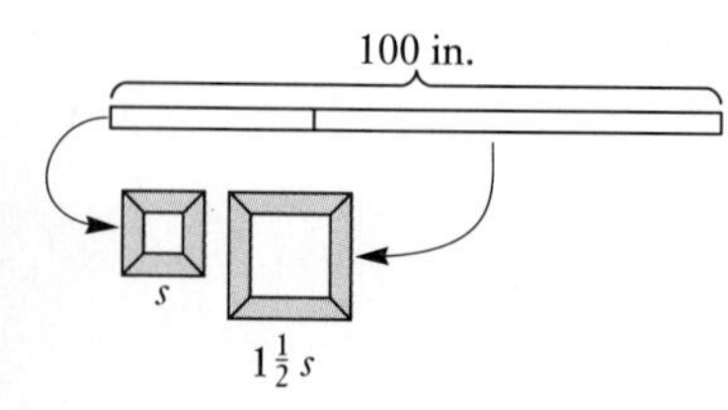

1. **Familiarize.** We note that the *perimeter* (distance around) of each square is four times the length of a side. Furthermore, if s is used to represent the length of a side in the smaller square, then $\left(1\frac{1}{2}\right)s$ will represent the length of a side in the larger square. Finally, note that the two perimeters must add up to 100 in.

$$\textit{Perimeter of a square} = 4 \cdot \textit{length of a side}$$

2. **Translate.**

$$\underbrace{\text{The perimeter of one square}}_{4s}\ \underbrace{\text{plus}}_{+}\ \underbrace{\text{the perimeter of the other}}_{4\left(1\frac{1}{2}s\right)}\ \underbrace{\text{is}}_{=}\ \underbrace{\text{100 in.}}_{100}$$

3. **Carry out.** We solve the equation:

$$4s + 4\left(1\tfrac{1}{2}s\right) = 100$$
$$4s + 6s = 100 \quad \text{Simplifying}$$
$$10s = 100 \quad \text{Collecting like terms}$$
$$s = \tfrac{1}{10} \cdot 100 \quad \text{Multiplying by } \tfrac{1}{10} \text{ on both sides}$$
$$s = 10. \quad \text{Simplifying}$$

4. **Check.** If 10 is the length of the smaller side, then $\left(1\frac{1}{2}\right)(10) = 15$ is the length of the larger side. The two perimeters would then be

$$4 \cdot 10 = 40 \quad \text{and} \quad 4 \cdot 15 = 60.$$

Since $40 + 60 = 100$, our answer checks.

5. **State.** The wood should be cut into two pieces, one 40 in. long and another 60 in. long. Each of these two pieces should then be quartered to form the frames. ■

EXAMPLE 5 Three numbers are such that the second is 6 less than 3 times the first and the third is 2 more than $\frac{2}{3}$ the first. The sum of the three numbers is 150. Find the largest of the three numbers.

Solution We proceed according to the five-step process.

1. **Familiarize.** Three numbers are involved, and we want to find the largest one. We list the information in a table, letting x represent the first number.

First Number	x
Second Number	6 less than 3 times the first
Third Number	2 more than $\frac{2}{3}$ the first

$$\text{First} + \text{Second} + \text{Third} = 150$$

The student is invited to make a guess at this point. We will proceed to the next step.

2. **Translate.** We can now name the second and third numbers by using x. (We often say that we name them "in terms of x.") We'll go back to the table and add another column:

First Number	x	x
Second Number	6 less than 3 times the first	$3x - 6$
Third Number	2 more than $\frac{2}{3}$ the first	$\frac{2}{3}x + 2$

We know that the sum of the three numbers is 150. Substituting, we obtain an equation:

$$\underbrace{\text{First}}_{\downarrow} + \underbrace{\text{Second}}_{\downarrow} + \underbrace{\text{Third}}_{\downarrow} = \underbrace{150.}_{\downarrow}$$
$$x + (3x - 6) + \left(\tfrac{2}{3}x + 2\right) = 150$$

3. **Carry out.** We solve the equation:

$$x + 3x - 6 + \tfrac{2}{3}x + 2 = 150 \quad \text{Leaving off unnecessary parentheses}$$
$$\left(4 + \tfrac{2}{3}\right)x - 4 = 150 \quad \text{Collecting like terms}$$
$$\tfrac{14}{3}x - 4 = 150$$
$$\tfrac{14}{3}x = 154 \quad \text{Adding 4 on both sides}$$
$$x = \tfrac{3}{14} \cdot 154 \quad \text{Multiplying on both sides by } \tfrac{3}{14}$$
$$x = 33.$$

We could check to see whether 33 is a solution of the equation, but we can skip that step because we will check later in the original problem.

Going back to the table, we can find the other two numbers:

Second: $3x - 6 = 3 \cdot 33 - 6 = 93$;
Third: $\frac{2}{3}x + 2 = \frac{2}{3} \cdot 33 + 2 = 24$.

4. **Check.** We go back to the original problem. We have three numbers: 33, 93, and 24. Is the second number 6 less than 3 times the first?

$$3 \times 33 - 6 = 99 - 6 = 93$$

The answer is *yes*.

Is the third number 2 more than $\frac{2}{3}$ the first?

$$\tfrac{2}{3} \times 33 + 2 = 22 + 2 = 24$$

The answer is *yes*.

Is the sum of the three numbers 150?

$$33 + 93 + 24 = 150$$

The answer is *yes*. The numbers do check.

5. **State.** The problem asks us to find the largest number, so the answer is: "The largest number is 93." ■

Note in Example 5 that although the equation $x = 33$ enabled us to find the largest number, 93, the number 33 was *not* the solution to the problem. By carefully labeling our variable in the first step, we may avoid the temptation of thinking that our variable always represents the solution to the problem.

EXERCISE SET 2.2

For each problem, familiarize yourself with the situation. Then translate to mathematical language. You need not actually solve the problem; just carry out the first two steps.

1. The sum of two numbers is 81. One of the numbers is 9 more than the other. What are the numbers?
2. The sum of two numbers is 95. One of the numbers is 11 more than the other. What are the numbers?
3. A person swims at a speed of 5 mph in still water. The current in a river is moving at 3.2 mph. How long will it take the person to swim 2.7 mi upriver?
4. A person swims at a speed of 4 mph in still water. The current in a river is moving at 1.5 mph. How long will it take the person to swim 3.75 mi upriver?
5. A paddleboat moves at a rate of 12 km/h in still water. How long will it take the boat to travel 35 km downriver if the river's current moves at a rate of 3 km/h?
6. A paddleboat moves at a rate of 14 km/h in still water. How long will it take the boat to travel 56 km downriver if the river's current moves at a rate of 7 km/h?
7. The degree measures of the angles in a triangle are three consecutive integers. Find the measures of the angles.
8. The degree measures of the angles in a triangle are three consecutive even integers. Find the measures of the angles.
9. A 12-ft piece of rope is to be cut into two pieces, one piece 4 ft longer than the other. How should the rope be cut?
10. A piece of wire 10 m long is to be cut into two pieces, one of them $\frac{2}{3}$ as long as the other. How should the wire be cut?
11. One angle of a triangle is three times as great as a second angle. The third angle measures 12° less than twice the second angle. Find the measures of the angles.
12. One angle of a triangle is four times as great as a second angle. The third angle measures 5° more than twice the second angle. Find the measures of the angles.
13. Find three consecutive odd integers such that the sum of the first, two times the second, and three times the third is 70.
14. Find two consecutive even integers such that two times the first plus three times the second is 76.
15. A piece of wire 100 cm long is to be cut into two pieces, each to be bent to make a square. The length of a side of one square is to be 2 cm greater than the length of a side of the other. How should the wire be cut?
16. A piece of wire 100 cm long is to be cut into two pieces, each to be bent to make a square. The area of one square is to be 144 cm^2 greater than that of the other. How should the wire be cut?

17. Three numbers are such that the second is six less than three times the first and the third is two more than two thirds of the second. The sum of the three numbers is 172. Find the largest number.

18. An appliance store is having a sale on 13 TV sets. They are displayed in order of increasing price from left to right. The price of each set differs by $20 from either set next to it. For the price of the set at the extreme right, a customer can buy both the second and seventh sets. What is the price of the least expensive set?

19. A student's scores on five tests are 93, 89, 72, 80, and 96. What must the score be on the next test so that the average will be 88?

20. The changes in population of a city for three consecutive years are, respectively, 20% increase, 30% increase, and 20% decrease. What is the percent of total change for those three years?

Solve the problem. Carry out all five problem-solving steps.

21. The product of two numbers is 12.3. If one of the factors is 3, find the other number.

22. The quotient of two numbers is 0.75. If the divisor is 50, find the other number.

23. The number 38.2 is less than some number by 12.1. What is the number?

24. The number 173.5 is greater than a certain number by 16.8. What is the number?

25. The number 128 is 0.4 of what number?

26. The number 456 is $\frac{1}{3}$ of what number?

27. One number is greater than another by 12. The sum of the numbers is 114. What is the larger number?

28. One number is less than another by 65. The sum of the numbers is 92. What is the smaller number?

29. One number is twice another number. The sum of the numbers is 495. What are the numbers?

30. One number is five times another. The sum of the numbers is 472. What are the numbers?

31. Solve the problem of Exercise 4.

32. Solve the problem of Exercise 3.

33. Solve the problem of Exercise 14.

34. Solve the problem of Exercise 13.

35. Solve the problem of Exercise 10.

36. Solve the problem of Exercise 9.

37. Solve the problem of Exercise 11.

38. Solve the problem of Exercise 15.

SKILL MAINTENANCE

Multiply.

39. $3a(2 - b)$

40. $-4(5x - 3y + 2)$

Factor.

41. $5x - 10$

42. $9a + 9b$

SYNTHESIS

43. The height and sides of a triangle are four consecutive integers. The height is the first integer, and the base is the fourth integer. The perimeter of the triangle is 42 in. Find the area of the triangle.

44. The salary of an employee is reduced n% during a time when a company was having financial difficulty. By what percent would the company then have to raise the salary in order to bring it back to where it was before the reduction?

45. A city's population grew 8% in 1989, 10% in 1990, and 11% in 1991. Find the population of the city at the start of 1989, if the population is 1,582,416 at the end of 1991.

46. A student's scores on four tests are 83, 91, 78, and 81. How many points above the average must the student score on the next test in order to raise the average 2 points?

2.3 Formulas and Problem Solving

A **formula** is a kind of recipe, or rule, for doing a certain kind of calculation and is often stated in the form of an equation. For example, in Section 2.2 we made use of the formula $P = 4s$, where P represents the perimeter of a square and s the length of a side. Other formulas that you may recall from a geometry course are $A = \pi r^2$ (for the area A of a circle of radius r), $C = \pi d$ (for the circumference C of a circle of diameter d), and $A = b \cdot h$ (for the area A of a parallelogram of height h and base length b).*

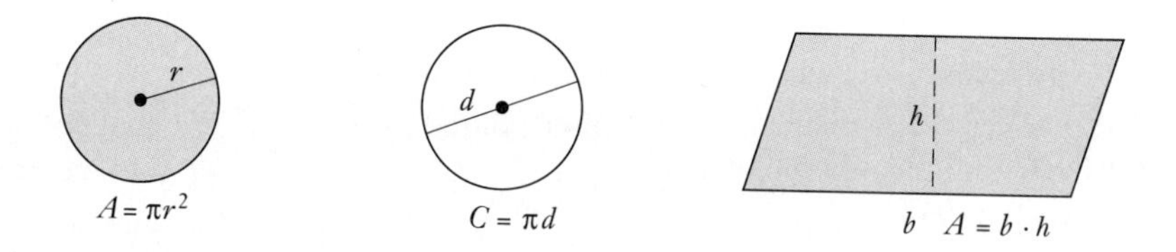

Additional geometric formulas are found in the table on page 616 of this text.

Solving Formulas

Suppose that we know the area and the width of a rectangular room and wish to find the length. To do so, we might begin with the equation $A = l \cdot w$, which expresses a rectangle's area A in terms of its length l and its width w. We then "solve" for l.

EXAMPLE 1 Solve the formula $A = l \cdot w$ for l.

Solution

$$A = l \cdot w \qquad \text{We want this letter alone.}$$

$$A \cdot \frac{1}{w} = l \cdot w \cdot \frac{1}{w} \qquad \text{Multiplying by } \frac{1}{w}$$

$$\frac{A}{w} = l \qquad \text{Simplifying}$$

■

Thus to find the length of a rectangular room, we can divide the area of the room by its width. Were we to do this calculation for a wide variety of rectangular rooms, the formula $l = A/w$ would be most useful.

When we solve a formula, we do the same things that we would do to solve any equation. The idea is to get a certain letter alone on one side of the equals sign.

EXAMPLE 2 The formula $I = Prt$ is used to determine the amount of simple interest I, earned on P dollars, when invested for t years at an interest rate r. Solve the above formula for t.

---

*The Greek letter π, read "pi," is *approximately* 3.14159265358979323846264. In this text, we will use 3.14 to approximate π unless otherwise noted.

Solution

$$I = Prt \quad \text{We want this letter alone.}$$

$$\frac{1}{Pr} \cdot I = \frac{1}{Pr} \cdot Prt \quad \text{Multiplying by } \frac{1}{Pr}$$

$$\frac{I}{Pr} = t \quad \text{Simplifying}$$

■

EXAMPLE 3 The formula $A = P + Prt$ tells how much a principal P, in dollars, will be worth when invested at a simple interest rate r in t years. Solve the formula for P.

Solution

$$A = P + Prt \quad \text{We want this letter alone.}$$

$$A = P(1 + rt) \quad \text{Factoring}$$

$$A \cdot \frac{1}{1 + rt} = P(1 + rt) \cdot \frac{1}{1 + rt} \quad \text{Multiplying by } \frac{1}{1 + rt}$$

$$\frac{A}{1 + rt} = P \quad \text{Simplifying}$$

■

Compare Example 3 with an equation in which only one letter appears. We will parallel Example 3 without simplifying.

EXAMPLE 3A Solve for x: $7 = x + 4 \cdot 5 \cdot x$.

Solution

$$7 = x + 4 \cdot 5 \cdot x$$

$$7 = x(1 + 4 \cdot 5) \quad \text{Factoring (or collecting like terms)}$$

$$7 \cdot \frac{1}{1 + 4 \cdot 5} = x(1 + 4 \cdot 5) \cdot \frac{1}{1 + 4 \cdot 5} \quad \text{Mulitplying by } \frac{1}{1 + 4 \cdot 5}$$

$$\frac{7}{1 + 4 \cdot 5} = x \quad \text{Simplifying}$$

In this case, we can simplify further, to $\frac{7}{21} = \frac{1}{3} = x$, whereas in Example 3 we could not simplify further. ■

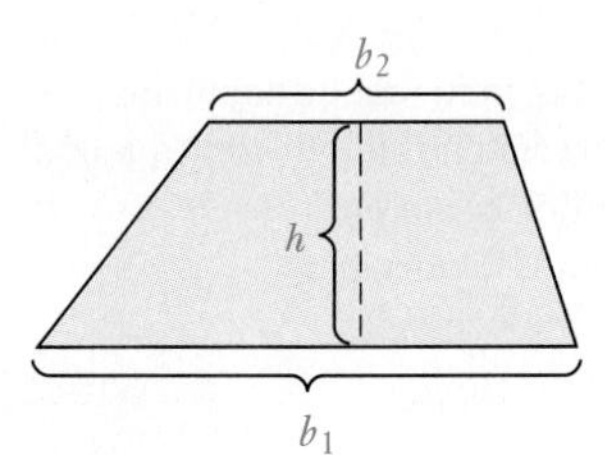

EXAMPLE 4 A trapezoid is a geometric shape with four sides, two of which, the bases, are parallel to each other. The formula for calculating the area A of a trapezoid with bases b_1 and b_2 (read "b sub one" and "b sub two") and height h is given by

$$A = \frac{h}{2}(b_1 + b_2),$$

where the *subscripts* 1 and 2 are used to distinguish the two bases from each other. Solve for b_1.

Solution

$$A = \frac{h}{2}(b_1 + b_2)$$

$$\frac{2}{h} \cdot A = \frac{2}{h} \cdot \frac{h}{2}(b_1 + b_2) \qquad \textbf{Multiplying by } \frac{2}{h}$$

$$\frac{2A}{h} = b_1 + b_2 \qquad \textbf{Simplifying}$$

$$\frac{2A}{h} - b_2 = b_1 \qquad \textbf{Adding } -b_2$$

■

Note that it is often more convenient to isolate the desired variable on the right. Compare Example 4 with an equation in which only one letter appears.

EXAMPLE 4A Solve for x: $7 = \frac{5}{2}(x + 3)$.

Solution

$$7 = \frac{5}{2}(x + 3)$$

$$\frac{2}{5} \cdot 7 = \frac{2}{5} \cdot \frac{5}{2}(x + 3) \qquad \textbf{Multiplying by } \frac{2}{5}$$

$$\frac{2 \cdot 7}{5} = x + 3 \qquad \textbf{Simplifying}$$

$$\frac{2 \cdot 7}{5} - 3 = x \qquad \textbf{Adding } -3$$

■

Again, we chose *not* to simplify so as to emphasize the similarities between Examples 4 and 4A. You may find the following summary useful.

To solve a formula for a given letter, identify the letter and:

1. Multiply on both sides to clear fractions or decimals, or to remove grouping symbols if that is needed.
2. Collect like terms on each side where convenient. This may require factoring.
3. Get all terms with the letter for which we are solving on one side of the equation and all other terms on the other side, using the addition principle.
4. Collect like terms again, if necessary. This may require factoring.
5. Solve for the letter in question, using the multiplication principle.

Formulas in Translating

The next example illustrates the use of formulas in translating problem situations to mathematical language and in carrying out the mathematical manipulations (steps 2 and 3 of the problem-solving process).

EXAMPLE 5 The density of gold is 7.72 grams per cubic centimeter (g/cm^3). A gold medallion has a mass of 38.6 g. Find the volume of the medallion.

Solution

2. **Translate.** If necessary, look up the formula

$$D = \frac{m}{V},$$

where D is the density, m is the mass, and V is the volume of a given material. Since we are interested in the volume, we solve the formula for V:

$$D = \frac{m}{V}$$

$$V \cdot D = V \cdot \frac{m}{V} \quad \text{Multiplying by } V$$

$$V \cdot D = m \quad \text{Simplifying}$$

$$V \cdot D \cdot \frac{1}{D} = m \cdot \frac{1}{D} \quad \text{Multiplying by } \frac{1}{D}$$

$$V = \frac{m}{D}. \quad \text{Simplifying}$$

We now have a formula that says that to find V, we divide m by D.

3. **Carry out.** We now put the numbers into the formula and calculate:

$$V = \frac{m}{D}$$

$$= \frac{38.6}{7.72} \quad \text{Substituting}$$

$$= 5.$$

We leave the check to the student. The solution is that the volume of the medallion is 5 cm^3. ■

In problems like Example 5, it is possible to first substitute the given values into the *original* formula and then solve. However, in situations in which calculations may be repeated with various sets of numbers, it is usually best to solve the original formula for the desired variable and *then* substitute.

EXERCISE SET 2.3

Solve.

1. $A = lw$, for w
2. $F = ma$, for a
3. $W = EI$, for I (an electricity formula)
4. $W = EI$, for E
5. $d = rt$, for r (a distance formula)
6. $d = rt$, for t
7. $V = lwh$, for l (a volume formula)
8. $I = Prt$, for r
9. $E = mc^2$, for m (a relativity formula)
10. $E = mc^2$, for c^2
11. $P = 2l + 2w$, for l (a perimeter formula)
12. $P = 2l + 2w$, for w
13. $c^2 = a^2 + b^2$, for a^2 (a geometry formula)
14. $c^2 = a^2 + b^2$, for b^2
15. $A = \pi r^2$, for r^2 (an area formula)
16. $A = \pi r^2$, for π
17. $W = \frac{11}{2}(h - 40)$, for h
18. $C = \frac{5}{9}(F - 32)$, for F (a temperature formula)
19. $V = \frac{4}{3}\pi r^3$, for r^3 (a volume formula)
20. $V = \frac{4}{3}\pi r^3$, for π
21. $A = \dfrac{h}{2}(b_1 + b_2)$, for h (an area formula)
22. $A = \dfrac{h}{2}(b_1 + b_2)$, for b_2
23. $F = \dfrac{mv^2}{r}$, for m (a physics formula)
24. $F = \dfrac{mv^2}{r}$, for v^2
25. $A = \dfrac{q_1 + q_2 + q_3}{n}$, for n
    (*Hint:* First clear the fraction.)
26. $r = \dfrac{s + t}{d}$, for d
27. $v = \dfrac{d_2 - d_1}{t}$, for t (a physics formula)
28. $v = \dfrac{s_2 - s_1}{m}$, for m
29. $v = \dfrac{d_2 - d_1}{t}$, for d_1
30. $v = \dfrac{s_2 - s_1}{m}$, for s_1
31. $r = m + mnp$, for m
32. $p = x - xyz$, for x
33. $y = ab - ac^2$, for a
34. $d = mn - mp^3$, for m

Problem Solving

35. The area of a parallelogram is 72 cm^2. The height of the figure is 6 cm. How long is the base?
36. The area of a parallelogram is 78 cm^2. The base of the figure is 13 cm. What is the height?
37. You are going to buy certificates of deposit at a bank. They pay 7% simple interest. You want your money to earn \$110 in a year. How much will you have to invest?
38. You have \$250 to invest for 6 months, and you expect it to earn at least \$8 in that time. What rate of simple interest will your money have to earn?
39. A garden is being constructed in the shape of a trapezoid. The dimensions are as shown in the figure. The unknown dimension is to be such that the area of the garden is 90 ft^2. Find that unknown dimension.

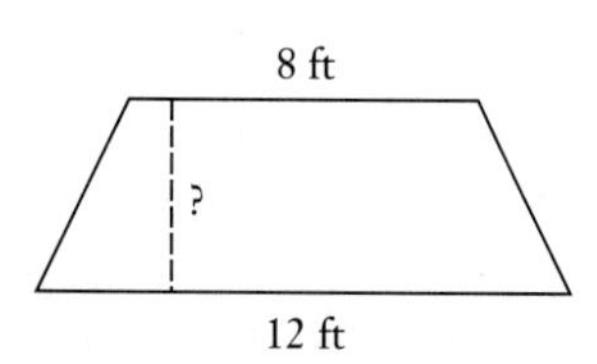

40. A rectangular garden is being constructed. There is 76 ft of fencing available, so the perimeter must be 76 ft at most. The width of the garden is to be 13 ft. What should the length be, in order to use just 76 ft of fence? (See the formula in Exercise 11.)

41. You are going to invest \$1600 at simple interest at 9%. How long will it take for your investment to be worth \$2608?

42. You are going to invest \$950 at simple interest at 7%. How long will it take for your investment to be worth \$1349?

43. The density of iron is 7.5 g/cm^3. A ship's anchor occupies a volume of 61.5 cm^3. Assuming that the anchor is pure iron, determine its mass.

44. The density of copper is 8.93 g/cm^3. A solid copper bar occupies a volume of 40.185 cm^3. Determine its mass.

SKILL MAINTENANCE

45. What percent of 5800 is 4176?

46. Simplify: $-5a + 9b - (3a - 4b)$.

47. Subtract: $-72.5 - (-14.06)$.

48. Simplify: $\frac{45x}{15x}$.

SYNTHESIS

Solve.

49. $s = v_i t + \frac{1}{2}at^2$, for a

50. $A = 4lw + w^2$, for l

51. $\frac{P_1V_1}{T_1} = \frac{P_2V_2}{T_2}$, for V_1

52. $\frac{P_1V_1}{T_1} = \frac{P_2V_2}{T_2}$, for T_2

53. $x = \frac{a}{b + c}$, for c

54. $m = \frac{(d/e)}{(e/f)}$, for d

55. $\frac{a}{b} = \frac{c}{d}$, for $\frac{a}{c}$

56. $m = \frac{(d/e)}{(e/f)}$, for f

57. $ab = c - bd$, for b

58. $mn - np = mp + nr$, for m

59. In Example 2, we solved the formula $I = Prt$ for t. Now use it to find how much time would be needed in order to earn \$6 on \$200 at 12% simple interest.

60. In Exercise 8, the formula $I = Prt$ was solved for r. Now use it to find what rate of interest would be required for a principal of \$120 to earn \$6.60 in half a year.

61. In Exercise 13, you solved for a^2. How might you solve for a?

62. In Exercise 19, you solved for r^3. How might you solve for r?

63. ▦ The density of copper is 8.93 g/cm^3. The mass of a roll of pennies is 177.6 g. If the diameter of a penny is 1.85 cm, how tall is a roll of pennies?

64. ▦ The density of copper is 8.93 g/cm^3. How long must a copper wire be if it is 1 cm thick and has a mass of 4280 g?

2.4 Solving Inequalities

An **inequality** is any sentence having one of the verbs $<$, $>$, $\leq$, or $\geq$. Examples are

$$-2 < 0, \quad x \leq 4, \quad x + 3 > 6, \quad \text{and} \quad 16 - 7y \geq 10y - 4.$$

Some replacements for an inequality make it true and some make it false. A replacement that makes it true is called a *solution*. The set of all solutions is called the *solution set.* When we have found the set of all solutions of an inequality, we say that we have *solved* the inequality.

EXAMPLES Determine whether the given number is a solution of the inequality.

1. $x + 3 < 6$; 5 We substitute and get $5 + 3 < 6$, or $8 < 6$, a false sentence. Therefore, 5 is not a solution.
2. $2x - 3 \geq -3$; 1 We substitute and get $2(1) - 3 \geq -3$, or $-1 \geq -3$, a true sentence. Therefore, 1 is a solution. ■

A *graph* of an inequality is a drawing that represents its solutions. An inequality in one variable can be graphed on a number line.

EXAMPLE 3 Graph $x < 4$ on a number line.

Solution The solutions are all real numbers less than 4, so we shade all numbers less than 4. Note that 4 is not a solution. We indicate this by using an open circle at 4.

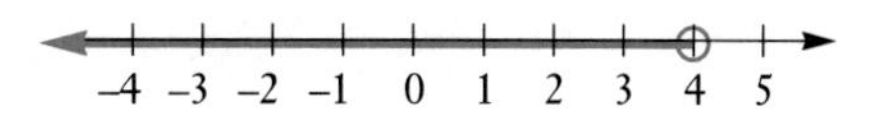

The solution set can be named as follows, using *set-builder notation* (see Chapter 1):

$$\{x | x < 4\}.$$

This is read

"The set of all x such that x is less than 4." ■

Interval Notation

Another way to list solutions to inequalities in one variable is to use **interval notation.** Pay special attention to the manner in which parentheses, ( ), and brackets, [ ], are used.

If a and b are real numbers such that $a < b$, we define the *open interval* (a, b) as follows:

$$(a, b) \text{ is the set of all numbers } x \text{ such that } a < x < b,$$

or

$$\{x | a < x < b\}.$$

Its graph excludes the endpoints:

The *closed interval* $[a, b]$ is defined as follows:

$$[a, b] \text{ is the set of all } x \text{ such that } a \leq x \leq b,$$

or

$$\{x | a \leq x \leq b\}.$$

Its graph is as follows.

Note that the endpoints are included. For example, the graph of $[-2, 3]$ is as follows:

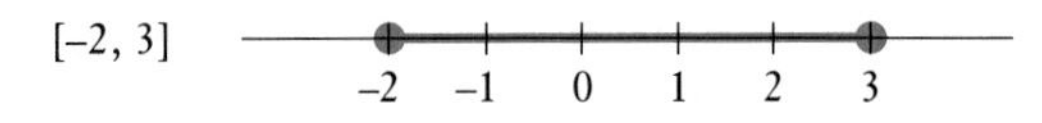

There are two kinds of *half-open intervals* defined as follows:

1. $(a, b] = \{x | a < x \leq b\}$. This is open on the left. Its graph is as follows:

2. $[a, b) = \{x | a \leq x < b\}$. This is open on the right. Its graph is as follows:

We use the symbols ∞ and $-\infty$ to represent positive and negative infinity, respectively. Thus the notation (a, ∞) represents the set of all real numbers greater than a, and $(-\infty, a)$ represents the set of all real numbers less than a.

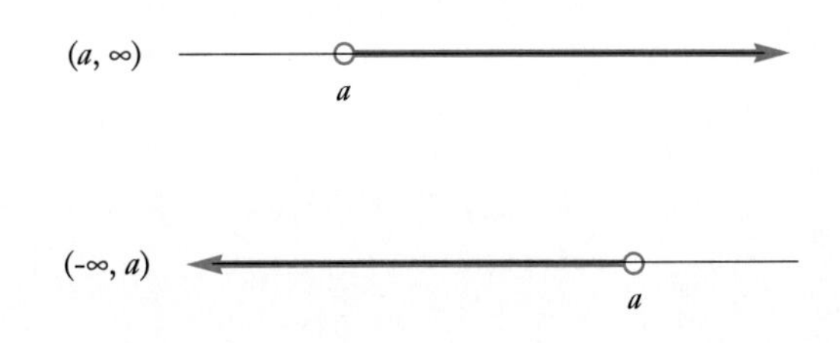

The notations $[a, \infty)$ and $(-\infty, a]$ are used when we wish to include the endpoint a.

EXAMPLE 4 Graph $y \geq -2$ on a number line and write the solution set using both set-builder and interval notations.

Solution Using set-builder notation, we write the solution set as $\{y | y \geq -2\}$.
Using interval notation, we write the solution set as $[-2, \infty)$.

To graph the solution, we shade all numbers to the right of -2 and use a solid circle at -2 to indicate that it is also a solution.

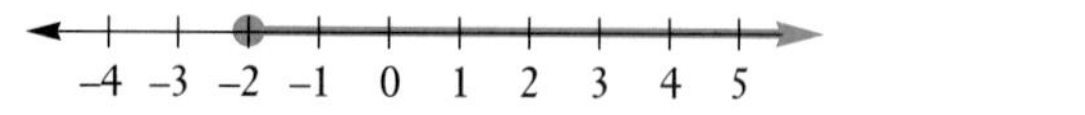

The Addition Principle

There is an addition principle for solving inequalities, similar to the one for solving equations.

> **THE ADDITION PRINCIPLE FOR INEQUALITIES**
>
> If the same number is added on both sides of a true inequality, another true inequality is obtained.
>
> If $a < b$ is true, then $a + c < b + c$ is true.
> If $a > b$ is true, then $a + c > b + c$ is true.
> If $a \leq b$ is true, then $a + c \leq b + c$ is true.
> If $a \geq b$ is true, then $a + c \geq b + c$ is true.

As with equations, we try to get the variable alone on one side in order to determine solutions easily.

EXAMPLE 5 Solve $x + 5 > 3$. Then graph.

Solution

$$x + 5 > 3$$
$$x + 5 + (-5) > 3 + (-5) \quad \text{Using the addition principle, add } -5$$
$$x > -2$$

Any number greater than -2 is a solution. Thus the solution set is

$$\{x \mid x > -2\}, \qquad \text{or} \qquad (-2, \infty),$$

and the graph of the inequality is as follows:

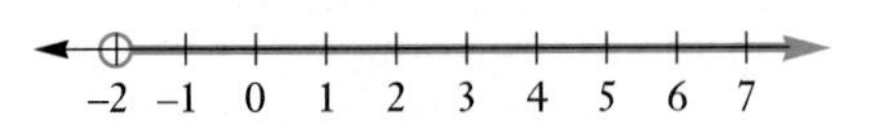

We cannot check all the solutions of an inequality by substitution, because there are too many of them. A partial check could be done by substituting a number greater than -2, say -1, into the original inequality:

$$\begin{array}{c|c} x + 5 > 3 & \\ \hline -1 + 5 & 3 \\ 4 & \end{array} \quad \text{TRUE}$$

Since $4 > 3$ is true, -1 is a solution. Any number greater than -2 is a solution. ■

EXAMPLE 6 Solve $4x - 1 \geq 5x - 2$. Then graph.

Solution

$$4x - 1 \geq 5x - 2$$
$$4x - 1 + 2 \geq 5x - 2 + 2 \quad \textbf{Adding 2}$$
$$4x + 1 \geq 5x \quad \textbf{Simplifying}$$
$$4x + 1 - 4x \geq 5x - 4x \quad \textbf{Adding } -4x$$
$$1 \geq x \quad \textbf{Simplifying}$$

We know that $1 \geq x$ has the same meaning as $x \leq 1$. Thus any number less than or equal to 1 is a solution. We can express the solution set as $\{x|1 \geq x\}$ or as $\{x|x \leq 1\}$. The latter is probably used most often. With interval notation, the solution is $(-\infty, 1]$. The graph is as follows:

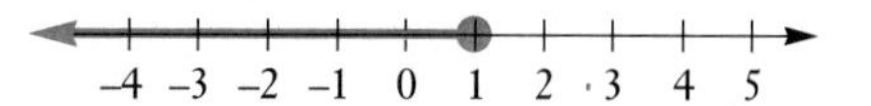

■

The Multiplication Principle

The multiplication principle for inequalities is somewhat different from the principle for equations.

Consider this true inequality:

$$-4 < 9.$$

If we multiply both numbers by 2, we get another true inequality:

$$-4(2) < 9(2), \quad \text{or} \quad -8 < 18.$$

If we multiply both numbers by *negative* two, we get a false inequality:

$$-4(-2) < 9(-2), \quad \text{or} \quad 8 < -18.$$

This is because negation reverses relative position on the number line. However, if we now reverse the inequality symbol, we get a true inequality:

$$8 > -18.$$

The $<$ symbol has been reversed!

> **THE MULTIPLICATION PRINCIPLE FOR INEQUALITIES**
>
> If we multiply on both sides of a true inequality by a positive number, we get another true inequality. If we multiply by a negative number and the inequality symbol is reversed, we get another true inequality.
>
> | For any positive real number c: | For any negative real number c: |
> |---|---|
> | If $a < b$ is true, then $ac < bc$ is true. | If $a < b$ is true, then $ac > bc$ is true. |
> | If $a > b$ is true, then $ac > bc$ is true. | If $a > b$ is true, then $ac < bc$ is true. |
> | If $a \leq b$ is true, then $ac \leq bc$ is true. | If $a \leq b$ is true, then $ac \geq bc$ is true. |
> | If $a \geq b$ is true, then $ac \geq bc$ is true. | If $a \geq b$ is true, then $ac \leq bc$ is true. |

The important thing to remember is that if you multiply by a negative number, you must reverse the inequality symbol.

When we solve an inequality using the multiplication principle, we can multiply by any number except zero.

EXAMPLE 7 Solve $3y < \frac{3}{4}$. Then graph.

Solution

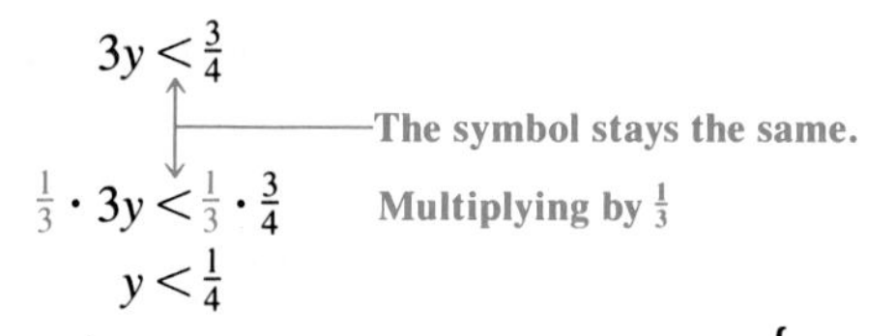

Any number less than $\frac{1}{4}$ is a solution. The solution set is $\left\{y \middle| y < \frac{1}{4}\right\}$, or $\left(-\infty, \frac{1}{4}\right)$. The graph is as follows:

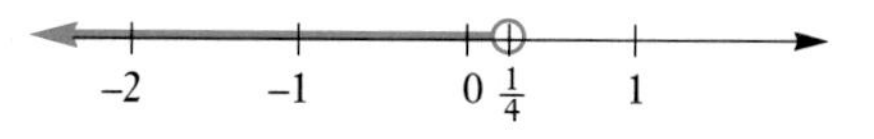

■

EXAMPLE 8 Solve $-4x \leq \frac{4}{5}$. Then graph.

Solution

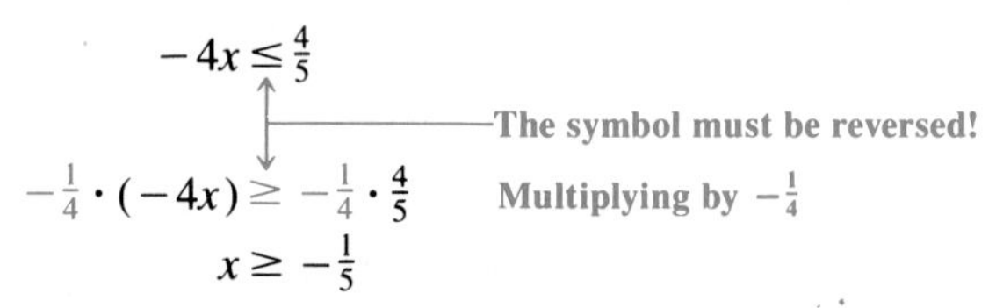

Any number greater than or equal to $-\frac{1}{5}$ is a solution. The solution set is $\left\{x \middle| x \geq -\frac{1}{5}\right\}$, or $\left[-\frac{1}{5}, \infty\right)$. The graph is as follows:

Using the Principles Together

We use the addition and multiplication principles together in solving inequalities in much the same way as in solving equations.

EXAMPLE 9 Solve: $-3(x + 8) - 5x > 4x - 9$.

Solution

$$-3(x + 8) - 5x > 4x - 9$$

$$-3x - 24 - 5x > 4x - 9 \quad \text{Using the distributive law}$$

$$-24 - 8x > 4x - 9$$

$$-24 - 8x + 8x > 4x - 9 + 8x \quad \text{Adding } 8x \text{ on both sides}$$

$$-24 > 12x - 9$$

$$-24 + 9 > 12x - 9 + 9 \quad \text{Adding 9}$$

$$-15 > 12x$$

The symbol stays the same.

$$-\frac{5}{4} > x \quad \text{Multiplying by } \frac{1}{12} \text{ and simplifying}$$

The solution set is $\left\{x \middle| -\frac{5}{4} > x\right\}$, or $\left\{x \middle| x < -\frac{5}{4}\right\}$, or $\left(-\infty, -\frac{5}{4}\right)$.

Problem Solving

EXAMPLE 10 In a business course, there will be three tests. You must get a total score of at least 270 to earn an A. You get scores of 91 and 86 on the first two tests. What scores on the last test will give you an A?

Solution

1. **Familiarize.** Making a table can help in the familiarization process.

Test 1	91
Test 2	86
Test 3	x
Total	270 or better

Note that we have let x = the score on the third test.

2. **Translate.** The words *at least* translate to $\geq$. We form an inequality:

$$91 + 86 + x \geq 270.$$

3. **Carry out.** We solve the inequality, obtaining

$$177 + x \geq 270 \quad \text{Collecting like terms}$$
$$x \geq 93. \quad \text{Adding } -177$$

4. **Check.** If you get 93 on the third test, then your total score will be $91 + 86 + 93$, which is 270. Any higher score will also give you an A.

5. **State.** A score of 93 or better will give you an A. ■

In translating phrases to inequalities, look for phrases like *is at least* 27 (≥ 27), *never exceeds* 90 (≤ 90), *is better than* 17 (> 17, or < 17, depending on the context), and *is at most* 32 (≤ 32). Of course, the phrases *is greater than* and *is less than* are also used.

EXERCISE SET 2.4

Determine whether the given numbers are solutions of the inequality.

1. $x - 2 \geq 6;\ -4, 0, 4, 8$
2. $3x + 5 \leq -10;\ -5, -10, 0, 27$
3. $t - 8 > 2t - 3;\ 0, -8, -9, -3$
4. $5y - 7 < 5 - y;\ 2, -3, 0, 3$

Graph each of the following inequalities, and write the solution sets using both set-builder and interval notation.

5. $x > 4$
6. $y < 5$
7. $t \leq 6$
8. $x \geq -4$
9. $y < -3$
10. $t > -2$
11. $x \geq -6$
12. $x \leq -5$

Solve. Then graph.

13. $x + 8 > 3$
14. $x + 5 > 2$
15. $y + 3 < 9$
16. $y + 4 < 10$
17. $a + 9 \leq -12$
18. $a + 7 \leq -13$
19. $t + 14 \geq 9$
20. $x - 9 \leq 10$
21. $y - 8 > -14$
22. $y - 9 > -18$
23. $x - 11 \leq -2$
24. $y - 18 \leq -4$
25. $8x \geq 24$
26. $9t < -81$
27. $0.3x < -18$
28. $0.5x < 25$
29. $-9x \geq -8.1$
30. $-8y \leq 3.2$
31. $-\frac{3}{4}x \geq -\frac{5}{8}$
32. $-\frac{5}{6}y \leq -\frac{3}{4}$
33. $2x + 7 < 19$
34. $5y + 13 > 28$
35. $5y + 2y \leq -21$
36. $-9x + 3x \geq -24$
37. $2y - 7 < 5y - 9$
38. $8x - 9 < 3x - 11$
39. $0.4x + 5 \leq 1.2x - 4$
40. $0.2y + 1 > 2.4y - 10$
41. $3x - \frac{1}{8} \leq \frac{3}{8} + 2x$
42. $2x - 3 < \frac{13}{4}x + 10 - 1.25x$

Solve.

43. $4(3y - 2) \geq 9(2y + 5)$
44. $4m + 5 \geq 14(m - 2)$
45. $3(2 - 5x) + 2x < 2(4 + 2x)$
46. $2(0.5 - 3y) + y > (4y - 0.2)8$
47. $5[3m - (m + 4)] > -2(m - 4)$
48. $[8x - 3(3x + 2)] - 5 \geq 3(x + 4) - 2x$
49. $3(r - 6) + 2 > 4(r + 2) - 21$
50. $5(t + 3) + 9 < 3(t - 2) + 6$
51. $19 - (2x + 3) \leq 2(x + 3) + x$
52. $13 - (2c + 2) \geq 2(c + 2) + 3c$
53. $\frac{1}{4}(8y + 4) - 17 < -\frac{1}{2}(4y - 8)$
54. $\frac{1}{3}(6x + 24) - 20 > -\frac{1}{4}(12x - 72)$
55. $2[4 - 2(3 - x)] - 1 \geq 4[2(4x - 3) + 7] - 25$
56. $5[3(7 - t) - 4(8 + 2t)] - 20 \leq -6[2(6 + 3t) - 4]$

57. $\frac{2}{3}(2x - 1) > 10$

58. $\frac{4}{5}(3x + 4) < 20$

59. $\frac{3}{4}(3 + 2x) + 1 \geq 13$

60. $\frac{7}{8}(5 - 4x) - 17 \geq 38$

61. $\frac{3}{4}\left(3x - \frac{1}{2}\right) - \frac{2}{3} < \frac{1}{3}$

62. $\frac{2}{3}\left(\frac{7}{8} - 4x\right) - \frac{5}{8} < \frac{3}{8}$

63. $0.7(3x + 6) \geq 1.1 - (x + 2)$

64. $0.9(2x + 8) < 20 - (x + 5)$

65. $a + (a - 3) \leq (a + 2) - (a + 1)$

66. $0.8 - 4(b - 1) > 0.2 + 3(4 - b)$

Solve.

67. A car rents for $30 per day plus 20¢ per mile. You are on a daily budget of $96. What mileages will allow you to stay within the budget?

68. A car can be rented for $35 per day with unlimited mileage, or for $28 per day plus 19¢ per mile. For what daily mileages would the unlimited mileage plan save you money?

69. You are taking a history course. There will be four tests. You have scores of 89, 92, and 95 on the first three tests. You must score a total of at least 360 to earn an A. What scores on the last test will give you an A?

70. You are taking a science course. There will be five tests, each worth 100 points. You have scores of 94, 90, and 89 on the first three. You must score a total of at least 450 to earn an A. What scores on your fourth test will still keep you eligible for an A?

71. You are going to invest $25,000, part at 14% and the rest at 16%. What is the most that can be invested at 14% in order to make at least $3600 interest per year?

72. You are going to invest $20,000, part at 12% and the rest at 16%. What is the most that can be invested at 12% in order to make at least $3000 interest per year?

73. In planning for a college dance, you find that one band will play for $250 plus 50% of the total ticket sales. Another band will play for a flat fee of $550. For the first band to produce more profit for the school than the other band, what is the highest price you can charge per ticket, assuming that 300 people will attend?

74. A bank offers two checking account plans. Plan A charges a base service charge of $2.00 per month plus 15¢ per check. Plan B charges a base service charge of $4.00 per month plus 9¢ per check. For what numbers of checks per month will plan B be better than plan A?

75. A medical insurance company offers two plans. With plan A, you pay the first $100 of your medical bills and the insuror pays 80% of the rest. With plan B, you pay the first $250 of your medical bills and the insurance company pays 90% of the rest. For what total amounts of medical bills will plan B save you money?

76. You can spend $3.50 at the laundromat washing your clothes, or you can have the laundromat do the laundry for 40¢ per pound. For what weights of clothes will it save you money to wash your clothes yourself?

SKILL MAINTENANCE

77. Simplify: $|-16|$.

78. Simplify: $-|-4|$.

SYNTHESIS

Solve. Assume that a, b, c, d, and m are positive constants.

79. $3ax + 2x \geq 5ax - 4$; assume $a > 1$

80. $6by - 4y \leq 7by + 10$

81. $a(by - 2) \geq b(2y + 5)$; assume $a > 2$

82. $c(6x - 4) < d(3 + 2x)$; assume $3c > d$

83. $c(2 - 5x) + dx > m(4 + 2x)$; assume $5c + 2m < d$

84. $a(3 - 4x) + cx < d(5x + 2)$; assume $c > 4a + 5d$

Determine whether each of the following statements is true or false. If false, give a counterexample.

85. For any real numbers a, b, c, and d, if $a < b$ and $c < d$, then $a - c < b - d$.

86. For any real numbers x and y, if $x < y$, then $x^2 < y^2$.

87. Determine whether the inequalities

$$x < 3 \qquad \text{and} \qquad x + \frac{1}{x} < 3 + \frac{1}{x}$$

are equivalent. Give reasons to support your answer.

88. Determine whether the inequalities

$$x < 3 \qquad \text{and} \qquad 0 \cdot x < 0 \cdot 3$$

are equivalent. Give reasons to support your answer.

2.5 Compound Inequalities

We now consider compound inequalities, that is, sentences formed by two or more inequalities, joined with the word *and* or the word *or*.

Intersections of Sets and Conjunctions of Sentences

The *intersection* of two sets A and B is the set of all members that are common to both A and B. We denote the intersection of sets A and B as

$$A \cap B.$$

The intersection of two sets is often pictured as follows:

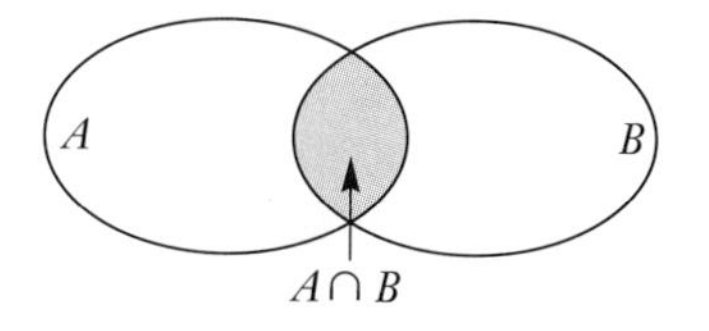

EXAMPLE 1 Find the intersection: $\{1, 2, 3, 4, 5\} \cap \{-2, -1, 0, 1, 2, 3\}$.

Solution The numbers 1, 2, and 3 are common to the two sets, so the intersection is $\{1, 2, 3\}$. ■

When two or more sentences are joined by the word *and* to make a compound sentence, the new sentence is called a **conjunction** of the sentences. The following is a conjunction of inequalities:

$$-2 < x \qquad \textit{and} \qquad x < 1.$$

For a conjunction to be true, all of the individual sentences must be true. The solution set of a conjunction is the intersection of the solution sets of the individual sentences. Let us consider the conjunction

$$-2 < x \qquad \textit{and} \qquad x < 1.$$

The graphs of each separate sentence are shown below, and the intersection is the last graph. We use both set-builder and interval notations.

$\{x | -2 < x\}$; $(-2, \infty)$

$\{x | x < 1\}$; $(-\infty, 1)$

$\{x | -2 < x\} \cap \{x | x < 1\}$; $(-2, 1)$

The conjunction $-2 < x$ and $x < 1$ can be abbreviated by $-2 < x < 1$. Thus the interval $(-2, 1)$ may be represented as $\{x | -2 < x < 1\}$, the set of all numbers that are simultaneously greater than -2 and less than 1.

> The conjunction $a < x$ *and* $x < b$ can be abbreviated as $a < x < b$.
> The conjunction $b > x$ *and* $x > a$ can be abbreviated as $b > x > a$.

If sets have no common members, we say that their intersection is the empty set, $\emptyset$. The following two sets have an empty intersection:

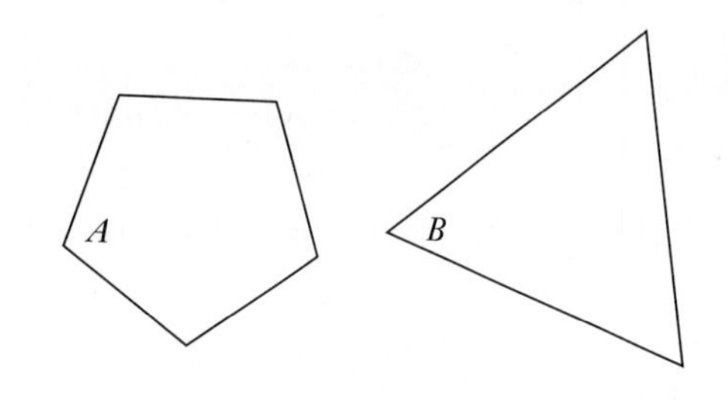

$$A \cap B = \emptyset.$$

EXAMPLE 2 Solve the conjunction $-2 \leq x + 1$ *and* $2x - 3 < 5$, and then graph the solution.

Solution We solve the two inequalities separately:

$$-2 \leq x + 1 \quad \textit{and} \quad 2x - 3 \leq 5;$$
$$-3 \leq x \quad \textit{and} \quad 2x < 8;$$
$$-3 \leq x \quad \textit{and} \quad x < 4.$$

The word *and* corresponds to set intersection. The solution set is thus the intersection of the solution set of $-3 \leq x$ and the solution set of $x < 4$:

$$\{x \mid -3 \leq x\} \cap \{x \mid x < 4\}, \qquad \text{or, in interval notation,} \qquad [-3, \infty) \cap (-\infty, 4).$$

The graph is the intersection of the individual graphs:

$\{x \mid -3 \leq x\}$; $[-3, \infty)$

–6 –5 –4 –3 –2 –1 0 1 2 3 4 5 6

$\{x \mid x < 4\}$; $(-\infty, 4)$

–6 –5 –4 –3 –2 –1 0 1 2 3 4 5 6

$$\{x \mid -3 \leq x\} \cap \{x \mid x < 4\} = \{x \mid -3 \leq x < 4\},$$

–6 –5 –4 –3 –2 –1 0 1 2 3 4 5 6

or, in interval notation, $[-3, \infty) \cap (-\infty, 4) = [-3, 4)$. ■

EXAMPLE 3 Solve: $-3 < 2x + 5 < 7$.

Solution

Method 1. We write the conjunction with the word *and:*

$$-3 < 2x + 5 \qquad \textit{and} \qquad 2x + 5 < 7.$$

Next we solve the individual inequalities separately.

$$\begin{aligned}
-3 &< 2x + 5 & \textit{and} \qquad 2x + 5 &< 7 \\
-3 + (-5) &< 2x + 5 + (-5) & \textit{and} \qquad 2x + 5 + (-5) &< 7 + (-5) \\
-8 &< 2x & \textit{and} \qquad 2x &< 2 \\
-4 &< x & \textit{and} \qquad x &< 1
\end{aligned}$$

We now abbreviate the answer:

$$-4 < x < 1.$$

The solution set is $\{x \mid -4 < x < 1\}$, or, in interval notation, $(-4, 1)$.

Method 2. With Method 1, we did the same thing to each inequality. We can shorten the writing as follows:

$$\begin{aligned}
-3 < \quad 2x + 5 \quad &< 7 \\
-3 + (-5) < 2x + 5 + (-5) &< 7 + (-5) \qquad \text{Adding } -5 \\
-8 < \quad 2x \quad &< 2 \\
-4 < \quad x \quad &< 1. \qquad \text{Multiplying by } \tfrac{1}{2}
\end{aligned}$$

The solution set is $\{x \mid -4 < x < 1\}$, or $(-4, 1)$. ■

EXAMPLE 4 Solve: $3 \le 5 - 2x < 7$.

Solution

$$3 \le 5 - 2x < 7$$

$$3 - 5 \le 5 - 2x - 5 < 7 - 5 \quad \text{Adding } -5$$

$$-2 \le -2x < 2 \quad \text{Simplifying}$$

Symbols must be reversed!

$$-\tfrac{1}{2}(-2) \ge -\tfrac{1}{2}(-2x) > -\tfrac{1}{2}(2) \quad \text{Multiplying by } -\tfrac{1}{2} \text{ and reversing the inequality signs}$$

$$1 \ge x > -1 \quad \text{Simplifying}$$

The solution set is $\{x \mid 1 \ge x > -1\}$, or $\{x \mid -1 < x \le 1\}$; in interval notation it is $(-1, 1]$. ■

Unions of Sets and Disjunctions of Sentences

The *union* of two sets A and B is formed by putting the sets together. We denote the union of sets A and B as

$$A \cup B.$$

The union of two sets is often pictured as follows:

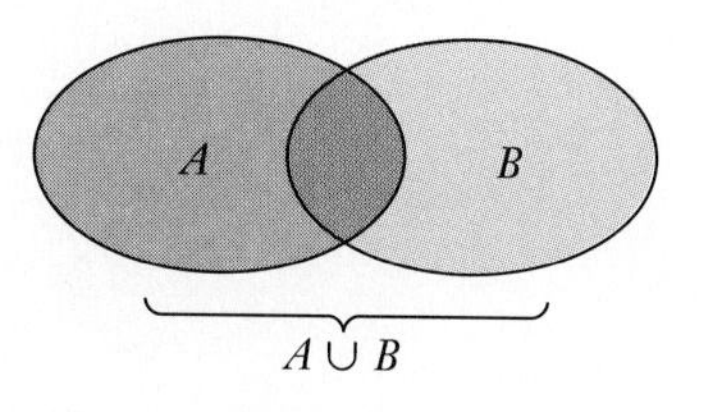

EXAMPLE 5 Find the union: $\{2, 3, 4\} \cup \{3, 5, 7\}$.

Solution The numbers in either or both sets are 2, 3, 4, 5, and 7, so the union is $\{2, 3, 4, 5, 7\}$. ■

When two or more sentences are joined by the word *or* to make a compound sentence, the new sentence is called a **disjunction** of the sentences. Here are three examples:

$$x < -3 \quad or \quad x > 3;$$

y is an odd number *or* y is a prime number;

$$x < 0 \quad or \quad x = 0 \quad or \quad x > 0.$$

For a disjunction to be true, at least one of the individual sentences must be true. The solution set of a disjunction is the union of the individual solution sets. Consider the disjunction

$$x < -3 \qquad or \qquad x > 3.$$

$\{x|x < -3\}$; $(-\infty, -3)$

$\{x|x > 3\}$; $(3, \infty)$

$\{x|x < -3\} \cup \{x|x > 3\} = \{x|x < -3 \text{ or } x > 3\}$;

$(-\infty, -3) \cup (3, \infty)$ cannot be simplified.

EXAMPLE 6 Graph: $x < -3 \quad or \quad x \geq 2.$

Solution We graph $x < -3$. We also graph $x \geq 2$. Then we take the union of the two graphs.

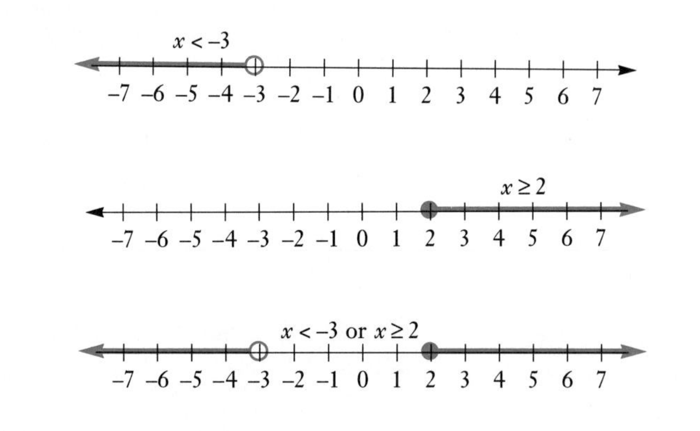

■

A compound inequality, such as

$$x < -3 \qquad or \qquad x \geq 2$$

as in Example 6, *cannot* be abbreviated to one like $2 \leq x < -3$ because to do so would be to say that x is simultaneously less than -3 and greater than or equal to 2. When the word *or* appears, you must keep that word. There is usually no short way to write a disjunction.

EXAMPLE 7 Solve: $-2x - 5 < -2 \quad or \quad x - 3 < -10$

Solution We solve the individual inequalities separately, but we continue to write the word *or.*

$$
\begin{array}{rcl}
-2x - 5 < -2 & or & x - 3 < -10 \\
-2x - 5 + 5 < -2 + 5 & or & x - 3 + 3 < -10 + 3 \\
-2x < 3 & or & x < -7 \\
x > -\frac{3}{2} & or & x < -7
\end{array}
$$

↑ Keep the word "or."

The solution set is $\left\{x \middle| x > -\frac{3}{2}\right\} \cup \{x|x < -7\}$, also written

$$\left\{x \middle| x > -\tfrac{3}{2} \text{ or } x < -7\right\}, \quad or \quad (-\infty, -7) \cup \left(-\tfrac{3}{2}, \infty\right).$$

■

EXERCISE SET 2.5

Find the intersection.

1. $\{5, 6, 7, 8\} \cap \{4, 6, 8, 10\}$
2. $\{9, 10, 27\} \cap \{8, 10, 38\}$
3. $\{2, 4, 6, 8\} \cap \{1, 3, 5\}$
4. $\emptyset \cap \{2, 4, 6\}$
5. $\{1, 2, 3, 4\} \cap \{1, 2, 3, 4\}$
6. $\{8, 9, 10\} \cap \emptyset$

Graph.

7. $1 < x < 6$
8. $0 \le y \le 3$
9. $-7 \le y \le -3$
10. $-9 \le x < -5$
11. $-4 \le -x < 3$
12. $x > -8 \quad and \quad x < -3$
13. $6 > -x \ge -2$
14. $x > -4 \quad and \quad x < 2$
15. $5 > x \ge -2$
16. $3 > x \ge 0$
17. $x < 5 \quad and \quad x \ge 1$
18. $x \ge -2 \quad and \quad x < 2$

Solve.

19. $-2 < x + 2 < 8$
20. $-1 < x + 1 \le 6$
21. $1 < 2y + 5 \quad and \quad y - 5 \le -3$
22. $3 \le 5x + 3 \quad and \quad 2x - 3 \le -1$
23. $-10 \le 3x - 5 \le -1$
24. $-18 \le -2x - 7 < 0$
25. $2 < x + 3 \quad and \quad 5 - 2x \ge -13$
26. $-6 \le x + 1 \quad and \quad 5 - 2x \ge -13$
27. $-6 \le 2x - 3 < 6$
28. $4 > -3m - 7 \ge 2$
29. $-\frac{1}{2} < \frac{1}{4}x - 3 \quad and \quad \frac{1}{2}x \le 7$
30. $-3 < \frac{3}{4}x - 9 \quad and \quad \frac{2}{3}x < 10$
31. $-3 < \frac{2x - 5}{4} < 8$
32. $-4 \le \frac{7 - 3x}{5} \le 4$

Find the union.

33. $\{4, 5, 6, 7, 8\} \cup \{1, 4, 6, 11\}$
34. $\{8, 9, 27\} \cup \{2, 8, 27\}$
35. $\{2, 4, 6, 8\} \cup \{1, 3, 5\}$
36. $\{8, 9, 10\} \cup \emptyset$
37. $\{4, 8, 11\} \cup \emptyset$
38. $\emptyset \cup \emptyset$

Graph.

39. $x < -1$ *or* $x > 2$

40. $x < -2$ *or* $x > 0$

41. $x \leq -3$ *or* $x > 1$

42. $x \leq -1$ *or* $x > 3$

43. $x < -8$ *or* $x > -2$

44. $t \leq -10$ *or* $t \geq -5$

Solve.

45. $x + 7 < -2$ *or* $x + 7 > 2$

46. $x + 9 < -4$ *or* $x + 9 > 4$

47. $2x - 8 \leq -3$ *or* $x - 8 \geq 3$

48. $x + 7 \leq -2$ *or* $3x - 7 \geq 2$

49. $3x - 9 < -5$ *or* $x - 9 > 6$

50. $4x - 4 < -8$ *or* $x - 4 > 12$

51. $7 > -4x + 5$ *or* $10 \leq -4x + 5$

52. $6 > 2x - 1$ *or* $-4 \leq 2x - 1$

53. $7 - x \leq -2$ *or* $7 - x > 2$

54. $9 - x < -4$ *or* $9 - x \geq 4$

55. $-2x - 2 < -6$ *or* $-2x - 2 > 6$

56. $-3m - 7 < -5$ *or* $-3m - 7 > 5$

57. $\frac{2}{3}x - 14 < -\frac{5}{6}$ *or* $\frac{2}{3}x - 14 > \frac{5}{6}$

58. $\frac{1}{4} - 3x \leq -3.7$ *or* $\frac{1}{4} - 5x \geq 4.8$

59. $\frac{2x - 5}{6} \leq -3$ *or* $\frac{2x - 5}{6} \geq 4$

60. $\frac{7 - 3x}{5} < -4$ *or* $\frac{7 - 3x}{5} > 4$

SKILL MAINTENANCE

Simplify.

61. $3\{2[x - 3(x + 4)] + 1\}$

62. $-19.2 - (-7.1)$

63. Which of the following numbers are rational?
$-7, 0, \frac{43}{10}, \sqrt{5}, 3.217, 29$

64. Which of the following are natural numbers?
$-4, 0, 2, \sqrt{7}, 9.2, \frac{11}{2}, 73$

SYNTHESIS

65. *Temperatures of liquids.* The formula

$$C = \tfrac{5}{9}(F - 32)$$

can be used to convert Fahrenheit temperatures F to Celsius temperatures C.

a) Gold is liquid for Celsius temperatures C such that $1063° \leq C < 2660°$. Find a similar such inequality for the corresponding Fahrenheit temperatures.

b) Silver is liquid for Celsius temperatures C such that $960.8° \leq C < 2180°$. Find a similar such inequality for the corresponding Fahrenheit temperatures.

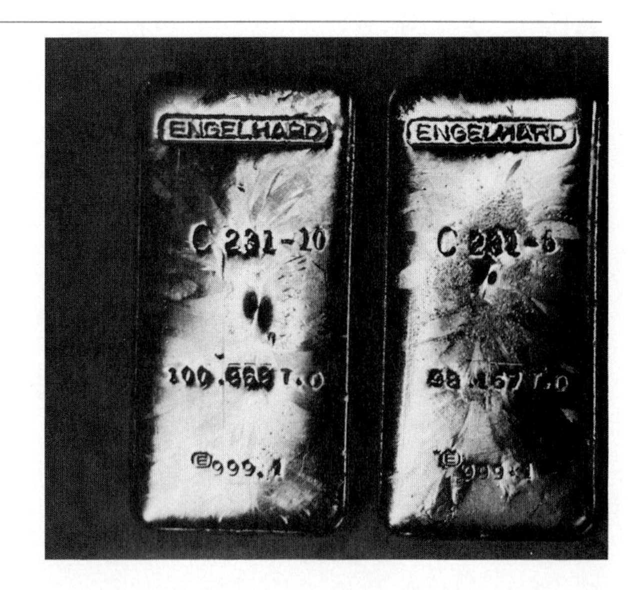

66. We say that x is *between* a and b if $a < x < b$. Find all the numbers on a number line from which you can subtract 3 and still be between -8 and 8.

67. *Women in the military ranks.* The percentage of the total active military duty force that are women has been steadily increasing. The number N of women in the active duty force t years since 1971 can be predicted by

$$N = 12{,}197.8t + 44{,}000.$$

For what years will the number of women always be at least 50,000 and at most 250,000? (Measure from the end of 1971 and make sure that $N \nless 50{,}000$ and $N \ngtr 250{,}000$ at any point in the time span).

68. *Records in the women's 100-m dash.* Florence Griffith Joyner set a world record of 10.49 sec in the women's 100-m dash in 1988. The formula

$$R = -0.03125t + 10.49$$

can be used to predict the world record in the women's 100-m dash t years after 1988. Predict (in terms of an inequality) those entire years for which the world record was between 11.5 and 10.8 sec. (Measure from the end of 1988.)

Solve and graph.

69. $4a - 2 \le a + 1 \le 3a + 4$

70. $4m - 8 > 6m + 5$ *or* $5m - 8 < -2$

71. $x - 10 < 5x + 6 \le x + 10$

72. $2[5(3 - y) - 2(y - 2)] > y + 4$

73. $-\frac{2}{15} \le \frac{2}{3}x - \frac{2}{5} \le \frac{2}{15}$

74. $2x - \frac{3}{4} < -\frac{1}{10}$ *or* $2x - \frac{3}{4} > \frac{1}{10}$

75. $3x < 4 - 5x < 5 + 3x$

76. $(x + 6)(x - 4) > (x + 1)(x - 3)$

Determine whether each of the following is true or false.

77. If $b > c$, then $b \nleq c$.

78. If $-b < -a$, then $a < b$.

79. If $c \neq a$, then $a < c$.

80. If $a < c$ and $c < b$, then $b \ngeq a$.

81. If $a < c$ and $b < c$, then $a < b$.

82. If $-a < c$ and $-c > b$, then $a < b$.

Solve.

83. $[4x - 2 < 8$ *or* $3(x - 1) < -2]$ *and* $-2 \le 5x \le 10$

84. $-2 \le 4m + 3 < 7$ *and* $[m - 5 \ge 4$ *or* $3 - m > 12]$

2.6 Equations and Inequalities with Absolute Value

Properties of Absolute Value

Intuitively, we can think of the absolute value of a number as its distance from zero on a number line. Let us recall the definition of absolute value.

The absolute value of x, denoted $|x|$, is defined as

$$|x| = \begin{cases} x & \text{if } x \geq 0, \\ -x & \text{if } x < 0. \end{cases}$$

In other words, the absolute value of a nonnegative number is the number itself. The absolute value of a negative number is the additive inverse, or opposite, of that number.

Some simple properties follow.

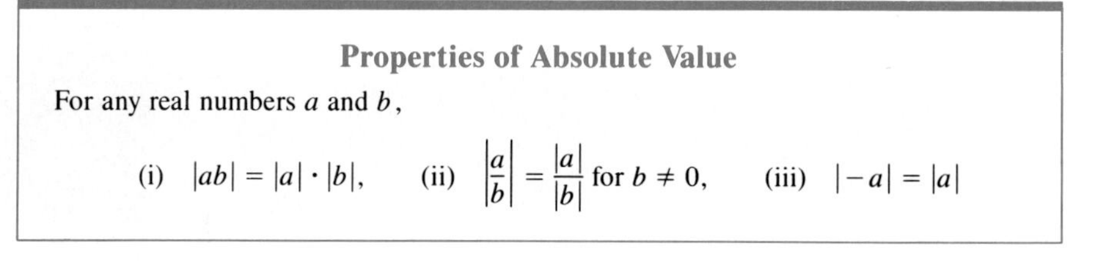

Properties of Absolute Value

For any real numbers a and b,

(i) $|ab| = |a| \cdot |b|$, (ii) $\left|\frac{a}{b}\right| = \frac{|a|}{|b|}$ for $b \neq 0$, (iii) $|-a| = |a|$

EXAMPLES Simplify, leaving as little as possible inside the absolute value signs.

1. $|5x| = |5| \cdot |x| = 5|x|$ Using property (i)

2. $|-3y| = |-3| \cdot |y| = 3|y|$

3. $|7x^2| = |7| \cdot |x^2| = 7|x^2|$

$= 7x^2$ Since x^2 is never negative for any real number x

4. $\left|\frac{6x}{-3x^2}\right| = \left|\frac{-2}{x}\right|$

$= \frac{|-2|}{|x|}$ Using property (ii)

$= \frac{2}{|x|}$ ■

Distance on a Number Line

The following number line shows that the distance between -3 and 2 is 5.

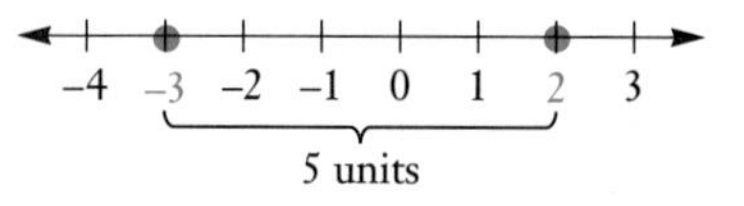

Another way to find the distance between two numbers on a number line is to take the absolute value of the difference, as follows:

$$|-3 - 2| = |-5| = 5, \quad \text{or} \quad |2 - (-3)| = |5| = 5.$$

Note that for any real numbers a and b, the distance between them is $|a - b|$ *or, equivalently,* $|b - a|$.

EXAMPLE 5 Find the distance between points having coordinates -8 and -92.

Solution

$$|-8-(-92)| = |84| = 84, \qquad \text{or} \qquad |-92-(-8)| = |-84| = 84$$ ■

EXAMPLE 6 Find the distance between x and 0.

Solution

$$|x - 0| = |x|$$ ■

Equations with Absolute Value

EXAMPLE 7 Solve $|x| = 4$. Then graph, using a number line.

Solution Since $|x| = |x - 0|$ is the distance from x to 0 (see Example 6), the solutions of the equation are those numbers x whose distance from 0 is 4. Those numbers are 4 and -4. The solution set is $\{4, -4\}$. The graph consists of just two points, as shown.

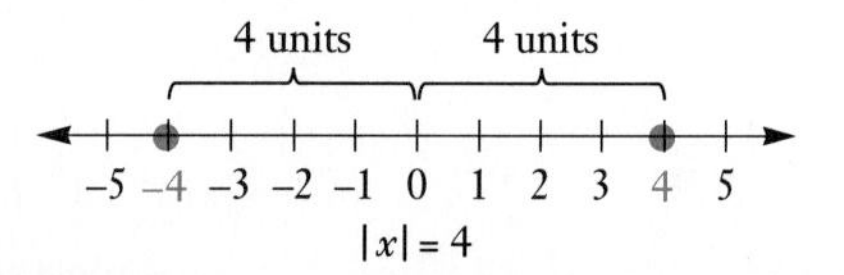

The following reasoning can also be used: The quantity inside the absolute value signs is either negative or nonnegative. If the quantity is negative it must be -4, and if it is nonnegative it must be 4, since only -4 and 4 have an absolute value of 4. The solution set is $\{4, -4\}$. ■

EXAMPLE 8 Solve: $|x| = 0$.

Solution The only number whose absolute value is 0 is 0 itself. Thus the solution is 0. The solution set is $\{0\}$. ■

EXAMPLE 9 Solve: $|x| = -7$.

Solution The absolute value of a number is always nonnegative. There is no number whose absolute value is -7. There is no solution. The solution set is $\emptyset$. ■

Examples 7–9 lead us to the following principles for solving equations with absolute value.

Absolute Value Principles for Equations

For any positive number a and any algebraic expression X,

a) If $|X| = a$, then $X = -a$ or $X = a$.
b) If $|X| = 0$, then $X = 0$.
c) If a is negative, $|X| = a$ has no solution.

Using absolute value principle (a), we can "trade in" one equation with the absolute value symbol for two equations with no absolute value symbol. Note that the expression inside the absolute value signs is often more than a single variable.

EXAMPLE 10 Solve: $|x - 2| = 3$.

Solution There are two methods we can use.

Method 1. This method uses absolute value principle (a), replacing X by $x - 2$ and a by 3. We form two equations and solve each separately.

$$|X| = a$$
$$|x - 2| = 3$$
$$x - 2 = -3 \quad or \quad x - 2 = 3 \qquad \text{Absolute value principle (a)}$$
$$x = -1 \quad or \quad x = 5$$

The solutions are -1 and 5.

Method 2. This method allows us to see the meaning of the solutions graphically. The solution set consists of those numbers that are 3 units from 2 on the number line.

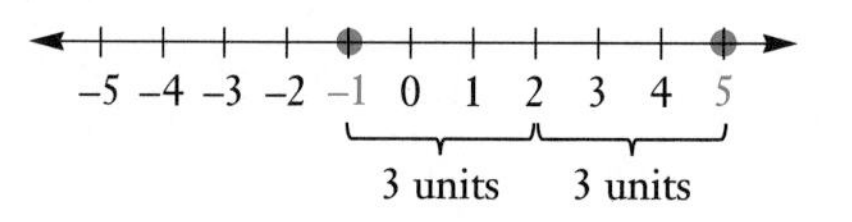

The solutions are -1 and 5.

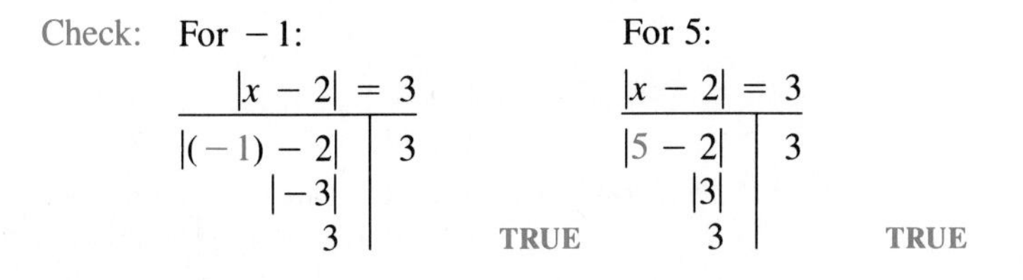

The solution set is $\{-1, 5\}$. ■

EXAMPLE 11 Solve: $|2x + 5| = 13$.

Solution We again use principle (a), this time replacing X by $2x + 5$ and a by 13:

$$|X| = a$$
$$|2x + 5| = 13$$
$$2x + 5 = -13 \quad \text{or} \quad 2x + 5 = 13 \qquad \text{Absolute value principle (a)}$$
$$2x = -18 \quad \text{or} \quad 2x = 8$$
$$x = -9 \quad \text{or} \quad x = 4.$$

The check is left to the student. The solutions are -9 and 4. The solution set is $\{-9, 4\}$. ■

Inequalities with Absolute Value

Our methods for solving equations with absolute value can be extended for solving inequalities.

EXAMPLE 12 Solve $|x| < 4$. Then graph.

Solution The solutions of $|x| < 4$ are the solutions of $|x - 0| < 4$ and are those numbers x whose distance from 0 is less than 4. By substituting or by looking at the number line, we can see that numbers like -3, -2, -1, $-\frac{1}{2}$, $-\frac{1}{4}$, 0, $\frac{1}{4}$, $\frac{1}{2}$, 1, 2, and 3 are all solutions. In fact, the solutions are all the numbers between -4 and 4. The solution set is $\{x|-4 < x < 4\}$ or the interval $(-4, 4)$. The graph is as follows:

$|x| < 4$ ■

EXAMPLE 13 Solve $|x| \geq 4$. Then graph.

Solution The solutions of $|x| \geq 4$ are solutions of $|x - 0| \geq 4$ and are those numbers whose distance from 0 is greater than or equal to 4—in other words, those numbers x such that $x \leq -4$ *or* $x \geq 4$. The solution set is $\{x|x \leq -4 \text{ or } x \geq 4\}$, or, in interval notation, $(-\infty, -4] \cup [4, \infty)$. We check with numbers like -4.1, -5, 4.1, and 5. The graph is as follows:

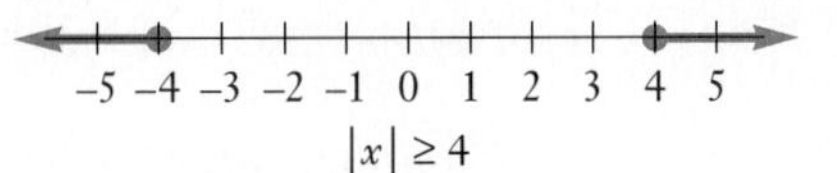

$|x| \geq 4$ ■

Again, the expression inside the absolute value signs can be something besides a single variable.

Absolute Value Principles for Inequalities

For any positive number a and any expression X:

a) The solutions of $|X| < a$ are those numbers that satisfy $-a < X < a$.
b) The solutions of $|X| > a$ are those numbers that satisfy $X < -a$ *or* $X > a$.

Similar statements can be made for $\leq$ and $\geq$.

a) As an example to illustrate part (a), the solutions of $|6x + 7| < 5$ are those numbers x for which

$$-5 < 6x + 7 < 5.$$

b) As an example to illustrate part (b), the solutions of $|2x - 9| > 4$ are those numbers x for which

$$2x - 9 < -4 \quad or \quad 2x - 9 > 4.$$

Note that an inequality of the form $|X| < a$ corresponds to a *con*junction, whereas an inequality of the form $|X| > a$ corresponds to a *dis*junction.

EXAMPLE 14 Solve $|3x - 2| < 4$. Then graph.

Solution We use part (a) of the absolute value principles for inequalities:

$$|X| < a$$
$$|3x - 2| < 4 \quad \text{Replacing } X \text{ by } 3x - 2 \text{ and } a \text{ by } 4$$
$$-4 < 3x - 2 < 4 \quad \text{Part (a) for inequalities}$$
$$-2 < 3x < 6 \quad \text{Adding 2}$$
$$-\tfrac{2}{3} < x < 2. \quad \text{Multiplying by } \tfrac{1}{3}$$

The solution set is $\{x \mid -\frac{2}{3} < x < 2\}$, or, in interval notation, $(-\frac{2}{3}, 2)$. The graph is as follows:

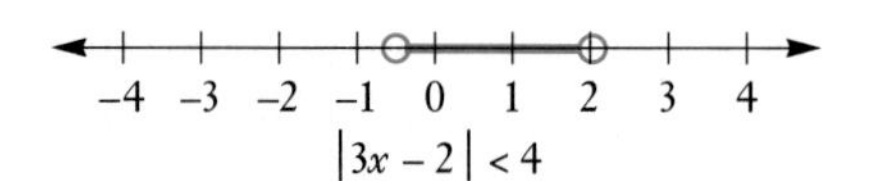

$|3x - 2| < 4$

■

EXAMPLE 15 Solve: $|8 - 4x| \leq 5$.

Solution We again use part (a) of the absolute value principles for inequalities:

$$|X| \le a$$

$$|8 - 4x| \le 5 \qquad \text{Replacing } X \text{ by } 8 - 4x \text{ and } a \text{ by } 5$$

$$-5 \le 8 - 4x \le 5 \qquad \text{Part (a) for inequalities}$$

$$-13 \le \ -4x \ \le -3 \qquad \text{Adding } -8$$

$$\tfrac{13}{4} \ge \ x \ \ge \tfrac{3}{4}. \qquad \text{Multiplying by } -\tfrac{1}{4} \text{ and reversing the inequality symbols}$$

The solution set is $\{x \mid \frac{13}{4} \ge x \ge \frac{3}{4}\}$, or $\{x \mid \frac{3}{4} \le x \le \frac{13}{4}\}$, or, in interval notation, $\left[\frac{3}{4}, \frac{13}{4}\right]$.

■

EXAMPLE 16 Solve $|4x + 2| \ge 6$. Then graph.

Solution We use part (b) of the absolute value principles for inequalities:

$$|X| \ge a$$

$$|4x + 2| \ge 6 \qquad \text{Replacing } X \text{ by } 4x + 2 \text{ and } a \text{ by } 6$$

$$4x + 2 \le -6 \quad \text{or} \quad 4x + 2 \ge 6 \qquad \text{Part (b)}$$

$$4x \le -8 \quad \text{or} \quad 4x \ge 4 \qquad \text{Adding } -2$$

$$x \le -2 \quad \text{or} \quad x \ge 1. \qquad \text{Multiplying by } \tfrac{1}{4}$$

The solution set is $\{x \mid x \le -2 \text{ or } x \ge 1\}$, or $(-\infty, -2] \cup [1, \infty)$, in interval notation. The graph is as follows:

−3 −2 −1 0 1 2 3

$|4x + 2| \ge 6$

■

EXERCISE SET 2.6

Simplify, leaving as little as possible inside the absolute value signs.

1. $|3x|$ **2.** $|17x|$ **3.** $|9x^2|$ **4.** $|6x^2|$ **5.** $|-4x^2|$

6. $|-10x^2|$ **7.** $|-8y|$ **8.** $|-13y|$ **9.** $\left|\frac{-4}{x}\right|$ **10.** $\left|\frac{y}{7}\right|$

11. $\left|\frac{x^2}{-y}\right|$ **12.** $\left|\frac{x^4}{-y}\right|$

Find the distance between the points having the given coordinates.

13. $-8, -42$ **14.** $-9, -36$ **15.** 26, 15 **16.** 54, 18

17. $-9, 24$ **18.** $-18, -37$

Solve. Then graph.

19. $|x| = 3$ **20.** $|x| = 5$ **21.** $|x| = -3$ **22.** $|x| = -5$

23. $|p| = 0$ **24.** $|y| = 8.6$ **25.** $|t| = 5.5$ **26.** $|m| = 0$

27. $|x - 3| = 12$ **28.** $|3x - 2| = 6$ **29.** $|2x - 3| = 4$ **30.** $|5x + 2| = 3$

Solve.

31. $|2y - 7| = 10$
32. $|3y - 4| = 8$
33. $|4x - 9| = 14$
34. $|9y - 2| = 17$
35. $|x| + 7 = 18$
36. $|x| - 2 = 6.3$
37. $|5x| = 40$
38. $|2y| = 18$
39. $5|q| - 2 = 9$
40. $7|z| + 2 = 16$
41. $\left|\frac{2x - 1}{3}\right| = 5$
42. $\left|\frac{4 - 5x}{6}\right| = 7$
43. $|m + 5| + 9 = 16$
44. $|t - 7| + 3 = 4$
45. $\left|\frac{2x - 1}{3}\right| = 1$
46. $\left|\frac{3x - 2}{5}\right| = 2$
47. $|3x - 4| = -2$
48. $|x - 6| = -8$

Solve. Then graph.

49. $|x| < 3.$
50. $|x| \leq 5$
51. $|x| \geq 2$
52. $|y| > 8$
53. $|t| \geq 5.5$
54. $|m| > 0$
55. $|x - 3| < 1$
56. $|x - 2| < 6$
57. $|x + 2| \leq 5$
58. $|x + 4| \leq 1$
59. $|x - 3| > 1$
60. $|x - 2| > 6$
61. $|2x - 3| \leq 4$
62. $|5x + 2| \leq 3$
63. $|2y - 7| > 10$
64. $|3y - 4| > 8$
65. $|4x - 9| \geq 14$
66. $|9y - 1| \geq 3$
67. $|y - 3| < 12$
68. $|p - 2| < 3$
69. $|2x + 3| \leq 4$
70. $|5x - 2| \leq 3$
71. $|4 - 3y| > 8$
72. $|7 - 2y| > 5$
73. $|9 - 4x| \geq 14$
74. $|2 - 9p| \geq 17$
75. $|3 - 4x| < 21$
76. $|-5 - 7x| \leq 30$
77. $\left|\frac{1}{2} + 3x\right| \geq 12$
78. $\left|\frac{1}{4}y - 6\right| > 24$
79. $\left|\frac{x - 7}{3}\right| < 4$
80. $\left|\frac{x + 5}{4}\right| \leq 2$
81. $\left|\frac{2 - 5x}{4}\right| \geq \frac{2}{3}$
82. $\left|\frac{1 + 3x}{5}\right| > \frac{7}{8}$

SKILL MAINTENANCE

Evaluate the given expression using the values provided.

83. $5ab - b^2$ if $a = 2$ and $b = -3$
84. $\frac{3 - x}{x^2}$ if $x = -6$

SYNTHESIS

Inequalities of the form $|X| = |Y|$ can be solved by forming the disjunction $X = Y$ *or* $X = -Y$. This simply states that the quantities X and Y either represent the same number or are additive inverses, or opposites, of each other. Use this approach to solve Exercises 85–90.

85. $|3x + 4| = |x - 7|$
86. $|2x - 8| = |x + 3|$
87. $|x + 5| = |x - 2|$
88. $|x - 7| = |x + 8|$
89. $\left|\frac{2x - 3}{6}\right| = \left|\frac{4 - 5x}{8}\right|$
90. $\left|\frac{6 - 8x}{5}\right| = \left|\frac{7 + 3x}{2}\right|$

Solve.

91. $|3x - 4| > -2$
92. $|x - 6| \leq -8$
93. $|x + 5| > x$
94. $2 \leq |x - 1| \leq 5$

Find an inequality with absolute value equivalent to each of the following.

95. $-3 < x < 3$
96. $-5 \leq y \leq 5$
97. $x \leq -6$ *or* $x \geq 6$
98. $p < -10$ *or* $p > 10$
99. $-5 < x < 1$
100. $-1 \leq x \leq 7$

101. $x < -8$ *or* $x > 2$

102. Pipe is being constructed so that it has a length of 5 ft with a tolerance of $\frac{1}{8}$ in. This means that the length of the pipe can be at most 5 ft plus $\frac{1}{8}$ in. and at least 5 ft minus $\frac{1}{8}$ in. Suppose p = the length of such a pipe. Find an inequality with absolute value whose solutions are all the possible lengths p.

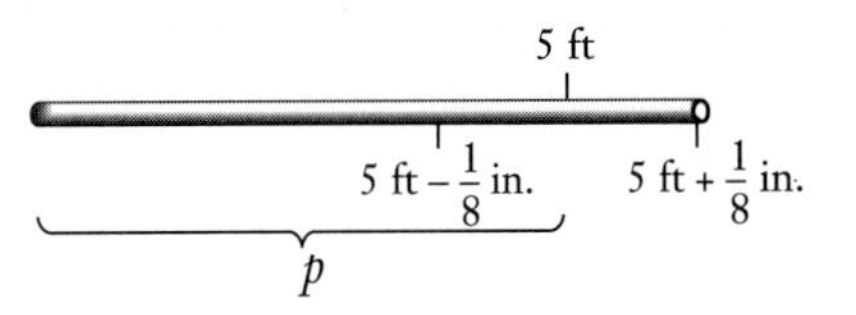

CHAPTER SUMMARY AND REVIEW 2

TERMS TO KNOW

Equivalent equations
Solution set
Empty set
Formula
Subscript

Inequality
Interval notation
Open interval
Closed interval
Half-open interval

Intersection
Conjunction
Union
Disjunction

IMPORTANT PROPERTIES AND FORMULAS

The addition principle for equations: If $a = b$ is true, then $a + c = b + c$ is true.
The multiplication principle for equations: If $a = b$ is true, then $a \cdot c = b \cdot c$ is true.

FIVE STEPS FOR PROBLEM SOLVING IN ALGEBRA

1. *Familiarize* yourself with the problem situation.
2. *Translate* to mathematical language.
3. *Carry out* some mathematical manipulation.
4. *Check* your possible answer in the original problem.
5. *State* the answer clearly.

To solve a formula for a given letter, identify the letter, and:

1. Multiply on both sides to clear fractions or decimals or to remove grouping symbols if that is needed.
2. Collect like terms on each side where convenient. This may require factoring.
3. Get all terms with the letter for which we are solving on one side of the equation and all other terms on the other side, using the addition principle.
4. Collect like terms again, if necessary. This may require factoring.
5. Solve for the letter in question, using the multiplication principle.

THE ADDITION PRINCIPLE FOR INEQUALITIES

If the same number is added on both sides of a true inequality, another true inequality is obtained.

THE MULTIPLICATION PRINCIPLE FOR INEQUALITIES

If we multiply on both sides of a true inequality by a positive number, we get another true inequality. If we multiply by a negative number and the inequality symbol is reversed, we get another true inequality.

Set intersection: $A \cap B = \{x \mid x \text{ is in } A \text{ and } x \text{ is in } B\}$
Set union: $A \cup B = \{x \mid x \text{ is in } A \text{ or in } B, \text{ or both}\}$
"$a < x$ *and* $x < b$" is equivalent to "$a < x < b$"

$|a| = a$ if $a \geq 0$; $|a| = -a$ if $a < 0$

PROPERTIES OF ABSOLUTE VALUE

$|ab| = |a| \cdot |b|$,

$\left|\frac{a}{b}\right| = \frac{|a|}{|b|}$,

$|-a| = |a|$

The distance between a and b is $|a - b|$ or, equivalently, $|b - a|$.

ABSOLUTE VALUE PRINCIPLES FOR EQUATIONS

For any positive number a and any algebraic expression X,

a) If $|X| = a$, then $X = -a$ or $X = a$.
b) If $|X| = 0$, then $X = 0$.
c) If a is negative, $|X| = a$ has no solution.

ABSOLUTE VALUE PRINCIPLES FOR INEQUALITIES

For any positive number a and any expression X:

a) The solutions of $|X| < a$ are those numbers that satisfy $-a < X < a$.
b) The solutions of $|X| > a$ are those numbers that satisfy $X < -a$ *or* $X > a$.

Similar statements can be made for $\leq$ and $\geq$.

REVIEW EXERCISES

The review sections to be tested in addition to the material in this chapter are Sections 1.1, 1.2, 1.3, and 1.4.

1. Determine whether the given pair of equations is equivalent:

$$6x - 1 = 3 \quad \text{and} \quad 3 = \frac{8}{x + 2}.$$

Solve. Don't forget to check.

2. $t - 3.7 = -5.5 - 2t$

3. $-5x + 4 = -51$

4. $\frac{3}{7} - \frac{3}{2}y = -3$

5. $4[3 - 2(x + 2)] = 2(5 - 4x)$

6. Find the solution set:

$$0.8(6x - 1) = 2.6 - (1 - 2x).$$

Solve the problem. Carry out all five problem-solving steps.

7. A person rows a boat at a speed of 4.2 mph in still water. The current in a river is moving at 2.2 mph. How long will it take the person to row 1.6 mi downriver?

8. A fence 250 ft long encloses a rectangular yard. The length of one side of the yard is $\frac{2}{3}$ the length of the adjacent side. What are the dimensions of the yard?

9. After an increase of 0.05, the population of a town is 4326. What was the original population of the town?

10. The total weight of three people is 443 lb. The first person weighs 0.6 as much as the second. The third person weighs 40 lb less than 3 times the weight of the second. Find the weights of the three people.

11. One number is 16 less than three times another number. The sum of the numbers is 68. What are the numbers?

12. Solve for m: $P = m/S$.

13. Solve for x: $c = mx - rx$.

14. The volume of a film canister is 62.8 cm^3. If the canister is 5 cm tall, determine its radius.

15. An investment of $750 is worth $850 after earning simple interest for 2 years. Determine the rate of simple interest.

16. Determine whether the given numbers are solutions of the inequality:

$$5x - 4 \geq 10 - 2x; \quad -6, 0, 2, 11.$$

17. Graph the inequality and write the solution set using both set-builder and interval notation:

$$t > -5.$$

18. Solve. Then graph.

$$1.3y - 3 \leq 2.7y + 1.2$$

Solve.

19. $\frac{3}{8}(2x - 4) > \frac{1}{4}(5x - 12)$

20. $0.6(5 - 3x) \leq -6 - (8.1 - 2x)$

21. You are going to invest $30,000, part at 13% and part at 15%. What is the most that can be invested at 13% in order to make at least $4300 interest per year?

22. One painter will paint your house, all materials and labor included, for $1235. A second painter will do the job for $105 per day of work plus $500 for materials. How many days can the second painter work and still save you money?

23. Find the intersection:

$$\{9, 18, 27, 36\} \cap \{6, 9, 12, 15, 18\}.$$

24. Find the union:

$$\{4, 8, 12, 16\} \cup \{12, 14, 16, 18\}.$$

Graph.

25. $x \geq -5$ *and* $x < 3$

26. $x < -4$ *or* $x \geq 2$

Solve.

27. $7 \le 2 - 3x$ *and* $\frac{3}{2}x - 3 \ge -9$

28. $5 \le -4x + 1 < 11$

29. $\frac{5}{3}x - 6 \ge 4$ *or* $\frac{5}{3}x - 6 \le -4$

30. Simplify, leaving as little as possible inside the absolute value signs:

$$\left|\frac{-5}{x^2}\right|.$$

Solve. Then graph.

31. $|x - 5| = 3$

32. $|2x - 1| = -5$

33. $\left|\frac{1 - 4x}{5}\right| = 3$

34. $|-3 - 6y| \le 11$

35. $\left|\frac{3x - 7}{2}\right| > 2$

SKILL MAINTENANCE

Evaluate.

36. $\frac{x + 6}{3}$, for $x = 2$

37. $\frac{x - y^2}{3 + 2x}$, for $x = 5$ and $y = 3$

38. Find $-x$ when $x = -23.4$.

39. Obtain an equivalent expression by factoring: $5x - 25y + 35$.

SYNTHESIS

Solve.

40. $3[x - 5(2x - 7)] - 4x - 2(3 - 5x) = 9 - 2(x + 7)$.

41. $|9 - 3x| = |7x + 12|$

42. Each of a student's test scores is three times as important as each of a student's quiz scores. If after four quizzes a student's average is 82.5, what score is needed on a test in order to raise the average to 85?

43. Fill in the following blank so as to assure an infinite number of solutions.

$$5x - 7(x + 3) - 4 = 2(7 - x) + \underline{\quad\quad}$$

44. Fill in the following blank so as to assure that no solution exists.

$$20 - 7[3(2x + 4) - 10] = 9 - 2(x - 5) + \underline{\quad\quad}$$

THINKING IT THROUGH

1. How does the word *solve* vary in meaning within this chapter?
2. Explain how the commutative, associative, and distributive laws are used in solving equations and inequalities.
3. Compare the solution sets of equations with those of inequalities.
4. Explain the differences in procedure between solving equations and solving inequalities.
5. Find the error or errors in the following:

$$\begin{aligned} 7 - 9x + 6x &< -9(x + 2) + 10x \\ 7 - 3x &< -9x - 27 + 10x \\ 7 - 3x &< x + 27 \\ -4x &< 20 \\ x &< -5 \end{aligned}$$

CHAPTER TEST 2

1. Determine whether the given pair of equations are equivalent:

$$7 = 5 - 4x \quad \text{and} \quad 5 + \frac{1}{x} = 2.$$

Solve.

2. $-4 = \frac{3}{8} - \frac{5}{4}x$

3. $5\left(-2 - \frac{3}{2}x\right) = -3[4 - (x - 3)]$

4. Find the solution set:

$$7(3 - 4x) = -1.4 - 2.1(1 - x).$$

Solve the problem.

5. After an increase of 0.12, a person's salary is \$28,168. What was her salary before the increase?

6. Twelve gallons of water is poured into three vats. The first vat contains $\frac{3}{4}$ as much water as the second vat. The third vat contains twice as much water as the second. How much water is in each vat?

7. Solve for h: $r = \dfrac{b + h}{h}$.

8. The density of gold is 7.72 grams per cubic centimeter (g/cm^3). A solid gold pendant occupies a volume of 3.50 cm^3. Determine its mass.

9. Determine whether the given numbers are solutions of the inequality:

$$-5 - 3x \leq 7x + 3; \quad -2, -\tfrac{4}{5}, 0, 3.$$

10. Solve. Then graph:

$$\frac{3}{8}t - 2 > -4 - \frac{1}{4}t.$$

Solve.

11. $\frac{5}{4}(3x - 8) \leq -5 - 2(3 - 2x)$

12. You can rent a car either for \$40 per day with unlimited mileage or for \$30 per day with an extra charge of 15¢ a mile. For what numbers of miles traveled would the unlimited mileage plan save you money?

13. Find the intersection:

$$\{3, 7, 11, 13\} \cap \{2, 4, 6, 8\}.$$

14. Find the union:

$$\{17, 19, 23, 29\} \cup \{3, 6, 12, 24\}.$$

Solve. Then graph.

15. $4x + 3 \geq 9$ *and* $2 - \frac{3}{5}x > -1$

16. $\frac{3}{4}x - 2 < -4$ *or* $2 + 3x > -3$

17. Simplify, leaving as little as possible inside the absolute value signs: $|-3x|$.

Solve. Then graph.

18. $\left|\dfrac{5 - 2x}{3}\right| = 2$

19. $|-2t + 5| < 9$

20. $\left|\dfrac{2 - 5x}{3}\right| \geq 1$

SKILL MAINTENANCE

21. Evaluate $2x - y$ for $x = 5$ and $y = 13$.

22. Simplify: $15 \div (1 - 2^2) \cdot 4$.

23. Subtract: $-3.9 - (-4.6)$.

24. Simplify: $9x - 4[2 - 3x]$.

SYNTHESIS

25. If the smell of gasoline is detectable at 3 parts per billion, what percent of the air is occupied by the gasoline when one can just barely smell it?

26. What's a better deal: a 13-in. diameter pizza for \$5 or a 17-in. diameter pizza for \$8? Explain.

27. The surface area of a cube is 486 cm^2. Find the volume of the cube.

28. Solve $m = \dfrac{x}{y - z}$ for z.

CHAPTER 3 Polynomials

FEATURE PROBLEM

Suppose fireworks are launched with an initial velocity of 80 ft/sec from a platform 22 ft high. The height of the fireworks after t sec is given by the polynomial

$$h = -16t^2 + 80t + 22.$$

How high are the fireworks after 2 sec?

THE MATHEMATICS

We find the height after 2 sec by evaluating the polynomial

$$-16t^2 + 80t + 22$$

for $t = 2$:

$$-16(2)^2 + 80(2) + 22.$$

A *polynomial* is a certain kind of algebraic expression with one or more terms. We will define polynomials more specifically in this chapter. Polynomials can be added, subtracted, multiplied, and divided, and can be used to form equations when problem solving.

The review sections to be tested in addition to the material in this chapter are Sections 1.2, 1.4, 2.2, and 2.4.

3.1 Polynomials: Addition and Subtraction

We have already seen how to evaluate certain kinds of algebraic expressions. We have also learned to collect like terms for certain kinds of algebraic expressions. Now we learn to evaluate polynomials and to collect like terms for polynomials.

Polynomial Expressions

The following are examples of *monomials:*

$$0, \quad -3, \quad z, \quad 8x, \quad -7y^2, \quad 4a^2b^3, \quad 1.3p^4q^5r^7.$$

Each expression is a constant or a constant times some variable or variables, each raised to a whole-number power.

> A *polynomial* is a monomial or a combination of sums and/or differences of monomials.

Expressions like these are called *polynomials in one variable*:

$$5x^2, \quad 8a, \quad 2, \quad 2x + 3, \quad -7x + 5, \quad 2y^2 + 5y - 3,$$
$$5a^4 - 3a^2 + \tfrac{1}{4}a - 8, \quad b^6 + 3b^5 - 8b + 7b^4 + \tfrac{1}{2}.$$

Expressions like these are called *polynomials in several variables*:

$$5x - xy^2 + 7y + 2, \quad 9xy^2z - 4x^3z + (-14x^4y^2) + 9, \quad 15x^3y^2.$$

The following are algebraic expressions that are not polynomials:

$$(1)\ \frac{y^2 - 3}{y^2 + 4}, \quad (2)\ 4x^{-3}b^4, \quad (3)\ \frac{2xy}{x^3 - y^3}.$$

Expressions (1) and (3) are not polynomials because they represent quotients. Expression (2) is not a polynomial because x is raised to a power that is negative.

The polynomial $5x^3y - 7xy^2 + 2$ has three terms:

$$5x^3y, \quad -7xy^2, \quad \text{and} \quad 2.$$

The **coefficients** of the terms are 5, -7, and 2.

The **degree of a term** is the sum of the exponents of any variables in that term. The degree of a constant term is 0, except when that constant happens to be 0 itself (the constant 0 has no degree). The **degree of a polynomial** is the same as the degree of its term of highest degree.

EXAMPLES For each polynomial, determine the degree of each term and the degree of the polynomial.

1. $2x^3 + 8x^2 - 17x - 3$

Term	$2x^3$	$8x^2$	$-17x$	-3
Degree	3	2	1	0
Degree of Polynomial	3			

2. $6x^2 + 8x^2y^3 - 17xy - 24xy^2z^4 + 2y + 3$

Term	$6x^2$	$8x^2y^3$	$-17xy$	$-24xy^2z^4$	$2y$	3
Degree	2	5	2	7	1	0
Degree of Polynomial	7					

■

We usually arrange polynomials in one variable so that the exponents either decrease (descending order) or increase (ascending order). The polynomial $x^2 - 7x - 12$ is in descending order. The polynomial $8 + 5x - x^2$ is in ascending order. Depending on the situation, you may see polynomials written in descending order, ascending order, or neither. Generally, if an exercise is written in one kind of order, we write the answer in that same order.

A polynomial of degree 0 or 1 is called **linear.** Polynomials in one variable are called **quadratic** if they are of degree 2 and are called **cubic** if they are of degree 3.

We can also arrange the terms of polynomials in several variables in ascending or descending order, with respect to the powers of one of the variables.

EXAMPLE 3 Arrange in ascending order: $x^4 + 2 - 5x^2 + 3x^3 + 7x$.

Solution

$$2 + 7x - 5x^2 + 3x^3 + x^4$$ ■

EXAMPLE 4 Arrange in descending powers of x:

$$y^4 + 2 - 5x^2 + 3x^3y + 7xy.$$

Solution

$$3x^3y - 5x^2 + 7xy + y^4 + 2$$ ■

The **leading term** of a polynomial in one variable is the first term when the polynomial has been written in descending order. Its coefficient is called the **leading coefficient.**

EXAMPLES For each polynomial find the leading term and the leading coefficient.

5. $5x^3 - 3x^2 - 7x + 4$

The leading term is $5x^3$. The leading coefficient is 5.

6. $6a^2 - 5a^7 - a^4 + 4a + 9$

The leading term is $-5a^7$ because, if this polynomial is written in descending order, we have $-5a^7 - a^4 + 6a^2 + 4a + 9$. The leading coefficient is -5. ■

Whereas a polynomial with a single term is called a **monomial,** a polynomial with two terms is a **binomial.** The following are examples of binomials:

$$2x - 7, \qquad a - 3b, \qquad 5x^2 - 7y^3.$$

A polynomial with three terms is a **trinomial.** The following are examples of trinomials:

$$x^2 - 7x - 12, \qquad a^2 - 2ab + b^2, \qquad 5x^3 - 6x^2 + 8x.$$

Evaluating Polynomials and Polynomials in Problem Solving

When we replace the variable in a polynomial by a number, the polynomial then represents a number called a **value** of the polynomial. Finding that number, or value, is called **evaluating the polynomial.**

EXAMPLE 7 Evaluate each polynomial for $x = 2$.

a) $3x + 5 = 3 \cdot 2 + 5 = 6 + 5 = 11$

b) $2x^2 - 7x + 3 = 2 \cdot 2^2 - 7 \cdot 2 + 3 = 2 \cdot 4 - 14 + 3$
$= 8 - 14 + 3 = -3$ ■

EXAMPLE 8 Evaluate each polynomial for $x = -5$.

a) $2 - x^3 = 2 - (-5)^3 = 2 - (-125) = 127$

b) $-x^2 - 3x + 1 = -(-5)^2 - 3(-5) + 1 = -25 + 15 + 1 = -9$ ■

Polynomials occur in many real-world situations and are used in problem solving. Although the next two examples are problem solving in nature, they involve only the evaluation of a polynomial. For that reason, we do not apply all five problem-solving steps.

EXAMPLE 9 *Games in a sports league.* In a sports league of n teams in which each team plays every other team twice, the total number of games to be played is given by the polynomial

$$n^2 - n.$$

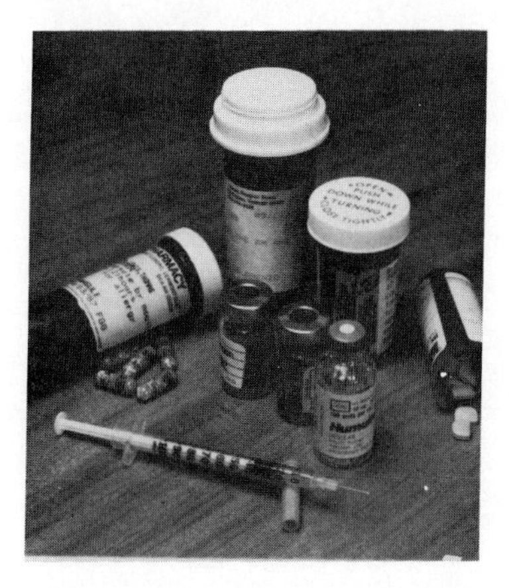

A women's slow-pitch softball league has 10 teams. What is the total number of games to be played?

Solution We evaluate the polynomial for $n = 10$:

$$n^2 - n = 10^2 - 10 = 100 - 10 = 90.$$

The league plays 90 games. ■

EXAMPLE 10 *Medical dosage.* The concentration, in parts per million, of a certain medication after time t, in hours, is given by the polynomial

$$-0.05t^2 + 2t + 2.$$

Find the concentration after 2 hr.

Solution We evaluate the polynomial for $t = 2$:

$$\begin{aligned} -0.05t^2 + 2t + 2 &= -0.05(2)^2 + 2(2) + 2 && \text{We carry out the calculation using the rules for order of operations.} \\ &= -0.05(4) + 2(2) + 2 \\ &= -0.2 + 4 + 2 \\ &= -0.2 + 6 \\ &= 5.8. \end{aligned}$$

The concentration after 2 hr is 5.8 parts per million. ■

Adding Polynomials

Recall from Section 1.4 that if two terms of a polynomial have the same variable(s) raised to the same power(s), they are **similar,** or **like, terms** and can be "combined" or "collected."

EXAMPLES Collect like terms.

11. $3x^2 - 4y + 2x^2 = 3x^2 + 2x^2 - 4y$ — Rearranging using the commutative law for addition

$= (3 + 2)x^2 - 4y$ — Using the distributive law

$= 5x^2 - 4y$

12. $9x^3 + 5x - 4x^2 - 2x^3 + 5x^2 = 7x^3 + x^2 + 5x$ — We normally perform the middle steps mentally and write just the answer.

13. $3x^2y + 5xy^2 - 3x^2y - xy^2 = 4xy^2$ ■

The sum of two polynomials can be found by writing a plus sign between them and then collecting like terms. Ordinarily, this can be done mentally.

EXAMPLE 14 Add: $(-3x^3 + 2x - 4) + (4x^3 + 3x^2 + 2)$.

Solution

$$(-3x^3 + 2x - 4) + (4x^3 + 3x^2 + 2) = x^3 + 3x^2 + 2x - 2$$ ■

Using columns is often helpful. To do so, we write the polynomials one under the other, writing like terms under one another and leaving spaces for missing terms. Let us do the addition in Example 14 using columns.

$$\begin{array}{rrrr} -3x^3 & & +2x & -4 \\ 4x^3 & +3x^2 & & +2 \\ \hline x^3 & +3x^2 & +2x & -2 \end{array}$$

EXAMPLE 15 Add: $4ax^2 + 4bx - 5$ and $-6ax^2 + 5bx + 8$.

Solution

$$\begin{array}{rrr} 4ax^2 & +4bx & -5 \\ -6ax^2 & +5bx & +8 \\ \hline -2ax^2 & +9bx & +3 \end{array}$$ ■

Although using columns is helpful for complicated examples, you should attempt, for the sake of working faster, to write only the answer whenever you can.

EXAMPLE 16 Add: $13x^3y + 3x^2y - 5y$ and $x^3y + 4x^2y - 3xy + 3y$.

Solution

$$(13x^3y + 3x^2y - 5y) + (x^3y + 4x^2y - 3xy + 3y)$$
$$= 14x^3y + 7x^2y - 3xy - 2y$$ ■

Additive Inverses and Subtraction

If the sum of two polynomials is 0, the polynomials are called *additive inverses,* or *opposites* of each other. For example,

$$(3x^2 - 5x + 2) + (-3x^2 + 5x - 2) = 0,$$

so the opposite of $(3x^2 - 5x + 2)$ is $(-3x^2 + 5x - 2)$. We can say the same thing using algebraic symbolism, as follows:

$$\underbrace{\text{The additive inverse or opposite of}}_{-} \underbrace{(3x^2 - 5x + 2)}_{(3x^2 - 5x + 2)} \underbrace{\text{is}}_{=} \underbrace{(-3x^2 + 5x - 2).}_{-3x^2 + 5x - 2}$$

Thus the additive inverse, or opposite, of a polynomial can be found by changing the sign of every term.

EXAMPLE 17 Write two equivalent expressions for the additive inverse, or opposite, of $7xy^2 - 6xy - 4y + 3$.

Solution

a) $-(7xy^2 - 6xy - 4y + 3)$ Writing an inverse sign in front
b) $-7xy^2 + 6xy + 4y - 3$ Changing the sign of every term ■

To subtract one polynomial from another, we add the opposite of the polynomial being subtracted.

EXAMPLE 18 Subtract: $(-9x^5 + 2x^2 + 4) - (2x^5 + 4x^3 - 3x^2)$.

Solution

$$(-9x^5 + 2x^2 + 4) - (2x^5 + 4x^3 - 3x^2)$$
$$= (-9x^5 + 2x^2 + 4) + (-2x^5 - 4x^3 + 3x^2)$$ Adding the opposite of the polynomial being subtracted
$$= -11x^5 - 4x^3 + 5x^2 + 4$$ ■

After some practice, you will find that you can skip some steps by mentally "changing the sign of each term" and then collecting like terms. Eventually, all you will write is the answer.

We can also use columns for subtraction. We change signs mentally and then add.

EXAMPLE 19 Subtract:

$$(4x^2y - 6x^3y^2 + x^2y^2) - (4x^2y + x^3y^2 + 3x^2y^3 - 8x^2y^2).$$

Solution

Think: (Subtract)

$$\begin{array}{l} 4x^2y - 6x^3y^2 \qquad\qquad + x^2y^2 \\ \underline{-(4x^2y + x^3y^2 + 3x^2y^3 - 8x^2y^2)} \end{array}$$

Write: (Add)

$$\begin{array}{r} 4x^2y - 6x^3y^2 \qquad\qquad + x^2y^2 \\ \underline{-4x^2y - x^3y^2 - 3x^2y^3 + 8x^2y^2} \\ -7x^3y^2 - 3x^2y^3 + 9x^2y^2 \end{array}$$ ■

As with addition, after sufficient experience, you should try to write only the answer.

EXERCISE SET 3.1

Determine the degree of each term and the degree of the polynomial.

1. $-11x^4 - x^3 + x^2 + 3x - 9$
2. $t^3 - 3t^2 + t + 1$
3. $y^3 + 2y^7 + x^2y^4 - 8$
4. $u^2 + 3v^5 - u^3v^4 - 7$
5. $a^5 + 4a^2b^4 + 6ab + 4a - 3$
6. $8p^6 + 2p^4t^4 - 7p^3t + 5p^2 - 14$

Arrange in descending order. Then find the leading term and the leading coefficient.

7. $23 - 4y^3 + 7y - 6y^2$

8. $5 - 8y + 6y^2 + 11y^3 - 18y^4$

9. $5x^2 + 3x^7 - x + 12$

10. $9 - 3x - 10x^4 + 7x^2$

11. $a + 5a^3 - a^7 - 19a^2 + 8a^5$

12. $a^3 - 7 + 11a^4 + a^9 - 5a^2$

Arrange in ascending powers of x.

13. $4x + 12 + 3x^4 - 5x^2$

14. $-5x^2 + 10x + 5$

15. $-9x^3y + 3xy^3 + x^2y^2 + 2x^4$

16. $5x^2y^2 - 9xy + 8x^3y^2 - 5x^4$

17. $4ax - 7ab + 4x^6 - 7ax^2$

18. $5xy^8 - 3ax^5 + 4ax^3 - 12a + 5x^6y$

Evaluate the polynomial for $x = 4$.

19. $-5x + 2$

20. $-3x + 1$

21. $2x^2 - 5x + 7$

22. $3x^2 + x + 7$

23. $x^3 - 5x^2 + x$

24. $7 - x + 3x^2$

Evaluate the polynomial for $x = -1$.

25. $3x + 5$

26. $6 - 2x$

27. $x^2 - 2x + 1$

28. $5x - 6 + x^2$

29. $-3x^3 + 7x^2 - 3x - 2$

30. $-2x^3 - 5x^2 + 4x + 3$

Daily accidents. The daily number of accidents (the average number of accidents per day) involving drivers of age a is approximated by the polynomial

$$0.4a^2 - 40a + 1039.$$

31. Evaluate the polynomial for $a = 18$ to find the number of daily accidents involving an 18-year-old driver.

32. Evaluate the polynomial for $a = 20$ to find the number of daily accidents involving a 20-year-old driver.

Falling distance. The distance s, in feet, traveled by a body falling freely from rest in t seconds is approximated by the monomial

$$16t^2.$$

33. A stone is dropped from a cliff and takes 8 sec to hit the ground. How high is the cliff?

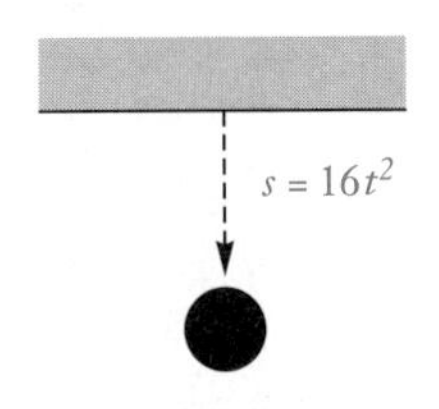

34. A brick is dropped from the top of a building and takes 3 sec to hit the ground. How high is the building?

Total revenue. An electronics firm is marketing a new kind of stereo. **Total revenue** is the total amount of money taken in. The firm determines that when it sells x stereos, it will take in

$$280x - 0.4x^2 \text{ dollars.}$$

35. What is the total revenue from the sale of 75 stereos?

36. What is the total revenue from the sale of 100 stereos?

Total cost. **Total cost** is the cost of producing x stereos. The electronics firm determines that the total cost, in dollars, of producing x stereos is given by

$$5000 + 0.6x^2.$$

37. What is the total cost of producing 75 stereos?

38. What is the total cost of producing 100 stereos?

Surface area of a right circular cylinder. The surface area of a right circular cylinder is given by the polynomial

$$2\pi rh + 2\pi r^2,$$

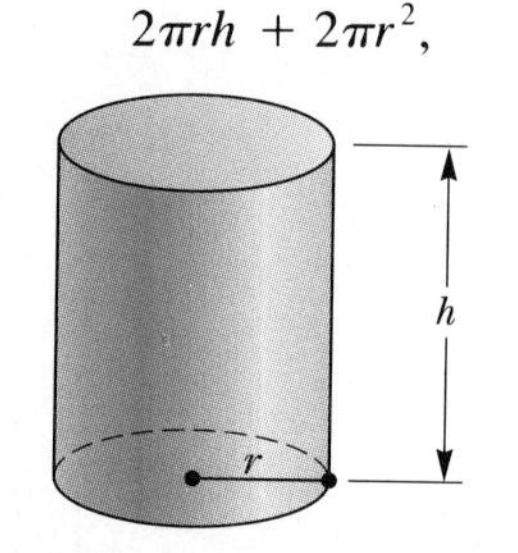

where h = the height and r = the radius of the base.

39. ▦ A 12-oz beverage can has height 4.7 in. and radius 1.2 in. Find the surface area of the can. (Use 3.14 as an approximation for π. Give your answer to the nearest hundredth.)

40. ▦ A 16-oz beverage can has height 6.3 in. and radius 1.2 in. Find the surface area of the can. (Use 3.14 as an approximation for π.)

Collect like terms.

41. $6x^2 - 7x^2 + 3x^2$

42. $-2y^2 - 7y^2 + 5y^2$

43. $5x - 4y - 2x + 5y$

44. $4a - 9b - 6a + 3b$

45. $5a + 7 - 4 + 2a - 6a + 3$

46. $9x + 12 - 8 - 7x + 5x + 10$

47. $3a^2b + 4b^2 - 9a^2b - 6b^2$

48. $5x^2y^2 + 4x^3 - 8x^2y^2 - 12x^3$

49. $8x^2 - 3xy + 12y^2 + x^2 - y^2 + 5xy + 4y^2$

50. $a^2 - 2ab + b^2 + 9a^2 + 5ab - 4b^2 + a^2$

51. $4x^2y - 3y + 2xy^2 - 5x^2y + 7y + 7xy^2$

52. $3xy^2 + 4xy - 7xy^2 + 7xy + x^2y$

Add.

53. $(3x^2 + 5y^2 + 6) + (2x^2 - 3y^2 - 1)$

54. $(9y^2 + 8y - 4) + (12y^2 - 5y + 8)$

55. $(2a + 3b - c) + (4a - 2b + 2c)$

56. $(5x - 4y + 2z) + (9x + 12y - 8z)$

57. $(a^2 - 3b^2 + 4c^2) + (-5a^2 + 2b^2 - c^2)$

58. $(x^2 - 5y^2 - 9z^2) + (-6x^2 + 9y^2 - 2z^2)$

59. $(x^2 + 2x - 3xy - 7) + (-3x^2 - x + 2xy + 6)$

60. $(3a^2 - 2b + ab + 6) + (-a^2 + 5b - 5ab - 2)$

61. $(7x^2y - 3xy^2 + 4xy) + (-2x^2y - xy^2 + xy)$

62. $(7ab - 3ac + 5bc) + (13ab - 15ac - 8bc)$

63. $(2r^2 + 12r - 11) + (6r^2 - 2r + 4) + (r^2 - r - 2)$

64. $(5x^2 + 19x - 23) + (-7x^2 - 11x + 12) + (-x^2 - 9x + 8)$

65. $\left(\frac{2}{3}xy + \frac{5}{6}xy^2 + 5.1x^2y\right) + \left(-\frac{4}{5}xy + \frac{3}{4}xy^2 - 3.4x^2y\right)$

66. $\left(\frac{1}{8}xy - \frac{3}{5}x^3y^2 + 4.3y^3\right) + \left(-\frac{1}{3}xy - \frac{3}{4}x^3y^2 - 2.9y^3\right)$

Write two equivalent expressions for the additive inverse, or opposite, of the polynomial.

67. $5x^3 - 7x^2 + 3x - 6$

68. $-8y^4 - 18y^3 + 4y - 9$

69. $-12y^5 + 4ay^4 - 7by^2$

70. $7ax^3y^2 - 8by^4 - 7abx - 12ay$

Subtract.

71. $(8x - 4) - (-5x + 2)$

72. $(9y + 3) - (-4y - 2)$

73. $(-3x^2 + 2x + 9) - (x^2 + 5x - 4)$

74. $(-9y^2 + 4y + 8) - (4y^2 + 2y - 3)$

75. $(5a - 2b + c) - (3a + 2b - 2c)$

76. $(8x - 4y + z) - (4x + 6y - 3z)$

77. $(3x^2 - 2x - x^3) - (5x^2 - 8x - x^3)$

78. $(8y^2 - 3y - 4y^3) - (3y^2 - 9y - 7y^3)$

79. $(5a^2 + 4ab - 3b^2) - (9a^2 - 4ab + 2b^2)$

80. $(9y^2 - 14yz - 8z^2) - (12y^2 - 8yz + 4z^2)$

81. $(6ab - 4a^2b + 6ab^2) - (3ab^2 - 10ab - 12a^2b)$

82. $(10xy - 4x^2y^2 - 3y^3) - (-9x^2y^2 + 4y^3 - 7xy)$

83. $(0.09y^4 - 0.052y^3 + 0.93) - (0.03y^4 - 0.084y^3 + 0.94y^2)$

84. $(1.23x^4 - 3.122x^3 + 1.11x) - (0.79x^4 - 8.734x^3 + 0.04x^2 + 6.71x)$

85. $\left(\frac{5}{8}x^4 - \frac{1}{4}x^2 - \frac{1}{2}\right) - \left(-\frac{3}{8}x^4 + \frac{3}{4}x^2 + \frac{1}{2}\right)$

86. $\left(\frac{5}{6}y^4 - \frac{1}{2}y^2 - 7.8y + \frac{1}{3}\right) - \left(-\frac{3}{8}y^4 + \frac{3}{4}y^2 + 3.4y - \frac{1}{5}\right)$

SKILL MAINTENANCE

87. Multiply: $3(y - 2)$.

88. Simplify: $3(x - 4) - 5(x + 16)$.

SYNTHESIS

89. Find a polynomial that gives the surface area of a box like this one, with an open top and dimensions as shown.

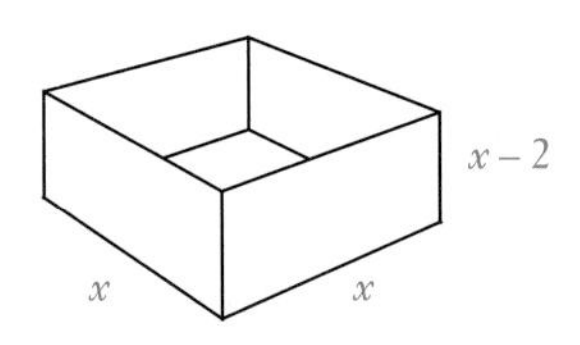

Add. Assume that the exponents are natural numbers.

90. $(2x^{2a} + 4x^a + 3) + (6x^{2a} + 3x^a + 4)$

91. $(47x^{4a} + 3x^{3a} + 22x^{2a} + x^a + 1) + (37x^{3a} + 8x^{2a} + 3)$

92. Find a polynomial in x and y with four terms, none of which are similar and all of which have degree 5. Write this polynomial so that it appears, at once, in both descending powers of x and ascending powers of y.

93. Is the sum of two nth degree polynomials in x always another nth degree polynomial in x? Why or why not?

3.2 Multiplication of Polynomials

We now multiply polynomials using techniques based, for the most part, on the distributive laws, but also on the associative and commutative laws. As we proceed in this chapter, we will develop special ways to find certain products.

Multiplying Monomials

We know that an exponential expression like a^3 means $a \cdot a \cdot a$. We also know that $a^1 = a$. Now consider multiplying powers with like bases:

$$a^3 \cdot a^2 = (a \cdot a \cdot a)(a \cdot a) = a \cdot a \cdot a \cdot a \cdot a = a^5.$$

Since an integer exponent greater than 1 tells us how many times we use a base as a factor, we have $(a \cdot a \cdot a)(a \cdot a) = a \cdot a \cdot a \cdot a \cdot a = a^5$ by the associative law. Note that the exponent in a^5 is the sum of the exponents in $a^3 \cdot a^2$. That is, $3 + 2 = 5$. Similarly,

$$b^4 \cdot b^3 = (b \cdot b \cdot b \cdot b)(b \cdot b \cdot b) = b^7, \qquad \text{where } 4 + 3 = 7.$$

Adding the exponents gives the correct result.

THE PRODUCT RULE FOR POWERS

For any real number a and any positive integers m and n,

$$a^m \cdot a^n = a^{m+n}.$$

(When multiplying with exponential notation, if the bases are the same, keep the base and add the exponents.)

EXAMPLES Multiply and simplify.

1. $8^4 \cdot 8^3 = 8^{4+3}$ Adding exponents: $a^m \cdot a^n = a^{m+n}$
$= 8^7$

2. $x^2 \cdot x^9 = x^{2+9}$
$= x^{11}$

3. $m^5 m^{10} m^3 = m^{5+10+3}$
$= m^{18}$

4. $x \cdot x^8 = x^1 \cdot x^8 = x^{1+8}$
$= x^9$

5. $(a^3b^2)(a^3b^5) = (a^3a^3)(b^2b^5)$
$= a^6b^7$ ■

When monomials contain coefficients and variables both, we multiply the coefficients first and then the variables, using the product rule if the bases are the same.

EXAMPLES Multiply and simplify.

6. $(-7x^5)(4x^3) = (-7 \cdot 4)(x^5 \cdot x^3)$ Using the associative and commutative laws
$= -28x^{5+3}$ Adding the exponents
$= -28x^8$ Simplifying

7. $(-8x^4y^7)(5x^3y^2) = (-8 \cdot 5)(x^4 \cdot x^3)(y^7 \cdot y^2)$ Using the associative and commutative laws
$= -40x^{4+3}y^{7+2}$ Adding exponents
$= -40x^7y^9$

8. $(3x^2yz^5)(-6x^5y^{10}z^2) = 3 \cdot (-6) \cdot x^2 \cdot x^5 \cdot y \cdot y^{10} \cdot z^5 \cdot z^2$
$= -18x^7y^{11}z^7$ ■

You should try to work mentally, writing only the answer.

Multiplying Monomials and Binomials

The distributive law is the basis for multiplying polynomials other than monomials. We first multiply a monomial and a binomial.

EXAMPLE 9 Multiply: $2x(3x - 5)$.

Solution

$$\begin{aligned} 2x \cdot (3x - 5) &= 2x \cdot (3x) - 2x \cdot (5) && \text{Using the distributive law} \\ &= 6x^2 - 10x && \text{Multiplying monomials} \end{aligned}$$

■

EXAMPLE 10 Multiply: $3a^2b(a^2 - b^2)$.

Solution

$$\begin{aligned} 3a^2b \cdot (a^2 - b^2) &= 3a^2b \cdot a^2 - 3a^2b \cdot b^2 && \text{Using the distributive law} \\ &= 3a^4b - 3a^2b^3 \end{aligned}$$

■

Now we multiply two binomials. To do so, we use the distributive law twice, first multiplying one of the binomials by each term of the other binomial.

EXAMPLE 11 Multiply: $(2y^3 + 1)(y^3 + 4)$.

Solution

$$(2y^3 + 1)(y^3 + 4) = \underbrace{(2y^3 + 1)y^3}_{\text{(a)}} + \underbrace{(2y^3 + 1)4}_{\text{(b)}}$$

We "distributed" the $2y^3 + 1$.

We consider parts (a) and (b) separately:

a) $(2y^3 + 1)y^3 = 2y^3y^3 + 1y^3$ **Using the distributive law**
$= 2y^6 + y^3$ **Multiplying the monomials**

b) $(2y^3 + 1)4 = 2y^3 \cdot 4 + 1 \cdot 4$ **Using the distributive law**
$= 8y^3 + 4.$ **Multiplying the monomials**

We replace parts (a) and (b) in the original expression with their answers and collect like terms, if we can:

$$\begin{aligned} (2y^3 + 1)(y^3 + 4) &= (2y^6 + y^3) + (8y^3 + 4) \\ &= 2y^6 + 9y^3 + 4. \end{aligned}$$

■

Multiplying Any Two Polynomials

To find a quick way to multiply any two polynomials, let us consider another example.

EXAMPLE 12 Multiply: $(p + 2)(p^4 - 2p^3 + 3)$.

Solution By the distributive law, we have

$(p + 2)(p^4 - 2p^3 + 3)$

$= (p + 2)(p^4) + (p + 2)(-2p^3) + (p + 2)(3)$ **Using the distributive law to "distribute" the $p + 2$**

$= p \cdot p^4 + 2 \cdot p^4 + p(-2p^3) + 2(-2p^3) + p \cdot 3 + 2 \cdot 3$ **Using the distributive law three more times**

$= p^5 + 2p^4 - 2p^4 - 4p^3 + 3p + 6$ **Multiplying the monomials**

$= p^5 - 4p^3 + 3p + 6.$ **Collecting like terms** ■

From the preceding examples we see the following:

> To multiply any two polynomials, we multiply each term of one polynomial by each term of the other polynomial and collect like terms.

For more complicated multiplications we can use columns. We multiply each term at the top by every term at the bottom, aligning like terms and listing all products in descending order. Then we add.

EXAMPLE 13 Multiply: $(4x^2 - 2x + 3)(x + 2)$.

Solution

$$\begin{array}{rrrrl} & 4x^2 & -\ 2x & +\ 3 & \\ & & x & +\ 2 & \\ \hline & 8x^2 & -\ 4x & +\ 6 & \text{Multiplying the top row by 2} \\ 4x^3 & -\ 2x^2 & +\ 3x & & \text{Multiplying the top row by } x \\ \hline 4x^3 & +\ 6x^2 & -\ x & +\ 6 & \text{Collecting like terms} \\ \uparrow & \uparrow & \uparrow & \uparrow & \text{Line up like terms in columns.} \end{array}$$

■

EXAMPLE 14 Multiply: $(5x^3 - 3x + 4)(-2x^2 - 3)$.

Solution When missing terms occur, it helps to leave spaces for them and align like terms as we multiply.

$$\begin{array}{rrrrrl} & 5x^3 & & -\ 3x & +\ 4 & \\ & & -\ 2x^2 & & -\ 3 & \\ \hline & -\ 15x^3 & & +\ 9x & -\ 12 & \text{Multiplying by } -3 \\ -10x^5 & +\ 6x^3 & -\ 8x^2 & & & \text{Multiplying by } -2x^2 \\ \hline -10x^5 & -\ 9x^3 & -\ 8x^2 & +\ 9x & -\ 12 & \text{Collecting like terms} \end{array}$$

■

Special Products: Products of Two Binomials

We now consider what are called *special products.* These are faster ways to multiply in certain situations.

Let us find a faster special product rule for the product of two binomials. Consider $(x + 7)(x + 4)$. We multiply each term of one binomial by every term of the other. Then we collect like terms:

$$(x + 7)(x + 4) = x \cdot x + 4 \cdot x + 7 \cdot x + 7 \cdot 4$$
$$= x^2 + 11x + 28.$$

We can rewrite the first line of this product to show a special product of two binomials:

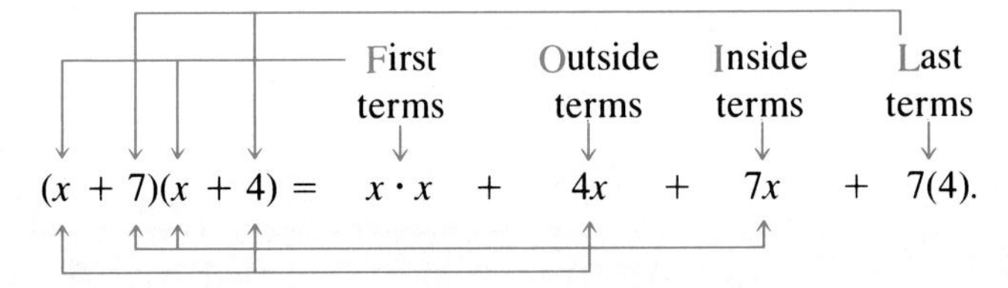

We use the mnemonic device FOIL to remember this method for multiplying.

EXAMPLES Multiply.

15. $(x + 5)(x - 8) = \overset{F}{x^2} \overset{O}{- 8x} \overset{I}{+ 5x} \overset{L}{- 40}$
$= x^2 - 3x - 40$ **Collecting like terms**

We write the result in descending order because the original binomials are in descending order.

16. $(2x - 3)(y + 2) = \overset{F}{2xy} \overset{O}{+ 4x} \overset{I}{- 3y} \overset{L}{- 6}$

17. $(2x + 3y)(x - 4y) = 2x^2 - 8xy + 3xy - 12y^2 = 2x^2 - 5xy - 12y^2$ ■

Squares of Binomials

We can use FOIL to develop a special product for the square of a binomial. Note the following:

$$(A + B)^2 = (A + B)(A + B)$$
$$= A^2 + AB + AB + B^2$$
$$= A^2 + 2AB + B^2$$

$$(A - B)^2 = (A - B)(A - B)$$
$$= A^2 - AB - AB + B^2$$
$$= A^2 - 2AB + B^2$$

THE SQUARE OF A BINOMIAL

The square of a sum or a difference of two terms is the square of the first term plus or minus twice the product of the two terms plus the square of the last term:

$$(A + B)^2 = A^2 + 2AB + B^2;$$
$$(A - B)^2 = A^2 - 2AB + B^2.$$

It can be helpful to memorize the words of the rule and say them while calculating.

EXAMPLES Multiply.

$$\begin{aligned}
(A - B)^2 &= A^2 - 2\,A\,B + B^2 \\
\textbf{18. } (y - 5)^2 &= y^2 - 2(y)(5) + 5^2 \\
&= y^2 - 10y + 25 \\
\textbf{19. } (2x + 3y)^2 &= (2x)^2 + 2(2x)(3y) + (3y)^2 \\
&= 4x^2 + 12xy + 9y^2
\end{aligned}$$

Note that $(2x)^2 = (2x)(2x)$. ■

Products of Sums and Differences

Another special case of a product of two binomials is the product of a sum and a difference. Note the following:

$$\begin{aligned}
(A + B)(A - B) &= A^2 - AB + AB - B^2 \\
&= A^2 - B^2.
\end{aligned}$$

MULTIPLYING THE SUM AND THE DIFFERENCE OF TWO TERMS

$$(A + B)(A - B) = A^2 - B^2$$

The product of the sum and the difference of two terms is the square of the first term minus the square of the second term.

EXAMPLES Multiply.

$$\begin{aligned}
(A + B)(A - B) &= A^2 - B^2 \\
\textbf{20. } (y + 5)(y - 5) &= y^2 - 5^2 \\
&= y^2 - 25 \\
\textbf{21. } (2xy + 3x)(2xy - 3x) &= (2xy)^2 - (3x)^2 \\
&= 4x^2y^2 - 9x^2 \\
\textbf{22. } (5y + 4 + 3x)(5y + 4 - 3x) &= (5y + 4)^2 - (3x)^2 \\
&= 25y^2 + 40y + 16 - 9x^2
\end{aligned}$$

Here we treated $(5y + 4)$ as the first expression A and $3x$ as B. This product could have been done by columns, but it is much faster to use the present rule. ■

Try to multiply polynomials mentally when possible, watching for any special products. As a precautionary note, *keep in mind the following*:

$(A + B)^2 \neq A^2 + B^2$ and, in general, $(A + B)^n \neq A^n + B^n$.

EXAMPLE 23 Multiply $(x + 4)^3$.

Solution

$$
\begin{aligned}
(x+4)^3 &= (x+4)(x+4)^2 && \text{Using the product rule} \\
&= (x+4)(x^2+8x+16) && \text{Squaring a binomial} \\
&= (x+4)x^2 + (x+4)8x + (x+4)16 && \text{Using the distributive law} \\
&= x^3 + 12x^2 + 48x + 64 && \text{Using the distributive law and collecting like terms}
\end{aligned}
$$

■

EXERCISE SET 3.2

Multiply.

1. $(6x^2)(7)$
2. $(5x^2)(-2)$
3. $(-x^3)(-x)$
4. $(-x^4)(x^2)$
5. $(-x^5)(x^3)$
6. $(-x^6)(-x^2)$
7. $2y^2 \cdot 5y$
8. $-3x^2 \cdot 2xy$
9. $5x(-4x^2y)$
10. $-3ab^2(2a^2b^2)$
11. $(2x^3y^2)(-5x^2y^4)$
12. $(7a^2bc^4)(-8ab^3c^2)$
13. $2x(3-x)$
14. $4a(a^2-5a)$
15. $3ab(a+b)$
16. $2xy(2x-3y)$
17. $5cd(3c^2d-5cd^2)$
18. $a^2(2a^2-5a^3)$
19. $(2x+3)(3x-4)$
20. $(2a-3b)(4a-b)$
21. $(s+3t)(s-3t)$
22. $(y+4)(y-4)$
23. $(x-y)(x-y)$
24. $(a+2b)(a+2b)$
25. $(y+8x)(2y-7x)$
26. $(x+y)(x-2y)$
27. $(a^2-2b^2)(a^2-3b^2)$
28. $(2m^2-n^2)(3m^2-5n^2)$
29. $(x-4)(x^2+4x+16)$
30. $(y+3)(y^2-3y+9)$
31. $(x+y)(x^2-xy+y^2)$
32. $(a-b)(a^2+ab+b^2)$
33. $(a^2+a-1)(a^2+4a-5)$
34. $(x^2-2x+1)(x^2+x+2)$
35. $(4a^2b-2ab+3b^2)(ab-2b+a)$
36. $(2x^2+y^2-2xy)(x^2-2y^2-xy)$
37. $\left(x-\frac{1}{2}\right)\left(x-\frac{1}{4}\right)$
38. $\left(b-\frac{1}{3}\right)\left(b-\frac{1}{3}\right)$
39. $(1.3x-4y)(2.5x+7y)$
40. $(40a-0.24b)(0.3a+10b)$

Multiply.

41. $(a+2)(a+3)$
42. $(x+5)(x+8)$
43. $(y+3)(y-2)$
44. $(y-4)(y+7)$
45. $(x+3)^2$
46. $(y-7)^2$
47. $(x-2y)^2$
48. $(2s+3t)^2$
49. $\left(b-\frac{1}{3}\right)\left(b-\frac{1}{2}\right)$
50. $\left(x-\frac{3}{2}\right)\left(x-\frac{2}{3}\right)$
51. $(2x+9)(x+2)$
52. $(3b+2)(2b-5)$
53. $(20a-0.16b)^2$
54. $(10p+2.3q)^2$
55. $(2x-3y)(2x+y)$
56. $(2a-3b)(2a-b)$
57. $\left(2a+\frac{1}{3}\right)^2$
58. $\left(3c-\frac{1}{2}\right)^2$

Multiply.

59. $(c+2)(c-2)$
60. $(x-3)(x+3)$
61. $(2a+1)(2a-1)$
62. $(3-2x)(3+2x)$
63. $(3m-2n)(3m+2n)$
64. $(3x+5y)(3x-5y)$
65. $(-mn+m^2)(mn+m^2)$
66. $(1.6+pq)(-1.6+pq)$
67. $(x+1)(x-1)(x^2+1)$
68. $(y-2)(y+2)(y^2+4)$
69. $(a-b)(a+b)(a^2-b^2)$
70. $(2x-y)(2x+y)(4x^2-y^2)$
71. $(a+b+1)(a+b-1)$
72. $(m+n+2)(m+n-2)$
73. $(2x+3y+4)(2x+3y-4)$
74. $(3a-2b+c)(3a-2b-c)$

75. Suppose P dollars is invested in a savings account at interest rate i, compounded annually, for 2 years. The amount A in the account after 2 years is given by

$$A = P(1 + i)^2.$$

Find an equivalent expression for A.

76. Suppose P dollars is invested in a savings account at interest rate i, compounded semiannually, for 1 year. The amount A in the account after 1 year is given by

$$A = P\left(1 + \frac{i}{2}\right)^2.$$

Find an equivalent expression for A.

SKILL MAINTENANCE

77. Divide: $\left(-\frac{3}{4}\right) \div \frac{4}{5}$.

78. Find an expression equivalent to $\frac{5x}{9}$ with a denominator of 27.

79. Use a commutative law to write an expression equivalent to $(x + y)^3$.

80. Use an associative law to write an expression equivalent to $(3 + 6x) + 4$.

SYNTHESIS

Multiply. Assume that variables in exponents represent natural numbers.

81. $(-8s^3t)(2s^5 - 3s^3t^4 + st^7 - t^{10})$

82. $y^3z^n(y^{3n}z^3 - 4yz^{2n})$

83. $[(2x - 1)^2 - 1]^2$

84. $[x + y + 1][x^2 - x(y + 1) + (y + 1)^2]$

85. $[(a + b)(a - b)][5 - (a + b)][5 + (a + b)]$

86. $(y - 1)^6(y + 1)^6$

87. $(r^2 + s^2)^2(r^2 + 2rs + s^2)(r^2 - 2rs + s^2)$

88. $\left(3x^5 - \frac{5}{11}\right)^2$

89. $(a - b + c - d)(a + b + c + d)$

90. $\left(\frac{2}{3}x + \frac{1}{3}y + 1\right)\left(\frac{2}{3}x - \frac{1}{3}y - 1\right)$

91. $[2(y - 3) - 6(x + 4)][5(y - 3) - 4(x + 4)]$

92. $\left(x - \frac{1}{7}\right)\left(x^2 + \frac{1}{7}x + \frac{1}{49}\right)$

93. $(4x^2 + 2xy + y^2)(4x^2 - 2xy + y^2)$

94. $(x^2 - 7x + 12)(x^2 + 7x + 12)$

95. $(x^a + y^b)(x^a - y^b)(x^{2a} + y^{2b})$

96. $\left[1 + \frac{1}{5}(x + 1)^2\right]^2$

97. $[a - (b - 1)][(b - 1)^2 + a(b - 1) + a^2]$

98. $(x - 1)(x^2 + x + 1)(x^3 + 1)$

99. $\left[\left(\frac{1}{3}x^3 - \frac{2}{3}y^2\right)\left(\frac{1}{3}x^3 + \frac{2}{3}y^2\right)\right]^2$

100. $10(0.1x^4y - 0.01xy^4)^2$

3.3 Factoring Polynomials

To **factor** a polynomial means to do the reverse of multiplying, that is, to find an equivalent expression that is a product. Factoring is an important algebraic skill, and in this section and the next we will study the types of factorizations that will commonly arise in your study of mathematics.

Terms with Common Factors

When factoring, you should always look first for factors common to all the terms of an expression.

EXAMPLE 1 Factor out a common factor: $4y^2 - 8$.

Solution

$$\begin{aligned} 4y^2 - 8 &= 4 \cdot y^2 - 4 \cdot 2 \quad \text{Noting that 4 is a common factor} \\ &= 4(y^2 - 2) \end{aligned}$$

■

In some cases, there is more than one common factor. In Example 2, for instance, 5 is a common factor, x^3 is a common factor, and $5x^3$ is a common factor. If there is more than one common factor, we usually choose the one with the largest coefficient and the greatest exponent, that is, the *largest common factor.* In Example 2, that factor is $5x^3$.

EXAMPLES Factor out a common factor.

2. $5x^4 - 20x^3 = 5x^3(x - 4)$ Try to write your answer directly.

3. $12x^2y - 20x^3y = 4x^2y(3 - 5x)$ Use the distributive law and the product rule for powers to check mentally.

4. $10p^6q^2 - 4p^5q^3 + 2p^4q^4 = 2p^4q^2(5p^2 - 2pq + q^2)$ ■

The polynomials in Examples 1–4 have been *factored completely.* They cannot be factored further. The factors in the resulting factorization are said to be **prime polynomials.**

Consider the polynomial

$$-3x^2 + 6.$$

We can factor this as

$$-3x^2 + 6 = 3(-x^2 + 2),$$

or as

$$-3x^2 + 6 = -3(x^2 - 2).$$

In certain situations, the latter factorization will be helpful. It allows the leading coefficient of the binomial to be 1.

EXAMPLES Factor out a common factor with a negative coefficient.

5. $-4x - 24 = -4(x + 6)$

6. $-2x^2 + 6x - 10 = -2(x^2 - 3x + 5)$ ■

Factoring by Grouping

In more complicated expressions, there may be a common *binomial* factor. We proceed as in the following examples.

EXAMPLE 7 Factor: $(a - b)(x + 5) + (a - b)(x - y^2)$.

Solution

$$\begin{aligned} (a - b)(x + 5) + (a - b)(x - y^2) &= (a - b)[(x + 5) + (x - y^2)] \\ &= (a - b)[2x + 5 - y^2] \end{aligned}$$

■

EXAMPLE 8 Factor: $y^3 + 3y^2 + 4y + 12$.

Solution

$$\begin{aligned} y^3 + 3y^2 + 4y + 12 &= (y^3 + 3y^2) + (4y + 12) && \text{Grouping} \\ &= y^2(y + 3) + 4(y + 3) && \text{Factoring out common factors} \\ &= (y + 3)(y^2 + 4) && \text{Factoring out } y + 3 \end{aligned}$$

In Example 8, we factor two parts of the expression. Then we factored as in Example 7. Sometimes trial and error must be used to see which terms are to be grouped. Not all polynomials with four terms can be factored into a product of two binomials by grouping—for example, $x^3 + x^2 + 3x - 3$.

Factoring Trinomials of the Type $x^2 + bx + c$

When trying to factor trinomials of the type $x^2 + bx + c$, we can use a trial-and-error procedure.

Recall the FOIL method of multiplying two binomials:

$$\begin{aligned} &\qquad\quad\ \ \text{F} \quad\ \text{O} \quad\ \ \text{I} \quad\ \ \text{L} \\ (x + 3)(x + 5) &= x^2 + \underbrace{5x + 3x}_{\downarrow} + 15 \\ &= x^2 + \quad 8x \quad + 15. \end{aligned}$$

The product is a trinomial. In this example, the leading term has a coefficient of 1. The constant term is positive. To factor $x^2 + 8x + 15$, we think of FOIL in reverse. We multiplied x times x to get the first term of the trinomial. So the first term of each binomial factor is x. We want to find numbers p and q such that

$$x^2 + 8x + 15 = (x + p)(x + q).$$

This is where trial and error is often used. To get the middle term and the last term of the trinomial, we look for two numbers whose product is 15 and whose sum is 8. In this case, we know from above that the numbers are 3 and 5. Thus the factorization is

$$(x + 3)(x + 5), \qquad \text{or} \qquad (x + 5)(x + 3).$$

EXAMPLE 9 Factor: $x^2 + 9x + 8$.

Solution We think of FOIL in reverse. The first term of each factor is x. We are looking for numbers a and b such that

$$x^2 + 9x + 8 = (x + a)(x + b).$$

Then we look for two numbers whose product is 8 and whose sum is 9. Since both 8 and 9 are positive, we need only consider positive factors.

Pairs of Factors	Sums of Factors
2, 4	6
1, 8	9 ← The numbers we need are 1 and 8.

The factorization is thus $(x + 1)(x + 8)$. We can check by multiplying to see whether we get the original trinomial. ■

When factoring trinomials with a leading coefficient of 1, it suffices to list all pairs of factors along with their sums, as we did above. At times, however, you may wish to simply form factors without calculating any sums. It is essential that you check any attempt made in this manner! For example, if we attempt the factorization

$$x^2 + 9x + 8 \stackrel{?}{=} (x + 2)(x + 4),$$

a check reveals that $(x + 2)(x + 4) = x^2 + 6x + 8 \neq x^2 + 9x + 8$. This type of trial-and-error procedure becomes easier to use with time. As you gain experience, you will find that many trials will be performed mentally.

Constant Term Positive. When the constant term of a trinomial is positive, we look for two numbers with the same sign. The sign is that of the middle term.

EXAMPLE 10 Factor: $y^2 - 9y + 20$.

Solution Since the constant term is positive and the coefficient of the middle term is negative, we look for a factorization of 20 in which both factors are negative. Their sum must be -9.

Pairs of Factors	Sums of Factors
$-1, -20$	-21
$-2, -10$	-12
$-4, -5$	-9 ← The numbers we need are -4 and -5.

The factorization is $(y - 4)(y - 5)$. ■

Constant Term Negative. When the constant term of a trinomial is negative, the middle term may be positive or negative. In these cases, we look for two factors whose product is negative. One of them must be positive and the other negative. Their sum must still be the coefficient of the middle term.

EXAMPLE 11 Factor: $x^3 - x^2 - 30x$.

Solution *Always* look first for a common factor! This time there is one, x. We first factor it out:

$$x^3 - x^2 - 30x = x(x^2 - x - 30).$$

Now we consider $x^2 - x - 30$. Since the constant term is negative, we look for a factorization of -30 in which one factor is positive and one factor is negative. Their sum must be -1, so the negative factor must have the larger absolute value. Thus we consider only pairs of factors in which the negative term has the larger absolute value.

Pairs of Factors	Sums of Factors
$1, -30$	-29
$3, -10$	-7
$5, -6$	-1 ← The numbers we want are 5 and -6.

The factorization of $x^2 - x - 30$ is

$$(x + 5)(x - 6).$$

Don't forget to include the factor that was factored out earlier! In this case, the factorization of the original trinomial is

$$x(x + 5)(x - 6).$$

■

The procedure considered here can also be applied to a trinomial with more than one variable.

EXAMPLE 12 Factor: $x^2 - 2xy - 48y^2$.

Solution We look for numbers p and q such that

$$x^2 - 2xy - 48y^2 = (x + py)(x + qy).$$

Our thinking is much the same as if we were factoring $x^2 - 2x - 48$. We look for factors of -48 whose sum is -2. Those factors are -8 and 6. Then

$$x^2 - 2xy - 48y^2 = (x + 6y)(x - 8y).$$

We check by multiplying:

$$\begin{aligned}(x + 6y)(x - 8y) &= x^2 - 8xy + 6xy - 48y^2\\ &= x^2 - 2xy - 48y^2.\end{aligned}$$

■

Sometimes a trinomial like $x^6 + 2x^3 - 15$ can be factored if we first make a substitution.

EXAMPLE 13 Factor: $x^6 + 2x^3 - 15$.

Solution We let $u = x^3$, because this way $u^2 = x^3 \cdot x^3 = x^6$. Then we consider $u^2 + 2u - 15$. Since

$$u^2 + 2u - 15 = (u - 3)(u + 5),$$

we can replace u by x^3 and factor the original trinomial:

$$x^6 + 2x^3 - 15 = (x^3 - 3)(x^3 + 5).$$

With practice, you will make such a substitution mentally. ■

Factoring Trinomials of the Type $ax^2 + bx + c$, $a \neq 1$

Now we consider trinomials where the leading coefficient is not 1. We consider two methods. Both methods involve trial and error. Use the method that works best for you, or you may use the one that your instructor chooses for you.

Method 1: The FOIL Method. We first consider the FOIL method for factoring trinomials of the type

$$ax^2 + bx + c, \qquad \text{where } a \neq 1.$$

Consider the following multiplication.

$$\begin{aligned} & \qquad\qquad\quad \text{F} \quad\;\; \text{O} \quad\;\; \text{I} \quad\;\; \text{L} \\ (2x + 5)(3x + 4) &= 6x^2 + 8x + 15x + 20 \\ &= 6x^2 + \quad 23x \quad + 20 \end{aligned}$$

Now to factor $6x^2 + 23x + 20$, we do the reverse of what we just did. We look for two binomials, $px + q$ and $rx + s$, whose product is this trinomial. The product of the first terms must be $6x^2$. The product of the outside terms plus the product of the inside terms must be $23x$. The product of the last terms must be 20. We know from the preceding discussion that the answer is

$$(2x + 5)(3x + 4).$$

Usually, however, finding such an answer is a trial-and-error process. It turns out that

$$(-2x - 5)(-3x - 4)$$

is also a correct answer, but we generally choose an answer in which the first coefficients are positive.

We will use the following method.

> To factor $ax^2 + bx + c$:
>
> **a)** First look for a common factor.
> **b)** Factor the first term, ax^2.
> **c)** Factor the last term, c.
> **d)** Look for factors in (b) and (c) such that the sum of their products is the middle term, bx.

EXAMPLE 14 Factor: $3x^2 + 10x - 8$.

Solution

a) First we look for a common factor. There is none (other than 1 or -1).

b) We factor the first term, $3x^2$. These factors are $3x, x$. We have this possibility:

$$(3x \quad)(x \quad).$$

c) We factor the last term, -8, which is negative. These factors are

$$8, -1 \text{ and } -8, 1 \qquad \text{and} \qquad -2, 4 \text{ and } 2, -4.$$

d) We look for factors in (b) and (c) such that the sum of the products (the "outside" and "inside" parts of FOIL) is the middle term, $10x$. We try some possibilities for factorization and check by multiplying:

$$(3x - 8)(x + 1) = 3x^2 - 5x - 8.$$

This gives us a middle term that is negative. We avoid such possibilities and look for possibilities that give a positive middle term:

$$(3x + 8)(x - 1) = 3x^2 + 5x - 8.$$

Notice that changing the signs in the binomials has the effect of changing the sign of the middle term. We try again:

$$(3x - 2)(x + 4) = 3x^2 + 10x - 8.$$

Thus the desired factorization is $(3x - 2)(x + 4)$. ■

EXAMPLE 15 Factor: $6x^6 - 19x^5 + 10x^4$.

a) First factor out the common factor x^4:

$$x^4(6x^2 - 19x + 10).$$

b) Next, consider that $6x^2 = 6x \cdot x$ and $6x^2 = 3x \cdot 2x$. Thus the possible factorizations of $6x^2 - 19x + 10$ are

$$(6x \quad)(x \quad) \qquad \text{and} \qquad (3x \quad)(2x \quad).$$

c) We factor the last term, 10, which is positive. These factors are

$$10, 1 \text{ and } -10, -1 \qquad \text{and} \qquad 5, 2 \text{ and } -5, -2.$$

d) We attempt to pair factors in (b) and (c) so that the sum of the products (the "outside" and "inside" parts of FOIL) is the middle term, $-19x$. We try a possible factorization and check by multiplying:

$$(3x - 10)(2x - 1) = 6x^2 - 23x + 10.$$

We try again:

$$(3x - 5)(2x - 2) = 6x^2 - 16x + 10.$$

Actually this last attempt could have been rejected by simply noting that $2x - 2$ has a common factor, 2. Since we removed the largest common factor in step (a), no other common factors can exist. We try again, reversing the -5 and -2:

$$(3x - 2)(2x - 5) = 6x^2 - 19x + 10.$$

The factorization of $6x^2 - 19x + 10$ is $(3x - 2)(2x - 5)$. But do not forget the common factor! We must include it to get a factorization of the original trinomial:

$$6x^6 - 19x^5 + 10x^4 = x^4(3x - 2)(2x - 5).$$ ■

Method 2: The Grouping Method. The second method for factoring trinomials of the type $ax^2 + bx + c$, $a \neq 1$, is known as the *grouping method*. It involves not only trial and error and FOIL but also factoring by grouping. We know how to factor the trinomial $x^2 + 7x + 10$. We look for factors of the constant term, 10, whose sum is the coefficient of the middle term, 7:

$$x^2 + 7x + 10$$

(1) Factor: $10 = 2 \cdot 5$.
(2) Sum of factors: $2 + 5 = 7$.

What happens when the leading coefficient is not 1? Consider the trinomial $6x^2 + 23x + 20$. The method we use is similar to what we used for the preceding trinomial, but we need two more steps. We first multiply the leading coefficient, 6, and the constant, 20, and get 120. Then we look for a factorization of 120 in which the sum of the factors is the coefficient of the middle term: 23. Next we split the middle term into a sum or difference using these factors.

$$6x^2 + 23x + 20$$

(1) Multiply 6 and 20: $6 \cdot 20 = 120$.
(2) Factor 120: $120 = 8 \cdot 15$, and $8 + 15 = 23$.
(3) Split the middle term: $23x = 8x + 15x$.
(4) Factor by grouping.

We factor by grouping as follows:

$$\begin{aligned} 6x^2 + 23x + 20 &= 6x^2 + 8x + 15x + 20 \\ &= 2x(3x + 4) + 5(3x + 4) \\ &= (3x + 4)(2x + 5). \end{aligned}$$

To factor $ax^2 + bx + c$, using the grouping method:

a) First look for a common factor.
b) Multiply the leading coefficient a and the constant c.
c) Try to factor the product ac such that the sum of the factors is b. That is, find integers p and q such that $pq = ac$ and $p + q = b$.
d) Split the middle term. That is, write it as a sum using the factors found in (c).
e) Factor by grouping.

EXAMPLE 16 Factor: $3x^2 + 10x - 8$.

Solution

a) First we look for a common factor. There is none (other than 1 or -1).

b) We multiply the leading coefficient and the constant, 3 and -8:

$$3(-8) = -24.$$

c) We try to factor -24 so that the sum of the factors is 10:

$$-24 = 12(-2) \quad \text{and} \quad 12 + (-2) = 10.$$

d) We split $10x$ using the results of part (c); that is, we split the middle term as follows:

$$10x = 12x - 2x.$$

e) We then factor by grouping:

$$\begin{aligned} 3x^2 + 10x - 8 &= 3x^2 + 12x - 2x - 8 \qquad \text{Substituting } 12x - 2x \text{ for the } 10x \\ &= 3x(x + 4) - 2(x + 4) \\ &= (x + 4)(3x - 2). \end{aligned}$$

■

EXERCISE SET 3.3

Factor.

1. $4a^2 + 2a$
2. $6y^2 + 3y$
3. $y^2 - 5y$
4. $x^2 + 9x$
5. $y^3 + 9y^2$
6. $x^3 + 8x^2$
7. $6x^2 - 3x^4$
8. $8y^2 + 4y^4$
9. $4x^2y - 12xy^2$
10. $5x^2y^3 + 15x^3y^2$
11. $3y^2 - 3y - 9$
12. $5x^2 - 5x + 15$
13. $4ab - 6ac + 12ad$
14. $8xy + 10xz - 14xw$
15. $10a^4 + 15a^2 - 25a - 30$
16. $12t^5 - 20t^4 + 8t^2 - 16$

Factor out a factor with a negative coefficient.

17. $-3x + 12$
18. $-5x - 40$
19. $-6y - 72$
20. $-8t + 72$
21. $-2x^2 + 4x - 12$
22. $-2x^2 + 12x + 40$

Factor.

23. $a(b - 2) + c(b - 2)$
24. $a(x^2 - 3) - 2(x^2 - 3)$
25. $(x - 2)(x + 5) + (x - 2)(x + 8)$
26. $(m - 4)(m + 3) + (m - 4)(m - 3)$
27. $a^2(x - y) + a^2(x - y)$
28. $3x^2(x - 6) + 3x^2(x - 6)$
29. $ac + ad + bc + bd$
30. $xy + xz + wy + wz$
31. $b^3 - b^2 + 2b - 2$
32. $y^3 - y^2 + 3y - 3$
33. $a^3 - 3a^2 + 2a - 6$
34. $t^3 + 6t^2 - 2t - 12$
35. $x^6 + x^5 - x^3 + x^2$
36. $y^4 - y^3 + y^2 + y$
37. $2y^4 + 6y^2 + 5y^2 + 15$
38. $2xy + x^2y - 6 - 3x$

Factor.

39. $x^2 + 9x + 20$
40. $x^2 + 8x + 15$
41. $t^2 - 8t + 15$
42. $y^2 - 12y + 27$
43. $x^2 - 27 - 6x$
44. $t^2 - 15 - 2t$

45. $2y^2 - 16y + 32$
46. $2a^2 - 20a + 50$
47. $p^2 + 3p - 54$
48. $m^2 + m - 72$
49. $14x + x^2 + 45$
50. $12y + y^2 + 32$
51. $y^2 + 2y - 63$
52. $p^2 + 3p - 40$
53. $t^2 - 11t + 28$
54. $y^2 - 14y + 45$
55. $3x + x^2 - 10$
56. $x + x^2 - 6$
57. $x^2 + 5x + 6$
58. $y^2 + 8y + 7$
59. $56 + x - x^2$
60. $32 + 4y - y^2$
61. $32y + 4y^2 - y^3$
62. $56x + x^2 - x^3$
63. $x^4 + 11x^2 - 80$
64. $y^4 + 5y^2 - 84$
65. $x^2 - 3x + 7$
66. $x^2 + 12x + 13$
67. $x^2 + 12xy + 27y^2$
68. $p^2 - 5pq - 24q^2$
69. $x^2 - 14x + 49$
70. $y^2 + 8y + 16$
71. $x^4 + 50x^2 + 49$
72. $p^4 - 80p^2 + 79$
73. $x^6 + 2x^3 - 63$
74. $x^8 - 7x^4 + 10$

Factor.

75. $3x^2 - 16x - 12$
76. $6x^2 - 5x - 25$
77. $6x^3 - 15x - x^2$
78. $10y^3 - 12y - 7y^2$
79. $3a^2 - 10a + 8$
80. $12a^2 - 7a + 1$
81. $35y^2 + 34y + 8$
82. $9a^2 + 18a + 8$
83. $4t + 10t^2 - 6$
84. $8x + 30x^2 - 6$
85. $8x^2 - 16 - 28x$
86. $18x^2 - 24 - 6x$
87. $12x^3 - 31x^2 + 20x$
88. $15x^3 - 19x^2 - 10x$
89. $14x^4 - 19x^3 - 3x^2$
90. $70x^4 - 68x^3 + 16x^2$
91. $3a^2 - a - 4$
92. $6a^2 - 7a - 10$
93. $9x^2 + 15x + 4$
94. $6y^2 - y - 2$
95. $3 + 35z - 12z^2$
96. $8 - 6a - 9a^2$
97. $-4t^2 - 4t + 15$
98. $-12a^2 + 7a - 1$
99. $3x^3 - 5x^2 - 2x$
100. $18y^3 - 3y^2 - 10y$
101. $24x^2 - 2 - 47x$
102. $15y^2 - 10 - 47y$
103. $21x^2 + 37x + 12$
104. $10y^2 + 23y + 12$
105. $40x^4 + 16x^2 - 12$
106. $24y^4 + 2y^2 - 15$
107. $12a^2 - 17ab + 6b^2$
108. $20p^2 - 23pq + 6q^2$
109. $2x^2 + xy - 6y^2$
110. $8m^2 - 6mn - 9n^2$
111. $6x^2 - 29xy + 28y^2$
112. $10p^2 + 7pq - 12q^2$
113. $9x^2 - 30xy + 25y^2$
114. $4p^2 + 12pq + 9q^2$
115. $6x^6 + x^3 - 2$
116. $2p^8 + 11p^4 + 15$

SKILL MAINTENANCE

117. A paddleboat moves at a rate of 9 km/h in still water. The current in a river is 3.7 km/h. How long will it take the boat to travel 31.75 km downriver?

118. Solve $2y - 9 > 5y + 18$. Then graph.

SYNTHESIS

Factor. Assume that all exponents are natural numbers.

119. $4y^{4a} + 12y^{2a} + 10y^{2a} + 30$
120. $2x^{4p} + 6x^{2p} + 5x^{2p} + 15$
121. $4x^{a+b} + 7x^{a-b}$
122. $7y^{2a+b} - 5y^{a+b} + 3y^{a+2b}$
123. $x^{2a} + 5x^a - 24$
124. $4x^{2a} - 4x^a - 3$
125. $a^2p^{2a} + a^2p^a - 2a^2$
126. $(x + 3)^2 - 2(x + 3) - 35$
127. $6(x - 7)^2 + 13(x - 7) - 5$

128. Find all integers m for which $x^2 + mx + 75$ can be factored.

129. Find all integers q for which $x^2 + qx - 32$ can be factored.

3.4 Special Factorizations

Rules for Exponents

Before learning quick methods for factoring certain types of polynomials, we will need to review some more rules for exponents.

Consider the product $x^3 \cdot x^3$. By the product rule of Section 3.2, we have $x^3 \cdot x^3 = x^{3+3} = x^6$. Note that $3 + 3 = 3 \cdot 2$ and $x^3 \cdot x^3 = (x^3)^2$, so we can write

$$(x^3)^2 = x^{3 \cdot 2} = x^6.$$

Similarly, $(y^8)^3 = (y^8)(y^8)(y^8) = y^{8+8+8} = y^{24}$. The same result is found by multiplying exponents:

$$(y^8)^3 = y^{8 \cdot 3} = y^{24}.$$

We have the following rule:

RAISING A POWER TO A POWER: THE POWER RULE

For any real number a and any natural numbers m and n,

$$(a^m)^n = a^{mn}.$$

(To raise a power to a power, we can multiply the exponents.)

EXAMPLES Simplify.

1. $(3^5)^4 = 3^{5 \cdot 4}$ Multiply exponents.
$= 3^{20}$

2. $(y^5)^7 = y^{5 \cdot 7}$
$= y^{35}$ ■

When an expression inside parentheses is raised to a power, the inside expression is the base. Let us compare $2a^3$ and $(2a)^3$.

$2a^3 = 2 \cdot a \cdot a \cdot a$;   The base is a.

$(2a)^3 = (2a)(2a)(2a)$ The base is $2a$.

$= (2 \cdot 2 \cdot 2)(a \cdot a \cdot a)$ Using the associative and commutative laws

$= 2^3a^3 = 8a^3$

We see that $2a^3$ and $(2a)^3$ are *not* equivalent. We also see that we can evaluate the power $(2a)^3$ by raising each factor to the power 3. This leads us to the following rule for raising a product to a power.

RAISING A PRODUCT TO A POWER

For any real numbers a and b and any natural number n,

$$(ab)^n = a^n b^n.$$

(To raise a product to the nth power, raise each factor to the nth power.)

EXAMPLES Simplify.

3. $(-2x)^3 = (-2)^3 \cdot x^3$ Raising each factor to the third power

$= -8x^3$

4. $(3x^4)^2 = 3^2 \cdot (x^4)^2$ Raising each factor to the second power

$= 9x^8$

There is a similar rule for raising a quotient to a power.

RAISING A QUOTIENT TO A POWER

For any real numbers a and b, $b \neq 0$, and any natural number n,

$$\left(\frac{a}{b}\right)^n = \frac{a^n}{b^n}.$$

(To raise a quotient to a power, raise the numerator to the power and divide by the denominator to the power.)

EXAMPLE 5 Simplify: $\left(\frac{x^2}{3}\right)^4$.

$$\left(\frac{x^2}{3}\right)^4 = \frac{(x^2)^4}{3^4} = \frac{x^8}{81}$$

Factoring a Difference of Squares

The following is a *difference of squares*:

$$x^2 - 9.$$

To factor a difference of two expressions that are squares, we can reverse a pattern that we used earlier for multiplying a sum and a difference.

$$A^2 - B^2 = (A + B)(A - B)$$

To factor a difference of two squares, write the product of the sum and the difference of the two quantities being squared.

EXAMPLE 6 Factor: $x^2 - 9$.

Solution

$$x^2 - 9 = x^2 - 3^2 = (x + 3)(x - 3)$$ ■

Sometimes the quantities being squared may be products with powers.

EXAMPLE 7 Factor: $25y^6 - 49x^2$.

Solution Note that $25y^6 = (5y^3)^2$ and $49x^2 = (7x)^2$.

$$\begin{array}{ccccccccc} A^2 & - & B^2 & = & (A & + & B)(A & - & B) \\ \downarrow & & \downarrow & & \downarrow & & \downarrow\ \ \downarrow & & \downarrow \\ 25y^6 & - & 49x^2 & = & (5y^3 & + & 7x)(5y^3 & - & 7x) \end{array}$$ ■

Common factors should always be considered first. This eases the factoring process because the type of factoring to be done becomes clearer.

EXAMPLE 8 Factor: $5 - 5x^2y^6$.

Solution There is a common factor, 5.

$$\begin{aligned} 5 - 5x^2y^6 &= 5(1 - x^2y^6) \\ &= 5[1^2 - (xy^3)^2] && \text{Rewriting } x^2y^6 \text{ as a quantity squared} \\ &= 5(1 + xy^3)(1 - xy^3) && \text{Factoring the difference of squares} \end{aligned}$$ ■

Trinomial Squares

Consider the trinomial

$$x^2 + 6x + 9.$$

To factor it, we could use a method considered in the preceding section. We would discover, after some work, that the factorization is

$$x^2 + 6x + 9 = (x + 3)(x + 3) = (x + 3)^2.$$

Note that the result is the square of a binomial. We also say that $x^2 + 6x + 9$ is a **trinomial square.** Now we can certainly use the trial-and-error procedure to factor trinomial squares, but we want to develop a faster procedure.

To do so, we must first be able to recognize when a trinomial is a square.

HOW TO RECOGNIZE A TRINOMIAL SQUARE

a) Two of the terms must be squares, such as A^2 and B^2.
b) There must be no minus sign before A^2 or B^2.
c) If we multiply A and B (which are the quantities that were squared) and double the result, we get the remaining term, or its opposite.

EXAMPLES Determine whether these are trinomial squares.

9. $x^2 + 10x + 25$

a) Two terms are squares: x^2 and 25.
b) There is no minus sign before either x^2 or 25.
c) If we multiply the quantities x and 5 and double the result, we get $10x$, the remaining term.

Thus the trinomial is a square.

10. $4x + 16 + 3x^2$

a) Only one term, 16, is a square. ($3x^2$ is not a square since 3 is not a perfect-square integer, and $4x$ is not a square since x is not a square.)

Therefore, the trinomial is not a square. ■

To factor trinomial squares, we use the same equations that we used to multiply out squares of binomials. They are as follows.

$$A^2 + 2AB + B^2 = (A + B)^2$$
$$A^2 - 2AB + B^2 = (A - B)^2$$

EXAMPLE 11 Factor: $x^2 - 10x + 25$.

Solution

$$x^2 - 10x + 25 = (x - 5)^2$$

Note the sign! (the minus sign in $x^2 - 10x$ and in $(x - 5)$)

We find the square terms and write the quantities that were squared with a minus sign between them.

To check, we multiply: $(x - 5)^2 = (x - 5)(x - 5) = x^2 - 10x + 25$. ■

EXAMPLE 12 Factor: $16y^2 + 49 + 56y$.

Solution

$$16y^2 + 49 + 56y = 16y^2 + 56y + 49 \quad \text{Using the commutative law}$$
$$= (4y + 7)^2 \quad \text{We find the square terms and write the quantities that were squared with a plus sign between them.}$$

The check is left to the student. ■

More Factoring by Grouping

Sometimes when factoring a polynomial with four terms, we may be able to factor by regrouping terms.

EXAMPLE 13 Factor: $x^3 + 3x^2 - 4x - 12$.

Solution

$$x^3 + 3x^2 - 4x - 12 = x^2(x + 3) - 4(x + 3)$$
$$= (x + 3)(x^2 - 4)$$
$$= (x + 3)(x + 2)(x - 2)$$ ■

A difference of squares can have more than two terms. For example, one of the squares may be a trinomial. We can factor by a type of grouping.

EXAMPLE 14 Factor: $x^2 + 6x + 9 - y^2$.

Solution

$$x^2 + 6x + 9 - y^2 = (x^2 + 6x + 9) - y^2 \quad \text{Grouping as a trinomial minus } y^2 \text{ to show a difference of squares}$$
$$= (x + 3)^2 - y^2$$
$$= (x + 3 + y)(x + 3 - y)$$ ■

Sums or Differences of Cubes

Consider the following products:

$$(A + B)(A^2 - AB + B^2) = A(A^2 - AB + B^2) + B(A^2 - AB + B^2)$$
$$= A^3 - A^2B + AB^2 + A^2B - AB^2 + B^3$$
$$= A^3 + B^3 \quad \text{Combining like terms}$$

and

$$(A - B)(A^2 + AB + B^2) = A(A^2 + AB + B^2) - B(A^2 + AB + B^2)$$
$$= A^3 + A^2B + AB^2 - A^2B - AB^2 - B^3$$
$$= A^3 - B^3. \quad \text{Combining like terms}$$

The above equations (reversed) show how we can factor a sum or a difference of two cubes.

$$A^3 + B^3 = (A + B)(A^2 - AB + B^2)$$
$$A^3 - B^3 = (A - B)(A^2 + AB + B^2)$$

The table of powers at the back of this text can help in the following examples and exercises.

EXAMPLE 15 Factor: $x^3 - 27$.

Solution

$$x^3 - 27 = x^3 - 3^3$$

In one set of parentheses, we write the first quantity that was cubed, x, and the second quantity that was cubed, -3. This gives us the expression $x - 3$:

$$(x - 3)(\qquad).$$

To get the next factor, we think of $x - 3$ and do the following:

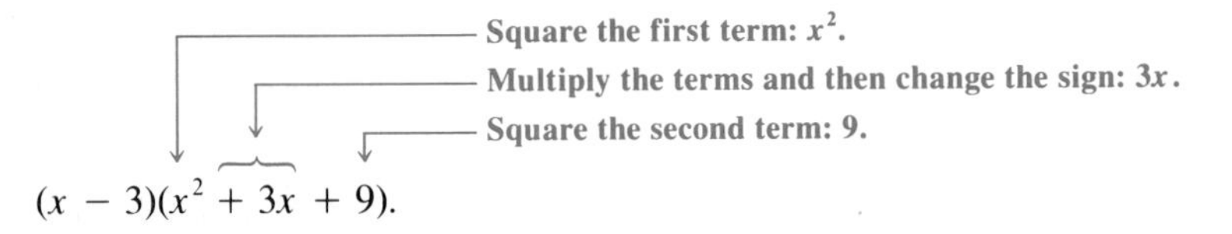

Note that we cannot factor $x^2 + 3x + 9$. (It is not a trinomial square nor can it be factored by trial.) ■

EXAMPLE 16 Factor: $125x^3 + y^3$.

Solution

$$125x^3 + y^3 = (5x)^3 + y^3$$

In one set of parentheses, we write the first quantity being cubed, $5x$, then a plus sign, and then the second quantity being cubed, y:

$$(5x + y)(\qquad).$$

To get the next factor, we think of $5x + y$ and do the following:

Square the first term: $(5x)^2$, or $25x^2$.
Multiply the terms and then change the sign: $-5xy$.
Square the second term: y^2.

$$(5x + y)(25x^2 - 5xy + y^2).$$

■

EXAMPLE 17 Factor: $128y^7 - 250x^6y$.

Solution We first look for a common factor:

$$\begin{aligned}128y^7 - 250x^6y &= 2y(64y^6 - 125x^6)\\ &= 2y[(4y^2)^3 - (5x^2)^3] \quad \textbf{Rewriting as quantities cubed}\\ &= 2y(4y^2 - 5x^2)(16y^4 + 20x^2y^2 + 25x^4).\end{aligned}$$

■

EXAMPLE 18 Factor: $a^6 - b^6$.

Solution We can express this polynomial as a difference of squares:

$$(a^3)^2 - (b^3)^2.$$

We factor as follows:

$$(a^3 + b^3)(a^3 - b^3).$$

One factor is a sum of two cubes, and the other factor is a difference of two cubes. We factor them:

$$(a + b)(a^2 - ab + b^2)(a - b)(a^2 + ab + b^2).$$

We have factored completely. ■

In Example 18, had we thought of factoring first as a difference of two cubes, we would have had

$$\begin{aligned}(a^2)^3 - (b^2)^3 &= (a^2 - b^2)(a^4 + a^2b^2 + b^4)\\ &= (a + b)(a - b)(a^4 + a^2b^2 + b^4).\end{aligned}$$

In this case, we might have missed some factors; $a^4 + a^2b^2 + b^4$ can be factored as $(a^2 - ab + b^2)(a^2 + ab + b^2)$, but we probably would not have known to do such factoring.

The following outline constitutes a strategy for factoring that you may find useful as you proceed through this text.

A. Always factor out the largest common factor that exists.
B. Look at the number of terms.

Two terms: Try factoring as a difference of squares first. Next, try factoring as a sum or a difference of cubes. Do *not* try to factor a *sum* of squares.

Three terms: Determine whether the trinomial is a square. If so, you know how to factor. If not, try trial and error, using the FOIL method or the grouping method.

Four or more terms: Try factoring by grouping and factoring out a common binomial factor. Next, try grouping into a difference of squares, one of which is a trinomial.

C. Always *factor completely*. If a factor with more than one term can be factored, you should factor it.

EXERCISE SET 3.4

Simplify.

1. $(2^4)^7$
2. $(3^4)^5$
3. $(a^5)^3$
4. $(x^3)^7$
5. $(3x)^2$
6. $(5a)^2$
7. $(-2m)^4$
8. $(-3x)^4$
9. $(5b^3)^2$
10. $(7y^6)^2$
11. $(-3a^2b^3)^3$
12. $(-5x^4y^3)^3$
13. $\left(\frac{x^5}{y^3}\right)^2$
14. $\left(\frac{a^4}{b^6}\right)^2$
15. $\left(\frac{3m^2}{n^7}\right)^3$
16. $\left(\frac{2x^7}{y^3}\right)^4$

Factor.

17. $x^2 - 16$
18. $y^2 - 9$
19. $p^2 - 49$
20. $m^2 - 64$
21. $p^2q^2 - 25$
22. $a^2b^2 - 81$
23. $6x^2 - 6y^2$
24. $8x^2 - 8y^2$
25. $4xy^4 - 4xz^4$
26. $25ab^4 - 25az^4$
27. $4a^3 - 49a$
28. $9x^3 - 25x$
29. $3x^8 - 3y^8$
30. $9a^4 - a^2b^2$
31. $9a^4 - 25a^2b^4$
32. $16x^6 - 121x^2y^4$
33. $\frac{1}{25} - x^2$
34. $\frac{1}{16} - y^2$
35. $0.04x^2 - 0.09y^2$
36. $0.01x^2 - 0.04y^2$

Factor.

37. $y^2 - 6y + 9$
38. $x^2 - 8x + 16$
39. $x^2 + 14x + 49$
40. $x^2 + 16x + 64$
41. $x^2 + 1 + 2x$
42. $x^2 + 1 - 2x$
43. $2a^2 + 8a + 8$
44. $4a^2 - 16a + 16$
45. $-18y^2 + y^3 + 81y$
46. $24a^2 + a^3 + 144a$
47. $12a^2 + 36a + 27$
48. $20y^2 + 100y + 125$
49. $2x^2 - 40x + 200$
50. $32x^2 + 48x + 18$
51. $y^4 + 8y^2 + 16$
52. $a^4 - 10a^2 + 25$
53. $p^2 - 2pq + q^2$
54. $m^2 + 2mn + n^2$
55. $a^2 + 4ab + 4b^2$
56. $49p^2 - 14pq + q^2$
57. $25a^2 - 30ab + 9b^2$
58. $49p^2 - 84pq + 36q^2$
59. $x^4 + 2x^2y^2 + y^4$
60. $p^8 + 2p^4q^4 + q^8$

Factor.

61. $m^3 - 7m^2 - 4m + 28$
62. $x^3 + 8x^2 - x - 8$
63. $a^3 - ab^2 - 2a^2 + 2b^2$
64. $p^2q - 25q + 3p^2 - 75$
65. $(a + b)^2 - 100$
66. $(p - 7)^2 - 144$
67. $a^2 + 2ab + b^2 - 9$
68. $x^2 - 2xy + y^2 - 25$
69. $r^2 - 2r + 1 - 4s^2$
70. $c^2 + 4cd + 4d^2 - 9p^2$
71. $9 - (a^2 + 2ab + b^2)$
72. $16 - (x^2 - 2xy + y^2)$

Factor.

73. $x^3 + 8$
74. $c^3 + 27$
75. $y^3 - 64$
76. $z^3 - 1$
77. $w^3 + 1$
78. $x^3 + 125$
79. $8a^3 + 1$
80. $27x^3 + 1$
81. $y^3 - 8$
82. $p^3 - 27$
83. $8 - 27b^3$
84. $64 - 125x^3$
85. $64y^3 + 1$
86. $125x^3 + 1$
87. $8x^3 + 27$
88. $27y^3 + 64$
89. $a^3 + \frac{1}{8}$
90. $b^3 + \frac{1}{27}$
91. $2y^3 - 128$
92. $3z^3 - 3$
93. $5x^3 - 40z^3$
94. $2y^3 - 54z^3$
95. $64x^6 - 8t^6$
96. $125c^6 - 8d^6$
97. $2y^4 - 128y$
98. $3z^5 - 3z^2$
99. $z^6 - 1$
100. $t^6 + 1$

SKILL MAINTENANCE

101. Add: $16.2 + (-19.3)$.
102. Solve: $\frac{4}{3}(2 - 3x) = 8$.

SYNTHESIS

Factor. Assume that variables in exponents represent natural numbers.

103. $27x^{6s} + 64y^{3t}$
104. $64a^{20} - 4b^{20}$
105. $4x^2 + 4xy + y^2 - r^2 + 6rs - 9s^2$
106. $(y + x)^3 + x^3$

107. $c^4d^4 - a^{16}$

108. $(1 - x)^3 - (x - 1)^6$

109. $c^{2w+1} - 2c^{w+1} + c$

110. $24x^{2a} - 6$

111. $y^9 - y$

112. $1 - \dfrac{x^{27}}{1000}$

113. $3a^2 + 3b^2 - 3c^2 - 3d^2 + 6ab - 6cd$

114. $3(x + 1)^2 + 9(x + 1) - 12$

115. $(m - 1)^3 + (m + 1)^3$

116. $3(a - 2)^2 - 30(a - 2) + 75$

117. Suppose that $\left(x + \dfrac{2}{x}\right)^2 = 6$. Find $x^3 + \dfrac{8}{x^3}$.

3.5 Using Factoring for Problem Solving

An equation like $2x^2 + 5x - 3 = 0$, with a polynomial on one side and 0 on the other, is called a **polynomial equation.** When the polynomial is quadratic we call the equation a *quadratic* equation and when the polynomial is linear we say that we have a *linear* equation. Any equation equivalent to a polynomial equation is also considered to be a polynomial equation. Polynomial equations occur often in applied problems, so the ability to solve them is an important skill.

One important method of finding solutions—also called *roots* or *zeros*—of polynomial equations involves factoring the polynomial and using a property of the real numbers.

The Principle of Zero Products

When we multiply two or more numbers, the product will be 0 if one of the factors is 0. Conversely, if a product is 0, then at least one of the factors must be 0. This property of real numbers gives us another principle for solving equations.

> **THE PRINCIPLE OF ZERO PRODUCTS**
>
> For any real numbers a and b:
>
> If $a \cdot b = 0$, then $a = 0$ or $b = 0$.
> If $a = 0$ or $b = 0$, then $a \cdot b = 0$.

To use this principle in solving equations, we make sure that there is a 0 on one side of the equation and a factorization on the other side.

EXAMPLE 1 Solve: $x^2 - x = 6$.

Solution To use the principle of zero products, we must have 0 on one side of the equation, so we add -6 on both sides:

$$x^2 - x - 6 = 0. \qquad \text{Getting 0 on one side}$$

We need a factorization on the other side, so we factor the polynomial:

$$(x - 3)(x + 2) = 0. \qquad \text{Factoring}$$

We now set each factor equal to 0 (this is a use of the principle of zero products):

$$x - 3 = 0 \quad \text{or} \quad x + 2 = 0. \qquad \text{Using the principle of zero products}$$

These are two linear equations. We solve them separately:

$$x = 3 \quad \text{or} \quad x = -2.$$

We check as follows:

Check:

$x^2 - x = 6$	
$3^2 - 3$	6
$9 - 3$	
6	

TRUE

$x^2 - x = 6$	
$(-2)^2 - (-2)$	6
$4 + 2$	
6	

TRUE

The numbers 3 and -2 are both solutions. The solution set is $\{3, -2\}$. ■

> To solve using the principle of zero products:
>
> **1.** Obtain a 0 on one side of the equation.
> **2.** Factor the other side.
> **3.** Set each factor equal to 0.
> **4.** Solve the resulting equations.

EXAMPLE 2 Solve: $7y + 3y^2 = -2$.

Solution There must be a 0 on one side of the equation. We add 2 to both sides to get 0 on one side and arrange in descending order. Then we factor and use the principle of zero products.

$$3y^2 + 7y + 2 = 0$$

$$(3y + 1)(y + 2) = 0 \qquad \text{Factoring}$$

$$3y + 1 = 0 \quad \text{or} \quad y + 2 = 0 \qquad \text{Using the principle of zero products}$$

$$y = -\tfrac{1}{3} \quad \text{or} \quad y = -2. \qquad \text{Solving both equations}$$

The check is left for the reader. The solutions are $-\frac{1}{3}$ and -2 and the solution set is $\{-\frac{1}{3}, -2\}$. ■

EXAMPLE 3 Solve: $5b^2 = 10b$.

Solution

$$5b^2 = 10b$$

$$5b^2 - 10b = 0 \qquad \text{Getting 0 on one side}$$

$$5b(b - 2) = 0 \qquad \text{Factoring}$$

$$5b = 0 \quad \text{or} \quad b - 2 = 0 \qquad \text{Using the principle of zero products}$$

$$b = 0 \quad \text{or} \quad b = 2$$

The solutions are 0 and 2. The solution set is $\{0, 2\}$. ■

EXAMPLE 4 Solve: $x^2 - 6x + 9 = 0$.

Solution

$$x^2 - 6x + 9 = 0$$

$$(x - 3)(x - 3) = 0 \quad \text{Factoring}$$

$$x - 3 = 0 \quad \text{or} \quad x - 3 = 0 \quad \text{Using the principle of zero products}$$

$$x = 3 \quad \text{or} \quad x = 3 \quad \text{We have a } \textit{repeated root} \text{ of 3.}$$

There is only one solution, 3. The solution set is $\{3\}$. ■

EXAMPLE 5 Solve: $3x^3 - 9x^2 = 30x$.

Solution

$$3x^3 - 9x^2 = 30x$$

$$3x^3 - 9x^2 - 30x = 0 \quad \text{Getting 0 on one side}$$

$$3x(x^2 - 3x - 10) = 0 \quad \text{Factoring out a common factor}$$

$$3x(x + 2)(x - 5) = 0 \quad \text{Factoring the trinomial}$$

$$3x = 0 \quad \text{or} \quad x + 2 = 0 \quad \text{or} \quad x - 5 = 0 \quad \text{Using the principle of zero products}$$

$$x = 0 \quad \text{or} \quad x = -2 \quad \text{or} \quad x = 5$$

The solutions are 0, -2, and 5. The solution set is $\{0, -2, 5\}$. ■

Problem Solving

Some problem situations can be translated to polynomial equations that we can now solve. The problem-solving process is the same as for other kinds of equations, except that we can now solve equations by factoring and using the principle of zero products.

EXAMPLE 6 A cabinet manufacturer determines that the revenue R, in thousands of dollars, from the sale of x cabinets is given by $R = 2x^2 + x$. If the cost C, in thousands of dollars, of producing x cabinets is given by $C = x^2 - 2x + 10$, how many cabinets must be produced and sold for the manufacturer to break even?

Solution

1. **Familiarize.** We note that for a firm to break even, its revenue must equal its cost of production. For this problem, x = the number of cabinets produced and sold.

2. **Translate.** Because revenue must *equal* cost for the firm to break even, we form the equation

$$2x^2 + x = x^2 - 2x + 10.$$

3. **Carry out.** We solve the equation:

$$2x^2 + x = x^2 - 2x + 10$$

$$x^2 + 3x - 10 = 0 \quad \text{Adding } -x^2 \text{ and } 2x \text{ and } -10 \text{ on both sides}$$

$$(x + 5)(x - 2) = 0 \quad \text{Factoring}$$

$$x + 5 = 0 \quad \text{or} \quad x - 2 = 0 \quad \text{Using the principle of zero products}$$

$$x = -5 \quad \text{or} \quad x = 2.$$

4. **Check.** We check the possible solutions in the original problem. The number -5 is not a solution because the number of cabinets cannot be negative. If 2 cabinets are produced, the revenue is $R = 2 \cdot 2^2 + 2 = 10$ thousand dollars, and the cost is $C = 2^2 - 2 \cdot 2 + 10 = 10$ thousand dollars. We have a solution.

5. **State.** The manufacturer breaks even when 2 cabinets are produced and sold. ■

The following problem involves the Pythagorean theorem, which relates the lengths of the sides of a right triangle. A right triangle has a 90° angle. The side opposite the 90° angle is called the **hypotenuse.** The other sides are called **legs.**

THE PYTHAGOREAN THEOREM

The sum of the squares of the legs of a right triangle is equal to the square of the hypotenuse:

$$a^2 + b^2 = c^2.$$

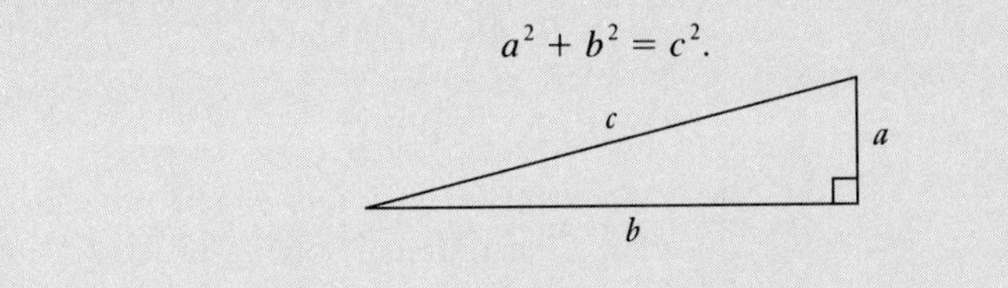

EXAMPLE 7 The lengths of the sides of a right triangle are three consecutive integers. Find these lengths.

Solution

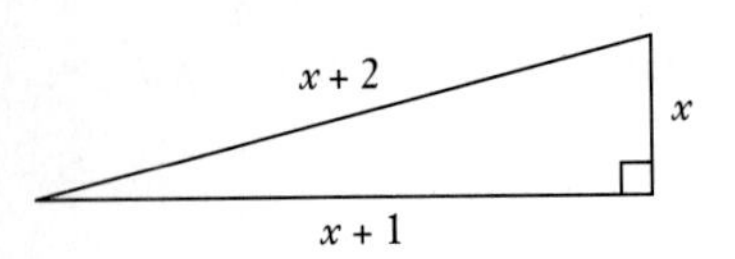

1. **Familiarize.** We first make a drawing. We let

$$x = \text{length of the first side.}$$

Since the lengths are consecutive integers, we know that

$$x + 1 = \text{length of the second side}$$

and

$$x + 2 = \text{length of the third side.}$$

Since the hypotenuse is laways the longest side of a right triangle, it would be the side that has length $x + 2$.

2. **Translate.** Applying the Pythagorean theorem, we obtain the following translation:

$$a^2 + b^2 = c^2$$
$$x^2 + (x + 1)^2 = (x + 2)^2.$$

3. **Carry out.** We solve the equation as follows:

$$x^2 + (x^2 + 2x + 1) = x^2 + 4x + 4 \quad \text{Squaring the binomials}$$
$$2x^2 + 2x + 1 = x^2 + 4x + 4 \quad \text{Collecting like terms}$$
$$x^2 - 2x - 3 = 0 \quad \text{Adding } -x^2 \text{ and } -4x \text{ and } -4 \text{ on both sides}$$
$$(x - 3)(x + 1) = 0$$
$$x - 3 = 0 \quad \text{or} \quad x + 1 = 0 \quad \text{Using the principle of zero products}$$
$$x = 3 \quad \text{or} \quad x = -1.$$

4. **Check.** The integer -1 cannot be a length of a side because it is negative. When $x = 3$, $x + 1 = 4$ and $x + 2 = 5$; and $3^2 + 4^2 = 5^2$. So 3, 4, and 5 check.

5. **State.** The integers 3, 4, and 5 are the lengths of the sides. ■

EXAMPLE 8 There are two square flower gardens on the campus mall. The length of a side of one garden is 4 ft less than the length of a side of the other. The area of the larger garden is 56 ft^2 greater than that of the smaller. Find the dimensions of each garden.

Solution

1. **Familiarize.** We draw a picture and label it. We let x = the length of a side of the larger garden. Then $x - 4$ = the length of the side of the smaller garden. We note that the larger garden has 56 ft^2 more area than the smaller garden.

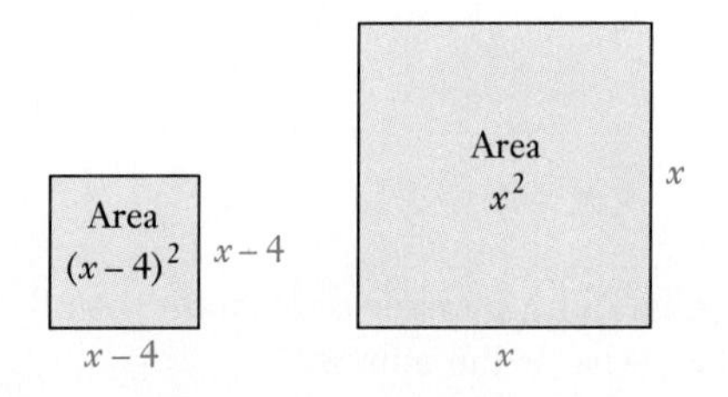

2. **Translate.** We use the following statement:

$$\underbrace{\text{Area of small garden}}_{(x-4)^2} \underset{+}{+} \underset{56}{56} \underset{=}{\text{is}} \underbrace{\text{area of large garden.}}_{x^2}$$

$$(x - 4)^2 + 56 = x^2$$

3. **Carry out.** We solve the equation:

$$(x - 4)^2 + 56 = x^2$$

$$x^2 - 8x + 16 + 56 = x^2 \quad \text{Squaring a binomial}$$

$$-8x + 72 = 0 \quad \text{Collecting like terms and simplifying}$$

$$x = 9.$$

4. **Check.** Because we used x to represent the length of a side of the larger garden, it follows that the smaller garden has sides of length $x - 4$, or 5. The areas of the gardens are then 5^2, or 25 ft^2, and 9^2, or 81 ft^2. The difference is 56 ft^2, so we do have a solution of the problem.

5. **State.** The smaller garden has sides of length 5 ft, and the larger garden has sides of length 9 ft. ■

Notice that in Example 8 the equation turned out to be linear. In problem solving it is usually best to keep an open mind when translating the words to mathematics. Sometimes the equation that you formulate may be easier (or harder) than you anticipated.

EXERCISE SET 3.5

Find the solution set for each of the following.

1. $x^2 + 3x = 28$

2. $y^2 - 4y = 45$

3. $x^2 - 12x + 36 = 0$

4. $y^2 + 16y + 64 = 0$

5. $9x + x^2 + 20 = 0$

6. $8y + y^2 + 15 = 0$

7. $x^2 + 8x = 0$

8. $t^2 + 9t = 0$

9. $x^2 - 9 = 0$

10. $p^2 - 16 = 0$

11. $z^2 = 36$

12. $y^2 = 81$

13. $x^2 + 14x + 45 = 0$

14. $y^2 + 12y + 32 = 0$

15. $p^2 - 11p = -28$

16. $x^2 - 14x = -45$

17. $8y^2 - 10y + 3 = 0$

18. $4x^2 + 11x + 6 = 0$

19. $6z - z^2 = 0$

20. $8y - y^2 = 0$

21. $5x^2 - 20 = 0$

22. $6y^2 - 54 = 0$

23. $21r^2 + r - 10 = 0$

24. $12a^2 - 5a - 28 = 0$

25. $2x^3 - 2x^2 = 12x$

26. $50y + 5y^3 = 35y^2$

27. $2x^3 = 128x$

28. $147y = 3y^3$

Solve.

29. If 4 times the square of a number is 21 more than 8 times the number, what is the number?

30. If 4 times the square of a number is 45 more than 8 times the number, what is the number?

31. The length of the top of a table is 5 cm more than the width. Find the length and the width if the area is 84 cm^2.

32. The length of the top of a workbench is 4 cm greater than the width. The area is 96 cm^2. Find the length and the width.

33. A student is planning a garden that is 25 m longer than it is wide. The garden will have an area of 7500 m^2. What will its dimensions be?

34. A flower bed is to be 3 m longer than it is wide. The flower bed will have an area of 108 m^2. What will its dimensions be?

35. The base of a triangle is 9 cm greater than the height. The area is 56 cm^2. Find the height and the base.

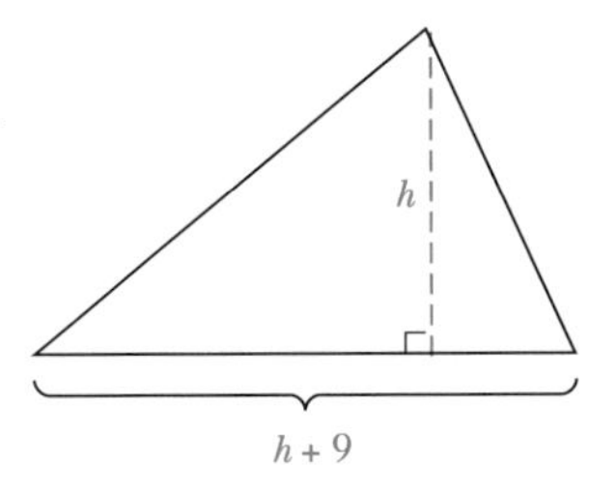

36. The base of a triangle is 5 cm less than the height. The area is 18 cm^2. Find the height and the base.

37. The perimeter of a square is 4 more than its area. Find the length of a side.

38. The area of a square is 12 more than its perimeter. Find the length of a side.

39. The revenue R from the sale of x clocks is given by $R = \frac{4}{9}x^2 + x$, where R is in hundreds of dollars. If the cost of producing x clocks is given by $C = \frac{1}{3}x^2 + x + 1$, where C is in hundreds of dollars, how many clocks must be produced and sold for the firm to break even?

40. Suppose that the cost C of making x video cameras is given by $C = \frac{1}{9}x^2 + 2x + 1$, where C is in thousands of dollars. If the revenue from the sale of x video cameras is given by $R = \frac{5}{36}x^2 + 2x$, where R is in thousands of dollars, how many cameras must be sold for the firm to break even?

41. One leg of a right triangle has length 9 m. The other sides have lengths that are consecutive integers. Find these lengths.

42. One leg of a right triangle is 10 cm. The other sides have lengths that are consecutive even integers. Find these lengths.

43. Suppose that an object is thrown upward with an initial velocity of 80 ft/sec from a height of 224 ft. Its height h after t seconds is given by

$$h = -16t^2 + 80t + 224.$$

After what amount of time will the object reach the ground? That is, for what positive value of t does $h = 0$?

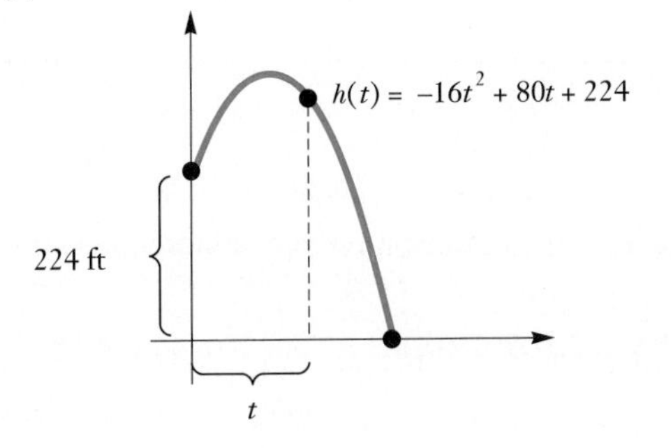

44. Suppose that an object is thrown upward with an initial velocity of 96 ft/sec from a height of 880 ft. Its height h after t seconds is given by

$$h = -16t^2 + 96t + 880.$$

After what amount of time will the object reach the ground? That is, for what positive value of t does $h = 0$?

SKILL MAINTENANCE

45. Obtain an equivalent expression by first removing parentheses: $5x - 9 - (2x - 3)$.

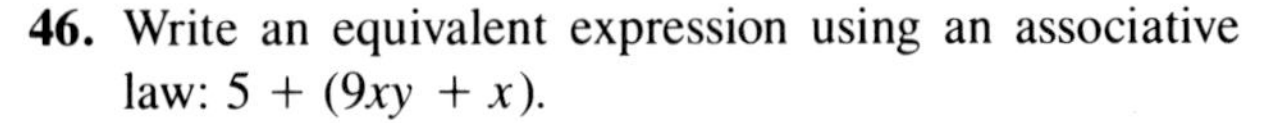

46. Write an equivalent expression using an associative law: $5 + (9xy + x)$.

47. One angle of a triangle is 10° more than a second angle. The third angle is 20° more than three times the second angle. Find the measures of the angles.

48. Solve: $2x - 14 + 9x > -8x + 16 + 10x$.

SYNTHESIS

Solve.

49. $(a - 5)^2 = 36$

50. $(x - 6)^2 = 81$

51. $(3x^2 - 7x - 20)(x - 5) = 0$

52. $(8x + 11)(12x^2 - 5x - 2) = 0$

53. $(x + 1)^3 = (x - 1)^3 + 26$

54. $(x - 2)^3 = x^3 - 2$

55. $3x^3 + 6x^2 - 27x - 54 = 0$

56. $2x^3 + 6x^2 = 8x + 24$

57. A square and an equilateral triangle have the same perimeter. The area of the square is 9 cm². What is the length of a side of the triangle?

58. The sum of two numbers is 17, and the sum of their squares is 205. Find the numbers.

59. A rectangular piece of tin is twice as long as it is wide. Squares 2 cm on a side are cut out of each corner, and the ends are turned up to make a box whose volume is 480 cm³. What are the dimensions of the piece of tin?

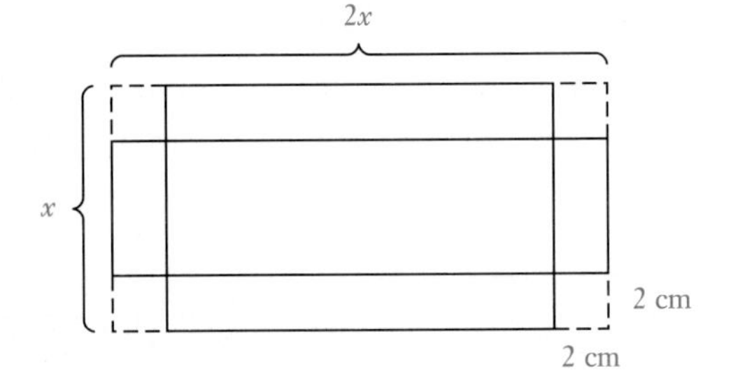

60. The hypotenuse of a right triangle is 3 cm longer than one of its legs and 6 cm more than the other leg. What is the area of the triangle?

3.6 Division of Polynomials

We now turn our attention to division of polynomials. As in division of real numbers, heavy use will be made of our multiplication and subtraction skills.

Dividing a Monomial by a Monomial

Expressions such as x^8/x^3, or equivalently $x^8 \div x^3$, represent the most basic type of polynomial division. Such an expression can be simplified by "removing" a factor of 1:

$$\frac{x^8}{x^3} = \frac{x \cdot x \cdot x \cdot x \cdot x \cdot x \cdot x \cdot x}{x \cdot x \cdot x}$$ Using the definition of exponential notation

$$= \frac{x \cdot x \cdot x}{x \cdot x \cdot x} \cdot x \cdot x \cdot x \cdot x \cdot x$$

$$= x \cdot x \cdot x \cdot x \cdot x$$ "Removing" a factor of 1

$$= x^5.$$

Notice that the exponent in the quotient, or the result of the division, can be found by subtracting the exponent of the denominator from the exponent of the numerator. This result can be generalized.

THE QUOTIENT RULE FOR POWERS

For any nonzero number a and any positive integers m and n, $m > n$,

$$\frac{a^m}{a^n} = a^{m-n}.$$

EXAMPLES Divide and simplify.

1. $\frac{r^9}{r^3} = r^{9-3} = r^6$ Subtracting exponents

2. $\frac{10x^{11}y^5}{2x^4y^3} = \frac{10}{2} \cdot x^{11-4} \cdot y^{5-3}$ The bases must be the same.

$= 5x^7y^2$ ■

Warning: Only expressions containing a *factor* of the form a^m/a^n can be simplified in this manner because doing so effectively "removes" a factor of 1. In general, *expressions like* $a^m + b^r/a^n + b^s$ *cannot be simplified.*

Dividing a Polynomial by a Monomial

Recall that $\frac{A+B}{C} = \frac{A}{C} + \frac{B}{C}$. We use this fact when dividing a polynomial with two or more terms by a monomial.

EXAMPLE 3 Divide and check: $(12x^3 + 8x^2 + x + 4) \div 4x$.

Solution

$$(12x^3 + 8x^2 + x + 4) \div 4x = \frac{12x^3 + 8x^2 + x + 4}{4x}$$ Writing as a fraction

$$= \frac{12x^3}{4x} + \frac{8x^2}{4x} + \frac{x}{4x} + \frac{4}{4x}$$ Note that $\frac{x}{x} = 1$ and $\frac{4}{4} = 1$

$$= 3x^2 + 2x + \frac{1}{4} + \frac{1}{x}$$ Doing four separate divisions

We can check by multiplying the quotient by $4x$:

$$4x\left(3x^2 + 2x + \frac{1}{4} + \frac{1}{x}\right) = 12x^3 + 8x^2 + x + 4.$$

Using the distributive law ■

Try to write only the answer.

EXAMPLE 4 Divide and check: $(15a^7b^9 - 18a^5b^3 + 12a^4b^2) \div 3a^4b$.

Solution

$$\frac{15a^7b^9 - 18a^5b^3 + 12a^4b^2}{3a^4b} = \frac{15a^7b^9}{3a^4b} - \frac{18a^5b^3}{3a^4b} + \frac{12a^4b^2}{3a^4b}$$

$$= 5a^3b^8 - 6ab^2 + 4b$$

Doing three separate divisions

Check: $3a^4b\,(5a^3b^8 - 6ab^2 + 4b) = 15a^7b^9 - 18a^5b^3 + 12a^4b^2$

Using the distributive law ■

> To divide a polynomial by a monomial, divide each term by the monomial.

Divisor Not a Monomial

When the divisor is not a monomial, we use a procedure very similar to long division in arithmetic.

EXAMPLE 5 Divide $2x^2 - 7x - 15$ by $x - 5$.

Solution

$$\begin{array}{r} 2x \phantom{{}-15} \\ x - 5\overline{)2x^2 - 7x - 15} \\ -(2x^2 - 10x) \phantom{{}-15} \\ \hline 3x \phantom{{}-15} \end{array}$$

Divide $2x^2$ by x: $2x^2/x = 2x$.

Multiply $x - 5$ by $2x$. Change signs mentally and subtract.

We now "bring down" the next term of the dividend, -15.

$$\begin{array}{r} 2x + 3 \\ x - 5\overline{)2x^2 - 7x - 15} \\ 2x^2 - 10x \phantom{{}-15} \\ \hline 3x - 15 \\ -(3x - 15) \\ \hline 0 \end{array}$$

Divide $3x$ by x: $3x/x = 3$.

Multiply $x - 5$ by 3. Change signs mentally and subtract.

The quotient is $2x + 3$.

Check: $(x - 5)(2x + 3) = 2x^2 - 7x - 15$. The answer checks. ■

To understand why we perform long division as we do, note that Example 5 amounted to "filling in" a missing polynomial:

$$(x - 5)(\quad ? \quad) = 2x^2 - 7x - 15.$$

We see that $2x$ must be in the unknown polynomial if we are to get the first term, $2x^2$, from the multiplication. To see what else is needed, note that

$$(x - 5)(2x \qquad) = 2x^2 - 10x \neq 2x^2 - 7x - 15.$$

The $2x^2 - 10x$ can be considered a (poor) approximation of $2x^2 - 7x - 15$. To see how far off our approximation is, we subtract:

$$\begin{array}{r} 2x^2 - 7x - 15 \\ -(2x^2 - 10x) \\ \hline 3x - 15 \end{array}$$

To get the needed terms, $3x - 15$, we include the term $+3$:

$$\begin{aligned}(x - 5)(2x + 3) &= 2x^2 - 10x + 3x - 15 \\ &= 2x^2 - 7x - 15.\end{aligned}$$

If we subtract now, we have a remainder of 0.

Often a nonzero remainder occurs.

EXAMPLE 6 Divide $x^2 + 5x + 8$ by $x + 3$.

Solution

$$\begin{array}{r} x \\ x + 3\overline{)x^2 + 5x + 8} \\ x^2 + 3x \\ \hline 2x \end{array}$$

$x \leftarrow$ Divide the first term of the dividend by the first term of the divisor: $x^2/x = x$.

$x^2 + 3x \leftarrow$ Multiply x above by $x + 3$.

$2x \leftarrow$ Subtract.

The subtraction we have done is $(x^2 + 5x) - (x^2 + 3x)$. Remember: To subtract, add the inverse (change the sign of every term, then add).

We now "bring down" the other terms of the dividend—in this case, 8:

$$\begin{array}{r} x + 2 \\ x + 3\overline{)x^2 + 5x + 8} \\ x^2 + 3x \\ \hline 2x + 8 \\ 2x + 6 \\ \hline 2 \end{array}$$

$x + 2 \leftarrow$ Divide the first term by the first term: $2x/x = 2$.

$2x + 8 \leftarrow$ The 8 has been "brought down."

$2x + 6 \leftarrow$ Multiply 2 by $x + 3$.

$2 \leftarrow$ Subtract: $(2x + 8) - (2x + 6)$.

The quotient is $x + 2$, and the remainder is 2.

Check: $(x + 3)(x + 2) + 2 = x^2 + 5x + 6 + 2$ — Add the remainder to the product.

$$= x^2 + 5x + 8$$

We can write our answer as $x + 2$, R2, or we can write our answer as follows:

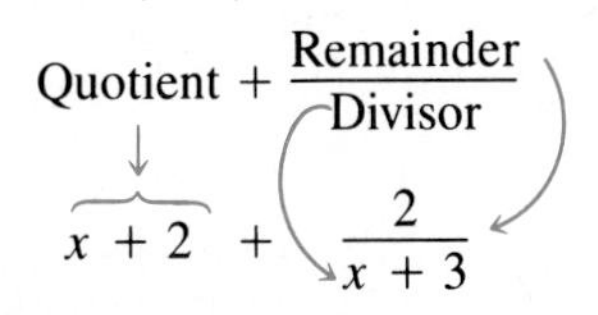

This is the way answers for the problems in this section will appear at the back of the book.

Check: When the answer is given in the preceding form, we can check by multiplying.

$$(x + 3)\left[(x + 2) + \frac{2}{x + 3}\right] = (x + 3)(x + 2) + (x + 3)\frac{2}{x + 3}$$ Using the distributive law

$$= x^2 + 5x + 6 + 2$$
$$= x^2 + 5x + 8$$ ■

You may have noticed that in each of our preceding examples all polynomials were written in descending order. When this is not the case, we rearrange terms before dividing.

Always remember the following:

1. Arrange polynomials in descending order.
2. If there are missing terms in the dividend, either write them with 0 coefficients or leave space for them.
3. Continue the long division process until the degree in the remainder is less than the degree of the divisor.

EXAMPLE 7 Divide: $(125y^3 - 8) \div (5y - 2)$.

Solution

$$\begin{array}{r} 25y^2 + 10y + 4 \\ 5y - 2\overline{)125y^3 + 0y^2 + 0y - 8} \\ \underline{125y^3 - 50y^2} \qquad\qquad \\ 50y^2 + 0y - 8 \\ \underline{50y^2 - 20y} \quad\; \\ 20y - 8 \\ \underline{20y - 8} \\ 0 \end{array}$$

When there are missing terms, we can write them in as in this example, or leave space as in Example 8.

This subtraction is $125y^3 - (125y^3 - 50y^2)$. We get $50y^2$.

The answer is $25y^2 + 10y + 4$. ■

EXAMPLE 8 Divide: $(9x^2 + x^3 - 5) \div (x^2 - 1)$.

Solution

$(x^3 + 9x^2 - 5) \div (x^2 - 1)$ **Rewriting in descending order**

$$
\begin{array}{r}
x + 9 \\
x^2 - 1 \overline{\smash{)} x^3 + 9x^2 - 5} \\
\underline{x^3 - x } \\
9x^2 + x - 5 \\
\underline{9x^2 - 9} \\
x + 4
\end{array}
$$

Here we've left space for the missing term.

The degree of the remainder is less than the degree of the divisor, so we are finished.

The answer is

$$x + 9 + \frac{x + 4}{x^2 - 1}.$$

This expression is the remainder over the divisor.

■

EXERCISE SET 3.6

Divide and simplify.

1. $\frac{a^9}{a^3}$
2. $\frac{x^{12}}{x^3}$
3. $\frac{8x^7}{4x^4}$
4. $\frac{20a^{20}}{5a^4}$
5. $\frac{m^7n^9}{m^2n^5}$
6. $\frac{m^{12}n^9}{m^4n^6}$
7. $\frac{35x^8y^5}{7x^2y}$
8. $\frac{45x^7y^8}{5xy^2}$
9. $\frac{-49a^5b^{12}}{7a^2b^2}$
10. $\frac{-42a^2b^7}{7ab^4}$
11. $\frac{9x^9y^7z}{3x^3y^2z}$
12. $\frac{10x^5y^8z^4}{5x^5y^2z^3}$

Divide and check.

13. $\frac{30x^8 - 15x^6 + 40x^4}{5x^4}$
14. $\frac{24y^6 + 18y^5 - 36y^2}{6y^2}$
15. $\frac{-14a^3 + 28a^2 - 21a}{7a}$
16. $\frac{-32x^4 - 24x^3 - 12x^2}{4x}$
17. $(9y^4 - 18y^3 + 27y^2) \div 9y$
18. $(24a^3 + 28a^2 - 20a) \div 2a$
19. $(36x^6 - 18x^4 - 12x^2) \div (-6x)$
20. $(18y^7 - 27y^4 - 3y^2) \div (-3y^2)$
21. $(a^2b - a^3b^3 - a^5b^5) \div a^2b$
22. $(x^3y^2 - x^3y^3 - x^4y^2) \div x^2y^2$
23. $(6p^2q^2 - 9p^2q + 12pq^2) \div (-3pq)$
24. $(16y^4z^2 - 8y^6z^4 + 12y^8z^3) \div 4y^4z$

Divide.

25. $(x^2 + 10x + 21) \div (x + 3)$
26. $(y^2 - 8y + 16) \div (y - 4)$
27. $(a^2 - 8a - 16) \div (a + 4)$
28. $(y^2 - 10y - 25) \div (y - 5)$
29. $(x^2 - 11x + 23) \div (x - 5)$
30. $(x^2 - 11x + 23) \div (x - 7)$
31. $(y^2 - 25) \div (y + 5)$
32. $(a^2 - 81) \div (a - 9)$
33. $(y^3 - 4y^2 + 3y - 6) \div (y - 2)$
34. $(x^3 - 5x^2 + 4x - 7) \div (x - 3)$
35. $(2x^3 + 3x^2 - x - 3) \div (x + 2)$
36. $(3x^3 - 5x^2 - 3x - 2) \div (x - 2)$

37. $(a^3 - a + 12) \div (a - 4)$

38. $(x^3 - x + 6) \div (x + 2)$

39. $(8x^3 + 27) \div (2x + 3)$

40. $(64y^3 - 8) \div (4y - 2)$

41. $(x^4 - x^2 - 42) \div (x^2 - 7)$

42. $(y^4 - y^2 - 54) \div (y^2 - 3)$

43. $(x^4 - x^2 - x + 2) \div (x - 1)$

44. $(y^4 - y^2 - y + 3) \div (y + 1)$

45. $(10y^3 + 6y^2 - 9y + 10) \div (5y - 2)$

46. $(6x^3 - 11x^2 + 11x - 2) \div (2x - 3)$

47. $(2x^4 - x^3 - 5x^2 + x - 6) \div (x^2 + 2)$

48. $(3x^4 + 2x^3 - 11x^2 - 2x + 5) \div (x^2 - 2)$

49. $(2x^5 + x^4 + 2x^3 + x) \div (x^2 + 1)$

50. $(2x^5 - 3x^3 + x^2 + 4) \div (x^2 - 1)$

SKILL MAINTENANCE

51. Subtract: $-3.1 - (-5.8)$.

52. Collect like terms: $3x - 5y + 9x - 2y$.

53. One number is larger than another number by 31. The sum of the numbers is 153. What is the larger number?

54. A van can be rented for \$40 per day plus 15¢ per mile or for \$58 with unlimited mileage. For what daily mileages would the unlimited plan save you money?

SYNTHESIS

Divide.

55. $(x^4 - x^3y + x^2y^2 + 2x^2y - 2xy^2 + 2y^3) \div (x^2 - xy + y^2)$

56. $(4a^3b + 5a^2b^2 + a^4 + 2ab^3) \div (a^2 + 2b^2 + 3ab)$

57. $(x^4 - y^4) \div (x - y)$

58. $(a^7 + b^7) \div (a + b)$

Solve.

59. Find k so that when $x^3 - kx^2 + 3x + 7k$ is divided by $x + 2$, the remainder will be 0.

60. When $x^2 - 3x + 2k$ is divided by $x + 2$, the remainder is 7. Find k.

CHAPTER SUMMARY AND REVIEW 3

TERMS TO KNOW

Polynomial
Monomial
Coefficient
Degree of a term
Degree of a polynomial
Linear
Quadratic
Cubic
Leading term
Leading coefficient
Binomial
Trinomial
Value of a polynomial
Evaluate a polynomial
Similar, or like, terms
FOIL
Square of a binomial
Factor
Prime polynomial
Factoring by grouping
Factoring trinomials
Difference of squares
Trinomial square
Sum or difference of cubes
Polynomial equation
Root
Zero of a polynomial
Pythagorean theorem
Hypotenuse
Legs

IMPORTANT PROPERTIES AND FORMULAS

Multiplying with like bases: $a^m \cdot a^n = a^{m+n}$ (Product rule for powers)

Dividing with like bases: $\frac{a^m}{a^n} = a^{m-n}$ (Quotient rule for powers)

Raising a product to a power: $(ab)^n = a^n b^n$

Raising a power to a power: $(a^m)^n = a^{mn}$ (Power rule)

Raising a quotient to a power: $\left(\frac{a}{b}\right)^n = \frac{a^n}{b^n}$

FACTORING FORMULAS

$$A^2 - B^2 = (A + B)(A - B)$$
$$A^2 + 2AB + B^2 = (A + B)^2$$
$$A^2 - 2AB + B^2 = (A - B)^2$$
$$A^3 + B^3 = (A + B)(A^2 - AB + B^2)$$
$$A^3 - B^3 = (A - B)(A^2 + AB + B^2)$$

To factor a polynomial:

A. Always factor out the largest common factor that exists.

B. Look at the number of terms.

Two terms: Determine first whether you have a difference of squares. Next, try to factor as a sum or a difference of cubes. Do *not* try to factor a sum of squares.

Three terms: Determine whether the trinomial is a square. If so, you know how to factor. If not, try trial and error, using the FOIL method or the grouping method.

Four terms: Try factoring by grouping and factoring out a common binomial factor. Next, try grouping into a difference of squares, one of which is a trinomial.

C. Always *factor completely*. If a factor with more than one term can still be factored, you should factor it.

THE PRINCIPLE OF ZERO PRODUCTS

For any real numbers a and b:

If $a \cdot b = 0$, then $a = 0$ or $b = 0$.
If $a = 0$ or $b = 0$, then $a \cdot b = 0$.

REVIEW EXERCISES

The review sections to be tested in addition to the material in this chapter are Sections 1.2, 1.4, 2.2, and 2.4.

Determine the degree of each term and the degree of the polynomial.

1. $-5x^6 + 3x^5 - x^3 + 4x^2 - 12$

2. $y^4 - 3x^2y^3 + 2xy^2 - y^2 + 7$

Arrange in descending order. Then find the leading term and the leading coefficient.

3. $-4b^2 - 21 + 2b^3 - 7b^4$

4. $17 + 3x^5 - 2x^3 + x^4 - 5x$

5. Arrange in ascending powers of x:

$$3x^6y - 7x^8y^3 + 2x^3 - 3x^2.$$

Evaluate the polynomial for the given values of x.

6. $-4x^3 + x^2 - 7x + 12$, for $x = -2$ and $x = 3$

7. $5x^2 - 11x + 8$, for $x = -1$ and $x = 0$

8. *Height of a thrown object.* Suppose that an object is thrown upward from ground level with an initial velocity of 80 ft/sec. Its height after t seconds is given by the polynomial

$$-16t^2 + 80t.$$

What is the height of the object after (a) 0 sec; (b) 1 sec; (c) 2.5 sec; (d) 4 sec; and (e) 5 sec?

Collect like terms.

9. $4x^2y - 3xy^2 - 5x^2y + xy^2$

10. $3ab - 10 + 5ab^2 - 2ab + 7ab^2 + 14$

Perform the indicated operation, and simplify, if possible.

11. $(-6x^3 - 4x^2 + 3x + 1) + (5x^3 + 2x + 6x^2 + 1)$

12. $(4x^3 - 2x^2 - 7x + 5) + (8x^2 - 3x^3 - 9 + 6x)$

13. $(-9xy^2 - xy + 6x^2y) + (-5x^2y - xy + 4xy^2) + (12x^2y - 3xy^2 + 6xy)$

14. $(4a - b + 3c) - (6a - 7b - 4c)$

15. $(8x^2 - 4xy + y^2) - (2x^2 + 3xy - 2y^2)$

16. $\left(\frac{5}{12}y^3 + \frac{4}{3}y^2 - 3.2y + \frac{1}{4}\right) - \left(-\frac{1}{8}y^3 + \frac{1}{2}y^2 - 2.3y + \frac{2}{7}\right)$

17. $(3x^2y)(-6xy^3)$

18. $7ab(2a^3b^2 - 3ab^4)$

19. $(4ab + 3c)(2ab - c)$

20. $(2x + 5y)(2x - 5y)$

21. $(2x - 5y)^2$

22. $(5x^2 - 7x + 3)(4x^2 + 2x - 9)$

23. $(x^2 + 4y^3)^2$

24. $\left(x - \frac{1}{3}\right)\left(x - \frac{1}{6}\right)$

25. $(y - 2)(2y + 3)(y - 5)$

26. $\dfrac{12m^5n^7 - 6m^2n^5}{9m^2n^3}$

27. $(y^2 - 20y + 64) \div (y - 6)$

28. $(6x^4 + 3x^2 + 5x + 4) \div (x^2 - 4)$

29. $(3x^4 + x^3 - 7x^2 + x - 10) \div (3x - 5)$

Factor.

30. $6x^2 + 5x$

31. $9y^4 - 3y^2$

32. $15x^4 - 18x^3 + 21x^2 - 9x$

33. $a^2 - 12a + 27$

34. $3m^2 + 14m + 8$

35. $25x^2 + 20x + 4$

36. $4y^2 - 16$

37. $a^2 - 81$
38. $ax + 2bx - ay - 2by$
39. $3y^3 + 6y^2 - 5y - 10$
40. $a^4 - 81$
41. $4x^4 + 24x^2 + 20$
42. $27x^3 - 8$
43. $\frac{1}{8}b^3 + \frac{1}{125}c^3$
44. $2z^8 - 16z^6$
45. $54x^6y - 2y$
46. $a^2 - 2ab + b^2 - 4t^2$
47. $6t^2 + 17pt + 5p^2$
48. $x^3 + 2x^2 - 9x - 18$

Solve.

49. $x^2 - 20x = -100$
50. $6b^2 - 13b + 6 = 0$
51. $8y^2 + 5 = 14y$
52. The area of a square is 5 more than 4 times the length of a side. What is the length of a side of the square?
53. The sum of the squares of three consecutive odd numbers is 83. Find the numbers.
54. A photograph is 3 in. longer than it is wide. When a 2-in. border is placed around the photograph, the total area of the mounted picture is 108 in^2. Find the dimensions of the photograph.

SKILL MAINTENANCE

55. Add: $5.78 + (-9.81)$.
56. Write an expression equivalent to $7 + mn$ by using a commutative law.
57. A 20-ft piece of rope is to be cut into two pieces, one piece 6 ft longer than the other. How long should each piece of rope be?
58. Solve: $\frac{3}{4}(-2x + 5) < 12$.

SYNTHESIS

Factor.

59. $128x^6 - 2y^6$
60. $(x - 1)^3 - (x + 1)^3$

THINKING IT THROUGH

Explain the error in each of the following.

1. $(a + 3)^2 = a^2 + 9$
2. $a^3 + b^3 = (a + b)(a^2 - 2ab + b^2)$
3. $(a - b)(a - b) = a^2 - b^2$
4. $(x + 3)(x - 4) = x^2 - 12$
5. $(p + 7)(p - 7) = p^2 + 49$
6. $(t - 3)^2 = t^2 - 9$
7. What law of real numbers is most important to our learning quick ways to multiply and factor polynomials?
8. In this chapter, we learned to solve equations that we could not have solved in Chapter 2. What new kinds of equations are we able to solve, how do the solutions differ from before, and how is the procedure different?

CHAPTER TEST 3

1. Determine the degree of each term and the degree of the polynomial: $3xy^3 - 4x^2y + 5x^5y^4 - 2x^4y$.
2. Arrange in descending order. Then find the leading term and the leading coefficient:
$$8a - 2 + a^2 - 4a^3.$$
3. Evaluate the polynomial $-5x^3 + 2x^2 - 23x + 17$ for $x = -4$ and $x = 2$.
4. Collect like terms: $5xy - 2xy^2 - 2xy + 5xy^2$.

Perform the indicated operation.

5. $(5m^3 - 4m^2n - 6mn^2 - 3n^3) + (9mn^2 - 4n^3 + 2m^3 + 6m^2n)$

6. $(9a - 4b - 5c) - (3a - 2b - 7c)$

7. $\left(\frac{3}{7}y^2 - 2.5y + \frac{5}{2}\right) - \left(\frac{1}{3}y^2 - 3.8y + \frac{1}{6}\right)$

8. $(-4x^3y)(-16x^2y^2)$

9. $(5x - 3yz)(2x + 4yz)$

10. $(x - 2y)(x + 2y)$

11. $(3m^2 + 4m - 2)(-m^2 - 3m + 5)$

12. $(7 - 4x)(11x + 5)$

13. $\dfrac{10x^3y^5 - 6x^5y^2}{2x^3y^2}$

14. $(x^3 + 5x^2 + 4x - 7) \div (x + 3)$

Factor.

15. $9x^2 + 7x$

16. $9x^2 + 25 - 30x$

17. $12m^2 + 20m + 3$

18. $3cy - dx - 3cx + dy$

19. $b^4 - 16$

20. $64x^3 - 27$

21. $3x + 24xy^3$

22. $x^3 - 3x^2 - 25x + 75$

Solve.

23. $18x^3 - 6x = 3x^2$

24. $9 + 49x^2 = 42x$

25. A photograph is 3 cm longer than it is wide. Its area is 40 cm^2. Find its length and its width.

SKILL MAINTENANCE

26. Subtract: $-3.2 - (-5.9)$.

27. Simplify: $-5(x + 2) + 3(4x + 7)$.

28. Solve: $-3(2x - 5) \geq 4x - 9$.

29. There are 70 questions on a test. The questions are either multiple-choice, true–false, or fill-in. There are twice as many true–false as fill-in and 5 more multiple-choice than true–false. How many of each type of question are there on the test?

SYNTHESIS

30. a) Multiply: $(x^2 + x + 1)(x^3 - x^2 + 1)$.
b) Factor: $x^5 + x + 1$.

31. Factor: $6x^{2n} - 7x^n - 20$.

CUMULATIVE REVIEW 1-3

1. Evaluate $5xy - z$ if $x = 3$, $y = 2$, and $z = 7$.

2. Use set-builder notation to name the following set: $\{7, 8, 9, 10, 11\}$.

3. Write an expression equivalent to $(2x + 5) + 7$ by using an associative law.

Simplify.

4. $7x - 4 + 2x - 5$

5. $3a - 2[a + 3(a - 4)]$

6. $8 \div 2 \cdot 2 + 3^2 - 5$

7. $5^2 - 3|4 - 3^2|$

8. Evaluate $4x^2$ for $x = -3$.

9. Find $-x$ if $x = -3.9$.

10. Use multiplication by 1 to find an expression equivalent to $\frac{4m}{9}$ with a denominator of 45.

11. Determine whether $2x - 5 = 8$ and $2x = 13$ are equivalent equations.

Solve.

12. $2x + 7.9 = 4.1$

13. $8x - 5 = 2x + 3(x - 7)$

14. $5x - 17 \geq 3x + 7$

15. $2(x - 4) + 7x < 3x - 7$

16. $-7 < 2x + 5 < 13$

17. $-4 \leq 2x - 7$ *and* $3x - 7 < 19$

18. $x + 7 < -5$ *or* $x - 3 > 8$

19. $|3x - 8| = 16$

20. $|x - 7| < 4$

21. $|3x + 5| > 17$

22. $x^2 - 9x + 14 = 0$

23. $6r^2 - 150 = 0$

24. $x^2 + 9 = -6x$

25. The sum of two numbers is 26. The larger is 2 more than twice the smaller. Find the numbers.

26. Find three consecutive odd numbers such that the sum of 4 times the first number and 5 times the third number is 47.

27. The sum of the squares of three consecutive positive integers is 50. Find the numbers.

28. The length of a rectangle is 3 ft longer than the width. The area is 54 ft^2. Find the perimeter of the rectangle.

29. Solve for r: $m = \frac{n}{3}(p + r)$.

30. The area of a triangle is 168 cm^2. The base of the figure is 14 cm. What is the height?

31. Determine the degree of each term and the degree of the polynomial: $7x^4 - 5x^2 + x^3y^5 + 17x - 20$.

32. Arrange in descending order and then find the leading term and the leading coefficient:

$$5x - 3x^8 + 4x^2 - 3x^4 - 17 + x^3.$$

33. Evaluate $7x^2 - 3x$ for $x = -2$.

Perform the indicated operation and simplify.

34. $(5x^3 - 4x^2 - 19) + (2x^3 - 3x^2 + 2x + 7)$

35. $(-7a^2 + 2a - 3) - (3a^2 - a + 5)$

36. $-7x^2y^3(-4xy^3)$

37. $(a + 2b)(3a - 4b)$

38. $\left(3x + \frac{1}{2}\right)^2$

39. $\frac{6x^{15}y^8}{3x^5y^2}$

40. $(36m^3 - 10m^2 + 4m) \div 2m$

41. $(3x^3 - 10x^2 + 10x - 1) \div (x - 2)$

Factor.

42. $9a^5 - 6a^2$

43. $x^3 + x^2 - 2x - 2$

44. $m^2 - 10m - 24$

45. $12x^6 - 2x^3 - 4$

46. $36a^2 - 49$

47. $m^2 + 10m + 25$

48. $x^2 - 6x + 9 - y^2$

49. $8z^3 - 1$

50. $3x^3 + 375$

SYNTHESIS

51. Simplify: $(5x^{a+3}y^{b-2})(-3x^{a-5}y^{b+5})$.

52. Solve: $5 \leq |4 - x| \leq 7$.

CHAPTER 4 Rational Expressions and Equations

FEATURE PROBLEM

A new photocopier works twice as fast as an old one. When the machines work together, a university can produce its staff manuals in 15 hr. How long would each machine take to do the job alone?

THE MATHEMATICS

Let t be the time required for the new machine to do the job alone. Then the problem translates to the *rational equation*

$$15\left(\frac{1}{t}\right) + 15\left(\frac{1}{2t}\right) = 1.$$

These are rational expressions.

A rational expression is an expression that indicates division, as the fractional symbols in arithmetic do. In this chapter, we will learn to add, subtract, multiply, and divide rational expressions, as well as to use them in equations. Then we will use rational expressions to solve problems that we could not have solved before.

The review sections to be tested in addition to the material in this chapter are Sections 1.3, 2.3, 2.5, and 2.6.

4.1 Multiplication and Division of Rational Expressions

Just as a rational number is a number that can be expressed as a quotient of two integers, a **rational expression** is an algebraic expression that can be expressed as a quotient, or ratio, of two polynomials. The following are rational expressions:

$$\frac{7}{8}, \quad \frac{a}{b}, \quad \frac{8}{y+5}, \quad \frac{x^2+7xy-4}{x^3-y^3}, \quad \frac{1+z^3}{1-z^6}.$$

Evaluating

Like polynomials, rational expressions can be evaluated by making a suitable substitution. This time, however, we must be careful to avoid division by zero.

EXAMPLE 1 Evaluate $\frac{x+y}{x-3}$ for (a) $x = 4$, $y = 3$; (b) $x = 3$, $y = 5$.

Solution

a) $\frac{4+3}{4-3} = \frac{7}{1} = 7$

b) $\frac{3+5}{3-3} = \frac{8}{0}$ is not defined. We say that 3 is not a *meaningful replacement* for x in this expression. ■

EXAMPLE 2 The rational expression

$$\frac{t^2+5t}{2t+5}$$

gives the number of hours required for two machines, working together, to complete a job that the first machine could do alone in t hours and the other machine could do in $t + 5$ hours. How long will it take the machines, working together, to complete a job that the first machine could do alone (a) in 1 hour; (b) in 3 hours?

Solution

a) $\frac{1^2+5(1)}{2(1)+5} = \frac{1+5}{2+5} = \frac{6}{7}$ hr

b) $\frac{3^2+5(3)}{2(3)+5} = \frac{9+15}{6+5} = \frac{24}{11}$ hr ■

Multiplying

Most of the calculations we do with rational expressions are very much like the calculations we do with fractional notation in arithmetic. By proceeding as in arithmetic, we will know that when we obtain one rational expression from another, the two expressions will be equivalent; that is, they will name the same number for all meaningful replacements.

For any rational expressions A/B and C/D, with $B \neq 0$ and $D \neq 0$,

$$\frac{A}{B} \cdot \frac{C}{D} = \frac{A \cdot C}{B \cdot D}.$$

(To multiply two rational expressions, multiply numerators and multiply denominators.)

EXAMPLE 3 Multiply.

$$\frac{x+3}{y-4} \cdot \frac{x^3}{y+5} = \frac{(x+3)x^3}{(y-4)(y+5)}$$

Multiplying numerators and multiplying denominators ■

Multiplying by 1

Any number multiplied by 1 is that same number. Any expression, other than 0/0, with the same numerator and denominator names the number 1:

$$\frac{y+5}{y+5}, \quad \frac{4x^2-5}{4x^2-5}, \quad \frac{-1}{-1}.$$

All name the number 1 for all meaningful replacements.

We can multiply by 1 to get equivalent expressions. For example, let us multiply $(x+y)/5$ by 1:

$$\frac{x+y}{5} \cdot \frac{x-y}{x-y} = \frac{(x+y)(x-y)}{5(x-y)}.$$

Multiplying by $\frac{x-y}{x-y}$, which is 1

We know that

$$\frac{x+y}{5} \quad \text{and} \quad \frac{(x+y)(x-y)}{5(x-y)}$$

are equivalent. This means that they will name the same number for all replacements that do not make a denominator zero.

EXAMPLES Multiply to obtain equivalent expressions.

4. $\frac{x^2+3}{x-1} \cdot \frac{x+1}{x+1} = \frac{(x^2+3)(x+1)}{(x-1)(x+1)}$

5. $\frac{-1}{-1} \cdot \frac{x-4}{x-y} = \frac{-1 \cdot (x-4)}{-1 \cdot (x-y)} = \frac{4-x}{y-x}$

Note that multiplication by −1 reverses the order of subtraction. ■

Simplifying

We can simplify rational expressions by reversing the procedure of multiplying by 1. We "remove" a factor of 1 much as we did in Section 3.6. First factor the numerator and the denominator and then factor the rational expression, finding a factor that is equal to 1. Use the largest factors common to the numerator and the denominator.

EXAMPLES Simplify by removing factors equal to 1.

6. $\dfrac{5x^2}{x} = \dfrac{5x \cdot x}{1 \cdot x}$ Factoring the numerator and the denominator

$= \dfrac{5x}{1} \cdot \dfrac{x}{x}$ Factoring the rational expression

$= 5x \cdot 1$ $\frac{x}{x} = 1$ for all meaningful replacements

$= 5x$ "Removing" a factor of 1

In Example 6, we supplied a 1 in the denominator. This can always be done, if necessary.

7. $\dfrac{4a + 8}{2} = \dfrac{2(2a + 4)}{2 \cdot 1}$ Factoring the numerator and the denominator

$= \dfrac{2}{2} \cdot \dfrac{2a + 4}{1}$ Factoring the rational expression

$= \dfrac{2a + 4}{1}$ "Removing" a factor of 1

$= 2a + 4$ ■

Warning: The following mistake is common on problems like Example 7:

$$\frac{4a + 8}{2} = \frac{4a}{2} + 8 = 2a + 8.$$

This is wrong! Such a shortcut ignores the fact that the entire quantity, $4a + 8$, is being divided. Only common *factors* of both the numerator and the denominator may be "removed."

EXAMPLES

8. $\dfrac{2x^2 + 4x}{6x^2 + 2x} = \dfrac{2x(x + 2)}{2x(3x + 1)}$ Factoring the numerator and the denominator

$= \dfrac{2x}{2x} \cdot \dfrac{x + 2}{3x + 1}$ Factoring the rational expression

$= \dfrac{x + 2}{3x + 1}$ "Removing" a factor of 1

9. $$\frac{x^2 - 1}{2x^2 - x - 1} = \frac{(x - 1)(x + 1)}{(2x + 1)(x - 1)}$$ Factoring the numerator and the denominator

$$= \frac{x - 1}{x - 1} \cdot \frac{x + 1}{2x + 1}$$ Factoring the rational expression

$$= \frac{x + 1}{2x + 1}$$ "Removing" a factor of 1

10. $$\frac{9x^2 + 6xy - 3y^2}{12x^2 - 12y^2} = \frac{3(x + y)(3x - y)}{12(x + y)(x - y)}$$ Factoring the numerator and the denominator

$$= \frac{3(x + y)}{3(x + y)} \cdot \frac{3x - y}{4(x - y)}$$ Factoring the rational expression

$$= \frac{3x - y}{4(x - y)}$$ "Removing" a factor of 1 ■

Multiplying and Simplifying

After multiplying, we ordinarily simplify, if possible. That is why we leave the numerator and the denominator in factored form. Even so, we might need to factor them further in order to simplify.

EXAMPLES Multiply and simplify.

11. $$\frac{x + 2}{x - 3} \cdot \frac{x^2 - 4}{x^2 + x - 2} = \frac{(x + 2)(x^2 - 4)}{(x - 3)(x^2 + x - 2)}$$ Multiplying numerators and also denominators

$$= \frac{(x + 2)(x - 2)(x + 2)}{(x - 3)(x + 2)(x - 1)}$$ Factoring numerators and denominators and looking for factors of 1

$$= \frac{x + 2}{x + 2} \cdot \frac{(x + 2)(x - 2)}{(x - 3)(x - 1)}$$ Factoring the rational expression

$$= \frac{(x + 2)(x - 2)}{(x - 3)(x - 1)},$$ "Removing" a factor of 1

or $$\frac{x^2 - 4}{x^2 - 4x + 3}$$ Multiplying out the numerator and the denominator

From here on, we will not multiply out the numerator and the denominator. This will ease many procedures that we do throughout this chapter.

12. $$\frac{a^3 - b^3}{a^2 - b^2} \cdot \frac{a^2 + 2ab + b^2}{a^2 + ab + b^2}$$

$$= \frac{(a^3 - b^3)(a^2 + 2ab + b^2)}{(a^2 - b^2)(a^2 + ab + b^2)}$$

$$= \frac{(a^3 - b^3)(a + b)(a + b)}{(a^2 - b^2)(a^2 + ab + b^2)}$$ Factoring the trinomial square

$$= \frac{(a - b)(a^2 + ab + b^2)(a + b)(a + b)}{(a - b)(a + b)(a^2 + ab + b^2)}$$ Factoring the difference of cubes and the difference of squares

$$= \frac{(a - b)(a^2 + ab + b^2)(a + b)}{(a - b)(a^2 + ab + b^2)(a + b)} \cdot \frac{a + b}{1}$$

$$= a + b$$ ■

Dividing and Simplifying

Two expressions are reciprocals of each other if their product is 1. As in arithmetic, to find the reciprocal of a rational expression, we interchange numerator and denominator.

The reciprocal of $\dfrac{x}{x^2 + 3}$ is $\dfrac{x^2 + 3}{x}$.

The reciprocal of $y - 8$ is $\dfrac{1}{y - 8}$.

It should be noted that any replacement that makes a numerator zero will not be a meaningful replacement in the expression for the reciprocal because it will make the denominator zero once the expression is inverted. For instance, x cannot be replaced by 3 in the reciprocal of $(x - 3)/4$.

For any rational expressions A/B and C/D, with $B, C, D \neq 0$,

$$\frac{A}{B} \div \frac{C}{D} = \frac{A}{B} \cdot \frac{D}{C}.$$

(To divide two rational expressions, multiply by the reciprocal of the divisor. We often refer to this as "*invert* and multiply.")

EXAMPLES Divide. Simplify if possible.

13. $\dfrac{x - 2}{x + 1} \div \dfrac{x + 5}{x - 3} = \dfrac{x - 2}{x + 1} \cdot \dfrac{x - 3}{x + 5}$ Multiplying by the reciprocal

$= \dfrac{(x - 2)(x - 3)}{(x + 1)(x + 5)}$ Multiplying numerators and denominators

14. $\dfrac{a^2 - 1}{a - 1} \div \dfrac{a^2 - 2a + 1}{a + 1} = \dfrac{a^2 - 1}{a - 1} \cdot \dfrac{a + 1}{a^2 - 2a + 1}$ Multiplying by the reciprocal

$= \dfrac{(a^2 - 1)(a + 1)}{(a - 1)(a^2 - 2a + 1)}$ Multiplying numerators and denominators

$= \dfrac{(a + 1)(a - 1)(a + 1)}{(a - 1)(a - 1)(a - 1)}$ Factoring the numerator and the denominator and looking for factors of 1

$= \dfrac{a - 1}{a - 1} \cdot \dfrac{(a + 1)(a + 1)}{(a - 1)(a - 1)}$ Factoring the rational expression

$= \dfrac{(a + 1)(a + 1)}{(a - 1)(a - 1)}$ "Removing" a factor of 1 ■

EXERCISE SET 4.1

Evaluate each rational expression, using the values provided. If a replacement is not meaningful, state this.

1. $\dfrac{4t^2 - 5t + 2}{t + 3}$; $t = 0$, $t = 3$, and $t = 7$

2. $\dfrac{5x^2 + 4x - 12}{6 - x}$; $x = 4$, $x = -1$, and $x = 3$

3. $\dfrac{3y^3 - 2y}{y - 5}$; $y = 0$, $y = 4$, and $y = 5$

4. $\dfrac{\pi r^2 + 2\pi r}{r - 1}$; $(\pi \approx 3.14)$ $r = 2$, $r = 5$, and $r = 1$

5. $\dfrac{2x^3 - 9}{x^2 - 4x + 4}$; $x = 0$, $x = 2$, and $x = -1$

6. $\dfrac{t^2 - 5t + 4}{t^2 - 9}$; $t = 1$, $t = 2$, and $t = -4$

7. $\dfrac{9 - t^2}{5 - 6t + t^2}$; $t = -3$, $t = 0$, and $t = 1$

8. $\dfrac{y^3 - 8}{y^2 - 8y + 16}$; $y = 2$, $y = 4$, and $y = -2$

9. $\dfrac{3x - 2y}{x^2 - 16}$; $x = 1$, $y = 2$; and $x = 4$, $y = 3$

10. $\dfrac{7x - 14y}{x^2 + 3}$; $x = 2$, $y = 1$; and $x = 3$, $y = -1$

11. $\dfrac{a^2 b}{a - b}$; $a = -3$, $b = 1$; and $a = 2$, $b = -1$

12. $\dfrac{a - b}{a^2 - 9}$; $a = 5$, $b = 1$; and $a = -3$, $b = 2$

13. $\dfrac{2r - t}{r + t}$; $r = 5$, $t = 3$; and $r = -2$, $t = -3$

14. $\dfrac{5 - rt}{2r - t}$; $r = 0$, $t = 3$; and $r = 4$, $t = 8$

15. $\dfrac{x^2 y}{3x - 2y}$; $x = 0$, $y = -1$; and $x = -2$, $y = -3$

16. $\dfrac{xy^2}{4x + 3y}$; $x = -1$, $y = -2$; and $x = 0$, $y = -5$

Multiply to obtain equivalent expressions. Do not simplify.

17. $\dfrac{3x}{3x} \cdot \dfrac{x + 1}{x + 3}$

18. $\dfrac{4 - y^2}{6 - y} \cdot \dfrac{-1}{-1}$

19. $\dfrac{t - 3}{t + 2} \cdot \dfrac{t + 3}{t + 3}$

20. $\dfrac{p - 4}{p - 5} \cdot \dfrac{p + 5}{p + 5}$

21. $\dfrac{x^2 - 3}{x - 6} \cdot \dfrac{x + 6}{x + 6}$

22. $\dfrac{t^2 - 9}{3 - t} \cdot \dfrac{t + 2}{t + 2}$

23. $\dfrac{t^2 - 3}{t^2 - 3} \cdot \dfrac{t^2 + 3}{t^2 - 4}$

24. $\dfrac{x^2 - 5}{x^2 - 5} \cdot \dfrac{x^2 - 3}{x^2 + 5}$

Simplify by removing a factor of 1.

25. $\dfrac{9y^2}{15y}$

26. $\dfrac{6x^3}{18x^2}$

27. $\dfrac{8t^3}{4t^7}$

28. $\dfrac{27y^7}{18y^9}$

29. $\dfrac{2a - 6}{2}$

30. $\dfrac{3a - 6}{3}$

31. $\dfrac{6x - 9}{12}$

32. $\dfrac{25a - 30}{15}$

33. $\dfrac{4y - 12}{4y + 12}$

34. $\dfrac{8x + 16}{8x - 16}$

35. $\dfrac{6x - 12}{5x - 10}$

36. $\dfrac{7x - 21}{3x - 9}$

37. $\dfrac{12 - 6x}{5x - 10}$

38. $\dfrac{21 - 7x}{3x - 9}$

39. $\dfrac{t^2 - 16}{t^2 - 8t + 16}$

40. $\dfrac{p^2 - 25}{p^2 + 10p + 25}$

41. $\dfrac{x^2 + 9x + 8}{x^2 - 3x - 4}$

42. $\dfrac{t^2 - 8t - 9}{t^2 + 5t + 4}$

43. $\dfrac{16 - t^2}{t^2 - 8t + 16}$

44. $\dfrac{25 - p^2}{p^2 + 10p + 25}$

Multiply and simplify.

45. $\dfrac{5x^2}{3t^5} \cdot \dfrac{9t^8}{25x}$

46. $\dfrac{7a^3}{10b^7} \cdot \dfrac{5b^3}{3a}$

47. $\dfrac{3x - 6}{5x} \cdot \dfrac{x^3}{5x - 10}$

48. $\dfrac{5t^3}{4t - 8} \cdot \dfrac{6t - 12}{10t}$

49. $\dfrac{y^2 - 16}{2y + 6} \cdot \dfrac{y + 3}{y - 4}$

50. $\dfrac{m^2 - n^2}{4m + 4n} \cdot \dfrac{m + n}{m - n}$

51. $\dfrac{x^2 - 16}{x^2} \cdot \dfrac{x^2 - 4x}{x^2 - x - 12}$

52. $\dfrac{y^2 + 10y + 25}{y^2 - 9} \cdot \dfrac{y^2 + 3y}{y + 5}$

53. $\dfrac{6 - 2t}{t^2 + 4t + 4} \cdot \dfrac{t^3 + 2t^2}{t^2 - 9}$

54. $\dfrac{x^2 - 6x + 9}{12 - 4x} \cdot \dfrac{x^2 - 9}{x^3 - 3x^2}$

55. $\dfrac{x^2 - 2x - 35}{2x^3 - 3x^2} \cdot \dfrac{4x^3 - 9x}{7x - 49}$

56. $\dfrac{y^2 - 10y + 9}{y^2 - 1} \cdot \dfrac{y + 4}{y^2 - 5y - 36}$

57. $\dfrac{c^3 + 8}{c^2 - 4} \cdot \dfrac{c^2 - 4c + 4}{c^2 - 2c + 4}$

58. $\dfrac{x^3 - 27}{x^2 - 9} \cdot \dfrac{x^2 - 6x + 9}{x^2 + 3x + 9}$

59. $\dfrac{a^3 - b^3}{3a^2 + 9ab + 6b^2} \cdot \dfrac{a^2 + 2ab + b^2}{a^2 - b^2}$

60. $\dfrac{x^3 + y^3}{x^2 + 2xy - 3y^2} \cdot \dfrac{x^2 - y^2}{3x^2 + 6xy + 3y^2}$

61. $\dfrac{4x^2 - 9y^2}{8x^3 - 27y^3} \cdot \dfrac{4x^2 + 6xy + 9y^2}{4x^2 + 12xy + 9y^2}$

62. $\dfrac{3x^2 - 3y^2}{27x^3 - 8y^3} \cdot \dfrac{6x^2 + 5xy - 6y^2}{6x^2 + 12xy + 6y^2}$

Divide and simplify.

63. $\dfrac{16a^7}{3b^5} \div \dfrac{8a^3}{6b}$

64. $\dfrac{9x^5}{8y^2} \div \dfrac{3x}{16y^9}$

65. $\dfrac{3y + 15}{y} \div \dfrac{y + 5}{y}$

66. $\dfrac{6x + 12}{x} \div \dfrac{x + 2}{x^3}$

67. $\dfrac{y^2 - 9}{y} \div \dfrac{y + 3}{y + 2}$

68. $\dfrac{x^2 - 4}{x} \div \dfrac{x - 2}{x + 4}$

69. $\dfrac{4a^2 - 1}{a^2 - 4} \div \dfrac{2a - 1}{a - 2}$

70. $\dfrac{25x^2 - 4}{x^2 - 9} \div \dfrac{5x - 2}{x + 3}$

71. $\dfrac{x^2 - y^2}{4x + 4y} \div \dfrac{3y - 3x}{12x^2}$

72. $\dfrac{5y - 5x}{15y^3} \div \dfrac{x^2 - y^2}{3x + 3y}$

73. $\dfrac{x^2 - 16}{x^2 - 10x + 25} \div \dfrac{3x - 12}{x^2 - 3x - 10}$

74. $\dfrac{y^2 - 36}{y^2 - 8y + 16} \div \dfrac{3y - 18}{y^2 - y - 12}$

75. $\dfrac{y^3 + 3y}{y^2 - 9} \div \dfrac{y^2 + 5y - 14}{y^2 + 4y - 21}$

76. $\dfrac{a^3 + 4a}{a^2 - 16} \div \dfrac{a^2 + 8a + 15}{a^2 + a - 20}$

77. $\dfrac{x^3 - 64}{x^3 + 64} \div \dfrac{x^2 - 16}{x^2 - 4x + 16}$

78. $\dfrac{8y^3 - 27}{64y^3 - 1} \div \dfrac{4y^2 - 9}{16y^2 + 4y + 1}$

79. $\dfrac{8a^3 + b^3}{2a^2 + 3ab + b^2} \div \dfrac{8a^2 - 4ab + 2b^2}{4a^2 + 4ab + b^2}$

80. $\dfrac{x^3 + 8y^3}{2x^2 + 5xy + 2y^2} \div \dfrac{x^3 - 2x^2y + 4xy^2}{8x^2 - 2y^2}$

SKILL MAINTENANCE

81. Calculate using the rules for order of operations:

$$\frac{3 - 2 \cdot 5^2}{2 + 7}$$

82. Solve: $2x^2 = 11x - 5$.

83. Solve the formula $g = mwr - s$ for w.

84. Solve: $|x| = 23$.

SYNTHESIS

Perform the indicated operations and simplify.

85. $\left[\dfrac{r^2 - 4s^2}{r + 2s} \div (r + 2s)\right] \cdot \dfrac{2s}{r - 2s}$

86. $\left[\dfrac{d^2 - d}{d^2 - 6d + 8} \cdot \dfrac{d - 2}{d^2 + 5d}\right] \div \dfrac{5d}{d^2 - 9d + 20}$

Simplify.

87. $\dfrac{x(x+1)-2(x+3)}{(x+1)(x+2)(x+3)}$

88. $\dfrac{2x-5(x+2)-(x-2)}{x^2-4}$

89. $\dfrac{m^2-t^2}{m^2+t^2+m+t+2mt}$

90. $\dfrac{a^3-2a^2+2a-4}{a^3-2a^2-3a+6}$

91. $\dfrac{x^3+x^2-y^3-y^2}{x^2-2xy+y^2}$

92. $\dfrac{u^6+v^6+2u^3v^3}{u^3-v^3+u^2v-uv^2}$

93. $\dfrac{x^5-x^3+x^2-1-(x^3-1)(x+1)^2}{(x^2-1)^2}$

94. Here is a number "trick." You tell a friend to think of a number, then add 3, double the result, then subtract 2, and finally divide by 2. Your friend tells you the answer. You then subtract 2 from that result and announce the original number. Explain why this "trick" works.

4.2 Addition and Subtraction of Rational Expressions

We add and subtract rational expressions as we do rational numbers. The reason for this is that when the variables in a rational expression are replaced with numbers (provided the denominator is nonzero), the rational expression becomes a number of the form p/q.

When Denominators Are the Same

When two or more rational numbers have the same denominator, we keep that denominator and add or subtract numerators as indicated. The same procedure is used for the addition and subtraction of rational expressions.

> To add or subtract when denominators are the same, add or subtract the numerators and keep the same denominator.
>
> $$\frac{A}{C}+\frac{B}{C}=\frac{A+B}{C} \quad \text{and} \quad \frac{A}{C}-\frac{B}{C}=\frac{A-B}{C}, \quad \text{where } C \neq 0.$$

EXAMPLE 1 Add:

$$\frac{3+x}{x}+\frac{4}{x}.$$

Solution

$$\frac{3+x}{x}+\frac{4}{x}=\frac{7+x}{x}$$

← This expression cannot be simplified further because x is not a factor of $7+x$. ■

EXAMPLE 2 Add:

$$\frac{4x^2 - 5xy}{x^2 - y^2} + \frac{2xy - y^2}{x^2 - y^2}.$$

Solution

$$\frac{4x^2 - 5xy}{x^2 - y^2} + \frac{2xy - y^2}{x^2 - y^2} = \frac{4x^2 - 3xy - y^2}{x^2 - y^2}$$ Adding numerators and combining like terms. The denominator is unchanged.

$$= \frac{(4x + y)(x - y)}{(x + y)(x - y)}$$ Factoring the numerator and the denominator and looking for factors of 1

$$= \frac{x - y}{x - y} \cdot \frac{4x + y}{x + y}$$ Factoring the rational expression

$$= \frac{4x + y}{x + y}$$ "Removing" a factor of 1 ■

Recall from Chapter 1 that a fraction bar acts as a grouping symbol. Thus, when a numerator containing a polynomial is subtracted, care must be taken to subtract, or change the sign of, each term in that polynomial.

EXAMPLE 3 Subtract:

$$\frac{4x + 5}{x + 3} - \frac{x - 2}{x + 3}.$$

Solution

$$\frac{4x + 5}{x + 3} - \frac{x - 2}{x + 3} = \frac{4x + 5 - (x - 2)}{x + 3}$$ The parentheses are important to make sure that you subtract the entire quantity.

$$= \frac{4x + 5 - x + 2}{x + 3}$$

$$= \frac{3x + 7}{x + 3}$$ ■

When Denominators Are Additive Inverses

When one denominator is the additive inverse, or opposite, of the other, we can multiply one of the rational expressions by $-1/-1$. That gives us a common denominator.

EXAMPLE 4 Add:

$$\frac{a}{2a} + \frac{a^3}{-2a}.$$

Solution

$$\frac{a}{2a} + \frac{a^3}{-2a} = \frac{a}{2a} + \frac{-1}{-1} \cdot \frac{a^3}{-2a} \qquad \text{Multiplying by } \frac{-1}{-1}\text{, which is 1}$$

$$= \frac{a}{2a} + \frac{-a^3}{2a} \qquad \text{Performing the multiplication}$$

$$= \frac{a - a^3}{2a} \qquad \text{Adding numerators}$$

$$= \frac{a(1 - a^2)}{2a} \qquad \text{Factoring the numerator and looking for factors of 1}$$

$$= \frac{a}{a} \cdot \frac{1 - a^2}{2} \qquad \text{Factoring the rational expression}$$

$$= \frac{1 - a^2}{2} \qquad \text{“Removing” a factor of 1}$$

■

EXAMPLE 5 Subtract:

$$\frac{5x}{x - 2y} - \frac{3y - 7}{2y - x}.$$

Solution

$$\frac{5x}{x - 2y} - \frac{3y - 7}{2y - x} = \frac{5x}{x - 2y} - \frac{-1}{-1} \cdot \frac{3y - 7}{2y - x}$$

$$= \frac{5x}{x - 2y} - \frac{7 - 3y}{x - 2y} \qquad \text{Performing the multiplication. } \textit{Note: } -1(2y - x) = -2y + x = x - 2y.$$

$$= \frac{5x - (7 - 3y)}{x - 2y} \qquad \text{Subtracting numerators}$$

$$= \frac{5x - 7 + 3y}{x - 2y}$$

■

In Example 5, you may have noticed that the expression $3y - 7$ was multiplied by -1 and then subtracted. This resulted in $-7 + 3y$, which is equivalent to $3y - 7$. Thus, instead of multiplying by -1 and then subtracting, we could have simply *added* $3y - 7$ to $5x$, while treating the denominators as we did.

When Denominators Are Different

To add rational expressions such as

$$\frac{7}{12xy^2} + \frac{8}{15x^3y} \qquad \text{or} \qquad \frac{x}{x^2 - y^2} + \frac{y}{x^2 - 4xy + 3y^2},$$

we must first find common denominators. As was the case with fractions, our work will be lessened if we can find the *least common multiple* (LCM) of the denominators involved. We do so by factoring each denominator and using each factor the greatest number of times that it occurs in any one factorization.

EXAMPLES Find the least common multiple for each pair of denominators.

6. $\dfrac{2}{21x}, \dfrac{7b}{3x^2}$

We factor the denominators:

$$\left.\begin{aligned} 21x &= 3 \cdot 7 \cdot x \\ 3x^2 &= 3 \cdot x \cdot x \end{aligned}\right\} \quad \text{The LCM is } 3 \cdot 7 \cdot x^2 = 21x^2$$

since the factors 3 and 7 each appear at most once in either denominator, while the factor x appears at most twice.

7. $\dfrac{5}{x^2 + x - 12}, \dfrac{7}{x^2 - 16}$

We factor the denominators:

$$\left.\begin{aligned} x^2 + x - 12 &= (x + 4)(x - 3) \\ x^2 - 16 &= (x + 4)(x - 4) \end{aligned}\right\} \quad \text{The LCM is } (x + 4)(x - 3)(x - 4)$$

since the factors $x + 4$, $x - 3$, and $x - 4$ each appear at most once in either denominator. ■

Once we have identified the least common multiple for the denominators, we multiply each rational expression by 1. We choose a symbol for 1 that will give us the LCM in each denominator. In this manner, we form equivalent rational expressions that have the common denominator. We then add or subtract numerators as indicated and simplify our answer, if possible.

EXAMPLE 8 Add:

$$\frac{2a}{15} + \frac{7b}{5a}.$$

Solution We first find the LCM of the denominators.

$$\left.\begin{aligned} 15 &= 3 \cdot 5 \\ 5a &= 5 \cdot a \end{aligned}\right\} \quad \text{The LCM is } 3 \cdot 5 \cdot a = 15a.$$

Now we multiply each expression by 1. We choose whatever symbol for 1 will give us the LCM in each denominator. In this case, we use a/a and $3/3$:

$$\begin{aligned} \frac{2a}{15} \cdot \frac{a}{a} + \frac{7b}{5a} \cdot \frac{3}{3} &= \frac{2a^2}{15a} + \frac{21b}{15a} \\ &= \frac{2a^2 + 21b}{15a}. \end{aligned}$$

Multiplying by a/a in the first expression gave us a denominator of $15a$. Multiplying by $3/3$ in the second expression also gave us a denominator of $15a$. ■

EXAMPLE 9 Add:

$$\frac{4}{x + 3} + \frac{2}{x - 2}.$$

Solution We first find the LCM of the denominators.

$$\left.\begin{matrix} x+3 \\ x-2 \end{matrix}\right\} \quad \text{The LCM is } (x+3)(x-2).$$

Next we multiply each expression by 1 to get the LCM in each denominator. Then we add and simplify if possible.

$$\frac{4}{x+3}+\frac{2}{x-2}=\frac{4}{x+3}\cdot\frac{x-2}{x-2}+\frac{2}{x-2}\cdot\frac{x+3}{x+3}$$ Multiplying by 1 to get the LCM in the denominator

$$=\frac{4(x-2)}{(x+3)(x-2)}+\frac{2(x+3)}{(x-2)(x+3)}$$

$$=\frac{4x-8}{(x+3)(x-2)}+\frac{2x+6}{(x-2)(x+3)}$$ Multiplying out the numerators

$$=\frac{4x-8+2x+6}{(x+3)(x-2)}$$ Adding numerators

$$=\frac{6x-2}{(x+3)(x-2)}$$ Combining like terms

We leave our answer in this form since factoring the numerator, in this case to $2(3x-1)$, will not enable us to simplify further. ■

EXAMPLE 10 Add:

$$\frac{x}{x^2-5x+6}+\frac{3}{x^2-7x+12}.$$

Solution We first find the LCM of the denominators:

$$\left.\begin{matrix} x^2-5x+6=(x-3)(x-2) \\ x^2-7x+12=(x-3)(x-4) \end{matrix}\right\} \quad \text{The LCM is } (x-3)(x-2)(x-4).$$

We now multiply by 1 to get the LCM in each expression.

$$\frac{x}{x^2-5x+6}+\frac{3}{x^2-7x+12}$$

$$=\frac{x}{(x-3)(x-2)}\cdot\frac{x-4}{x-4}+\frac{3}{(x-3)(x-4)}\cdot\frac{x-2}{x-2}$$ Multiplying by 1, using the factors of the LCM that were absent from each denominator

$$=\frac{x(x-4)}{(x-3)(x-2)(x-4)}+\frac{3(x-2)}{(x-3)(x-4)(x-2)}$$

$$=\frac{x^2-4x}{(x-3)(x-2)(x-4)}+\frac{3x-6}{(x-3)(x-4)(x-2)}$$ Multiplying out the numerators

$$= \frac{x^2 - x - 6}{(x - 3)(x - 2)(x - 4)}$$ Adding numerators and combining like terms

$$= \frac{(x - 3)(x + 2)}{(x - 3)(x - 2)(x - 4)}$$ Factoring the numerator

$$= \frac{x - 3}{x - 3} \cdot \frac{x + 2}{(x - 2)(x - 4)} = \frac{x + 2}{(x - 2)(x - 4)}$$ Simplifying by "removing" a factor of 1 ■

EXAMPLE 11 Subtract:

$$\frac{2y + 1}{y^2 - 7y + 6} - \frac{y + 3}{y^2 - 5y - 6}.$$

Solution

$$\frac{2y + 1}{y^2 - 7y + 6} - \frac{y + 3}{y^2 - 5y - 6}$$

$$= \frac{2y + 1}{(y - 6)(y - 1)} - \frac{y + 3}{(y - 6)(y + 1)}$$ The LCM is $(y - 6)(y - 1)(y + 1)$

$$= \frac{2y + 1}{(y - 6)(y - 1)} \cdot \frac{y + 1}{y + 1} - \frac{y + 3}{(y - 6)(y + 1)} \cdot \frac{y - 1}{y - 1}$$

$$= \frac{(2y + 1)(y + 1) - (y + 3)(y - 1)}{(y - 6)(y - 1)(y + 1)}$$

$$= \frac{2y^2 + 3y + 1 - (y^2 + 2y - 3)}{(y - 6)(y - 1)(y + 1)}$$ The parentheses are important.

$$= \frac{2y^2 + 3y + 1 - y^2 - 2y + 3}{(y - 6)(y - 1)(y + 1)}$$

$$= \frac{y^2 + y + 4}{(y - 6)(y - 1)(y + 1)}$$ ■

EXAMPLE 12 Combine:

$$\frac{2x}{x^2 - 4} + \frac{5}{2 - x} - \frac{1}{2 + x}.$$

Solution

$$\frac{2x}{x^2 - 4} + \frac{5}{2 - x} - \frac{1}{2 + x}$$

$$= \frac{2x}{(x - 2)(x + 2)} + \frac{5}{2 - x} - \frac{1}{2 + x}$$ Note that $2 - x$ is the opposite of $x - 2$

$$= \frac{2x}{(x - 2)(x + 2)} + \frac{-1}{-1} \cdot \frac{5}{(2 - x)} - \frac{1}{x + 2}$$

$$= \frac{2x}{(x - 2)(x + 2)} + \frac{-5}{x - 2} - \frac{1}{x + 2}$$ The LCM is $(x - 2)(x + 2)$.

$$= \frac{2x}{(x-2)(x+2)} + \frac{-5}{x-2} \cdot \frac{x+2}{x+2} - \frac{1}{x+2} \cdot \frac{x-2}{x-2}$$

$$= \frac{2x - 5(x+2) - (x-2)}{(x-2)(x+2)} = \frac{2x - 5x - 10 - x + 2}{(x-2)(x+2)}$$ Removing parentheses twice

$$= \frac{-4x - 8}{(x-2)(x+2)} = \frac{-4(x+2)}{(x-2)(x+2)}$$

$$= \frac{-4}{x-2} \cdot \frac{x+2}{x+2} = \frac{-4}{x-2}$$ "Removing" a factor of 1 ■

EXERCISE SET 4.2

Perform the indicated operations. Simplify when possible.

1. $\frac{3}{2a} + \frac{5}{2a}$
2. $\frac{4}{3y} + \frac{8}{3y}$
3. $\frac{3}{4a^2b} - \frac{7}{4a^2b}$
4. $\frac{5}{3m^2n^2} - \frac{4}{3m^2n^2}$
5. $\frac{a-3b}{a+b} + \frac{a+5b}{a+b}$
6. $\frac{x-5y}{x+y} + \frac{x+7y}{x+y}$
7. $\frac{4y+2}{y-2} - \frac{y-3}{y-2}$
8. $\frac{3t+2}{t-4} - \frac{t-2}{t-4}$
9. $\frac{3x-4}{x^2-5x+4} + \frac{3-2x}{x^2-5x+4}$
10. $\frac{5x-4}{x^2-6x-7} + \frac{5-4x}{x^2-6x-7}$
11. $\frac{3a-8}{a^2-9} - \frac{2a-5}{a^2-9}$
12. $\frac{4a-7}{a^2-25} - \frac{3a-2}{a^2-25}$
13. $\frac{a^2}{a-b} + \frac{b^2}{b-a}$
14. $\frac{s^2}{r-s} + \frac{r^2}{s-r}$
15. $\frac{3}{x} - \frac{8}{-x}$
16. $\frac{2}{a} - \frac{5}{-a}$
17. $\frac{2x-9}{x^2-25} - \frac{4-x}{25-x^2}$
18. $\frac{y-9}{y^2-16} - \frac{7-y}{16-y^2}$
19. $\frac{t^2+3}{t^4-16} + \frac{7}{16-t^4}$
20. $\frac{y^2-5}{y^4-81} + \frac{4}{81-y^4}$
21. $\frac{m-3n}{m^3-n^3} - \frac{2n}{n^3-m^3}$
22. $\frac{r-6s}{r^3-s^3} - \frac{5s}{s^3-r^3}$
23. $\frac{y-2}{y+4} + \frac{y+3}{y-5}$
24. $\frac{x-2}{x+3} + \frac{x+2}{x-4}$
25. $2 + \frac{x-3}{x+1}$
26. $3 + \frac{y+2}{y-5}$
27. $\frac{4xy}{x^2-y^2} + \frac{x-y}{x+y}$
28. $\frac{5ab}{a^2-b^2} + \frac{a+b}{a-b}$
29. $\frac{9x+2}{3x^2-2x-8} + \frac{7}{3x^2+x-4}$
30. $\frac{3y+2}{2y^2-y-10} + \frac{8}{2y^2-7y+5}$
31. $\frac{4}{x+1} + \frac{x+2}{x^2-1} + \frac{3}{x-1}$
32. $\frac{-2}{y+2} + \frac{5}{y-2} + \frac{y+3}{y^2-4}$
33. $\frac{x-1}{3x+15} - \frac{x+3}{5x+25}$
34. $\frac{y-2}{4y+8} - \frac{y+6}{5y+10}$

35. $\dfrac{5ab}{a^2 - b^2} - \dfrac{a - b}{a + b}$

36. $\dfrac{6xy}{x^2 - y^2} - \dfrac{x + y}{x - y}$

37. $\dfrac{x}{x^2 + 9x + 20} - \dfrac{4}{x^2 + 7x + 12}$

38. $\dfrac{x}{x^2 + 11x + 30} - \dfrac{5}{x^2 + 9x + 20}$

39. $\dfrac{3y}{y^2 - 7y + 10} - \dfrac{2y}{y^2 - 8y + 15}$

40. $\dfrac{5x}{x^2 - 6x + 8} - \dfrac{3x}{x^2 - x - 12}$

41. $\dfrac{y}{y^2 - y - 20} + \dfrac{2}{y + 4}$

42. $\dfrac{2t + 9}{t^2 - t - 6} + \dfrac{1}{t + 2}$

43. $\dfrac{3y + 2}{y^2 + 5y - 24} + \dfrac{7}{y^2 + 4y - 32}$

44. $\dfrac{3x + 2}{x^2 - 7x + 10} + \dfrac{2x}{x^2 - 8x + 15}$

45. $\dfrac{3x - 1}{x^2 + 2x - 3} - \dfrac{x + 4}{x^2 - 9}$

46. $\dfrac{3p - 2}{p^2 + 2p - 24} - \dfrac{p - 3}{p^2 - 16}$

47. $\dfrac{2}{a^2 - 5a + 4} + \dfrac{-2}{a^2 - 4}$

48. $\dfrac{3}{a^2 - 7a + 6} + \dfrac{-3}{a^2 - 9}$

49. $3 + \dfrac{t}{t + 2} - \dfrac{2}{t^2 - 4}$

50. $2 + \dfrac{t}{t - 3} - \dfrac{3}{t^2 - 9}$

51. $\dfrac{1}{x + 1} - \dfrac{x}{x - 2} + \dfrac{x^2 + 2}{x^2 - x - 2}$

52. $\dfrac{2}{y + 3} - \dfrac{y}{y - 1} + \dfrac{y^2 + 2}{y^2 + 2y - 3}$

53. $\dfrac{4x}{x^2 - 1} + \dfrac{3x}{1 - x} - \dfrac{4}{x - 1}$

54. $\dfrac{5y}{1 - 2y} - \dfrac{2y}{2y + 1} + \dfrac{3}{4y^2 - 1}$

55. $\dfrac{1}{t^2 + 5t + 6} - \dfrac{2}{t^2 + 3t + 2} + \dfrac{1}{t^2 - 3t - 4}$

56. $\dfrac{2}{x^2 - 5x + 6} - \dfrac{4}{x^2 - 2x - 3} + \dfrac{2}{x^2 + 4x + 3}$

SKILL MAINTENANCE

57. Use the commutative law for addition to write an expression equivalent to $5a + 7bc$.

58. Find the intersection:

$$\{2, 5, 10, 12\} \cap \{-3, 5, 12, 15\}.$$

59. Solve: $7 < 12 - 3x < 15$.

60. Simplify, leaving as little as possible inside the absolute value signs: $|-7x^4|$.

SYNTHESIS

Perform the indicated operations and simplify.

61. $\dfrac{x^2 - 7x + 12}{x^2 - x - 29/3} \cdot \left(\dfrac{3x + 2}{x^2 + 5x - 24} + \dfrac{7}{x^2 + 4x - 32}\right)$

62. $\dfrac{8t^5}{2t^2 - 10t + 12} \div \left(\dfrac{2t}{t^2 - 8t + 15} - \dfrac{3t}{t^2 - 7t + 10}\right)$

63. $\dfrac{9t^3}{3t^3 - 12t^2 + 9t} \div \left(\dfrac{t + 4}{t^2 - 9} - \dfrac{3t - 1}{t^2 + 2t - 3}\right)$

64. $\dfrac{x}{x^2 + 11x + 30} - \dfrac{5x + 20}{7x^3 + 63x^2 + 140x} \div \dfrac{x^2 + x - 12}{7x^2 - 21x}$

65. The LCM of two expressions is $8a^4b^7$. One of the expressions is $2a^3b^7$. List all the possibilities for the other expression.

Planet orbits and LCMs. Earth, Jupiter, Saturn, and Uranus all revolve around the sun. Earth takes 1 year, Jupiter takes 12 years, Saturn takes 30 years, and Uranus takes 84 years.

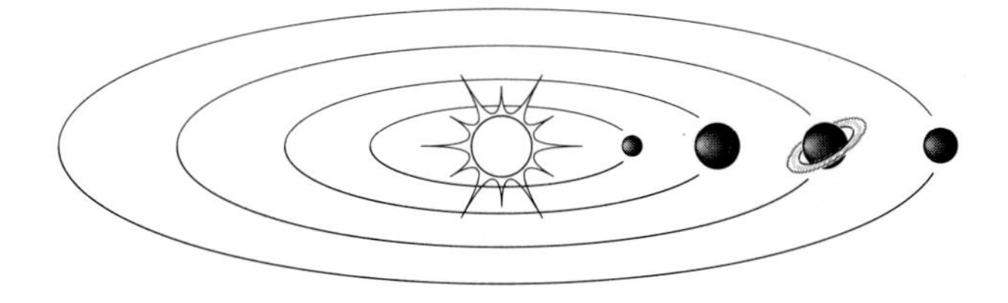

66. How often will Earth, Jupiter, and Saturn line up with each other?

67. In how many years will these four planets align themselves exactly as they are this evening?

4.3 Complex Rational Expressions

A **complex rational expression** is a rational expression that has one or more rational expressions within its numerator or its denominator. Here are some examples:

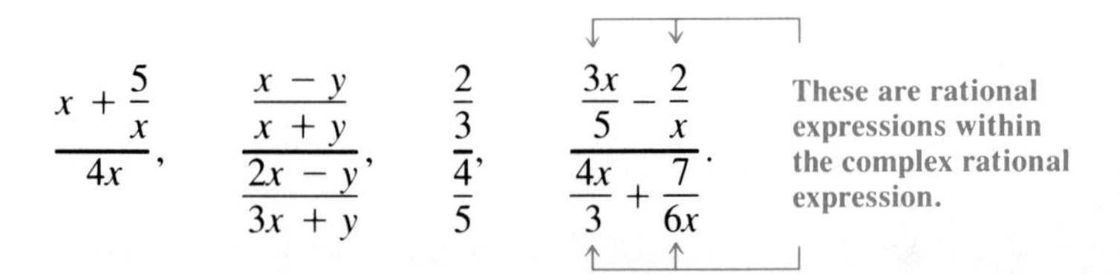

$$\frac{x + \frac{5}{x}}{4x}, \quad \frac{\frac{x-y}{x+y}}{\frac{2x-y}{3x+y}}, \quad \frac{\frac{2}{3}}{\frac{4}{5}}, \quad \frac{\frac{3x}{5} - \frac{2}{x}}{\frac{4x}{3} + \frac{7}{6x}}.$$

We will consider two methods that can be used to simplify complex rational expressions.

Adding or Subtracting Within the Complex Rational Expression

One method of simplifying complex rational expressions involves first adding or subtracting, as necessary, to get a single rational expression in the numerator and a single rational expression in the denominator. The problem is then simplified to one involving the division of two expressions.

EXAMPLE 1 Simplify:

$$\frac{\frac{3}{x} - \frac{2}{x^2}}{\frac{3}{x-2} + \frac{1}{x^2}}.$$

Solution

$$\frac{\frac{3}{x}-\frac{2}{x^2}}{\frac{3}{x-2}+\frac{1}{x^2}} = \frac{\frac{3}{x}\cdot\frac{x}{x}-\frac{2}{x^2}}{\frac{3}{x-2}\cdot\frac{x^2}{x^2}+\frac{1}{x^2}\cdot\frac{x-2}{x-2}}$$

Multiplying $3/x$ by 1 to get a common denominator

Multiplying $3/(x-2)$ and $1/x^2$ by 1 to get a common denominator

$$= \frac{\frac{3x}{x^2}-\frac{2}{x^2}}{\frac{3x^2}{(x-2)x^2}+\frac{x-2}{x^2(x-2)}}$$

$$= \frac{\frac{3x-2}{x^2}}{\frac{3x^2+x-2}{(x-2)x^2}}$$

Subtracting in the numerator and adding in the denominator. We now have a rational expression divided by a rational expression.

$$= \frac{3x-2}{x^2}\cdot\frac{(x-2)x^2}{3x^2+x-2}$$

To divide, multiply by the reciprocal of the divisor.

$$= \frac{3x-2}{x^2}\cdot\frac{(x-2)x^2}{(3x-2)(x+1)}$$

Factoring

$$= \frac{x^2(3x-2)}{x^2(3x-2)}\cdot\frac{x-2}{x+1}$$

"Removing" a factor of 1

$$= \frac{x-2}{x+1}$$

Simplifying

■

We outline the procedure just used.

To simplify a complex rational expression by first adding or subtracting:

1. Add or subtract, as necessary, to get a single rational expression in the numerator.
2. Add or subtract, as necessary, to get a single rational expression in the denominator.
3. Divide the rational expression in the numerator by the rational expression in the denominator (invert and multiply).
4. Simplify, if possible, by removing a factor of 1.

EXAMPLE 2 Simplify:

$$\frac{1+\frac{1}{x}}{1-\frac{1}{x^2}}.$$

Solution

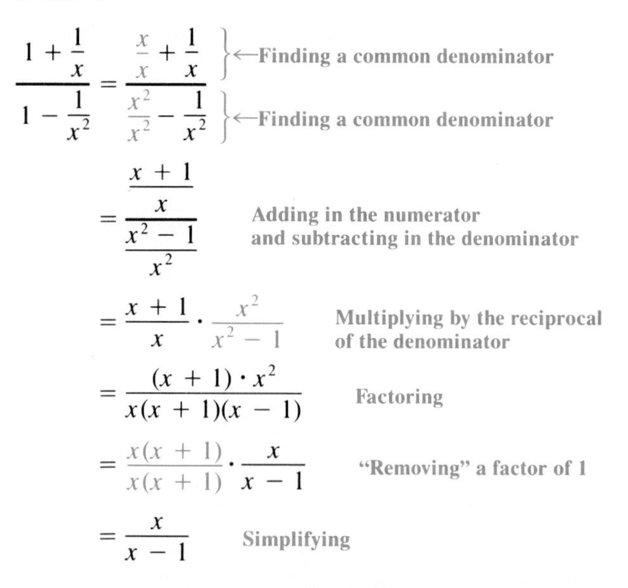

$$\frac{1+\frac{1}{x}}{1-\frac{1}{x^2}} = \frac{\frac{x}{x}+\frac{1}{x}}{\frac{x^2}{x^2}-\frac{1}{x^2}} = \frac{\frac{x+1}{x}}{\frac{x^2-1}{x^2}} = \frac{x+1}{x}\cdot\frac{x^2}{x^2-1} = \frac{(x+1)\cdot x^2}{x(x+1)(x-1)} = \frac{x(x+1)}{x(x+1)}\cdot\frac{x}{x-1} = \frac{x}{x-1}$$

Multiplying by the LCM of All the Denominators

Another method of simplifying complex rational expressions involves multiplying the entire complex rational expression by 1. To write 1, we use the least common multiple of the denominators found in the top and bottom of the complex rational expression.

EXAMPLE 3 Simplify:

$$\frac{\frac{1}{a}+\frac{1}{b}}{\frac{1}{a^3}+\frac{1}{b^3}}.$$

Solution

$$\frac{\frac{1}{a}+\frac{1}{b}}{\frac{1}{a^3}+\frac{1}{b^3}}$$

The denominators in the top and the bottom of the complex rational expression are a, b, a^3, and b^3. The LCM of these denominators is a^3b^3. We multiply by 1 using a^3b^3/a^3b^3.

$$= \frac{\frac{1}{a}+\frac{1}{b}}{\frac{1}{a^3}+\frac{1}{b^3}}\cdot 1 = \frac{\frac{1}{a}+\frac{1}{b}}{\frac{1}{a^3}+\frac{1}{b^3}}\cdot\frac{a^3b^3}{a^3b^3}$$ Multiplying by 1

$$= \frac{\left(\frac{1}{a}+\frac{1}{b}\right)a^3b^3}{\left(\frac{1}{a^3}+\frac{1}{b^3}\right)a^3b^3}$$ Multiplying by a^3b^3. Remember to use parentheses.

$$= \frac{\frac{1}{a} \cdot a^3b^3 + \frac{1}{b} \cdot a^3b^3}{\frac{1}{a^3} \cdot a^3b^3 + \frac{1}{b^3} \cdot a^3b^3}$$ Using the distributive law to carry out the multiplications

$$= \frac{a^2b^3 + a^3b^2}{b^3 + a^3}$$ Simplifying. Note that we have *cleared* all rational expressions from the numerator and denominator.

$$= \frac{a^2b^2(b + a)}{(b + a)(b^2 - ab + a^2)}$$ Factoring and looking for a factor of 1

$$= \frac{b + a}{b + a} \cdot \frac{a^2b^2}{b^2 - ab + a^2}$$ Factoring the rational expression

$$= \frac{a^2b^2}{b^2 - ab + a^2}$$ "Removing" a factor of 1 ■

We outline the procedure just used.

> To simplify a complex rational expression by using the LCM:
>
> 1. Find the LCM of all the denominators *within* the complex rational expression.
> 2. Multiply the complex rational expression by 1, using the LCM to form the expression for 1.
> 3. Distribute and simplify. No rational expressions should remain within the numerator or the denominator of the complex rational expression.
> 4. Factor and simplify, if possible.

Note that we choose the LCM to form the number 1 in order to clear all rational expressions in the top and bottom of the complex rational expression.

EXAMPLE 4 Simplify:

$$\frac{\frac{3}{2x - 2} - \frac{1}{x + 1}}{\frac{1}{x - 1} + \frac{x}{x^2 - 1}}.$$

Solution Note that to find the LCM, we may have to factor first:

$$\frac{\frac{3}{2x - 2} - \frac{1}{x + 1}}{\frac{1}{x - 1} + \frac{x}{x^2 - 1}} = \frac{\frac{3}{2(x - 1)} - \frac{1}{x + 1}}{\frac{1}{x - 1} + \frac{x}{(x - 1)(x + 1)}}$$ The LCM is $2(x - 1)(x + 1)$.

$$= \frac{\frac{3}{2(x - 1)} - \frac{1}{x + 1}}{\frac{1}{x - 1} + \frac{x}{(x - 1)(x + 1)}} \cdot \frac{2(x - 1)(x + 1)}{2(x - 1)(x + 1)}$$ Multiplying by 1

$$= \frac{\frac{3}{2(x-1)} \cdot 2(x-1)(x+1) - \frac{1}{x+1} \cdot 2(x-1)(x+1)}{\frac{1}{x-1} \cdot 2(x-1)(x+1) + \frac{x}{(x-1)(x+1)} \cdot 2(x-1)(x+1)}$$ Using the distributive law

$$= \frac{3(x+1) - 2(x-1)}{2(x+1) + 2x}$$ Simplifying. This is why we left the denominators in factored form. Study this step carefully.

$$= \frac{3x + 3 - 2x + 2}{2x + 2 + 2x}$$ Using the distributive law

$$= \frac{x+5}{4x+2}.$$

It is difficult to say which method is best to use. To simplify expressions such as

$$\frac{\frac{3x+1}{x-5}}{\frac{2-x}{x+3}} \quad \text{or} \quad \frac{\frac{3}{x} - \frac{2}{x}}{\frac{1}{x+1} + \frac{5}{x+1}},$$

the first method discussed is probably easier to use since it is little or no work to write the expression as a quotient of two rational expressions. We then invert and multiply.

On the other hand, expressions such as

$$\frac{\frac{3}{a^2b} - \frac{4}{bc^3}}{\frac{1}{b^3c} + \frac{2}{ac^4}} \quad \text{or} \quad \frac{\frac{5}{a^2 - b^2} + \frac{2}{a^2 + 2ab + b^2}}{\frac{1}{a-b} + \frac{4}{a+b}}$$

may be more easily solved using the second method. Either method will work on any complex rational expression.

EXERCISE SET 4.3

Simplify.

1. $\dfrac{\frac{1}{x} + 4}{\frac{1}{x} - 3}$

2. $\dfrac{\frac{1}{y} + 7}{\frac{1}{y} - 5}$

3. $\dfrac{x - \frac{1}{x}}{x + \frac{1}{x}}$

4. $\dfrac{y + \frac{1}{y}}{y - \frac{1}{y}}$

5. $\dfrac{\frac{3}{x} + \frac{4}{y}}{\frac{4}{x} - \frac{3}{y}}$

6. $\dfrac{\frac{2}{y} + \frac{5}{z}}{\frac{1}{y} - \frac{4}{z}}$

7. $\dfrac{\frac{x^2 - y^2}{xy}}{\frac{x - y}{y}}$

8. $\dfrac{\frac{a^2 - b^2}{ab}}{\frac{a - b}{b}}$

9. $\dfrac{a - \frac{3a}{b}}{b - \frac{b}{a}}$

10. $\dfrac{1 - \frac{2}{3x}}{x - \frac{4}{9x}}$

11. $\dfrac{\frac{1}{a} + \frac{1}{b}}{\frac{a^2 - b^2}{ab}}$

12. $\dfrac{\frac{1}{x} + \frac{1}{y}}{\frac{x^2 - y^2}{xy}}$

13. $\dfrac{\dfrac{1}{x+h}-\dfrac{1}{x}}{h}$

14. $\dfrac{\dfrac{1}{a-h}-\dfrac{1}{a}}{h}$

15. $\dfrac{\dfrac{y^2-y-6}{y^2-5y-14}}{\dfrac{y^2+6y+5}{y^2-6y-7}}$

16. $\dfrac{\dfrac{x^2-x-12}{x^2-2x-15}}{\dfrac{x^2+8x+12}{x^2-5x-14}}$

17. $\dfrac{\dfrac{1}{x-2}+\dfrac{3}{x-1}}{\dfrac{2}{x-1}+\dfrac{5}{x-2}}$

18. $\dfrac{\dfrac{2}{y-3}+\dfrac{1}{y+1}}{\dfrac{3}{y+1}+\dfrac{4}{y-3}}$

19. $\dfrac{\dfrac{a}{a+3}-\dfrac{2}{a-1}}{\dfrac{a}{a+3}-\dfrac{1}{a-1}}$

20. $\dfrac{\dfrac{a}{a+2}-\dfrac{3}{a-3}}{\dfrac{a}{a+2}-\dfrac{1}{a-3}}$

21. $\dfrac{\dfrac{x}{x^2+3x-4}-\dfrac{1}{x^2+3x-4}}{\dfrac{x}{x^2+6x+8}+\dfrac{3}{x^2+6x+8}}$

22. $\dfrac{\dfrac{x}{x^2+5x-6}+\dfrac{6}{x^2+5x-6}}{\dfrac{x}{x^2-5x+4}-\dfrac{2}{x^2-5x+4}}$

23. $\dfrac{\dfrac{y}{y^2-1}+\dfrac{3}{1-y^2}}{\dfrac{y^2}{y^2-1}+\dfrac{9}{1-y^2}}$

24. $\dfrac{\dfrac{y}{y^2-4}+\dfrac{5}{4-y^2}}{\dfrac{y^2}{y^2-4}+\dfrac{25}{4-y^2}}$

25. $\dfrac{\dfrac{2}{a^2-1}+\dfrac{1}{a+1}}{\dfrac{3}{a^2-1}+\dfrac{2}{a-1}}$

26. $\dfrac{\dfrac{3}{a^2-9}+\dfrac{2}{a+3}}{\dfrac{4}{a^2-9}+\dfrac{1}{a+3}}$

27. $\dfrac{\dfrac{5}{x^2-4}-\dfrac{3}{x-2}}{\dfrac{4}{x^2-4}-\dfrac{2}{x+2}}$

28. $\dfrac{\dfrac{4}{x^2-1}-\dfrac{3}{x+1}}{\dfrac{5}{x^2-1}-\dfrac{2}{x-1}}$

29. $\dfrac{\dfrac{y^2}{y^2-9}-\dfrac{y}{y+3}}{\dfrac{y}{y^2-9}-\dfrac{1}{y-3}}$

30. $\dfrac{\dfrac{y^2}{y^2-25}-\dfrac{y}{y-5}}{\dfrac{y}{y^2-25}-\dfrac{1}{y+5}}$

31. $\dfrac{\dfrac{a}{a+3}+\dfrac{4}{5a}}{\dfrac{a}{2a+6}+\dfrac{3}{a}}$

32. $\dfrac{\dfrac{a}{a+2}+\dfrac{5}{a}}{\dfrac{a}{2a+4}+\dfrac{1}{3a}}$

33. $\dfrac{\dfrac{x}{x+y}+\dfrac{x}{y}}{\dfrac{x}{3x+3y}+\dfrac{y}{x}}$

34. $\dfrac{\dfrac{x}{x+y}+\dfrac{y}{x}}{\dfrac{x}{5x+5y}+\dfrac{x}{y}}$

35. $\dfrac{\dfrac{1}{x^2-1}+\dfrac{1}{x^2+4x+3}}{\dfrac{1}{x^2-1}+\dfrac{1}{x^2-3x+2}}$

36. $\dfrac{\dfrac{1}{x^2-4}+\dfrac{1}{x^2+3x+2}}{\dfrac{1}{x^2-4}+\dfrac{1}{x^2-4x+4}}$

37. $\dfrac{\dfrac{y}{y^2-4}-\dfrac{2y}{y^2+y-6}}{\dfrac{2y}{y^2-4}-\dfrac{y}{y^2+5y+6}}$

38. $\dfrac{\dfrac{y}{y^2-1}-\dfrac{3y}{y^2+5y+4}}{\dfrac{3y}{y^2-1}-\dfrac{y}{y^2-4y+3}}$

39. $\dfrac{\dfrac{1}{a^2+7a+12}+\dfrac{1}{a^2+a-6}}{\dfrac{1}{a^2+2a-8}+\dfrac{1}{a^2+5a+4}}$

40. $\dfrac{\dfrac{1}{a^2-5a+6}+\dfrac{1}{a^2-4a+3}}{\dfrac{1}{a^2-3a+2}+\dfrac{1}{a^2+3a-10}}$

41. $\dfrac{\dfrac{2}{x^2-7x+12}-\dfrac{1}{x^2+7x+10}}{\dfrac{2}{x^2-x-6}-\dfrac{1}{x^2+x-20}}$

42. $\dfrac{\dfrac{3}{x^2+2x-3}-\dfrac{1}{x^2-3x-10}}{\dfrac{3}{x^2-6x+5}-\dfrac{1}{x^2+5x+6}}$

SKILL MAINTENANCE

43. Solve for x: $\dfrac{a}{x+y} = b$.

44. Evaluate: $\dfrac{x^2 - 3}{2x}$, for $x = -4$.

45. A concert committee needs to take in $4000 from ticket sales to break even. If a total of 400 tickets are to be sold at full price and 200 tickets sold at half price, how should the tickets be priced?

46. A waitress received $14 in tips on Monday, $11 in tips on Tuesday, and $18 in tips on Wednesday. How much will she have to earn in tips on Thursday if her average tips for the four days is to be $15?

SYNTHESIS

Simplify.

47. $2 + \cfrac{2}{2 + \cfrac{2}{2 + \cfrac{2}{2 + \frac{2}{x}}}}$

48. $\left[\dfrac{\dfrac{x+3}{x-3} + 1}{\dfrac{x+3}{x-3} - 1}\right]^4$

Find the reciprocal and simplify.

49. $x^2 - \dfrac{1}{x}$

50. $\dfrac{1 - \frac{1}{a}}{a - 1}$

51. $\dfrac{a^6 - b^6}{a + b}$

52. $1 + \cfrac{1}{1 + \cfrac{1}{1 + \cfrac{1}{1 + \frac{1}{x}}}}$

4.4 Rational Equations

A **rational equation** is an equation that contains one or more rational expressions. Here are some examples:

$$\frac{2}{3} - \frac{5}{6} = \frac{1}{x}, \qquad \frac{x-1}{x-5} = \frac{4}{x-5}, \qquad x + \frac{6}{x} = 5.$$

As you will see in Section 4.5, equations of this type occur frequently in applications. To solve rational equations, we borrow a technique we used in Section 4.3: When we multiply by the least common multiple of the denominators, the problem is *cleared of fractions.* In Section 4.3, we used the LCM over itself to form 1. Here we multiply by the LCM on both sides of an equation.

> To solve a rational equation, we multiply on both sides by the LCM of all the denominators. This is called *clearing of fractions.*

EXAMPLE 1 Solve:

$$\frac{2}{3} - \frac{5}{6} = \frac{1}{x}.$$

Solution The LCM of all denominators is $6x$, or $2 \cdot 3 \cdot x$, which, for convenience, we leave in factored form. We multiply on both sides by the LCM:

$$(2 \cdot 3 \cdot x) \cdot \left(\frac{2}{3} - \frac{5}{6}\right) = (2 \cdot 3 \cdot x) \cdot \frac{1}{x} \quad \text{Multiplying by the LCM}$$

$$2 \cdot 3 \cdot x \cdot \frac{2}{3} - 2 \cdot 3 \cdot x \cdot \frac{5}{6} = 2 \cdot 3 \cdot x \cdot \frac{1}{x} \quad \text{Using the distributive law}$$

$$\frac{2 \cdot 3 \cdot x \cdot 2}{3} - \frac{2 \cdot 3 \cdot x \cdot 5}{6} = \frac{2 \cdot 3 \cdot x}{x} \quad \text{Locating factors of 1}$$

$$4x - 5x = 6 \quad \text{Simplifying}$$

$$-x = 6$$

$$x = -6.$$

Check: $\frac{2}{3} - \frac{5}{6} = \frac{1}{x}$

$\frac{2}{3} - \frac{5}{6}$	$\frac{1}{-6}$
$\frac{4}{6} - \frac{5}{6}$	$-\frac{1}{6}$
$-\frac{1}{6}$	

TRUE

The number -6 is the solution. ■

When clearing of fractions, as we did in Example 1, be sure to use the distributive law carefully when multiplying on both sides by the LCM.

Note that when we clear of fractions, all the denominators disappear. Then we have an equation without rational expressions, which we know how to solve.

EXAMPLE 2 Solve:

$$\frac{x+4}{3x} + \frac{x+8}{5x} = 2.$$

Solution The LCM of the denominators is $3 \cdot 5 \cdot x$, which we multiply by on both sides:

$$3 \cdot 5 \cdot x \left(\frac{x+4}{3x} + \frac{x+8}{5x}\right) = 3 \cdot 5 \cdot x \cdot 2 \quad \text{Multiplying by the LCM}$$

$$3 \cdot 5 \cdot x \cdot \frac{x+4}{3x} + 3 \cdot 5 \cdot x \cdot \frac{x+8}{5x} = 3 \cdot 5 \cdot x \cdot 2 \quad \text{Using the distributive law}$$

$$\frac{3 \cdot 5 \cdot x\,(x+4)}{3x} + \frac{3 \cdot 5 \cdot x \cdot (x+8)}{5x} = 30x \qquad \text{Locating factors of 1}$$

$$5(x+4) + 3(x+8) = 30x \qquad \text{Simplifying}$$

$$5x + 20 + 3x + 24 = 30x$$

$$8x + 44 = 30x$$

$$44 = 22x$$

$$2 = x.$$

Check:

$$\frac{x+4}{3x} + \frac{x+8}{5x} = 2$$

$\frac{2+4}{3 \cdot 2} + \frac{2+8}{5 \cdot 2}$	2
$\frac{6}{6} + \frac{10}{10}$	
2	

TRUE

The number 2 is the solution. ■

When solving rational equations, it is extremely important to check possible solutions in the original equation. They may not check, even though we have made no error.

EXAMPLE 3 Solve:

$$\frac{x-1}{x-5} = \frac{4}{x-5}.$$

Solution The LCM of the denominators is $x - 5$. We multiply by $x - 5$:

$$(x-5) \cdot \frac{x-1}{x-5} = (x-5) \cdot \frac{4}{x-5}$$

$$x - 1 = 4$$

$$x = 5.$$

Check:

$$\frac{x-1}{x-5} = \frac{4}{x-5}$$

$\frac{5-1}{5-5}$	$\frac{4}{5-5}$
$\frac{4}{0}$	$\frac{4}{0}$

Division by 0 is undefined.

We know that 5 is *not* a solution of the original equation because it results in division by 0. In fact, the equation has no solution, and we say that 5 is an *extraneous* root of the equation. ■

To help see why 5 is not a solution to Example 3, consider the fact that the multiplication principle for equations requires that we multiply by a *nonzero* number on both sides if we are to form an equivalent equation. When both sides of an equation are multiplied by an expression containing variables, it is possible that certain replacements will make that expression equal to 0. Thus it is safe to say that, *if* a solution to

$$\frac{x-1}{x-5} = \frac{4}{x-5}$$

exists, then the solution will be a solution of $x - 1 = 4$. We *cannot* conclude that every solution to $x - 1 = 4$ will be a solution of the original equation.

> In solving an equation, when we multiply by an expression containing a variable, we may get an equation having solutions that are not solutions of the original equation.

EXAMPLE 4 Solve:

$$\frac{x^2}{x-3} = \frac{9}{x-3}.$$

Solution The LCM of the denominators is $x - 3$. We multiply by $x - 3$:

$$(x-3)\cdot\frac{x^2}{x-3} = (x-3)\cdot\frac{9}{x-3}$$

$$x^2 = 9 \quad \text{Simplifying}$$

$$x^2 - 9 = 0 \quad \text{Getting 0 on one side}$$

$$(x+3)(x-3) = 0 \quad \text{Factoring}$$

$$x = 3 \quad \text{or} \quad x = -3. \quad \text{Using the principle of zero products}$$

Check: For 3:

$$\frac{x^2}{x-3} = \frac{9}{x-3}$$

$\frac{3^2}{3-3}$	$\frac{9}{3-3}$
$\frac{9}{0}$	$\frac{9}{0}$

UNDEFINED

For -3:

$$\frac{x^2}{x-3} = \frac{9}{x-3}$$

$\frac{(-3)^2}{-3-3}$	$\frac{9}{-3-3}$
$\frac{9}{-6}$	$\frac{9}{-6}$

TRUE

The number -3 is a solution, but 3 is not (it is extraneous since it results in division by 0). ■

EXAMPLE 5 Solve:

$$x + \frac{6}{x} = 5.$$

Solution The LCM of the denominators is x. We multiply on both sides by x:

$$x\left(x + \frac{6}{x}\right) = 5 \cdot x \qquad \text{Multiplying on both sides by } x$$

$$x^2 + x \cdot \frac{6}{x} = 5x$$

$$x^2 + 6 = 5x \qquad \text{Simplifying}$$

$$x^2 - 5x + 6 = 0 \qquad \text{Getting 0 on one side}$$

$$(x - 3)(x - 2) = 0 \qquad \text{Factoring}$$

$$x = 3 \quad \text{or} \quad x = 2. \qquad \text{Using the principle of zero products}$$

The solutions are 2 and 3. ■

EXAMPLE 6 Solve:

$$\frac{2}{x - 1} = \frac{3}{x + 1}.$$

Solution The LCM of the denominators is $(x - 1)(x + 1)$.

$$(x - 1)(x + 1) \cdot \frac{2}{(x - 1)} = (x - 1)(x + 1) \cdot \frac{3}{x + 1} \qquad \text{Multiplying}$$

$$2(x + 1) = 3(x - 1) \qquad \text{Simplifying}$$

$$2x + 2 = 3x - 3$$

$$5 = x$$

As the reader should confirm, 5 checks in the original equation. The number 5 is the solution. ■

EXAMPLE 7 Solve:

$$\frac{2}{x + 5} + \frac{1}{x - 5} = \frac{16}{x^2 - 25}.$$

Solution The LCM is $(x + 5)(x - 5)$. We multiply by $(x + 5)(x - 5)$:

$$(x + 5)(x - 5) \cdot \left[\frac{2}{x + 5} + \frac{1}{x - 5}\right] = (x + 5)(x - 5) \cdot \frac{16}{x^2 - 25}$$

$$(x + 5)(x - 5) \cdot \frac{2}{x + 5} + (x + 5)(x - 5) \cdot \frac{1}{x - 5} = (x + 5)(x - 5) \cdot \frac{16}{x^2 - 25}$$

Using the distributive law

$$2(x - 5) + (x + 5) = 16$$
$$2x - 10 + x + 5 = 16$$
$$3x - 5 = 16$$
$$3x = 21$$
$$x = 7.$$

The check is left for the reader. The solution is 7. ■

EXERCISE SET 4.4

Solve.

1. $\frac{2}{5} + \frac{7}{8} = \frac{y}{20}$
2. $\frac{4}{5} + \frac{1}{3} = \frac{t}{9}$
3. $\frac{x}{3} - \frac{x}{4} = 12$
4. $\frac{y}{5} - \frac{y}{3} = 15$
5. $\frac{1}{3} - \frac{5}{6} = \frac{1}{x}$
6. $\frac{5}{8} - \frac{2}{5} = \frac{1}{y}$
7. $\frac{2}{3} - \frac{1}{5} = \frac{7}{3x}$
8. $\frac{1}{2} - \frac{2}{7} = \frac{3}{2x}$
9. $\frac{2}{6} + \frac{1}{2x} = \frac{1}{3}$
10. $\frac{12}{15} - \frac{1}{3x} = \frac{4}{5}$
11. $\frac{4}{z} + \frac{2}{z} = 3$
12. $\frac{4}{3y} - \frac{3}{y} = \frac{10}{3}$
13. $y + \frac{5}{y} = -6$
14. $x + \frac{4}{x} = -5$
15. $2x - \frac{6}{x} = 1$
16. $2x - \frac{15}{x} = 1$
17. $\frac{y - 1}{y - 3} = \frac{2}{y - 3}$
18. $\frac{x - 2}{x - 4} = \frac{2}{x - 4}$
19. $\frac{x + 1}{x} = \frac{3}{2}$
20. $\frac{y + 2}{y} = \frac{5}{3}$
21. $\frac{x - 3}{x + 2} = \frac{1}{5}$
22. $\frac{y - 5}{y + 1} = \frac{3}{5}$
23. $\frac{3}{y + 1} = \frac{2}{y - 3}$
24. $\frac{4}{x - 1} = \frac{3}{x + 2}$
25. $\frac{7}{5x - 2} = \frac{5}{4x}$
26. $\frac{5}{y + 4} = \frac{3}{y - 2}$
27. $\frac{2}{x} - \frac{3}{x} + \frac{4}{x} = 5$
28. $\frac{4}{y} - \frac{6}{y} + \frac{8}{y} = 8$
29. $\frac{1}{2} - \frac{4}{9x} = \frac{4}{9} - \frac{1}{6x}$
30. $-\frac{1}{3} - \frac{5}{4y} = \frac{3}{4} - \frac{1}{6y}$
31. $\frac{z}{z - 1} = \frac{6}{z + 1}$
32. $\frac{2y}{y + 3} = \frac{-4}{y - 7}$
33. $\frac{60}{x} - \frac{60}{x - 5} = \frac{2}{x}$
34. $\frac{50}{y} - \frac{50}{y - 2} = \frac{4}{y}$
35. $\frac{x}{x - 2} + \frac{x}{x^2 - 4} = \frac{x + 3}{x + 2}$
36. $\frac{3}{y - 2} + \frac{2y}{4 - y^2} = \frac{5}{y + 2}$
37. $\frac{a}{2a - 6} - \frac{3}{a^2 - 6a + 9} = \frac{a - 2}{3a - 9}$

38. $\dfrac{2}{x+4}+\dfrac{2x-1}{x^2+2x-8}=\dfrac{1}{x-2}$

39. $\dfrac{2x+3}{x-1}=\dfrac{10}{x^2-1}+\dfrac{2x-3}{x+1}$

40. $\dfrac{5}{y+1}+\dfrac{3y+5}{y^2+4y+3}=\dfrac{2}{y+3}$

SKILL MAINTENANCE

41. Factor completely: $81x^4-y^4$.

42. Solve: $|3x-4|<9$.

43. There are 50 questions on a test. The questions are either multiple-choice, true–false, or fill-in. There are two and a half times as many multiple-choice as fill-in and 5 fewer fill-in than true–false. How many of each type of question are there on the test?

44. Find two consecutive even numbers whose product is 288.

SYNTHESIS

Solve.

45. $\left(\dfrac{1}{1+x}+\dfrac{x}{1-x}\right)\div\left(\dfrac{x}{1+x}-\dfrac{1}{1-x}\right)=-1$

46. $\dfrac{x+3}{x+2}-\dfrac{x+4}{x+3}=\dfrac{x+5}{x+4}-\dfrac{x+6}{x+5}$

47. ▦ $\dfrac{2.315}{y}-\dfrac{12.6}{17.4}=\dfrac{6.71}{7}+0.763$

48. ▦ $\dfrac{6.034}{x}-43.17=\dfrac{0.793}{x}+18.15$

49. $\dfrac{x^3+8}{x+2}=x^2-2x+4$

50. $\dfrac{(x-3)^2}{x-3}=x-3$

Equations that are true for all meaningful replacements of the variables are called *identities.* Determine whether each of the following equations is an identity.

51. $\dfrac{x^2+6x-16}{x-2}=x+8$

52. $\dfrac{x^3+8}{x^2-4}=\dfrac{x^2-2x+4}{x-2}$

Solve and check.

53. $\dfrac{x-\frac{3}{2}}{x+\frac{2}{3}}=\dfrac{x+\frac{1}{2}}{x-\frac{2}{3}}$

54. $\dfrac{2+\frac{x}{2}}{2-\frac{x}{4}}=\dfrac{2}{\frac{x}{2}-2}$

4.5 Applications of Rational Equations

Now that we have considered equations containing rational expressions, we can use that skill in solving certain kinds of problems that we could not have handled before. The problem-solving steps and strategy are the same.

Problems Using Rational Expressions

EXAMPLE 1 If a certain number is added to 5 times the reciprocal of 2 more than that number, the result is 4. Find the number.

Solution

1. **Familiarize.** We let y equal the number in question.
2. **Translate.**

A certain number	added to	five	times	the reciprocal of 2 more than the number	is	4.
y	$+$	5	$\cdot$	$\dfrac{1}{y+2}$	$=$	4

3. **Carry out.** We solve as follows:

$$y + 5 \cdot \frac{1}{y+2} = 4$$

$$(y+2)\cdot y + (y+2)\cdot 5 \cdot \frac{1}{y+2} = (y+2)\cdot 4 \qquad \text{Multiplying by the LCM}$$

$$y^2 + 2y + 5 = 4y + 8 \qquad \text{Simplifying}$$

$$y^2 - 2y - 3 = 0 \qquad \text{Collecting like terms all on one side}$$

$$(y-3)(y+1) = 0 \qquad \text{Factoring}$$

$$y - 3 = 0 \quad \text{or} \quad y + 1 = 0 \qquad \text{Principle of zero products}$$

$$y = 3 \quad \text{or} \quad y = -1.$$

4. **Check.** The possible solutions are 3 and -1. We check 3 in the conditions of the problem.

   Number: 3. Two more than the number: 5.

   Reciprocal of 2 more than the number: $\frac{1}{5}$.

   5 times the reciprocal of 2 more than the number: $5 \cdot \frac{1}{5} = 1$.

   Sum of the number and 5 times the reciprocal of 2 more than the number: $3 + 1 = 4$.

   The number 3 checks. So does the number -1, but we leave that check to you.

5. **State.** The answer is that two numbers satisfy the conditions of the problem. They are 3 and -1. ■

Problems Involving Work

EXAMPLE 2 Pam can mow a lawn in 4 hr. Jack can mow the same lawn in 5 hr. How long would it take them, working together with two mowers, to mow the lawn?

Solution

1. **Familiarize.** We first consider two common, *but incorrect,* approaches to the problem.

a) One *incorrect* approach is to simply add the two times:

$$4 \text{ hr} + 5 \text{ hr} = 9 \text{ hr}.$$

This cannot be correct since either Pam or Jack can do the job alone in less than 9 hr.

b) Another *incorrect* approach is to split up the job so that Pam does half and Jack does half. But, in this case, Pam would finish her half in 2 hr and Jack would complete his half in $2\frac{1}{2}$ hr. Since Pam would be idle for $\frac{1}{2}$ hr, they would not be working *together* the whole time. We can now see that by working together, Pam and Jack will finish the lawn in a time between 2 and $2\frac{1}{2}$ hr.

Let's consider how much of the job is done in 1 hr, 2 hr, 3 hr, and so on. Since Pam can do the whole lawn in 4 hr, she can do $\frac{1}{4}$ of the job in 1 hr. Jack can do the whole lawn in 5 hr, so he can do $\frac{1}{5}$ of the job in 1 hr. Together, they can do

$$\frac{1}{4} + \frac{1}{5} = \frac{9}{20} \text{ of the job in one hour,} \qquad \text{(I)}$$
$$2\left(\frac{1}{4}\right) + 2\left(\frac{1}{5}\right) = \frac{18}{20} \text{ of the job in two hours,} \qquad \text{(II)}$$
$$3\left(\frac{1}{4}\right) + 3\left(\frac{1}{5}\right) = \frac{27}{20} \text{ of the job in three hours.} \qquad \text{(III)}$$

But $\frac{27}{20}$ represents mowing more than one lawn. Let's let t represent the number of hours required for Pam and Jack, working together, to mow the lawn.

2. **Translate.** From equations (II) and (III), we see that the time we want is some number t for which

$$t\left(\frac{1}{4}\right) + t\left(\frac{1}{5}\right) = 1, \quad \text{or} \quad \frac{t}{4} + \frac{t}{5} = 1,$$

where 1 represents the idea that one entire job is completed in time t.

3. **Carry out.** We solve the equation:

$$\frac{t}{4} + \frac{t}{5} = 1$$
$$20\left(\frac{t}{4} + \frac{t}{5}\right) = 20 \cdot 1 \qquad \text{Multiplying by the LCM, 20}$$
$$20 \cdot \frac{t}{4} + 20 \cdot \frac{t}{5} = 20 \qquad \text{Using the distributive law}$$
$$5t + 4t = 20$$
$$9t = 20$$
$$t = \frac{20}{9}.$$

4. **Check.** The possible solution is $\frac{20}{9}$, or $2\frac{2}{9}$ hr. If Pam works $\frac{20}{9}$ hr, she will do $\frac{20}{9} \cdot \frac{1}{4}$ of the work, or $\frac{5}{9}$ of it. If Jack works $\frac{20}{9}$ hr, he will do $\frac{20}{9} \cdot \frac{1}{5}$ of the work, or $\frac{4}{9}$ of it. Altogether, they will do $\frac{5}{9} + \frac{4}{9}$ of the work, or one complete job.

5. **State.** It will take $2\frac{2}{9}$ hr for Pam and Jack, together, to mow the lawn. ■

EXAMPLE 3 It takes Red 9 hr longer to build a wall than it takes Mort. If they work together, they can build the wall in 20 hr. How long would it take each, working alone, to build the wall?

Solution

1. **Familiarize.** Unlike Example 2, this problem does not provide us with the times required by the individuals to do the job alone. Let's have t = the amount of time it would take Mort working alone and $t + 9$ = the amount of time it would take Red working alone.

2. **Translate.** Using the same reasoning that we employed in Example 2, we see that Mort can build $1/t$ of a wall in 1 hr and Red can build $1/(t + 9)$ of a wall in 1 hr. In 20 hr, Mort does $20(1/t)$ of the job and Red does $20[1/(t + 9)]$ of the job. If we add these fractional parts, we get the entire job, represented by 1. This gives us the following equation:

$$20\left(\frac{1}{t}\right) + 20\left(\frac{1}{t + 9}\right) = 1.$$

3. **Carry out.** We solve the equation.

$$\frac{20}{t} + \frac{20}{t + 9} = 1$$

$$t(t + 9)\left(\frac{20}{t} + \frac{20}{t + 9}\right) = t(t + 9)1 \quad \text{Multiplying by the LCM}$$

$$(t + 9)20 + t \cdot 20 = t(t + 9) \quad \text{Distributing and simplifying}$$

$$40t + 180 = t^2 + 9t$$

$$0 = t^2 - 31t - 180 \quad \text{Getting 0 on one side}$$

$$0 = (t - 36)(t + 5) \quad \text{Factoring}$$

$$t - 36 = 0 \quad \text{or} \quad t + 5 = 0 \quad \text{Principle of zero products}$$

$$t = 36 \quad \text{or} \quad t = -5$$

4. **Check.** Since negative time has no meaning in the problem, -5 is not a solution of the original problem. The number 36 checks since, if Mort took 36 hr alone and Red took $36 + 9 = 45$ hr alone, in 20 hr they would have completed

$$\frac{20}{36} + \frac{20}{45} = \frac{5}{9} + \frac{4}{9} = 1 \text{ wall.}$$

5. **State.** It would take Mort 36 hr to build the wall alone, and Red 45 hr. ■

Problems Involving Motion

When a problem deals with distance, speed (rate), and time, we need to recall the following:

If r represents rate, t represents time, and d represents distance, then:

$$d = rt, \qquad r = \frac{d}{t}, \qquad \text{and} \qquad t = \frac{d}{r}.$$

You should remember at least one of these equations and obtain the others by using algebraic manipulations as needed in a problem situation.

EXAMPLE 4 A bicycle racer is traveling 15 km/h faster than a person on a touring bike. In the time it takes the racer to travel 80 km, the person on the touring bike has gone 50 km. Find the speed of each biker.

Solution

1. **Familiarize.** We let x represent the speed, in kilometers per hour, of the slower biker. Then the racer's speed is $x + 15$. We make a drawing and label it.

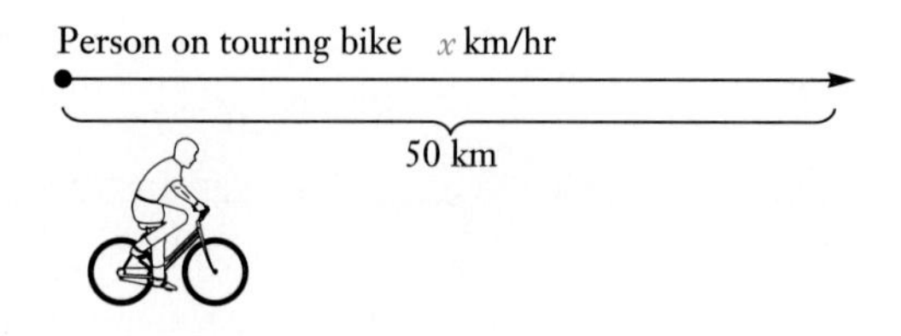

Using a table, we list the information.

	Distance	Speed	Time
Fast Bike	80	$x + 15$	t
Slower Bike	50	x	t

Both distances are covered in the same amount of time.

2. **Translate.** The table suggests that we use

$$\text{time} = \frac{\text{distance}}{\text{rate}}$$

since the times for the two bikers are equal and can be expressed in terms of x. We get the equations

$$t = \frac{80}{x + 15} \qquad \text{and} \qquad t = \frac{50}{x}.$$

3. **Carry out.** Since both rational expressions represent the same time, t, we set them equal to each other:

$$\frac{80}{x+15} = \frac{50}{x}. \quad \text{Both expressions represent the same time.}$$

Multiplying by the LCM, simplifying, and solving, we get $x = 25$.

4. **Check.** If our answer checks, the touring bike is going 25 km/h and the racing bike is going $25 + 15 = 40$ km/h.

   Traveling 80 km at 40 km/h, the racer is on the road for $\frac{80}{40} = 2$ hr. Traveling 50 km at 25 km/h, the person on the touring bike is on the road for $\frac{50}{25} = 2$ hr. Our answer checks since the two times are the same.

5. **State.** The racer's speed is 40 km/h, and the person on the touring bike is traveling at a speed of 25 km/h. ■

In the following example, although the distance is the same in both directions, the key to the translation lies in an additional piece of given information.

EXAMPLE 5 The speed of a boat in still water is 10 mph. It travels 24 mi upstream and 24 mi downstream in a total time of 5 hr. What is the speed of the current?

Solution

1. **Familiarize.** We first make a drawing and write out the pertinent information. We know that the speed of the boat is 10 mph, but we do not know the speed of the current. Let's call it c.

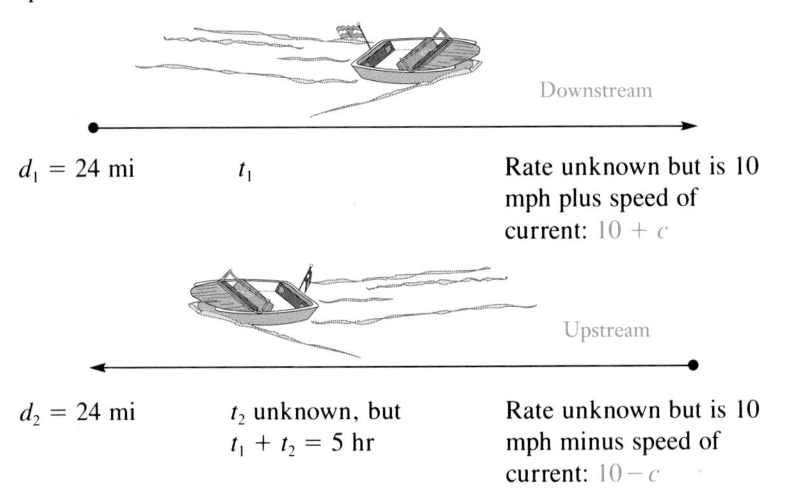

We can also organize the information in a table.

	Distance	Speed	Time
Downstream	24	$10 + c$	t_1
Upstream	24	$10 - c$	t_2

2. **Translate.** This time, the basis of our translation is the equation involving the times:

$$t_1 + t_2 = 5.$$

Since $t = d/r$, we have $t_1 = 24/(10 + c)$ and $t_2 = 24/(10 - c)$. Substituting, we have

$$\frac{24}{10 + c} + \frac{24}{10 - c} = 5.$$

3. **Carry out.** Solving for c, we get $c = -2$ or $c = 2$.
4. **Check.** Since speed cannot be negative in this problem, -2 cannot be a solution. But 2 checks.
5. **State.** The speed of the current is 2 mph. ■

EXERCISE SET 4.5

Solve.

1. The reciprocal of 5 plus the reciprocal of 7 is the reciprocal of what number?
2. The reciprocal of 3 plus the reciprocal of 6 is the reciprocal of what number?
3. The sum of a number and 6 times its reciprocal is -5. Find the number.
4. The sum of a number and 21 times its reciprocal is -10. Find the number.
5. The reciprocal of the product of two consecutive integers is $\frac{1}{72}$. Find the two integers.
6. The reciprocal of the product of two consecutive integers is $\frac{1}{42}$. Find the two integers.
7. Sam, an experienced shipping clerk, can fill a certain order in 5 hr. Willy, a new clerk, needs 9 hr to do the same job. Working together, how long will it take them to fill the order?
8. Paul can paint a room in 4 hr. Sally can paint the same room in 3 hr. Working together, how long will it take them to paint the room?
9. A swimming pool can be filled in 12 hr if water enters through a pipe alone or in 30 hr if water enters through a hose alone. If water is entering through both the pipe and the hose, how long will it take to fill the pool?
10. A tank can be filled in 18 hr by pipe A alone and in 22 hr by pipe B alone. How long will it take to fill the tank if both pipes are working?
11. Bill can clear a lot in 5.5 hr. His partner can do the same job in 7.5 hr. How long will it take them to clear the lot working together?
12. One printing press can print an order of booklets in 4.5 hr. Another press can do the same job in 5.5 hr. How long will it take if both presses are used?
13. Person A can paint a neighbor's house 4 times as fast as person B. The year they worked together it took them 8 days. How long would it take each to paint the house alone?
14. Person A can deliver papers 3 times as fast as person B. If they work together, it takes them 1 hr. How long would it take each to deliver the papers alone?
15. Rosita can wax her car in 2 hr. When she works together with Helga, they can wax the car in 45 min. How long would it take Helga, working by herself, to wax the car?
16. Hannah can sand the living room floor in 3 hr. When she works together with Henri, the job takes 2 hr. How long would it take Henri, working by himself, to sand the floor?

17. Jake can cut and split a cord of firewood in 6 fewer hr than Skyler can. When they work together, it takes them 4 hr. How long would it take each of them to do the job alone?

18. Sara takes 3 hr longer to paint a floor than it takes Kate. When they work together, it takes them 2 hr. How long would each take to do the job alone?

19. Together it takes John and Deb 2 hr 55 min to sort recyclables. Alone, John would require 2 more hr than Deb. How long would it take Deb to do the job alone? (*Hint:* Convert minutes to hours or hours to minutes.)

20. Together, Larry and Mo require 4 hr 48 min to pave a driveway. Alone Larry would require 4 hr more than Mo. How long would it take Mo to do the job alone? (*Hint:* Convert minutes to hours.)

21. A new photocopier works twice as fast as an old one. When the machines work together, a university can produce all its staff manuals in 15 hr. Find the time it would take each machine, working alone, to complete the same job.

22. Working together, two people can do a job in 1.5 hr. Person A takes 4 hr longer, working alone, than person B alone. How long would it take person B to do the job?

23. The speed of a stream is 3 mph. A boat travels 4 mi upstream in the same time it takes to travel 10 mi downstream. What is the speed of the boat in still water?

24. The speed of a stream is 4 mph. A boat travels 6 mi upstream in the same time it takes to travel 12 mi downstream. What is the speed of the boat in still water?

25. The speed of a moving sidewalk at an airport is 7 ft/sec. A person can walk 80 ft forward on the moving sidewalk in the same time it takes to walk 15 ft in the opposite direction. At what rate would the person walk on a nonmoving sidewalk?

26. The speed of a moving sidewalk at an airport is 6 ft/sec. A person can walk 70 ft forward on the moving sidewalk in the same time it takes to walk 20 ft in the opposite direction. At what rate would the person walk on a nonmoving sidewalk?

27. Rosanna walks 2 mph slower than Simone. In the time it takes Simone to walk 8 mi, Rosanna walks 5 mi. Find the speed of each person.

28. A local bus travels 7 mph slower than the express. If the express travels 90 mi in the time it takes the local to travel 75 mi, find the speed of each bus.

29. The speed of train A is 12 mph less than the speed of train B. Train A travels 230 mi in the same time it takes train B to travel 290 mi. Find the speed of each train.

30. The speed of a passenger train is 14 mph faster than the speed of a freight train. The passenger train travels 400 mi in the same time it takes the freight train to travel 330 mi. Find the speed of each train.

31. Motorboat A travels 10 km/h faster than motorboat B. Motorboat A travels 75 km in the same time it takes motorboat B to travel 50 km. Find the speed of each boat.

32. Jaime's moped travels 8 km/h faster than Mara's. Jaime travels 69 km in the same time it takes Mara to travel 45 km. Find the speed of each person's moped.

33. Suzie has a boat that can move at a speed of 15 km/h in still water. She rides 140 km downriver in the same time it takes to ride 35 km upstream. What is the speed of the river?

34. A paddleboat can move at a speed of 2 km/h in still water. The boat is paddled 4 km downstream in a river in the same time it takes to go 1 km upstream. What is the speed of the river?

35. A barge moves 7 km/h in still water. It travels 45 km upriver and 45 km downriver in a total time of 14 hr. What is the speed of the current?

36. Janet bicycles 9 mph with no wind. She bikes 24 mi against the wind and 24 mi with the wind in a total time of 6 hr. Find the wind speed.

37. A plane travels 100 mph in still air. It travels 240 mi with a headwind and 240 mi with a tailwind in a total time of 5 hr. Find the wind speed.

38. Al swims 55 m per minute in still water. He swims 150 m upstream and 150 m downstream in a total time of 5.5 min. What is the speed of the current?

39. A car traveled 120 mi at a certain speed. If the speed had been 10 mph faster, the trip could have been made in 2 hr less time. Find the speed.

40. A boat travels 45 mi upstream and 45 mi back. The time required for the round trip is 8 hr. The speed of the stream is 3 mph. Find the speed of the boat in still water.

SKILL MAINTENANCE

41. 8% of what number is 480?

42. Solve for x: $3xy - 5x = 9y$.

43. Solve: $|x - 2| = 9$.

44. Combine similar terms:

$$4y - 5xy^2 + 6xy - 3xy^2 - 2y.$$

SYNTHESIS

45. At what time after 4:00 will the minute hand and the hour hand of a clock first be in the same position?

46. At what time after 10:30 will the hands of a clock first be perpendicular?

47. A boat travels 96 km downstream in 4 hr. It travels 28 km upstream in 7 hr. Find the speed of the boat and the speed of the stream.

48. An airplane carries enough fuel for 6 hr of flight time, and its speed in still air is 240 mph. It leaves an airport against a wind of 40 mph and returns to the same airport with a wind of 40 mph. How far can it fly under those conditions without refueling?

49. A motor boat travels 3 times as fast as the current. A trip up the river and back takes 10 hr, and the total distance of the trip is 100 km. Find the speed of the current.

50. An employee drives to work at 50 mph and arrives 1 min late. The employee drives to work at 60 mph and arrives 5 min early. How far does the employee live from work?

51. A tank can be filled in 9 hr and drained in 11 hr. How long will it take to fill the tank if the drain is left open?

52. A tub can be filled in 10 min and drained in 8 min. How long will it take to empty a full tub if the water is left on?

Average speed is defined as *total distance divided by total time.*

53. A driver went 200 km. For the first 100 km of the trip, the driver traveled at a speed of 40 km/h. For the second half of the trip, the driver traveled at a speed of 60 km/h. What was the average speed for the entire trip? (It was *not* 50 km/h.)

54. For the first 50 mi of a 100-mi trip, a driver travels at 40 mph. What speed would the driver have to travel for the last half of the trip so that the average speed for the entire trip would be 45 mph?

CHAPTER SUMMARY AND REVIEW 4

TERMS TO KNOW

Rational expression
Meaningful replacement
Least common multiple
Complex rational expression
Rational equation
Clearing of fractions
Extraneous root

IMPORTANT PROPERTIES AND FORMULAS

Addition: $\frac{A}{C} + \frac{B}{C} = \frac{A + B}{C}$

Subtraction: $\frac{A}{C} - \frac{B}{C} = \frac{A - B}{C}$

Multiplication: $\frac{A}{B} \cdot \frac{C}{D} = \frac{AC}{BD}$

Division: $\frac{A}{B} \div \frac{C}{D} = \frac{A}{B} \cdot \frac{D}{C}$

To find the least common multiple, LCM, use each factor the greatest number of times that it occurs in any one factorization.

SIMPLIFYING COMPLEX EXPRESSIONS

I: By first adding or subtracting

1. Add or subtract, as necessary, to get a single rational expression in the numerator.
2. Add or subtract, as necessary, to get a single rational expression in the denominator.
3. Invert and multiply.
4. Simplify, if possible.

II: By using the LCM

1. Find the LCM of all the denominators *within* the complex rational expression.
2. Multiply the complex rational expression by 1, using the LCM to form the expression for 1.
3. Distribute and simplify.
4. Factor and simplify, if possible.

TO SOLVE AN EQUATION INVOLVING RATIONAL EXPRESSIONS

1. Multiply on both sides by the LCM of all the denominators.
2. Solve the resulting equation.
3. Check the answer in the original problem.

If r represents rate, t represents time, and d represents distance, then:

$$d = rt, \qquad r = \frac{d}{t}, \qquad \text{and} \qquad t = \frac{d}{r}.$$

REVIEW EXERCISES

The review sections to be tested in addition to the material in this chapter are Sections 1.3, 2.3, 2.5, and 2.6.

1. Is -1 a meaningful replacement in

$$\frac{x^2 - 6x - 7}{x^2 - 5x - 6}?$$

Why or why not?

2. Evaluate

$$\frac{t^2 - 16}{t^2 - 4t + 3},$$

using the following values.
a) $t = -1$ **b)** $t = 4$ **c)** $t = 2$

Find the LCM of the denominators.

3. $\frac{5}{6x^3}, \frac{y}{16x^2}$

4. $\frac{1}{x^2 + x - 20}, \frac{7}{x^2 + 3x - 10}$

Perform the indicated operations and simplify.

5. $\frac{x^3}{x + 2} + \frac{8}{x + 2}$

6. $\frac{4x - 2}{x^2 - 5x + 4} - \frac{3x - 1}{x^2 - 5x + 4}$

7. $\frac{3a^2b^3}{5c^3d^2} \cdot \frac{15c^9d^4}{9a^7b}$

8. $\frac{1}{6m^2n^3p} + \frac{2}{9mn^4p^2}$

9. $\frac{y^2 - 64}{2y + 10} \cdot \frac{y + 5}{y + 8}$

10. $\frac{x^3 - 8}{x^2 - 25} \cdot \frac{x^2 + 10x + 25}{x^2 + 2x + 4}$

11. $\frac{9a^2 - 1}{a^2 - 9} \div \frac{3a + 1}{a + 3}$

12. $\frac{x^3 - 64}{x^2 - 16} \div \frac{x^2 + 5x + 6}{x^2 - 3x - 18}$

13. $\frac{x}{x^2 + 5x + 6} - \frac{2}{x^2 + 3x + 2}$

14. $\frac{9xy}{x^2 - y^2} + \frac{x + y}{x - y}$

15. $\frac{2x^2}{x - y} + \frac{2y^2}{x + y}$

16. $\frac{3}{y + 4} - \frac{y}{y - 1} + \frac{y^2 + 3}{y^2 + 3y - 4}$

17. Simplify: $\dfrac{3 + \frac{3}{y}}{4 + \frac{4}{y}}$.

18. Simplify: $\dfrac{\frac{2}{a} + \frac{2}{b}}{\frac{4}{a^3} + \frac{4}{b^3}}$.

19. Simplify: $\dfrac{\dfrac{y^2+4y-77}{y^2-10y+25}}{\dfrac{y^2-5y-14}{y^2-25}}$.

20. Simplify: $\dfrac{\dfrac{5}{x^2-9}-\dfrac{3}{x+3}}{\dfrac{4}{x^2+6x+9}+\dfrac{2}{x-3}}$.

Solve.

21. $\dfrac{6}{x}+\dfrac{4}{x}=5$

22. $\dfrac{x}{7}+\dfrac{x}{4}=1$

23. $\dfrac{5}{3x+2}=\dfrac{3}{2x}$

24. $\dfrac{4x}{x+1}+\dfrac{4}{x}+9=\dfrac{4}{x^2+x}$

25. $\dfrac{90}{x^2-3x+9}-\dfrac{5x}{x+3}=\dfrac{405}{x^3+27}$

Solve.

26. Joan can paint a house in 12 hr. Kelly can do the same job in 9 hr. How long would it take them working together to paint the house?

27. A river's current is 6 mph. A boat travels 50 mi downriver in the same time that it takes to travel 30 mi upstream. What is the speed of the boat in still water?

28. A car and a motorcycle leave a rest area at the same time, with the car traveling 8 mph faster than the motorcycle. The car then travels 105 mi in the time it takes the motorcycle to travel 93 mi. Find the speed of each vehicle.

SKILL MAINTENANCE

29. Simplify: $24 \div 2^3 - 3 \cdot 5^2$.

30. Solve: $x = \dfrac{t+4q}{n}$ for q.

31. Solve: $2x - 7 < 9$ *or* $20 - 3x < -13$.

32. Solve: $|3x - 5| = 7$.

SYNTHESIS

Solve.

33. $\dfrac{5}{x-13}-\dfrac{5}{x}=\dfrac{65}{x^2-13x}$

34. $\dfrac{\dfrac{x}{x^2-25}+\dfrac{2}{x-5}}{\dfrac{3}{x-5}-\dfrac{4}{x^2-10x+25}}=1$

35. One summer Sara sold 4 sweepers for every 3 sweepers sold by her brother Stephen. Together they sold 98 sweepers. How many did each sell?

THINKING IT THROUGH

1. Explain at least three different uses of the LCM studied in this chapter.

2. You have learned to solve new kinds of equations in this chapter. Explain how the equations differ from those you have studied previously and how the equation-solving process differs.

3. Explain the difference between a rational expression and a rational equation.

4. Explain why it is necessary to check all answers to rational equations found using the procedure discussed in this chapter.

5. Explain why it is necessary, when simplifying complex rational expressions, to combine the fractions in the numerator and those in the denominator before carrying out the division.

CHAPTER TEST 4

Perform the indicated operation and simplify.

1. $\dfrac{4y^2 - 4}{3y + 9} \cdot \dfrac{y + 3}{y + 1}$

2. $\dfrac{x^3 + 27}{x^2 - 16} \div \dfrac{x^2 + 8x + 15}{x^2 + x - 20}$

Find the LCM of the denominators.

3. $\dfrac{1}{x^2 - 16}, \dfrac{7}{x^3 - 64}$

4. $\dfrac{3x}{x^2 + 8x - 33}, \dfrac{x + 1}{x^2 - 12x + 27}$

Perform the indicated operation and simplify, if possible.

5. $\dfrac{25x}{x + 5} + \dfrac{x^3}{x + 5}$

6. $\dfrac{3a^2}{a - b} - \dfrac{3b^2 - 6ab}{b - a}$

7. $\dfrac{4ab}{a^2 - b^2} + \dfrac{a^2 + b^2}{a + b}$

8. $\dfrac{6}{x^3 - 64} - \dfrac{4}{x^2 - 16}$

9. $\dfrac{4}{y + 3} - \dfrac{y}{y - 2} + \dfrac{y^2 + 4}{y^2 + y - 6}$

Simplify.

10. $\dfrac{\dfrac{5}{x} - \dfrac{3}{y}}{\dfrac{2}{x} + \dfrac{3}{y}}$

11. $\dfrac{\dfrac{x^2 - 5x - 36}{x^2 - 36}}{\dfrac{x^2 + x - 12}{x^2 - 12x + 36}}$

12. $\dfrac{\dfrac{4}{x + 3} - \dfrac{2}{x^2 - 3x + 2}}{\dfrac{3}{x - 2} + \dfrac{1}{x^2 + 2x - 3}}$

Solve.

13. $\dfrac{1}{x} + \dfrac{3}{x} = 4$

14. $\dfrac{6}{5a + 3} = \dfrac{4}{2a - 5}$

15. Tom can mow the yard in 3.5 hr. Larry can mow the yard in 4.5 hr. How long will it take them to mow it together?

16. The product of the reciprocals of two consecutive integers is $\frac{1}{30}$. Find both integers.

17. A biker can travel 12 mph with no wind. The same rider can bicycle 8 mi against the wind in the same time that it takes to bicycle 14 mi with the wind. What is the speed of the wind?

18. Working together, Phil and Jill can paint a house in 3 days. When Jill works alone she can paint the house in 5 days. How long would it take Phil, working alone, to paint the house?

SKILL MAINTENANCE

19. Evaluate $3 + x^2 - y$ for $x = -4$ and $y = -7$.

20. Solve and then graph the solutions of $|2x - 7| < 9$.

21. Find the union: $\{-2, 1, 5\} \cup \{1, 3, 7\}$.

22. Solve: $z = xy - x$ for x.

SYNTHESIS

23. Solve: $\dfrac{6}{x - 15} - \dfrac{6}{x} = \dfrac{90}{x^2 - 15x}$.

24. Find the LCM: $1 - t^6,\ 1 + t,\ 1 + t^3,\ 1 + t + t^2$.

CHAPTER 5

Exponents and Radicals

FEATURE PROBLEM

We see road pavement messages of various sizes. If the letters are too small, they cannot be read in time to heed their warnings. If the letters are too large, they take lots of paint to maintain and can be a hazard when the road gets wet. What length L will make a message most readable from a distance d when the eyes are a height h from the surface of the road?

THE MATHEMATICS

In a psychological study, it was determined that the proper length L of letters painted on pavement is given by

$$L = \frac{0.000169d^{2.27}}{h},$$

← This is a fractional exponent.

where d is the distance, in feet, of the lettering from the car and h is the height of the eye above the road.

In this chapter we study square roots, cube roots, fourth roots, and so on. We study these in connection with radical expressions that involve these roots, and solve problems involving such roots. We also define and use fractional exponents.

The review sections to be tested in addition to the material in this chapter are Sections 3.2, 3.5, 4.1, and 4.4.

5.1 Simplifying and Multiplying Radical Expressions

In this section we consider roots, such as square roots and cube roots. We look at the symbolism used for them and the ways we can manipulate symbols to get equivalent expressions.

Square Roots

When a number is raised to the second power, the number is squared. Often we need to know what number was squared in order to produce some value a. If such a number can be found, we call that number a *square root* of a.

> The number c is a *square root* of a if $c^2 = a$.

For example,

5 is a square root of 25 because $5 \cdot 5 = 25$;

-5 is a square root of 25 because $(-5)(-5) = 25$;

-4 does not have a real number square root because there is no real number c such that $c^2 = -4$.

In Section 5.6 we will see that there is a number system, different from the real number system, in which negative numbers do have square roots.

> Every positive real number has two real number square roots. The number 0 has just one square root, 0 itself. Negative numbers do not have real number square roots.

EXAMPLE 1 Find the two square roots of 64.

Solution The square roots are 8 and -8 because $8^2 = 64$ and $(-8)^2 = 64$. ■

> The *principal square root* of a nonnegative number is its nonnegative square root. The symbol $\sqrt{a}$ represents the principal square root of a. To name the negative square root of a, we write $-\sqrt{a}$.

EXAMPLES Simplify. Remember, $\sqrt{}$ indicates the principal square root.

2. $\sqrt{25} = 5$
3. $-\sqrt{64} = -8$ Since $\sqrt{64} = 8$, $-\sqrt{64} = -8$.
4. $\sqrt{\frac{25}{64}} = \frac{5}{8}$
5. $\sqrt{0.0049} = 0.07$ ■

> The symbol $\sqrt{}$ is called a *radical sign*. An expression written with a radical sign is called a *radical expression*. The expression written under the radical sign is called the *radicand*.

The following are radical expressions:

$$\sqrt{5}, \quad \sqrt{a}, \quad -\sqrt{5x}, \quad \sqrt{\frac{y^2+7}{\sqrt{x}}}.$$

An expression like $\sqrt{5}$ may be read as "radical 5," "the square root of 5," or simply "root 5."

Evaluating Expressions of the Form $\sqrt{a^2}$

We will often have reason to evaluate radical expressions. Notice that when 5 or -5 is substituted into the expression $\sqrt{x^2}$, the result is the same.

EXAMPLE 6 Evaluate the expression $\sqrt{x^2}$ for the following values: (a) 5; (b) 0; (c) -5.

Solution

a) $\sqrt{5^2} = \sqrt{25} = 5$
b) $\sqrt{0^2} = \sqrt{0} = 0$
c) $\sqrt{(-5)^2} = \sqrt{25} = 5$ ■

We have the following general rule that we can use in simplifying radical expressions.

> For any real number a,
>
> $$\sqrt{a^2} = |a|.$$
>
> (The principal square root of a^2 is the absolute value of a.)

When an expression contains perfect squares, like 25 or $(m-3)^2$, in the radicand, we need to use absolute value signs when simplifying unless we know that the quantities being squared do not represent negative numbers.

EXAMPLES Simplify. Assume that the variable can represent any real number.

7. $\sqrt{(x+1)^2} = |x+1|$ Since $x+1$ might be negative (for example, if $x=-3$), absolute value notation is necessary.

8. $\sqrt{x^2-8x+16} = \sqrt{(x-4)^2} = |x-4|$ Since $x-4$ might be negative, absolute value notation is necessary.

EXAMPLES Simplify. Assume that only nonnegatives have been squared to form any perfect square radicands.

9. $\sqrt{y^2} = y$ We are assuming that y is nonnegative, so no absolute value notation is necessary. When y *is* negative, $\sqrt{y^2} \neq y$.

10. $\sqrt{(5x+2)^2} = 5x+2$ Assuming that $5x+2$ is nonnegative

11. $\sqrt{x^2-2x+1} = \sqrt{(x-1)^2} = x-1$ Assuming that $x-1$ is nonnegative

Cube Roots

We often need to know what number was cubed in order to produce a certain value. When such a number is found, we say that we have found a *cube root.*

> The number c is the *cube root* of a if $c^3 = a$. In symbols, we write $\sqrt[3]{a}$ to denote the cube root of a.

For example,

2 is the cube root of 8 because $2^3 = 2 \cdot 2 \cdot 2 = 8$;

-4 is the cube root of -64 because $(-4)^3 = (-4)(-4)(-4) = -64$.

In the real number system, every number has exactly one cube root. The cube root of a positive number is positive, and the cube root of a negative number is negative. In simplifying expressions involving cube roots, we need not use absolute value signs.

EXAMPLES Simplify the following.

12. $\sqrt[3]{-27} = -3$ Since $(-3)(-3)(-3) = -27$

13. $\sqrt[3]{8y^3} = 2y$ Since $(2y)(2y)(2y) = 8y^3$

Odd and Even kth Roots

The fifth root of a number a is the number c for which $c^5 = a$ and, in general, the kth root of a is the number c for which $c^k = a$. We denote the kth root of a by $\sqrt[k]{a}$. The number k is called the *index*. When the *index* is 2, we do not write it. Whenever the index is an odd number, we say that we are taking an *odd* root.

Every number has just one root for an odd k. If the number is positive, its root is positive. If the number is negative, its root is negative.

EXAMPLES Find each of the following.

14. $\sqrt[5]{32} = 2$ Since $2^5 = 32$

15. $\sqrt[5]{-32} = -2$ Since $(-2)^5 = -32$

16. $-\sqrt[5]{32} = -2$ Taking the opposite of $\sqrt[5]{32}$

17. $-\sqrt[5]{-32} = -(-2) = 2$ Taking the opposite of $\sqrt[5]{-32}$

18. $\sqrt[7]{x^7} = x$

19. $\sqrt[9]{(x-1)^9} = x - 1$ ■

Absolute value signs are never needed when finding odd roots.

When the index k in $\sqrt[k]{a}$ is an even number, we say that we are taking an *even* root. Every positive real number has two kth roots when k is even. One of those roots is positive and one is negative. The notation $\sqrt[k]{a}$ indicates the positive kth root when k is even. Negative real numbers do not have kth roots when k is even. When finding even kth roots, absolute value signs are sometimes necessary, as with square roots.

EXAMPLES Find the following. Assume that variables can represent any real numbers.

20. $\sqrt[4]{16} = 2$ Since $2^4 = 16$

21. $-\sqrt[4]{16} = -2$ Taking the opposite of $\sqrt[4]{16}$

22. $\sqrt[4]{-16}$ No real number even root exists.

23. $\sqrt[4]{81x^4} = 3|x|$ Use absolute value since x could represent a negative number.

24. $\sqrt[6]{(y+7)^6} = |y + 7|$ Use absolute value since $y + 7$ could be negative. ■

For any real number a:

a) $\sqrt[k]{a^k} = |a|$ when k is even. We use absolute value when k is even unless a is known to be nonnegative.

b) $\sqrt[k]{a^k} = a$ when k is odd. We do not use absolute value when k is odd.

Meaningful Replacements

Since negative numbers do not have even roots in the system of real numbers, any replacement that makes a radicand negative when the index is even is not meaningful.

EXAMPLE 25 Determine whether 0 and 3 are meaningful replacements in $\sqrt{5x - 4}$.

Solution We substitute 0 for x in the radicand $5x - 4$:

$$5(0) - 4 = 0 - 4 = -4.$$

Since the radicand is negative, 0 is not a meaningful replacement. We substitute 3 for x in $5x - 4$:

$$5(3) - 4 = 15 - 4 = 11.$$

Since the radicand is not negative, 3 is a meaningful replacement. ■

Multiplying

Note that

$$\sqrt{4}\,\sqrt{25} = 2 \cdot 5 = 10 \quad \text{and} \quad \sqrt{4 \cdot 25} = \sqrt{100} = 10.$$

Likewise,

$$\sqrt[3]{27}\,\sqrt[3]{8} = 3 \cdot 2 = 6 \quad \text{and} \quad \sqrt[3]{27 \cdot 8} = \sqrt[3]{216} = 6.$$

These examples suggest the following rule.

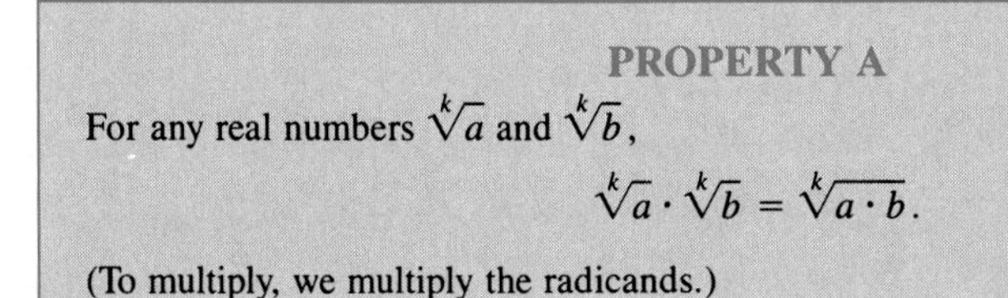

PROPERTY A

For any real numbers $\sqrt[k]{a}$ and $\sqrt[k]{b}$,

$$\sqrt[k]{a} \cdot \sqrt[k]{b} = \sqrt[k]{a \cdot b}.$$

(To multiply, we multiply the radicands.)

EXAMPLES Multiply.

26. $\sqrt{3} \cdot \sqrt{5} = \sqrt{3 \cdot 5} = \sqrt{15}$
27. $\sqrt{x+2} \cdot \sqrt{x-2} = \sqrt{(x+2)(x-2)} = \sqrt{x^2-4}$
28. $\sqrt[3]{4} \cdot \sqrt[3]{5} = \sqrt[3]{4 \cdot 5} = \sqrt[3]{20}$ ■

Read from right to left, property A also enables us to simplify radical expressions. The idea is to factor the radicand, obtaining factors that are perfect kth powers. We then use property A to form two or more radical expressions.

In the examples that follow, we will assume that no radicands were formed by raising negative quantities to even powers. This assumption eliminates the need for absolute value signs.

EXAMPLES Simplify.

29. $\sqrt{50} = \sqrt{25 \cdot 2} = \sqrt{25} \cdot \sqrt{2} = 5\sqrt{2}$
30. $\sqrt{5x^2} = \sqrt{x^2 \cdot 5} = \sqrt{x^2} \cdot \sqrt{5} = x\sqrt{5}$ **Assuming that x is nonnegative**
31. $\sqrt[3]{32} = \sqrt[3]{8 \cdot 4} = \sqrt[3]{8} \cdot \sqrt[3]{4} = 2\sqrt[3]{4}$ **Note that 8 is a perfect cube.**
32. $\sqrt{2x^2 - 4x + 2} = \sqrt{\ (x-1)^2}$ **Factoring the radicand**
$= \sqrt{(x-1)^2} \cdot \sqrt{2}$ **Factoring into two radicals**
$= (x-1)\sqrt{2}$ **Assuming that $x - 1$ is nonnegative** ■

Note that, in simplified form, a radical expression that represents a kth root never contains any factors that are perfect kth powers in a radicand. For instance $3\sqrt{20x}$ would *not* be considered simplified form since 4, a perfect square, is a factor of the radicand, $20x$.

EXAMPLES Simplify.

33. $\sqrt{216x^5y^3} = \sqrt{36 \cdot 6 \cdot x^4 \cdot x \cdot y^2 \cdot y}$
$= \sqrt{36 \cdot x^4 \cdot y^2 \cdot 6 \cdot x \cdot y}$ Factoring the radicand
$= \sqrt{36}\,\sqrt{x^4}\sqrt{y^2}\,\sqrt{6xy}$ Factoring into several radicals
$= 6x^2y\sqrt{6xy}$

34. $\sqrt[3]{16x^7y^{11}} = \sqrt[3]{8 \cdot 2 \cdot x^6 \cdot x \cdot y^9 \cdot y^2}$
$= \sqrt[3]{8 \cdot x^6 \cdot y^9 \cdot 2 \cdot x \cdot y^2}$ Factoring the radicand. Because we're taking the cube, or third root, we form powers that are multiples of 3. Note that 6 is the largest power less than 7 that is a multiple of 3 and 9 is the largest power less than 11 that is a multiple of 3.
$= \sqrt[3]{8}\,\sqrt[3]{x^6}\,\sqrt[3]{y^9}\,\sqrt[3]{2xy^2}$ Factoring into several radicals
$= 2x^2y^3\sqrt[3]{2xy^2}$

35. $\sqrt[3]{-81(x+y)^5} = \sqrt[3]{-27 \cdot 3 \cdot (x+y)^3(x+y)^2}$
$= \sqrt[3]{-27 \cdot (x+y)^3 \cdot 3 \cdot (x+y)^2}$ Factoring the radicand. Look for powers that are multiples of 3.
$= \sqrt[3]{-27}\,\sqrt[3]{(x+y)^3}\,\sqrt[3]{3(x+y)^2}$ Factoring into several radicals
$= -3(x+y)\sqrt[3]{3(x+y)^2}$ ■

Sometimes after multiplying, we can simplify.

EXAMPLES Multiply and then simplify by factoring.

36. $\sqrt{15}\,\sqrt{6} = \sqrt{15 \cdot 6} = \sqrt{90} = \sqrt{9 \cdot 10} = 3\sqrt{10}$

37. $3\sqrt[3]{25} \cdot 2\sqrt[3]{5} = 6 \cdot \sqrt[3]{25 \cdot 5}$
$= 6 \cdot \sqrt[3]{125}$ Multiplying radicands
$= 6 \cdot 5$, or 30 Taking the cube root of 125

38. $\sqrt[3]{18y^3}\,\sqrt[3]{4x^2} = \sqrt[3]{18y^3 \cdot 4x^2}$
$= \sqrt[3]{72y^3x^2}$ Multiplying radicands
$= \sqrt[3]{8y^3 \cdot 9x^2}$ Factoring the radicand
$= \sqrt[3]{8y^3}\,\sqrt[3]{9x^2}$ Factoring into two radicals
$= 2y\sqrt[3]{9x^2}$ Taking the cube root of $8y^3$ ■

EXERCISE SET 5.1

Find the square roots of the number.

1. 16 **2.** 225 **3.** 144 **4.** 9
5. 400 **6.** 81 **7.** 49 **8.** 900

Simplify.

9. $-\sqrt{\frac{49}{36}}$ 10. $-\sqrt{\frac{361}{9}}$ 11. $\sqrt{196}$ 12. $\sqrt{441}$ 13. $-\sqrt{\frac{16}{81}}$

14. $-\sqrt{\frac{81}{144}}$ 15. $\sqrt{0.09}$ 16. $\sqrt{0.36}$ 17. $-\sqrt{0.0049}$ 18. $\sqrt{0.0144}$

Identify the radicand in the expression.

19. $5\sqrt{p^2 + 4}$ 20. $-7\sqrt{y^2 - 8}$ 21. $x^2y^2\sqrt{\frac{x}{y + 4}}$ 22. $a^2b^3\sqrt{\frac{a}{a^2 - b}}$

Simplify. Assume that variables can represent any real number.

23. $\sqrt{16x^2}$ 24. $\sqrt{25t^2}$ 25. $\sqrt{(-7c)^2}$ 26. $\sqrt{(-6b)^2}$
27. $\sqrt{(a + 1)^2}$ 28. $\sqrt{(5 - b)^2}$ 29. $\sqrt{x^2 - 4x + 4}$
30. $\sqrt{y^2 + 16y + 64}$ 31. $\sqrt{4x^2 + 28x + 49}$ 32. $\sqrt{9x^2 - 30x + 25}$

Simplify. Assume that only nonnegatives have been squared to form any perfect square radicands.

33. $\sqrt{16x^2}$ 34. $\sqrt{25t^2}$ 35. $\sqrt{(-6b)^2}$ 36. $\sqrt{(-7c)^2}$
37. $\sqrt{(a + 1)^2}$ 38. $\sqrt{(5 + b)^2}$ 39. $\sqrt{4x^2 + 8x + 4}$
40. $\sqrt{9x^2 + 36x + 36}$ 41. $\sqrt{9t^2 - 12t + 4}$ 42. $\sqrt{25t^2 - 20t + 4}$

Simplify.

43. $\sqrt[3]{27}$ 44. $-\sqrt[3]{64}$ 45. $\sqrt[3]{-64x^3}$ 46. $\sqrt[3]{-125y^3}$
47. $\sqrt[3]{-216}$ 48. $-\sqrt[3]{-1000}$ 49. $-\sqrt[3]{-125y^3}$ 50. $-\sqrt[3]{-64x^3}$
51. $\sqrt[3]{0.343(x + 1)^3}$ 52. $\sqrt[3]{0.000008(y - 2)^3}$

Find each of the following. Assume that variables can represent any real number.

53. $\sqrt[4]{625}$ 54. $-\sqrt[4]{256}$ 55. $\sqrt[5]{-1}$ 56. $-\sqrt[5]{-32}$
57. $\sqrt[5]{-\frac{32}{243}}$ 58. $\sqrt[5]{-\frac{1}{32}}$ 59. $\sqrt[6]{x^6}$ 60. $\sqrt[8]{y^8}$
61. $\sqrt[4]{(5a)^4}$ 62. $\sqrt[4]{(7b)^4}$ 63. $\sqrt[10]{(-6)^{10}}$ 64. $\sqrt[12]{(-10)^{12}}$
65. $\sqrt[414]{(a + b)^{414}}$ 66. $\sqrt[1976]{(2a + b)^{1976}}$ 67. $\sqrt[7]{y^7}$ 68. $\sqrt[3]{(-6)^3}$
69. $\sqrt[5]{(x - 2)^5}$ 70. $\sqrt[9]{(2xy)^9}$

Determine whether the given numbers are meaningful replacements in the expression.

71. $\sqrt{x - 3}$; $-2, 5$ 72. $\sqrt{2x - 5}$; $3, 2$ 73. $\sqrt{3 - 4x}$; $-1, 1$ 74. $\sqrt{x^2 + 3}$; $0, 4.3$
75. $\sqrt{1 - x^2}$; $1, 3$ 76. $\sqrt{x^2 + 2x + 1}$; $-3, 4$ 77. $\sqrt[3]{2x + 7}$; $-4, 5$ 78. $\sqrt[4]{3 - 5x}$; $1, 2$

For exercises 79–146, assume that no radicands were formed by raising negative quantities to even powers.

Multiply.

79. $\sqrt{3}\sqrt{2}$ 80. $\sqrt{5}\sqrt{7}$ 81. $\sqrt[3]{2}\sqrt[3]{5}$ 82. $\sqrt[3]{7}\sqrt[3]{2}$
83. $\sqrt[4]{8}\sqrt[4]{9}$ 84. $\sqrt[4]{6}\sqrt[4]{3}$ 85. $\sqrt{3a}\sqrt{10b}$ 86. $\sqrt{2x}\sqrt{13y}$
87. $\sqrt[5]{9t^2}\sqrt[5]{2t}$ 88. $\sqrt[5]{8y^3}\sqrt[5]{10y}$ 89. $\sqrt{x - a}\sqrt{x + a}$
90. $\sqrt{y - b}\sqrt{y + b}$ 91. $\sqrt[3]{0.3x}\sqrt[3]{0.2x}$ 92. $\sqrt[3]{0.7y}\sqrt[3]{0.3y}$
93. $\sqrt[4]{x - 1}\sqrt[4]{x^2 + x + 1}$ 94. $\sqrt[5]{x - 2}\sqrt[5]{(x - 2)^2}$

Simplify by factoring.

95. $\sqrt{27}$ **96.** $\sqrt{28}$ **97.** $\sqrt{45}$ **98.** $\sqrt{12}$

99. $\sqrt{8}$ **100.** $\sqrt{18}$ **101.** $\sqrt{24}$ **102.** $\sqrt{20}$

103. $\sqrt{180x^4}$ **104.** $\sqrt{175y^6}$ **105.** $\sqrt[3]{800}$ **106.** $\sqrt[3]{270}$

107. $\sqrt[3]{-16x^6}$ **108.** $\sqrt[3]{-32a^6}$ **109.** $\sqrt[3]{54x^8}$ **110.** $\sqrt[3]{40y^3}$

111. $\sqrt[3]{80x^8}$ **112.** $\sqrt[3]{108m^5}$ **113.** $\sqrt[4]{32}$ **114.** $\sqrt[4]{80}$

115. $\sqrt[4]{810}$ **116.** $\sqrt[4]{160}$ **117.** $\sqrt[4]{96a^8}$ **118.** $\sqrt[4]{240x^8}$

119. $\sqrt[4]{162c^4d^6}$ **120.** $\sqrt[4]{243x^8y^{10}}$ **121.** $\sqrt[3]{(x+y)^4}$ **122.** $\sqrt[3]{(a-b)^5}$

123. $\sqrt[3]{8000(m+n)^8}$ **124.** $\sqrt[3]{-1000(x+y)^{10}}$

125. $\sqrt[5]{-a^6b^{11}c^{17}}$ **126.** $\sqrt[5]{x^{13}y^8z^{22}}$

Multiply and simplify by factoring.

127. $\sqrt{3}\sqrt{6}$ **128.** $\sqrt{5}\sqrt{10}$ **129.** $\sqrt{15}\sqrt{12}$ **130.** $\sqrt{2}\sqrt{32}$

131. $\sqrt{6}\sqrt{8}$ **132.** $\sqrt{18}\sqrt{14}$ **133.** $\sqrt[3]{3}\sqrt[3]{18}$ **134.** $\sqrt{45}\sqrt{60}$

135. $\sqrt{5b^3}\sqrt{10c^4}$ **136.** $\sqrt[3]{-6a}\sqrt[3]{20a^4}$ **137.** $\sqrt[3]{10x^5}\sqrt[3]{-75x^2}$ **138.** $\sqrt{2x^3y}\sqrt{12xy}$

139. $\sqrt[3]{y^4}\sqrt[3]{16y^5}$ **140.** $\sqrt[3]{5^2t^4}\sqrt[3]{5^4t^6}$

141. $\sqrt[3]{(b+3)^4}\sqrt[3]{(b+3)^2}$ **142.** $\sqrt[3]{(x+y)^3}\sqrt[3]{(x+y)^5}$

143. $\sqrt{12a^3b}\sqrt{8a^4b^2}$ **144.** $\sqrt{18a^2b^5}\sqrt{30a^3b^4}$

145. $\sqrt[5]{a^2(b+c)^4}\sqrt[5]{a^4(b+c)^7}$ **146.** $\sqrt[5]{x^3(y-z)^7}\sqrt[5]{x^6(y-z)^9}$

SKILL MAINTENANCE

147. Use the commutative law for addition to write an equivalent expression: $5x(y + z)$.

148. Solve: $5(2 - x) = 4x - 7$.

Multiply and simplify.

149. $x^3 \cdot x^5$

150. $(-3x^2y^5)(7x^3y^6)$

SYNTHESIS

151. *An application: Speed of a skidding car.* After an accident, police can estimate the speed at which a car was traveling by measuring its skid marks. The formula

$$r = 2\sqrt{5L}$$

can be used, where r is the speed in miles per hour and L is the length of the skid marks in feet. Estimate (to the nearest tenth mile per hour) the speed of a car that left skid marks (a) 20 ft long; (b) 70 ft long; (c) 90 ft long.

152. *An application: Wind chill temperature.* In cold weather we feel colder if there is wind than if there is not. *Wind chill temperature* is the temperature at which, without wind, we would feel as cold in an actual situation with wind. Here is a formula for finding wind chill temperature:

$$T_w = 33 - \frac{(10.45 + 10\sqrt{v} - v)(33 - T)}{22},$$

where T is the actual temperature given in degrees Celsius and v is the wind speed in meters per second. Find the wind chill temperature to the nearest tenth of a degree for the given actual temperatures and wind speeds.

a) $T = 7°\text{C}$, $v = 8$ m/sec
b) $T = 0°\text{C}$, $v = 12$ m/sec
c) $T = -5°\text{C}$, $v = 14$ m/sec
d) $T = -23°\text{C}$, $v = 15$ m/sec

153. Solve for k if $\sqrt[3]{5x^{k+1}}\sqrt[3]{25x^k} = 5x^7$.

154. Solve for k if $\sqrt[5]{4a^{3k+2}}\sqrt[5]{8a^{6-k}} = 2a^4$.

155. What assumption do we make about x if $\sqrt{(2x+3)^2} = 2x + 3$?

5.2 Dividing and Simplifying with Radical Notation

In this section we will manipulate radical expressions using division rather than multiplication.

Roots of Quotients

Note that

$$\sqrt[3]{\frac{27}{8}} = \frac{3}{2} \quad \text{and} \quad \frac{\sqrt[3]{27}}{\sqrt[3]{8}} = \frac{3}{2}.$$

This example suggests that we can take the root of a quotient by taking the roots of the numerator and the denominator separately.

PROPERTY B

For any real numbers $\sqrt[k]{a}$ and $\sqrt[k]{b}$, $b \neq 0$,

$$\sqrt[k]{\frac{a}{b}} = \frac{\sqrt[k]{a}}{\sqrt[k]{b}}.$$

(We can take the kth roots of the numerator and the denominator separately.)

To help understand property B, note that both $\sqrt[k]{a/b}$ and $\sqrt[k]{a}/\sqrt[k]{b}$, when raised to the kth power, are equal to a/b.

EXAMPLES Simplify by taking roots of the numerator and the denominator.

1. $\sqrt[3]{\frac{27}{125}} = \frac{\sqrt[3]{27}}{\sqrt[3]{125}} = \frac{3}{5}$ Taking the cube roots of the numerator and the denominator

2. $\sqrt{\frac{25}{y^2}} = \frac{\sqrt{25}}{\sqrt{y^2}} = \frac{5}{y}$ Taking the square roots of the numerator and the denominator. Assume $y > 0$.

3. $\sqrt{\dfrac{16x^3}{y^8}} = \dfrac{\sqrt{16x^3}}{\sqrt{y^8}} = \dfrac{\sqrt{16x^2 \cdot x}}{\sqrt{y^8}} = \dfrac{4x\sqrt{x}}{y^4}$ Assume $x \geq 0$, $y \neq 0$.

4. $\sqrt[3]{\dfrac{27y^{14}}{343x^3}} = \dfrac{\sqrt[3]{27y^{14}}}{\sqrt[3]{343x^3}} = \dfrac{\sqrt[3]{27y^{12}y^2}}{\sqrt[3]{343x^3}} = \dfrac{\sqrt[3]{27y^{12}}\sqrt[3]{y^2}}{\sqrt[3]{343x^3}} = \dfrac{3y^4\sqrt[3]{y^2}}{7x}$ Assume $x \neq 0$. ■

Dividing Radical Expressions

Reading property B from right to left gives us a way to divide radical expressions.

> To divide, we divide the radicands, provided they have the same index:
>
> $$\frac{\sqrt[k]{a}}{\sqrt[k]{b}} = \sqrt[k]{\frac{a}{b}}.$$
>
> After doing so, we can sometimes simplify by taking roots.

EXAMPLES Divide. Then simplify by taking roots, if possible.

5. $\dfrac{\sqrt{80}}{\sqrt{5}} = \sqrt{\dfrac{80}{5}} = \sqrt{16} = 4$

6. $\dfrac{5\sqrt[3]{32}}{\sqrt[3]{2}} = 5\sqrt[3]{\dfrac{32}{2}} = 5\sqrt[3]{16} = 5\sqrt[3]{8 \cdot 2} = 5\sqrt[3]{8}\sqrt[3]{2} = 5 \cdot 2\sqrt[3]{2} = 10\sqrt[3]{2}$

7. $\dfrac{\sqrt{128xy}}{3\sqrt{2}} = \dfrac{1}{3}\dfrac{\sqrt{128xy}}{\sqrt{2}} = \dfrac{1}{3}\sqrt{\dfrac{128xy}{2}} = \dfrac{1}{3}\sqrt{64xy} = \dfrac{1}{3}\sqrt{64}\sqrt{xy}$

$= \dfrac{1}{3} \cdot 8\sqrt{xy} = \dfrac{8}{3}\sqrt{xy}$ Assume $xy \geq 0$.

8. $\dfrac{\sqrt[4]{33a^9b^5}}{\sqrt[4]{2b}} = \sqrt[4]{\dfrac{33a^9b^5}{2b}} = \sqrt[4]{\dfrac{33}{2}a^9b^4}$

$= \sqrt[4]{a^8b^4}\sqrt[4]{\dfrac{33}{2}a} = a^2b\sqrt[4]{\dfrac{33}{2}a}$ Note that 8 is the largest power less than 9 that is a multiple of the index 4. Assume $a \geq 0$, $b > 0$. ■

It is important to remember that when we divide radical expressions by dividing the radicands, both radicals must have the same index.

Powers and Roots Combined

Consider the following:

$$\sqrt[3]{8^2} = \sqrt[3]{64} = 4,$$
$$(\sqrt[3]{8})^2 = (2)^2 = 4.$$

We can either square and then take the root or take the root first and then square. This suggests another important property of radical expressions.

PROPERTY C

For any nonnegative number a, any index k, and any natural number m,

$$\sqrt[k]{a^m} = \left(\sqrt[k]{a}\right)^m.$$

(We can raise to a power and then take a root or we can take a root and then raise to a power.)

In some cases one way of calculating is easier than the other.

EXAMPLES Calculate as shown. Then use property C to calculate in another way.

9. a) $\sqrt[3]{27^2} = \sqrt[3]{729} = 9$ — Finding 27^2 and then taking the cube root

b) $\left(\sqrt[3]{27}\right)^2 = (3)^2 = 9$ — Taking the cube root and then squaring

10. a) $\sqrt[3]{2^6} = \sqrt[3]{64} = 4$ — Finding 2^6 and then taking the cube root

b) $\left(\sqrt[3]{2}\right)^6 = \sqrt[3]{2}\,\sqrt[3]{2}\,\sqrt[3]{2}\,\sqrt[3]{2}\,\sqrt[3]{2}\,\sqrt[3]{2} = 2 \cdot 2 = 4$

Note that in part (a), we could have easily used the fact that $2^6 = (2^2)^3$. Then $\sqrt[3]{2^6} = \sqrt[3]{(2^2)^3} = 2^2$, or 4. In part (b), we could have used the fact that $\left(\sqrt[3]{2}\right)^3 = 2$. Then $\left(\sqrt[3]{2}\right)^6 = \left(\sqrt[3]{2}\right)^3 \cdot \left(\sqrt[3]{2}\right)^3 = 2 \cdot 2$, or 4.

11. a) $\left(\sqrt{5x}\right)^3 = \sqrt{5x}\,\sqrt{5x}\,\sqrt{5x} = 5x\,\sqrt{5x}$

b) $\sqrt{(5x)^3} = \sqrt{5^3x^3} = \sqrt{5^2x^2}\,\sqrt{5x} = 5x\sqrt{5x}$

Assume $x \geq 0$.

12. a) $\left(\sqrt{5a^2b^3}\right)^3 = \sqrt{5a^2b^3} \cdot \sqrt{5a^2b^3} \cdot \sqrt{5a^2b^3}$

$= 5a^2b^3\,\sqrt{5a^2b^3}$

$= 5a^2b^3\,\sqrt{a^2b^2 \cdot 5b} = 5a^2b^3\,\sqrt{a^2b^2}\,\sqrt{5b}$

$= 5a^2b^3 \cdot ab\,\sqrt{5b} = 5a^3b^4\,\sqrt{5b}$

b) $\sqrt{(5a^2b^3)^3} = \sqrt{125a^6b^9}$

$= \sqrt{25a^6b^8}\,\sqrt{5b}$

$= 5a^3b^4\,\sqrt{5b}$

Assume $a, b \geq 0$.

13. a) $\left(\sqrt[3]{16x^3y^2}\right)^2 = \left(\sqrt[3]{8x^3 \cdot 2y^2}\right)^2$

$= \left(2x\,\sqrt[3]{2y^2}\right)^2$ — Simplifying the cube root

$= 2x\,\sqrt[3]{2y^2} \cdot 2x\,\sqrt[3]{2y^2}$

$= 4x^2\,\sqrt[3]{4y^4}$ — Multiplying and combining radicals

$= 4x^2\,\sqrt[3]{y^3 4y}$

$= 4x^2y\,\sqrt[3]{4y}$

b) $\sqrt[3]{(16x^3y^2)^2} = \sqrt[3]{256x^6y^4}$

$= \sqrt[3]{64x^6y^3 4y}$ — Factoring the radicand

$= \sqrt[3]{64x^6y^3}\,\sqrt[3]{4y}$

$= 4x^2y\,\sqrt[3]{4y}$

■

EXERCISE SET 5.2

Simplify by taking roots of the numerator and the denominator. Assume that all radicands represent positive numbers.

1. $\sqrt{\frac{16}{25}}$ **2.** $\sqrt{\frac{100}{81}}$ **3.** $\sqrt[3]{\frac{64}{27}}$ **4.** $\sqrt[3]{\frac{343}{512}}$

5. $\sqrt{\frac{49}{y^2}}$ **6.** $\sqrt{\frac{121}{x^2}}$ **7.** $\sqrt{\frac{25y^3}{x^4}}$ **8.** $\sqrt{\frac{36a^5}{b^6}}$

9. $\sqrt[3]{\frac{8x^5}{27y^3}}$ **10.** $\sqrt[3]{\frac{64x^7}{216y^6}}$ **11.** $\sqrt[4]{\frac{16a^4}{81}}$ **12.** $\sqrt[4]{\frac{81x^4}{y^8}}$

13. $\sqrt[4]{\frac{a^5b^8}{c^{10}}}$ **14.** $\sqrt[4]{\frac{x^9y^{12}}{z^6}}$ **15.** $\sqrt[5]{\frac{32x^6}{y^{11}}}$ **16.** $\sqrt[5]{\frac{243a^9}{b^{13}}}$

17. $\sqrt[6]{\frac{x^6y^8}{z^{15}}}$ **18.** $\sqrt[6]{\frac{a^9b^{12}}{c^{13}}}$

Divide. Then simplify by taking roots, if possible. Assume that all radicands represent positive numbers.

19. $\frac{\sqrt{21a}}{\sqrt{3a}}$ **20.** $\frac{\sqrt{28y}}{\sqrt{4y}}$ **21.** $\frac{\sqrt[3]{54}}{\sqrt[3]{2}}$ **22.** $\frac{\sqrt[3]{40}}{\sqrt[3]{5}}$

23. $\frac{\sqrt{40xy^3}}{\sqrt{8x}}$ **24.** $\frac{\sqrt{56ab^3}}{\sqrt{7a}}$ **25.** $\frac{\sqrt[3]{96a^4b^2}}{\sqrt[3]{12a^2b}}$ **26.** $\frac{\sqrt[3]{189x^5y^7}}{\sqrt[3]{7x^2y^2}}$

27. $\frac{\sqrt{72xy}}{2\sqrt{2}}$ **28.** $\frac{\sqrt{75ab}}{3\sqrt{3}}$ **29.** $\frac{\sqrt[4]{48x^5y^{11}}}{\sqrt[4]{3xy^2}}$ **30.** $\frac{\sqrt[4]{2a^9b^5}}{\sqrt[4]{162a^2b}}$

Calculate as shown. Then use property C to calculate in another way. Assume that all radicands represent nonnegative numbers.

31. $\sqrt{(6a)^3}$ **32.** $\sqrt{(7y)^3}$ **33.** $(\sqrt{16b^2})^3$ **34.** $(\sqrt{25r^2})^3$

35. $\sqrt{(18a^2b)^3}$ **36.** $\sqrt{(12x^2y)^3}$ **37.** $(\sqrt[3]{3c^2d})^4$ **38.** $(\sqrt[3]{2x^2y})^4$

39. $\sqrt[3]{(5x^2y)^2}$ **40.** $\sqrt[3]{(6ab^2)^2}$ **41.** $\sqrt[4]{(x^2y)^3}$ **42.** $\sqrt[4]{(2a^3)^3}$

43. $(\sqrt[3]{8a^4b})^2$ **44.** $(\sqrt[3]{27xy^5})^2$ **45.** $(\sqrt[4]{16x^2y^3})^2$ **46.** $(\sqrt[4]{16xy^5})^3$

SKILL MAINTENANCE

Solve.

47. $\frac{12x}{x-4} - \frac{3x^2}{x+4} = \frac{384}{x^2-16}$

48. $\frac{2}{3} + \frac{1}{t} = \frac{4}{5}$

49. The width of a rectangle is one fourth the length. The area is twice the perimeter. Find the dimensions of the rectangle.

SYNTHESIS

50. Pendulums. The *period* of a pendulum is the time it takes to complete one cycle, swinging to and fro. If a pendulum consists of a weight on a string, the period T is given by the formula

$$T = 2\pi\sqrt{\frac{L}{980}},$$

where T is in seconds and L is the length of the pendulum in centimeters. Find to the nearest hundredth of a second the period of a pendulum of length (a) 65 cm; (b) 98 cm; (c) 120 cm. Use 3.14 for π.

Divide and simplify.

51. $\dfrac{7\sqrt{a^9b^5}\,\sqrt{25x^5y^8}}{5\sqrt{a^3b^3}\,\sqrt{49x^3y^4}}$

52. $\dfrac{(\sqrt[3]{81mn^2})^2}{(\sqrt[3]{mn})^2}$

53. $\dfrac{\sqrt{44x^2y^9z}\,\sqrt{22y^9z^6}}{(\sqrt{11xy^8z^2})^2}$

54. $\dfrac{\sqrt{x^3 - y^3}}{\sqrt{x - y}}$

55. Explain why no assumptions were made regarding the numbers that x and y represent in Example 13.

56. Explain why $\sqrt[3]{x^6} = x^2$ for any value x, whereas $\sqrt[2]{x^6} = x^3$ only when $x \geq 0$.

5.3 Addition, Subtraction, and More Multiplication with Radicals

Any two real numbers can be added. For instance, the sum of 7 and $\sqrt{3}$ can be expressed as

$$7 + \sqrt{3}.$$

We cannot simplify this name for the sum unless we use a rational approximation of $\sqrt{3}$. However, when we have *like radicals* (radicals having the same index and radicand), we can use the distributive law to simplify by collecting like radical terms.

EXAMPLES Add or subtract. Simplify by collecting like radical terms, if possible.

1. $6\sqrt{7} + 4\sqrt{7} = (6 + 4)\sqrt{7}$ **Using the distributive law (factoring out $\sqrt{7}$)**
$= 10\sqrt{7}$

2. $8\sqrt[3]{2} - 7x\sqrt[3]{2} + 5\sqrt[3]{2} = (8 - 7x + 5)\sqrt[3]{2}$ **Factoring out $\sqrt[3]{2}$**
$= (13 - 7x)\sqrt[3]{2}$ **These parentheses *are* necessary!**

3. $6\sqrt[5]{4x} + 4\sqrt[5]{4x} - \sqrt[3]{4x} = (6 + 4)\sqrt[5]{4x} - \sqrt[3]{4x}$ Try to do this step mentally.

$= 10\sqrt[5]{4x} - \sqrt[3]{4x}$ ■

Note that these expressions have the *same* radicand but are *not* like radicals because they have *different* indices!

One way to think of problems like Example 1 is as follows: When 4 square roots of 7 are added to 6 square roots of 7, the result is 10 square roots of 7.

Sometimes we need to factor one or more of the radicals in order to have terms with like radicals.

EXAMPLES Add or subtract. Simplify by collecting like radical terms, if possible.

4. $3\sqrt{8} - 5\sqrt{2} = 3\sqrt{4 \cdot 2} - 5\sqrt{2}$ Factoring 8

$= 3\sqrt{4} \cdot \sqrt{2} - 5\sqrt{2}$ Factoring $\sqrt{4 \cdot 2}$ into two radicals

$= 3 \cdot 2\sqrt{2} - 5\sqrt{2}$ Taking the square root of 4

$= 6\sqrt{2} - 5\sqrt{2}$

$= (6 - 5)\sqrt{2}$ Factoring out $\sqrt{2}$

$= \sqrt{2}$

5. $5\sqrt{2} - 4\sqrt{3}$ No simplification possible

6. $5\sqrt[3]{16y^4} + 7\sqrt[3]{2y} = 5\sqrt[3]{8y^3 \cdot 2y} + 7\sqrt[3]{2y}$

$= 5\sqrt[3]{8y^3} \cdot \sqrt[3]{2y} + 7\sqrt[3]{2y}$ } Factoring the first radical

$= 5 \cdot 2y \cdot \sqrt[3]{2y} + 7\sqrt[3]{2y}$ Taking the cube root

$= 10y\sqrt[3]{2y} + 7\sqrt[3]{2y}$

$= (10y + 7)\sqrt[3]{2y}$ Factoring ■

More About Multiplication with Radicals

We now consider multiplication in which some factors contain more than one term. To do that, we use the procedures for multiplying polynomials, which are based on the distributive law.

EXAMPLES Multiply.

7. $\sqrt{3}(x - \sqrt{5}) = \sqrt{3} \cdot x - \sqrt{3} \cdot \sqrt{5}$ Using the distributive law

$= x\sqrt{3} - \sqrt{15}$ Multiplying radicals; using property A of Section 5.1

8. $\sqrt[3]{y}(\sqrt[3]{y^2} + \sqrt[3]{2}) = \sqrt[3]{y} \cdot \sqrt[3]{y^2} + \sqrt[3]{y} \cdot \sqrt[3]{2}$ Using the distributive law

$= \sqrt[3]{y^3} + \sqrt[3]{2y}$ Multiplying radicals; using property A

$= y + \sqrt[3]{2y}$ Simplifying $\sqrt[3]{y^3}$ ■

When each factor contains two terms, we multiply much like we did with binomials.

EXAMPLE 9 Multiply: $(\sqrt{a} + \sqrt{3})(\sqrt{b} + \sqrt{3})$. Assume $a, b \geq 0$.

Solution

$$(\sqrt{a} + \sqrt{3})(\sqrt{b} + \sqrt{3}) = \sqrt{a}\sqrt{b} + \sqrt{a}\sqrt{3} + \sqrt{3}\sqrt{b} + \sqrt{3}\sqrt{3} \quad \text{FOIL}$$
$$= \sqrt{ab} + \sqrt{3a} + \sqrt{3b} + 3$$

EXAMPLE 10 Multiply: $(4\sqrt{3} + \sqrt{2})(\sqrt{3} - 5\sqrt{2})$.

Solution

$$(4\sqrt{3} + \sqrt{2})(\sqrt{3} - 5\sqrt{2}) = 4(\sqrt{3})^2 - 20\sqrt{3} \cdot \sqrt{2} + \sqrt{2} \cdot \sqrt{3} - 5(\sqrt{2})^2$$
$$= 4 \cdot 3 - 20\sqrt{6} + \sqrt{6} - 5 \cdot 2$$
$$= 12 - 20\sqrt{6} + \sqrt{6} - 10$$
$$= 2 - 19\sqrt{6} \quad \text{Simplifying}$$

EXAMPLE 11 Multiply: $(\sqrt{3} + x)^2$.

Solution

$$(\sqrt{3} + x)^2 = (\sqrt{3})^2 + 2x\sqrt{3} + x^2 \quad \text{Squaring a binomial}$$
$$= 3 + 2x\sqrt{3} + x^2$$

EXAMPLE 12 Multiply: $(\sqrt{5} + \sqrt{7})(\sqrt{5} - \sqrt{7})$.

Solution

$$(\sqrt{5} + \sqrt{7})(\sqrt{5} - \sqrt{7}) = (\sqrt{5})^2 - (\sqrt{7})^2 \quad \text{This is now a difference of two squares.}$$
$$= 5 - 7$$
$$= -2$$

Rationalizing Denominators or Numerators

Fractional expressions are often considered simpler when the denominator is free of radicals. Thus, in simplifying, we generally remove the radicals in a denominator. This is called **rationalizing the denominator,** and it can be done by multiplying by 1 in such a way as to obtain a perfect power in the denominator. On occasion, we prefer to rationalize the numerator. In either case, we can accomplish the rationalization by multiplying by 1, as in the following examples. Note that we can write the expression for 1 using two radicals (as in Examples 15 and 16) or by working under one radical (as in Examples 13, 14, and 17).

EXAMPLES Simplify by rationalizing the denominator.

13. $\sqrt{\frac{1}{2}} = \sqrt{\frac{1}{2} \cdot \frac{2}{2}} = \sqrt{\frac{2}{4}} = \frac{\sqrt{2}}{\sqrt{4}} = \frac{\sqrt{2}}{2}$ We multiply by $\frac{2}{2}$ so that we have a perfect square in the denominator.

14. $\sqrt[3]{\frac{7}{9}} = \sqrt[3]{\frac{7}{9} \cdot \frac{3}{3}} = \sqrt[3]{\frac{21}{27}} = \frac{\sqrt[3]{21}}{\sqrt[3]{27}} = \frac{\sqrt[3]{21}}{3}$ We multiply by $\frac{3}{3}$ so that we have a perfect cube in the denominator.

15. $\frac{\sqrt{7}}{\sqrt{5}} = \frac{\sqrt{7}}{\sqrt{5}} \cdot \frac{\sqrt{5}}{\sqrt{5}} = \frac{\sqrt{35}}{\sqrt{25}} = \frac{\sqrt{35}}{5}$ Using $\frac{\sqrt{5}}{\sqrt{5}}$ for 1

16. $\frac{\sqrt{2a}}{\sqrt{5b}} = \frac{\sqrt{2a}}{\sqrt{5b}} \cdot \frac{\sqrt{5b}}{\sqrt{5b}} = \frac{\sqrt{10ab}}{\sqrt{(5b)^2}} = \frac{\sqrt{10ab}}{5b}$ Note that for the original expression to make sense, we must have $a \geq 0$ and $b > 0$.

17. $\frac{\sqrt[3]{54x^3}}{\sqrt[3]{4y^5}} = \sqrt[3]{\frac{54x^3}{4y^5} \cdot \frac{2y}{2y}}$ We multiply by $\frac{2y}{2y}$ so that we have a perfect cube in the denominator.

$$= \sqrt[3]{\frac{27x^3 \cdot 4y}{8y^6}}$$

$$= \frac{\sqrt[3]{27x^3} \cdot \sqrt[3]{4y}}{\sqrt[3]{8y^6}}$$

$$= \frac{3x \cdot \sqrt[3]{4y}}{2y^2}$$ Assume $y \neq 0$.

When a denominator or a numerator to be rationalized has an expression like $a + \sqrt{b}$ or $\sqrt{c} - \sqrt{d}$, we choose the symbol for 1 in such a way that the product in the part being rationalized becomes a difference of squares. The symbol for 1 will have two terms in its numerator and denominator. The following examples illustrate this process.

EXAMPLES Rationalize the denominator. Assume that all letters represent positive numbers.

18. $\frac{1}{\sqrt{2} + \sqrt{3}} = \frac{1}{\sqrt{2} + \sqrt{3}} \cdot \frac{\sqrt{2} - \sqrt{3}}{\sqrt{2} - \sqrt{3}}$ The number $\sqrt{2} - \sqrt{3}$ is called the *conjugate* of $\sqrt{2} + \sqrt{3}$. It is found by changing the middle sign. We use the conjugate to form the symbol for 1.

$$= \frac{\sqrt{2} - \sqrt{3}}{(\sqrt{2} + \sqrt{3})(\sqrt{2} - \sqrt{3})}$$

$$= \frac{\sqrt{2} - \sqrt{3}}{(\sqrt{2})^2 - (\sqrt{3})^2}$$

$$= \frac{\sqrt{2} - \sqrt{3}}{2 - 3} = \frac{\sqrt{2} - \sqrt{3}}{-1}$$

$$= \sqrt{3} - \sqrt{2}$$

19.

$$\frac{\sqrt{x} + \sqrt{y}}{\sqrt{x} - \sqrt{y}} = \frac{\sqrt{x} + \sqrt{y}}{\sqrt{x} - \sqrt{y}} \cdot \frac{\sqrt{x} + \sqrt{y}}{\sqrt{x} + \sqrt{y}}$$ The conjugate of $\sqrt{x} - \sqrt{y}$ is $\sqrt{x} + \sqrt{y}$.

$$= \frac{(\sqrt{x} + \sqrt{y})^2}{(\sqrt{x})^2 - (\sqrt{y})^2}$$

$$= \frac{x + 2\sqrt{xy} + y}{x - y}$$

EXAMPLES Rationalize the numerator. Assume that all letters represent positive numbers.

$$20.\quad \frac{1-\sqrt{2}}{5} = \frac{1-\sqrt{2}}{5}\cdot\frac{1+\sqrt{2}}{1+\sqrt{2}}$$ Using the conjugate of the numerator to form 1

$$= \frac{(1-\sqrt{2})(1+\sqrt{2})}{5(1+\sqrt{2})}$$

$$= \frac{1-2}{5(1+\sqrt{2})}$$

$$= \frac{-1}{5+5\sqrt{2}}$$

$$21.\quad \frac{\sqrt{x+h}-\sqrt{x}}{h} = \frac{\sqrt{x+h}-\sqrt{x}}{h}\cdot\frac{\sqrt{x+h}+\sqrt{x}}{\sqrt{x+h}+\sqrt{x}}$$

$$= \frac{(x+h)-x}{h(\sqrt{x+h}+\sqrt{x})}$$

$$= \frac{h}{h(\sqrt{x+h}+\sqrt{x})}$$

$$= \frac{1}{\sqrt{x+h}+\sqrt{x}}$$

Example 21 is important in the study of calculus. ■

EXERCISE SET 5.3

Add or subtract. Simplify by collecting like radical terms, if possible. Assume that all variables and radicands represent nonnegative numbers.

1. $6\sqrt{3} + 2\sqrt{3}$
2. $8\sqrt{5} + 9\sqrt{5}$
3. $9\sqrt[3]{5} - 6\sqrt[3]{5}$
4. $14\sqrt[5]{2} - 6\sqrt[5]{2}$
5. $4\sqrt[3]{y} + 9\sqrt[3]{y}$
6. $6\sqrt[4]{t} - 3\sqrt[4]{t}$
7. $8\sqrt{2} - 6\sqrt{2} + 5\sqrt{2}$
8. $2\sqrt{6} + 8\sqrt{6} - 3\sqrt{6}$
9. $4\sqrt[3]{3} - \sqrt{5} + 2\sqrt[3]{3} + \sqrt{5}$
10. $5\sqrt{7} - 8\sqrt[4]{11} + \sqrt{7} + 9\sqrt[4]{11}$
11. $8\sqrt{27} - 3\sqrt{3}$
12. $9\sqrt{50} - 4\sqrt{2}$
13. $8\sqrt{45} + 7\sqrt{20}$
14. $9\sqrt{12} + 16\sqrt{27}$
15. $18\sqrt{72} + 2\sqrt{98}$
16. $12\sqrt{45} - 8\sqrt{80}$
17. $3\sqrt[3]{16} + \sqrt[3]{54}$
18. $\sqrt[3]{27} - 5\sqrt[3]{8}$
19. $\sqrt{5a} + 2\sqrt{45a^3}$
20. $4\sqrt{3x^3} - \sqrt{12x}$
21. $\sqrt[3]{24x} - \sqrt[3]{3x^4}$
22. $\sqrt[3]{54x} - \sqrt[3]{2x^4}$
23. $\sqrt{8y-8} + \sqrt{2y-2}$
24. $\sqrt{12t+12} + \sqrt{3t+3}$
25. $\sqrt{x^3-x^2} + \sqrt{9x-9}$
26. $\sqrt{4x-4} - \sqrt{x^3-x^2}$
27. $5\sqrt[3]{32} - \sqrt[3]{108} + 2\sqrt[3]{256}$
28. $3\sqrt[3]{8x} - 4\sqrt[3]{27x} + 2\sqrt[3]{64x}$
29. $\sqrt{x^3+x^2} + \sqrt{4x^3+4x^2} - \sqrt{9x^3+9x^2}$
30. $\sqrt{5x^2+4} - 5\sqrt{45x^2+36} + 3\sqrt{20x^2+16}$
31. $\sqrt[4]{x^5-x^4} + 3\sqrt[4]{x^9-x^8}$
32. $\sqrt[4]{16a^4+16a^5} - 2\sqrt[4]{a^8+a^9}$

Multiply. Assume that all variables represent nonnegative real numbers.

33. $\sqrt{6}(2 - 3\sqrt{6})$
34. $\sqrt{3}(4 + \sqrt{3})$
35. $\sqrt{2}(\sqrt{3} - \sqrt{5})$
36. $\sqrt{5}(\sqrt{5} - \sqrt{2})$
37. $\sqrt{3}(2\sqrt{5} - 3\sqrt{4})$
38. $\sqrt{2}(3\sqrt{10} - 2\sqrt{2})$
39. $\sqrt[3]{2}(\sqrt[3]{4} - 2\sqrt[3]{32})$
40. $\sqrt[3]{3}(\sqrt[3]{9} - 4\sqrt[3]{21})$
41. $\sqrt[3]{a}(\sqrt[3]{2a^2} + \sqrt[3]{16a^2})$
42. $\sqrt[3]{x}(\sqrt[3]{3x^2} - \sqrt[3]{81x^2})$
43. $\sqrt[4]{x}(\sqrt[4]{x^7} + \sqrt[4]{3x^2})$
44. $\sqrt[4]{a}(\sqrt[4]{2a} - \sqrt[4]{a^{11}})$
45. $(5 - \sqrt{7})(5 + \sqrt{7})$
46. $(3 + \sqrt{5})(3 - \sqrt{5})$
47. $(\sqrt{5} + \sqrt{8})(\sqrt{5} - \sqrt{8})$
48. $(\sqrt{3} - \sqrt{5})(\sqrt{3} + \sqrt{5})$
49. $(3 - 2\sqrt{7})(3 + 2\sqrt{7})$
50. $(4 - 3\sqrt{2})(4 + 3\sqrt{2})$
51. $(\sqrt{a} + \sqrt{b})(\sqrt{a} - \sqrt{b})$
52. $(\sqrt{x} - \sqrt{y})(\sqrt{x} + \sqrt{y})$
53. $(3 - \sqrt{5})(2 + \sqrt{5})$
54. $(2 + \sqrt{6})(4 - \sqrt{6})$
55. $(2\sqrt{7} - 4\sqrt{2})(3\sqrt{7} + 6\sqrt{2})$
56. $(4\sqrt{5} + 3\sqrt{3})(3\sqrt{5} - 4\sqrt{3})$
57. $(\sqrt{a} + \sqrt{2})(\sqrt{a} + \sqrt{3})$
58. $(2 - \sqrt{x})(1 - \sqrt{x})$
59. $(2 + \sqrt{3})^2$
60. $(\sqrt{5} + 1)^2$
61. $(a + \sqrt{b})^2$
62. $(x - \sqrt{y})^2$
63. $(2x - \sqrt{y})^2$
64. $(3a + \sqrt{b})^2$
65. $(\sqrt{m} + \sqrt{n})^2$
66. $(\sqrt{r} - \sqrt{s})^2$

Rationalize the denominator. Assume that all variables represent positive numbers.

67. $\sqrt{\frac{6}{5}}$
68. $\sqrt{\frac{11}{6}}$
69. $\sqrt[3]{\frac{16}{9}}$
70. $\sqrt[3]{\frac{2}{9}}$
71. $\frac{\sqrt[3]{5y^4}}{\sqrt[3]{6x^4}}$
72. $\frac{\sqrt[3]{3a^4}}{\sqrt[3]{7b^2}}$
73. $\sqrt{\frac{9}{20x^2y}}$
74. $\sqrt{\frac{5}{32ab^2}}$

Rationalize the numerator. Assume that all variables represent nonnegative numbers.

75. $\sqrt{\frac{14}{21}}$
76. $\sqrt{\frac{12}{15}}$
77. $\frac{\sqrt[3]{7}}{\sqrt[3]{2}}$
78. $\frac{\sqrt[3]{5}}{\sqrt[3]{4}}$
79. $\frac{\sqrt{ab}}{3}$
80. $\frac{\sqrt{xy}}{5}$

Rationalize the denominator. Assume that all variables represent nonnegative numbers and that no denominators are 0.

81. $\frac{5}{8 - \sqrt{6}}$
82. $\frac{7}{9 + \sqrt{10}}$
83. $\frac{\sqrt{x} - \sqrt{y}}{\sqrt{x} + \sqrt{y}}$
84. $\frac{\sqrt{a} + \sqrt{b}}{\sqrt{a} - \sqrt{b}}$
85. $\frac{3\sqrt{x} + \sqrt{y}}{2\sqrt{x} + 3\sqrt{y}}$
86. $\frac{2\sqrt{a} - \sqrt{b}}{3\sqrt{a} + 2\sqrt{b}}$

Rationalize the numerator. Assume that all variables represent nonnegative numbers and that no denominators are 0.

87. $\frac{\sqrt{3} + 5}{8}$
88. $\frac{3 - \sqrt{2}}{5}$
89. $\frac{4\sqrt{6} - 5\sqrt{3}}{2\sqrt{3} + 7\sqrt{6}}$
90. $\frac{8\sqrt{2} + 5\sqrt{3}}{5\sqrt{3} - 7\sqrt{2}}$
91. $\frac{\sqrt{3} + 2\sqrt{x}}{\sqrt{3} - \sqrt{x}}$
92. $\frac{\sqrt{5} - 3\sqrt{x}}{\sqrt{5} + \sqrt{x}}$

SKILL MAINTENANCE

93. Multiply: $(4x + 2)(x - 5)$.

94. A kitchen floor is 3 ft longer than it is wide. The area of the floor is 180 ft^2. Find the dimensions of the kitchen.

Add.

95. $\dfrac{3}{x - 2} + \dfrac{2}{x + 2}$

96. $\dfrac{7}{x^2 - 5x + 4} + \dfrac{3}{x^2 + 2x - 3}$

SYNTHESIS

Add or subtract by collecting like terms. Assume that all variables represent nonnegative real numbers unless otherwise indicated.

97. $\sqrt{432} - \sqrt{6125} + \sqrt{845} - \sqrt{4800}$

98. $\sqrt{1250x^3y} - \sqrt{1800xy^3} - \sqrt{162x^3y^3}$

99. $\frac{1}{2}\sqrt{36a^5bc^4} - \frac{1}{2}\sqrt[3]{64a^4b^6} + \frac{1}{6}\sqrt{144a^3bc^2}$

100. $7x\sqrt{(x + y)^3} - 5xy\sqrt{x + y} - 2y\sqrt{(x + y)^3}$ (Assume $x + y \geq 0$.)

For Exercises 101–104 assume that all radicands are positive and that no denominators are 0. Rationalize the denominator.

101. $\dfrac{a - \sqrt{a + b}}{\sqrt{a + b} - b}$

102. $\dfrac{3\sqrt{y} + 4\sqrt{yz}}{5\sqrt{y} - 2\sqrt{z + y}}$

103. $\dfrac{b + \sqrt{b}}{1 + b + \sqrt{b}}$

104. $\dfrac{a - b}{a^2 + a\sqrt{a - b}}$

Simplify.

105. $\sqrt{1 + x^2} + \dfrac{1}{\sqrt{1 + x^2}}$

106. $\sqrt{1 - x^2} - \dfrac{x^2}{2\sqrt{1 - x^2}}$

107. ▦ *Water flow from a faucet.* As water flows at a velocity v_0 from a faucet with internal diameter d, the width w of the stream at a distance h below the outlet is given by

$$w = d\sqrt{\frac{v_0}{\sqrt{v_0^2 + 19.6h}}}.$$

Find the width of a stream 0.1 m below a faucet of internal diameter 0.03 m if the water is flowing at a rate of 0.6 m/sec.

108. ▦ Select the smallest positive number from the following list:

$10\sqrt{26} - 51,\quad 51 - 10\sqrt{26},\quad 18 - 5\sqrt{13},$
$3\sqrt{11} - 10,\quad 10 - 3\sqrt{11}.$

5.4 Rational Numbers as Exponents

Zero and Negative Exponents

Although we have not yet settled on the meaning of such expressions as 8^0 or 5^{-4}, we have developed several rules for natural number exponents. We list them here for convenience, along with the section in which the rule first appeared.

Rules for Natural Number Exponents

For any natural numbers m and n,

Multiplying with like bases: $a^m \cdot a^n = a^{m+n}$ (Product Rule for Powers, Section 3.2)

Dividing with like bases: $\frac{a^m}{a^n} = a^{m-n}$; $a \neq 0$, $m > n$ (Quotient Rule for Powers, Section 3.6)

Raising a product to a power: $(ab)^n = a^n b^n$ (Section 3.4)

Raising a power to a power: $(a^m)^n = a^{mn}$ (Power Rule, Section 3.4)

Raising a quotient to a power: $\left(\frac{a}{b}\right)^n = \frac{a^n}{b^n}$ (Section 3.4)

Let us focus now on the quotient rule, with $m \not> n$. Suppose now that the bases in the numerator and the denominator are both raised to the *same* power:

$$\frac{x^5}{x^5} = 1 \quad \text{or} \quad \frac{8^3}{8^3} = 1.$$

These results follow from the fact that any (nonzero) expression divided by itself is equal to 1. On the other hand, were we to subtract exponents, we would obtain

$$\frac{x^5}{x^5} = x^{5-5} = x^0 \quad \text{or} \quad \frac{8^3}{8^3} = 8^{3-3} = 8^0.$$

Thus to continue subtracting exponents when dividing with like bases, we must have the following:

For any real number a, $a \neq 0$,

$$a^0 = 1.$$

(Any nonzero number raised to the zero power is 1.)

Defining the zero exponent in this manner preserves the following pattern:

$4^3 = 4 \cdot 4 \cdot 4,$

$4^2 = 4 \cdot 4,$ Dividing by 4 on both sides

$4^1 = 4,$ Dividing by 4 on both sides

$4^0 = 1.$ Dividing by 4 on both sides

EXAMPLE 1 Evaluate x^0 when $x = -7$ and $x = 1.379$.

Solution When $x = -7$, $x^0 = (-7)^0 = 1$. Similarly, when $x = 1.379$, $x^0 = (1.379)^0 = 1$. ■

We can use the rule for dividing powers with like bases to lead us to a definition of exponential notation when the exponent is a negative integer. Consider $5^3/5^7$ and first simplify it using procedures we have learned for working with fractions:

$$\frac{5^3}{5^7} = \frac{5 \cdot 5 \cdot 5}{5 \cdot 5 \cdot 5 \cdot 5 \cdot 5 \cdot 5 \cdot 5} = \frac{5 \cdot 5 \cdot 5 \cdot 1}{5 \cdot 5 \cdot 5 \cdot 5 \cdot 5 \cdot 5 \cdot 5}$$

$$= \frac{5 \cdot 5 \cdot 5}{5 \cdot 5 \cdot 5} \cdot \frac{1}{5 \cdot 5 \cdot 5 \cdot 5}$$

$$= \frac{1}{5^4}. \qquad \text{Removing a factor of 1}$$

Now suppose we apply the rule for dividing powers with the same bases. Then

$$\frac{5^3}{5^7} = 5^{3-7} = 5^{-4}.$$

From these two expressions for $5^3/5^7$, it follows that

$$5^{-4} = \frac{1}{5^4}.$$

This leads to our definition of negative exponents.

For any real number a that is nonzero and any integer n,

$$a^{-n} = \frac{1}{a^n}.$$

(The numbers a^{-n} and a^n are reciprocals.)

EXAMPLES Express using positive exponents. Then simplify.

2. $3^{-2} = \frac{1}{3^2} = \frac{1}{9}$

3. $3x^{-4} = 3\left(\frac{1}{x^4}\right) = \frac{3}{x^4}$

4. $a^{-2}b^3c^{-1} = \left(\frac{1}{a^2}\right)(b^3)\left(\frac{1}{c^1}\right) = \frac{b^3}{a^2c}$

5. $\frac{1}{5^{-2}} = \frac{1}{\frac{1}{5^2}} = 1 \cdot \frac{5^2}{1} = 25$

Example 5 reveals that when a factor of the numerator or the denominator is raised to any power, the factor can be moved to the other side of the fraction bar provided the sign of the exponent is changed. Thus, for example,

$$\frac{a^{-2}b^3}{c^{-4}} = \frac{c^4}{a^2b^{-3}}.$$

Because of our definitions of zero and negative powers, all of the preceding rules still hold.

EXAMPLES Simplify. Use only positive exponents in the answer.

6. $7^{-3} \cdot 7^{8} = 7^{-3+8}$ Adding exponents
$= 7^{5}$

7. $\frac{5^4}{5^{-2}} = 5^{4-(-2)}$ Subtracting exponents
$= 5^{6}$

8. $(5y)^{-3} = 5^{-3}y^{-3}$ Raising each factor to the power
$= \frac{1}{125y^3}$

9. $(-2x^3y^{-1})^{-4} = (-2)^{-4}(x^3)^{-4}(y^{-1})^{-4}$ Raising each factor to the negative fourth power
$= \frac{1}{(-2)^4} \cdot x^{-12}y^{4}$ Multiplying powers
$= \frac{y^4}{16x^{12}}$

■

Fractional Exponents

To preserve the rules for exponents when fractions appear as the powers, a suitable definition must be made. Consider $a^{1/2} \cdot a^{1/2}$. If we still want to multiply by adding exponents, it must follow that $a^{1/2} \cdot a^{1/2} = a^{1/2+1/2}$ or a^1. Thus we should define $a^{1/2}$ to be a square root of a. Similarly, $a^{1/3} \cdot a^{1/3} \cdot a^{1/3} = a^{1/3+1/3+1/3}$ or a^1. Thus $a^{1/3}$ should be defined to mean $\sqrt[3]{a}$.

> For any nonnegative number a and any index n, $a^{1/n}$ means $\sqrt[n]{a}$ (the principal nth root of a). If a is negative, then n must be odd.

Whenever we use fractional exponents, we assume that no even roots of negative numbers appear.

EXAMPLES Rewrite without fractional exponents.

10. $x^{1/2} = \sqrt{x}$
11. $27^{1/3} = \sqrt[3]{27}$, or 3
12. $(abc)^{1/5} = \sqrt[5]{abc}$

■

EXAMPLES Rewrite with fractional exponents.

13. $\sqrt[5]{7xy} = (7xy)^{1/5}$ **Parentheses are required.**

14. $\sqrt[7]{\dfrac{x^3y}{9}} = \left(\dfrac{x^3y}{9}\right)^{1/7}$ ■

How should we define $a^{2/3}$? If the usual properties of exponents are to hold, we have $a^{2/3} = (a^{1/3})^2$, or $(\sqrt[3]{a})^2$. Furthermore,

$$(\sqrt[3]{a})^2 = \sqrt[3]{a} \cdot \sqrt[3]{a} = \sqrt[3]{a^2}.$$

We make our definition accordingly.

> For any natural numbers m and n ($n \neq 1$) and any real number a for which $\sqrt[n]{a}$ exists,
>
> $$a^{m/n} \quad \text{means} \quad (\sqrt[n]{a})^m, \quad \text{or} \quad \sqrt[n]{a^m}.$$

EXAMPLES Rewrite without fractional exponents.

15. $(27)^{2/3} = (\sqrt[3]{27})^2$, or $\sqrt[3]{27^2}$

$= 3^2$, or 9 **This is most easily computed by first taking the cube root and then squaring.**

16. $4^{3/2} = (\sqrt{4})^3$, or $\sqrt{4^3}$

$= 2^3$, or 8 **Taking the square root and then cubing** ■

EXAMPLES Rewrite with fractional exponents.

17. $\sqrt[3]{9^4} = 9^{4/3}$

18. $(\sqrt[4]{7xy})^5 = (7xy)^{5/4}$

The index of the radical is the denominator of the exponent. ■

Just as a^{-n} and a^n are reciprocals of one another, so too are $a^{-m/n}$ and $a^{m/n}$.

EXAMPLES Rewrite with positive exponents.

19. $9^{-1/2} = \dfrac{1}{9^{1/2}}$ **$9^{-1/2}$ and $9^{1/2}$ are reciprocals.**

Since $9^{1/2} = \sqrt{9} = 3$, the answer simplifies to $\frac{1}{3}$.

20. $\left(\dfrac{8x}{y}\right)^{-4/5} = \dfrac{(8x)^{-4/5}}{y^{-4/5}}$

$= \dfrac{y^{4/5}}{(8x)^{4/5}}$ **Try to do these steps mentally.**

$= \left(\dfrac{y}{8x}\right)^{4/5}$ ■

The rules for exponents that we listed earlier can now be extended to all rational exponents, as indicated in the following box.

For any real numbers a and b and any rational exponents m and n for which a^m, a^n, and b^m are defined:

1. $a^m \cdot a^n = a^{m+n}$ In multiplying, we can add exponents if the bases are the same.
2. $\dfrac{a^m}{a^n} = a^{m-n}$ In dividing, we can subtract exponents if the bases are the same. (Assume $a \neq 0$.)
3. $(a^m)^n = a^{m \cdot n}$ To raise a power to a power, we can multiply the exponents.
4. $(ab)^m = a^m b^m$ To raise a product to a power, we can raise each factor to the power and multiply.

EXAMPLES Use the laws of exponents to simplify.

21. $3^{1/5} \cdot 3^{3/5} = 3^{1/5+3/5} = 3^{4/5}$ Adding exponents

22. $\dfrac{7^{1/4}}{7^{1/2}} = 7^{1/4-1/2} = 7^{1/4-2/4} = 7^{-1/4}$ Subtracting exponents after finding a common denominator

23. $(7.2^{2/3})^{3/4} = 7.2^{2/3 \cdot 3/4} = 7.2^{6/12}$ Multiplying exponents

$= 7.2^{1/2}$ Using arithmetic to simplify the exponent

24. $(a^{-1/3}b^{2/5})^{1/2} = a^{-1/3 \cdot 1/2} \cdot b^{2/5 \cdot 1/2}$ Raising a product to a power and multiplying exponents

$= a^{-1/6}b^{1/5}$ ■

Simplifying Radical Expressions

Fractional exponents can be used to simplify some radical expressions. The procedure is as follows.

1. Convert radical expressions to exponential expressions.
2. Use arithmetic and the laws of exponents to simplify.
3. Convert back to radical notation and simplify when appropriate.

EXAMPLES Use fractional exponents to simplify.

25. $\sqrt[6]{x^3} = x^{3/6}$ Converting to an exponential expression

$= x^{1/2}$ Using arithmetic to simplify the exponent

$= \sqrt{x}$ Converting back to radical notation

26. $\sqrt[6]{4} = 4^{1/6}$ Converting to an exponential expression

$= (2^2)^{1/6}$ Recognizing that 4 is 2^2

$= 2^{2/6}$ Multiplying exponents

$= 2^{1/3}$ Simplifying the exponent

$= \sqrt[3]{2}$ Converting back to radical notation

27. $\sqrt[8]{a^{12}b^4} = (a^{12}b^4)^{1/8}$ Converting to exponential notation

$= a^{12/8} \cdot b^{4/8}$ Raising a product to a power and multiplying exponents

$= a^{3/2} \cdot b^{1/2}$ Simplifying exponents

$= (a^3b)^{1/2}$ Rewriting as a product raised to a power. Study this step carefully.

$= \sqrt{a^3b} = a\sqrt{ab}$ Converting back to radical notation and simplifying

28. $\sqrt[3]{5} \cdot \sqrt{2} = 5^{1/3} \cdot 2^{1/2}$ Converting to exponential notation

$= 5^{2/6} \cdot 2^{3/6}$ Rewriting so that exponents have a common denominator

$= (5^2 \cdot 2^3)^{1/6}$ Rewriting as a product raised to a power

$= \sqrt[6]{5^2 \cdot 2^3}$ Converting back to radical notation

$= \sqrt[6]{200}$ Multiplying under the radical

29. $\sqrt{x-2} \cdot \sqrt[4]{3y} = (x-2)^{1/2}(3y)^{1/4}$ Converting to exponential notation

$= (x-2)^{2/4}(3y)^{1/4}$ Writing exponents with a common denominator

$= [(x-2)^2(3y)]^{1/4}$ Rewriting as a product raised to a power

$= \sqrt[4]{(x^2-4x+4) \cdot 3y}$ Converting back to radical notation

$= \sqrt[4]{3x^2y - 12xy + 12y}$ Multiplying under the radical

30. $\dfrac{\sqrt[4]{(x+y)^3}}{\sqrt{x+y}} = \dfrac{(x+y)^{3/4}}{(x+y)^{1/2}}$ Converting to exponential notation

$= (x+y)^{3/4 - 1/2}$ Subtracting exponents

$= (x+y)^{1/4}$ Simplifying

$= \sqrt[4]{x+y}$ Converting back to radical notation ■

We have now seen several different methods of simplifying radical expressions. We list them.

Some Ways to Simplify Radical Expressions

1. *Simplifying by factoring.* We factor the radicand, looking for factors raised to powers that are multiples of the index.

   Example: $\sqrt[3]{16} = \sqrt[3]{2^3}\sqrt[3]{2}$
   $= 2\sqrt[3]{2}$

2. *Rationalizing denominators.* Radical expressions are often considered simpler if there are no radicals in the denominator.

   Example: $\dfrac{1}{\sqrt{2}} = \dfrac{1}{\sqrt{2}} \cdot \dfrac{\sqrt{2}}{\sqrt{2}}$
   $= \dfrac{\sqrt{2}}{2}$

3. *Collecting like radical terms.*

Example: $\sqrt{8} + 3\sqrt{2} = \sqrt{4} \cdot \sqrt{2} + 3\sqrt{2}$
$= 2\sqrt{2} + 3\sqrt{2}$
$= 5\sqrt{2}$

4. *Using fractional exponents to simplify.* We convert to exponential notation and then use arithmetic and the laws of exponents to simplify the exponents. Then we convert back to radical notation.

Example: $\sqrt[3]{p} \cdot \sqrt[4]{q^3} = p^{1/3} \cdot q^{3/4} = p^{4/12} \cdot q^{9/12}$
$= \sqrt[12]{p^4 q^9}$

EXERCISE SET 5.4

Note: Assume for all exercises that even roots are of nonnegative quantities and that all denominators are nonzero.

Simplify.

1. 10^0 **2.** 9^0 **3.** $(-5)^0$ **4.** $(-7)^0$

5. x^0 when $x = -12$ **6.** y^0 when $y = 23$ **7.** $5x^0$ when $x = -4$

8. $7m^0$ when $m = 1.7$ **9.** n^0, $n \neq 0$ **10.** t^0, $t \neq 0$

Write equivalent expressions without negative exponents.

11. 6^{-3} **12.** 8^{-4} **13.** $\left(\frac{2}{3}\right)^{-1}$ **14.** $\left(\frac{1}{4}\right)^{-2}$

15. $(-11)^{-1}$ **16.** $(-4)^{-3}$ **17.** $(5x)^{-3}$ **18.** $(4xy)^{-5}$

19. x^2y^{-3} **20.** $2a^2b^{-5}$ **21.** $x^{-2}y^5$ **22.** $a^2b^{-3}c^4d^{-5}$

23. $\frac{x^3}{y^{-2}}$ **24.** $\frac{y^4z^3}{x^{-1}}$ **25.** $\frac{y^{-5}}{x^2}$ **26.** $\frac{z^{-4}}{3x^5}$

Write equivalent expressions with negative exponents.

27. $\frac{1}{3^4}$ **28.** $\frac{1}{9^2}$ **29.** $\frac{1}{(-16)^2}$ **30.** $\frac{1}{(-8)^6}$ **31.** 6^4 **32.** 8^5

33. $6x^2$ **34.** $-4y^5$ **35.** $\frac{1}{(5y)^3}$ **36.** $\frac{1}{(5x)^5}$ **37.** $\frac{1}{3y^4}$ **38.** $\frac{1}{4b^3}$

Simplify. When negative exponents appear in the answer, write a second answer using only positive exponents.

39. $8^{-6} \cdot 8^2$ **40.** $9^{-5} \cdot 9^3$ **41.** $b^2 \cdot b^{-5}$

42. $a^4 \cdot a^{-3}$ **43.** $a^{-3} \cdot a^4 \cdot a^2$ **44.** $x^{-8} \cdot x^5 \cdot x^3$

45. $(3x^{-5})(2x^{-3})$ **46.** $(-4a^7)(3a^{-9})$ **47.** $(-2x^{-3})(7x^{-8})$

48. $(6x^{-4}y^3)(-4x^{-8}y^{-2})$ **49.** $(6^{-4})^{-3}$ **50.** $(7^{-8})^{-5}$

51. $(-2x^3y^{-4})^{-2}$ **52.** $(-3a^2b^{-5})^{-3}$ **53.** $\frac{a^3}{a^{-2}}$

54. $\frac{y^4}{y^{-5}}$ **55.** $\frac{-24x^6y^7}{18x^{-3}y^9}$ **56.** $\frac{14a^4b^{-3}}{-8a^8b^{-5}}$

Rewrite without fractional exponents.

57. $x^{1/4}$
58. $y^{1/5}$
59. $8^{1/3}$
60. $16^{1/2}$
61. $(xyz)^{1/3}$
62. $(ab)^{1/4}$
63. $(a^2b^2)^{1/5}$
64. $(x^3y^3)^{1/4}$
65. $16^{3/4}$
66. $4^{7/2}$
67. $9^{5/2}$
68. $81^{3/2}$
69. $(81x)^{3/4}$
70. $(125a)^{2/3}$
71. $(25x^4)^{3/2}$
72. $(9y^6)^{3/2}$

Rewrite with fractional exponents.

73. $\sqrt[3]{20}$
74. $\sqrt[3]{19}$
75. $\sqrt{17}$
76. $\sqrt{6}$
77. $\sqrt{x^3}$
78. $\sqrt{a^5}$
79. $\sqrt[5]{m^2}$
80. $\sqrt[5]{n^4}$
81. $\sqrt[4]{cd^3}$
82. $\sqrt[5]{xy^2}$
83. $(\sqrt{3mn})^3$
84. $(\sqrt[3]{7xy})^4$

Use the properties of exponents to simplify.

85. $5^{3/4} \cdot 5^{1/8}$
86. $11^{2/3} \cdot 11^{1/2}$
87. $\dfrac{7^{5/8}}{7^{3/8}}$
88. $\dfrac{9^{9/11}}{9^{7/11}}$
89. $a^{2/3} \cdot a^{5/4}$
90. $x^{3/4} \cdot x^{2/3}$
91. $(x^{2/3})^{3/7}$
92. $(a^{3/2})^{2/5}$
93. $(m^{2/3}n^{1/2})^{1/4}$
94. $(x^{1/3}y^{2/5})^{1/4}$
95. $(a^{-2/3}b^{-1/4})^{-6}$
96. $(m^{-1/5}n^{-5/6})^{-10}$

Use fractional exponents to simplify.

97. $\sqrt[6]{a^4}$
98. $\sqrt[6]{y^2}$
99. $\sqrt[3]{8y^6}$
100. $\sqrt{x^4y^6}$
101. $\sqrt[6]{4x^2}$
102. $\sqrt[4]{16x^4y^2}$
103. $\sqrt[5]{32c^{10}d^{15}}$
104. $\sqrt[4]{16x^{12}y^{16}}$
105. $\sqrt[6]{\dfrac{m^{12}n^{24}}{64}}$
106. $\sqrt[5]{\dfrac{x^{15}y^{20}}{32}}$
107. $\sqrt[8]{r^4s^2}$
108. $\sqrt[12]{64t^6s^6}$

Use fractional exponents to write a single radical expression.

109. $\sqrt[3]{7} \cdot \sqrt{2}$
110. $\sqrt[3]{7} \cdot \sqrt[4]{5}$
111. $\sqrt{x}\sqrt[3]{2x}$
112. $\sqrt[3]{y}\sqrt[5]{3y}$
113. $\sqrt{x}\sqrt[3]{x-2}$
114. $\sqrt[4]{3x}\sqrt{y+4}$
115. $\dfrac{\sqrt[3]{(a+b)^2}}{\sqrt{a+b}}$
116. $\dfrac{\sqrt[3]{(x+y)^2}}{\sqrt[4]{(x+y)^3}}$
117. $\sqrt[5]{yx^2}\sqrt{xy}$
118. $\sqrt[5]{a^3b}\sqrt{ab}$
119. $\sqrt{a^3bc^4}\sqrt[3]{ab^2c}$
120. $\sqrt[3]{x^2yz}\sqrt{xy^3z^2}$

SKILL MAINTENANCE

Solve.

121. $x^2 - 1 = 8$
122. $\dfrac{1}{x} + 2 = 5$
123. Multiply: $(3x^3 - 1)(2x^3 + 3)$.

124. For homes selling under \$100,000, the real-estate transfer tax in Vermont is 0.5% of the selling price. Find the selling price of a home that had a transfer tax of \$467.50.

SYNTHESIS

Use fractional exponents to write a single radical expression and simplify.

125. $\sqrt[5]{yx^2\sqrt{xy}}$
126. $\sqrt{x^5\sqrt[3]{x^4}}$
127. $\dfrac{\sqrt{(a+b)^3}\sqrt[3]{(a+b)^2}}{\sqrt[4]{a+b}}$
128. $\sqrt[4]{\sqrt[3]{8x^3y^6}}$

Simplify.

129. $\dfrac{1}{\sqrt[3]{3}-\sqrt[3]{2}}$

130. $[\sqrt[10]{\sqrt[5]{x^{15}}}]^5\,[\sqrt[5]{\sqrt[10]{x^{15}}}]^5$

131. $\sqrt[p]{x^{5p}y^{7p+1}z^{p+3}}$

132. *An application: Road pavement messages.* In a psychological study, it was determined that the proper length L of the letters of a word printed on pavement is given by

$$L = \frac{(0.00252)d^{2.27}}{h},$$

where d is the distance of a car from the lettering and h is the height of the eye above the surface of the road. All units are in meters. This formula says that if a person is h meters above the surface of the road and is to be able to recognize a message d meters away, that message will be the most recognizable if the length of the letters is L. Find L to the nearest tenth of a meter, given d and h.

a) $h = 1$ m, $d = 60$ m
b) $h = 0.9906$ m, $d = 75$ m
c) $h = 2.4$ m, $d = 80$ m
d) $h = 1.1$ m, $d = 100$ m

5.5 Radical Equations and Quadratic Form

Solving Radical Equations

Equations like $\sqrt[3]{2x} + 1 = 5$ and $\sqrt{x} + \sqrt{4x - 2} = 7$ are known as **radical equations.** A radical equation has variables in one or more radicands. To solve such equations, we need a new principle. Suppose an equation of the form $a = b$ is true. If we square both sides, we get another true equation, this one of the form $a^2 = b^2$. This can be generalized.

THE PRINCIPLE OF POWERS

If an equation $a = b$ is true, then $a^n = b^n$ is true for any rational number n for which a^n and b^n exist.

Note that the principle of powers is an "if–then" statement. The statement which is obtained by interchanging the sentence parts—"if $a^n = b^n$ is true for some rational number n, then $a = b$ is true"—*is not always true*. For example, $3^2 = (-3)^2$ is true, but $3 = -3$ is *not* true. This means that we must always check our solution in the original problem when we use the principle of powers, because $a = b$ and $a^n = b^n$ are not equivalent equations.

EXAMPLE 1 Solve: $\sqrt{x} - 3 = 4$.

Solution

$$\sqrt{x} - 3 = 4$$

$$\sqrt{x} = 7 \quad \text{Adding to isolate the radical}$$

$$(\sqrt{x})^2 = 7^2 \quad \text{Using the principle of powers}$$

$$x = 49$$

Check:

$\sqrt{x} - 3 = 4$	
$\sqrt{49} - 3$	4
$7 - 3$	
4	

TRUE

The solution is 49. ■

We will see in the next example that the principle of powers does not always give equivalent equations.

EXAMPLE 2 Solve: $\sqrt{x} = -3$.

Solution We might observe at the outset that this equation has no real number solution because the principal square root of a number is never negative. Let us continue as above, for comparison.

$$(\sqrt{x})^2 = (-3)^2 \quad \text{Principle of powers (squaring)}$$

$$x = 9$$

Check:

$\sqrt{x} = -3$	
$\sqrt{9}$	-3
3	

FALSE

The number 9 is extraneous since it does not check. Hence the equation has no solution. ■

> In solving radical equations, possible solutions found using the principle of powers *must* be checked!

In Example 1, we first added on both sides to isolate the radical before using the principle of powers. We use this procedure on more complicated problems as well.

EXAMPLE 3 Solve: $x = \sqrt{x+7} + 5$.

Solution

$$\begin{aligned} x &= \sqrt{x+7} + 5 \\ x - 5 &= \sqrt{x+7} && \text{Adding } -5 \text{ to isolate the radical term} \\ (x-5)^2 &= \left(\sqrt{x+7}\right)^2 && \text{Principle of powers; squaring both sides} \\ x^2 - 10x + 25 &= x + 7 \\ x^2 - 11x + 18 &= 0 \\ (x-9)(x-2) &= 0 && \text{Factoring} \\ x = 9 \quad &\text{or} \quad x = 2 && \text{Using the principle of zero products} \end{aligned}$$

The possible solutions are 9 and 2. Let us check.

Check: For 9:

$$\begin{array}{c|l} x & \sqrt{x+7}+5 \\ \hline 9 & \sqrt{9+7}+5 \\ & 9 \end{array} \quad \text{TRUE}$$

For 2:

$$\begin{array}{c|l} x & \sqrt{x+7}+5 \\ \hline 2 & \sqrt{2+7}+5 \\ & 8 \end{array} \quad \text{FALSE}$$

Since 9 checks but 2 does not, the solution is 9. ■

Suppose, in Example 3, that we had used the principle of powers *before* we added -5 to each side. We then would have had the expression $\left(\sqrt{x+7}+5\right)^2$ or $x + 7 + 10\sqrt{x+7} + 25$ on the right side, and the radical would still have been in the problem.

EXAMPLE 4 Solve: $\sqrt[3]{2x+1} + 5 = 0$.

Solution

$$\begin{aligned} \sqrt[3]{2x+1} + 5 &= 0 \\ \sqrt[3]{2x+1} &= -5 && \text{Adding } -5\text{; this isolates the radical term.} \\ \left(\sqrt[3]{2x+1}\right)^3 &= (-5)^3 && \text{Principle of powers; raising to the third power} \\ 2x + 1 &= -125 \\ 2x &= -126 && \text{Adding } -1 \\ x &= -63 \end{aligned}$$

Check:

$$\begin{array}{r|l} \sqrt[3]{2x+1}+5 & 0 \\ \hline \sqrt[3]{2\cdot(-63)+1}+5 & 0 \\ \sqrt[3]{-125}+5 & 0 \\ -5+5 & \\ 0 & \end{array} \quad \text{TRUE}$$

The solution is -63. ■

When two radical terms appear in an equation, we often need to use the principle of powers twice.

EXAMPLE 5 Solve: $\sqrt{x-3} + \sqrt{x+5} = 4$.

Solution

$$\sqrt{x-3} + \sqrt{x+5} = 4$$

$$\sqrt{x-3} = 4 - \sqrt{x+5} \quad \text{Adding } -\sqrt{x+5}\text{; this isolates one of the radical terms.}$$

$$\left(\sqrt{x-3}\right)^2 = \left(4 - \sqrt{x+5}\right)^2 \quad \text{Principle of powers; squaring both sides}$$

To simplify the right side, we square 4, then find twice the product of 4 and $\sqrt{x+5}$, and then square $\sqrt{x+5}$.

$$x - 3 = 16 - 8\sqrt{x+5} + (x+5)$$

$$-3 = 21 - 8\sqrt{x+5} \quad \text{Adding } -x \text{ and collecting like terms}$$

$$-24 = -8\sqrt{x+5} \quad \text{Isolating the remaining radical term}$$

$$3 = \sqrt{x+5} \quad \text{Dividing by } -8$$

$$3^2 = \left(\sqrt{x+5}\right)^2 \quad \text{Squaring}$$

$$9 = x + 5$$

$$4 = x$$

The number 4 checks and is the solution. ■

In some cases we can solve by raising both sides of an equation to a fractional power.

EXAMPLE 6 Solve: $(x-4)^3 = 7$.

Solution

$$(x-4)^3 = 7$$

$$\left[(x-4)^3\right]^{1/3} = 7^{1/3} \quad \text{Principle of powers}$$

$$x - 4 = \sqrt[3]{7} \quad \text{Multiplying exponents and writing radical form}$$

$$x = 4 + \sqrt[3]{7} \quad \text{Adding 4 on both sides}$$

The number $4 + \sqrt[3]{7}$ checks and is the solution. ■

Note that in Example 6, 1/3 was the reciprocal of 3. When a fractional exponent with an *even* denominator is used, we must consider both positive and negative roots.

EXAMPLE 7 Solve: $(x+3)^2 = 5$.

Solution

$$(x + 3)^2 = 5$$

$$\left[(x + 3)^2\right]^{1/2} = \pm 5^{1/2}$$ Because 2 is even, we must consider both $5^{1/2}$ and $-5^{1/2}$. The notation $\pm 5^{1/2}$ represents both $5^{1/2}$ and $-5^{1/2}$.

$$x + 3 = \pm\sqrt{5}$$ The notation $\pm\sqrt{5}$ represents both $\sqrt{5}$ and $-\sqrt{5}$.

$$x = -3 \pm \sqrt{5}$$ Adding -3 on both sides

The solutions are $-3 + \sqrt{5}$ and $-3 - \sqrt{5}$. The check is left for the reader. ■

In Example 7, had we not considered $-\sqrt{5}$ as well as $\sqrt{5}$, one of the solutions would have eluded us. We will refer to this use of the principle of powers as the *principle of positive and negative roots.*

PRINCIPLE OF POSITIVE AND NEGATIVE ROOTS

For any nonnegative number a,

$$\text{if} \quad x^2 = a, \quad \text{then} \quad x = \pm\sqrt{a}.$$

Equations in Quadratic Form

We have already solved polynomial equations like $x^2 - 11x + 18 = 0$ and $5x^2 = 4x + 1$. These are called **quadratic equations** because they contain polynomials of second degree. Often, equations that are not really quadratic may be written in *quadratic form* after a suitable substitution is made. For example, consider this fourth-degree equation:

$$x^4 - 9x^2 + 8 = 0$$

$$\downarrow \quad \downarrow \quad \downarrow \quad \downarrow$$

$$(x^2)^2 - 9(x^2) + 8 = 0$$ Thinking of x^4 as $(x^2)^2$

$$\downarrow \quad \downarrow \quad \downarrow \quad \downarrow$$

$$u^2 - 9u + 8 = 0$$ To make this clearer, write u instead of x^2.

The equation $u^2 - 9u + 8 = 0$ can be solved by factoring. We can then find x by remembering that $x^2 = u$. Equations that may be written in quadratic form are said to be *reducible to quadratic.*

EXAMPLE 8 Solve: $x^4 - 9x^2 + 8 = 0$.

Solution Let $u = x^2$. Then we solve the equation by substituting u for x^2:

$$u^2 - 9u + 8 = 0$$

$$(u - 8)(u - 1) = 0$$ Factoring

$$u - 8 = 0 \quad \text{or} \quad u - 1 = 0$$ Principle of zero products

$$u = 8 \quad \text{or} \quad u = 1.$$

Now we substitute x^2 for u and solve these equations:

$$x^2 = 8 \quad \text{or} \quad x^2 = 1$$
$$x = \pm\sqrt{8} \quad \text{or} \quad x = \pm 1 \qquad \text{Principle of positive and negative roots}$$
$$x = \pm 2\sqrt{2} \quad \text{or} \quad x = \pm 1. \qquad \text{Simplifying}$$

To check, note that for $x = 2\sqrt{2}$ or $-2\sqrt{2}$, $x^2 = 8$ and $x^4 = 64$. Similarly, for $x = 1$ or -1, $x^2 = 1$ and $x^4 = 1$. Thus, instead of making four checks, we need make only two.

Check: For $\pm 2\sqrt{2}$:

$x^4 - 9x^2 + 8$	$= 0$
$(\pm 2\sqrt{2})^4 - 9(\pm 2\sqrt{2})^2 + 8$	0
$64 - 9 \cdot 8 + 8$	
0	TRUE

For ± 1:

$x^4 - 9x^2 + 8$	$= 0$
$(\pm 1)^4 - 9(\pm 1)^2 + 8$	0
$1 - 9 + 8$	
0	TRUE

The solutions are 1, -1, $2\sqrt{2}$, and $-2\sqrt{2}$. ■

Caution: A common error is to solve for u and then forget to solve for x. Remember that you must find values for the *original* variable!

Sometimes great care must be taken in deciding what substitution to make.

EXAMPLE 9 Solve: $y^{-2} - y^{-1} - 6 = 0$.

Solution We rewrite the equation using positive exponents:

$$\frac{1}{y^2} - \frac{1}{y} - 6 = 0.$$

Note that if we let $u = 1/y$, then $u^2 = 1/y^2$. The equation can then be written as a quadratic:

$$u^2 - u - 6 = 0$$
$$(u - 3)(u + 2) = 0$$
$$u = 3 \quad \text{or} \quad u = -2.$$

Now we substitute $1/y$ for u and solve these equations:

$$\frac{1}{y} = 3 \quad \text{or} \quad \frac{1}{y} = -2.$$

Solving, we get

$$y = \frac{1}{3} \quad \text{or} \quad y = \frac{1}{(-2)} = -\frac{1}{2}.$$

The numbers $\frac{1}{3}$ and $-\frac{1}{2}$ both check. They are the solutions. ■

EXAMPLE 10 Solve: $t^{2/5} - t^{1/5} - 2 = 0$.

Solution Let $u = t^{1/5}$. Then it follows that $u^2 = t^{2/5}$. We solve the resulting equation:

$$u^2 - u - 2 = 0$$
$$(u - 2)(u + 1) = 0$$
$$u = 2 \quad \text{or} \quad u = -1.$$

Now we substitute $t^{1/5}$ for u and solve:

$$t^{1/5} = 2 \quad \text{or} \quad t^{1/5} = -1$$
$$t = 32 \quad \text{or} \quad t = -1.$$

Principle of powers; raising to the 5th power

The solutions are 32 and -1.

EXERCISE SET 5.5

Solve.

1. $\sqrt{2x - 3} = 1$
2. $\sqrt{x + 3} = 6$
3. $\sqrt{3x} + 1 = 7$
4. $\sqrt{2x} - 1 = 7$
5. $\sqrt{y + 1} - 5 = 8$
6. $\sqrt{x - 2} - 7 = -4$
7. $\sqrt{y - 3} + 4 = 2$
8. $\sqrt{y + 4} + 6 = 7$
9. $\sqrt[3]{x + 5} = 2$
10. $\sqrt[3]{x - 2} = 3$
11. $\sqrt[4]{y - 3} = 2$
12. $\sqrt[4]{x + 3} = 3$
13. $\sqrt{y + 3} - 20 = 0$
14. $\sqrt{x + 4} - 11 = 0$
15. $\sqrt{x + 2} = -4$
16. $\sqrt{y - 3} = -2$
17. $\sqrt{2x + 3} - 5 = -2$
18. $\sqrt{3x + 1} - 4 = -1$
19. $\sqrt[3]{6x + 9} + 8 = 5$
20. $\sqrt[3]{3y + 6} + 2 = 3$
21. $\sqrt{3y + 1} = \sqrt{2y + 6}$
22. $\sqrt{5x - 3} = \sqrt{2x + 3}$
23. $2\sqrt{t - 1} = \sqrt{3t - 1}$
24. $\sqrt{y + 10} = 3\sqrt{2y + 3}$
25. $\sqrt{y - 5} + \sqrt{y} = 5$
26. $\sqrt{x - 9} + \sqrt{x} = 1$
27. $3 + \sqrt{z - 6} = \sqrt{z + 9}$
28. $\sqrt{4x - 3} = 2 + \sqrt{2x - 5}$
29. $\sqrt{x + 2} + \sqrt{3x + 4} = 2$
30. $\sqrt{6x + 7} - \sqrt{3x + 3} = 1$
31. $\sqrt{4y + 1} - \sqrt{y - 2} = 3$
32. $\sqrt{y + 15} - \sqrt{2y + 7} = 1$
33. $x^2 = 10$
34. $x^2 = 14$
35. $(x + 2)^3 = 5$
36. $(x - 2)^3 = 9$
37. $(y - 5)^2 = 6$
38. $(y + 6)^2 = 13$
39. $(x - 7)^2 + 3 = 10$
40. $(x + 5)^2 - 4 = 10$
41. $(t + 3)^4 = 12$
42. $(t - 2)^4 = 8$
43. $x^4 - 10x^2 + 25 = 0$
44. $x^4 - 3x^2 + 2 = 0$
45. $x^4 - 12x^2 + 27 = 0$
46. $x^4 - 9x^2 + 20 = 0$
47. $9x^4 - 14x^2 + 5 = 0$
48. $4x^4 - 19x^2 + 12 = 0$
49. $x^{-2} - x^{-1} - 6 = 0$
50. $4x^{-2} - x^{-1} - 5 = 0$
51. $2x^{-2} + x^{-1} - 1 = 0$
52. $m^{-2} + 9m^{-1} - 10 = 0$
53. $t^{2/3} + t^{1/3} - 6 = 0$ (*Hint:* Let $u = t^{1/3}$.)
54. $w^{2/3} - 2w^{1/3} - 8 = 0$
55. $z^{1/2} - z^{1/4} - 2 = 0$
56. $m^{1/3} - m^{1/6} - 6 = 0$
57. $x^{2/5} + x^{1/5} - 6 = 0$
58. $x^{1/2} - x^{1/4} - 6 = 0$
59. $t^{1/3} + 2t^{1/6} = 3$
60. $m^{1/2} + 6 = 5m^{1/4}$

SKILL MAINTENANCE

61. Solve:

$$\frac{3}{2x} + \frac{1}{x} = \frac{2x + 3.5}{3x}.$$

62. The base of a triangle is 2 in. longer than the height. The area is $31\frac{1}{2}$ in^2. Find the height and the base.

SYNTHESIS

An application: Sighting to the horizon. The formula $V = 1.2\sqrt{h}$ can be used to approximate the distance V, in miles, that a person can see to the horizon from a height h, in feet.

63. 🖩 How far can you see to the horizon through an airplane window at a height of 30,000 ft?

64. 🖩 How high above sea level must a sailor climb to see 10.2 mi out to sea?

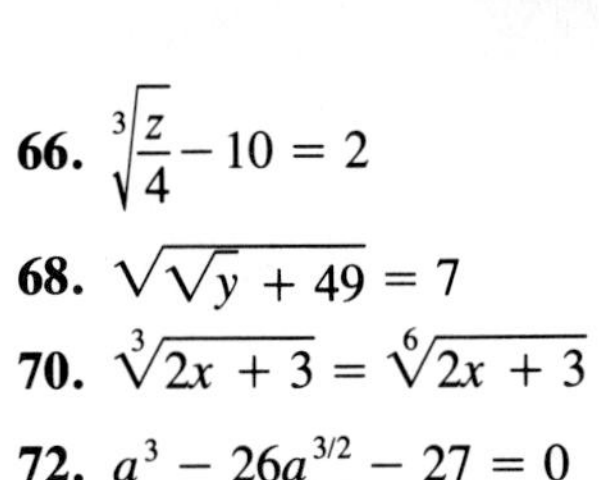

Solve.

65. $\dfrac{x + \sqrt{x+1}}{x - \sqrt{x+1}} = \dfrac{5}{11}$

66. $\sqrt[3]{\dfrac{z}{4}} - 10 = 2$

67. $\sqrt[4]{z^2 + 17} = 3$

68. $\sqrt{\sqrt{y} + 49} = 7$

69. $9x^{3/2} - 8 = x^3$

70. $\sqrt[3]{2x + 3} = \sqrt[6]{2x + 3}$

71. $\sqrt{x - 3} - \sqrt[4]{x - 3} = 2$

72. $a^3 - 26a^{3/2} - 27 = 0$

5.6 Complex Numbers

Imaginary and Complex Numbers

In the real number system, negative numbers do not have square roots. Mathematicians have invented a larger number system that contains the real number system but is such that negative numbers do have square roots. That system is called the **complex number system** and makes use of the number i.

> We define the number i so that $i^2 = -1$. Thus $i = \sqrt{-1}$.

EXAMPLE 1 Express each of the following in terms of i.

a) $\sqrt{-5} = \sqrt{-1 \cdot 5} = \sqrt{-1} \cdot \sqrt{5} = i\sqrt{5}$, or $\sqrt{5}i$ ← **i is *not* under the radical.**

b) $\sqrt{-16} = \sqrt{-1 \cdot 16} = \sqrt{-1} \cdot \sqrt{16} = i \cdot 4 = 4i$

c) $-\sqrt{-13} = -\sqrt{-1 \cdot 13} = -\sqrt{-1} \cdot \sqrt{13} = -i\sqrt{13}$, or $-\sqrt{13}i$

d) $-\sqrt{-64} = -\sqrt{-1 \cdot 64} = -\sqrt{-1} \cdot \sqrt{64} = -i \cdot 8 = -8i$

e) $\sqrt{-48} = \sqrt{-1 \cdot 48} = \sqrt{-1} \cdot \sqrt{48} = i\sqrt{48} = i \cdot 4\sqrt{3} = 4\sqrt{3}i = 4i\sqrt{3}$ ■

> An *imaginary number* is a number that can be written bi, where b is some real number and $b \neq 0$.

To form the system of complex numbers, we take the imaginary* numbers and the real numbers, as well as all possible sums of real and imaginary numbers. These are complex numbers:

$$7 - 4i, \qquad -\pi + 19i, \qquad 37, \qquad i\sqrt{8}.$$

> A *complex number* is any number that can be written $a + bi$, where a and b are any real numbers. (Note that a and/or b can be 0.)

Note that because complex numbers include numbers of the form $0 + bi = bi$, every imaginary number is considered a complex number. Similarly, since every number of the form $a + 0i = a$ is complex, we see that every real number is considered a complex number. Complex numbers of the form $a + bi$ with $a, b \neq 0$ are neither imaginary nor real.

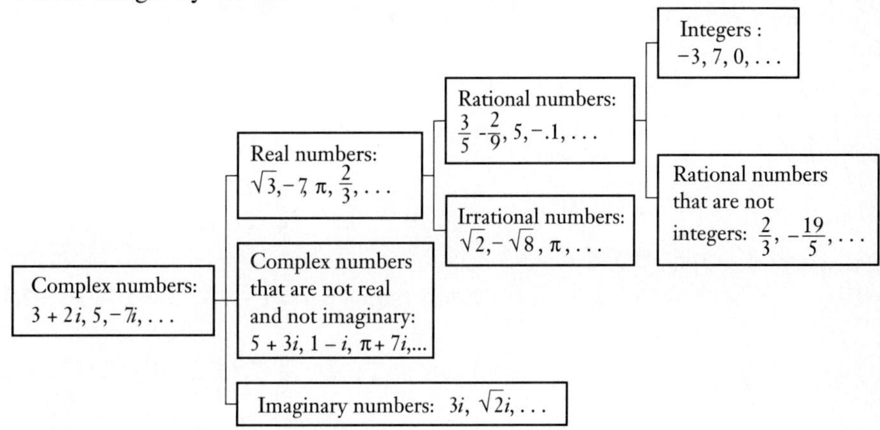

Addition and Subtraction

Complex numbers follow the commutative and associative laws of addition. Thus we can add and subtract them as we do binomials in real numbers.

EXAMPLES Add or subtract.

2. $(8 + 6i) + (3 + 2i) = (8 + 3) + (6 + 2)i$ Collecting the real parts and the imaginary parts

$= 11 + 8i$

3. $(3 + 2i) - (5 - 2i) = (3 - 5) + [2 - (-2)]i$ Note that the 5 and the $-2i$ are both being subtracted

$= -2 + 4i$ ■

*Don't let the name *imaginary* fool you. The imaginary numbers are very important in such fields as engineering and the physical sciences.

Multiplying

Complex numbers also obey the commutative and associative laws of multiplication. However, the property $\sqrt{a}\sqrt{b} = \sqrt{ab}$ does *not* hold for imaginary numbers. That is, all square roots of negatives must be expressed in terms of i before we multiply. For example,

$$\sqrt{-2}\cdot\sqrt{-5} = i\sqrt{2}\cdot i\sqrt{5} = i^2\sqrt{10} = -\sqrt{10} \text{ is correct,}$$

but

$$\sqrt{-2}\cdot\sqrt{-5} = \sqrt{(-2)(-5)} = \sqrt{10} \text{ is wrong!}$$

Keeping this in mind, we multiply in much the same way as we did with polynomials.

EXAMPLES Multiply.

4. $\sqrt{-16}\sqrt{-25} = \sqrt{-1}\cdot\sqrt{16}\cdot\sqrt{-1}\cdot\sqrt{25}$
$= i\cdot 4\cdot i\cdot 5$
$= i^2\cdot 20$
$= -1\cdot 20$ $\quad i^2 = -1$
$= -20$

5. $\sqrt{-5}\sqrt{-7} = \sqrt{-1}\cdot\sqrt{5}\cdot\sqrt{-1}\cdot\sqrt{7}$
$= i\cdot\sqrt{5}\cdot i\cdot\sqrt{7}$
$= i^2\cdot\sqrt{35}$
$= -1\cdot\sqrt{35}$
$= -\sqrt{35}$

6. $-3i\cdot 8i = -24\cdot i^2$
$= -24\cdot(-1)$
$= 24$

7. $(-4i)(3-5i) = (-4i)\cdot 3 + (-4i)(-5i)$ $\quad$ Using the distributive law
$= -12i + 20i^2$
$= -12i - 20$ $\quad i^2 = -1$
$= -20 - 12i$ $\quad$ Writing in the form $a + bi$

8. $(1 + 2i)(1 + 3i) = 1 + 3i + 2i + 6i^2$ $\quad$ Multiplying each term of one number by every term of the other (FOIL)
$= 1 + 3i + 2i - 6$ $\quad i^2 = -1$
$= -5 + 5i$ $\quad$ Collecting like terms ■

Powers of *i*

We now want to simplify certain expressions involving higher powers of i. To do so, we recall that -1 raised to an *even* power is 1, and -1 raised to an *odd* power is -1. Simplifying powers of i can then be done by using the fact that $i^2 = -1$ and expressing the given power of i in terms of i^2. Consider the following:

i, or $\sqrt{-1}$,
$i^2 = -1$,
$i^3 = i^2\cdot i = (-1)i = -i$,

$$i^4 = (i^2)^2 = (-1)^2 = 1,$$
$$i^5 = i^4 \cdot i = (i^2)^2 \cdot i = (-1)^2 \cdot i = i, \text{ or } \sqrt{-1},$$
$$i^6 = (i^2)^3 = (-1)^3 = -1.$$

Note that the powers of i cycle themselves through the values i, -1, $-i$, and 1.

EXAMPLES Simplify.

9. $i^{37} = i^{36} \cdot i = (i^2)^{18} \cdot i = (-1)^{18} \cdot i = 1 \cdot i = i$
10. $i^{58} = (i^2)^{29} = (-1)^{29} = -1$
11. $i^{75} = i^{74} \cdot i = (i^2)^{37} \cdot i = (-1)^{37} \cdot i = -i$
12. $i^{80} = (i^2)^{40} = (-1)^{40} = 1$ ■

Now let us simplify other expressions.

EXAMPLES Simplify to the form $a + bi$.

13. $8 - i^2 = 8 - (-1) = 8 + 1 = 9$
14. $17 + 6i^3 = 17 + 6 \cdot i^2 \cdot i = 17 + 6(-1)i = 17 - 6i$
15. $i^{22} - 67i^2 = (i^2)^{11} + 67 = (-1)^{11} + 67 = -1 + 67 = 66$
16. $i^{23} + i^{48} = (i^{22}) \cdot i + (i^2)^{24} = (i^2)^{11} \cdot i + (-1)^{24} = (-1)^{11} \cdot i + (-1)^{24}$
$= -i + 1 = 1 - i$ ■

Complex Conjugates and Division

Consider the following multiplication:

$$\begin{aligned}(3 - 2i)(3 + 2i) &= 9 + 6i - 6i - 4i^2 \\ &= 9 - 4(-1) = 13.\end{aligned}$$

Note that the imaginary terms $6i$ and $-6i$ added to 0, so our answer was a real number. This will happen any time numbers of the form $a + bi$ and $a - bi$ are multiplied. Pairs of numbers like $3 - 2i$ and $3 + 2i$ are known as **complex conjugates,** or simply, conjugates.

> The *complex conjugate* of a complex number $a + bi$ is $a - bi$.
> The *complex conjugate* of a complex number $a - bi$ is $a + bi$.

EXAMPLES Find the conjugate.

17. $5 + 7i$ — The conjugate is $5 - 7i$.
18. $14 - 3i$ — The conjugate is $14 + 3i$.
19. $-3 - 9i$ — The conjugate is $-3 + 9i$.
20. $4i$ — The conjugate is $-4i$. ■

EXAMPLES Multiply.

21. $(5 + 7i)(5 - 7i) = 5^2 - (7i)^2$
$= 25 - 49i^2$
$= 25 + 49 \qquad i^2 = -1$
$= 74$

22. $(2 - 3i)(2 + 3i) = 2^2 - (3i)^2$
$= 4 - 9i^2$
$= 4 + 9$
$= 13$ ■

We use conjugates in dividing complex numbers.

EXAMPLE 23 Divide:

$$\frac{-5 + 9i}{1 - 2i}.$$

Solution

$$\frac{-5 + 9i}{1 - 2i} \cdot \frac{1 + 2i}{1 + 2i} = \frac{(-5 + 9i)(1 + 2i)}{(1 - 2i)(1 + 2i)}$$ **Multiplying by 1**

$$= \frac{-23 - i}{5}$$ **Performing the multiplication. The student should check this.**

$$= -\frac{23}{5} - \frac{1}{5}i$$ **Writing in the form $a + bi$** ■

Note the similarity between this example and rationalizing denominators. The symbol for the number 1 was formed using the conjugate of the divisor.

EXAMPLE 24 Divide:

$$\frac{3 + 5i}{4 + 3i}.$$

Solution The divisor, $4 + 3i$, tells us to write 1 as $(4 - 3i)/(4 - 3i)$:

$$\frac{3 + 5i}{4 + 3i} \cdot \frac{4 - 3i}{4 - 3i} = \frac{(3 + 5i)(4 - 3i)}{(4 + 3i)(4 - 3i)}$$ **Multiplying by 1**

$$= \frac{27 + 11i}{25}$$

$$= \frac{27}{25} + \frac{11}{25}i.$$ ■

EXERCISE SET 5.6

Express in terms of i.

1. $\sqrt{-15}$
2. $\sqrt{-17}$
3. $\sqrt{-16}$
4. $\sqrt{-25}$
5. $-\sqrt{-36}$
6. $-\sqrt{-49}$
7. $-\sqrt{-12}$
8. $-\sqrt{-20}$
9. $\sqrt{-250}$
10. $\sqrt{-180}$

Add or subtract and simplify.

11. $(3 + 2i) + (5 - i)$
12. $(-2 + 3i) + (7 + 8i)$
13. $(4 - 3i) + (5 - 2i)$
14. $(-2 - 5i) + (1 - 3i)$
15. $(9 - i) + (-2 + 5i)$
16. $(6 + 4i) + (2 - 3i)$
17. $(3 - i) - (5 + 2i)$
18. $(-2 + 8i) - (7 + 3i)$
19. $(4 - 2i) - (5 - 3i)$
20. $(-2 - 3i) - (1 - 5i)$
21. $(9 + 5i) - (-2 - i)$
22. $(6 - 3i) - (2 + 4i)$
23. $(-5 - 2i) + (-7 - 4i)$
24. $(-3 - 7i) + (-4 - 5i)$
25. $(2 + 3i) + (1 + 2i) + (4 + i)$
26. $(3 + 2i) + (4 + i) + (5 + 3i)$
27. $(5 - 2i) + (3 + 4i) - (2 + 7i)$
28. $(3 + 5i) - (7 + 9i) + (4 - 3i)$
29. $(5 - 9i) - (3 - 4i) - (-2 + i)$
30. $(9 - 7i) - (5 - 6i) - (-3 + 4i)$

Multiply. Write the answer in the form $a + bi$.

31. $\sqrt{-25}\sqrt{-36}$
32. $\sqrt{-81}\sqrt{-49}$
33. $\sqrt{-6}\sqrt{-5}$
34. $\sqrt{-7}\sqrt{-10}$
35. $\sqrt{-50}\sqrt{-3}$
36. $\sqrt{-72}\sqrt{-3}$
37. $\sqrt{-48}\sqrt{-6}$
38. $\sqrt{-15}\sqrt{-75}$
39. $5i \cdot 8i$
40. $6i \cdot 9i$
41. $5i \cdot (-7i)$
42. $7i \cdot (-4i)$
43. $5i(3 - 2i)$
44. $4i(5 - 6i)$
45. $-3i(7 - 4i)$
46. $-7i(9 - 3i)$
47. $(3 + 2i)(1 + i)$
48. $(4 + 3i)(2 + 5i)$
49. $(2 + 3i)(6 - 2i)$
50. $(5 + 6i)(2 - i)$
51. $(6 - 5i)(3 + 4i)$
52. $(5 - 6i)(2 + 5i)$
53. $(7 - 2i)(2 - 6i)$
54. $(-4 + 5i)(3 - 4i)$
55. $(5 - 3i)(4 - 5i)$
56. $(7 - 3i)(4 - 7i)$
57. $(-2 + 3i)(-2 + 5i)$
58. $(-3 + 6i)(-3 + 4i)$
59. $(-5 - 4i)(3 + 7i)$
60. $(2 + 9i)(-3 - 5i)$
61. $(3 - 2i)^2$
62. $(5 - 2i)^2$
63. $(2 + 3i)^2$
64. $(4 + 2i)^2$
65. $(-2 + 3i)^2$
66. $(-5 - 2i)^2$

Simplify.

67. i^7
68. i^9
69. i^{40}
70. i^{42}
71. i^{53}
72. i^{32}
73. i^{62}
74. i^{83}
75. $5 - i^{22}$
76. $9 + i^{18}$
77. $9i^2 + 23i^{32}$
78. $3i^{60} - 15i^6$
79. $i^{35} - i^{48}$
80. $i^{50} - i^{45}$
81. $-5i^{17} + i^{25}$
82. $-7i^{29} + i^{33}$

Multiply.

83. $(3 + 7i)(3 - 7i)$
84. $(5 - 2i)(5 + 2i)$
85. $(8 - 5i)(8 + 5i)$
86. $(2 + 5i)(2 - 5i)$
87. $(-3 + 4i)(-3 - 4i)$
88. $(-5 - 3i)(-5 + 3i)$

Divide.

89. $\dfrac{5}{3 - i}$
90. $\dfrac{3}{5 + i}$
91. $\dfrac{2i}{7 + 3i}$
92. $\dfrac{4i}{2 - 5i}$
93. $\dfrac{7}{6i}$
94. $\dfrac{3}{10i}$
95. $\dfrac{8 - 3i}{7i}$
96. $\dfrac{3 + 8i}{5i}$

97. $\frac{3+2i}{2+i}$

98. $\frac{4+5i}{5-i}$

99. $\frac{5-2i}{2+5i}$

100. $\frac{3-2i}{4+3i}$

101. $\frac{3-5i}{3-2i}$

102. $\frac{2-7i}{5-4i}$

SKILL MAINTENANCE

Solve.

103. $\frac{196}{x^2-7x+49} - \frac{2x}{x+7} = \frac{2058}{x^3+343}$

104. $\frac{5}{t} - \frac{3}{2} = \frac{4}{7}$

105. Simplify: $(x^2+2x+1)(3x-1)$.

SYNTHESIS

106. Evaluate $\frac{z^4-z^2}{z-1}$ when $z = 2i - 1$.

107. Evaluate $\frac{1}{w-w^2}$ when $w = \frac{1-i}{10}$.

Express in terms of i.

108. $\frac{1}{8}\left(-24 - \sqrt{-1024}\right)$

109. $12\sqrt{-\frac{1}{32}}$

110. $7\sqrt{-64} - 9\sqrt{-256}$

Simplify.

111. $\frac{i^5+i^6+i^7+i^8}{(1-i)^4}$

112. $(1-i)^3(1+i)^3$

113. $\frac{5-\sqrt{5}i}{\sqrt{5}i}$

114. $\frac{6}{1+\frac{3}{i}}$

115. $\left(\frac{1}{2}-\frac{1}{3}i\right)^2 - \left(\frac{1}{2}+\frac{1}{3}i\right)^2$

116. $\frac{i-i^{38}}{1+i}$

CHAPTER SUMMARY AND REVIEW 5

TERMS TO KNOW

Square root
Principal square root
Radical sign
Radical expression
Radicand
Cube root
Index
kth root
Odd root
Even root
Like radicals
Rationalize the denominator
Conjugate
Radical equation
Quadratic equation
Reducible to quadratic
Imaginary number
Complex number
Complex conjugate

IMPORTANT PROPERTIES AND FORMULAS

The number c is a square root of a if $c^2 = a$.
The number c is a cube root of a if $c^3 = a$.

For any real number a:

a) $\sqrt[k]{a^k} = |a|$ when k is even. We use absolute value when k is even unless a is known to be nonnegative.
b) $\sqrt[k]{a^k} = a$ when k is odd. We do not use absolute value when k is odd.

For any real numbers $\sqrt[k]{a}$ and $\sqrt[k]{b}$,

$$\sqrt[k]{a} \cdot \sqrt[k]{b} = \sqrt[k]{a \cdot b}.$$

(To multiply, we multiply the radicands.)

For any real numbers $\sqrt[k]{a}$ and $\sqrt[k]{b}$, $b \neq 0$,

$$\sqrt[k]{\frac{a}{b}} = \frac{\sqrt[k]{a}}{\sqrt[k]{b}}.$$

(We can take the kth roots of the numerator and of the denominator separately.)

For any real number a, $a \neq 0$,

$$a^0 = 1.$$

(Any nonzero number raised to the zero power is 1.)

For any real number a that is nonzero and any integer n,

$$a^{-n} = \frac{1}{a^n}.$$

(The numbers a^{-n} and a^n are reciprocals.)

For any nonnegative number a and any index n,

$$a^{1/n} \quad \text{means} \quad \sqrt[n]{a}.$$

If a is negative, then n must be odd.

For any natural numbers m and n ($n \neq 1$), and any real number a for which $\sqrt[n]{a}$ exists,

$$a^{m/n} \quad \text{means} \quad \left(\sqrt[n]{a}\right)^m \quad \text{or } \sqrt[n]{a^m}.$$

For any real numbers a and b and any rational exponents m and n for which a^m, a^n, and b^m are defined:

1. $a^m \cdot a^n = a^{m+n}$ In multiplying, we can add exponents if the bases are the same.
2. $\dfrac{a^m}{a^n} = a^{m-n}$ In dividing, we can subtract exponents if the bases are the same. (Assume $a \neq 0$.)
3. $(a^m)^n = a^{m \cdot n}$ To raise a power to a power, we can multiply the exponents.
4. $(ab)^m = a^m b^m$ To raise a product to a power, we can raise each factor to the power and multiply.

SOME WAYS TO SIMPLIFY RADICAL EXPRESSIONS

1. *Simplifying by factoring.* We factor the radicand, looking for factors raised to powers that are multiples of the index.

   Example: $\sqrt[3]{16} = \sqrt[3]{8}\sqrt[3]{2} = 2\sqrt[3]{2}$

2. *Rationalizing denominators.* Radical expressions are often considered simpler if there are no radicals in the denominator.

   Example: $\dfrac{1}{\sqrt{2}} = \dfrac{1}{\sqrt{2}} \cdot \dfrac{\sqrt{2}}{\sqrt{2}} = \dfrac{\sqrt{2}}{2}$

3. *Collecting like radical terms.*

   Example: $\sqrt{8} + 3\sqrt{2} = \sqrt{4} \cdot \sqrt{2} + 3\sqrt{2} = 2\sqrt{2} + 3\sqrt{2} = 5\sqrt{2}$

4. *Using fractional exponents to simplify.* We convert to exponential notation and then use arithmetic and the laws of exponents to simplify the exponents. Then we convert back to radical notation.

   Example: $\sqrt[3]{p} \cdot \sqrt[4]{q^3} = p^{1/3} \cdot q^{3/4} = p^{4/12} \cdot q^{9/12}$
   $= \sqrt[12]{p^4 q^9}$

THE PRINCIPLE OF POWERS

If an equation $a = b$ is true, then $a^n = b^n$ is true for any rational number n for which a^n and b^n exist.

PRINCIPLE OF POSITIVE AND NEGATIVE ROOTS

For any nonnegative number a, if $x^2 = a$, then $x = \pm\sqrt{a}$.

A complex number is any number that can be written $a + bi$, where a and b are any real numbers and $i = \sqrt{-1}$.

REVIEW EXERCISES

The review sections to be tested in addition to the material in this chapter are Sections 3.2, 3.5, 4.1, and 4.4.

Simplify.

1. $-\sqrt{\frac{36}{81}}$

2. $\sqrt{0.0049}$

Simplify. Assume that letters can represent *any* real number.

3. $\sqrt{81a^2}$

4. $\sqrt{(c+8)^2}$

5. $\sqrt{x^2-6x+9}$

6. $\sqrt{4x^2+4x+1}$

7. $\sqrt[5]{-32}$

8. $\sqrt[3]{-\frac{1}{27}}$

9. $\sqrt[10]{x^{10}}$

10. $-\sqrt[13]{(-3)^{13}}$

11. Determine whether -2 and 6 are meaningful replacements in $\sqrt{4-2x}$.

For Exercises 12–17 assume that all variables are nonnegative.
Multiply and simplify by factoring.

12. $\sqrt{3x^2}\sqrt{6y^3}$

13. $\sqrt[3]{a^5b}\sqrt[3]{27b}$

Divide. Then simplify by taking roots, if possible.

14. $\dfrac{\sqrt[3]{60xy^3}}{\sqrt[3]{10x}}$

15. $\dfrac{\sqrt{75x}}{2\sqrt{3}}$

Simplify.

16. $\left(\sqrt{8xy^2}\right)^2$

17. $\left(\sqrt[3]{4a^2b}\right)^2$

Perform the indicated operation. Simplify by collecting like radical terms, if possible.

18. $12\sqrt[3]{135}-3\sqrt[3]{40}$

19. $\sqrt{50}+2\sqrt{18}+\sqrt{32}$

20. $\left(\sqrt[3]{27}-\sqrt[3]{2}\right)\left(\sqrt[3]{27}+\sqrt[3]{2}\right)$

21. $\left(\sqrt{5}-3\sqrt{8}\right)\left(\sqrt{5}+2\sqrt{8}\right)$

22. $\left(1-\sqrt{7}\right)^2$

23. Rationalize the denominator. Assume that a and b represent positive numbers.

$$\frac{5\sqrt{12a}}{\sqrt{a}+\sqrt{b}}$$

24. Rationalize the numerator of the expression in Exercise 23.

Simplify. When negative exponents appear in the answer, write a second answer using only positive exponents.

25. $(-3)^0$

26. $\dfrac{-8x^3}{6x^{-2}}$

27. $(2x^{-3})^{-2}$

28. Rewrite without negative exponents: $-2x^{-3}y^2$.

29. Rewrite with negative exponents: $\dfrac{1}{5x^6}$.

30. Rewrite with fractional exponents: $\left(\sqrt[5]{8x^6y^2}\right)^4$.

31. Rewrite without fractional exponents: $(5a)^{3/4}$.

Use fractional exponents to write a single radical expression.

32. $x^{1/3}\cdot y^{1/4}$

33. $\sqrt[4]{x}\sqrt[3]{x-3}$

Solve.

34. $\sqrt{3x-3}=1+\sqrt{x}$

35. $\sqrt[4]{x+3}=2$

36. $(x-3)^2=19$

37. $x^{-2}+5x^{-1}+6=0$

38. Express in terms of i and simplify: $-\sqrt{-8}$.

39. Add: $(-4 + 3i) + (2 - 12i)$.

40. Subtract: $(4 - 7i) - (3 - 8i)$.

Multiply.

41. $(2 + 5i)(2 - 5i)$

42. i^{13}

43. $(6 - 3i)(2 - i)$

Divide.

44. $\dfrac{-3 + 2i}{5i}$

45. $\dfrac{6 - 3i}{2 + i}$

SKILL MAINTENANCE

46. Multiply: $(3m + 2n)(3m - 2n)$.

47. Solve:
$$\frac{7}{x + 2} + \frac{5}{x^2 - 2x + 4} = \frac{84}{x^3 + 8}.$$

48. Solve: $2x^2 + 3x - 27 = 0$.

49. Multiply and simplify:
$$\frac{x^2 + 3x}{x^2 - y^2} \cdot \frac{x^2 - xy + 2x - 2y}{x^2 - 9}.$$

SYNTHESIS

50. Solve: $\sqrt{11x + \sqrt{6 + x}} = 6$.

51. Simplify: $\dfrac{2}{1 - 3i} - \dfrac{3}{4 + 2i}$.

THINKING IT THROUGH

1. We learned to solve a new kind of equation in this chapter. Explain how the procedure for solving this kind of equation differs from others we have solved.

2. Explain why $\sqrt{x^2} = |x|$, when x is considered to be an arbitrary real number.

3. Explain the difference between a complex number and a real number. Give two examples of complex numbers that are not real numbers.

4. Compare the procedure of rationalizing a denominator with that of dividing complex numbers. Explain why the procedures are similar.

CHAPTER TEST 5

In Questions 1–5, assume that letters can represent *any* real number. Simplify.

1. $\sqrt{\dfrac{100}{49}}$

2. $\sqrt{36y^2}$

3. $\sqrt{x^2 + 10x + 25}$

4. $\sqrt[3]{-8}$

5. $\sqrt[10]{(-4)^{10}}$

6. Determine whether 0 and 3 are meaningful replacements in $\sqrt{2x - 1}$.

7. Multiply and simplify by factoring:
$$\sqrt[3]{x^4}\sqrt[3]{8x^5}.$$

8. Divide. Then simplify by taking roots, if possible. (Assume $x, y > 0$.)
$$\sqrt{\frac{20x^3y}{4y}}$$

9. Simplify: $\left(\sqrt[3]{16a^2b}\right)^2$.

10. Add. Then simplify by collecting like terms.

$$3\sqrt{128} + 2\sqrt{18} + 2\sqrt{32}$$

11. Multiply and simplify:

$$(\sqrt{20} + 2\sqrt{5})(\sqrt{20} - 3\sqrt{5}).$$

12. Rationalize the denominator:

$$\frac{1 + \sqrt{2}}{3 - 5\sqrt{2}}.$$

13. Rewrite with fractional exponents:

$$(\sqrt{5xy^2})^5.$$

14. Simplify. Use only positive exponents in the answer.

$$(-2x^{-5})(8x^3)$$

15. Use fractional exponents to write a single radical expression: $\sqrt[4]{2y}\sqrt{x - 3}$.

16. Solve: $\sqrt{y - 6} = \sqrt{y + 9} - 3$.

17. Solve: $t^{2/3} + 6t^{1/3} - 16 = 0$.

18. Express in terms of i and simplify: $\sqrt{-18}$.

19. Subtract: $(5 + 8i) - (-2 + 3i)$.

20. Multiply: $(1 - i)^2$.

21. Divide: $\dfrac{-7 + 14i}{6 - 8i}$.

22. Simplify: $5i^{25} - i^{12}$.

SKILL MAINTENANCE

23. Solve: $6x^2 = 13x + 5$.

24. Divide and simplify:

$$\frac{x^3 - 27}{x^2 - 16} \div \frac{x^2 + 3x + 9}{x + 4}.$$

25. Solve:

$$\frac{11x}{x + 3} + \frac{33}{x} + 12 = \frac{99}{x^2 + 3x}.$$

26. Multiply: $(x^2 + 3)(x^2 + 3)$.

SYNTHESIS

27. Solve:

$$\sqrt{2x - 2} + \sqrt{7x + 4} = \sqrt{13x + 10}.$$

28. Simplify:

$$\frac{1 - 4i}{4i(1 - 4i)^{-1}}.$$

CHAPTER 6

Graphs, Functions, and Linear Equations

FEATURE PROBLEM

A person buys a video cassette recorder on which there is a revolution counter. Estimate the time elapsed when the counter has reached 600.

THE MATHEMATICS

Using a booklet that lists the counter readings for 0, 1, 2, 3, and 4 hours, we can make a graph and then approximate how much time has elapsed when the counter reaches 600.

Graphs are useful because they allow us to see relationships pictorially. For example, a graph of an equation in two variables helps us to see how those two variables are related. Graphs are useful in problem solving and in other ways as well. In this chapter you will take a good look at graphs of equations.

A certain kind of relationship between two variables is known as a *function*. Functions are very important in mathematics in general, and in problem solving in particular. You will learn what we mean by a function and then begin to use functions to solve problems.

The review sections to be tested in addition to the material in this chapter are Sections 2.1, 3.4, 4.2, and 5.3.

6.1 Graphs

It has often been said that a picture is worth a thousand words. As we turn our attention to the study of graphs, we discover that in mathematics this is quite literally the case. Our study of graphs will provide us with a pictorial means of solving problems.

Points and Ordered Pairs

On a number line each point is the graph of a number. On a plane each point is the graph of a pair of numbers. The idea of using two perpendicular number lines, called **axes,** to identify points in a plane is commonly attributed to the great French mathematician and philosopher René Descartes (1596–1650). Because the variables x and y are often used on the horizontal and vertical axes, respectively, we frequently refer to the **x-, y-coordinate system.** In honor of Descartes, this representation is also called the **Cartesian coordinate system.**

Plotting Points. Note that on the following figure (2, 3) and (3, 2) give different points. These are called **ordered pairs** of numbers because it makes a difference which number comes first. The ordered pair (0, 0) is called the **origin.**

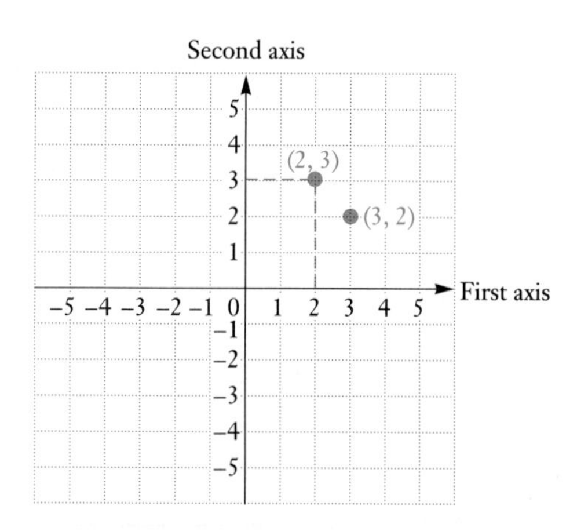

EXAMPLE 1 Plot the point $(-4, 3)$.

Solution The first number, -4, tells us the distance in the first, or horizontal, direction. We move 4 units *left.* The second number tells us the distance in the second, or vertical, direction. We move 3 units *up.*

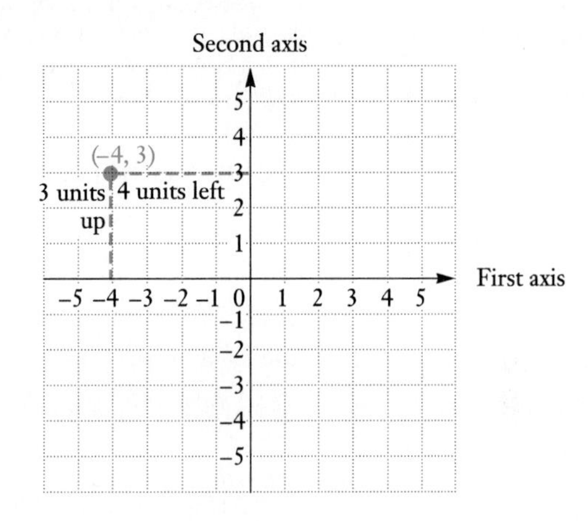

EXAMPLE 2 Plot the points $(0, -3)$ and $(2.5, 0)$.

Solution To graph the point $(0, -3)$, we move 0 units in the horizontal direction and 3 units *down*. To graph $(2.5, 0)$ we move 2.5 units to the *right* and 0 units in the vertical direction.

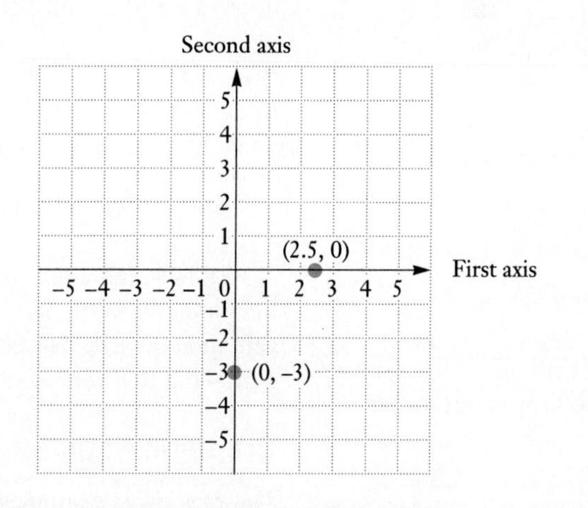

The numbers in an ordered pair are called **coordinates.** In $(-4, 3)$ the **first coordinate** is -4 and the **second coordinate*** is 3.

*The first coordinate is sometimes called the *abscissa* and the second coordinate is called the *ordinate*.

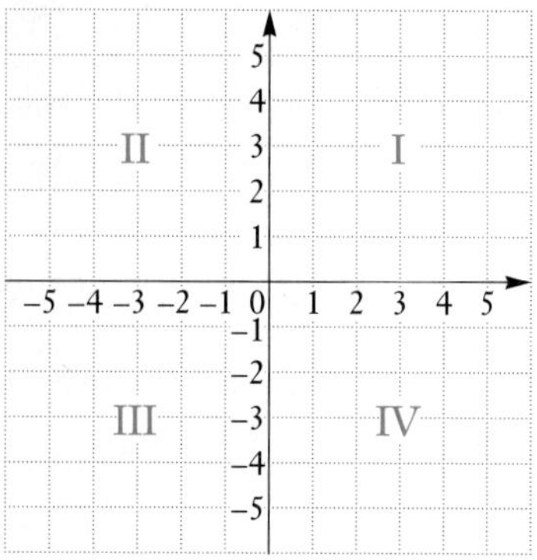

Quadrants. The axes divide the plane into four regions, as shown, called *quadrants*. In region I (the *first* quadrant), both coordinates of a point are positive. In region II (the *second* quadrant), the first coordinate is negative and the second coordinate is positive. In region III (the *third* quadrant), both coordinates are negative. Finally, in region IV (the fourth quadrant), the first coordinate is positive and the second coordinate is negative.

Note that the axes themselves serve as boundary lines and do not belong to any quadrant.

Solutions of Equations

If an equation has two variables, its solutions are pairs of numbers. We usually take the variables in alphabetical order; hence we get ordered pairs of numbers for solutions.

EXAMPLE 3 Determine whether the pairs (4, 2), $(-1, -4)$, and (2, 5) are solutions of the equation $y = 3x - 1$.

Solution

$$\begin{array}{c|l} y = & 3x - 1 \\ \hline 2 & 3 \cdot 4 - 1 \\ & 12 - 1 \\ & 11 \end{array}$$

We substitute 4 for x and 2 for y (alphabetical order of variables).

FALSE

Because $2 \neq 11$, the pair (4, 2) is *not* a solution.

$$\begin{array}{c|l} y = & 3x - 1 \\ \hline -4 & 3(-1) - 1 \\ & -3 - 1 \\ & -4 \end{array}$$

TRUE

Because $-4 = -4$, the pair $(-1, -4)$ *is* a solution.

The reader can confirm that the pair (2, 5) is also a solution. ■

In fact, there are infinitely many solutions to the equation $y = 3x - 1$. Rather than attempt to list all of these solutions, we will use a graph as a convenient representation of such a large set of solutions. Thus to *graph* an equation means to make a drawing that represents its solutions.

Graphing Equations of the Type $y = mx$

The next three examples are equations of the form $y = mx$, where m is some fixed number. We will find that an equation of this type has a graph that is a straight line.

EXAMPLE 4 Graph the equation $y = x$.

Solution We will use alphabetical order. Thus the first axis is the x-axis and the second axis is the y-axis.

Next, we find some ordered pairs that are solutions of the equation. In this case it is easy. Here are a few pairs that satisfy the equation $y = x$:

$$(0, 0), \quad (1, 1), \quad (5, 5), \quad (-1, -1), \quad (-6, -6).$$

Now we plot these points. We can see that if we were to plot a million solutions, the dots that we drew would merge into a solid line. We see the pattern, so we can draw the line with a ruler. The line is the graph of the equation $y = x$. We label the line $y = x$ on the graph paper.

Note that the coordinates of *any* point on the line—for example, (2.5, 2.5)—will satisfy the equation $y = x$.

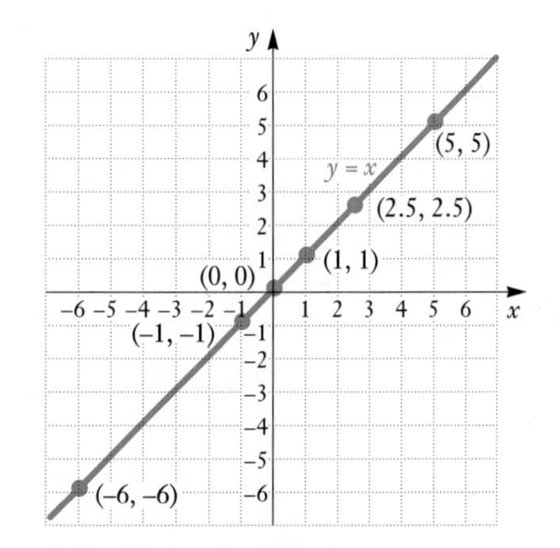

EXAMPLE 5 Graph the equation $y = 2x$.

Solution We find some ordered pairs that are solutions. This time we will list these pairs in a table. To find an ordered pair, we can choose *any* number for x and then determine y. For example, if we choose 3 for x, then $y = 2 \cdot 3$ (substituting into the equation of the line), or 6. We make some negative choices for x, as well as some positive ones. If a number takes us off the graph paper, we generally do not use it. Next, we plot these points. If we had enough of them, they would make a solid line. We can draw the line with a ruler, and we label it $y = 2x$.

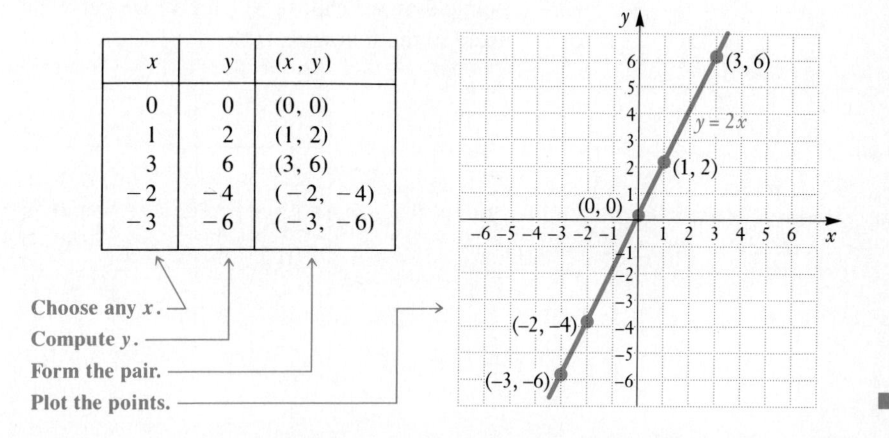

x	y	(x, y)
0	0	(0, 0)
1	2	(1, 2)
3	6	(3, 6)
−2	−4	(−2, −4)
−3	−6	(−3, −6)

EXAMPLE 6 Graph the equation $y = -\frac{1}{2}x$.

Solution To find an ordered pair, we choose any convenient number for x and then determine y. For example, if we choose 4 for x, we get $y = \left(-\frac{1}{2}\right)(4)$, or -2. When we choose -6 for x, we get $y = \left(-\frac{1}{2}\right)(-6)$, or 3. We find several ordered pairs, plot them, and draw the line.

x	y	(x, y)
4	-2	$(4, -2)$
-6	3	$(-6, 3)$
0	0	$(0, 0)$
2	-1	$(2, -1)$

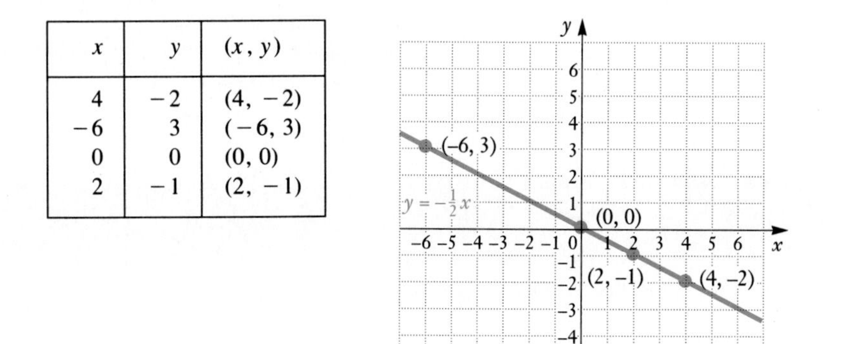

As you can see, the graphs in Examples 4–6 are straight lines. We will refer to any equation whose graph is a straight line as a **linear equation.** Linear equations will be discussed in more detail in Sections 6.3–6.5.

Nonlinear Equations

There are many equations whose graphs are not straight lines. Let's look at some of these **nonlinear equations**.

EXAMPLE 7 Graph: $y = |x|$.

Solution We select numbers for x and find the corresponding values for y. For example, if we choose -1 for x, we get $y = |-1| = 1$. Several ordered pairs are listed in the following table.

x	y
-3	3
-2	2
-1	1
0	0
1	1
2	2
3	3

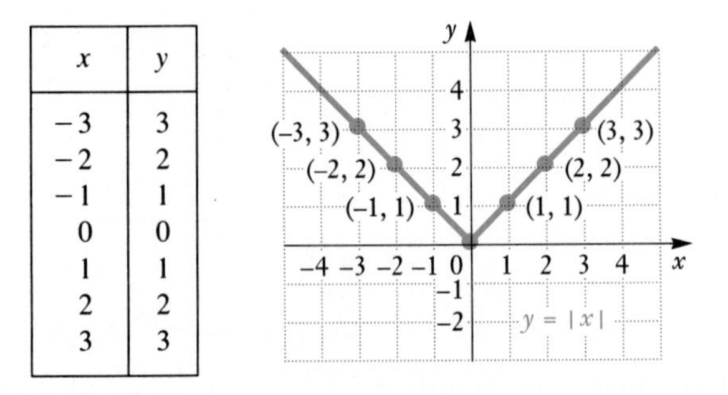

We plot these points, noting that the absolute value of a positive number is the same as the absolute value of its opposite. Thus the x-values 3 and -3 both are paired with the y-value 3. We see that the graph is V-shaped, centered at the origin. ■

EXAMPLE 8 Graph: $y = \frac{1}{x}$.

Solution We select x-values and find the corresponding y-values. The table lists the ordered pairs $\left(2, \frac{1}{2}\right)$, $\left(-2, -\frac{1}{2}\right)$, $\left(\frac{1}{2}, 2\right)$, and so on.

x	y
3	$\frac{1}{3}$
2	$\frac{1}{2}$
1	1
$\frac{1}{2}$	2
$-\frac{1}{2}$	-2
-1	-1
-2	$-\frac{1}{2}$
-3	$-\frac{1}{3}$

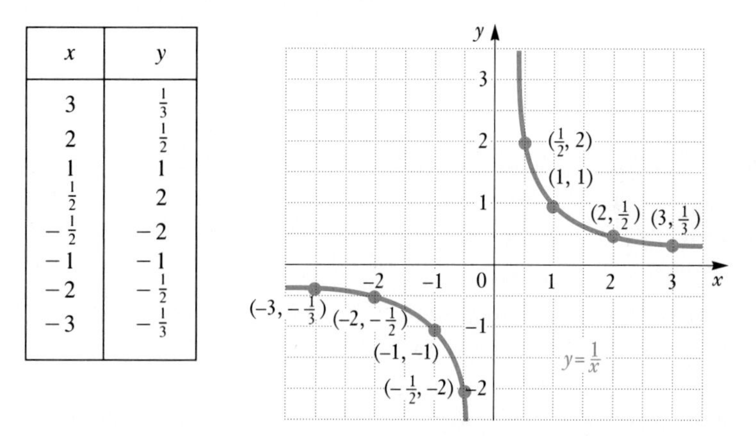

We plot these points, noting that each first coordinate is paired with its reciprocal. Because 1/0 is undefined, we cannot use 0 as a first coordinate. Thus this graph has two "branches"—one on either side of the y-axis. Note that for x-values far to the right or far to the left of 0, the graph will approach, but not touch, the x-axis. ■

EXAMPLE 9 Graph: $y = x^2 - 5$.

Solution We select numbers for x and find the corresponding values for y. For example, if we choose -2 for x, we get $y = (-2)^2 - 5 = 4 - 5 = -1$. The table lists several ordered pairs.

x	y
0	-5
-1	-4
1	-4
-2	-1
2	-1
-3	4
3	4

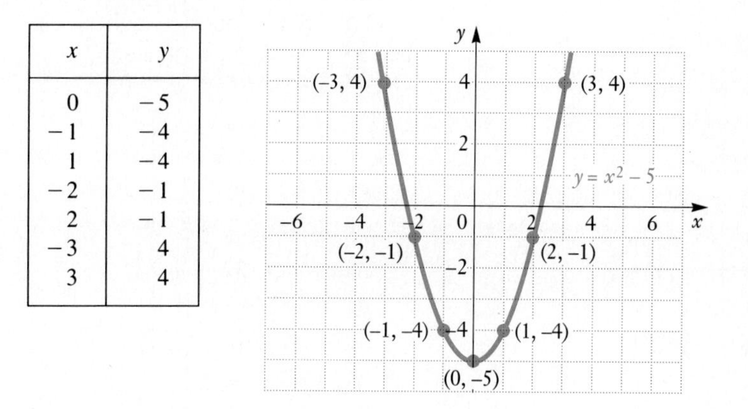

Next, we plot these points. We note that as the absolute value of x increases, $x^2 - 5$ also increases. Thus the graph is a curve that rises on either side of the y-axis, as shown in the figure. ■

You can always use a calculator to find as many values as desired. This can be especially helpful when you are uncertain about the shape of a graph. Very complicated graphs are often drawn with the aid of a computer or a graphing calculator. As you may have discovered already, determining just a few ordered pairs that solve an equation can be quite time-consuming. Computers can calculate large numbers of ordered pairs in very little time (and without complaining!) when properly programmed. Many software packages and graphing calculators are currently on the market.

EXERCISE SET 6.1

Plot the following points.

1. $A(5, 3)$, $B(2, 4)$, $C(0, 2)$, $D(0, -6)$, $E(3, 0)$, $F(-2, 0)$, $G(1, -3)$, $H(-5, 3)$, $J(-4, 4)$
2. $A(3, 5)$, $B(1, 5)$, $C(0, 4)$, $D(0, -4)$, $E(5, 0)$, $F(-5, 0)$, $G(1, -5)$, $H(-7, 4)$, $J(-5, 5)$
3. $A(3, 0)$, $B(4, 2)$, $C(5, 4)$, $D(6, 6)$, $E(3, -4)$, $F(3, -3)$, $G(3, -2)$, $H(3, -1)$
4. $A(1, 1)$, $B(2, 3)$, $C(3, 5)$, $D(4, 7)$, $E(-2, 1)$, $F(-2, 2)$, $G(-2, 3)$, $H(-2, 4)$, $J(-2, 5)$, $K(-2, 6)$
5. Plot the points $M(2, 3)$, $N(5, -3)$, and $P(-2, -3)$. Draw $\overline{MN}$, $\overline{NP}$, and $\overline{MP}$. ($\overline{MN}$ means the line segment from M to N.) What kind of geometric figure is formed? What is its area?
6. Plot the points $Q(-4, 3)$, $R(5, 3)$, $S(2, -1)$, and $T(-7, -1)$. Draw $\overline{QR}$, $\overline{RS}$, $\overline{ST}$, and $\overline{TQ}$. What kind of figure is formed? What is its area?

In which quadrant is each of the following points found?

7. $(-3, -5)$
8. $(2, 17)$
9. $(-6, 1)$
10. $(4, -8)$
11. $\left(3, \frac{1}{2}\right)$
12. $(-1, -8)$
13. $(7, -0.2)$
14. $(-4, 31)$

Determine whether the ordered pair is a solution of the indicated equation.

15. $(1, -1)$; $y = 2x - 3$
16. $(2, 5)$; $y = 3x - 1$
17. $(3, 4)$; $3s + t = 4$
18. $(2, 3)$; $2p + q = 5$
19. $(3, 5)$; $4x - y = 7$
20. $(2, 7)$; $5x - y = 3$
21. $\left(0, \frac{3}{5}\right)$; $2a + 5b = 3$
22. $\left(0, \frac{3}{2}\right)$; $3f + 4g = 6$
23. $(2, -1)$; $4r + 3s = 5$
24. $(2, -4)$; $5w + 2z = 2$
25. $(3, 2)$; $3x - 2y = -4$
26. $(1, 2)$; $2x - 5y = -6$
27. $(-1, 3)$; $y = 3x^2$
28. $(2, 4)$; $2r^2 - s = 5$
29. $(2, 3)$; $5s^2 - t = 7$
30. $(2, 3)$; $y = x^3 - 5$

Graph.

31. $y = -2x$
32. $y = -\frac{1}{2}x$
33. $y = x + 3$
34. $y = x - 2$
35. $y = 3x - 2$
36. $y = -4x + 1$
37. $y = -2x + 3$
38. $y = -3x + 1$
39. $y = \frac{2}{3}x + 1$
40. $y = \frac{1}{3}x + 2$
41. $y = -\frac{3}{2}x + 1$
42. $y = -\frac{2}{3}x - 2$
43. $y = \frac{3}{4}x + 1$
44. $y = x^2$
45. $y = -x^2$
46. $y = x^2 + 2$
47. $y = x^2 - 2$
48. $x = y^2 + 2$

49. $y = 3 - x^2$

50. $y = x^3 - 2$

51. $y = -\frac{1}{x}$

52. $y = \frac{3}{x}$

53. $y = |x| + 2$

54. $y = -|x|$

SKILL MAINTENANCE

55. A garden is being constructed in the shape of a triangle. One side is 12 ft long. How tall should the triangle be to make the area of the garden 156 ft^2?

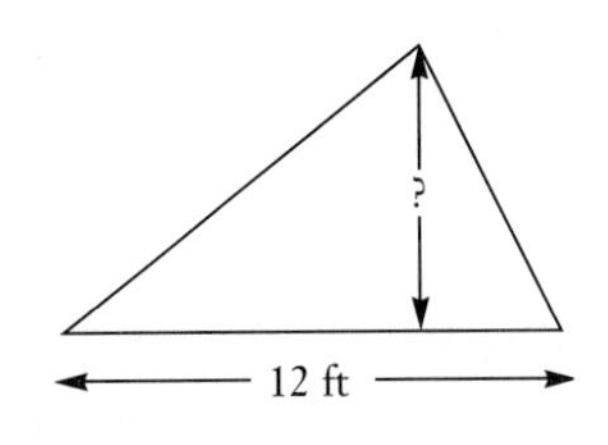

56. What rate of interest is required for a principal of \$320 to earn \$17.60 in half a year?

57. Subtract: $-3.9 - (-2.5)$.

58. Solve: $3(x - 2) = 7x - 8$.

SYNTHESIS

59. Using the same set of axes, graph $y = 6x$, $y = 3x$, $y = \frac{1}{2}x$, $y = -6x$, $y = -3x$, and $y = -\frac{1}{2}x$, and compare the slants of the lines. What does the number in front of the x tell you about the slant of the line?

60. If $(-10, -2)$, $(-3, 4)$, and $(6, 4)$ are the coordinates of three consecutive vertices of a parallelogram, what are the coordinates of the fourth vertex?

61. Which of the following equations have $\left(-\frac{1}{3}, \frac{1}{4}\right)$ as a solution?

a) $-\frac{3}{2}x - 3y = -\frac{1}{4}$

b) $8y - 15x = \frac{7}{2}$

c) $0.16y = -0.09x + 0.1$

d) $2(-y + 2) - \frac{1}{4}(3x - 1) = 4$

▦ Use a calculator. Find and graph at least 10 solutions for the equation. Then graph the equation.

62. $y = x^3 + 3x^2 + 3x + 1$; use values of x from -3 to 1

63. $y = x^3 - 6x^2 + 12x - 8$; use values of x from 0 to 4

64. $y = 1/x^2$; use values of x from -3 to 3

65. $y = -1/x^2$; use values of x from -3 to 3

66. $y = |x - 2.3|$; use values of x from -2 to 5

67. $y = 1/(x - 2)$; use values of x from -1 to 5

68. $y = 1/(x - 1)^2$; use values of x from -2 to 4

69. Using the same set of axes, graph $y = 2x$, $y = 2x - 3$, and $y = 2x + 3$. What does the sign of the number added to $2x$ tell you about the location of the line with respect to the graph of $y = 2x$?

70. If $(-1, 1)$ and $(4, -4)$ are the endpoints of a diagonal of a square, what are the coordinates of the other two vertices of the square?

71. Plot the following points: $A(-5, 2)$, $B(-3, 2)$, $C\left(-2, \frac{1}{2}\right)$, $D(-1, 2)$, $E(1, 2)$, $F(-1, -1)$, $G(1, -4)$, $H(-1, -4)$, $I\left(-2, -2\frac{1}{2}\right)$, $J(-3, -4)$, $K(-5, -4)$, $L(-3, -1)$. Draw $\overline{AB}$, $\overline{BC}$, $\overline{CD}$, $\overline{DE}$, $\overline{EF}$, $\overline{FG}$, $\overline{GH}$, $\overline{HI}$, $\overline{IJ}$, $\overline{JK}$, $\overline{KL}$, and $\overline{LA}$ in order to find an "unknown."

COMPUTER–CALCULATOR EXERCISES

Use a computer software package or a graphing calculator to graph the equation.

72. $y = x^3 - 3x + 2$

73. $y = x^3 - 3x^2 + 2$

74. $y = \dfrac{x^2 - x - 2}{x - 1}$

75. $y = 280x - 0.4x^2$

6.2 Functions

In this section we consider a mathematical concept that is highly useful in problem solving—that is, a *function*. A function is a certain kind of correspondence from one set to another. For example:

To each person in a math class . . . there corresponds . . . his or her age.

To each item in a store . . . there corresponds . . . its price.

To each real number . . . there corresponds . . . the cube of that number.

In each example, the first set is called the **domain.** The second set is called the **range.** Given a member of the domain, there is *just one* member of the range to which it corresponds. This kind of correspondence is called a **function.**

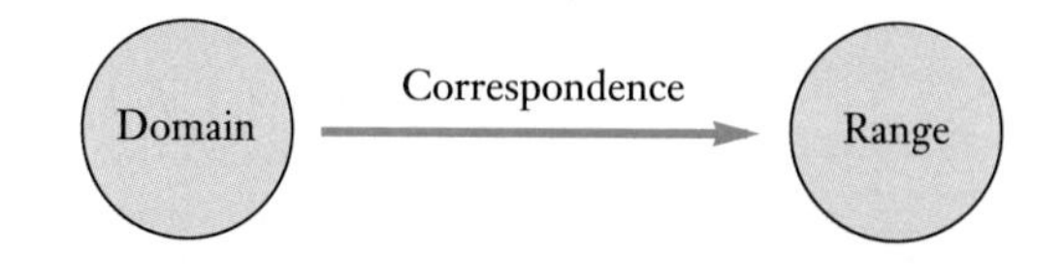

EXAMPLES Which of the following correspondences are functions?

1.

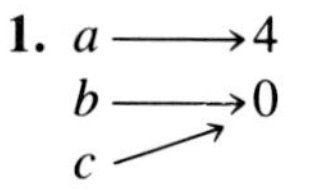

The correspondence is a function because for each member of the domain, there corresponds just one member of the range.

2. San Francisco → Giants
New York → Giants
New York → Mets
Cleveland → Browns

The correspondence is not a function because a member of the domain (New York) corresponds to more than one member of the range. ■

> A *function* is a correspondence between a first set, called the *domain*, and a second set, called the *range*, such that to each member of the domain, there corresponds *exactly one* member of the range.

EXAMPLES Which of the following are functions?

	Domain	*Correspondence*	*Range*
3.	A family	Each person's weight	A set of positive numbers

This correspondence is a function because each person has *only one* weight.

4.	The natural numbers	The square of each number	A set of natural numbers

This correspondence is a function because each natural number has *only one* square.

5.	The set of all states	The state's members of the U.S. Senate	The set of all U.S. senators

This correspondence is not a function because every state has two U.S. senators. ■

When a correspondence between two sets is not a function, it is still an example of a **relation.**

> A *relation* is a correspondence between a first set, called the *domain,* and a second set, called the *range,* such that to any member of the domain, there corresponds *at least one* member of the range.

Thus, although Examples 1–5 do not all represent functions, they do represent relations. A function can be thought of as a special type of relation—one in which each member of the domain is paired up with only *one* member of the range.

Notation for Functions

To understand function notation it helps to imagine a "function machine." Think of putting a member of the domain (an *input*) into the machine. The machine knows the correspondence and gives you a member of the range (the *output*).

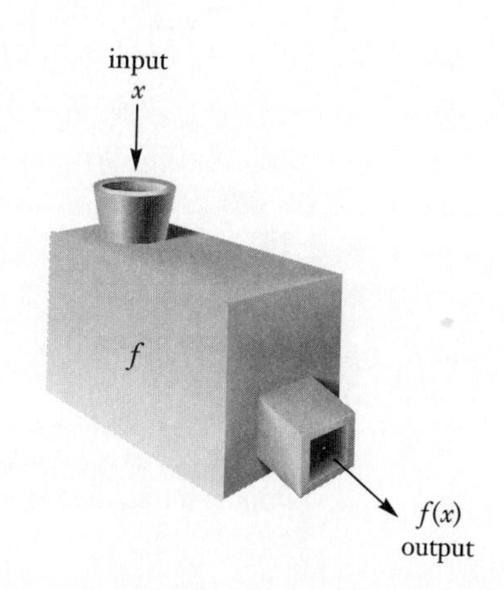

The function has been named f. We call the input x, and its output $f(x)$. This is read "f of x," or "f at x," or "the value of f at x." Note that $f(x)$ does *not* mean "f times x."

Some functions can be described by formulas or equations. For example, $f(x) = 2x + 3$ describes the function that takes an input x, multiplies it by 2, and then adds 3.

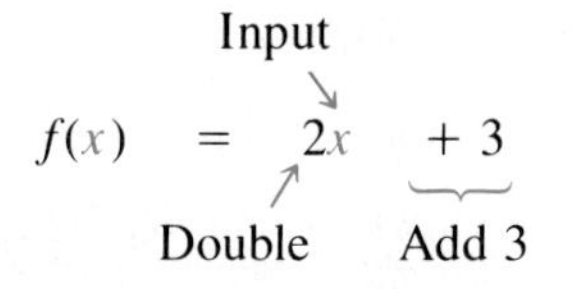

To find the output $f(4)$, we take the input, 4, double it, and add 3 to get 11. That is, we substitute 4 into the formula for $f(x)$:

$$f(4) = 2 \cdot 4 + 3$$
$$= 11.$$

Sometimes, instead of writing $f(x) = 2x + 3$, we might write $y = 2x + 3$, where it is understood that the value of y, the *dependent variable*, is calculated after first choosing a value for x, the *independent variable*. To understand why $f(x)$ notation is so useful, consider two equivalent statements:

a) If $f(x) = 2x + 3$, then $f(4) = 11$.
b) If $y = 2x + 3$, then the value of y is 11 when x is 4.

Note that the notation used in part (a) is far more concise.

EXAMPLES Find these function values.

6. For $f(x) = 3x + 2$, find $f(5)$.

$$f(5) = 3 \cdot 5 + 2 = 17$$

7. For $g(z) = 5z^2 - 4$, find $g(3)$.

$$g(3) = 5(3)^2 - 4 = 41$$

8. For $A(r) = 3r^2 + 2r$, find $A(-2)$.

$$A(-2) = 3(-2)^2 + 2(-2) = 8$$

9. For $h(x) = 5$, find $h(3)$ and $h(-1)$.

$h(3) = 5$

$h(-1) = 5$ **Regardless of input, a *constant function*, like h, always has the same output.**

10. For $f(x) = 3x + 2$, find $f(a + 1)$.

$$f(a + 1) = 3(a + 1) + 2 = 3a + 3 + 2 = 3a + 5$$

■

Notice that whether we write $f(x) = 3x + 2$, or $f(t) = 3t + 2$, or $f(\ \ ) = 3(\ \ ) + 2$, we still have $f(5) = 17$. Thus the independent variable can be thought of as a *dummy* variable. The letter chosen for the dummy variable is not as important as the algebraic manipulations to which it is subjected.

Functions and Formulas

Functions are often described by formulas and are used extensively in solving problems. You are probably already familiar with this, except that you may not have seen function notation in that connection. For example, a formula for finding the area A of a circle with radius r is

$$A = \pi r^2.$$

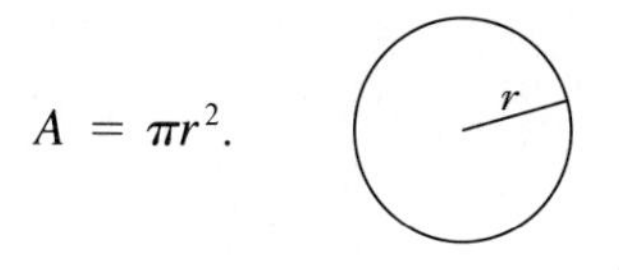

The area A is thus a function of r. To emphasize that fact we often write

$$A(r) = \pi r^2.$$

EXAMPLE 11 Find the area of a circle having a radius of 12 cm.

Solution We use the area function $A(r) = \pi r^2$:

$$\begin{aligned} A(12) &= \pi(12)^2 \quad \text{Substituting for } r \\ &= \pi \times 144 \\ &\approx 452 \text{ cm}^2. \end{aligned}$$

■

Functions and Tables

Tables are often used in problem solving and frequently provide examples of functions. To use tables in problem solving, we generally look up some function value in them.

EXAMPLE 12 The cost of replacing a defective tire is a function of tread depth and can be found from the following table.

Input	Tread Depth (in millimeters)	9	8	7	6	5	4	3	2	1
Output	Percent Charged	No charge	20	30	40	55	70	80	90	100

Find the cost of replacing a tire whose original price was \$64.50 and whose tread depth is 4 mm.

Solution

Translate. The problem translates to

$$\text{Cost} = ?\% \times \text{original cost}.$$

To find the percent charged, we treat the table as a function. When the tread depth, 4, is used as an input, the output (percent charged) is 70%. This gives us the equation

$$\text{Cost} = 70\% \times \$64.50.$$

Carry out. We calculate:

$$\begin{aligned}\text{Cost} &= 0.70 \times 64.50 \qquad \textbf{Converting to decimal notation}\\ &= \$45.15.\end{aligned}$$

■

Functions and Graphs

Functions are often described by graphs. To use a graph in problem solving, we do about what we would do with a table except that we find a function value by reading a graph instead of a table. In what follows we first draw a graph and then estimate a function value from it.

EXAMPLE 13 A person buys a video cassette recorder on which there is a revolution counter. There is also a booklet with a table that relates the counter reading and the time for which the tape has run.

Counter Reading	Time of Tape (in hours)
000	0
400	1
700	3
800	4

a) Use the data in the table to draw a graph of the time that a tape has run as a function of the counter reading.

b) Use the graph to estimate the time elapsed when the counter has reached 600.

Solution

a) We draw a graph first. We plot the points and connect them with a smooth curve.

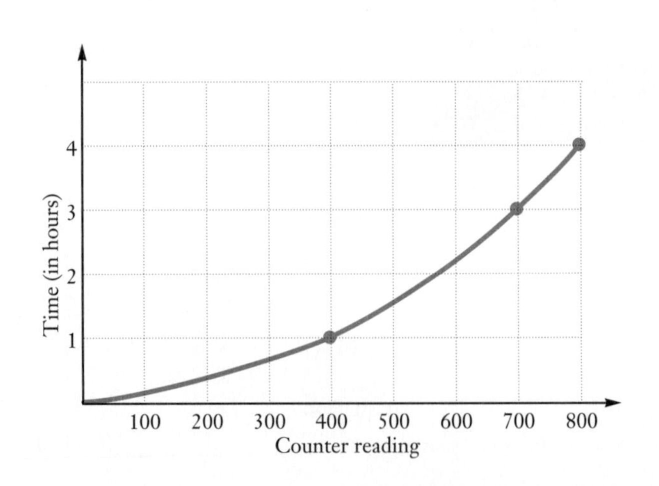

b) We then find the number 600 on the first (horizontal) axis. Then we move along the vertical line to the curve, and from there to the left, to the second (vertical) axis. There, we read an approximate function value. The time elapsed is about 2.25 hours.

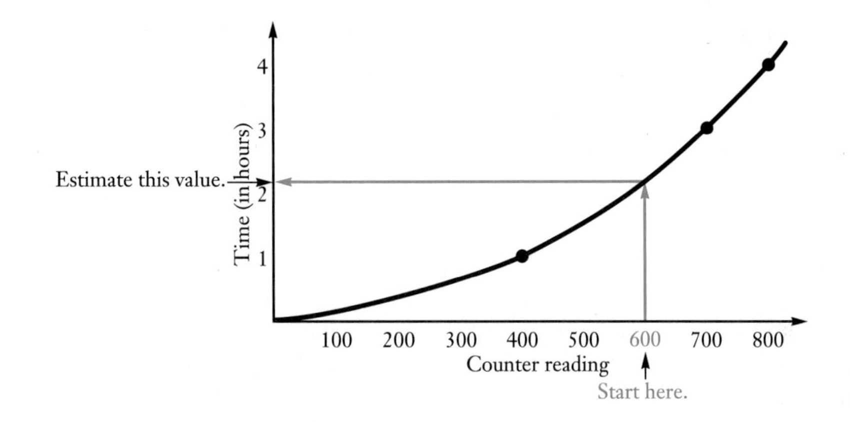

■

The Vertical Line Test

The members of the domain are those values on the horizontal axis that serve as a first coordinate for a point on the graph. The members of the range are those values on the vertical axis that serve as a second coordinate for a point on the graph. A graph cannot represent a function if a value on the horizontal axis appears as the first coordinate of more than one point (otherwise, to one member of the domain there would correspond more than one member of the range). Thus, if a vertical line drawn anywhere on the graph intersects the graph at more than one point, the graph is not one of a function.

> **THE VERTICAL LINE TEST**
>
> If it is possible for a vertical line to intersect a graph more than once, the graph is not a graph of a function.

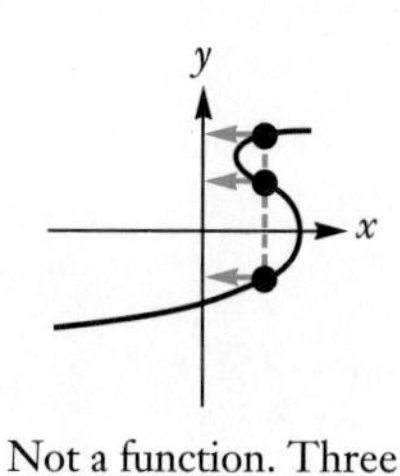

Not a function. Three y-values correspond to one x-value.

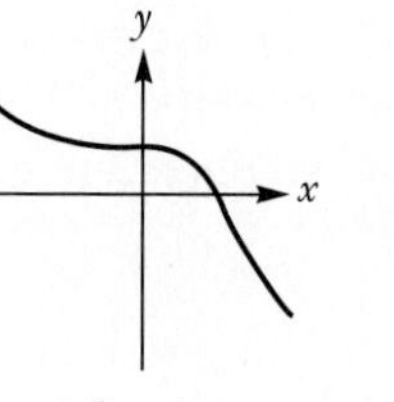

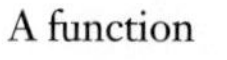

A function

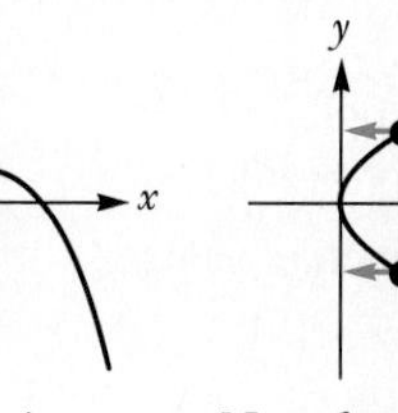

A function

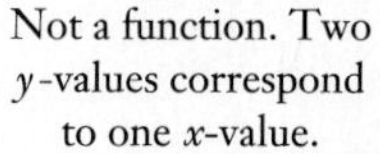

Not a function. Two y-values correspond to one x-value.

EXERCISE SET 6.2

Determine whether the correspondence is a function.

1.

a	P
b	Q
c	R
d	S
	T

2.

1	a
2	b
3	c
4	d
5	

3.

*Firm**	*Number of Partners*
Deloitte Haskins & Sells	850
Arthur Young & Co.	850
Peat, Marwick Main & Co.	1900
Coopers & Lybrand	1270

4.

*Firm**	*Female Partners*
Arthur Anderson	38
Arthur Young & Co.	44
Touche Ross	23
Ernst & Whinney	27
Coopers & Lybrand	54

5.

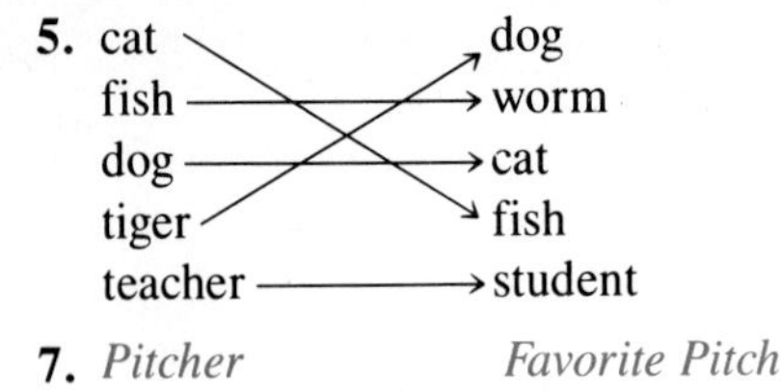

6.

A	a
B	b
C	c
D	d
	e

7.

Pitcher	*Favorite Pitch*
Viola	Slider
Hammaker	Fastball
Worrell	Curveball
Morris	

8.

Site	*Year*
Lake Placid	1980
Oslo	1976
Squaw Valley	1960
Innsbruck	1952
	1932

9.

a	M
b	N
c	O
d	P
e	

10.

Jack	John
Kate	James
Marnie	Katherine
Jim	Margaret
Peggy	

Determine whether each of the following is a function. Identify any relations that are not functions.

	Domain	*Correspondence*	*Range*
11.	A math class	Each person's seat number	A set of numbers
12.	A set of numbers	Square each number and then add 4.	A set of numbers
13.	A set of geometric figures	Find the area of each figure.	A set of numbers
14.	A family	Each person's eye color	A set of colors
15.	The people in a town	Each person's aunt	A set of females
16.	A set of avenues	Find an intersecting road.	A set of cross streets

Find the function values.

17. $g(x) = x + 1$

a) $g(0)$ **b)** $g(-4)$ **c)** $g(-7)$
d) $g(8)$ **e)** $g(a + 2)$

18. $h(x) = x - 4$

a) $h(4)$ **b)** $h(8)$ **c)** $h(-3)$
d) $h(-4)$ **e)** $h(a - 1)$

19. $f(n) = 5n^2 + 4$
a) $f(0)$ **b)** $f(-1)$ **c)** $f(3)$
d) $f(t)$ **e)** $f(2a)$

20. $g(n) = 3n^2 - 2$
a) $g(0)$ **b)** $g(-1)$ **c)** $g(3)$
d) $g(t)$ **e)** $g(2a)$

21. $g(r) = 3r^2 + 2r - 1$
a) $g(2)$ **b)** $g(3)$ **c)** $g(-3)$
d) $g(1)$ **e)** $g(3r)$

22. $h(r) = 4r^2 - r + 2$
a) $h(3)$ **b)** $h(0)$ **c)** $h(-1)$
d) $h(-2)$ **e)** $h(3r)$

23. $f(x) = \dfrac{x - 3}{2x - 5}$
a) $f(0)$ **b)** $f(4)$ **c)** $f(-1)$
d) $f(3)$ **e)** $f(x + 2)$

24. $s(x) = \dfrac{3x - 4}{2x + 5}$
a) $s(10)$ **b)** $s(2)$ **c)** $s\left(-\frac{5}{2}\right)$
d) $s(-1)$ **e)** $s(x + 3)$

Consider a function g described by $g(x) = -2x - 4$ and a function h described by $h(x) = 3x^2$. Find the following.

25. $g(5) + h(-2)$

26. $g(-4) \cdot h(4)$

27. $2g(-3) - 5h(12)$

28. $g(a + 4) - g(a)$

The function A described by $A(s) = s^2\dfrac{\sqrt{3}}{4}$ gives the area of an equilateral triangle with side s.

29. Find the area when a side measures 4 cm.

30. Find the area when a side measures 6 in.

The function V described by $V(r) = 4\pi r^2$ gives the surface area of a sphere with radius r.

31. Find the area when the radius is 3 in.

32. Find the area when the radius is 5 cm.

Temperature conversion. The function F described by $F(C) = \frac{9}{5}C + 32$ gives the Fahrenheit temperature corresponding to the Celsius temperature C.

33. Find the Fahrenheit temperature equivalent to -10°C.

34. Find the Fahrenheit temperature equivalent to 5°C.

Estimating height. The function H described by $H(x) = 2.75x + 71.48$ can be used to predict the height, in centimeters, of a woman whose *humerus* (the bone from the elbow to the shoulder) is x cm long. Predict the height of a woman whose humerus is the length given.

35. 40 cm

36. 43 cm

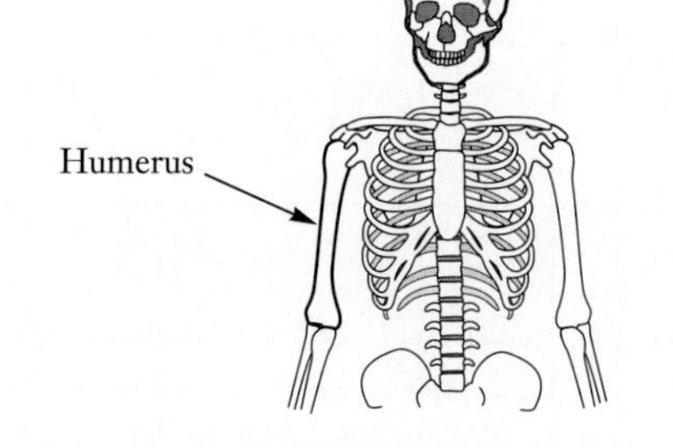

37. Using the table in Example 12, find the cost of replacing a tire with a tread depth of 7 mm and an original price of \$78.50.

38. Using the table in Example 12, find the cost of replacing a tire with a tread depth of 3 mm and an original price of \$72.40.

Point of legal intoxication. The following table can be used to predict the number of drinks required for a person of a specified weight to be legally intoxicated (blood alcohol level of 0.10 or above). One 12-oz glass of beer, a 5-oz glass of wine, or a cocktail containing $1\frac{1}{2}$ oz of a distilled liquor all count as one drink.

Input	Body Weight (in pounds)	100	140	160	180	200	220
Output	Number of Drinks	3	4	5	5.5	6	6.5

39. Use the preceding table to draw a graph and to estimate the number of drinks that a 240-lb person would have to drink to be considered intoxicated.

40. Use the preceding table to draw a graph and to estimate the number of drinks an 80-lb person would have to drink to be considered intoxicated.

For Exercises 41 and 42 use the following graph, which shows the cost of manufacturing screwdrivers as a function of the number produced.

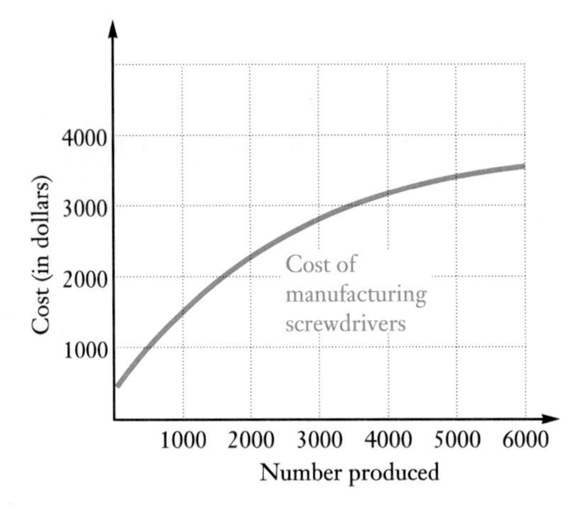

41. Approximate the cost of producing 1200 screwdrivers.

42. Approximate the cost of producing 5100 screwdrivers.

For Exercises 43 and 44 use the following graph, which shows the annual heart attack rate per 10,000 men as a function of blood cholesterol level.*

43. Approximate the annual heart attack rate per 10,000 men for those whose blood cholesterol level is 225 mg/dl.

44. Approximate the annual heart attack rate per 10,000 men for those whose blood cholesterol level is 275 mg/dl.

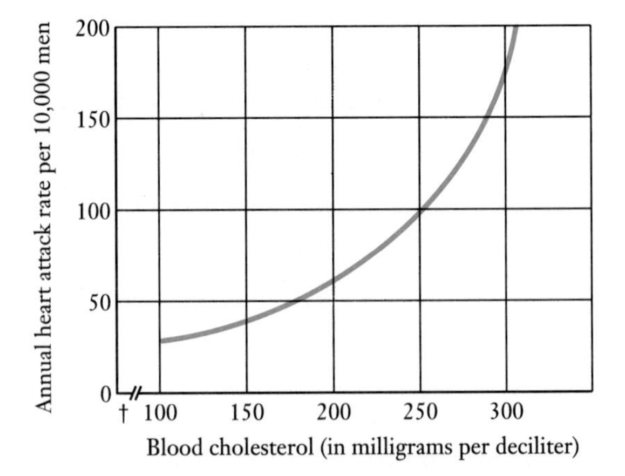

Nutrition Action Healthletter, Jan./Feb. 1988, Center for Science in the Public Interest, 1501 16th St. N.W., Washington, D.C. 20036-1499. Copyright 1989, Center for Science in the Public Interest.

†The slash marks on the horizontal axis indicate that the axis has been truncated for convenience.

A city experiencing rapid growth recorded the following dates and populations.

Input	Year	1985	1987	1989	1991
Output	Population (in Tens of Thousands)	5.8	6	7	10

45. Use the data in the table to draw a graph of the population as a function of time.

46. Use the graph in Exercise 45 to predict the city's population in the year 1993.

47. Use the graph in Exercise 45 to estimate what the population was in 1988.

48. Use the graph in Exercise 45 to estimate what the population was in 1990.

Determine whether each of the following is the graph of a function.

49.

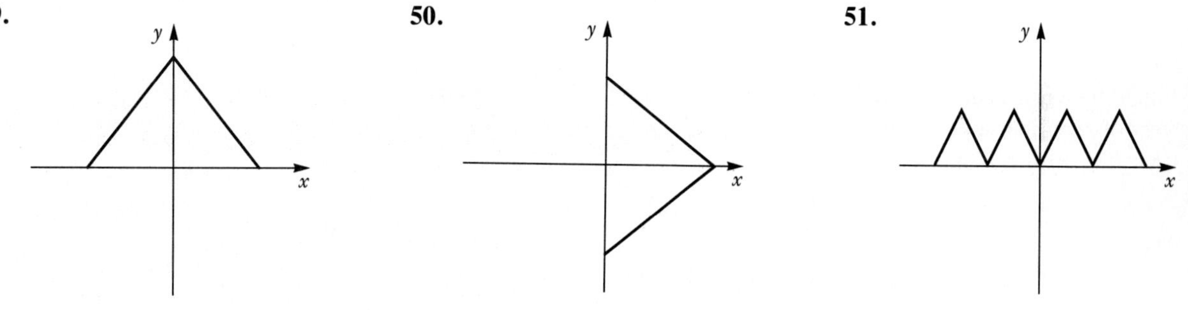

52.

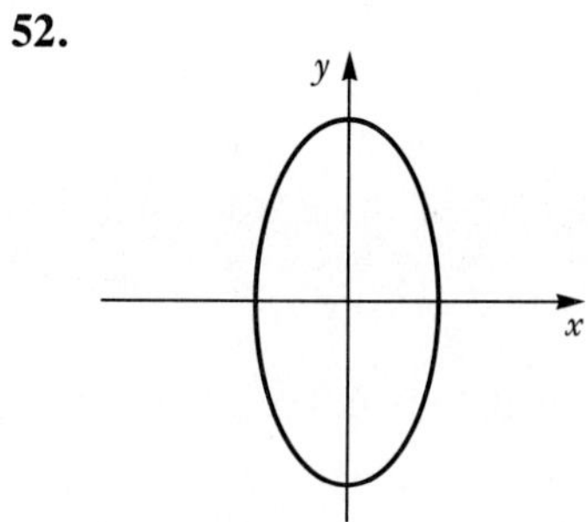

53.

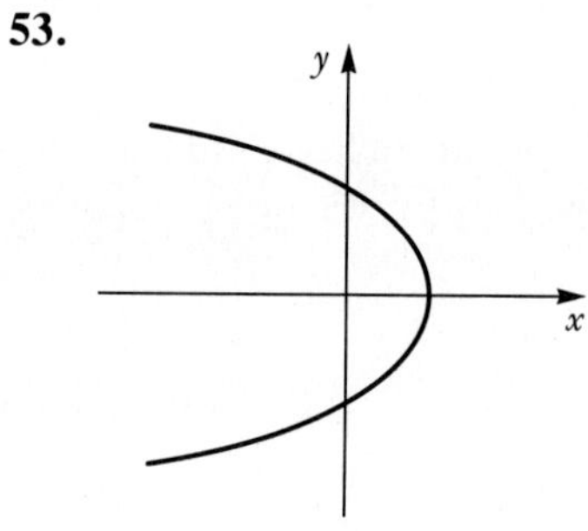

54.

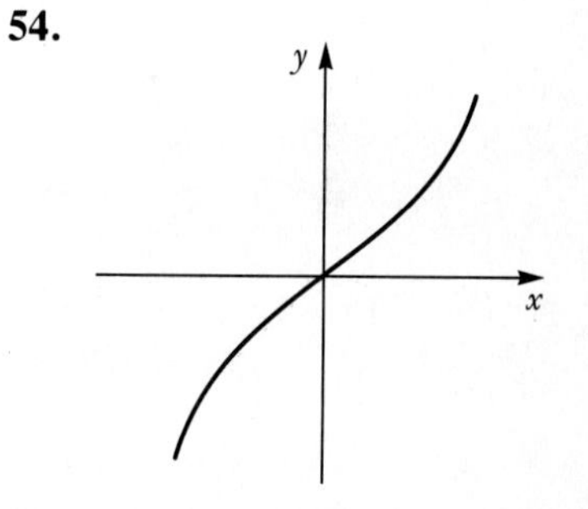

55.

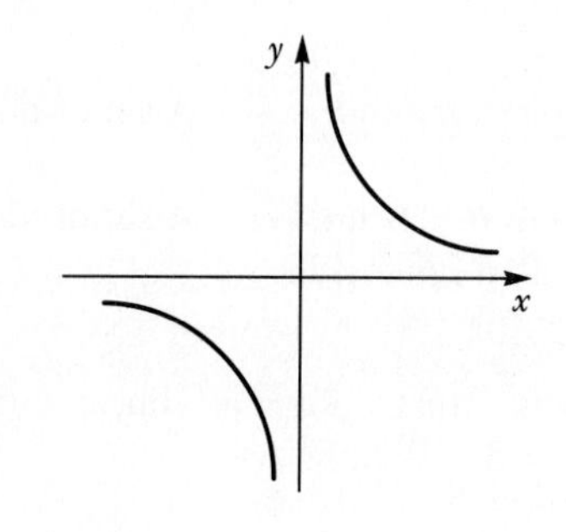

56.

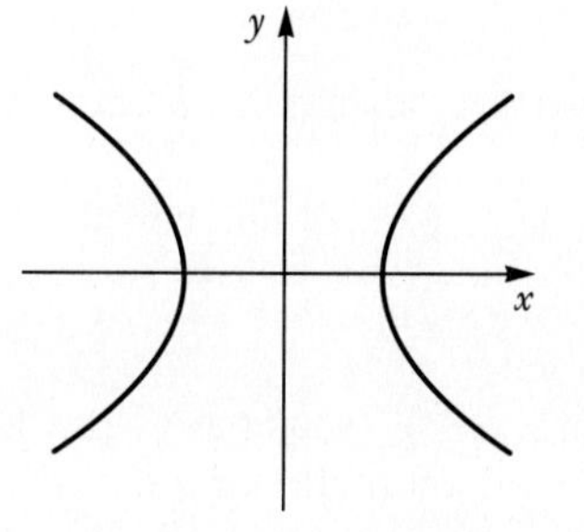

57.

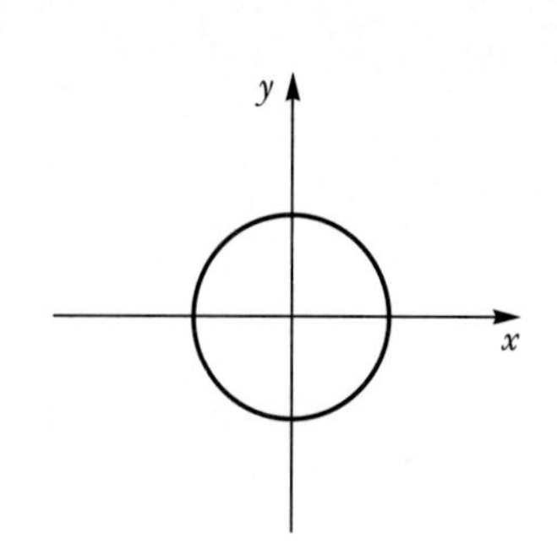

58.

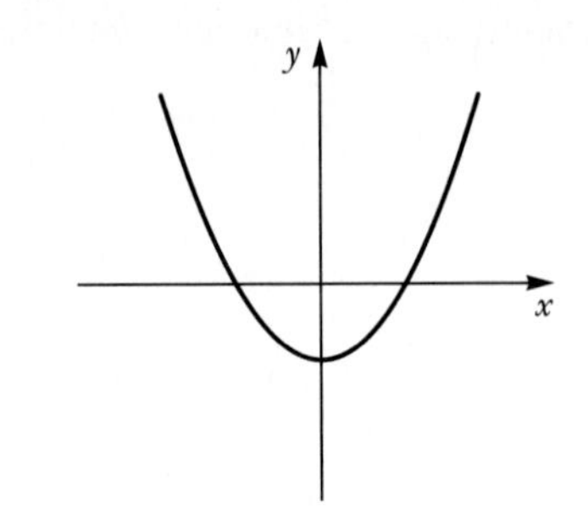

SKILL MAINTENANCE

59. Find three consecutive even integers such that the sum of the first, two times the second, and three times the third is 124.

60. Simplify: $\left(\frac{x^3}{y^5}\right)^2$.

61. The surface area of a rectangular solid of length l, width w, and height h is given by $S = 2lh + 2lw + 2wh$. Solve for l.

62. Add: $\frac{2x - 1}{x^2 + 1} + \frac{3x - 4}{x^2 + 1}$.

SYNTHESIS

For each of these functions, use a calculator to find the indicated function values.

63. ▦ $f(x) = 4.3x^2 - 1.4x$

a) $f(1.034)$ b) $f(-3.441)$

c) $f(27.35)$ d) $f(-16.31)$

64. ▦ $g(x) = 2.2x^3 + 3.5$

a) $g(17.3)$ b) $g(-64.2)$

c) $g(0.095)$ d) $g(-6.33)$

Determine whether each of the following is a function.

	Domain	*Correspondence*	*Range*
65.	A set of pairs of numbers (a, b), $a \neq b$	Take the larger of a and b.	A set of numbers
66.	A set of ordered pairs of numbers (a, b), $b \neq 2a$	Subtract a from b. Then take the larger of a and $b - a$.	A set of numbers

67. For a function g you have $g(-1) = -7$ and $g(3) = 8$. Find a formula for g if $g(x)$ is of the form $g(x) = mx + b$, where m and b are constants.

68. You know that for a linear function f, $f(x - 1) = 5x$. What is $f(6)$?

69. Does the chart constitute a function? Why or why not?

APPROXIMATE ENERGY EXPENDITURE BY A 150-POUND PERSON IN VARIOUS ACTIVITIES	
Activity	**Calories per Hour**
Lying down or sleeping	80
Sitting	100
Driving an automobile	120
Standing	140
Domestic work	180
Walking, $2\frac{1}{2}$ mph	210
Bicycling, $5\frac{1}{2}$ mph	210
Gardening	220
Golf; lawn mowing, power mower	250
Bowling	270
Walking, $3\frac{3}{4}$ mph	300
Swimming, $\frac{1}{4}$ mph	300
Square dancing, volleyball, roller skating	350
Wood chopping or sawing	400
Tennis	420
Skiing, 10 mph	600
Squash and handball	600
Bicycling, 13 mph	660
Running, 10 mph	900

Source: Based on material prepared by Robert E. Johnson, M.D., Ph.D., and colleagues, University of Illinois.

For Exercises 70–73, use the following graph of a woman's "nonstress test." This graph shows the size of a pregnant woman's contractions as a function of time.

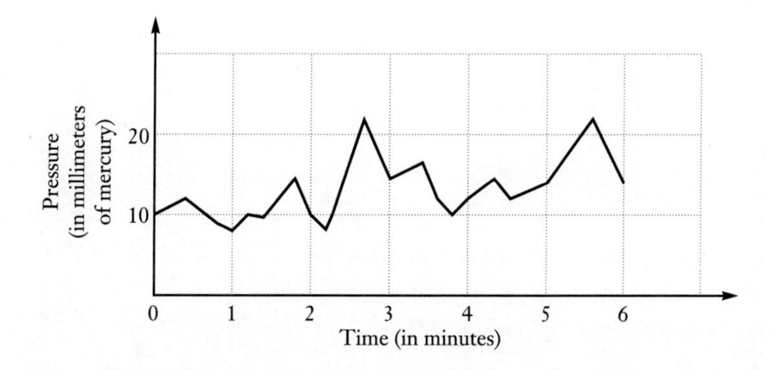

70. How large is the largest contraction that occurred during the test?

71. At what time during the test did the largest contraction occur?

72. On the basis of the information provided, how large a contraction would you expect 60 seconds from the end of the test?

73. What is the frequency of the woman's largest contractions?

74. *The greatest integer function* $f(x) = [\![x]\!]$ is defined as follows: $[\![x]\!]$ is the greatest integer that is less than or equal to x. For example, if $x = 3.74$, then $[\![x]\!] = 3$; and if $x = -0.98$, then $[\![x]\!] = -1$. Graph the greatest integer function for values of x such that $-5 \leq x \leq 5$. (The notation $f(x) = \text{INT}[x]$ is used in many computer programs for the greatest integer function.)

6.3 Graphs of Linear Functions

Functions can have different kinds of graphs. In this section we are interested in functions whose graphs are straight lines. Such functions and their graphs are called *linear.*

Graphs of Equations of the Type $y = mx + b$ or $f(x) = mx + b$

We have seen that for any number m, the graph of $y = mx$ is a straight line passing through the origin. What will happen if we add a number b on the right side to get the equation $y = mx + b$?

EXAMPLE 1 Graph $y = 2x$ and $y = 2x + 3$, using the same set of axes.

Solution We first make a table containing values for both equations.

We then plot these points. Drawing a solid line for $y = 2x + 3$ and a dashed line for $y = 2x$, we see that the graph of $y = 2x + 3$ can be obtained by moving, or *translating*, the graph of $y = 2x$ three units up.

x	y	y
	$y = 2x$	$y = 2x + 3$
0	0	3
1	2	5
-1	-2	1
2	4	7
-2	-4	-1

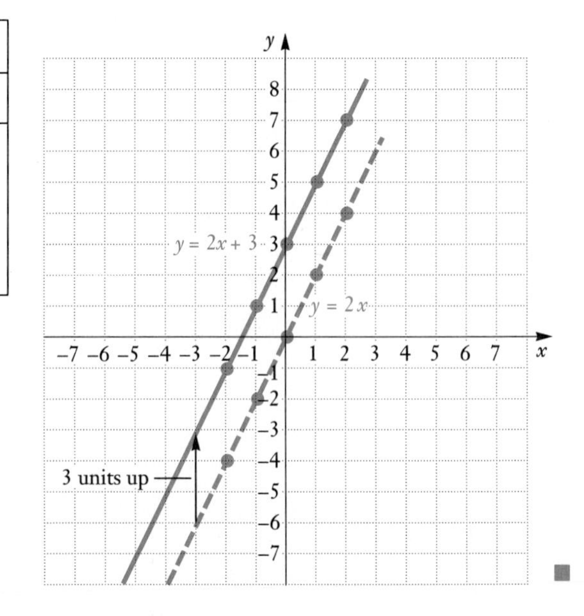

■

EXAMPLE 2 Graph $f(x) = \frac{1}{3}x$ and $g(x) = \frac{1}{3}x - 2$, using the same set of axes.

Solution We first make a table containing values for both equations. By choosing multiples of 3, we avoid fractions.

x	$f(x)$	$g(x)$
	$f(x) = \frac{1}{3}x$	$g(x) = \frac{1}{3}x - 2$
0	0	-2
3	1	-1
-3	-1	-3
6	2	0

We then plot these points. Drawing a solid line for $g(x) = \frac{1}{3}x - 2$ and a dashed line for $f(x) = \frac{1}{3}x$, we see that the graph of $g(x) = \frac{1}{3}x - 2$ looks just like the graph of $f(x) = \frac{1}{3}x$ but is moved, or translated, 2 units down.

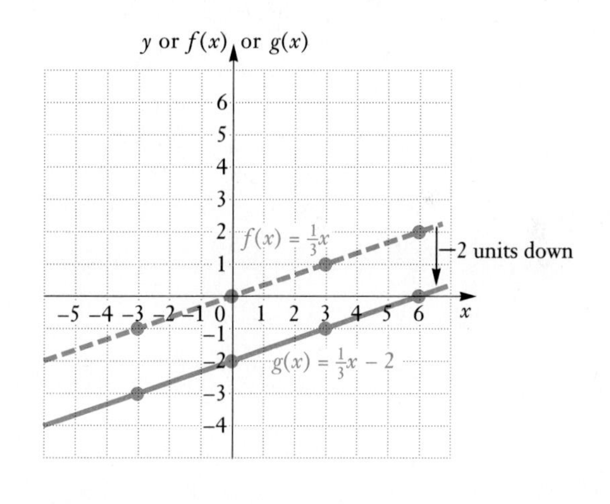

■

Intercepts and Slopes. Note that the graph of $y = 2x + 3$ passed through the point (0, 3) and the graph of $y = \frac{1}{3}x - 2$ passed through the point $(0, -2)$. In general, the graph of $y = mx + b$ is a line parallel to $y = mx$, passing through the point $(0, b)$. The point $(0, b)$ is called the **y-intercept.** To save time, we sometimes will simply refer to the number b as the y-intercept.

EXAMPLE 3 Find the y-intercept: $y = -5x + 4$.

Solution The y-intercept is (0, 4), or simply 4. ■

EXAMPLE 4 Find the y-intercept: $f(x) = 5.3x - 12$.

Solution The y-intercept is $(0, -12)$, or simply -12. Using function notation, we write that the y-intercept is $f(0)$, or -12. ■

In examining the preceding two graphs, we note that in each case the slant of the dotted line appears to be the same as the slant of the solid line. This leads us to suspect that it is the number m, in the equation $y = mx + b$, that is responsible for the slant of the line. In Section 6.4 we will prove that this is indeed the case, but for now we simply state that the number m is called the *slope* of the line $y = mx + b$. Thus the slope of the lines in Example 1 is 2 and the slope of the lines in Example 2 is $\frac{1}{3}$. Note that $2 = \frac{2}{1}$ and that from any point on either line in Example 1, if we go *up* 2 units and *to the right* 1 unit (or *down* 2 units and *to the left* 1 unit), we return to the line. Similarly, in Example 2, if we go *up* 1 unit and *to the right* 3 units (or *down* 1 unit and *to the left* 3 units), we return to the line.

EXAMPLE 5 Determine the slope of the line given by $y = \frac{2}{3}x + 4$, and draw its graph.

Solution Here $m = \frac{2}{3}$, so the slope is $\frac{2}{3}$. This means that from *any* point on the graph, we can find a second point by simply going up 2 units and to the right 3 units. Since the y-intercept, $(0, 4)$, is a point known to be on the graph, we calculate that $(0 + 3, 4 + 2)$, or $(3, 6)$, is also on the graph. Knowing two points, we can now draw the graph.

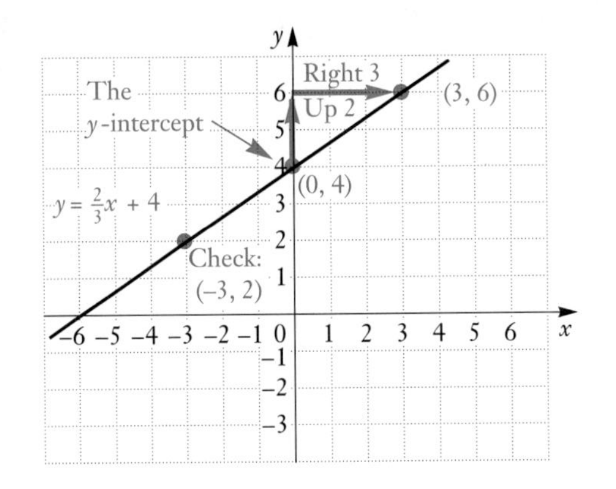

Important: As a check, we choose some other value for x, say -3, and determine y (in this case, 2). We plot that point and see whether it is on the line. If it is not, there has been some error. ■

In Example 5, the slope is $\frac{2}{3}$. We can find some other names for $\frac{2}{3}$ by multiplying by 1 and then find other points on the graph.

$$\frac{2}{3} = \frac{2}{3} \cdot \frac{2}{2} = \frac{4}{6}$$

Thus from any point on the graph, we can find another point by going *up* 4 units and *to the right* 6.

$$\frac{2}{3} = \frac{2}{3} \cdot \frac{3}{3} = \frac{6}{9}$$

Thus we can find a point by going *up* 6 and *to the right* 9 from any point on the graph.

When numbers are negative, we reverse directions.

$$\frac{2}{3} = \frac{2}{3} \cdot \frac{-1}{-1} = \frac{-2}{-3}$$

Thus from any point on the graph, we can find another point by going *down* 2 units and then *to the left* 3 units.

If the slope of a line is negative, it slants downward from left to right.

EXAMPLE 6 Graph: $y = -\frac{1}{2}x + 5$.

Solution The y-intercept is (0, 5). The slope is $-\frac{1}{2}$, or $\frac{-1}{2}$. From the y-intercept we go *down* 1 unit and *to the right* 2 units. That gives us the point (2, 4). We can now draw the graph.

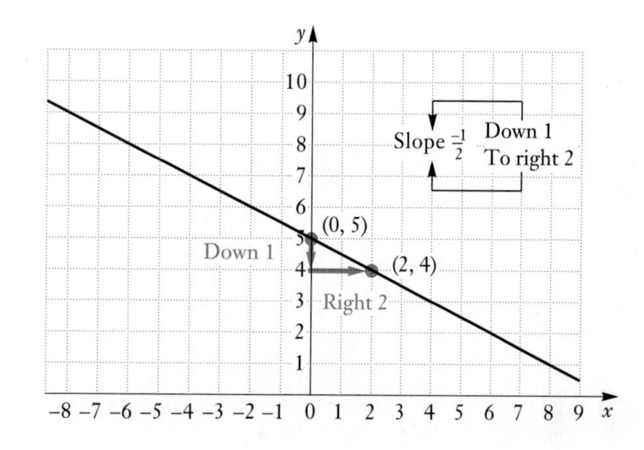

As a new type of check, we rename the slope and find another point:

$$\frac{-1}{2} = \frac{-1}{2} \cdot \frac{-3}{-3} = \frac{3}{-6}.$$

Thus we can go *up* 3 units and then *to the left* 6 units. This gives the point (− 6, 8).

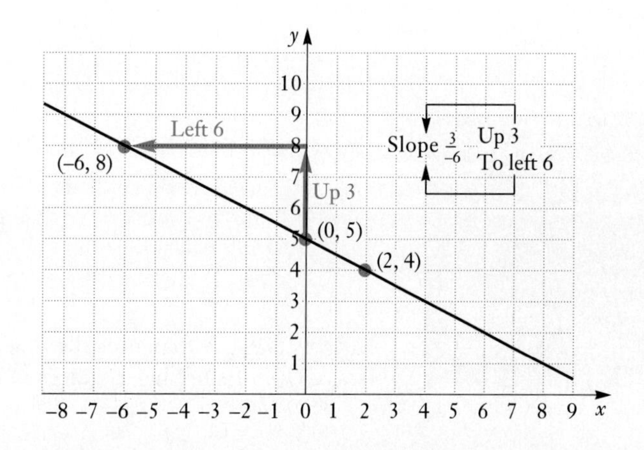

■

We also could have checked by choosing a number for x and finding y. In either type of check, we find a third point not already known to be on the graph. If the third point does not line up with the other two, we must inspect our calculations for an error.

Because an equation of the form $y = mx + b$ enables us to make use of the slope and the y-intercept when graphing, it is called the *slope–intercept* form of a linear equation.

THE SLOPE–INTERCEPT EQUATION

Any equation $y = mx + b$ has a graph that is a straight line. It goes through the y-intercept $(0, b)$ and has slope m. Any equation of the form $y = mx + b$ is said to be a *slope–intercept equation*.

EXAMPLE 7 Determine an equation for a line with slope $-\frac{2}{3}$ and y-intercept $(0, 4)$.

Solution We use the slope–intercept form, $y = mx + b$:

$$y = -\frac{2}{3}x + 4. \quad \text{Substituting } -\tfrac{2}{3} \text{ for } m \text{ and 4 for } b$$ ■

Linear Functions. Consider the function $f(x) = 3x + 2$. If we think of it as $y = 3x + 2$, we realize that the graph is a line and that we already know how to graph it. Because the function's domain is not specified, we will assume the domain to be all numbers that could work as inputs. Thus the domain of f is the set of all real numbers.

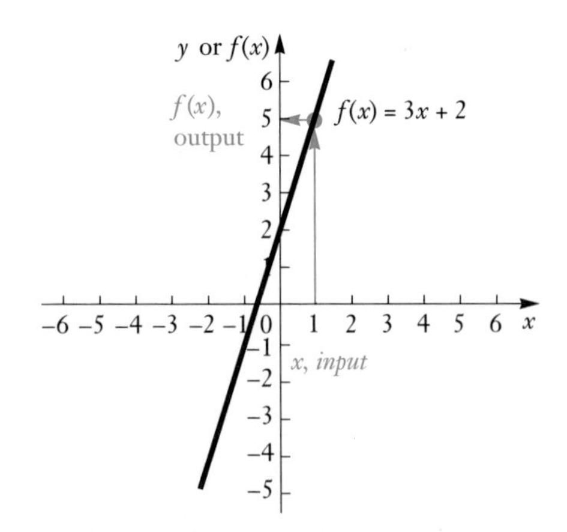

EXAMPLE 8 A firm uses the function $C(x) = \frac{3}{5}x + 2$ to calculate the cost in dollars, $C(x)$, of shipping x ounces of chocolates.

a) Using the horizontal axis for values of x and the vertical axis for values of $C(x)$, graph the equation $C(x) = \frac{3}{5}x + 2$.

b) Use the graph to estimate the cost of shipping $6\frac{1}{2}$ oz of chocolates. Compare this value to the one obtained from the function formula.

Solution

a) We locate the y-intercept (0, 2) and from there count *up* 3 units and *to the right* 5 units. That gives us the point (5, 5). We can now draw the line. As a check, we note that the pair (10, 8) also satisfies the equation.

Note that for this function, all inputs must be nonnegative.

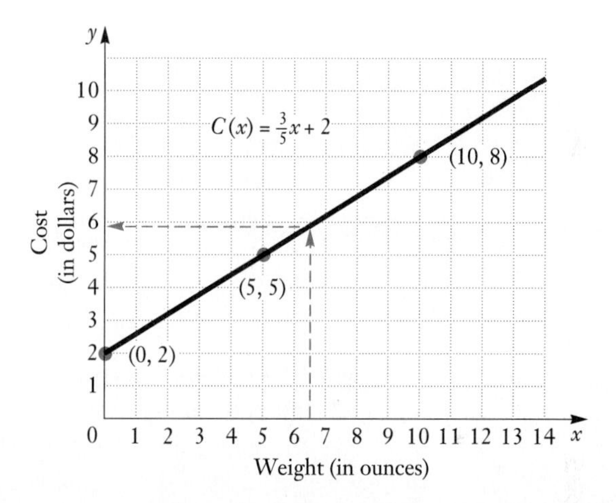

b) To estimate the cost of shipping $6\frac{1}{2}$ oz, we determine the coordinate on the vertical (cost) axis that appears to be paired with the coordinate $6\frac{1}{2}$ on the horizontal (weight) axis. We do so by drawing a vertical line segment from $6\frac{1}{2}$ up to the line and then a horizontal segment from the line over to the cost axis.

We estimate that it will cost approximately \$6.00 to ship $6\frac{1}{2}$ oz of chocolates.

A more precise calculation is made by substitution into the function formula:

$$C\left(6\tfrac{1}{2}\right) = C\left(\tfrac{13}{2}\right) = \tfrac{3}{5} \cdot \tfrac{13}{2} + 2$$
$$= \$5.90.$$

■

Graphs That Are Parallel to an Axis

Some equations have graphs that are parallel to one of the axes. These equations have a missing variable.

EXAMPLE 9 Graph $y = 3$. Determine whether the graph is that of a function.

Solution Because x is missing, the pairs $(-1, 3)$, $(0, 3)$, and $(2, 3)$ all satisfy the equation. In fact, any pair of the form $(x, 3)$ will work. The graph is a line parallel to the x-axis.

x	y
-1	3
0	3
2	3

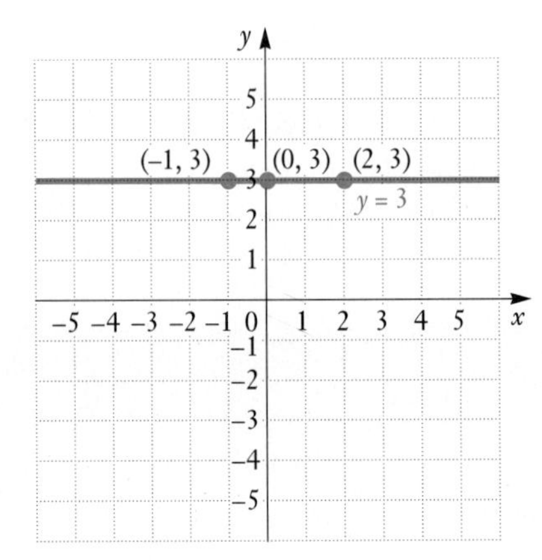

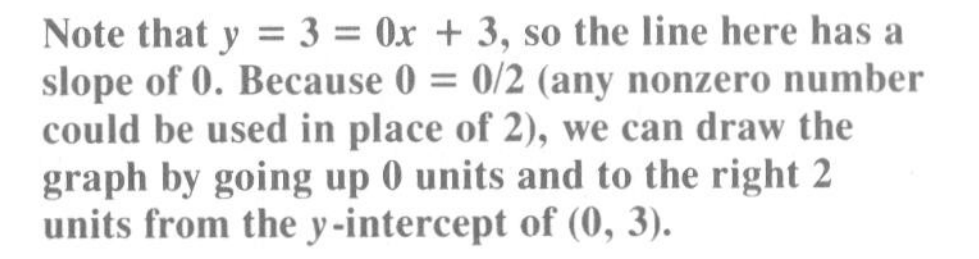

Note that $y = 3 = 0x + 3$, so the line here has a slope of 0. Because $0 = 0/2$ (any nonzero number could be used in place of 2), we can draw the graph by going up 0 units and to the right 2 units from the y-intercept of $(0, 3)$.

Since the graph satisfies the vertical line test, the graph is a function. ■

EXAMPLE 10 Graph $x = -2$. Determine whether the graph is that of a function.

Solution With y missing, any number for y will do. Thus we know that all ordered pairs of the form $(-2, y)$ are solutions. The graph is a line parallel to the y-axis. Among other pairs, it contains the pairs $(-2, -4)$, $(-2, 0)$, and $(-2, 3)$.

x	y
-2	3
-2	0
-2	-4

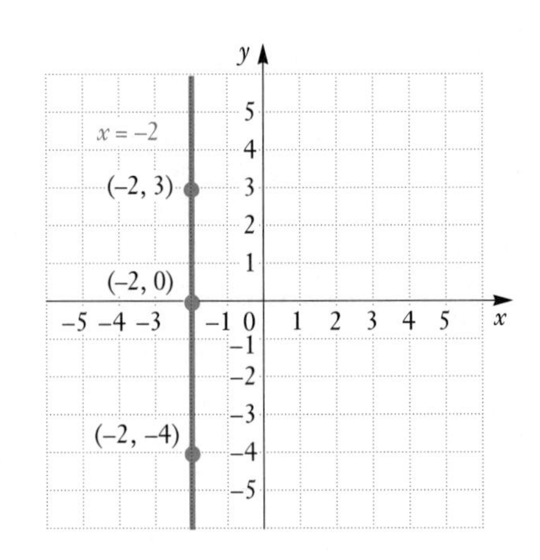

Because y is absent from this equation, it is impossible to write it in the form $y = mx + b$.

Recalling our definition of a function, we note that since the first coordinate, -2, is paired with more than one second coordinate, the graph is *not* that of a function. The graph *fails* the vertical line test. ■

Which Equations Are Linear?

Let us now consider an equation of the form $Ax + By = C$, where A, B, and C are real numbers. Suppose that A and B are both nonzero and solve for y:

$$Ax + By = C$$

$$By = -Ax + C \qquad \text{Adding } -Ax \text{ on both sides}$$

$$y = -\frac{A}{B}x + \frac{C}{B}. \qquad \text{Dividing by } B \text{ on both sides}$$

Since the last equation is a slope–intercept equation, we see that $Ax + By = C$ is a linear equation when $A \neq 0$ and $B \neq 0$.

Suppose next that A or B (but not both) is 0. If $A = 0$, then $By = C$ and $y = C/B$. If $B = 0$, then $Ax = C$ and $x = C/A$. In the first case the graph is a horizontal line, and in the second case the line is vertical. Thus $Ax + By = C$ is a linear equation when A or B (but not both) is 0. We have now justified the following result.

THE STANDARD FORM OF A LINEAR EQUATION

Any equation of the form $Ax + By = C$, where A, B, and C are real numbers and A and B are not both 0, is linear.

Any equation in the form $Ax + By = C$ is said to be a linear equation in *standard form*.

EXAMPLE 11 Determine whether the equation $y = x^2 - 5$ is linear.

Solution We attempt to put the equation in standard form.

$$y = x^2 - 5$$

$$-x^2 + y = -5 \qquad \text{Adding } -x^2 \text{ on both sides}$$

This last equation is not linear because it has an x^2-term. From Example 9 in Section 6.1, we see that the graph is not a straight line. ■

EXAMPLE 12 Determine the slope and the y-intercept for the linear equation $2x - 3y = 6$.

Solution We convert to a slope–intercept equation:

$$2x - 3y = 6$$

$$-3y = -2x + 6 \qquad \text{Adding } -2x$$

$$y = -\tfrac{1}{3}(-2x + 6) \qquad \text{Multiplying by } -\tfrac{1}{3}$$

$$y = \tfrac{2}{3}x - 2. \qquad \text{Using the distributive law}$$

Because we have an equation $y = mx + b$, we know that the slope is $\frac{2}{3}$ and the y-intercept is $(0, -2)$. ■

EXERCISE SET 6.3

Determine the slope and the y-intercept.

1. $y = 4x + 5$
2. $y = 5x + 3$
3. $f(x) = -2x - 6$
4. $g(x) = -5x + 7$
5. $y = -\frac{3}{8}x - 0.2$
6. $y = \frac{15}{7}x + 2.2$
7. $g(x) = 0.5x - 9$
8. $f(x) = -3.1x + 5$
9. $y = 7$
10. $y = -2$
11. $f(x) = 3.7$
12. $g(x) = 5.2$

Give an equation for a line that will have the indicated slope and y-intercept.

13. Slope $\frac{2}{3}$, y-intercept -7
14. Slope $-\frac{3}{4}$, y-intercept 5
15. Slope -4, y-intercept 2
16. Slope 2, y-intercept -1
17. Slope $-\frac{7}{9}$, y-intercept 3
18. Slope $-\frac{4}{11}$, y-intercept 9
19. Slope 5, y-intercept $\frac{1}{2}$
20. Slope 6, y-intercept $\frac{2}{3}$
21. Slope 0.7, y-intercept 3.8
22. Slope 1.7, y-intercept -4.3

Determine the slope and the y-intercept. Then draw a graph. Be sure to use a third point as a check.

23. $y = \frac{5}{2}x + 1$
24. $y = \frac{2}{5}x + 4$
25. $f(x) = -\frac{5}{2}x + 4$
26. $f(x) = -\frac{2}{5}x + 3$
27. $y = 2x - 5$
28. $y = -2x + 4$
29. $f(x) = \frac{1}{3}x + 6$
30. $f(x) = -3x + 6$
31. $f(x) = -0.25x + 2$
32. $f(x) = 1.5x - 3$
33. $y = \frac{4}{5}x - 2$
34. $y = -\frac{5}{4}x + 1$
35. $f(x) = \frac{5}{4}x - 2$
36. $f(x) = \frac{4}{3}x + 2$
37. $f(x) = 4$
38. $f(x) = -1$

39. *Cost of a taxi ride.* The cost, in dollars, of a taxi ride in Pelham is given by $C(m) = 0.25m + 2$, where m is the number of miles traveled.

a) Graph $C(m) = 0.25m + 2$.
b) Use the graph to estimate the cost of a $5\frac{1}{2}$-mi taxi ride.

40. *Cost of renting a car.* The cost, in dollars, of renting a car for a day is given by $C(m) = 0.3m + 10$, where m is the number of miles driven.

a) Graph $C(m) = 0.3m + 10$.
b) Use the graph to estimate the cost of driving the rental car 40 mi.

41. *Natural gas demand.* The demand, in quadrillions of joules, for natural gas can be approximated by $D(t) = \frac{1}{5}t + 20$, where t is the number of years after 1960.

a) Graph the equation $D(t) = \frac{1}{5}t + 20$.
b) Use the graph to predict the demand for gas in 1995.

42. *Cricket chirps per minute.* The number of cricket chirps per minute is given by $N(t) = 7.2t - 32$, where t is the temperature in degrees Celsius.

a) Graph the equation $N(t) = 7.2t - 32$.
b) Use the graph to predict the number of cricket chirps per minute when it is 5°C.

43. *Cost of a telephone call.* The cost, in dollars, of a long distance telephone call is approximated by $C(m) = 0.80m + 1$, where m is the length of the call in minutes.

a) Graph $C(m) = 0.80m + 1$.
b) Use the graph to approximate the cost of a 4-min call.
c) Use the graph to determine how long a phone call can be made for $7.40.

44. *Life expectancy of American women.* The life expectancy of American women is given by $A(t) = \frac{3}{20}t + 72$, where t is the number of years since 1950.

a) Graph $A(t) = \frac{3}{20}t + 72$.
b) Use the graph to predict the life expectancy of American women in the year 2000.
c) Use the graph to determine the year in which the life expectancy of American women is 78 yr.

Graph.

45. $3y = 9$
46. $6g(x) + 24 = 0$
47. $3x = -15$
48. $2x = 10$

49. $4g(x) + 3x = 12 + 3x$

50. $6x - 4y + 12 = -4y$

51. $6y + 3x = -2(4 - 3y)$

52. $2x + 5y = 7 - 2y + 2x$

Determine whether the equation is linear. Find the slope of any nonvertical lines.

53. $3x + 5f(x) + 15 = 0$

54. $5x - 3f(x) = 15$

55. $3x - 12 = 0$

56. $16 + 4y = 0$

57. $2x + 4g(x) = 19$

58. $3g(x) = 5x^2 + 4$

59. $5x - 4xy = 12$

60. $3y = 7xy - 5$

61. $\frac{3y}{4x} = 5y + 2$

62. $6y - \frac{4}{y} = 0$

63. $f(x) = x^3$

64. $f(x) = \frac{1}{2}x^2$

SKILL MAINTENANCE

65. Multiply:

$$\sqrt{2}(\sqrt{7x} - \sqrt{5}).$$

66. A piece of wire 32.8 ft long is to be cut into two pieces, and those pieces are each to be bent to make a square. The length of a side of one square is to be 2.2 ft greater than the length of a side of the other. How should the wire be cut?

67. Simplify:

$$9\{2x - 3[5x + 2(-3x + y^0 - 2)]\}.$$

68. Solve:

$$3[7x - 2(4 + 5x)] = 5(x + 2) - 7.$$

SYNTHESIS

In Exercises 69–78 assume that a, b, and c are constants and that x and y are variables. Determine the slope and the y-intercept.

69. $ay = -5x + b$

70. $bx + 2 = 4y - 9a$

71. $ax + by = c$

72. $ax + by = c - ay$

Determine whether the equation is linear.

73. $ax + 3y = b - c$

74. $by = cx - dy + 2$

75. $a^2x = by + 5$

76. $axy + cy = 2x$

77. $\frac{x}{a} - by = 17$

78. $axy + \frac{b}{x} = -axy + 2$

79. The graph of a linear function passes through the points (0, 3.1) and (2, 7.8). Write an equation for the function.

80. The graph of a linear function passes through the points (1, 3) and (4, −2). Write an equation for the function. (*Hint:* Find m and then solve for b in $y = mx + b$.)

6.4 Another Look at Linear Graphs

Graphing Using Intercepts

Once we have determined that an equation is linear, we can sometimes draw its graph very quickly by plotting the y-intercept and the x-intercept and then drawing a straight line through those points. Note that to find the y-intercept we replace x with 0 and solve for y. To find the x-intercept we replace y with 0 and solve for x.

> The x-intercept is $(a, 0)$. To find a let $y = 0$ and solve the original equation for x.
> The y-intercept is $(0, b)$. To find b let $x = 0$ and solve the original equation for y.

EXAMPLE 1 Graph the equation $3x + 2y = 12$ by using intercepts.

Solution This equation is linear. *To find the y-intercept, we let $x = 0$.* Then we solve for y. We have

$$3 \cdot 0 + 2y = 12$$
$$2y = 12, \qquad \text{or } y = 6.$$

The y-intercept is $(0, 6)$.

To find the x-intercept, we let $y = 0$. Then we solve for x. We have

$$3x + 2 \cdot 0 = 12$$
$$3x = 12, \qquad \text{or } x = 4.$$

The x-intercept is $(4, 0)$.

We plot the two intercepts and draw the line. A third point could be calculated and used as a check.

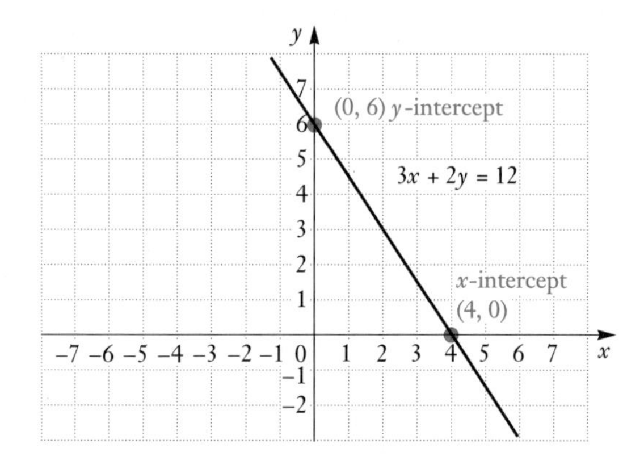

Finding the Slope

Knowing the coordinates of *any* two points on a line, we can calculate the slope of the line. To do this we divide the vertical distance between the points, the *rise,* by the horizontal distance, the *run*. That is, "slope = rise/run." Equivalently, we often say "slope = change in y/change in x."

In the following definition (x_1, y_1) and (x_2, y_2)—read "x sub-one, y sub-one and x sub-two, y sub-two"—represent any two distinct points on a line. The letter m is traditionally used and has its roots in the French verb *monter,* to climb.

> The *slope* of a line containing points (x_1, y_1) and (x_2, y_2) is given by
>
> $$m = \frac{\text{rise}}{\text{run}} = \frac{\text{the change in } y}{\text{the change in } x} = \frac{y_2 - y_1}{x_2 - x_1}.$$

EXAMPLE 2 Two points of a line are (1, 3) and (5, 6). Find the slope of the line.

Solution We graph the points and draw the line. Then we mark the rise (the distance *up*) and the run (the distance *over*) by completing a right triangle. The rise is 3 (from 3 to 6) and the run is 4 (from 1 to 5).

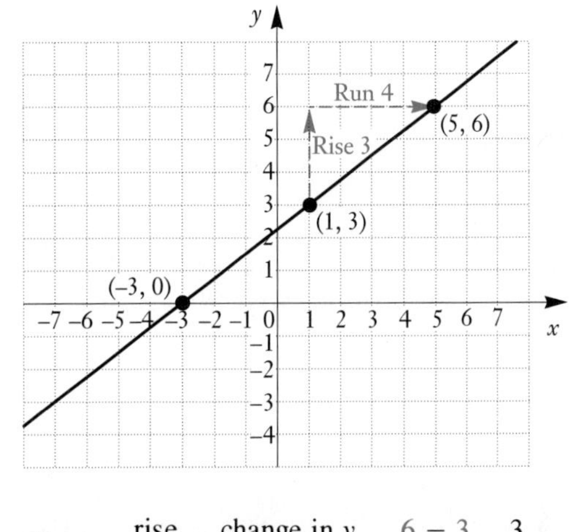

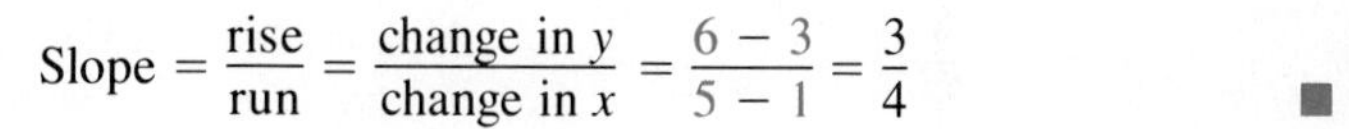

$$\text{Slope} = \frac{\text{rise}}{\text{run}} = \frac{\text{change in } y}{\text{change in } x} = \frac{6-3}{5-1} = \frac{3}{4}$$

■

EXAMPLE 3 Find the slope of the line in Example 2 by using a different pair of points.

Solution We can use *any* two points. Let's choose $(-3, 0)$ and (5, 6). The rise is 6 (from 0 to 6), and the run is 8 (from -3 to 5).

$$\text{Slope} = \frac{6-0}{5-(-3)} = \frac{6}{8} = \frac{3}{4}$$

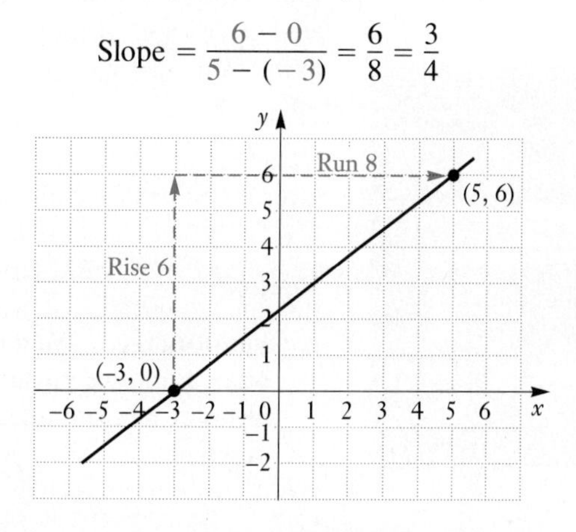

■

EXAMPLE 4 Graph the line containing the points $(1, -3)$ and $(-2, 2)$, and find the slope.

Solution The graph reveals that from $(1, -3)$ to $(-2, 2)$, the change in y, or the rise, is $2 - (-3)$, or 5. The change in x, or the run, is $-2 - 1$, or -3.

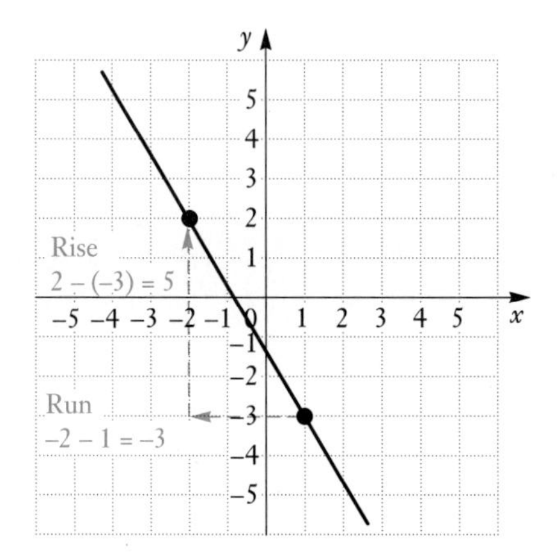

$$\text{Slope} = \frac{5}{-3} = -\frac{5}{3}$$

Suppose that, in Example 4, we subtracted coordinates in the opposite order. For the change in y we would get $-3 - 2$, or -5. For the change in x we would get $1 - (-2)$, or 3. For the slope, we get

$$\text{Slope} = \frac{-5}{3} = -\frac{5}{3}.$$

This is the same answer that we got before. It does not matter which point is considered (x_1, y_1) and which point is considered (x_2, y_2) as long as we subtract coordinates in the same order in both the numerator and the denominator.

EXAMPLE 5 A racer bikes 3 km in 4 min and 9 km in 12 min. Assuming that the racer maintains a steady pace, determine the rate of travel.

Solution

1. **Familiarize.** We might familiarize ourselves with this problem in one of two ways. First, we might note that $d = r \cdot t$ is a formula that expresses distance in terms of rate and time. Second, we might use a graph to give a "picture" of the problem.

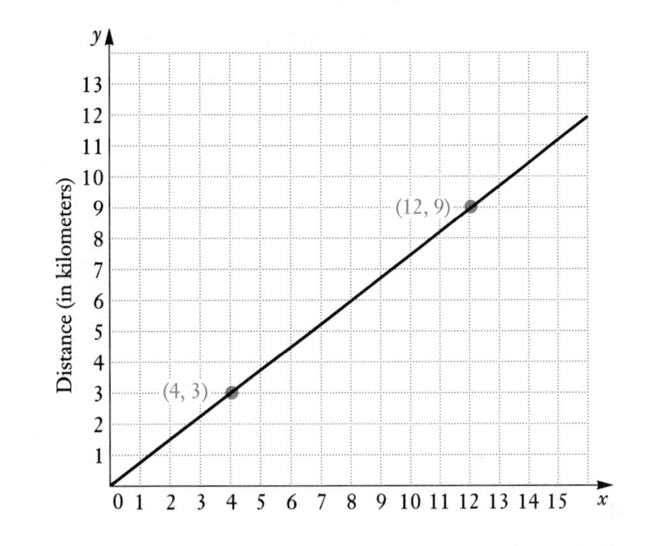

2. **Translate.** Using the first approach, we could substitute either 4 for t and 3 for d, or 12 for t and 9 for d. Thus we could form the equation

$$3 \text{ km} = r \cdot 4 \text{ min} \qquad \text{or} \qquad 9 \text{ km} = r \cdot 12 \text{ min}.$$

Using the second approach, we could form the equation

$$\text{Rate} = \text{change in distance/change in time.}$$

Thus we could form the equation

$$r = (9 \text{ km} - 3 \text{ km})/(12 \text{ min} - 4 \text{ min}).$$

Note that this is exactly the formula for the slope of the line drawn in the graph.

3. **Carry out.** Using the first approach, we solve

$$3 = r \cdot 4 \qquad \text{or} \qquad 9 = r \cdot 12$$

to obtain $r = \frac{3}{4}$ km/min.

Using the second approach, we solve

$$r = \frac{9 - 3}{12 - 4} = \frac{6}{8} = \frac{3}{4} \text{ km/min.}$$

4. **Check.** If the rate is $\frac{3}{4}$ km/min, in 4 min the racer has biked $\frac{3}{4} \cdot 4 = 3$ km, and in 12 minutes, $\frac{3}{4} \cdot 12 = 9$ km. Our answer checks. The fact that both approaches produced the same answer serves as another check.

5. **State.** The racer bikes at a rate of $\frac{3}{4}$ km/min. ■

In general, the slope of a line can be regarded as a rate of change—that is, the amount of vertical change corresponding to one unit of horizontal change. Thus a slope of $\frac{3}{4}$ can be interpreted as a rise of $\frac{3}{4}$ unit for each horizontal change of 1 unit.

Zero Slope and Lines with Undefined Slope

If two different points have the same second coordinate, what about the slope of the line joining them? Since, in this case, $y_2 = y_1$, we have

$$\frac{y_2 - y_1}{x_2 - x_1} = \frac{0}{x_2 - x_1} = 0.$$

Every horizontal line has a slope of 0.

Suppose that two different points are on a vertical line. Then they have the same first coordinate. In this case we have $x_2 = x_1$, so

$$\frac{y_2 - y_1}{x_2 - x_1} = \frac{y_2 - y_1}{0}.$$

Since we cannot divide by 0, this is undefined.

The slope of a vertical line is undefined.

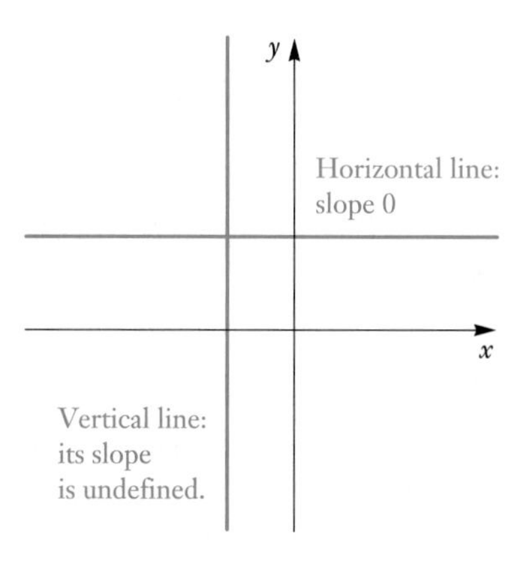

We conclude this section by proving that m in the equation $y = mx + b$ does indeed represent the slope of the line.

Proof. Observe that $(0, b)$ and $(1, m + b)$ are solutions to any equation of the form $y = mx + b$. Using the slope formula, we see that the slope of the line is given by

$$\text{Slope} = \frac{y_2 - y_1}{x_2 - x_1} = \frac{(m + b) - b}{1 - 0}$$
$$= m.$$

■

EXERCISE SET 6.4

Find the intercepts. Then graph, using the intercepts and a third point as a check.

1. $x - 2 = y$

2. $x - 4 = y$

3. $3x - 1 = y$

4. $3x - 4 = y$

5. $5x - 4y = 20$

6. $3x + 5y = 15$

7. $y = -5 - 5x$

8. $y = -2 - 2x$

9. $5y = -15 + 3x$

10. $7x = 3y - 21$

11. $6x - 7 + 3y = 9x - 2y + 8$

12. $7x - 8 + 4y = 8y + 4x + 4$

13. $1.4y - 3.5x = -9.8$

14. $3.6x - 2.1y = 22.68$

15. $5x + 2y = 7$

16. $3x - 4y = 10$

For each pair of points, find the slope, if it exists, of the line containing them.

17. $(6, 9)$ and $(4, 5)$

18. $(8, 7)$ and $(2, -1)$

19. $(3, 8)$ and $(9, -4)$

20. $(17, -12)$ and $(-9, -15)$

21. $(-8, -7)$ and $(-9, -12)$

22. $(14, 3)$ and $(2, 12)$

23. $(-16.3, 12.4)$ and $(-5.2, 8.7)$

24. $(14.4, -7.8)$ and $(-12.5, -17.6)$

25. $(3.2, -12.8)$ and $(3.2, 2.4)$

26. $(-1.5, 7.6)$ and $(-1.5, 8.8)$

27. $(7, 3.4)$ and $(-1, 3.4)$

28. $(-3, 4.2)$ and $(5.1, 4.2)$

29. $(0, 9.1)$ and $(9.1, 0)$

30. $(4.3, 0)$ and $(0, -4.3)$

31. *Running rate.* An Olympic marathoner passes the 5-km point of a race after 30 min and reaches the 25-km point after 2.5 hr. Find the speed of the marathoner.

32. *Skiing rate.* A cross-country skier travels 3 km in 15 min and 12 km in 1 hr. Find the speed of the skier.

33. *Rate of production.* At the beginning of a production run, 4.5 tons of sugar had already been refined. Six hours later, the total amount of refined sugar reached 8.1 tons. Calculate the rate of production.

34. *Work rate.* As a painter begins work, one fourth of a house has already been painted. Eight hours later, the house is two-thirds done. Calculate the painter's work rate.

35. *Rate of descent.* A plane descends to sea level from 12,000 ft after being airborne for $1\frac{1}{2}$ hr. The entire flight time is 2 hr and 10 min. Determine the plane's average rate of descent.

36. *Rate of descent.* A climber leaves a 6200-ft peak at 1:40 P.M. and reaches the base (elevation 950 ft) at 6:10 P.M. Find the climber's average rate of descent.

37. *Sales rate of Avon Products, Co.* In 1985, the total sales of Avon Products, Co., were about \$2.0 million. In 1987 they were about \$2.8 million. Determine the rate at which sales are increasing.

38. In 1985, the number of U.S. visitors overseas was about 7.5 million. In 1988 the number grew to 11.6 million. Determine the rate at which the number of U.S. visitors overseas is growing.

For each equation find the slope, if it exists. If the slope is undefined, state this.

39. $3x = 12 + y$

40. $5y - 12 = 3x$

41. $5x - 6 = 15$

42. $-12 = 4x - 7$

43. $5y = 6$

44. $19 = -6y$

45. $y - 6 = 14$

46. $3y - 5 = 8$

47. $12 - 4x = 9 + x$

48. $15 + 7x = 3x - 5$

49. $2y - 4 = 35 + x$

50. $2x - 17 + y = 0$

51. $3y + x = 3y + 2$

52. $x - 4y = 12 - 4y$

53. $3y - 2x = 5 + 9y - 2x$

54. $17y + 4x + 3 = 7 + 4x$

SKILL MAINTENANCE

55. Factor:

$$7x^2 - 175.$$

56. Rationalize the denominator:

$$\frac{3}{4 - \sqrt{5}}.$$

57. The formula

$$f = \frac{F(c - v_0)}{c - v_s}$$

is used when studying the physics of sound. Solve for F.

58. Solve for y:

$$m = \frac{y - y_1}{x - x_1}.$$

SYNTHESIS

59. A line contains the points $(-100, 4)$ and $(0, 0)$. List four more points of the line.

60. Give an equation, in standard form, for the line whose x-intercept is 5 and whose y-intercept is -4.

61. Find the x-intercept of $y = mx + b$, assuming $m \neq 0$.

62. Determine a so that the slope of the line through this pair of points has the value m.

$$(-2, 3a), (4, -a); \quad m = -\tfrac{5}{12}$$

63. Find the slope of the line that contains the pair of points.

a) $(5b, -6c), (b, -c)$
b) $(b, d), (b, d + e)$
c) $(c + f, a + d), (c - f, -a - d)$

64. Suppose that two linear equations have the same y-intercept but that equation A has an x-intercept that is half the x-intercept of equation B. How do the slopes compare?

6.5 Other Equations of Lines

Specifying a line's slope and one point through which the line passes enables us to draw the line. In this section we study how this same information can be used to produce an *equation* for the line.

Point–Slope Equations

EXAMPLE 1 Find an equation for a line of slope $\frac{3}{4}$ that passes through the point $(1, 2)$.

Solution Because any other point (x, y) on the line satisfies the slope formula, we must have

$$\frac{y - 2}{x - 1} = \frac{3}{4}.$$

Thus any pair (x, y) that is a solution of the preceding equation lies on the graph. Unfortunately, the point $(1, 2)$ itself cannot be substituted because we would obtain 0/0. We can avoid this difficulty as follows:

$$\frac{y-2}{x-1} = \frac{3}{4}$$

$$(x-1)\cdot\frac{y-2}{x-1} = \frac{3}{4}(x-1) \qquad \text{Multiplying by } x-1$$

$$y - 2 = \frac{3}{4}(x-1). \qquad \text{Simplifying}$$

The pair $(1, 2)$ *is* a solution of this last equation. ■

As a generalization of Example 1, let's now suppose that a line of slope m passes through the point (x_1, y_1). For any other point (x, y) to lie on this line, we must have

$$\frac{y - y_1}{x - x_1} = m.$$

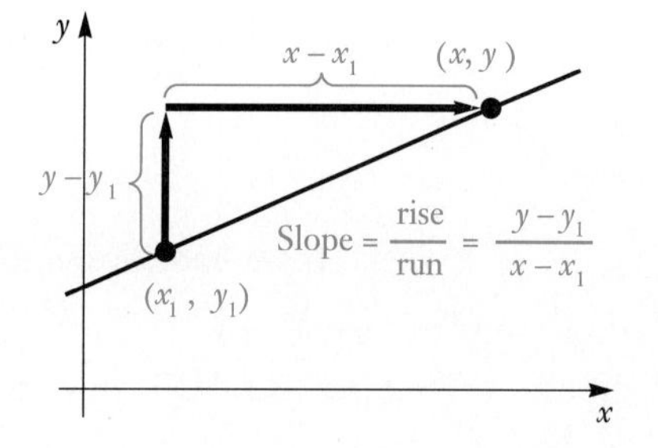

Multiplying by $x - x_1$ gives us the equation we are looking for.

> The *point–slope equation* of a line is
> $$y - y_1 = m(x - x_1).$$

EXAMPLE 2 Find an equation of the line containing the point $(5, -1)$ with slope $-\frac{1}{2}$. Write the equation using function notation.

Solution We substitute in the equation $y - y_1 = m(x - x_1)$:

$$
\begin{aligned}
y - y_1 &= m(x - x_1) \\
y - (-1) &= -\tfrac{1}{2}(x - 5) && \text{Substituting} \\
y + 1 &= -\tfrac{1}{2}x + \tfrac{5}{2} && \text{Simplifying} \\
y &= -\tfrac{1}{2}x + \tfrac{3}{2} && \text{Subtracting 1 on both sides} \\
f(x) &= -\tfrac{1}{2}x + \tfrac{3}{2}. && \text{Using function notation}
\end{aligned}
$$

■

Because the slope of a vertical line is undefined, vertical lines do not have point–slope equations. All other lines do.

Sometimes some preliminary work must be done before we can find a point–slope equation. Such is the case in the next example.

EXAMPLE 3 Find an equation for the line passing through the points (1, 4) and (3, 7).

Solution We first determine the slope of the line and then use the point–slope equation. Note that

$$m = \frac{7 - 4}{3 - 1} = \frac{3}{2}.$$

Since the line passes through (1, 4), we have

$$y - 4 = \frac{3}{2}(x - 1). \qquad \textbf{(1)}$$

If we had chosen to use the point (3, 7), we would have obtained

$$y - 7 = \frac{3}{2}(x - 3). \qquad \textbf{(2)}$$

Equations (1) and (2) may be rewritten as $y = \frac{3}{2}x + \frac{5}{2}$ by distributing the $\frac{3}{2}$ and solving for y. In this manner we can show that Eqs. (1) and (2) are equivalent. ■

EXAMPLE 4 A car dealership's sales figures indicate that 160 new cars were sold in 1986 and 200 new cars were sold in 1990. Assuming constant growth since 1985, how many new cars can the dealership expect to sell in 1997?

Solution

1. **Familiarize.** We form the pairs (1, 160) and (5, 200) and plot them after choosing suitable scales on the two axes. A constant growth rate means that a linear relationship exists between years and number of cars sold. We will let n represent the number of cars sold and t represent the number of years since 1985.

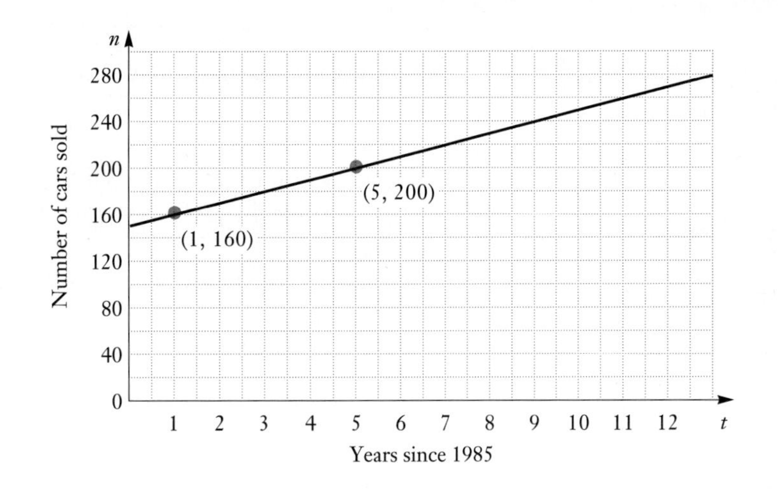

2. **Translate.** We seek an equation relating n to t. To accomplish this we find the slope and use the point–slope form:

$$m = \frac{200 - 160}{5 - 1} = \frac{40}{4} = 10.$$

Thus,

$$n - 160 = 10(t - 1)$$

$$n - 160 = 10t - 10 \qquad \text{Using the distributive law}$$

$$n = 10t + 150. \qquad \text{Adding 160 on both sides}$$

3. **Carry out.** Using function notation, we have

$$n(t) = 10t + 150.$$

To predict sales in 1997, we find

$$n(12) = 10 \cdot 12 + 150 \qquad \text{1997 is 12 years from 1985.}$$

$$= 270.$$

4. **Check.** To check we could duplicate our calculations. Our number does appear to be reasonable, and the pair (12, 270) appears to lie on the graph.

5. **State.** The dealership can expect to sell about 270 cars in 1997. ■

Note that in Example 4 because of the scale used on the two axes, a slope of 10 does not appear to be as steep as it otherwise might be.

Parallel and Perpendicular Lines

If two lines are vertical, they are parallel. How can we tell whether nonvertical lines are parallel? The answer is simple: We look at their slopes.

> Two nonvertical lines are parallel if they have the same slope.

EXAMPLE 5 Determine whether the line passing through the points (1, 7) and (4, −2) is parallel to the line $f(x) = -3x + 4.2$.

Solution The slope of the line passing through (1, 7) and (4, −2) is given by

$$m = \frac{7-(-2)}{1-4} = \frac{9}{-3} = -3.$$

Because the function given by $f(x) = -3x + 4.2$ also has a slope of −3, we conclude that the lines are parallel. ■

If one line is vertical and another is horizontal, they are perpendicular. There are other instances in which two lines are perpendicular.

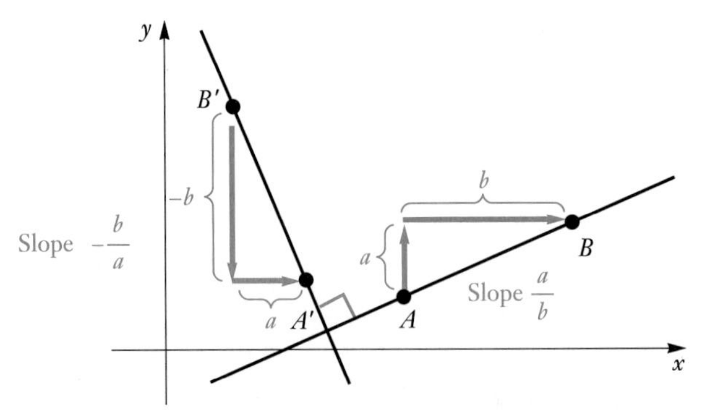

Consider a line $\overleftrightarrow{AB}$ as shown, with slope a/b. Then think of rotating the figure 90° to get a line $\overleftrightarrow{A'B'}$ perpendicular to $\overleftrightarrow{AB}$. For the new line the rise and the run are interchanged, but the run is now negative. Thus the slope of the new line is $-b/a$. Let us multiply the slopes:

$$\frac{a}{b}\left(-\frac{b}{a}\right) = -1.$$

This is the condition under which lines will be perpendicular.

> Two lines are perpendicular if the product of their slopes is −1. (If one line has slope m, the slope of a line perpendicular to it is $-1/m$. That is, we take the reciprocal and change the sign.) Lines are also perpendicular if one of them is vertical and the other is horizontal.

EXAMPLE 6 Given the line $8y = 7x - 24$ and the point $(-1, 2)$,

Solution

a) Find an equation for the line containing the point and parallel to the line.
We find the slope:

$$y = \tfrac{7}{8}x - 3. \quad \text{Solving for } y. \text{ The slope is } \tfrac{7}{8}.$$

Now we use the point–slope equation:

$$y - 2 = \tfrac{7}{8}[x - (-1)], \quad \text{or} \quad y = \tfrac{7}{8}x + \tfrac{23}{8}. \quad \text{Substituting } \tfrac{7}{8} \text{ for the slope and } (-1, 2) \text{ for the point}$$

b) Find an equation for the line containing the point and perpendicular to the line.
To find the slope, we take the reciprocal of $\frac{7}{8}$ and then change the sign. The slope is $-\frac{8}{7}$.

$$y - 2 = -\tfrac{8}{7}[x - (-1)], \quad \text{or} \quad y = -\tfrac{8}{7}x + \tfrac{6}{7}. \quad \text{Substituting } -\tfrac{8}{7} \text{ for the slope and } (-1, 2) \text{ for the point}$$

EXERCISE SET 6.5

Find an equation of the line having the specified slope and containing the point indicated. Leave your answer in point–slope form.

1. $m = 4$, $(3, 2)$
2. $m = 5$, $(5, 4)$
3. $m = -2$, $(4, 7)$
4. $m = -3$, $(7, 3)$
5. $m = 3$, $(-2, -4)$
6. $m = 1$, $(-5, -7)$
7. $m = -2$, $(8, 0)$
8. $m = -3$, $(-2, 0)$
9. $m = 0$, $(0, -7)$
10. $m = 0$, $(0, 4)$
11. $m = \frac{3}{4}$, $(5, -1)$
12. $m = \frac{2}{5}$, $(-3, 7)$

Find an equation of the line having the specified slope and containing the indicated point. Write your final answer in slope–intercept form.

13. $m = 5$, $(2, -3)$
14. $m = -4$, $(-1, 5)$
15. $m = -\frac{2}{3}$, $(4, -7)$
16. $m = \frac{3}{7}$, $(-2, 1)$
17. $m = -0.6$, $(-3, -4)$
18. $m = -3.1$, $(5, -2)$

Find a function for the line containing the pair of points.

19. $(1, 4)$ and $(5, 6)$
20. $(2, 6)$ and $(4, 1)$
21. $(-1, -1)$ and $(2, 2)$
22. $(-3, -3)$ and $(6, 6)$
23. $(-2, 0)$ and $(0, 5)$
24. $(6, 0)$ and $(0, -3)$
25. $(3, 5)$ and $(-5, 3)$
26. $(4, 6)$ and $(-6, 4)$
27. $(0, 0)$ and $(5, 2)$
28. $(0, 0)$ and $(7, 3)$
29. $(-4, -7)$ and $(-2, -1)$
30. $(-2, -3)$ and $(-4, -6)$

31. *Records in the 400-meter run.* In 1930 the record for the 400-m run was 46.8 sec. In 1970 it was 43.8 sec. Let R represent the record in the 400-m run and t the number of years since 1930.

a) Fit a linear function to the data points.
b) Use the function of part (a) to predict the record in 1990; in 2000.
c) When will the record be 40 seconds?

32. *Records in the 1500-meter run.* In 1930 the record for the 1500-m run was 3.85 min. In 1950 it was 3.70 min. Let R represent the record in the 1500-m run and t the number of years since 1930.

a) Fit a linear function to the data points.
b) Use the function of part (a) to predict the record in 1984; in 1989.
c) When will the record be 3.3 minutes?

33. *Consumer's demand.* Suppose that when the price of coffee is \$4 per pound, 6.5 million lb are sold; and when the price is \$5 per pound, 4.0 million lb are sold.

a) Fit a linear function to the data points.
b) Use the function of part (a) to predict consumer demand if the price were to drop to \$2 per lb.

34. *Life expectancy of males in the United States.* In 1950 the life expectancy of males was 65 years. In 1970 it was 68 years. Let E represent life expectancy and t the number of years since 1950.

a) Fit a linear function to the data points.
b) Use the function of part (a) to predict the life expectancy of males in 1988; in 1995.

35. *Seller's supply.* Suppose that when the price of coffee is \$4 per pound, suppliers are willing to sell 5.0 million lb; and when the price is \$5 per pound, they will supply 7.0 million lb.

a) Fit a linear function to the data points.
b) Use the function of part (a) to predict the amount that suppliers are willing to sell when the price drops to \$2 per pound.

36. *Temperature as a function of depth.* The temperature 5 km beneath the earth's surface is 70°C and the temperature 10 km beneath the surface is 120°C.

a) Fit a linear function to the data points.
b) Use the function of part (a) to determine the temperature 17 km beneath the earth's surface.

37. *Items in a supermarket.* In 1981 the average supermarket contained 12,877 items. In 1985 that figure had grown to 17,459 items. Assuming constant growth, predict the average number of items to be found in a supermarket in the year 1995.

38. *Pressure at sea depth.* The pressure 100 ft beneath the ocean's surface is approximately 4 atm (atmospheres), whereas at a depth of 200 ft, the pressure is about 7 atm.

a) Fit a linear function to the data points.
b) Use the function of part (a) to determine the pressure at a depth of 690 ft.

39. *Country radio stations.* The number of country radio stations grew from 1534 in 1980 to 2275 in 1986. Assuming constant growth, predict the number of country radio stations in the year 1994.

40. *Advertising.* An automobile dealer discovers that when \$1000 is spent on radio advertising, weekly sales increase by \$101,000. When \$1250 is spent on radio advertising, sales increase by \$126,000. Suppose that the sales increase is a linear function of the amount of radio advertising. Calculate the increase in sales when (a) the amount spent on radio advertising is \$1500; (b) when it is \$2000.

Without graphing, tell whether the graphs of each pair of equations are parallel.

41. $x + 6 = y$, $y - x = -2$

42. $2x - 7 = y$, $y - 2x = 8$

43. $y + 3 = 5x$, $3x - y = -2$

44. $y + 8 = -6x$, $-2x + y = 5$

45. $y = 3x + 9$, $2y = 6x - 2$

46. $y = -7x - 9$, $-3y = 21x + 7$

Write an equation of the line containing the specified point and parallel to the indicated line.

47. $(3, 7)$, $x + 2y = 6$

48. $(0, 3)$, $3x - y = 7$

49. $(2, -1)$, $5x - 7y = 8$

50. $(-4, -5)$, $2x + y = -3$

51. $(-6, 2)$, $3x - 9y = 2$

52. $(-7, 0)$, $5x + 2y = 6$

53. $(-3, -2)$, $3x + 2y = -7$

54. $(-4, 3)$, $6x - 5y = 4$

Without graphing, tell whether the graphs of each pair of equations are perpendicular.

55. $y = 4x - 5$, $4y = 8 - x$

56. $2x - 5y = -3$, $2x + 5y = 4$

57. $x + 2y = 5$, $2x + 4y = 8$

58. $y = -x + 7$, $y = x + 3$

Write an equation of the line containing the specified point and perpendicular to the indicated line.

59. $(2, 5)$, $2x + y = -3$

60. $(4, 0)$, $x - 3y = 0$

61. $(3, -2)$, $3x + 4y = 5$

62. $(-3, -5)$, $5x - 2y = 4$

63. $(0, 9)$, $2x + 5y = 7$

64. $(-3, -4)$, $-3x + 6y = 2$

65. $(-4, -7)$, $3x - 5y = 6$

66. $(-4, 5)$, $7x - 2y = 1$

SKILL MAINTENANCE

67. Find the solution set:

$$3x - 12 = 2(x - 6) + x.$$

68. A basketball team increases its score by 7 points in each of three consecutive games. If the team scored a total of 228 points in all three games, what was its score in the first game?

69. Add: $\frac{3}{x^2 - 9} + \frac{2}{x - 3}$.

70. Subtract: $-\frac{7}{4} - \left(-\frac{2}{3}\right)$.

SYNTHESIS

Solve each problem, assuming that a linear equation fits the situation.

71. A piece of copper pipe has a length of 100 cm at 18°C. At 20°C the length of the pipe changes to 100.00356 cm. Find the length of the pipe at 40°C and at 0°C.

72. The value of a copying machine is \$5200 when it is purchased. After 2 years its value is \$4225. Find its value after 8 years.

73. *Sales commissions.* A person applying for a sales position is offered alternative salary plans.

Plan A: A base salary of \$600 per month plus a commission of 4% of the gross sales for the month

Plan B: A base salary of \$700 per month plus a commission of 6% of the gross sales for the month in excess of \$10,000

a) For each plan, formulate a function that expresses monthly earnings as a function of gross sales, x.

b) For what gross sales values is plan B preferable?

74. Water freezes at 32° Fahrenheit and at 0° Celsius. Water boils at 212°F and at 100°C. What Celsius temperature corresponds to a room temperature of 70°F?

75. For a linear function f, $f(-1) = 3$ and $f(2) = 4$.

a) Find an equation for f.
b) Find $f(3)$.
c) Find a such that $f(a) = 100$.

76. For a linear function g, $g(3) = -5$ and $g(7) = -1$.

a) Find an equation for g.
b) Find $g(-2)$.
c) Find a such that $g(a) = 75$.

77. Find the value of k so that $5y - kx = 7$ and the line containing the points $(7, -3)$ and $(-2, 5)$ are parallel.

78. Find the value of k so that $7y - kx = 9$ and the line containing the points $(2, -1)$ and $(-4, 5)$ are perpendicular.

6.6 The Algebra of Functions

Functions, like real numbers, can be added, subtracted, multiplied, and divided.

Forming New Functions

Let's return to the idea of a function as a machine. Suppose that some number a is in the domain of two functions. If the functions are defined by formulas for $f(x)$ and $g(x)$, the number a will be paired with the number $f(a)$ by one function and the number $g(a)$ by the other function. These values can then be added to obtain $f(a) + g(a)$.

EXAMPLE 1 Let $f(x) = x + 5$ and $g(x) = x^2$. Find $f(2) + g(2)$.

Solution We visualize two function machines. Because 2 is in the domain of each function, we can compute $f(2)$ and $g(2)$.

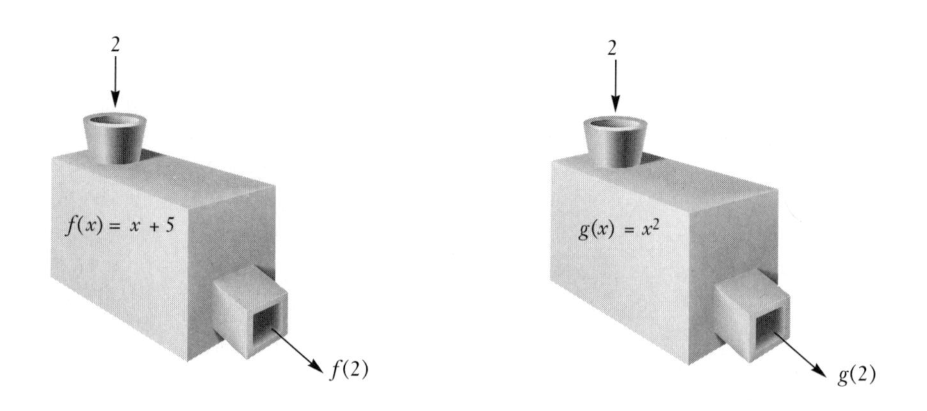

Since

$$f(2) = 2 + 5 = 7 \quad \text{and} \quad g(2) = 2^2 = 4,$$

we have

$$f(2) + g(2) = 7 + 4 = 11.$$

In Example 1, suppose that we first added the functions together. We would then obtain $f(x) + g(x) = x^2 + x + 5$, which can be regarded as a "new" function. The notation $(f + g)(x)$ is generally used to denote a function formed in this manner. A similar notation exists for the subtraction, multiplication, and division of functions.

THE ALGEBRA OF FUNCTIONS

If f and g are functions and x is in the domain of both functions, then:

1. $(f + g)(x) = f(x) + g(x)$;
2. $(f - g)(x) = f(x) - g(x)$;
3. $(f \cdot g)(x) = f(x) \cdot g(x)$;
4. $(f/g)(x) = f(x)/g(x)$, provided $g(x) \neq 0$.

EXAMPLES For the following, let $f(x) = x^2 - 1$ and $g(x) = x + 2$.

2. Find $(f + g)(3)$.

Since $f(3) = 3^2 - 1 = 8$ and $g(3) = 3 + 2 = 5$, we have

$$\begin{aligned}(f + g)(3) &= f(3) + g(3)\\ &= 8 + 5 \quad \text{Substituting}\\ &= 13.\end{aligned}$$

Alternatively, we could first find $(f + g)(x)$:

$$\begin{aligned}(f + g)(x) &= f(x) + g(x)\\ &= x^2 - 1 + x + 2\\ &= x^2 + x + 1. \quad \text{Combining like terms}\end{aligned}$$

Thus $(f + g)(3) = 3^2 + 3 + 1 = 13$.

3. Find $(f - g)(x)$ and $(f - g)(-1)$.

$$\begin{aligned}(f - g)(x) &= f(x) - g(x)\\ &= x^2 - 1 - (x + 2) \quad \text{Substituting}\\ &= x^2 - x - 3. \quad \text{Removing parentheses and combining like terms}\end{aligned}$$

$$\begin{aligned}\text{Thus } (f - g)(-1) &= (-1)^2 - (-1) - 3\\ &= -1. \quad \text{Simplifying}\end{aligned}$$

4. Find $(f/g)(x)$ and $(f/g)(-4)$.

$$(f/g)(x) = f(x)/g(x)$$
$$= \frac{x^2 - 1}{x + 2}$$
$$(f/g)(-4) = \frac{(-4)^2 - 1}{-4 + 2} \qquad \text{Substituting}$$
$$= \frac{15}{-2} = -7.5$$

5. Find $(f \cdot g)(x)$ and $(f \cdot g)(-3)$.

$$(f \cdot g)(x) = f(x) \cdot g(x)$$
$$= (x^2 - 1)(x + 2)$$
$$= x^3 + 2x^2 - x - 2$$
$$(f \cdot g)(-3) = (-3)^3 + 2(-3)^2 - (-3) - 2$$
$$= -8$$

■

Although it is usually difficult to visualize the product or quotient of two or more functions, sums and differences frequently occur in applications. In the graph shown here the total number of airline passengers, $F(t)$, in the New York area is regarded as a function of time. The number of passengers using Kennedy Airport is denoted by $k(t)$, the number of passengers using LaGuardia Airport is $l(t)$, and the number of passengers using Newark Airport is $n(t)$. Although separate graphs for k, l, and n have not been drawn, we can see that

$$F(t) = k(t) + l(t) + n(t).$$

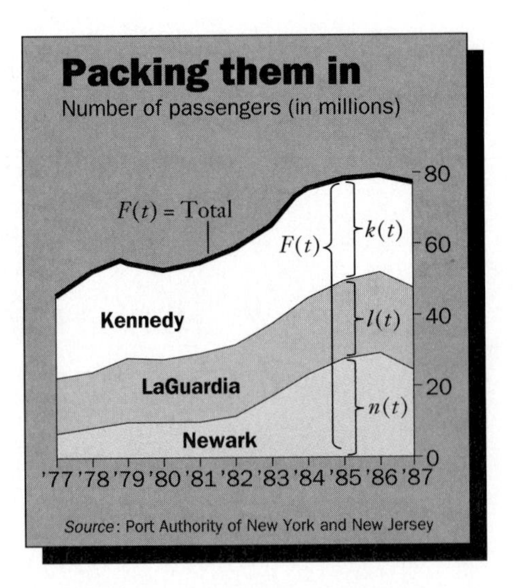

Two graphs that share the same axes can be added by adding the two function values associated with any given input.

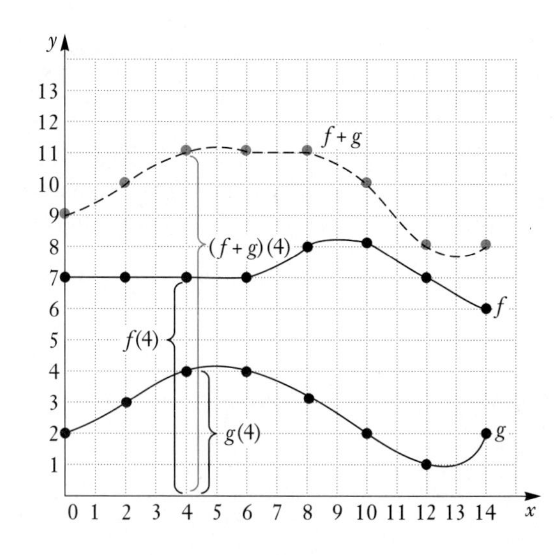

Note that $(f + g)(4)$ can be calculated by first finding the separate values $f(4)$ and $g(4)$ and then adding the two values together. Thus

$$(f + g)(4) = f(4) + g(4) = 7 + 4 = 11.$$

The Domain of a Sum, Difference or Product of Two Functions

To find $(f + g)(a)$, $(f - g)(a)$, or $(f \cdot g)(a)$, it makes sense that we must first be able to find $f(a)$ and $g(a)$. Thus we need to determine whether a is in the domains of f and g.

EXAMPLE 6 Let

$$f(x) = \frac{5}{x} \quad \text{and} \quad g(x) = \frac{2x - 6}{x + 1}.$$

Find the domain of $f + g$, the domain of $f - g$, and the domain of $f \cdot g$.

Solution Note that because division by 0 is undefined, we have

$$\text{Domain of } f = \{x \mid x \text{ is a real number and } x \neq 0\}$$

and

$$\text{Domain of } g = \{x \mid x \text{ is a real number and } x \neq -1\}.$$

To find $f(a) + g(a)$, $f(a) - g(a)$, or $f(a) \cdot g(a)$, we must know that a is in *both* of the preceding domains. Thus

$$\begin{aligned} \text{Domain of } f + g &= \text{Domain of } f - g = \text{Domain of } f \cdot g \\ &= \{x \mid x \text{ is a real number and } x \neq 0 \text{ and } x \neq -1\}. \end{aligned}$$

■

The Domain of a Quotient of Two Functions

Suppose that in Example 4 we had wanted to find $(f/g)(-2)$. Finding $f(-2) = (-2)^2 - 1 = 3$ and $g(-2) = -2 + 2 = 0$ poses no particular problem. However, when we find

$$\begin{aligned}(f/g)(-2) &= f(-2)/g(-2)\\ &= 3/0,\end{aligned}$$

we see that $(f/g)(-2)$ is undefined. Thus, although -2 is in the domain of both f and g, it is not in the domain of f/g.

EXAMPLE 7 Let $F(x) = x^3$ and $G(x) = 4x - 3$. Find the domain of F/G.

Solution The domain of F and G is all real numbers. However, the domain of

$$(F/G)(x) = \frac{x^3}{4x - 3}$$

is the set of all real numbers such that $4x - 3 \neq 0$. Because $4x - 3 = 0$ when $x = \frac{3}{4}$, we conclude that the domain of F/G is the set of all real numbers except $\frac{3}{4}$. In set-builder notation,

$$\text{Domain of } F/G = \{x \mid x \text{ is a real number and } x \neq \tfrac{3}{4}\}.$$ ■

EXAMPLE 8 Let

$$f(x) = \frac{3}{x - 4} \quad \text{and} \quad g(x) = \frac{6}{x + 2}.$$

Find the domain of $(f/g)(x)$.

Solution Since the domain of $f = \{x \mid x \text{ is a real number and } x \neq 4\}$ and the domain of $g = \{x \mid x \text{ is a real number and } x \neq -2\}$, we conclude that the domain of f/g is the set of all real numbers except -2, 4, and any x-values such that $g(x) = 0$. Because $6/(x + 2)$ is never 0, there is no x-value such that $g(x) = 0$. Thus

$$\text{Domain of } f/g = \{x \mid x \text{ is a real number and } x \neq 4 \text{ and } x \neq -2\}.$$ ■

Comment: It is tempting to write

$$\begin{aligned}(f/g)(x) &= \frac{\frac{3}{x-4}}{\frac{6}{x+2}} = \frac{3}{x - 4} \div \frac{6}{x + 2}\\ &= \frac{3}{x - 4} \cdot \frac{x + 2}{6}\\ &= \frac{x + 2}{2(x - 4)},\end{aligned}$$

in which case the domain of f/g would exclude only 4. Because of the fact that $(f/g)(x)$ is defined as $f(x)/g(x)$, we can use only x-values that are in the domains of *both* functions. Thus we stipulate

$$(f/g)(x) = \frac{x+2}{2(x-4)}, \qquad \textit{provided } x \neq -2.$$

EXAMPLE 9 Let

$$p(x) = \frac{5}{x} \quad \text{and} \quad q(x) = \frac{2x-6}{x+1}.$$

Find the domain of p/q.

Solution

$$\text{Domain of } p = \{x \mid x \text{ is a real number and } x \neq 0\}$$
$$\text{Domain of } q = \{x \mid x \text{ is a real number and } x \neq -1\}$$

Since $q(x) = 0$ when $2x - 6 = 0$, we have $q(x) = 0$ when $x = 3$. We conclude that

$$\text{Domain of } p/q = \{x \mid x \text{ is a real number and } x \neq 0,\ x \neq -1, \text{ and } x \neq 3\}.$$ ■

To find the domain of a sum, difference, product, or quotient of two functions:

1. Determine the domain of each function.
2. The domain of the sum, difference, or product is the set of all values common to both domains.
3. The domain of the quotient is the set of all values common to both domains, excluding any value that would lead to division by 0.

Domains and Graphs

There is an interesting visual interpretation to the algebra of functions. Consider f and g as sketched in the figure.

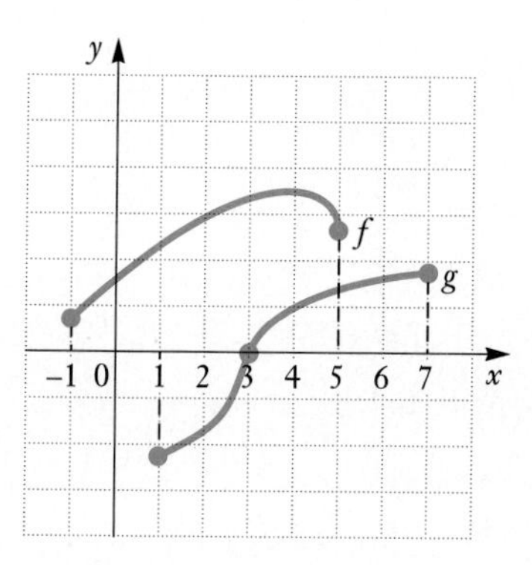

Note that

$$\text{the domain of } f = \{x \mid -1 \leq x \leq 5\}$$

and

$$\text{the domain of } g = \{x \mid 1 \leq x \leq 7\}$$

can be regarded as the *projections*, or "shadows," of f and g on the x-axis. Thus for $f+g$, $f-g$, or $f \cdot g$, the domain is $\{x \mid 1 \leq x \leq 5\}$, or what is common to the domains of f and g. Since $g(3) = 0$, the domain of f/g is

$$\{x \mid 1 \leq x \leq 5 \text{ and } x \neq 3\}.$$

EXERCISE SET 6.6

Let $f(x) = -3x + 1$ and $g(x) = x^2 + 2$. Find the following.

1. $f(1) + g(1)$ **2.** $f(2) + g(2)$ **3.** $f(-1) + g(-1)$ **4.** $f(-2) + g(-2)$

5. $f(-7) - g(-7)$ **6.** $f(-5) - g(-5)$ **7.** $f(5) - g(5)$ **8.** $f(4) - g(4)$

9. $f(2) \cdot g(2)$ **10.** $f(3) \cdot g(3)$ **11.** $f(-3) \cdot g(-3)$ **12.** $f(-4) \cdot g(-4)$

13. $f(0)/g(0)$ **14.** $f(1)/g(1)$ **15.** $f(-3)/g(-3)$ **16.** $f(-4)/g(-4)$

Let $F(x) = x^2 - 3$ and $G(x) = 4 - x$. Find the following.

17. $(F + G)(-3)$ **18.** $(F + G)(-2)$ **19.** $(F + G)(x)$ **20.** $(F + G)(a)$

21. $(F - G)(-4)$ **22.** $(F - G)(-5)$ **23.** $(F \cdot G)(2)$ **24.** $(F \cdot G)(3)$

25. $(F \cdot G)(-3)$ **26.** $(F \cdot G)(-4)$ **27.** $(F/G)(0)$ **28.** $(F/G)(1)$

29. $(F/G)(-2)$ **30.** $(F/G)(-1)$

For each pair of functions f and g, determine the domain of the sum, difference, and product of the two functions.

31. $f(x) = x^2$, $g(x) = 3x - 4$

32. $f(x) = 5x - 1$, $g(x) = 2x^2$

33. $f(x) = \dfrac{1}{x - 2}$, $g(x) = 4x^3$

34. $f(x) = 3x^2$, $g(x) = \dfrac{1}{x - 4}$

35. $f(x) = \dfrac{2}{x}$, $g(x) = x^2 - 4$

36. $f(x) = x^3 + 1$, $g(x) = \dfrac{5}{x}$

37. $f(x) = 4x + \dfrac{2}{x - 1}$, $g(x) = 3x^3$

38. $f(x) = 9 - x^2$, $g(x) = \dfrac{3}{x - 5} + 2x$

39. $f(x) = \dfrac{3}{x - 2}$, $g(x) = \dfrac{5}{4 - x}$

40. $f(x) = \dfrac{5}{x - 3}$, $g(x) = \dfrac{1}{x - 2}$

41. $f(x) = \dfrac{3}{x + 2}$, $g(x) = \dfrac{x}{x - 4}$

42. $f(x) = \dfrac{2x}{3 - x}$, $g(x) = \dfrac{4}{x - 5}$

For each pair of functions f and g, determine the domain of f/g.

43. $f(x) = x^4$, $g(x) = x - 3$

44. $f(x) = 2x^3$, $g(x) = 5 - x$

45. $f(x) = 3x - 2$, $g(x) = 2x - 8$

46. $f(x) = 5 + x$, $g(x) = 6 - 2x$

47. $f(x) = \dfrac{2}{x}$, $g(x) = \dfrac{3}{x - 4}$

48. $f(x) = \dfrac{5}{x - 3}$, $g(x) = \dfrac{4}{x}$

49. $f(x) = \dfrac{3}{x - 4}$, $g(x) = 5 - x$

50. $f(x) = \dfrac{1}{2 - x}$, $g(x) = 7 - x$

51. $f(x) = \dfrac{2x}{x + 1}$, $g(x) = 2x + 5$

52. $f(x) = \dfrac{7x}{x - 2}$, $g(x) = 3x + 7$

53. $f(x) = \dfrac{x - 1}{x - 4}$, $g(x) = 3x^2$

54. $f(x) = \dfrac{x + 2}{x + 3}$, $g(x) = 4x^3$

55. $f(x) = 3x^2$, $g(x) = \dfrac{x-1}{x-4}$

56. $f(x) = 4x^3$, $g(x) = \dfrac{x+2}{x+3}$

57. $f(x) = \dfrac{x-1}{x-2}$, $g(x) = \dfrac{x-3}{x-4}$

58. $f(x) = \dfrac{x+4}{x+3}$, $g(x) = \dfrac{x+2}{x+1}$

59. $f(x) = \dfrac{x-3}{x+2}$, $g(x) = \dfrac{3x-5}{x+1}$

60. $f(x) = \dfrac{2x-1}{x-3}$, $g(x) = \dfrac{5x-1}{x+6}$

61. $f(x) = x + \dfrac{1}{2x-5}$, $g(x) = \dfrac{x}{3x-1}$

62. $f(x) = x^2 - \dfrac{1}{x}$, $g(x) = \dfrac{2x}{6-5x}$

SKILL MAINTENANCE

63. Solve: $5x - 7 = 0$.

64. Evaluate $3x - y^2$ if $x = -4$ and $y = 5$.

65. Factor: $24 - 3x^3$

66. Rationalize the numerator:

$$\frac{\sqrt{5} + \sqrt{x}}{\sqrt{5}}.$$

SYNTHESIS

67. Let

$$p(x) = \begin{cases} 2x, & \text{if } x > 1, \\ x^2, & \text{if } x < 1 \end{cases}$$

and

$$q(x) = \frac{x-3}{x-2}.$$

Find the domain of p/q.

68. Let

$$m(x) = 3x \text{ for } -1 < x < 5$$

and

$$n(x) = 2x - 3.$$

Find the domain of m/n.

69. Let

$f = \{(-2, 1), (-1, 2), (0, 3), (1, 4), (2, 5)\}$

and

$g = \{(-4, 4), (-3, 3), (-2, 4), (-1, 0), (0, 5), (1, 6)\}$.

Find the domains of $f + g$, $f - g$, $f \cdot g$, and f/g.

70. For f and g as defined in Exercise 69, find $(f + g)(-2)$, $(f \cdot g)(0)$, and $(f/g)(1)$.

71. Let

$$F(x) = \sqrt{x-3} \quad \text{and} \quad G(x) = \sqrt{x-5}.$$

Find the domain of F/G.

72. Let

$$f(x) = \frac{1}{x+3} \quad \text{and} \quad g(x) = \sqrt{x+5}.$$

Find the domain of f/g.

73. The graphs of f and g are labeled in the graph. Find the domains of $f + g, f - g, f \cdot g$, and f/g.

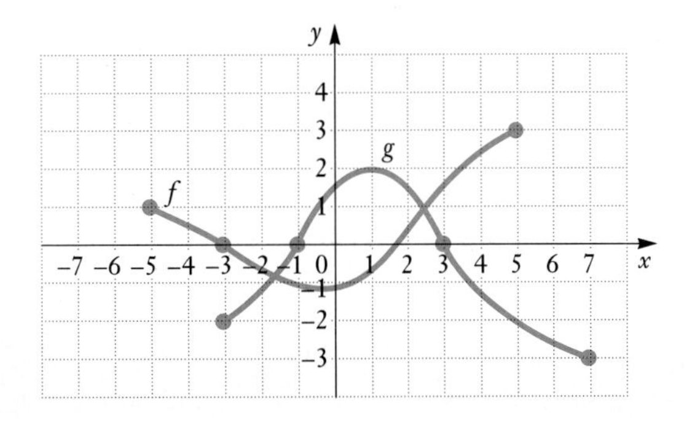

74. Write equations for two functions f and g such that the domain of $f + g$ is

$$\{x \mid x \text{ is a real number and } x \neq -2 \text{ and } x \neq 5\}.$$

75. Sketch the graph of two functions f and g such that the domain of f/g is

$$\{x \mid -2 \leq x \leq 3 \text{ and } x \neq 1\}.$$

76. Use the graph of the number of airline passengers that appears after Example 5 to estimate each of the following.

a) $(k + l + n)(1983)$
b) $(l + n)(1986)$
c) $(l + k)(1986)$

6.7 Variation and Problem Solving

We extend our study of formulas and functions by examining three situations that frequently arise in problem solving: direct variation, inverse variation, and mixed variation.

Direct Variation

Let's say that a worker earns \$18 per hour. In 1 hr \$18 is earned. In 2 hr \$36 is earned. In 3 hr \$54 is earned, and so on. This gives rise to a set of ordered pairs of numbers, all having the same ratio:

$$(1, 18), (2, 36), (3, 54), (4, 72), \quad \text{and so on.}$$

The ratio of earnings to time is $\frac{18}{1}$ in every case.

Whenever a situation gives rise to pairs of numbers in which the ratio is constant, we say that there is **direct variation.** Here the earnings *vary directly* as the time:

$$E = 18t \qquad \text{or, using function notation,} \qquad E(t) = 18t.$$

Note that since 18 is positive, as one variable increases, so does the other. Similarly, as one variable decreases, so does the other.

> Whenever a situation gives rise to a linear function $f(x) = kx$, or $y = kx$, where k is a nonzero constant, we say that there is *direct variation,* that y *varies directly as* x, or that y is *proportional to* x. The number k is called the *variation constant,* or *constant of proportionality.*

EXAMPLE 1 Find the variation constant and an equation of variation in which y varies directly as x, and $y = 32$ when $x = 2$.

Solution We know that (2, 32) is a solution of $y = kx$. Therefore,

$$32 = k \cdot 2 \qquad \text{Substituting}$$

$$\frac{32}{2} = k, \qquad \text{or} \qquad k = 16. \qquad \text{Solving for } k$$

The variation constant is 16. The equation of variation is $y = 16x$. The notation $y(x) = 16x$ or $f(x) = 16x$ is also used. ■

EXAMPLE 2 *Water from melting snow.* The number of centimeters W of water produced from melting snow varies directly as S, the number of centimeters of snow. Meteorologists have found that under certain conditions, 150 cm of snow will melt to 16.8 cm of water. To how many centimeters of water will 200 cm of snow melt under these conditions?

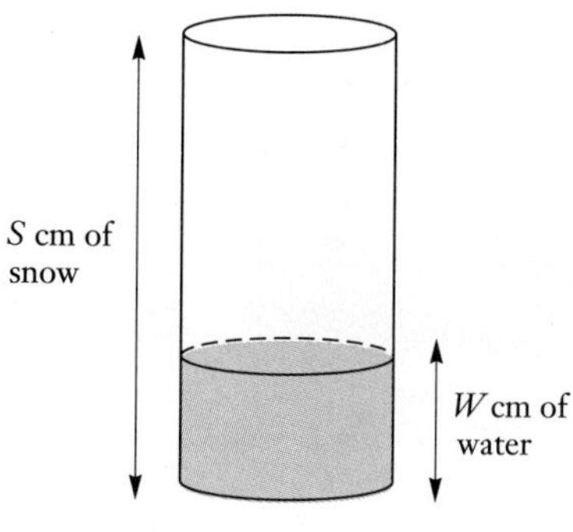

Solution

1. **Familiarize.** Because of the phrase "W . . . varies directly as S," we decide to express the amount of water as a function of the amount of snow. Thus $W(S) = kS$, where k is the variation constant. From the information provided, we know that $W(150) = 16.8$. That is, 150 cm of snow melts to 16.8 cm of water.

2. **Translate.** We find the variation constant using the data and then find the equation of variation:

$$W(S) = kS$$

$$W(150) = k \quad 150 \qquad \text{Replacing } S \text{ with } 150$$

$$16.8 = k \cdot 150 \qquad \text{Substituting}$$

$$\frac{16.8}{150} = k \qquad \text{Solving for } k$$

$$0.112 = k. \qquad \text{This is the variation constant.}$$

The equation of variation is $W(S) = 0.112S$. This is the translation.

3. **Carry out.** To find how much water 200 cm of snow will melt to, we compute $W(200)$:

$$\begin{aligned} W(S) &= 0.112S \\ W(200) &= 0.112(200) \qquad \text{Replacing } S \text{ with } 200 \\ &= 22.4. \end{aligned}$$

4. **Check.** To check, we could reexamine all our calculations. Note that our answer seems reasonable since 200/22.4 and 150/16.8 are equal. In fact, this type of check can be used on any direct-variation problem.

5. **State.** 200 cm of snow will melt into 22.4 cm of water. ■

Inverse Variation

To see what we mean by inverse variation, consider the following situation.

A bus is traveling a distance of 20 mi. At a speed of 20 mph, the trip will take 1 hr. At 40 mph, it will take $\frac{1}{2}$ hr. At 60 mph, it will take $\frac{1}{3}$ hr, and so on. This gives rise to a set of pairs of numbers, all having the same product:

$$(20, 1), \left(40, \tfrac{1}{2}\right), \left(60, \tfrac{1}{3}\right), \left(80, \tfrac{1}{4}\right), \qquad \text{and so on.}$$

Whenever a situation gives rise to pairs of numbers whose product is constant, we say that there is **inverse variation.** The time t required for the bus to travel 20 mi at rate r is given by

$$t = \frac{20}{r} \qquad \text{or, using function notation,} \qquad t(r) = \frac{20}{r}.$$

Note that since 20 is positive, as one variable increases, the other decreases.

> Whenever a situation gives rise to a function $f(x) = k/x$, or $y = k/x$, where k is a nonzero constant, we say that there is *inverse variation*, that *y varies inversely as x*, or that y is *inversely proportional to x*. The number k is called the *variation constant*, or *constant of proportionality.*

EXAMPLE 3 Find the variation constant and an equation of variation in which y varies inversely as x, and $y = 32$ when $x = 0.2$.

Solution We know that (0.2, 32) is a solution of

$$y = \frac{k}{x}.$$

Therefore,

$$32 = \frac{k}{0.2} \quad \textbf{Substituting}$$

$$(0.2)32 = k$$

$$6.4 = k. \quad \textbf{Solving for } k$$

The variation constant is 6.4. The equation of variation is

$$y = \frac{6.4}{x}.$$

■

There are many problems that translate to an equation of inverse variation.

EXAMPLE 4 The time t required to do a certain job varies inversely as the number of people P who work on the job (assuming that all do the same amount of work). It takes 4 hr for 12 people to build a woodshed. How long would it take 3 people to do the same job?

Solution

1. **Familiarize.** Because of the phrase "t . . . varies inversely as P," we decide to express the amount of time required, in hours, as a function of the number of people working. Thus we have $t(P) = k/P$. From the information provided we know that $t(12) = 4$. That is, it takes 4 hr for 12 people to do the job.

2. **Translate.** We find the variation constant using the data and then find the equation of variation:

$$t(P) = \frac{k}{P}$$

$$t(12) = \frac{k}{12} \quad \text{Replacing } P \text{ with } 12$$

$$4 = \frac{k}{12} \quad \text{Substituting}$$

$$48 = k. \quad \text{Solving for } k\text{, the variation constant}$$

The equation of variation is $t(P) = 48/P$. This is the translation.

3. **Carry out.** To find how long it would take 3 people to do the job, we solve for $t(3)$:

$$t(P) = \frac{48}{P}$$

$$t(3) = \frac{48}{3} \quad \text{Replacing } P \text{ with } 3$$

$$t = 16.$$

4. **Check.** We could now recheck each step. Note that, as expected, as the number of people working goes *down*, the time required for the job goes *up*.

5. **State.** It will take 3 people 16 hr to build a woodshed. ■

Combined Variation

Often one variable varies directly or inversely with more than one other variable. For example, in the formula for the volume of a right circular cylinder, $V = \pi r^2 h$, we would say that V varies *jointly* as h and the square of r.

> y varies *jointly* as x and z if there is some nonzero constant k such that $y = kxz$.

EXAMPLE 5 Find an equation of variation in which y varies *jointly* as x and z and inversely as the square of w, and $y = 105$ when $x = 3$, $z = 20$, and $w = 2$.

Solution The equation of variation is of the form

$$y = k \cdot \frac{xz}{w^2},$$

so

$$105 = k \cdot \frac{3 \cdot 20}{2^2} \quad \text{and} \quad k = 7.$$

Thus $y = 7 \cdot \dfrac{xz}{w^2}$. ■

EXAMPLE 6 *The volume of a tree trunk.* The volume of wood V in a tree trunk varies jointly as the height h and the square of the girth g (girth is distance around). If the volume is 35 ft^3 when the height is 20 ft and the girth is 5 ft, what is the height when the volume is 85.75 ft^3 and the girth is 7 ft?

Solution

1. **Familiarize.** We'll make a table, including the data from the problem and the data we need to find.

	Volume of Wood	Height of Tree	Girth of Tree
Smaller Tree	35 ft^3	20 ft	5 ft
Larger Tree	85.75 ft^3	h	7 ft

h g

Let h, g, and V represent the height, girth, and volume of a tree, respectively. We wish to determine h when V is 85.75 ft^3 and g is 7 ft.

We know from the statement of the problem that in this situation the volume varies jointly as the height and the square of the girth.

2. **Translate.** First we find k using the first set of data. Then we solve for g using the second set of data:

$$V = khg^2$$
$$35 = k \cdot 20 \cdot 5^2$$
$$0.07 = k. \quad \text{This is the variation constant.}$$

The equation of variation is $V = 0.07hg^2$.

3. **Carry out.** The translation is $V = 0.07hg^2$. We substitute and solve for g:

$$85.75 = 0.07 \cdot h \cdot 7^2$$
$$85.75 = 3.43\,h$$
$$25 \text{ ft} = h. \quad \text{Dividing by 3.43 on both sides}$$

We could have first solved the formula for h and then substituted; either approach is valid.

4. **Check.** We should now recheck all our calculations and perhaps make an estimate to see whether our answer is reasonable. We leave this for the student.

5. **State.** The answer is that the height of the tree is 25 ft. ■

EXERCISE SET 6.7

Find the variation constant and an equation of variation in which y varies directly as x and the following conditions exist.

1. $y = 24$ when $x = 3$

2. $y = 5$ when $x = 12$

3. $y = 3.6$ when $x = 1$

4. $y = 2$ when $x = 5$

5. $y = 15$ when $x = 3$

6. $y = 1$ when $x = 2$

7. $y = 30$ when $x = 8$

8. $y = 1$ when $x = \frac{1}{3}$

9. $y = 0.8$ when $x = 0.5$

10. $y = 0.6$ when $x = 0.4$

Solve.

11. *Electric current and voltage.* The electric current I, in amperes, in a circuit varies directly as the voltage V. When 12 volts is applied, the current is 4 amperes. What is the current when 18 volts is applied?

12. *Hooke's law.* Hooke's law states that the distance d that a spring is stretched by a hanging object varies directly as the mass m of the object. If the distance is 40 cm when the mass is 3 kg, what is the distance when the mass is 5 kg?

13. The number N of plastic straws produced by a machine varies directly as the amount of time t the machine is operating. If the machine produces 20,000 straws in 8 hr, how many straws can it produce in 50 hr?

14. The number N of aluminum cans used each year varies directly as the number of people using the cans. If 250 people use 60,000 cans in one year, how many cans are used each year in a city with a population of 850,000?

15. The amount of pollution A entering the atmosphere varies directly as the number of people N living in an area. If 60,000 people result in 42,600 tons of pollutants entering the atmosphere, how many tons enter the atmosphere in a city with a population of 750,000?

16. *Weight on the moon.* The weight M of an object on the moon varies directly as its weight E on Earth. A person who weighs 95 lb on Earth weighs 15.2 lb on the moon. How much would a 105-lb person weigh on the moon?

17. *Weight on Mars.* The weight M of an object on Mars varies directly as its weight E on Earth. A person who weighs 95 lb on Earth weighs 38 lb on Mars. How much would a 100-lb person weigh on Mars?

18. *Mass of water in body.* The number of kilograms W of water in a human body varies directly as the total weight. A person weighing 96 kg contains 64 kg of water. How many kilograms of water are in a 75-kg person?

19. *Ohm's law.* Ohm's law states that the voltage V in an electric circuit varies directly as the number of amperes I of electric current in the circuit. If the voltage is 10 volts when the current is 3 amperes, what is the voltage when the current is 15 amperes?

20. *Relative aperture.* The relative aperture, or f-number, of a 23.5-mm lens is directly proportional to the focal length F of the lens. If a 150-mm focal length has an f-number of 6.3, find the f-number of a 23.5-mm lens with a focal length of 80 mm.

Find the variation constant and an equation of variation in which y varies inversely as x, and the following conditions exist.

21. $y = 6$ when $x = 10$
22. $y = 16$ when $x = 4$
23. $y = 4$ when $x = 3$
24. $y = 4$ when $x = 9$
25. $y = 12$ when $x = 3$
26. $y = 9$ when $x = 5$
27. $y = 27$ when $x = \frac{1}{3}$
28. $y = 81$ when $x = \frac{1}{9}$

Solve.

29. *Current and resistance.* The current I in an electrical conductor varies inversely as the resistance R of the conductor. If the current is $\frac{1}{2}$ ampere when the resistance is 240 ohms, what is the current when the resistance is 540 ohms?

30. *Pumping rate.* The time t required to empty a tank varies inversely as the rate r of pumping. If a pump can empty a tank in 45 min at the rate of 600 kL/min, how long will it take the pump to empty the same tank at the rate of 1000 kL/min?

31. *Volume and pressure.* The volume V of a gas varies inversely as the pressure P upon it. The volume of a gas is 200 cm^3 under a pressure of 32 kg/cm^2. What will be its volume under a pressure of 40 kg/cm^2?

32. *Work rate.* The time T required to do a job varies inversely as the number of people P working. It takes 5 hr for 7 bricklayers to complete a certain job. How long will it take 10 bricklayers to complete the job?

33. *Rate of travel.* The time t required to drive a fixed distance varies inversely as the speed r. It takes 5 hr at a speed of 80 km/h to drive a fixed distance. How long will it take to drive the fixed distance at a speed of 60 km/h?

34. *Wavelength and frequency.* The wavelength W of a radio wave varies inversely as its frequency F. A wave with a frequency of 1200 kilohertz has a length of 300 meters. What is the length of a wave with a frequency of 800 kilohertz?

Find an equation of variation in which:

35. y varies directly as the square of x, and $y = 0.15$ when $x = 0.1$.

36. y varies directly as the square of x, and $y = 6$ when $x = 3$.

37. y varies inversely as the square of x, and $y = 0.15$ when $x = 0.1$.

38. y varies inversely as the square of x, and $y = 6$ when $x = 3$.

39. y varies jointly as x and z, and $y = 56$ when $x = 7$ and $z = 8$.

40. y varies directly as x and inversely as z, and $y = 4$ when $x = 12$ and $z = 15$.

41. y varies jointly as x and the square of z, and $y = 105$ when $x = 14$ and $z = 5$.

42. y varies jointly as x and z and inversely as w, and $y = \frac{3}{2}$ when $x = 2$, $z = 3$, and $w = 4$.

43. y varies directly as x and inversely as the square of z, and $y = 1.2$ when $x = 14$ and $z = 5$.

44. y varies directly as the square of x and inversely as z, and $y = 64$ when $x = 8$ and $z = 3$.

45. y varies jointly as w and the square of x and inversely as z, and $y = 49$ when $w = 3$, $x = 7$, and $z = 12$.

46. y varies directly as x and inversely as w and the square of z, and $y = 4.5$ when $x = 15$, $w = 5$, and $z = 2$.

47. y varies jointly as x and z and inversely as the product of w and p, and $y = \frac{3}{28}$ when $x = 3$, $z = 10$, $w = 7$, and $p = 8$.

48. y varies jointly as x and z and inversely as the square of w, and $y = \frac{12}{5}$ when $x = 16$, $z = 3$, and $w = 5$.

Problem Solving

49. *Stopping distance of a car.* The stopping distance d of a car after the brakes have been applied varies directly as the square of the speed r. If a car traveling 60 mph can stop in 200 ft, how fast can a car go and still stop in 72 ft?

50. *Area of a cube.* The area of a cube varies directly as the square of the length of a side. If a cube has an area of 168.54 in^2 when the length of a side is 5.3 in., how long is a side when the area of the cube is 294 in^2?

51. *Intensity of a signal.* The intensity I of a television signal varies inversely as the square of the distance d from the transmitter. If the intensity is 25 watts per square meter (W/m^2) at a distance of 2 km, how far from the transmitter are you when the intensity is 2.56 W/m^2?

52. *Distance of a fall.* The distance d that an object falls varies directly as the square of the amount of time t that it is falling. If an object falls 64 ft in 2 sec, how long will it take the object to fall 400 ft?

53. *Weight of an astronaut.* The weight W of an object varies inversely as the square of the distance d from the center of the earth. At sea level (6400 km from the center of the earth), an astronaut weighs 100 lb. Find her weight when she is 200 km above the surface of the earth and the spacecraft is not in motion.

54. *Intensity of light.* The intensity I of light from a light bulb varies inversely as the square of the distance d from the bulb. Suppose I is 90 W/m^2 when the distance is 5 m. Find the intensity at a distance of 10 m.

55. *Earned run average.* A pitcher's earned run average A varies directly as the number R of earned runs allowed and inversely as the number I of innings pitched. Suppose that a pitcher had an earned run average of 2.92 and gave up 85 earned runs in 262 innings. How many earned runs would be given up had the pitcher pitched 300 innings with the same average? Round to the nearest whole number.

56. *Volume of a gas.* The volume V of a given mass of a gas varies directly as the temperature T and inversely as the pressure P. If $V = 231\ \text{cm}^3$ when $T = 42°$ and $P = 20\ \text{kg/cm}^2$, what is the volume when $T = 30°$ and $P = 15\ \text{kg/cm}^2$?

57. *Electrical resistance.* At a fixed temperature the resistance R of a wire varies directly as the length l and inversely as the square of its diameter d. If the resistance is 0.1 ohm when the diameter is 1 mm and the length is 50 cm, what is the diameter when the resistance is 1 ohm and the length is 2000 cm?

58. *Volume of a can.* The volume V of a can varies jointly as its height h and the square of its radius r. If a 12-fluid-ounce soda comes in a can that is 12 cm high with a 3.2-cm radius, what is the radius of a 9-fluid-ounce can that is 4 cm high?

59. ▦ *Drag force.* The drag force F on a boat varies jointly as the wet surface area A and the square of the boat's velocity. If a boat going 6.5 mph experiences a drag force of 86 N (Newtons) when the wet surface area is 41.2 ft^2, how fast must a boat with 28.5 ft^2 of wet surface area go to experience a drag force of 94 N?

60. ▦ *Atmospheric drag.* Wind resistance, or atmospheric drag, tends to slow down moving objects. Atmospheric drag varies jointly as an object's surface area A and velocity v. If a car traveling at a speed of 40 mph with a surface area of 37.8 ft^2 experiences a drag of 222 N, how fast must a car with 51 ft^2 of surface area travel to experience a drag force of 430 N?

SKILL MAINTENANCE

61. Rationalize the denominator:

$$\frac{7}{\sqrt[3]{5x^2}}.$$

62. Solve:

$$3(2 - 5x) = 4x - 7(2x + 3).$$

63. Factor: $1 - 16x^2$.

64. Simplify: $(1 - \sqrt{2})^2$.

SYNTHESIS

65. Suppose that y varies directly as x and x is tripled. What is the effect on y?

66. Suppose that y varies inversely as x and x is tripled. What is the effect on y?

67. Suppose that y varies inversely as the square of x and x is multiplied by n. What is the effect on y?

68. Suppose that y varies directly as the square of x and x is multiplied by n. What is the effect on y?

69. The area of a circle varies directly as the square of the length of a diameter. What is the variation constant?

70. A peanut butter jar in the shape of a right circular cylinder is 4 in. high and 3 in. in diameter and sells for \$1.20. Assuming the same ratio of volume of peanut butter to cost, how much should a jar 6 in. high and 6 in. in diameter cost?

71. If y varies inversely as the cube of x and x is multiplied by 0.5, what is the effect on y?

72. *The gravity model.* It has been determined that the average number of telephone calls in a day N, between two cities, is directly proportional to the populations P_1 and P_2 of the cities and inversely proportional to the square of the distance between the cities. This model is called the "gravity model" because the equation of variation resembles the equation that applies to Newton's law of gravity.

a) The population of Indianapolis is 744,624, the population of Cincinnati is 452,524, and the distance between the cities is 174 km. The average number of daily phone calls between the two cities is 11,153. Find the value k and write the equation of variation.

b) The average number of daily phone calls between Indianapolis and New York is 4270, and the population of New York is 7,895,563. Find the distance between Indianapolis and New York.

73. *Tension of a stringed instrument.* The tension T on a string in a musical instrument varies jointly as the string's mass per unit length m, the square of its length l, and the square of its fundamental frequency f. A 2-m-long string weighing 5 gm/m with a fundamental frequency of 80 has a tension of 100 N. How long should the same string be if its tension is going to be changed to 72 N?

74. *Golf distance finder.* A device used in golf to estimate the distance d to a hole measures the size s that the 7-ft pin *appears* to be in a viewfinder. The viewfinder uses the principle, diagrammed here, that s gets bigger when d gets smaller. If $s = 0.56$ in. when $d = 50$ yd, find an equation of variation that expresses d as a function of s. What is d when $s = 0.40$ in.?

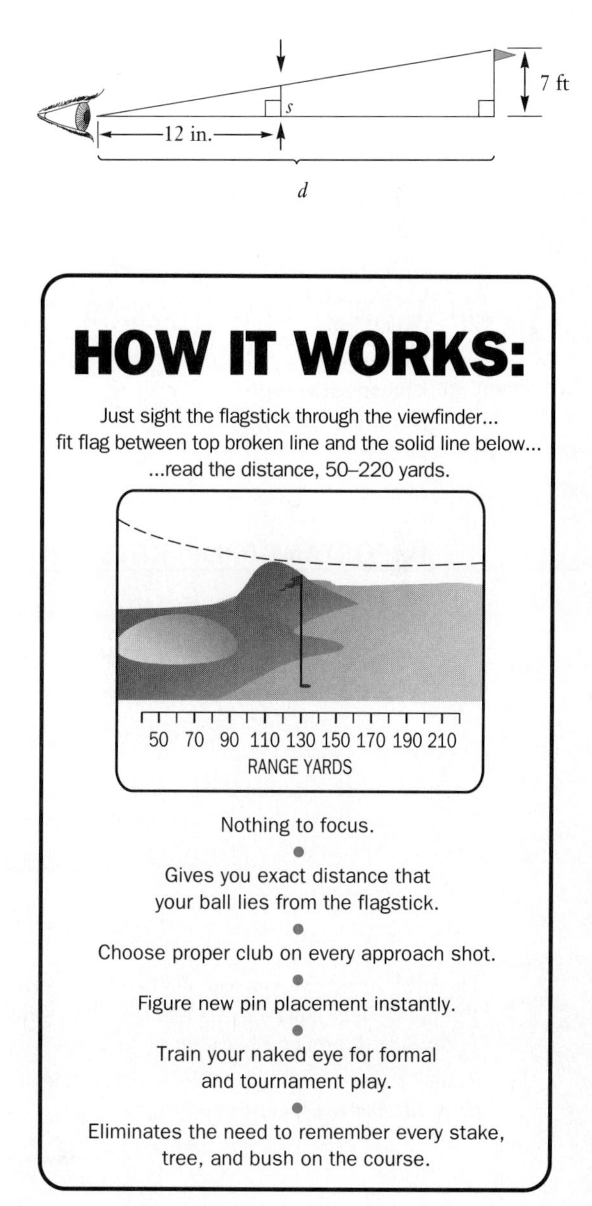

75. Suppose that the number of customer complaints is inversely proportional to the number of employees hired. Will a firm benefit more, in terms of customer complaints, by expanding from 5 to 10 employees, or from 20 to 25? Explain. Consider using a graph to help justify your answer.

CHAPTER SUMMARY AND REVIEW 6

TERMS TO KNOW

Axes
x-, y-coordinate system
Cartesian coordinate system
Ordered pair
Origin
Coordinates
Quadrant
Graph
Linear equation
Nonlinear equation
Domain
Range
Function
Relation
Input
Output
Dependent variable
Independent variable
Constant function
Dummy variable
Linear function
y-intercept
Slope
x-intercept
Rise
Run
Zero slope
Undefined slope
Varies directly
Variation constant
Varies inversely
Varies jointly

IMPORTANT PROPERTIES AND FORMULAS

THE VERTICAL LINE TEST

If it is possible for a vertical line to intersect a graph more than once, the graph is not that of a function.

The graph of an equation $x = a$, where a is constant, is a vertical line through the point $(a, 0)$.
The graph of an equation $y = b$, where b is constant, is a horizontal line through the point $(0, b)$.

$$\text{Slope} = m = \frac{\text{rise}}{\text{run}} = \frac{\text{change in } y}{\text{change in } x} = \frac{y_2 - y_1}{x_2 - x_1}$$

The slope–intercept equation of a line is $y = mx + b$.
The point–slope equation of a line is $y - y_1 = m(x - x_1)$.
The standard form of a linear equation is $Ax + By = C$.

Parallel lines: slopes equal, y-intercepts different.
Perpendicular lines: product of slopes $= -1$.

THE ALGEBRA OF FUNCTIONS

1. $(f + g)(x) = f(x) + g(x)$
2. $(f - g)(x) = f(x) - g(x)$
3. $(f \cdot g)(x) = f(x) \cdot g(x)$
4. $(f/g)(x) = f(x)/g(x)$, provided $g(x) \neq 0$

TO FIND THE DOMAIN OF A SUM, DIFFERENCE, PRODUCT, OR QUOTIENT OF TWO FUNCTIONS

1. Determine the domain of each function.
2. The domain of the sum, difference, or product is the set of all values common to both domains.
3. The domain of the quotient is the set of all values common to both domains, excluding any value that would lead to division by 0.

y varies directly as x if there is some nonzero constant k such that $y = kx$.
y varies inversely as x if there is some nonzero constant k such that $y = k/x$.
y varies jointly as x and z if there is some nonzero constant k such that $y = kxz$.

REVIEW EXERCISES

The review sections to be tested in addition to the material in this chapter are Sections 2.1, 3.4, 4.2, and 5.3.
Determine whether the ordered pair is a solution.

1. $(3, 7)$, $4p - q = 5$
2. $(-2, 4)$, $x - 3y = 10$
3. $\left(0, -\frac{1}{2}\right)$, $3a + 4b = 2$
4. $(8, -2)$, $3c - 2d = 28$

Graph.

5. $y = -3x + 2$
6. $f(x) = -x^2 + 1$
7. $8x + 32 = 0$

Find the slope and the y-intercept.

8. $g(x) = -4x - 9$
9. $-6y + 2x = 7$

Determine whether each is a linear equation.

10. $2x - 7 = 0$
11. $3x - 8f(x) = 7$
12. $2a + 7b^2 = 3$
13. $2p - \dfrac{7}{q} = 1$

Graph, using intercepts.

14. $-2x + 4y = 7$

Find the slope of the line (if it is defined).

15. Containing the points $(4, 5)$ and $(-3, 1)$
16. Containing $(-16.4, 2.8)$ and $(-16.4, 3.5)$

Find an equation of the line.

17. With slope -2 and containing $(-3, 4)$
18. Containing $(2, 5)$ and $(-4, -3)$

Determine whether the lines are parallel or perpendicular.

19. $y + 5 = -x$, $x - y = 2$
20. $3x - 5 = 7y$, $7y - 3x = 7$
21. $4y + x = 3$, $2x + 8y = 5$

Find an equation of the line.

22. Containing the point $(2, -5)$ and parallel to the line $3x - 5y = 9$

23. Containing the point $(2, -5)$ and perpendicular to the line $3x - 5y = 9$

Let $g(x) = 2x - 5$ and $h(x) = 3x + 7$. Find the following.

24. $g(0)$

25. $h(-5)$

26. $(g \cdot h)(4)$

27. $(g - h)(-2)$

28. $(g/h)(-1)$

29. $g(a + 1)$

30. The domain of $g + h$ and $g \cdot h$

31. The domain of h/g

32. Find an equation of variation in which y varies jointly as x and z and inversely as w and $y = 1$ when $x = 3$, $z = 2$, and $w = 5$.

33. The power P expended by heat in an electric circuit of fixed resistance varies directly as the square of the current I in the circuit. A circuit expends 180 watts when a current of 6 amperes is flowing. What is the heat expended when the current is 10 amperes?

SKILL MAINTENANCE

34. Multiply: $(2 + \sqrt{3})(2 - \sqrt{3})$.

35. Simplify: $(5a^3b)^2$.

36. Solve:

$$3(x - 2) + x = 5(x + 4).$$

37. Subtract and simplify:

$$\frac{3x}{x + 2} - \frac{24}{x^2 - 4}.$$

SYNTHESIS

38. Find the y-intercept of the function given by

$$f(x) + 3 = 0.17x^2 + (5 - 2x)^x - 7.$$

39. Find the value of a so that the lines $3x - 4y = 12$ and $ax + 6y = -9$ are parallel.

THINKING IT THROUGH

1. Explain the usefulness of the concept of slope when describing a straight line.

2. Explain why the slope of a vertical line is undefined but the slope of a horizontal line is 0.

3. Explain and compare the situations in which one would use the slope–intercept equation rather than the point–slope equation.

4. Explain the idea of a function in as many ways as possible.

CHAPTER TEST 6

Determine whether the ordered pair is a solution of the equation.

1. $(2, 5)$, $3y - 4z = -14$

2. $(-6, 8)$, $2s - t = -4$

3. $(0, -5)$, $x - 4y = -20$

4. $(1, -4)$, $-2p + 5q = 18$

Graph.

5. $y = -5x + 4$

6. $y = -2x^2 + 3$

Find the slope and y-intercept.

7. $y = 3x - 5$

8. $-3y + 4x = 9$

9. Graph: $3x - 18 = 0$.

10. Which of these are linear equations?

a) $8x - 7 = 0$ **b)** $4b - 9a^2 = 2$ **c)** $2x - 5y = 3$

11. Graph: $-5x + 2y = -12$.

Find the slope, if it exists, of the line containing the following points.

12. $(-2, -2)$ and $(6, 3)$

13. $(-3.1, 5.2)$ and $(-4.4, 5.2)$

14. Which of these equations has zero slope and which has an undefined slope?

a) $2y = 7$ **b)** $3x - 4y = 6x - 4y$

15. Find an equation of the line with slope 4 and containing $(-2, -4)$.

16. Find an equation of the line containing $(3, -1)$ and $(4, -2)$.

Determine without graphing whether the pair of lines is parallel or perpendicular.

17. $4y + 2 = 3x$,
$-3x + 4y = -12$

18. $y = -2x + 5$,
$2y - x = 6$

Find an equation of the line.

19. Containing $(-3, 2)$ and parallel to the line $2x - 5y = 8$

20. Containing $(-3, 2)$ and perpendicular to the line $2x - 5y = 8$

21. Find the following function values, given that $g(x) = -3x - 4$ and $h(x) = x^2 + 1$.

a) $g(0)$ **b)** $h(-2)$
c) $(g/h)(2)$ **d)** $(g - h)(-3)$

22. *The cost of renting a car.* If you rent a car for one day and drive it 250 miles, the cost is \$100. If you drive it 300 miles, the cost is \$115.

a) Fit a linear function to the data points.

b) Use the function to find how much it will cost to rent the car for one day and drive it 500 miles.

23. The area of a balloon varies directly as the square of its radius. The area is 3.4 cm^2 when the radius is 5 cm. What is the area when the radius is 7 cm?

SKILL MAINTENANCE

24. Solve: $3(2x - 4) = 5x - (12 - x)$.

25. Add: $2 + \frac{y+3}{y-5}$.

26. Rationalize the denominator: $\frac{2}{\sqrt[4]{8x^2}}$.

27. Factor: $y^2 + 16 - 8y$.

SYNTHESIS

28. The function $f(t) = 5 + 15t$ can be used to determine a bicycle racer's position, in miles from the starting line, measured in hours since passing the 5-mile mark.

a) How far from the start will the racer be 1 hr and 40 min after passing the 5-mi mark?

b) At what rate is the racer traveling?

29. The graph of the function $f(x) = mx + b$ contains the points $(r, 3)$ and $(7, s)$. Express s in terms of r if the graph is parallel to the line $3x - 2y = 7$.

CUMULATIVE REVIEW 1-6

1. Subtract: $-\frac{1}{5} - \frac{1}{2}$.

2. Multiply: $\left(-\frac{1}{5}\right)\left(-\frac{1}{2}\right)$.

3. Calculate: $\frac{2 + 4 \cdot 5 - 3^2}{(6-4)^2}$.

4. Write an equivalent expression using a commutative law: $xy + 3$.

Solve.

5. $0.3(2x - 4) = 10 - (x - 1)$

6. $2m - (3m - 4) = 10m + 2 + m$

7. $4 - 5x \leq 14$

8. $-3 \leq 2x + 5 \leq -2$

9. $3 - 2x < -2$ *or* $3 - 2x > 2$

10. $\left|\frac{1}{4} + 2x\right| \geq 8$

11. $\left|\frac{x + 5}{3}\right| = 6$

12. $t^2 = 3t$

13. $x^2 = 20$

14. $6y^2 + 14 + 25y = 0$

15. $\frac{x - 3}{x - 2} = \frac{1}{8}$

16. $\frac{6}{x - 2} - \frac{2}{x} = \frac{3}{x^2 - 2x}$

17. $\sqrt{x - 3} + 5 = 8$

18. $A = \frac{a_1 + a_2 + a_3}{n}$ for a_2

19. Evaluate for $x = -2$: $3x - x^2 + 4$.

20. Simplify: $\left(\frac{2x^2}{y^3}\right)^4$.

Perform the indicated operations and simplify if possible.

21. $(8x^2 - 6x + 3) + (3x^2 + 9x - 7)$

22. $(3y^2 + 2y + 1) - (y^2 - 3y + 1)$

23. $(2x - y)(3x^2 + 2xy - y^2)$

24. $(2x + 7)(3x - 9)$

25. $(2x^4y - 3x^3y^2 + x^2y^3) \div x^2y$

26. $(x^4 + 3x^2 + 2) \div (x^2 - 1)$

27. $\frac{x^2 + 2x - 8}{x^2 - 9} \cdot \frac{x^2 + 5x + 6}{x^2 + 3x - 4}$

28. $\frac{2x + 8}{x^2 - 1} \div \frac{x^2 + 3x - 4}{x^2 + 3x + 2}$

29. $\frac{x^2}{x - y} + \frac{y^2}{y - x}$

30. $\frac{2x}{x - 1} - \frac{x + 1}{x}$

Factor.

31. $4x^2 + 9 - 12x$

32. $27y^4 + 8y$

33. $27y^4 - 12y^2$

34. $x^3 + 2x^2 - x - 2$

35. $x^2 + x + 5$

36. $y^4 + 49y^2 + 48$

37. Simplify: $\dfrac{\frac{2}{x + 1} + \frac{3}{x - 2}}{\frac{5}{x - 2} - \frac{2}{x + 1}}$.

38. Simplify. Assume that variables can represent any real number.

$$\sqrt{(6 - x)^2}$$

Simplify. Assume that all variables represent positive numbers

39. $-\sqrt[5]{-1}$

40. $(\sqrt[3]{2xy^2})^4$

41. $\sqrt[3]{80}$

Perform the indicated operations and simplify if possible. Assume that all radicands represent positive numbers.

42. $(3 - \sqrt{x})(1 + \sqrt{x})$

43. $6\sqrt{50} - 3\sqrt{18}$

44. Rewrite without fractional exponents: $(9x)^{3/2}$.

45. Rewrite with negative exponents: $\frac{1}{5x^2}$.

46. Find an equation of the line containing $(2, 1)$ and $(6, -3)$.

47. Find an equation of the line containing $(2, 1)$ and perpendicular to $y + 2x = 7$.

Graph.

48. $y + 3 = 6$

49. $2x + 3y = 8$

50. $y = -x + 4$

Solve.

51. The time t required to do a job varies inversely as the number of people P working. It takes 2 hr for 3 carpenters to complete a certain job. How long will it take 4 carpenters to complete the job?

52. The base of a triangle is 3 cm greater than the height. The area is 35 cm^2. Find the height and the base.

53. A person swims at a speed of 4 mph in still water. The current in a river is moving at 3.2 mph. How long will it take the person to swim 2.4 mi upriver?

54. The speed of a boat in still water is 10 mph. The boat travels 40 mi downriver in the same time it takes to go 10 mi upstream. What is the speed of the river?

55. Multiply: $(-2 + i)(-2 - 2i)$.

56. For $f(x) = 2x + 1$ and $g(x) = x^2$, find the domain of f/g.

SYNTHESIS

Solve.

57. $x^4 - 3x^2 - 10 = 0$

58. $\frac{18}{x - 9} + \frac{10}{x + 5} = \frac{28x}{x^2 - 4x - 45}$

59. $\sqrt[3]{\frac{x}{6}} - 8 = -18$

60. $x^4 + 5x = 5x^3 + x^2$

CHAPTER 7

Quadratic Equations and Functions

FEATURE PROBLEM

A plastics manufacturer plans to produce a one-compartment vertical file by bending an 8 in. by 14 in. piece of plastic along two lines to form a U shape. How tall should the file be to maximize the volume that the file can hold?

THE MATHEMATICS

Let x represent the height of the file, in inches. When we translate, we see that we must find the maximum value attained by the function

$$f(x) = 8 \cdot x \cdot (14 - 2x)$$

or

$$\underbrace{f(x) = -16x^2 + 112x.}_{\text{This is a } \textit{quadratic function.}}$$

When a problem situation is translated to mathematical language, we sometimes obtain a function or equation containing a second-degree polynomial. Such equations and functions are called *quadratic*. In this chapter we will learn to solve quadratic equations and graph quadratic functions. Then we will have the opportunity to solve problems for which the translation contains a quadratic equation or function.

The review sections to be tested in addition to the material in this chapter are Sections 3.6, 5.1, 5.4, and 6.5.

7.1 Quadratic Equations

In Section 3.5 we began a study of *quadratic equations*. These equations, which contain second-degree polynomials, were solved by factoring and using the principle of zero products. We also saw, in Section 5.5, that certain equations could be solved by using the principle of positive and negative roots. It is to this principle that we will turn as we endeavor to find a method for solving *any* quadratic equation.

A quadratic equation like

$$5x^2 + 8x - 2 = 0$$

is said to be in *standard form*. The quadratic equation

$$5x^2 = 2 - 8x$$

is equivalent to the preceding one, but it is not in standard form.

> An equation of the type $ax^2 + bx + c = 0$, where a, b, and c are real number constants and $a > 0$, is called the *standard form of a quadratic equation*.

An equation like $-3x^2 + 4x - 7 = 0$ can be rewritten as an equivalent equation that is in standard form by multiplying on both sides by -1:

$$-1(-3x^2 + 4x - 7) = -1(0)$$
$$3x^2 - 4x + 7 = 0.$$

Equations of the Type $ax^2 + c = 0$

We now consider equations in which $b = 0$ — that is, equations of the form $ax^2 + c = 0$ or $ax^2 = -c$.

EXAMPLE 1 Solve: $3x^2 = 6$.

Solution

$$3x^2 = 6$$
$$x^2 = 2 \quad \text{Multiplying by } \tfrac{1}{3}$$
$$x = \pm\sqrt{2} \quad \text{Using the principle of positive and negative roots}$$

In this equation we can check both numbers at once.

Check:

$3x^2 = 6$	
$3(\pm\sqrt{2})^2$	6
$\cdot 2$	
6	

TRUE

The solutions are $\sqrt{2}$ and $-\sqrt{2}$, or $\pm\sqrt{2}$. ■

Sometimes we get solutions that are not real numbers but complex numbers instead.

EXAMPLE 2 Solve: $4x^2 + 9 = 0$.

Solution

$$4x^2 + 9 = 0$$
$$x^2 = -\frac{9}{4} \quad \text{Adding } -9 \text{ and multiplying by } \tfrac{1}{4}$$
$$x = \sqrt{-\frac{9}{4}} \quad \text{or} \quad x = -\sqrt{-\frac{9}{4}} \quad \text{Taking square roots (principle of positive and negative roots)}$$
$$x = \frac{3}{2}i \quad \text{or} \quad x = -\frac{3}{2}i \quad \text{Simplifying}$$

The check is left to the student. The solutions are $\frac{3}{2}i$ and $-\frac{3}{2}i$, or $\pm\frac{3}{2}i$. ■

Equations of the Type $ax^2 + bx = 0$

When c is 0 but b is not, we can factor and use the principle of zero products.

EXAMPLE 3 Solve: $3x^2 + 5x = 0$.

Solution

$$3x^2 + 5x = 0$$
$$x(3x + 5) = 0 \quad \text{Factoring}$$
$$x = 0 \quad \text{or} \quad 3x + 5 = 0 \quad \text{Principle of zero products}$$
$$x = 0 \quad \text{or} \quad 3x = -5$$
$$x = 0 \quad \text{or} \quad x = -\tfrac{5}{3}$$

This time, the checks cannot be done together.

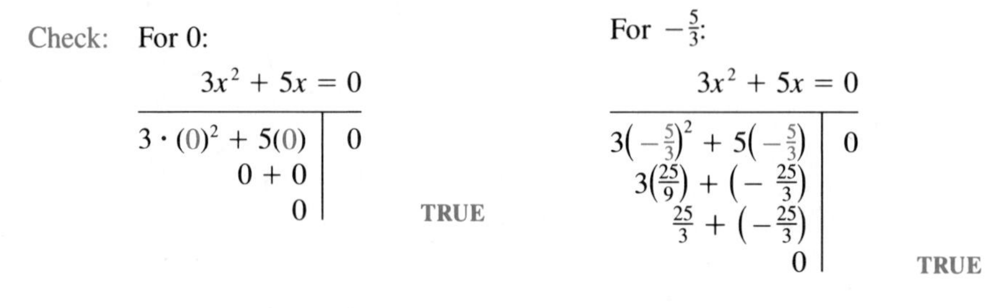

The solutions are 0 and $-\frac{5}{3}$.

Equations of the Type $ax^2 + bx + c = 0$

When none of the constants a, b, or c is 0, we can try factoring, much as we did in Section 3.5. Unfortunately, many quadratic equations are extremely difficult to solve by factoring. The procedure used in the next example enables us to solve an equation for which factoring would not work.

EXAMPLE 4 Solve: $x^2 + 6x + 4 = 0$.

Solution

$$x^2 + 6x + 4 = 0$$

$$x^2 + 6x = -4 \quad \text{Adding } -4 \text{ on both sides}$$

$$x^2 + 6x + 9 = -4 + 9 \quad \text{Adding 9 on both sides. We explain this shortly.}$$

$$(x + 3)^2 = 5 \quad \text{Factoring the trinomial square}$$

$$x + 3 = \pm\sqrt{5} \quad \text{Using the principle of positive and negative roots}$$

$$x = -3 \pm \sqrt{5} \quad \text{Adding } -3 \text{ on both sides}$$

The check is left to the student. The solutions are $-3 + \sqrt{5}$ and $-3 - \sqrt{5}$, or $-3 \pm \sqrt{5}$.

Let's examine how the above procedure works.

Completing the Square

The decision to add 9 on both sides in Example 4 was not made arbitrarily. We chose 9 because it made the left side a trinomial square. The 9 was obtained by taking half of the coefficient of x and squaring it—that is,

$$\left(\tfrac{1}{2} \cdot 6\right)^2 = 3^2, \quad \text{or} \quad 9.$$

To help see why this procedure works, examine the following drawings.

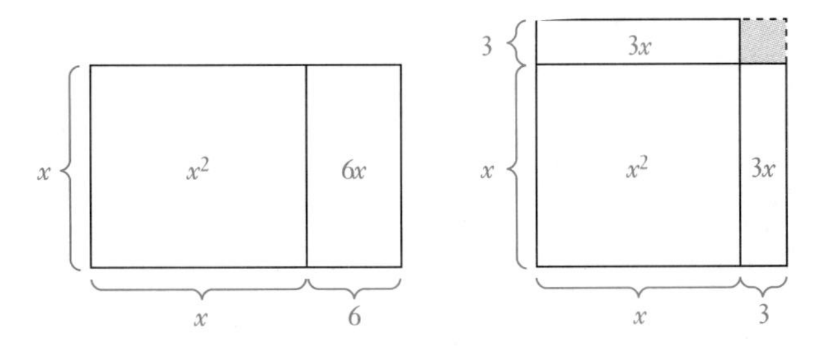

Note that both figures represent the same area, $x^2 + 6x$. However, only the figure on the right can be converted into a square with the addition of a constant term. The constant term, 9, can be interpreted as the area of the "missing" piece of the diagram on the right. It *completes* the square.

EXAMPLE 5 Complete the square for $x^2 - 5x$.

Solution We take half of -5 and get $-\frac{5}{2}$. We then square $-\frac{5}{2}$: $\left(-\frac{5}{2}\right)^2 = \frac{25}{4}$. We add the result to get $x^2 - 5x + \frac{25}{4}$. This trinomial square is equivalent to $\left(x - \frac{5}{2}\right)^2$. Note that for purposes of factoring, it is best *not* to write $\frac{25}{4}$ as a mixed number or a decimal. ■

EXAMPLE 6 Complete the square for $x^2 + \frac{3}{4}x$.

Solution We take half of $\frac{3}{4}$ and get $\frac{1}{2} \cdot \frac{3}{4} = \frac{3}{8}$. We then square $\frac{3}{8}$: $\left(\frac{3}{8}\right)^2 = \frac{9}{64}$. We add $\frac{9}{64}$ to get $x^2 + \frac{3}{4}x + \frac{9}{64}$, which is equivalent to $\left(x + \frac{3}{8}\right)^2$. ■

Solving Quadratics by Completing the Square

We can now use the method of completing the square to solve quadratic equations similar to Example 4.

EXAMPLE 7 Solve: $x^2 - 8x - 7 = 0$.

Solution

$$x^2 - 8x - 7 = 0$$

$$x^2 - 8x = 7 \qquad \text{Adding 7 on both sides}$$

$$x^2 - 8x + 16 = 7 + 16 \qquad \text{Adding 16 on both sides to complete the square}$$

$$(x - 4)^2 = 23 \qquad \text{Factoring}$$

$$x - 4 = \pm\sqrt{23} \qquad \text{Using the principle of positive and negative roots}$$

$$x = 4 \pm \sqrt{23} \qquad \text{Adding 4 on both sides}$$

The check is left to the student. The solutions are $4 - \sqrt{23}$ and $4 + \sqrt{23}$, or $4 \pm \sqrt{23}$. ■

EXAMPLE 8 Solve: $x^2 + 5x - 3 = 0$.

Solution

$$\begin{aligned} x^2 + 5x - 3 &= 0 \\ x^2 + 5x &= 3 && \text{Adding 3} \\ x^2 + 5x + \frac{25}{4} &= 3 + \frac{25}{4} && \text{Completing the square: } \left(\frac{5}{2}\right)^2 = \frac{25}{4} \\ \left(x + \frac{5}{2}\right)^2 &= \frac{37}{4} && \text{Factoring} \\ x + \frac{5}{2} &= \pm\frac{\sqrt{37}}{2} \\ x &= \frac{-5 \pm \sqrt{37}}{2} && \text{Adding } -\frac{5}{2} \end{aligned}$$

The checking of possible solutions such as these is quite cumbersome. When we use the method of completing the square or the quadratic formula of Section 7.2, we will never obtain any numbers that are not solutions of the original quadratic equation unless we have made a mistake. Thus it is usually advisable to check our work rather than to check by substituting into the original equation. The solutions are $(-5 - \sqrt{37})/2$ and $(-5 + \sqrt{37})/2$, or $(-5 \pm \sqrt{37})/2$. ■

Sometimes we need to divide on both sides before completing the square. We do this to have 1 as the coefficient of x^2.

EXAMPLE 9 Solve: $3x^2 + 7x - 2 = 0$.

Solution

$$\begin{aligned} 3x^2 + 7x - 2 &= 0 \\ 3x^2 + 7x &= 2 && \text{Adding 2} \\ x^2 + \frac{7}{3}x &= \frac{2}{3} && \text{Dividing by 3 on both sides} \\ x^2 + \frac{7}{3}x + \frac{49}{36} &= \frac{2}{3} + \frac{49}{36} && \text{Completing the square} \\ \left(x + \frac{7}{6}\right)^2 &= \frac{73}{36} && \text{Factoring and simplifying} \\ x + \frac{7}{6} &= \pm\frac{\sqrt{73}}{6} && \text{Using the principle of positive and negative roots} \\ x &= \frac{-7 \pm \sqrt{73}}{6} && \text{Adding } -\frac{7}{6} \end{aligned}$$

The solutions are $\dfrac{-7 - \sqrt{73}}{6}$ and $\dfrac{-7 + \sqrt{73}}{6}$, or $\dfrac{-7 \pm \sqrt{73}}{6}$. ■

We summarize the procedure used in Example 9. This can be used to solve *any* quadratic equation.

To solve a quadratic equation in x by completing the square:

1. Isolate the terms with variables on one side of the equation, and arrange them in descending order.
2. Divide by the coefficient of x^2 on both sides if that coefficient is not 1.
3. Complete the square by taking half of the coefficient of x and adding its square on both sides.
4. Factor one side. Find a common denominator on the other side and simplify.
5. Use the principle of positive and negative roots.
6. Solve for x by adding appropriately on both sides.

EXERCISE SET 7.1

Solve.

1. $4x^2 = 20$
2. $3x^2 = 21$
3. $10x^2 = 0$
4. $9x^2 = 0$
5. $16x^2 = 1$
6. $25x^2 = 9$
7. $2x^2 - 18 = 0$
8. $3x^2 - 75 = 0$
9. $2x^2 - 3 = 0$
10. $3x^2 - 7 = 0$
11. $-3x^2 + 5 = 0$
12. $-2x^2 + 1 = 0$
13. $x^2 + 100 = 0$
14. $x^2 + 81 = 0$
15. $x^2 + 5 = 0$
16. $x^2 + 6 = 0$
17. $0 = 4 + 25x^2$
18. $0 = 16 + 9x^2$
19. $2x^2 + 14 = 0$
20. $3x^2 + 15 = 0$
21. $\frac{4}{9}x^2 = 1$
22. $\frac{16}{25}x^2 = 1$

Solve.

23. $x^2 - 5x = 0$
24. $x^2 - 6x = 0$
25. $5x^2 + 10x = 0$
26. $3x^2 + 12x = 0$
27. $3x^2 - 2x = 0$
28. $7x^2 - 3x = 0$
29. $14x^2 + 9x = 0$
30. $19x^2 + 8x = 0$
31. $9x^2 - 11x = 0$
32. $7x^2 - 13x = 0$

Complete the square. Then write the trinomial square in factored form.

33. $x^2 + 10x$
34. $x^2 + 16x$
35. $x^2 - 8x$
36. $x^2 - 6x$
37. $x^2 - 24x$
38. $x^2 - 18x$
39. $x^2 + 9x$
40. $x^2 + 3x$
41. $x^2 - 7x$
42. $x^2 - 11x$
43. $x^2 + \frac{2}{3}x$
44. $x^2 + \frac{2}{5}x$
45. $x^2 - \frac{5}{6}x$
46. $x^2 - \frac{5}{3}x$
47. $x^2 + \frac{9}{5}x$
48. $x^2 + \frac{9}{4}x$

Solve. Use the method of completing the square. Show your work.

49. $x^2 + 8x = -7$
50. $x^2 + 6x = 7$
51. $x^2 - 10x = 22$
52. $x^2 - 8x = -9$
53. $x^2 + 6x + 5 = 0$
54. $x^2 + 10x + 9 = 0$

55. $x^2 - 10x + 21 = 0$
56. $x^2 - 10x + 24 = 0$
57. $x^2 + 5x + 6 = 0$
58. $x^2 + 7x + 12 = 0$
59. $x^2 + 4x + 1 = 0$
60. $x^2 + 6x + 7 = 0$
61. $x^2 - 10x + 23 = 0$
62. $x^2 - 6x + 4 = 0$
63. $x^2 + 6x + 13 = 0$
64. $x^2 + 8x + 25 = 0$
65. $2x^2 - 5x - 3 = 0$
66. $3x^2 + 5x - 2 = 0$
67. $4x^2 + 8x + 3 = 0$
68. $9x^2 + 18x + 8 = 0$
69. $6x^2 - x - 2 = 0$
70. $6x^2 - x - 15 = 0$
71. $2x^2 + 4x + 1 = 0$
72. $2x^2 + 5x + 2 = 0$
73. $3x^2 - 5x - 3 = 0$
74. $4x^2 - 6x - 1 = 0$

SKILL MAINTENANCE

75. Graph: $y = 2x + 1$.
76. Simplify: $\sqrt{88}$.
77. Approximate to the nearest tenth:

$$14 - \sqrt{88}.$$

78. Rationalize the denominator: $\sqrt{\frac{2}{5}}$.

SYNTHESIS

Solve.

79. 🖩 $25.55x^2 - 1635.2 = 0$
80. $(3x^2 - 7x - 20)(2x - 5) = 0$
81. $x(2x^2 + 9x - 56)(3x + 10) = 0$
82. $\left(x - \frac{1}{3}\right)\left(x - \frac{1}{3}\right) + \left(x - \frac{1}{3}\right)\left(x + \frac{2}{9}\right) = 0$

Solve for x.

83. $ax^2 - bx = 0$
84. $ax^2 - b = 0$

7.2 The Quadratic Formula

We studied completing the square for two reasons. The most important is that it is a useful tool in other places in mathematics. The second reason is that it can be used to derive a general formula for solving quadratic equations, called the **quadratic formula.** To obtain the quadratic formula, consider $ax^2 + bx + c = 0$, where a, b, and c can be any real numbers, provided $a > 0$. (For $a < 0$, multiply by -1 first on both sides.) Using the same steps as in Example 9 in Section 7.1, we solve by completing the square:

$$ax^2 + bx + c = 0$$

$$ax^2 + bx = -c \qquad \text{Adding } -c \text{ on both sides}$$

$$x^2 + \frac{b}{a}x = -\frac{c}{a} \qquad \text{Dividing by } a \text{ on both sides}$$

$$x^2 + \frac{b}{a}x + \frac{b^2}{4a^2} = \frac{b^2}{4a^2} - \frac{c}{a} \qquad \text{Completing the square: } \left(\frac{1}{2} \cdot \frac{b}{a}\right)^2 = \frac{b^2}{4a^2}$$

$$\left(x + \frac{b}{2a}\right)^2 = \frac{b^2 - 4ac}{4a^2} \qquad \text{Factoring and simplifying}$$

$$x + \frac{b}{2a} = \pm\frac{\sqrt{b^2 - 4ac}}{2a} \qquad \text{Using the principle of positive and negative roots. Since } a > 0,\ \sqrt{4a^2} = 2a.$$

$$x = \frac{-b \pm \sqrt{b^2 - 4ac}}{2a} \qquad \text{Adding } -\frac{b}{2a} \text{ on both sides}$$

It is important for you to remember the quadratic formula and to know how to use it.

THE QUADRATIC FORMULA

The solutions of a quadratic equation $ax^2 + bx + c = 0$ are given by

$$x = \frac{-b \pm \sqrt{b^2 - 4ac}}{2a}.$$

EXAMPLE 1 Use the quadratic formula to solve $5x^2 - 8x = 3$.

Solution First we find standard form and determine a, b, and c:

$$5x^2 - 8x - 3 = 0; \qquad a = 5, \quad b = -8, \quad c = -3.$$

We then substitute into the quadratic formula:

$$x = \frac{-(-8) \pm \sqrt{(-8)^2 - 4 \cdot 5 \cdot (-3)}}{2 \cdot 5}$$

$$x = \frac{8 \pm \sqrt{64 + 60}}{10} = \frac{8 \pm \sqrt{124}}{10}$$

$$x = \frac{8 \pm \sqrt{4 \cdot 31}}{10} = \frac{8 \pm 2\sqrt{31}}{10}$$

$$x = \frac{2(4 \pm \sqrt{31})}{2 \cdot 5} \qquad \text{Factoring}$$

$$x = \frac{4 \pm \sqrt{31}}{5}. \qquad \text{Removing a factor of 1}$$

The solutions are

$$\frac{4 + \sqrt{31}}{5} \quad \text{and} \quad \frac{4 - \sqrt{31}}{5}.$$

■

Some quadratic equations have solutions that are not real numbers.

EXAMPLE 2 Solve: $x^2 + x + 1 = 0$.

Solution

$$a = 1, \qquad b = 1, \qquad c = 1$$

$$x = \frac{-b \pm \sqrt{b^2 - 4ac}}{2a}$$

$$x = \frac{-1 \pm \sqrt{1^2 - 4 \cdot 1 \cdot 1}}{2 \cdot 1} = \frac{-1 \pm \sqrt{1 - 4}}{2}$$

$$x = \frac{-1 \pm \sqrt{-3}}{2} = \frac{-1 \pm i\sqrt{3}}{2}$$

The solutions are

$$\frac{-1 + i\sqrt{3}}{2} \quad \text{and} \quad \frac{-1 - i\sqrt{3}}{2}.$$

■

EXAMPLE 3 Solve:

$$2 + \frac{7}{x} = \frac{4}{x^2}.$$

Solution First we find standard form.

$$2x^2 + 7x = 4 \quad \text{Multiplying by } x^2\text{, the LCM of the denominators}$$

$$2x^2 + 7x - 4 = 0 \quad \text{Adding } -4$$

$$a = 2, \quad b = 7, \quad c = -4$$

$$x = \frac{-7 \pm \sqrt{7^2 - 4 \cdot 2 \cdot (-4)}}{2 \cdot 2}$$

$$x = \frac{-7 \pm \sqrt{49 + 32}}{4}$$

$$x = \frac{-7 \pm \sqrt{81}}{4} = \frac{-7 \pm 9}{4}$$

$$x = \frac{-7 + 9}{4} = \frac{1}{2} \quad \text{or} \quad x = \frac{-7 - 9}{4} = -4$$

In this case, since we started with a fractional equation, we *do* need to check. The quadratic formula always gives correct results when we start with the standard form. In this example, however, we cleared fractions before obtaining standard form, and this step could possibly introduce numbers that do not check in the original equation.

We should at least check that neither of the numbers makes a denominator 0. Since neither of them does, the solutions are $\frac{1}{2}$ and -4. ■

In Example 3 you may have noticed that the equation $2x^2 + 7x - 4 = 0$ could have been solved by factoring. We chose not to factor in order to further illustrate the use of the quadratic formula.

The solutions of a quadratic equation can *always* be found by using the quadratic formula. They are not always easy to find by factoring. A general strategy for solving quadratic equations follows.

To solve a quadratic equation:

a) Try factoring.
b) If factoring seems difficult, use the quadratic formula; it *always works*!

Approximating Solutions

Many solutions of quadratic equations are irrational numbers because they involve square roots. In such cases we may wish to find rational number approximations to the solutions, using a calculator.

EXAMPLE 4 Approximate, to the nearest tenth, the solutions of the equation in Example 1.

Solution Using a calculator, we find that $\sqrt{31} \approx 5.568$. Thus,

$$\frac{4+\sqrt{31}}{5} \approx \frac{4+5.568}{5} \quad \text{and} \quad \frac{4-\sqrt{31}}{5} \approx \frac{4-5.568}{5}$$

$$\approx \frac{9.568}{5} \qquad \text{Adding or subtracting} \qquad \approx \frac{-1.568}{5}$$

$$\approx 1.914 \qquad \text{Dividing} \qquad \approx -0.314$$

$$\approx 1.9 \qquad \text{Rounding to the nearest tenth} \qquad \approx -0.3$$

The solutions, approximated to the nearest tenth, are -0.3 and 1.9. ■

EXERCISE SET 7.2

Solve. Use factoring or the quadratic formula.

1. $x^2 + 6x + 4 = 0$
2. $x^2 - 6x - 4 = 0$
3. $x^2 + 4x - 5 = 0$
4. $x^2 - 2x - 15 = 0$
5. $y^2 + 7y = 30$
6. $y^2 - 7y = 30$
7. $2t^2 - 3t - 2 = 0$
8. $5m^2 + 3m - 2 = 0$
9. $3p^2 = -8p - 5$
10. $3u^2 = 18u - 6$
11. $x^2 - x + 1 = 0$
12. $x^2 + x + 2 = 0$
13. $x^2 + 13 = 4x$
14. $x^2 + 13 = 6x$
15. $z^2 + 5 = 0$
16. $t^2 + 3 = 0$
17. $r^2 + 3r = 8$
18. $h^2 + 4 = 6h$
19. $1 + \frac{2}{x} + \frac{5}{x^2} = 0$
20. $1 + \frac{5}{x^2} = \frac{2}{x}$
21. $x^2 - 2x + 5 = 0$
22. $x^2 - 4x + 5 = 0$
23. $5 = 2x^2$
24. $2 = 3x^2$
25. $x(x - 2) = -3x$
26. $x(x - 3) = -4x$
27. $2x + 1 = -5x^2$
28. $x + 2 = -3x^2$
29. $(2t - 3)^2 + 17t = 15$
30. $2y^2 - (y + 2)(y - 3) = 12$
31. $(x - 2)^2 + (x + 1)^2 = 0$
32. $(x + 3)^2 + (x - 1)^2 = 0$
33. $x + \frac{1}{x} = \frac{13}{6}$
34. $\frac{3}{x} + \frac{x}{3} = \frac{5}{2}$

Approximate solutions to the nearest tenth. Use a calculator or Table 1 in the appendix of this text.

35. $x^2 + 4x - 7 = 0$
36. $x^2 + 6x + 4 = 0$
37. $x^2 - 6x + 4 = 0$
38. $x^2 - 4x + 1 = 0$
39. $2x^2 - 3x - 7 = 0$
40. $3x^2 - 3x - 2 = 0$

SKILL MAINTENANCE

41. A store has cereal worth $1.50 per pound and granola worth $2.50 per pound. It wants to mix the two to obtain a 50-lb mixture worth $1.90 per pound. How much cereal and how much granola should be used?

42. Solve: $\frac{1}{2x} + \frac{1}{6x} = 2$.

SYNTHESIS

Solve.

43. ▦ $2.2x^2 + 0.5x - 1 = 0$

44. ▦ $5.33x^2 - 8.23x - 3.24 = 0$

45. ▦ $t^2 + 0.2t - 0.3 = 0$

46. ▦ $p^2 + 0.3p - 0.2 = 0$

47. ▦ $x^2 - 0.75x - 0.5 = 0$

48. ▦ $z^2 + 0.84z - 0.4 = 0$

49. $x^2 + x - \sqrt{2} = 0$

50. $x^2 - x - \sqrt{3} = 0$

51. $x^2 + \sqrt{5}x - \sqrt{3} = 0$

52. $\sqrt{2}x^2 + 5x + \sqrt{2} = 0$

53. $x^2 + 3x + i = 0$

54. $ix^2 - 2x + 1 = 0$

55. Solve for x in terms of y: $3x^2 + xy + 4y^2 = 9$.

56. Let $f(x) = x^2 + x - 8$.

a) Find the x-intercepts of the graph of the function.

b) Find all x-values for which $f(x) = 9$.

57. One solution of $kx^2 + 3x - k = 0$ is -2. Find the other.

7.3 Solutions of Quadratic Equations

We now examine the relationship between the solutions of a quadratic equation and the quadratic equation itself.

The Discriminant

From the quadratic formula, we know that the solutions x_1 and x_2 of a quadratic equation are given by

$$x_1 = \frac{-b + \sqrt{b^2 - 4ac}}{2a} \quad \text{and} \quad x_2 = \frac{-b - \sqrt{b^2 - 4ac}}{2a}.$$

The expression $b^2 - 4ac$ determines the nature of the solutions—that is, whether the solutions are real numbers and whether the numbers are different. This expression is called the *discriminant.* If it is 0, then whether we choose the plus or minus sign in the formula doesn't matter. There is just one solution and it is a real number. If the discriminant is positive, there will be two real solutions. If the discriminant is negative, we will be taking the square root of a negative number; hence there will be two nonreal solutions, and they will be complex conjugates. In this manner we gain information about the solutions of a quadratic equation without solving the equation. We summarize:

Discriminant $b^2 - 4ac$	Nature of Solutions
0	Only one solution; it is a real number.
Positive	Two different real number solutions.
Negative	Two different nonreal solutions (complex conjugates).

Since all real numbers are considered complex (written as $a + 0i$), we have used the word *nonreal* to describe numbers that are complex but not real.

EXAMPLE 1 Determine the nature of the solutions of $9x^2 - 12x + 4 = 0$.

Solution We have

$$a = 9, \quad b = -12, \quad \text{and} \quad c = 4.$$

We compute the discriminant:

$$\begin{aligned} b^2 - 4ac &= (-12)^2 - 4 \cdot 9 \cdot 4 \\ &= 144 - 144 = 0. \end{aligned}$$

There is just one solution, and it is a real number. ■

EXAMPLE 2 Determine the nature of the solutions of $x^2 + 5x + 8 = 0$.

Solution We have

$$a = 1, \quad b = 5, \quad \text{and} \quad c = 8.$$

We compute the discriminant:

$$\begin{aligned} b^2 - 4ac &= 5^2 - 4 \cdot 1 \cdot 8 \\ &= 25 - 32 = -7. \end{aligned}$$

Because the discriminant is negative, there are two nonreal solutions. ■

EXAMPLE 3 Determine the nature of the solutions of $x^2 + 5x + 6 = 0$.

Solution

$$a = 1, \quad b = 5, \quad \text{and} \quad c = 6$$

$$b^2 - 4ac = 5^2 - 4 \cdot 1 \cdot 6 = 1$$

Because the discriminant is positive, there are two solutions and they are real numbers. ■

Writing Equations from Solutions

We know by the principle of zero products that $(x - 2)(x + 3) = 0$ has solutions 2 and -3. If we know the solutions of an equation, we can write an equation by using the principle in reverse.

EXAMPLE 4 Find a quadratic equation whose solutions are $\frac{1}{3}$ and $-\frac{2}{5}$.

Solution

$$x = \tfrac{1}{3} \quad \text{or} \quad x = -\tfrac{2}{5}$$

$$x - \tfrac{1}{3} = 0 \quad \text{or} \quad x + \tfrac{2}{5} = 0 \qquad \text{Getting the 0's on one side}$$

$$\left(x - \tfrac{1}{3}\right)\left(x + \tfrac{2}{5}\right) = 0 \qquad \text{Principle of zero products (multiplying)}$$

$$3 \cdot \left(x - \tfrac{1}{3}\right)\left(x + \tfrac{2}{5}\right) \cdot 5 = 0 \cdot 3 \cdot 5 \qquad \text{Multiplying by } 3 \cdot 5 \text{ on both sides to clear fractions}$$

$$(3x - 1)(5x + 2) = 0 \qquad \text{Using the distributive law}$$

$$15x^2 + 6x - 5x - 2 = 0 \qquad \text{Using FOIL}$$

$$15x^2 + x - 2 = 0 \qquad \text{Collecting like terms}$$

■

EXAMPLE 5 Write a quadratic equation whose solutions are $2i$ and $-2i$.

Solution

$$x = 2i \quad \text{or} \quad x = -2i$$

$$x - 2i = 0 \quad \text{or} \quad x + 2i = 0 \qquad \text{Getting the 0's on one side}$$

$$(x - 2i)(x + 2i) = 0 \qquad \text{Principle of zero products}$$

$$x^2 + 2ix - 2ix - (2i)^2 = x^2 - (2i)^2 = 0 \qquad \text{Using FOIL}$$

$$x^2 + 4 = 0$$

■

EXAMPLE 6 Write a quadratic equation whose solutions are $\sqrt{3}$ and $-2\sqrt{3}$.

Solution

$$x = \sqrt{3} \quad \text{or} \quad x = -2\sqrt{3}$$

$$x - \sqrt{3} = 0 \quad \text{or} \quad x + 2\sqrt{3} = 0 \qquad \text{Getting the 0's on one side}$$

$$\left(x - \sqrt{3}\right)\left(x + 2\sqrt{3}\right) = 0 \qquad \text{Principle of zero products}$$

$$x^2 + 2\sqrt{3}x - \sqrt{3}x - 2\left(\sqrt{3}\right)^2 = 0 \qquad \text{Using FOIL}$$

$$x^2 + \sqrt{3}x - 6 = 0 \qquad \text{Collecting like terms}$$

■

Note that in Example 4 we multiplied on both sides to clear the fractions. We normally perform this step so that the equation has no fractional coefficients.

EXERCISE SET 7.3

Determine the nature of the solutions of the equation.

1. $x^2 - 6x + 9 = 0$
2. $x^2 + 10x + 25 = 0$
3. $x^2 + 7 = 0$
4. $x^2 + 2 = 0$
5. $x^2 - 2 = 0$
6. $x^2 - 5 = 0$
7. $4x^2 - 12x + 9 = 0$
8. $4x^2 + 8x - 5 = 0$
9. $x^2 - 2x + 4 = 0$
10. $x^2 + 3x + 4 = 0$
11. $a^2 - 10a + 21 = 0$
12. $t^2 - 8t + 16 = 0$
13. $6x^2 + 5x - 4 = 0$
14. $10x^2 - x - 2 = 0$
15. $9t^2 - 3t = 0$
16. $4m^2 + 7m = 0$
17. $x^2 + 5x = 7$
18. $x^2 + 4x = -6$

19. $y^2 = \frac{1}{2}y - \frac{3}{5}$
20. $y^2 + \frac{9}{4} = 4y$
21. $4x^2 - 4\sqrt{3}x + 3 = 0$
22. $6y^2 - 2\sqrt{3}y - 1 = 0$
23. $3t^2 - 5\sqrt{2}t + 7 = 0$
24. $5x^2 - 3\sqrt{5}x + 4 = 0$

Write a quadratic equation having the given numbers as solutions.

25. $-11, 9$
26. $-4, 4$
27. 7, only solution (*Hint:* It must be a double solution.)
28. -5, only solution
29. $-3, -5$
30. $-2, -7$
31. $4, \frac{2}{3}$
32. $5, \frac{3}{4}$
33. $\frac{1}{2}, \frac{1}{3}$
34. $-\frac{1}{4}, -\frac{1}{2}$
35. $-\frac{2}{5}, \frac{6}{5}$
36. $\frac{2}{7}, -\frac{3}{7}$
37. $\sqrt{2}, 3\sqrt{2}$
38. $-\sqrt{3}, 2\sqrt{3}$
39. $-\sqrt{5}, -2\sqrt{5}$
40. $-\sqrt{6}, -3\sqrt{6}$
41. $3i, -3i$
42. $4i, -4i$
43. $-1 + i, -1 - i$
44. $-3 + i, -3 - i$
45. $5 - 2i, 5 + 2i$
46. $2 - 7i, 2 + 7i$
47. $\frac{1 + 3i}{2}, \frac{1 - 3i}{2}$
48. $\frac{2 - i}{3}, \frac{2 + i}{3}$

SKILL MAINTENANCE

49. During a one-hour television show, there were 12 commercials. Some of the commercials were 30 sec long and the others were 60 sec long. The amount of time for 30-sec commercials was 6 min less than the total number of minutes of commercial time during the show. How many 30-sec commercials were used? How many 60-sec commercials were used?

SYNTHESIS

For each equation under the given condition, (a) find k, and (b) find the other solution.

50. $kx^2 - 17x + 33 = 0$; one solution is 3
51. $kx^2 - 2x + k = 0$; one solution is -3
52. $x^2 - kx + 2 = 0$; one solution is $1 + i$
53. $x^2 - (6 + 3i)x + k = 0$; one solution is 3
54. Find k for which

$$kx^2 - 4x + (2k - 1) = 0$$

and the product of the solutions is 3.

55. Find a quadratic equation for which the sum of the solutions is $\sqrt{3}$ and the product is 8.

56. The graph of an equation of the form

$$y = ax^2 + bx + c$$

is a curve similar to the one shown below. Determine a, b, and c from the information given.

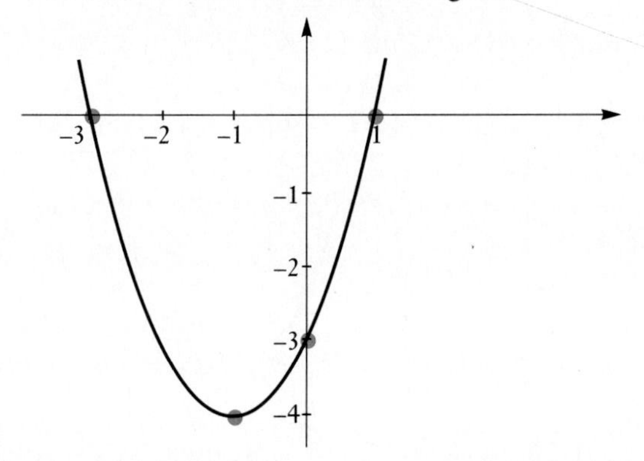

57. Prove each of the following.

a) The sum of the solutions of

$$ax^2 + bx + c = 0$$

is $-b/a$.

b) The product of the solutions of

$$ax^2 + bx + c = 0$$

is c/a.

7.4 Quadratic Functions and Their Graphs

The following bar graph shows the fall and rise of the rate of unemployment during a recent year. A curve drawn along the graph would approximate the graph of a *quadratic function*. We now consider such graphs.

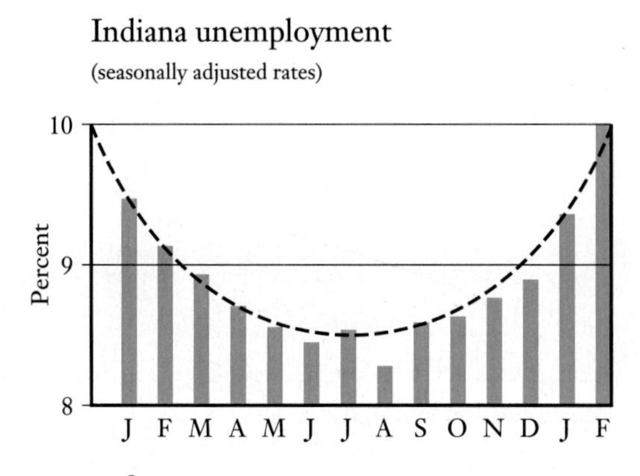

Graphs of $f(x) = ax^2$

In Chapter 6 the notion of a function was introduced. We graphed equations such as $f(x) = 3x + 2$ or $y = 3x + 2$. The equations studied in Chapter 6 were linear for the most part—that is, their graphs were straight lines. We now consider equations (or functions) in which the right-hand side is a quadratic polynomial:

$$f(x) = ax^2 + bx + c.$$

A function of this type is referred to as a **quadratic function.**

EXAMPLE 1 Graph: $f(x) = x^2$.

Solution We choose numbers for x, some positive and some negative, and for each number we compute $f(x)$.

We plot the ordered pairs and connect them with a smooth curve.

x	$f(x) = x^2$	$(x, f(x))$
-3	9	$(-3, 9)$
-2	4	$(-2, 4)$
-1	1	$(-1, 1)$
3	9	$(3, 9)$
2	4	$(2, 4)$
1	1	$(1, 1)$
0	0	$(0, 0)$

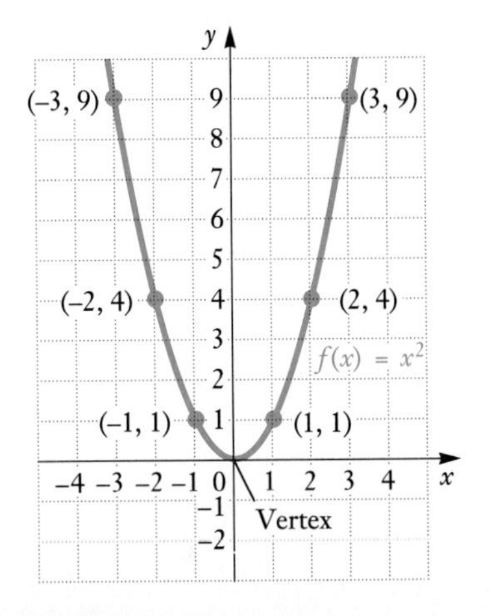

■

All quadratic functions have graphs similar to the one in Example 1. Such curves are called *parabolas*. They are smooth, cup-shaped curves that are symmetric with respect to a vertical line known as the parabola's *line of symmetry*. In the graph of $f(x) = x^2$, the y-axis is the line of symmetry. If the paper were folded on this line, the two halves of the curve would match. The point (0, 0) is known as the *vertex* of this parabola.

By plotting points, we can see how the graphs of $g(x) = \frac{1}{2}x^2$ and $h(x) = 2x^2$ compare with the graph of $f(x) = x^2$.

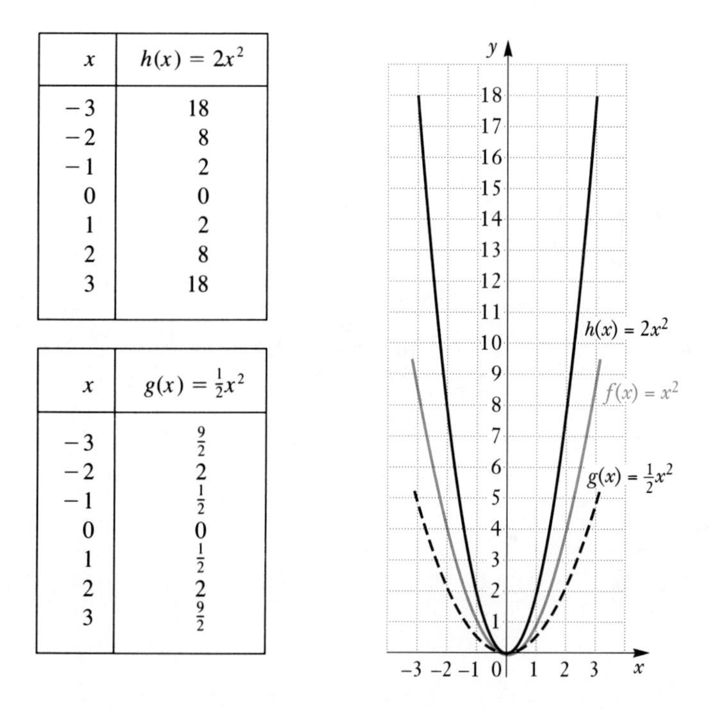

x	$h(x) = 2x^2$
-3	18
-2	8
-1	2
0	0
1	2
2	8
3	18

x	$g(x) = \frac{1}{2}x^2$
-3	$\frac{9}{2}$
-2	2
-1	$\frac{1}{2}$
0	0
1	$\frac{1}{2}$
2	2
3	$\frac{9}{2}$

Note that the graph of $g(x) = \frac{1}{2}x^2$ is a flatter parabola than the graph of $f(x) = x^2$, and the graph of $h(x) = 2x^2$ is narrower. The vertex and the line of symmetry have not changed.

When we consider the graph of $k(x) = -\frac{1}{2}x^2$, we see that the parabola opens downward and is the same shape as the graph of $g(x) = \frac{1}{2}x^2$.

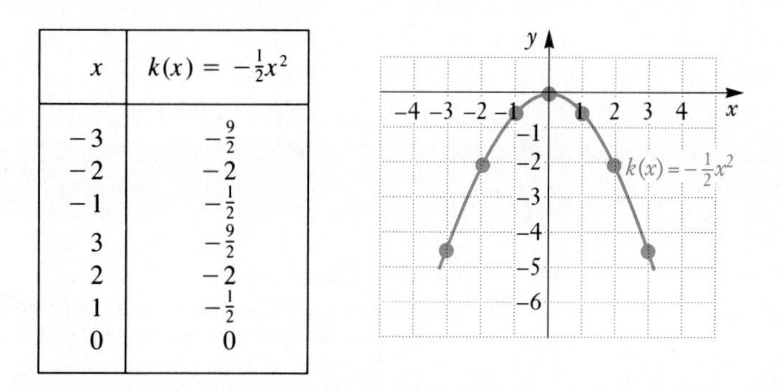

x	$k(x) = -\frac{1}{2}x^2$
-3	$-\frac{9}{2}$
-2	-2
-1	$-\frac{1}{2}$
3	$-\frac{9}{2}$
2	-2
1	$-\frac{1}{2}$
0	0

> The graph of $g(x) = ax^2$ is a parabola with the vertical axis as its line of symmetry and its vertex at the origin.
>
> If $|a|$ is greater than 1, the parabola is narrower than $f(x) = x^2$.
>
> If $|a|$ is between 0 and 1, the parabola is flatter than $f(x) = x^2$.
>
> If a is positive, the parabola opens upward; if a is negative, the parabola opens downward.

Graphs of $f(x) = a(x - h)^2$

Why not now consider graphs of $f(x) = ax^2 + bx + c$, where b and c are not both 0? In effect, we will do that, but in a disguised form. It turns out to be convenient to consider functions $f(x) = a(x - h)^2$, that is, where we start with ax^2 but then replace x by $x - h$, where h is some constant.*

EXAMPLE 2 Graph: $f(x) = 2(x - 3)^2$.

Solution We choose some values of x and compute $f(x)$. Then we plot the points and draw the curve.

Compare these values of x and $f(x)$ to the values used earlier when graphing $h(x) = 2x^2$.

x	$f(x) = 2(x - 3)^2$
0	18
1	8
2	2
3	0
4	2
5	8
6	18

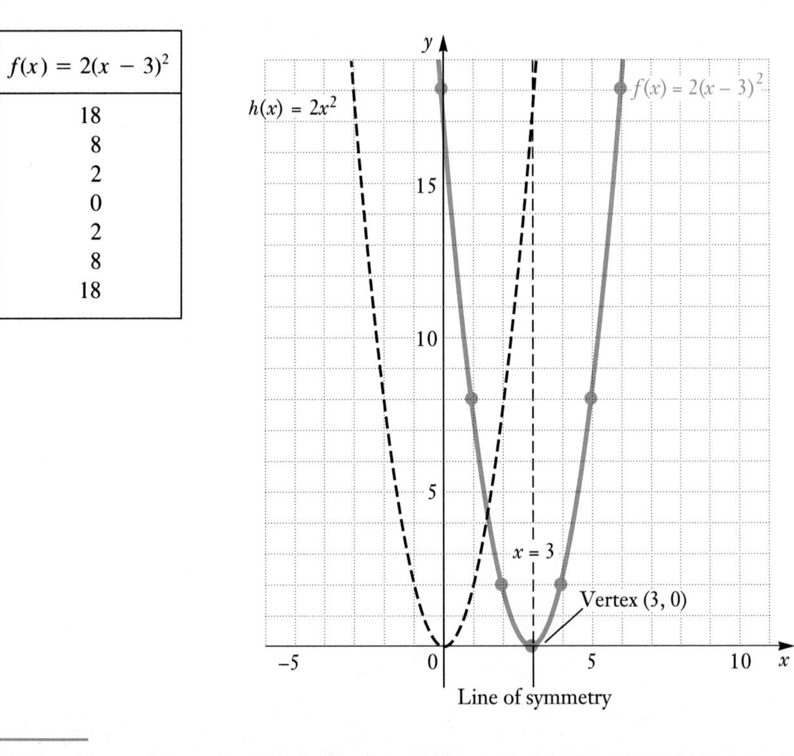

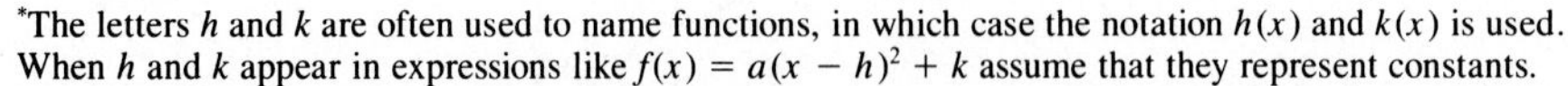
*The letters h and k are often used to name functions, in which case the notation $h(x)$ and $k(x)$ is used. When h and k appear in expressions like $f(x) = a(x - h)^2 + k$ assume that they represent constants.

The graph of $f(x) = 2(x - 3)^2$ looks just like the graph of $h(x) = 2x^2$ except that it is moved three units to the right. The line of symmetry is now the line $x = 3$, and the vertex is the point (3, 0). ■

The graph of $g(x) = a(x - h)^2$ looks like the graph of $f(x) = ax^2$ except it is moved to the right or left h units. If h is positive, it is moved to the right. If h is negative, it is moved to the left. The vertex is now $(h, 0)$, and the line of symmetry is $x = h$.

EXAMPLE 3 Graph: $g(x) = 2(x + 3)^2$.

Solution We know that the graph looks like that of $h(x) = 2x^2$ but moved right or left. Think of $x + 3$ as $x - (-3)$. The number we are *subtracting* is negative, so we move *left*. We can draw $f(x) = 2x^2$ lightly, then move the graph three units to the left. We should compute a few values as a check.

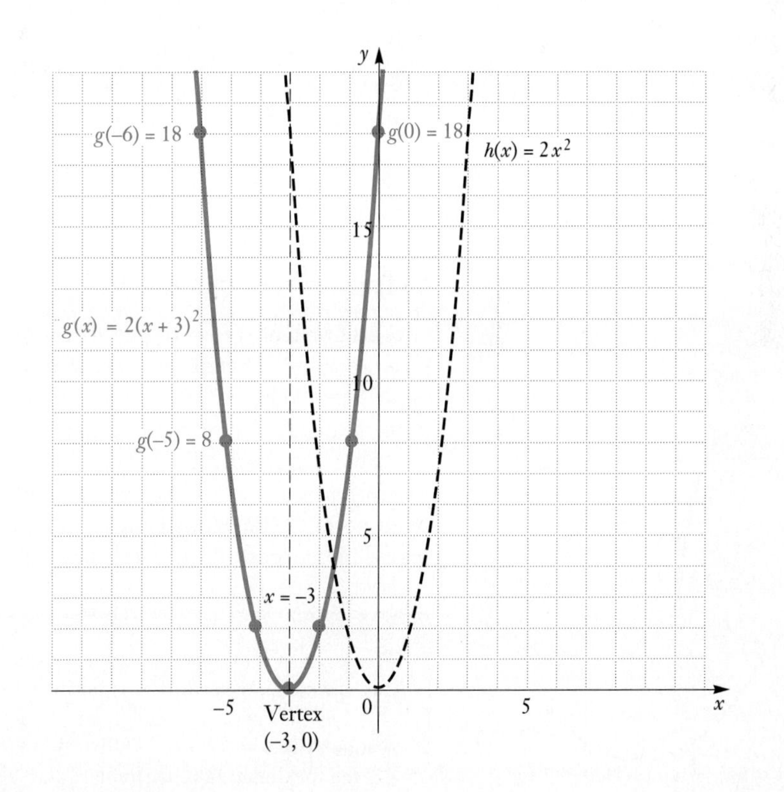

■

EXAMPLE 4 Graph: $k(x) = -2(x + 2)^2$.

Solution We know that the graph looks like that of $h(x) = 2x^2$ but moved to the left two units, and it will also open downward because of the negative coefficient, -2. (See the figure at the top of the next page.)

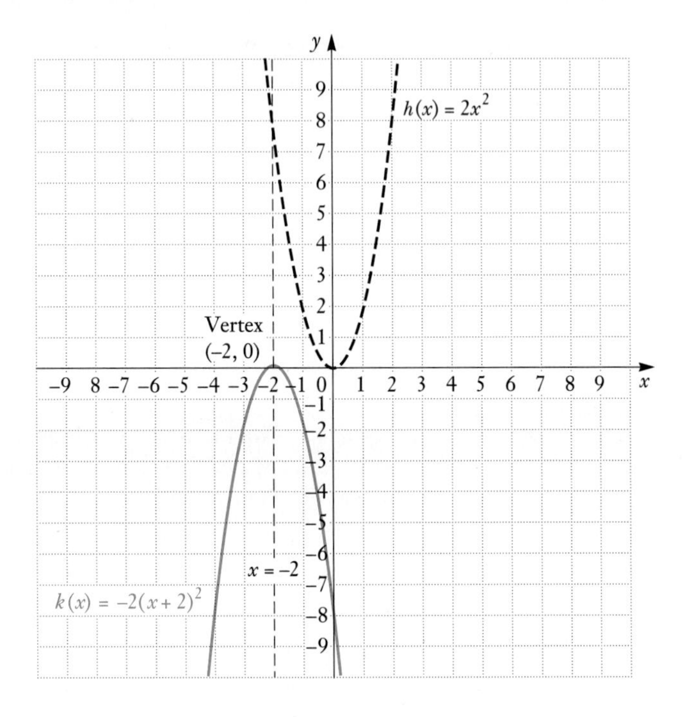

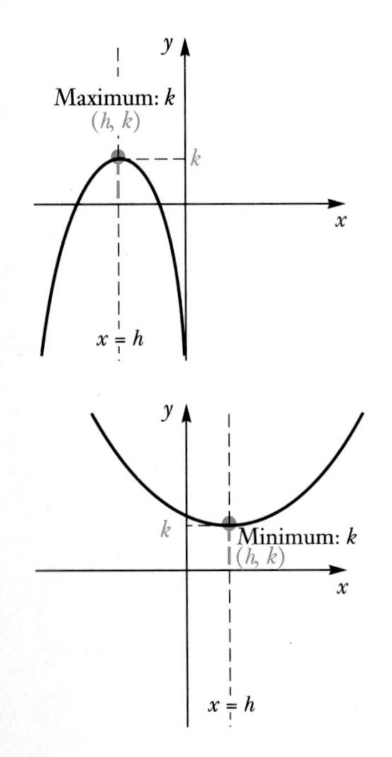

Graphs of $f(x) = a(x - h)^2 + k$

Given a graph of $f(x) = a(x - h)^2$, what happens to it if we add a constant k? Suppose we add 2. This increases each function value $f(x)$ by 2, so the curve is moved up. If k should be -3, the curve is moved down. The vertex of the parabola will be at the point (h, k) and the line of symmetry will be $x = h$.

Note that if a parabola opens upward ($a > 0$), the function value, or y-value, at the vertex is a least, or *minimum,* value. That is, it is less than the y-value at any other point. If the parabola opens downward ($a < 0$), the function value at the vertex will be a greatest, or *maximum,* value.

> The graph of $f(x) = a(x - h)^2 + k$ looks like that of $g(x) = a(x - h)^2$ except that it is moved up or down. If k is positive, the curve is moved up k units. If k is negative, the curve is moved down k units. The vertex is at (h, k), and the line of symmetry is $x = h$. If $a > 0$, then k is the minimum function value. If $a < 0$, then k is the maximum function value.

EXAMPLE 5 Graph $f(x) = 2(x + 3)^2 - 5$, and find the minimum function value.

Solution We know that the graph looks like that of $g(x) = 2(x + 3)^2$ (see Example 3) but moved down five units. You should confirm this by plotting some points. For instance, $f(-4) = -3$, while $g(-4) = 2$.

The vertex is now $(-3, -5)$, and the minimum function value is -5.

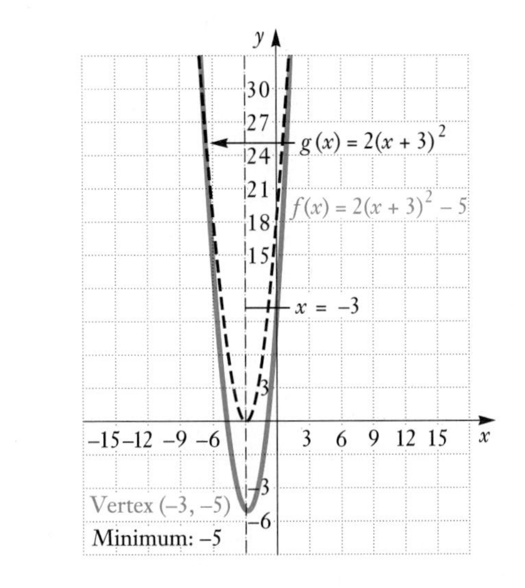

EXAMPLE 6 Graph $h(x) = \frac{1}{2}(x - 3)^2 + 5$, and find the minimum function value.

Solution We know that the graph looks just like that of $f(x) = \frac{1}{2}x^2$ but moved to the right three units and up five units. The vertex is (3, 5), and the line of symmetry is the line $x = 3$. We draw $f(x) = \frac{1}{2}x^2$ and then move the curve over and up. We plot a few points as a check. The minimum function value is 5.

x	$h(x)$
0	$9\frac{1}{2}$
1	7
2	$5\frac{1}{2}$
6	$9\frac{1}{2}$

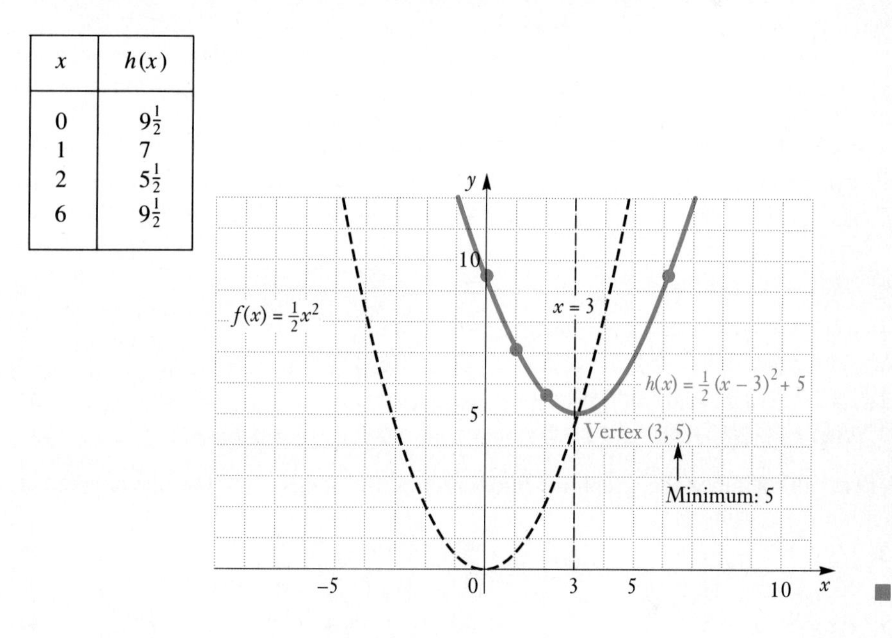

EXAMPLE 7 Graph $y = -2(x+3)^2 + 5$, and find the maximum function value.

Solution We know that the graph looks like that of $y = 2x^2$ but moved to the left three units, up five units, and turned upside down. The vertex is $(-3, 5)$. This time there is a *greatest* (maximum) function value, 5.

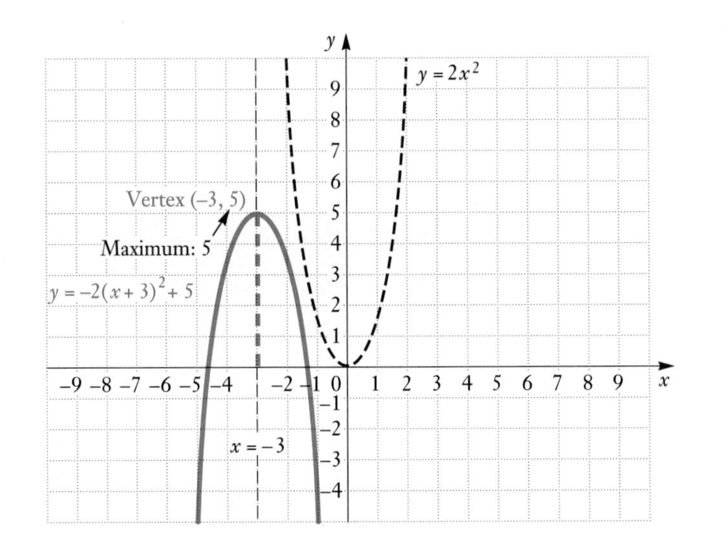

EXERCISE SET 7.4

Graph.

1. $f(x) = x^2$
2. $f(x) = -x^2$
3. $f(x) = -4x^2$
4. $f(x) = -3x^2$
5. $g(x) = \frac{1}{4}x^2$
6. $g(x) = \frac{1}{3}x^2$
7. $h(x) = -\frac{1}{3}x^2$
8. $h(x) = -\frac{1}{4}x^2$
9. $f(x) = \frac{3}{2}x^2$
10. $f(x) = \frac{5}{2}x^2$

For each of the following, graph the function, label the vertex, and draw the line of symmetry.

11. $g(x) = (x+1)^2$
12. $g(x) = (x+4)^2$
13. $f(x) = (x-4)^2$
14. $f(x) = (x-1)^2$
15. $h(x) = (x-3)^2$
16. $h(x) = (x-7)^2$
17. $f(x) = -(x+4)^2$
18. $f(x) = -(x-2)^2$
19. $g(x) = -(x-1)^2$
20. $g(x) = -(x+5)^2$
21. $f(x) = 2(x-1)^2$
22. $f(x) = 2(x+4)^2$
23. $h(x) = -\frac{1}{2}(x-3)^2$
24. $h(x) = -\frac{3}{2}(x-2)^2$
25. $f(x) = \frac{1}{2}(x+1)^2$
26. $f(x) = \frac{1}{3}(x+2)^2$
27. $g(x) = -3(x-2)^2$
28. $g(x) = -4(x-7)^2$
29. $f(x) = -2(x+9)^2$
30. $f(x) = 2(x+7)^2$
31. $h(x) = -3\left(x-\frac{1}{2}\right)^2$
32. $h(x) = -2\left(x+\frac{1}{2}\right)^2$

For each of the following, graph the function, find the vertex, find the line of symmetry, and find the maximum value or the minimum value.

33. $f(x) = (x-3)^2 + 1$
34. $f(x) = (x+2)^2 - 3$
35. $f(x) = (x+1)^2 - 2$
36. $f(x) = (x-1)^2 + 2$
37. $g(x) = (x+4)^2 + 1$
38. $g(x) = -(x-2)^2 - 4$
39. $f(x) = \frac{1}{2}(x-5)^2 + 2$
40. $f(x) = \frac{1}{2}(x+1)^2 - 2$
41. $h(x) = -2(x-1)^2 - 3$

42. $h(x) = -2(x + 1)^2 + 4$
43. $f(x) = -3(x + 4)^2 + 1$
44. $f(x) = -2(x - 5)^2 - 3$
45. $g(x) = -\frac{3}{2}(x - 1)^2 + 2$
46. $g(x) = \frac{3}{2}(x + 2)^2 - 1$

Without graphing, find the vertex, find the line of symmetry, and find the maximum value or the minimum value.

47. $f(x) = 8(x - 9)^2 + 5$
48. $f(x) = 10(x + 5)^2 - 8$
49. $h(x) = -\frac{2}{7}(x + 6)^2 + 11$
50. $h(x) = -\frac{3}{11}(x - 7)^2 - 9$
51. $f(x) = 5\left(x + \frac{1}{4}\right)^2 - 13$
52. $f(x) = 6\left(x - \frac{1}{4}\right)^2 + 19$
53. $f(x) = -7(x - 10)^2 - 20$
54. $f(x) = -9(x + 12)^2 + 23$
55. $f(x) = \sqrt{2}(x + 4.58)^2 + 65\pi$
56. $f(x) = 4\pi(x - 38.2)^2 - \sqrt{34}$

SKILL MAINTENANCE

57. Divide and simplify:

$$\frac{21a^9b^7}{3a^3b}.$$

58. Solve: $6x^2 - 13x + 2 = 0$.

SYNTHESIS

For each of the following, write the equation of the parabola that has the shape of $f(x) = 2x^2$ or $f(x) = -2x^2$ and has a maximum or minimum value at the specified point.

59. Maximum: $(0, 4)$
60. Minimum: $(2, 0)$
61. Minimum: $(6, 0)$
62. Maximum: $(0, 3)$
63. Maximum: $(3, 8)$
64. Minimum: $(-2, 3)$
65. Minimum: $(-3, 6)$
66. Maximum: $(-4, -3)$
67. Minimum: $(2, -3)$

For Exercises 68 and 69, write an equation of the parabola.

68. The parabola has a minimum value at the same point as $f(x) = 3(x - 4)^2$, but for all x the function values are twice the values obtained from $f(x) = 3(x - 4)^2$.

69. The parabola is the same shape as

$$f(x) = -\tfrac{1}{2}(x - 2)^2 + 4$$

and has a maximum value at the same point as

$$f(x) = -2(x - 1)^2 - 6.$$

A shift of a graph in which we do not change the size or shape of the graph and in which there is no rotation is called a *translation*. What we have seen for parabolas holds for other functions: For a function $f(x)$, if we replace x by $x - h$, where h is constant, the graph will be moved horizontally. If we add a constant k to a function $f(x)$, the graph will be moved vertically.

Use the graph of the function $y = g(x)$, and draw a graph for Exercises 70–75.

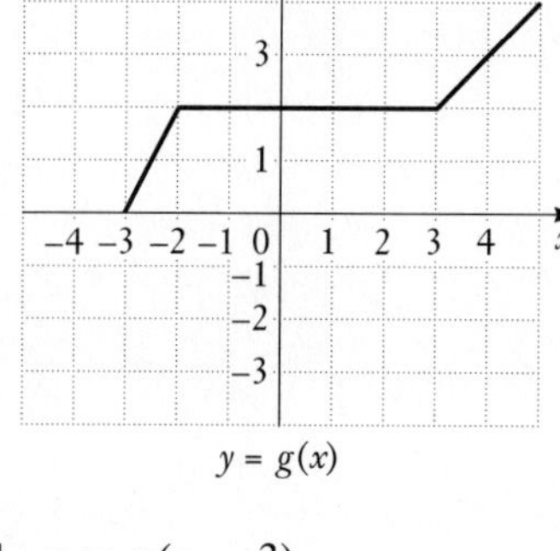

$y = g(x)$

70. $y = g(x + 2)$
71. $y = g(x - 3)$
72. $y - 3 = g(x)$
73. $y = g(x) + 4$
74. $y + 2 = g(x - 4)$
75. $y = g(x - 2) + 3$

7.5 More About Graphing Quadratic Functions

Completing the Square

The procedures discussed in Section 7.4 enable us to graph any quadratic function $f(x) = ax^2 + bx + c$. By *completing the square* (see Section 7.1), we can always rewrite the polynomial $ax^2 + bx + c$ in the form

$$a(x - h)^2 + k.$$

EXAMPLE 1 Graph: $f(x) = x^2 - 6x + 4$.

Solution

$$f(x) = x^2 - 6x + 4$$
$$= (x^2 - 6x) + 4$$

We complete the square inside the parentheses by taking half the x-coefficient,

$$\tfrac{1}{2} \cdot (-6) = -3,$$

and squaring it:

$$(-3)^2 = 9.$$

Then we add 9 inside the parentheses. However, we will no longer have an equivalent expression if we just add 9 inside the parentheses. Thus we must at the same time *subtract* 9 outside the parentheses. In this way, we form an equivalent expression, since the overall effect is that of adding 0:

$$f(x) = (x^2 - 6x + 9) + 4 - 9 \qquad \text{Adding and subtracting 9}$$
$$= (x - 3)^2 - 5. \qquad \text{Factoring and simplifying}$$

The vertex is $(3, -5)$. The line of symmetry is $x = 3$. We plot a few points as a check and draw the curve.

x	$f(x)$
0	4
1	−1
5	−1
6	4

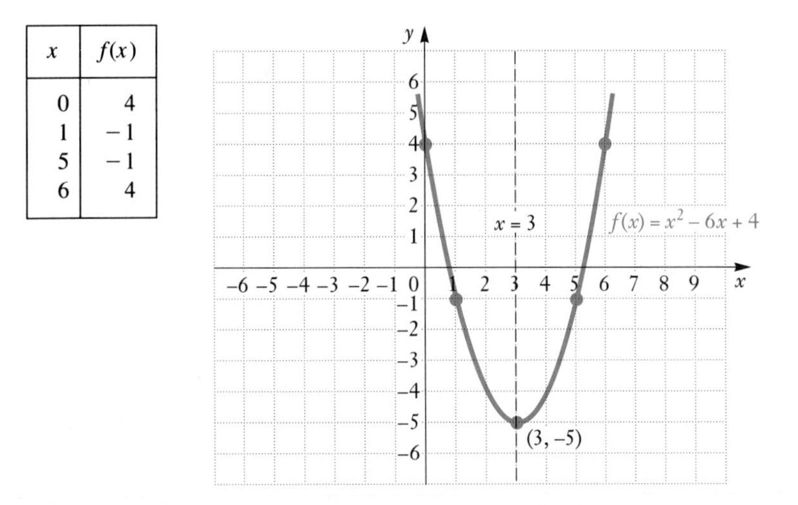

■

When the leading coefficient is not 1, we use the distributive law to factor out that number from the first two terms. Then we complete the square.

EXAMPLE 2 Graph: $g(x) = 3x^2 + 12x + 13$.

Solution Since the coefficient of x^2 is not 1, we need to factor that number—in this case, 3—from the first two terms. Remember that we want the form $g(x) = a(x - h)^2 + k$:

$$\begin{aligned} g(x) &= 3x^2 + 12x + 13 \\ &= 3(x^2 + 4x) + 13. \end{aligned}$$

To complete the square inside the parentheses, we take half of 4 and square it, to get 4. Thus we plan to add 4 inside the parentheses. However, since the parentheses are preceded by a factor of 3, we must subtract $3 \cdot 4$, or 12, from 13 to form an equivalent expression:

$$\begin{aligned} g(x) &= 3(x^2 + 4x + 4) + 13 - 12 && \text{Adding and subtracting 12} \\ &= 3(x + 2)^2 + 1. && \text{This is the desired form.} \end{aligned}$$

The vertex is $(-2, 1)$. The line of symmetry is $x = -2$. The coefficient 3 is positive, so the graph opens upward.

We plot a few points as a check and draw the curve.

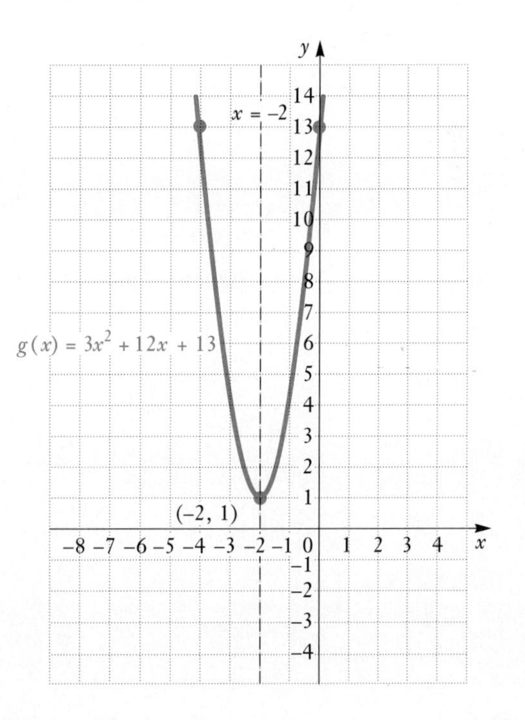

■

EXAMPLE 3 Graph: $f(x) = -2x^2 + 10x - 7$.

Solution We factor -2 from the expression $-2x^2 + 10x$. This makes the coefficient of x^2 inside the parentheses 1:

$$\begin{aligned} f(x) &= -2x^2 + 10x - 7 \\ &= -2(x^2 - 5x) - 7. \end{aligned}$$

We take half the x-coefficient and square it, to get $\frac{25}{4}$. Since we next plan to add $\frac{25}{4}$ *inside* the parentheses, we must also subtract $(-2)(\frac{25}{4})$ *outside* the parentheses:

$$\begin{aligned} f(x) &= -2\left(x^2 - 5x + \tfrac{25}{4}\right) - 7 - (-2)\left(\tfrac{25}{4}\right) \\ &= -2\left(x - \tfrac{5}{2}\right)^2 + \tfrac{11}{2}. \end{aligned}$$

The vertex is $\left(\frac{5}{2}, \frac{11}{2}\right)$. The line of symmetry is $x = \frac{5}{2}$. The coefficient -2 is negative, so the graph opens downward.

We plot a few points as a check and draw the curve.

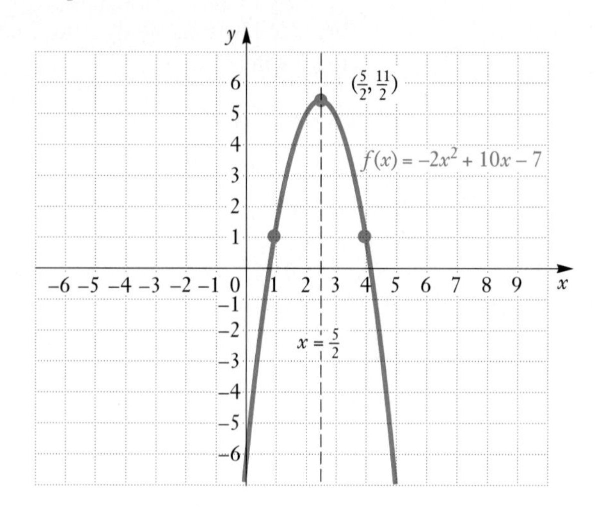

■

Intercepts

The points at which a graph crosses the x-axis are called its **x-intercepts.** These are, of course, the points at which $y = 0$.

To find the x-intercepts of a quadratic function $f(x) = ax^2 + bx + c$, we solve the equation

$$0 = ax^2 + bx + c.$$

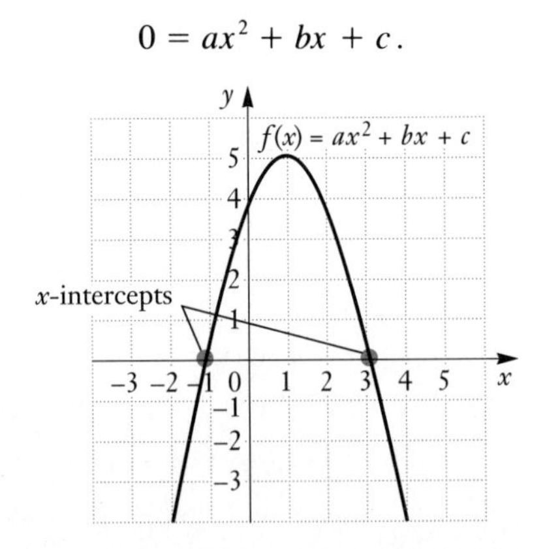

EXAMPLE 4 Find the x-intercepts of the graph of $f(x) = x^2 - 2x - 2$.

Solution We solve the equation

$$0 = x^2 - 2x - 2.$$

The equation is difficult to factor, so we use the quadratic formula and get $x = 1 \pm \sqrt{3}$. Thus the x-intercepts are $(1 - \sqrt{3}, 0)$ and $(1 + \sqrt{3}, 0)$.

For plotting, we approximate, to get $(-0.7, 0)$ and $(2.7, 0)$. We sometimes refer to the x-coordinates as *intercepts*. It is useful to have these points when graphing a function. ■

The discriminant, $b^2 - 4ac$, tells us how many real number solutions the equation $0 = ax^2 + bx + c$ has, so it also indicates how many intercepts there are. Compare the following graphs.

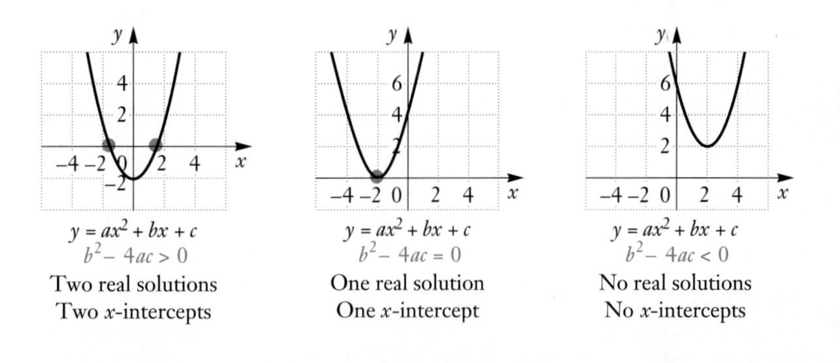

EXERCISE SET 7.5

Complete the square. Then graph.

1. $f(x) = x^2 - 2x - 3$
2. $f(x) = x^2 + 2x - 5$
3. $g(x) = x^2 + 6x + 13$
4. $g(x) = x^2 - 4x + 5$
5. $f(x) = x^2 + 4x - 1$
6. $f(x) = x^2 - 10x + 21$
7. $h(x) = 2x^2 + 16x + 25$
8. $h(x) = 2x^2 - 16x + 23$
9. $f(x) = -x^2 + 4x + 6$
10. $f(x) = -x^2 - 4x + 3$
11. $g(x) = x^2 + 3x - 10$
12. $g(x) = x^2 + 5x + 4$
13. $f(x) = 3x^2 - 24x + 50$
14. $f(x) = 4x^2 + 8x - 3$
15. $h(x) = x^2 - 9x$
16. $h(x) = x^2 + x$
17. $f(x) = -2x^2 - 4x - 6$
18. $f(x) = -3x^2 + 6x + 2$
19. $g(x) = 2x^2 - 10x + 14$
20. $g(x) = 2x^2 + 6x + 8$
21. $f(x) = -3x^2 - 3x + 1$
22. $f(x) = -2x^2 + 2x + 1$
23. $h(x) = \frac{1}{2}x^2 + 4x + \frac{19}{3}$
24. $h(x) = \frac{1}{2}x^2 - 3x + 2$

Find the x-intercepts. If no x-intercepts exist, state this.

25. $f(x) = x^2 - 4x + 1$
26. $f(x) = x^2 + 6x + 10$
27. $g(x) = -x^2 + 2x + 3$
28. $g(x) = x^2 - 2x - 5$
29. $f(x) = x^2 - 3x - 4$
30. $f(x) = x^2 - 8x + 5$
31. $h(x) = -x^2 + 3x - 2$
32. $h(x) = 2x^2 - 4x + 6$
33. $f(x) = 2x^2 + 4x - 1$
34. $f(x) = x^2 - x + 2$
35. $g(x) = x^2 - x + 1$
36. $g(x) = 4x^2 + 12x + 9$
37. $f(x) = -x^2 - 3x - 3$
38. $f(x) = 3x^2 - 5x - 4$

SKILL MAINTENANCE

Divide.

39. $(x^2 - 8x + 17) \div (x - 4)$

40. $(125y^3 - 8) \div (5y - 2)$

SYNTHESIS

Find the maximum or the minimum value.

41. $f(x) = 2.31x^2 - 3.135x - 5.89$

42. $f(x) = -18.8x^2 + 7.92x + 6.18$

Find the x-intercepts. If none exist, state this.

43. $f(x) = 0.05x^2 - 4.735x + 100.23$

44. $f(x) = 1.13x^2 + 2.809x - 7.114$

45. $f(x) = 2.12x^2 + 3.21x + 9.73$

46. $f(x) = 0.13x^2 - 0.071x - 0.12$

47. Graph the function

$$f(x) = x^2 - x - 6.$$

Then use the graph to approximate solutions to the following equations.

a) $x^2 - x - 6 = 2$
b) $x^2 - x - 6 = -3$

48. Graph the function

$$f(x) = \frac{x^2}{8} + \frac{x}{4} - \frac{3}{8}.$$

Use the graph to approximate solutions to the following equations.

a) $\frac{x^2}{8} + \frac{x}{4} - \frac{3}{8} = 0$
b) $\frac{x^2}{8} + \frac{x}{4} - \frac{3}{8} = 1$
c) $\frac{x^2}{8} + \frac{x}{4} - \frac{3}{8} = 2$

Find an equivalent equation of the type $f(x) = a(x - h)^2 + k$.

49. $f(x) = ax^2 + bx + c$

50. $f(x) = 3x^2 + mx + m^2$

Graph.

51. $f(x) = |x^2 - 1|$

52. $f(x) = |3 - 2x - x^2|$

53. The graph of a quadratic function has a vertex at $(1, -8)$, and $(-3, 0)$ and $(5, 0)$ as its x-intercepts. Find an equation for the function.

7.6 Problem Solving and Quadratic Functions

Let's look now at some of the many situations in which quadratic functions are used for problem solving.

Maximum and Minimum Problems

For a quadratic function the value $f(x)$ at the vertex will be either greater than any other $f(x)$ or less than any other $f(x)$. If the graph opens upward, f will achieve a minimum. If the graph opens downward, f will achieve a maximum. In certain problems we want to find a maximum or a minimum. If the problem situation translates to a quadratic function, we can solve by finding $f(x)$ at the vertex.

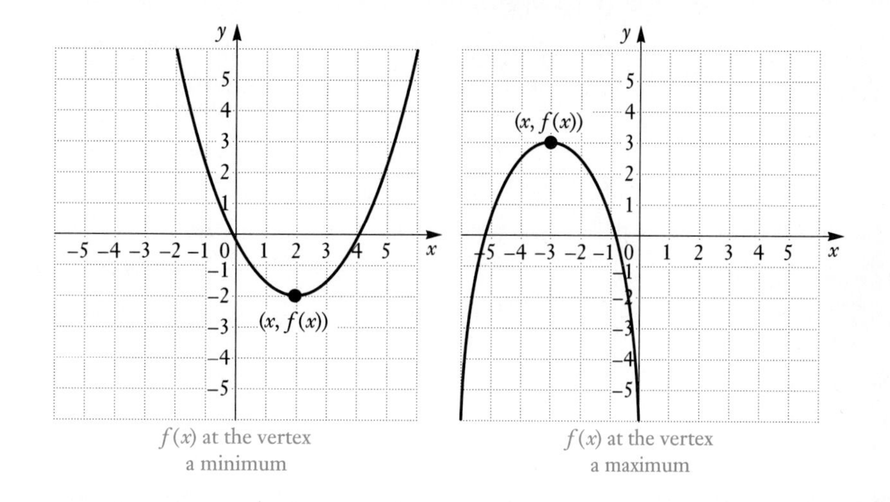

$f(x)$ at the vertex a minimum

$f(x)$ at the vertex a maximum

EXAMPLE 1 What are the dimensions of the largest rectangular pen that can be enclosed with 64 m of fence?

Solution

1. **Familiarize.** We make a drawing and label it.

   Perimeter: $2w + 2l = 64$ m

   Area: $A = l \cdot w$

   We may need to look up definitions or formulas in step (1). Also, we may be able to get a better feel for the problem by looking at some possible dimensions for a rectangular pen that can be enclosed with 64 m of fence.

l	w	A
22	10	220
20	12	240
18	14	252

2. **Translate.** We have two equations, one of which expresses area as a function of length and width:

$$2w + 2l = 64$$
$$A = l \cdot w.$$

3. **Carry out.** We need to express A as a function of l or w but not both. To do so, we solve for l in the first equation to obtain $l = 32 - w$. Substituting for l in the second equation, we get a quadratic function, $A(w)$, or just A*:

   $A = (32 - w)w$ **Substituting for l**

   $A = -w^2 + 32w.$ **This is a parabola opening downward, so a maximum exists.**

   Completing the square, we get

   $$A = -(w - 16)^2 + 256.$$

   The maximum function value of 256 occurs when $w = 16$ and $l = 32 - 16$, or 16.

4. **Check.** Note that 256 is greater than any of the values for A found in the *Familiarize* step. To be more certain, we check function values for $w = 15$ and $w = 17$. This is not a foolproof check, but it is fairly close:

   $$A(15) = -(15 - 16)^2 + 256 = -1 + 256 = 255$$
   $$A(17) = -(17 - 16)^2 + 256 = -1 + 256 = 255.$$

   Since 256 is greater than 255, it looks as though we have a maximum.

5. **State.** To maximize the area of the pen, the dimensions should be 16 m by 16 m. ■

EXAMPLE 2 What is the minimum product of two numbers whose difference is 5? What are the numbers?

Solution

1. **Familiarize.** We try some pairs of numbers that differ by 5 and compute their products:

   $$1 \cdot 6 = 6$$
   $$0 \cdot 5 = 0$$
   $$(-1) \cdot 4 = -4.$$

   We suspect that one of the two numbers will be negative and the other positive. Let x represent the larger number and $x - 5$ the other number.

2. **Translate.** We represent the product of the two numbers by $p = x(x - 5)$, or $p(x) = x^2 - 5x$.

3. **Carry out.** The function $p(x) = x^2 - 5x$ represents a parabola opening upward. Completing the square, we get

   $$p(x) = \left(x - \tfrac{5}{2}\right)^2 - \tfrac{25}{4}.$$

   The minimum function value of $-\frac{25}{4}$ occurs when x is $\frac{5}{2}$ and $x - 5$ is $-\frac{5}{2}$.

4. **Check.** Note that if $x = 2$, then $x - 5 = -3$ and $2(-3) = -6$. Also note that if $x = 3$, then $x - 5 = -2$ and $3(-2) = -6$. Thus, since $-\frac{25}{4} < -6$, it appears that we have a minimum.

---

*We often use function notation, such as $A(w)$, when we wish to emphasize that the value of a variable (in this case A) depends on some input (in this case w).

5. **State.** The numbers $-\frac{5}{2}$ and $\frac{5}{2}$ differ by 5 and yield a minimum product of $-\frac{25}{4}$. ■

Solving Formulas

Recall that to solve a formula for a certain letter, we use the principles for solving equations to get that letter alone on one side. When square roots appear, we can usually eliminate the radical signs by squaring both sides.

EXAMPLE 3 *A pendulum formula.* Solve $T = 2\pi\sqrt{l/g}$ for l. (Assume $l, g > 0$.)

Solution

$$T = 2\pi\sqrt{\frac{l}{g}}$$

$$T^2 = \left(2\pi\sqrt{\frac{l}{g}}\right)^2 \qquad \text{Principle of powers (squaring)}$$

$$T^2 = 2^2\pi^2\frac{l}{g}$$

$$gT^2 = 4\pi^2 l \qquad \text{Clearing fractions}$$

$$\frac{gT^2}{4\pi^2} = l \qquad \text{Multiplying by } \frac{1}{4\pi^2}$$

We now have l alone on one side and l does not appear on the other side, so the formula is solved for l. ■

EXAMPLE 4 *A right triangle formula.* Solve $c^2 = a^2 + b^2$ for a. Assume a, b, $c > 0$.

Solution

$$c^2 = a^2 + b^2$$

$$c^2 - b^2 = a^2 \qquad \text{Adding } -b^2 \text{ to get } a^2 \text{ alone}$$

$$\sqrt{c^2 - b^2} = a \qquad \text{Taking the square root}$$

■

EXAMPLE 5 *A motion formula.* Solve $s = vt + 16t^2$ for t. (Assume $v, t \geq 0$.)

Solution This time we use the quadratic formula to get t alone on one side of the equation:

$$16t^2 + vt - s = 0. \qquad \text{Writing the equation in standard form.}$$

Note that since t is the variable we're solving for, we treat v and s as constants. Thus, to use the quadratic formula, we have

$$a = 16, \quad b = v, \quad c = -s,$$

$$t = \frac{-v \pm \sqrt{v^2 - 4 \cdot 16 \cdot (-s)}}{2 \cdot 16}. \qquad \text{Using the quadratic formula}$$

Since taking the negative square root would result in a negative value for t, we use only the positive root:

$$t = \frac{-v + \sqrt{v^2 + 64s}}{32}.$$

■

EXERCISE SET 7.6

1. A rancher is fencing off a rectangular field with a fixed perimeter of 76 ft. What dimensions will yield the maximum area? What is the maximum area?

2. A carpenter is building a rectangular room with a fixed perimeter of 68 ft. What dimensions will yield the maximum area? What is the maximum area?

3. What is the maximum product of two numbers whose sum is 16? What numbers yield this product?

4. What is the maximum product of two numbers whose sum is 28? What numbers yield this product?

5. What is the maximum product of two numbers whose sum is 22? What numbers yield this product?

6. What is the maximum product of two numbers whose sum is 45? What numbers yield this product?

7. What is the minimum product of two numbers whose difference is 4? What are the numbers?

8. What is the minimum product of two numbers whose difference is 10? What are the numbers?

9. What is the minimum product of two numbers whose difference is 9? What are the numbers?

10. What is the minimum product of two numbers whose difference is 7? What are the numbers?

11. What is the maximum product of two numbers whose sum is -7? What numbers yield this product?

12. What is the maximum product of two numbers whose sum is -9? What numbers yield this product?

13. A farmer decides to enclose a rectangular garden, using the side of a barn as one side of the rectangle. What is the maximum area that the farmer can enclose with 40 ft of fence? What should the dimensions of the garden be to yield this area?

14. A stone mason has enough stones to enclose a rectangular patio with 60 ft of perimeter, assuming that the attached house forms one side of the rectangle. What is the maximum area that the mason can enclose? What should the dimensions of the patio be to yield this area?

15. A rectangular compost container is to be formed in a corner of a fenced yard, with 8 ft of chicken wire completing the other two sides of the rectangle. If the chicken wire is 3 ft high, what dimensions of the base will maximize the container's volume?

16. A plastics manufacturer plans to produce a one-compartment vertical file by bending the long side of an 8 in. by 14 in. sheet of plastic along two lines to form a U shape. How tall should the file be to maximize the volume that the file can hold?

Solve the formula for the indicated letter. Assume that all variables represent nonnegative numbers.

17. $A = 6s^2$, for s
(Area of a cube)

18. $A = 4\pi r^2$, for r
(Area of a sphere)

19. $F = \dfrac{Gm_1m_2}{r^2}$, for r
(Law of gravity)

20. $N = \dfrac{kQ_1Q_2}{s^2}$, for s
(Number of phone calls between two cities)

21. $E = mc^2$, for c
(Energy–mass relationship)

22. $A = \pi r^2$, for r
(Area of a circle)

23. $a^2 + b^2 = c^2$, for b
(Pythagorean formula in two dimensions)

24. $a^2 + b^2 + c^2 = d^2$, for c
(Pythagorean formula in three dimensions)

25. $N = \dfrac{k^2 - 3k}{2}$, for k
(Number of diagonals of a polygon)

26. $s = v_0t + \dfrac{gt^2}{2}$, for t
(A motion formula)

27. $A = 2\pi r^2 + 2\pi rh$, for r
(Area of a right cylindrical solid)

28. $A = \pi r^2 + \pi rs$, for r
(Area of a cone)

29. $N = \frac{1}{2}(n^2 - n)$, for n
(Number of games in a league of n teams)

30. $A = A_0(1 - r)^2$, for r
(A business formula)

31. $A = 2w^2 + 4lw$, for w
(Area of a rectangular solid)

32. $A = 4\pi r^2 + 2\pi rh$, for r
(Area of a capsule)

33. $T = 2\pi\sqrt{\dfrac{l}{g}}$, for g
(A pendulum formula)

34. $W = \sqrt{\dfrac{1}{LC}}$, for L
(An electricity formula)

35. $P_1 - P_2 = \dfrac{32LV}{gD^2}$, for D

36. $N + p = \dfrac{6.2A^2}{pR^2}$, for R

37. $m = \dfrac{m_0}{\sqrt{1 - \dfrac{v^2}{c^2}}}$, for v
(A relativity formula)

38. Solve the formula given in Exercise 37 for c.

SKILL MAINTENANCE

Multiply and simplify.

39. $\sqrt[4]{5x^3y^5}\,\sqrt[4]{125x^2y^3}$

40. $\sqrt{9a^3}\,\sqrt{16ab^4}$

SYNTHESIS

41. The sum of the base and the height of a triangle is 38 cm. Find the dimensions for which the area is a maximum, and find the maximum area.

42. The perimeter of a rectangle is 44 ft. Find the least possible length of a diagonal.

43. *Maximizing revenue.* When a theater owner charges \$2 for admission, she averages 100 people attending. For each 10¢ increase in admission price, the average number attending decreases by 1. What should the owner charge to make the most money?

44. *Maximizing yield.* An orange grower finds that she gets an average yield of 40 bu per tree when she plants 20 trees on an acre of ground. Each time she adds a tree to an acre, the yield per tree decreases by 1 bu, due to congestion. How many trees per acre should she plant for maximum yield?

Recall that total profit P is the difference between total revenue R and total cost C. Given the following total revenue and total cost functions, find the total profit function, the maximum value of the total profit function, and the value of x at which it occurs.

45. $R(x) = 200x - x^2$,
$C(x) = 5000 + 8x$

46. $R(x) = 300x - x^2$,
$C(x) = 50 + 80x$

47. *Maximizing height.* The height above the ground of a launched object is a quadratic function of the time that it is in the air. Suppose that a flare is launched from a cliff 64 ft above sea level. If 3 sec after being launched the flare is again level with the cliff, and if 2 sec after that it lands in the sea, what is the maximum height that the flare will reach?

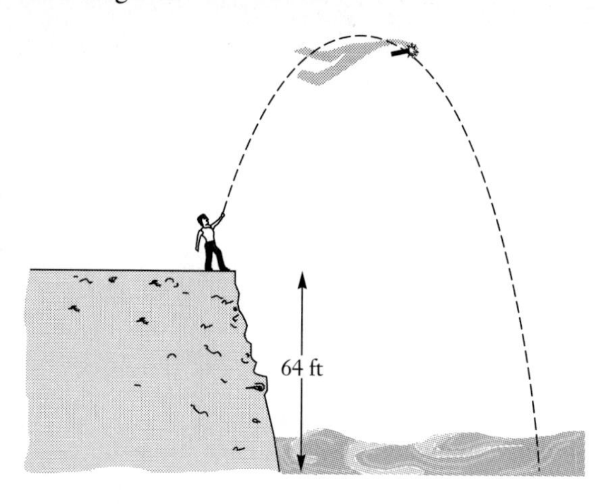

48. The cables supporting a straight-line suspension bridge are parabolic in shape. Suppose that a suspension bridge is being designed to cross a river that is 160 ft wide and that the vertical cables are 30 ft above road level at the bridge's midpoint and are 80 ft above road level at a point 50 ft from the bridge's midpoint. How long are the longest vertical cables?

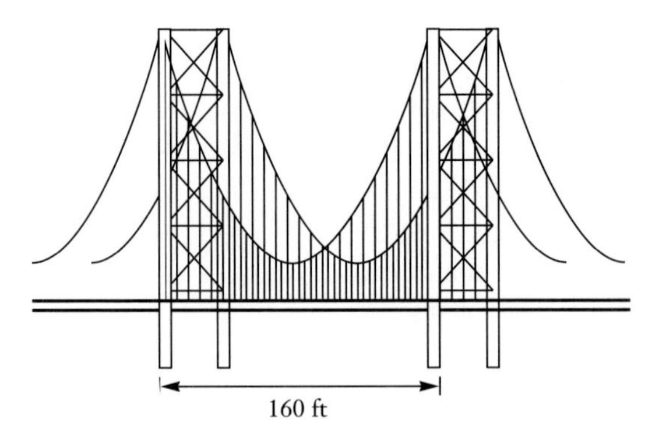

49. Solve for n:

$$mn^4 - r^2pm^3 - r^2n^2 + p = 0.$$

50. Solve for t:

$$rt^2 - rt - st^2 + s^2r - st = 0.$$

CHAPTER SUMMARY AND REVIEW 7

TERMS TO KNOW

Quadratic equation	Discriminant	Vertex
Standard form	Quadratic function	Minimum value
Completing the square	Parabola	Maximum value
The quadratic formula	Line of symmetry	x-intercept

IMPORTANT PROPERTIES AND FORMULAS

To solve a quadratic equation in x by completing the square:

1. Isolate the terms with variables on one side of the equation and arrange them in descending order.
2. Divide by the coefficient of x^2 on both sides if that coefficient is not 1.
3. Complete the square by taking half of the coefficient of x and adding its square to both sides.
4. Factor one side. Find a common denominator on the other side and simplify.
5. Use the principle of positive and negative roots.
6. Solve for x by adding appropriately on both sides.

THE QUADRATIC FORMULA

$$x = \frac{-b \pm \sqrt{b^2 - 4ac}}{2a}$$

Discriminant $b^2 - 4ac$	Nature of Solutions
0	Only one solution; it is a real number
Positive	Two different real number solutions
Negative	Two different nonreal number solutions (complex conjugates)

The graph of $g(x) = ax^2$ is a parabola with the vertical axis as its line of symmetry and its vertex at the origin.

If $|a|$ is greater than 1, the parabola is narrower than $f(x) = x^2$.

If $|a|$ is between 0 and 1, the parabola is flatter than $f(x) = x^2$.

If a is positive, the parabola opens upward; if a is negative, the parabola opens downward.

The graph of $g(x) = a(x - h)^2$ looks just like the graph of $f(x) = ax^2$ except that it is moved to the right or left h units. If h is positive, it is moved to the right. If h is negative, it is moved to the left. The vertex is now $(h, 0)$ and the line of symmetry is $x = h$.

The graph of $f(x) = a(x - h)^2 + k$ looks just like the graph of $g(x) = a(x - h)^2$, except that it is moved up or down. If k is positive, the curve is moved up k units. If k is negative, the curve is moved down k units. The vertex is at (h, k) and the line of symmetry is $x = h$. If $a > 0$, then k is the minimum function value. If $a < 0$, then k is the maximum function value.

REVIEW EXERCISES

The review sections to be tested in addition to the material in this chapter are Sections 3.6, 5.1, 5.4, and 6.5.

Solve.

1. $2x^2 - 7 = 0$

2. $14x^2 + 5x = 0$

3. $x^2 - 12x + 27 = 0$

4. $4x^2 + 3x + 1 = 0$

5. $x^2 - 7x + 13 = 0$

6. $4x(x - 1) + 15 = x(3x + 4)$

7. $x^2 + 4x + 1 = 0$. Approximate the solutions to the nearest tenth.

Find the number that completes the square.

8. $x^2 - 12x$

9. $x^2 + \frac{3}{5}x$

Solve by completing the square. Show your work.

10. $x^2 - 2x - 8 = 0$

11. $x^2 - 6x + 1 = 0$

Determine the nature of the solutions of each equation.

12. $x^2 + 3x - 6 = 0$

13. $x^2 + 2x + 5 = 0$

14. Write a quadratic equation having the solutions $\frac{1}{5}$, $-\frac{3}{5}$.

15. Write a quadratic equation having -4 as its only solution.

16. **a)** Graph: $f(x) = \frac{1}{2}(x - 1)^2$.
b) Label the vertex.
c) Draw the line of symmetry.

17. **a)** Graph: $f(x) = -3(x + 2)^2 + 4$.
b) Label the vertex.
c) Draw the line of symmetry.
d) Find the maximum or the minimum value.

18. For the function $f(x) = 2x^2 - 12x + 23$:
a) find an equivalent equation of the type $f(x) = a(x - h)^2 + k$;
b) find the vertex and the line of symmetry;
c) graph the function.

19. Complete the square and then graph:

$$f(x) = -x^2 - 4x - 5.$$

Find the x-intercepts.

20. $f(x) = x^2 - 9x + 14$

21. $g(x) = -x^2 + x - 2$

22. Solve $N = 3\pi\sqrt{1/p}$, for p.

23. Solve $2A + T = 3T^2$, for T.

24. What is the minimum product of two numbers whose difference is 22? What numbers yield this product?

25. A carpenter is building a rectangular room with a fixed perimeter of 80 ft. What dimensions will yield the maximum area? What is the maximum area?

SKILL MAINTENANCE

26. Divide and simplify: $(x^3 + 2x - 3) \div (x + 1)$.

27. Multiply and simplify: $\sqrt[3]{9t^6}\sqrt[3]{3s^4t^9}$.

28. Use fractional exponents to simplify: $\sqrt[5]{243x^{15}y^{20}}$.

29. Find an equation of the line containing $(3, -5)$ and $(-1, -2)$.

SYNTHESIS

30. Write the equation of the parabola that has the shape $f(x) = -2x^2$ and a maximum value at $(0, 10)$.

31. Find h and k if, for $3x^2 - hx + 4k = 0$, the sum of the solutions is 20 and the product is 80.

32. The average of two positive integers is 171. One of the numbers is the square root of the other. Find the integers.

THINKING IT THROUGH

1. Explain as many characteristics as you can of the graph of a parabola $y = ax^2 + bx + c$.
2. Give a general strategy for solving a quadratic equation.
3. Explain why there will never be only one complex root of a quadratic equation with real coefficients.
4. Describe the new types of problems we can solve by using the techniques in this chapter.

CHAPTER TEST 7

Solve.

1. $3x^2 - 4 = 0$
2. $8x^2 = 7x$
3. $4x(x - 2) - 3x(x + 1) = -18$
4. $x^2 + x + 1 = 0$
5. $3x^2 - 5x = 1$
6. $x^2 + 4x = 2$. Approximate the solutions to the nearest tenth, using a calculator or Table 1 in the appendix of this text.

Complete the square.

7. $x^2 + 14x$
8. $x^2 - \frac{2}{7}x$

Solve by completing the square. Show your work.

9. $x^2 + 3x - 18 = 0$
10. $x^2 + 10x + 15 = 0$
11. Determine the nature of the solutions of the equation $x^2 + 5x + 17 = 0$.
12. Write a quadratic equation having solutions $\sqrt{3}$ and $3\sqrt{3}$.
13. Solve: $V = \frac{1}{3}\pi(R^2 + r^2)$, for r.
14. Find the x-intercepts of $f(x) = x^2 - x - 6$.
15. **a)** Graph: $f(x) = 4(x - 3)^2 + 5$.
    b) Label the vertex.
    c) Draw the line of symmetry.
    d) Find the maximum or the minimum function value.
16. For the function $f(x) = 2x^2 + 4x - 6$:
    a) Find an equivalent equation of the type $f(x) = a(x - h)^2 + k$;
    b) Find the vertex and the line of symmetry;
    c) Graph the function.
17. What is the minimum product of two numbers having a difference of 8?
18. A farmer is enclosing a rectangular field, using the side of a barn as one side of the rectangle. What is the maximum area that the farmer can enclose with 60 ft of fence? What should the dimensions of the field be to yield this area?

SKILL MAINTENANCE

19. Divide and simplify: $\dfrac{-16a^5b^2}{2ab^2}$.
20. Find an equation of the line containing (1, 2) and parallel to $y + x = 5$.
21. Simplify, assuming that variables can represent any real number: $\sqrt{(16xy)^2}$.
22. Use fractional exponents to write a single radical expression:
$$\frac{\sqrt[4]{(x + y)^5}}{\sqrt{x + y}}.$$

SYNTHESIS

23. Find a quadratic function for which $f(3) = 0$, $f(-1) = 0$, and $f(0) = -6$.
24. Solve: $3x^2 - 4x - \sqrt{5} = 0$.

CHAPTER 8 Polynomial and Rational Functions

FEATURE PROBLEM

Weight Above Earth. A person's weight at height h miles above sea level satisfies the equation

$$W = \left(\frac{r}{r+h}\right)^2 W_0,$$

where W_0 is a person's weight at sea level and r is the radius of the earth, in miles. Suppose a person weighs 150 lb at sea level. Express W as a function of h. Assume that the radius of the earth is 3965 miles.

THE MATHEMATICS

Substituting 150 for W_0 and 3965 for r, we obtain the function W, described as follows:

$$W(h) = 150\left(\frac{3965}{3965 + h}\right)^2.$$

This is a *rational function.*

In Sections 3.5, 7.1, and 7.2 we developed methods for solving any equation, $ax^2 + bx + c = 0$, where a, b, and c are real numbers. We now turn our attention to finding solutions of polynomial equations of degree three or more. After developing a method for solving polynomial equations, we will graph polynomial functions. A *polynomial function* is a function that can be described by a polynomial in one variable, like $f(x) = 5x^7 + 3x^6 - 4x^2 - 5$. The chapter closes with a study of rational functions and inequalities.

The review sections to be tested in addition to the material in this section are Sections 5.6, 6.2, 6.6, and 7.3.

8.1 Synthetic Division

We will find that in solving polynomial equations it is often necessary to divide a polynomial by a binomial of the type $x - a$. To do this division it is possible to streamline the procedure used in Section 3.6. The streamlined process is referred to as *synthetic division*.

Compare the following. In each stage, we attempt to write a bit less than in the previous stage, while retaining enough essentials to solve the problem. At the end, we will return to usual polynomial notation.

Stage 1. We note that the coefficients provide the essential information, provided that the dividend is written in descending order.

$$\begin{array}{r} 4x^2 + 5x + 11 \\ x - 2\overline{)4x^3 - 3x^2 + x + 7} \\ \underline{4x^3 - 8x^2} \qquad\qquad \\ 5x^2 + x \qquad \\ \underline{5x^2 - 10x} \qquad \\ 11x + 7 \\ \underline{11x - 22} \\ 29 \end{array} \qquad \begin{array}{r} 4 + 5 + 11 \\ 1 - 2\overline{)4 - 3 + 1 + 7} \\ \underline{4 - 8} \qquad\quad \\ 5 + 1 \qquad \\ \underline{5 - 10} \qquad \\ 11 + 7 \\ \underline{11 - 22} \\ 29 \end{array}$$

Note that, because the coefficient of x is 1 in the divisor, all the coefficients in our answer—after the first term, 4—will be the remainder found in each stage of the long-division process. In the next stage we don't bother to duplicate these numbers. We also show where -2 is used in the problem.

Stage 2.

$$\begin{array}{r} 4x^2 + 5x + 11 \\ x - 2\overline{)4x^3 - 3x^2 + x + 7} \\ \underline{4x^3 - 8x^2} \qquad\qquad \\ 5x^2 + x \qquad \\ \underline{5x^2 - 10x} \qquad \\ 11x + 7 \\ \underline{11x - 22} \\ 29 \end{array} \qquad \begin{array}{rl} 4 + 5 + 11 & \\ 1 - 2\overline{)4 - 3 + 1 + 7} & \\ \underline{(-2)4} \qquad\qquad & \leftarrow \textbf{Subtract: } -3 - (-2)4 = 5. \\ 5 + 1 \qquad & \\ \underline{(-2)5} \qquad & \leftarrow \textbf{Subtract: } 1 - (-2)5 = 11. \\ 11 + 7 & \\ \underline{(-2)11} & \leftarrow \textbf{Subtract: } 7 - (-2)11 = 29. \\ 29 & \end{array}$$

To simplify computations, we now change the sign of the -2 in the divisor and, in exchange, *add* at each step in the long division. We stop writing the 1 in the divisor.

Stage 3.

$$
\begin{array}{r|l}
 & 4x^2 + 5x + 11 \\ \hline
x - 2 & 4x^3 - 3x^2 + x + 7 \\
 & \underline{4x^3 - 8x^2} \\
 & \quad 5x^2 + x \\
 & \quad \underline{5x^2 - 10x} \\
 & \qquad 11x + 7 \\
 & \qquad \underline{11x - 22} \\
 & \qquad\qquad 29
\end{array}
\qquad
\begin{array}{r|l}
 & 4 + 5 + 11 \\ \hline
2 & 4 - 3 + 1 + 7 \quad \text{Change the } 1 - 2 \text{ to } 2. \\
 & \quad \underline{8} \longleftarrow \text{Add: } -3 + 8 = 5. \\
 & \quad 5 + 1 \\
 & \qquad \underline{10} \longleftarrow \text{Add: } 1 + 10 = 11. \\
 & \qquad 11 + 7 \\
 & \qquad\quad \underline{22} \longleftarrow \text{Add: } 7 + 22 = 29. \\
 & \qquad\quad 29
\end{array}
$$

As you can see from the highlighted numbers, there is still some duplication that we can eliminate.

Stage 4.

$$
\begin{array}{r|l}
 & 4x^2 + 5x + 11 \\ \hline
x - 2 & 4x^3 - 3x^2 + x + 7 \\
 & \underline{4x^3 - 8x^2} \\
 & \quad 5x^2 + x \\
 & \quad \underline{5x^2 - 10x} \\
 & \qquad 11x + 7 \\
 & \qquad \underline{11x - 22} \\
 & \qquad\qquad 29
\end{array}
\qquad
\begin{array}{r|rrrr}
 & 4 & 5 & 11 & \\ \hline
2 & 4 & -3 & 1 & 7 \\
 & & \underline{8} & \underline{10} & \underline{22} \\
 & & 5 & 11 & 29
\end{array}
$$

Don't lose sight of how the products 8, 10, and 22 are found. In our final stage we streamline still further. This final stage is what is commonly referred to as **synthetic division.**

Stage 5.

$$
\begin{array}{r|l}
 & 4x^2 + 5x + 11 \\ \hline
x - 2 & 4x^3 - 3x^2 + x + 7 \\
 & \underline{4x^3 - 8x^2} \\
 & \quad 5x^2 + x \\
 & \quad \underline{5x^2 - 10x} \\
 & \qquad 11x + 7 \\
 & \qquad \underline{11x - 22} \\
 & \qquad\qquad 29
\end{array}
\qquad
\begin{array}{r|rrr|r}
2 & 4 & -3 & 1 & 7 \\
 & & 8 & 10 & 22 \\ \hline
 & 4 & 5 & 11 & 29 \longleftarrow \text{This is the remainder.}
\end{array}
$$

- 11: This is the zero-degree coefficient.
- 5: This is the first-degree coefficient.
- 4: This is the second-degree coefficient.

The quotient is $4x^2 + 5x + 11$. The remainder is 29.

EXAMPLE 1 Use synthetic division to divide:

$$(x^3 + 6x^2 - x - 30) \div (x - 2).$$

Solution

a)
$$\begin{array}{r|rrrr} 2 & 1 & 6 & -1 & -30 \\ & & & & \\ \hline & 1 & & & \end{array}$$

Write the 2 of $x - 2$ and the coefficients of the dividend. Bring down the first coefficient.

b)
$$\begin{array}{r|rrrr} 2 & 1 & 6 & -1 & -30 \\ & & 2 & & \\ \hline & 1 & 8 & & \end{array}$$

Multiply 1 by 2 to get 2. Add 6 and 2.

c)
$$\begin{array}{r|rrrr} 2 & 1 & 6 & -1 & -30 \\ & & 2 & 16 & \\ \hline & 1 & 8 & 15 & \end{array}$$

Multiply 8 by 2 and then add.

d)
$$\begin{array}{r|rrr|r} 2 & 1 & 6 & -1 & -30 \\ & & 2 & 16 & 30 \\ \hline & 1 & 8 & 15 & 0 \end{array}$$

Multiply 15 by 2 and add.

The quotient is $x^2 + 8x + 15$. The remainder is 0. ■

Always remember the following:

1. Arrange polynomials in descending order.
2. When there are missing terms, be sure to write 0's.
3. The divisor must be of the form $x - a$—that is, a variable minus a constant, and the coefficient of the variable must be 1.

EXAMPLES Use synthetic division to find the quotient and the remainder.

2. $(2x^3 + 7x^2 - 5) \div (x + 3)$
There is no x-term, so we must write a 0 for the coefficient. Note that $x + 3 = x - (-3)$.

$$\begin{array}{r|rrr|r} -3 & 2 & 7 & 0 & -5 \\ & & -6 & -3 & 9 \\ \hline & 2 & 1 & -3 & 4 \end{array}$$

The quotient is $2x^2 + x - 3$. The remainder is 4.

3. $(x^3 + 4x^2 - x - 4) \div (x + 4)$

$$\begin{array}{r|rrr|r} -4 & 1 & 4 & -1 & -4 \\ & & -4 & 0 & 4 \\ \hline & 1 & 0 & -1 & 0 \end{array}$$

The quotient is $x^2 - 1$. The remainder is 0.

4. $(8x^5 - 6x^3 + x - 8) \div (x + 2)$

$$\begin{array}{r|rrrrr|r} -2 & 8 & 0 & -6 & 0 & 1 & -8 \\ & & -16 & 32 & -52 & 104 & -210 \\ \hline & 8 & -16 & 26 & -52 & 105 & -218 \end{array}$$

The quotient is $8x^4 - 16x^3 + 26x^2 - 52x + 105$. The remainder is -218. ■

Rather than use fractional form in some quotients as in Section 3.6, we will indicate the remainder when writing the answers to the problems in this section. This is done as preparation for later work in this chapter (see Exercise 27).

EXERCISE SET 8.1

Use synthetic division to find the quotient and the remainder.

1. $(x^3 - 2x^2 + 2x - 5) \div (x - 1)$

2. $(x^3 - 2x^2 + 2x - 5) \div (x + 1)$

3. $(a^2 + 11a - 19) \div (a + 4)$

4. $(a^2 + 11a - 19) \div (a - 4)$

5. $(x^3 - 7x^2 - 13x + 3) \div (x - 2)$

6. $(x^3 - 7x^2 - 13x + 3) \div (x + 2)$

7. $(3x^3 + 7x^2 - 4x + 3) \div (x + 3)$

8. $(3x^3 + 7x^2 - 4x + 3) \div (x - 3)$

9. $(y^3 - 3y + 10) \div (y - 2)$

10. $(x^3 - 2x^2 + 8) \div (x + 2)$

11. $(3x^4 - 25x^2 - 18) \div (x - 3)$

12. $(6y^4 + 15y^3 + 28y + 6) \div (y + 3)$

13. $(x^3 - 27) \div (x - 3)$

14. $(y^3 + 27) \div (y + 3)$

15. $(y^5 - 1) \div (y - 1)$

16. $(x^5 - 32) \div (x - 2)$

17. $(3x^4 + 8x^3 + 2x^2 - 7x - 4) \div (x + 2)$

18. $(2x^4 - x^3 - 5x^2 + x + 7) \div (x + 1)$

19. $(3x^3 + 7x^2 - x + 1) \div \left(x + \frac{1}{3}\right)$

20. $(8x^3 - 6x^2 + 7x - 1) \div \left(x - \frac{1}{2}\right)$

SKILL MAINTENANCE

Graph on a plane.

21. $2x - 3y = 6$

22. $5x + 3y = 15$

23. $y = 4$

24. $x = -2$

SYNTHESIS

Divide, using synthetic division.

25. ▦ $(3.41x^4 - 24.25x^2 - 13.47) \div (x - 2.41)$

26. ▦ $(5.032x^5 + 11.414x^3 + 217.3) \div (x + 17.07)$

27. Let $f(x) = x^3 - 5x^2 + 5x - 4$.

a) Use synthetic division to show that $x - 4$ is a factor of $x^3 - 5x^2 + 5x - 4$.

b) Why does the result in part (a) indicate that $f(4) = 0$?

c) Check that $f(4) = 0$ by substituting 4 into the given polynomial function.

28. Devise a way to use synthetic division when the divisor is a linear polynomial $ax + b$, with leading coefficient different from 1.

8.2 Symmetry

In Sections 8.5 and 8.6 we will consider some methods for graphing functions that are more complex than those studied thus far. While the study of symmetry has important uses throughout mathematics, our present discussion is included to better prepare us for the later sections of this chapter.

Even and Odd Functions

Consider the following two graphs and tables of values.

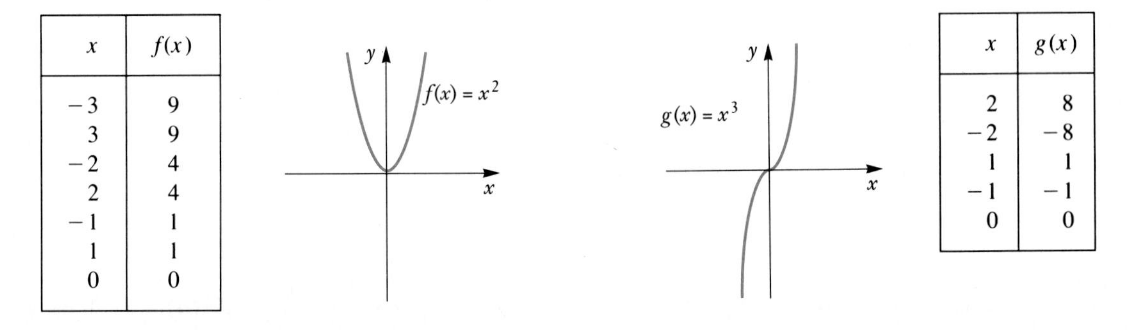

x	$f(x)$
-3	9
3	9
-2	4
2	4
-1	1
1	1
0	0

x	$g(x)$
2	8
-2	-8
1	1
-1	-1
0	0

Since for any number a, $a^2 = (-a)^2$, we see that for the preceding function, $f(x) = f(-x)$. In a somewhat different vein, notice that for any number a, $(-a)^3 = -a^3$ (when opposites are cubed, the results are opposites). Thus we see that $g(-x) = -g(x)$. The functions $f(x) = x^2$ and $g(x) = x^3$ are examples of *even* and *odd functions,* respectively.

> A function f is an *even function* if $f(x) = f(-x)$ for all x in the domain of f.
> A function f is an *odd function* if $f(-x) = -f(x)$ for all x in the domain of f.

EXAMPLES Classify each of the following functions as even, odd, or neither.

1. $f(x) = 5x^7 + 2x$

Since
$$\begin{aligned} f(-x) &= 5(-x)^7 + 2(-x) \\ &= 5(-x)^7 - 2x \qquad (-x^7) = (-1 \cdot x)^7 = (-1)^7x^7 = -1x^7 \\ &= -5x^7 - 2x, \end{aligned}$$

we see that $f(-x) \neq f(x)$, so f is not even. Since $f(-x) = -5x^7 - 2x = -(5x^7 + 2x) = -f(x)$, we see that f is odd. This is not surprising when we consider that both terms involve odd powers of x.

2. $h(x) = 6x^4 - 2x$.

Since
$$\begin{aligned} h(-x) &= 6(-x)^4 - 2(-x) \\ &= 6x^4 + 2x, \qquad (-x)^4 = (-1 \cdot x)^4 = (-1)^4x^4 = 1x^4 \end{aligned}$$

we see that $h(-x) \neq h(x)$, so h is not even. Since $h(-x) \neq -h(x)$, we see that h is not odd. We conclude that h is neither even nor odd.

3. $f(x) = 5x^6 - 3x^2 + 2$

$$\text{Since} \quad f(-x) = 5(-x)^6 - 3(-x)^2 + 2$$
$$= 5x^6 - 3x^2 + 2$$
$$= f(x),$$

we conclude that f is an even function. The student can check to see that $f(-x) \neq -f(x)$, so f is not odd. ■

Symmetry and Graphing

Observe that in the preceding graph of $f(x) = x^2$, the points $(-3, 9)$, $(2, 4)$, and $(-1, 1)$ had counterparts $(3, 9)$, $(-2, 4)$, and $(1, 1)$, respectively, that acted like **reflections across the *y*-axis.** Because the reflection of a point across the y-axis is a point with the same y-value but an opposite x-value, we can say that the points $(a, f(a))$ and $(-a, f(-a))$ are reflections of each other only if $f(a) = f(-a)$. Thus, we see that if f is an even function and $(a, f(a))$ is in the graph of f, then the point's reflection $(-a, f(a))$ is also in the graph. Any graph that contains the reflection of every point across the y-axis is said to be **symmetric with respect to the *y*-axis.** Some examples follow.

Symmetry with Respect to the y-axis

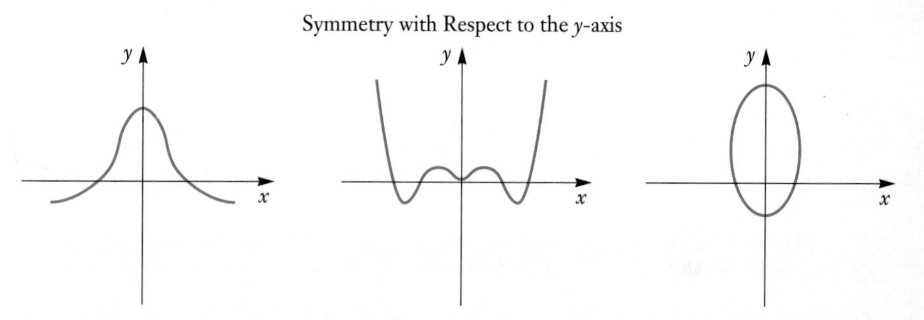

In Section 8.5 it will be helpful to keep in mind that all even functions are symmetric with respect to the y-axis.

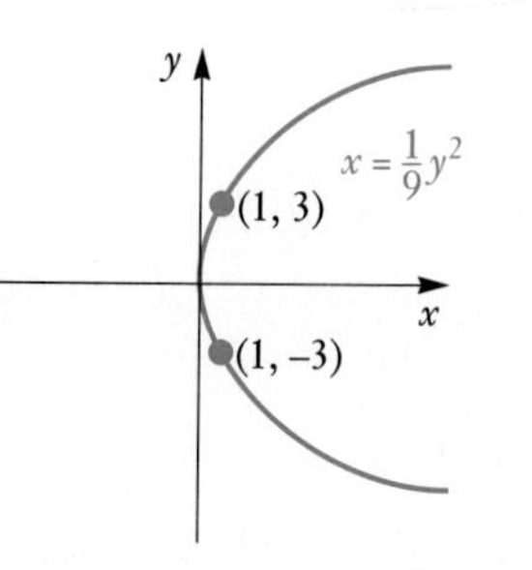

Consider now points like $(1, 3)$ and $(1, -3)$, which appear in the graph of $x = \frac{1}{9}y^2$. Points like these have the same x-values but opposite y-values and are **reflections across the *x*-axis** of each other. Thus the reflection across the x-axis of some point (x_1, y_1) is the point $(x_1, -y_1)$. Any graph that contains the reflection of every point across the x-axis is said to be **symmetric with respect to the *x*-axis.** Some examples follow.

Symmetry with Respect to the x-axis

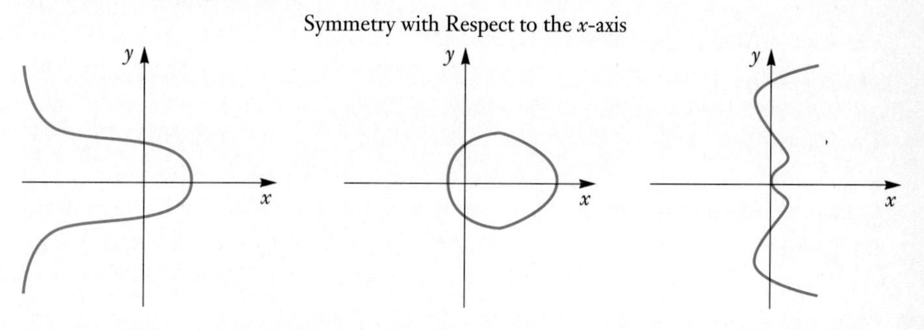

At the start of this section we graphed the function $g(x) = x^3$. Observe that the points $(-2, -8)$ and $(1, 1)$ had counterparts $(2, 8)$ and $(-1, -1)$, respectively. One way to obtain a point like $(2, 8)$ from the point $(-2, -8)$ is to first reflect $(-2, -8)$ across the y-axis to obtain the point $(2, -8)$ and then reflect *that* point across the x-axis to obtain the point $(2, 8)$. Any pair of points for which one point can be obtained from the other by making two successive reflections across the axes is said to be **symmetric with respect to the origin.** Note that the points $(a, f(a))$ and $(-a, f(-a))$ are symmetric with respect to the origin only if $f(-a) = -f(a)$. In other words, a graph is symmetric with respect to the origin if x-values that are opposites of each other are paired with y-values that are opposites of each other. Some examples follow.

Symmetry with Respect to the origin

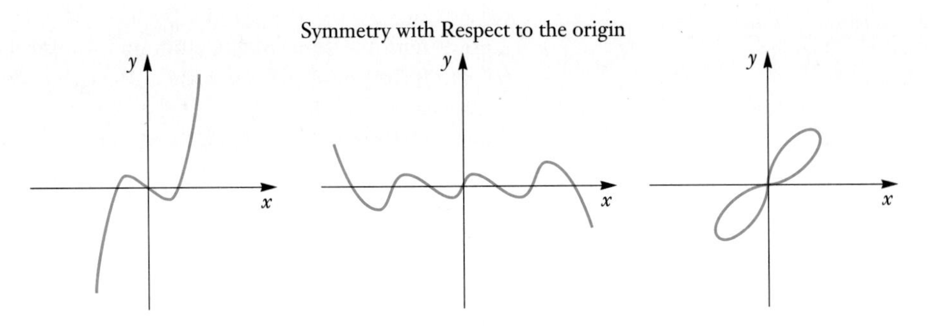

In Section 8.5 it will be helpful to keep in mind that all odd functions are symmetric with respect to the origin.

EXERCISE SET 8.2

Determine whether the function is even, odd, both even and odd, or neither even nor odd.

1. $f(x) = 2x^2 + 4x$

2. $f(x) = -3x^3 + 2x$

3. $f(x) = 3x^4 - 4x^2$

4. $f(x) = 5x^2 + 2x^4 - 1$

5. $f(x) = 7x^3 + 4x - 2$

6. $f(x) = 4x$

7. $f(x) = |3x|$

8. $f(x) = x^{24}$

9. $f(x) = x^{17}$

10. $f(x) = x + \dfrac{1}{x}$

11. $f(x) = x - |x|$

12. $f(x) = \sqrt{x}$

13. $f(x) = \sqrt[3]{x}$

14. $f(x) = 7$

15. $f(x) = 0$

16. $f(x) = \sqrt[3]{x - 2}$

17. $f(x) = \sqrt{x^2 + 1}$

18. $f(x) = \dfrac{x^2 + 1}{x^3 - x}$

Determine whether the graph is symmetric with respect to the x-axis, the y-axis, or the origin. If more than one type of symmetry exists or if none exist, state this.

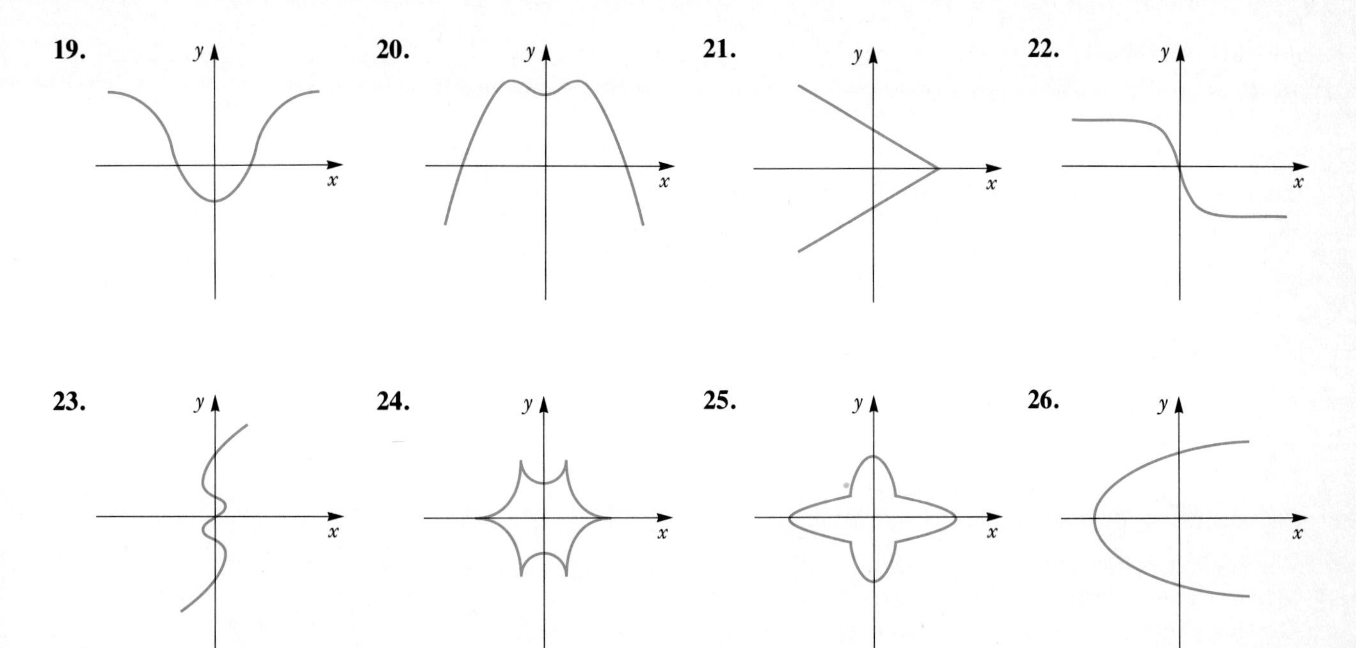

Determine whether the function is even, odd, or neither. Then list any symmetries that exist.

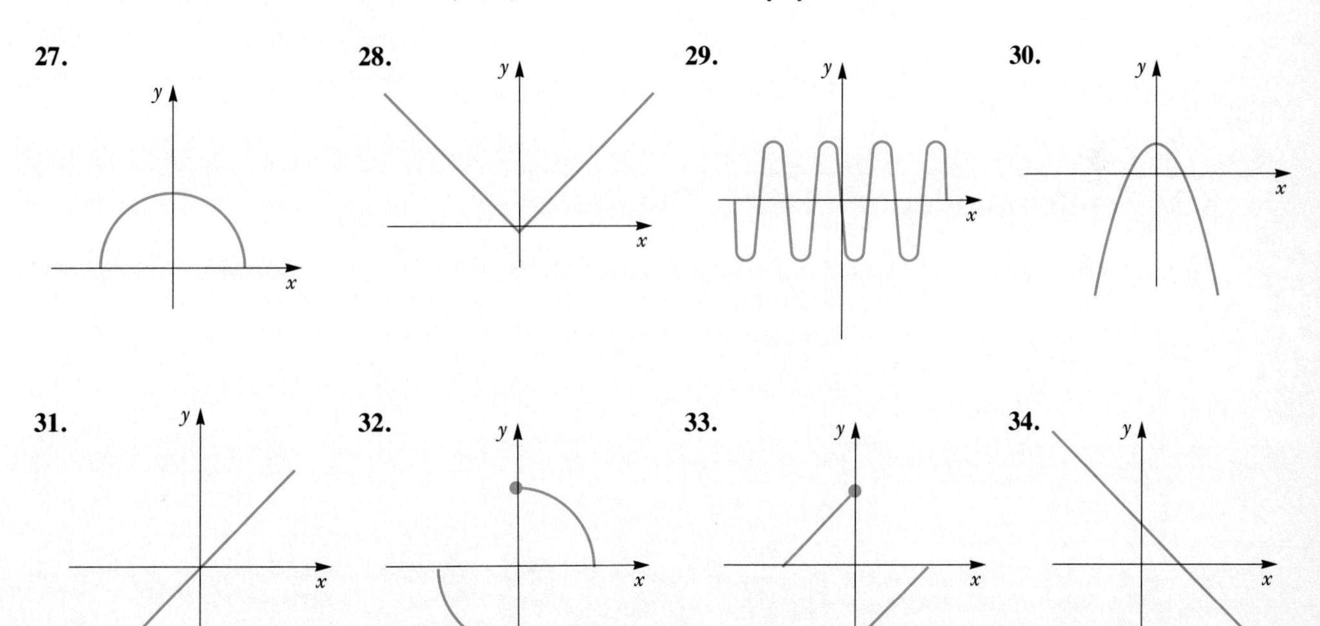

SKILL MAINTENANCE

35. Express $-\sqrt{-3600}$ in terms of i.

36. Determine the domain of f if

$$f(x) = \frac{9 - x^2}{x^2 - 25}.$$

Determine whether each of the following is the graph of a function.

37.

38.

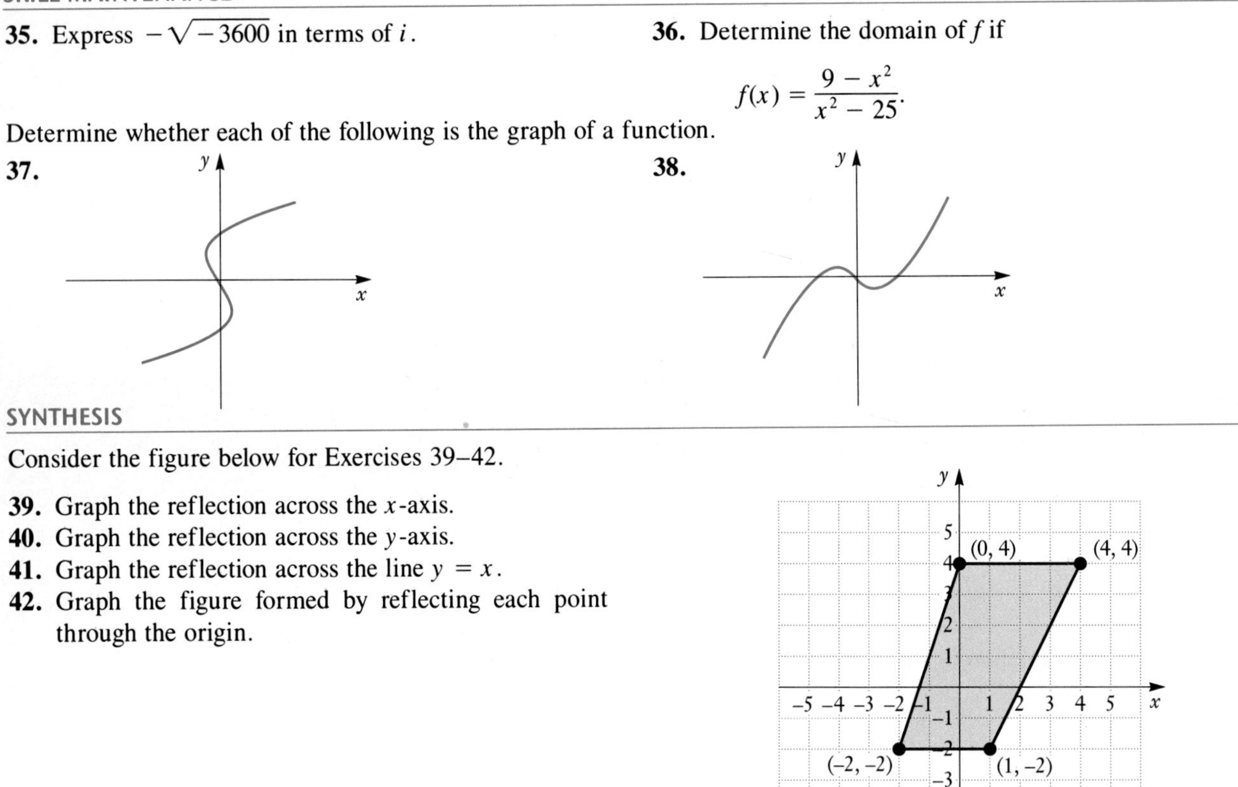

SYNTHESIS

Consider the figure below for Exercises 39–42.

39. Graph the reflection across the x-axis.

40. Graph the reflection across the y-axis.

41. Graph the reflection across the line $y = x$.

42. Graph the figure formed by reflecting each point through the origin.

8.3 The Remainder and Factor Theorems

There is an interesting connection between evaluating a polynomial function and polynomial division. Let's evaluate the function $f(x) = 2x^4 - 4x^3 - 5x + 6$ for $x = 3$ and divide the polynomial $2x^4 - 4x^3 - 5x + 6$ by $x - 3$.

To evaluate, we substitute.

$$f(3) = 2 \cdot 3^4 - 4 \cdot 3^3 - 5 \cdot 3 + 6$$
$$f(3) = 45$$

To divide, we use synthetic division.

$$\begin{array}{r|rrrrr} 3 & 2 & -4 & 0 & -5 & 6 \\ & & 6 & 6 & 18 & 39 \\ \hline & 2 & 2 & 6 & 13 & \big| \; 45 \end{array}$$

Notice that upon division by $x - 3$ the remainder is the same as $f(3)$. It is natural to ask if, for any polynomial $P(x)$ and any real number r, the remainder upon division by $x - r$ has the same value as $P(r)$.

To answer this question, note that any polynomial $P(x)$ can be rewritten as $(x - r) \cdot Q(x) + R$, where $Q(x)$ is the quotient polynomial that arises upon division of $P(x)$ by $x - r$, and R is some constant (the remainder). Thus we have, for any real number x,

$$P(x) = (x - r)Q(x) + R,$$

and in particular,

$$\begin{aligned} P(r) &= (r - r)Q(r) + R \\ &= 0 \cdot Q(r) + R \\ &= R. \end{aligned}$$

This tells us that the function value $P(r)$ is the remainder obtained when $P(x)$ is divided by $x - r$. We commonly refer to this result as the remainder theorem.

THE REMAINDER THEOREM

The remainder obtained by dividing $P(x)$ by $x - r$ is $P(r)$.

EXAMPLE 1 Given that $P(x) = 2x^5 - 3x^4 + x^3 - 2x^2 + x - 8$, find $P(10)$.

Solution By the remainder theorem, $P(10)$ is the remainder when $P(x)$ is divided by $x - 10$. We use synthetic division to find that remainder.

$$\begin{array}{r|rrrrrr} 10 & 2 & -3 & 1 & -2 & 1 & -8 \\ & & 20 & 170 & 1710 & 17{,}080 & 170{,}810 \\ \hline & 2 & 17 & 171 & 1708 & 17{,}081 & 170{,}802 \end{array}$$

Thus $P(10) = 170{,}802$. ■

A **root**, or **zero**, of a polynomial $P(x)$ is an x-value for which $P(x) = 0$. The remainder theorem gives us a method which is usually faster than direct substitution for checking possible roots.

EXAMPLE 2 Determine whether -4 is a zero, or root, of $P(x)$, where $P(x) = x^3 + 8x^2 + 8x - 32$.

Solution We use synthetic division and the remainder theorem to find $P(-4)$.

$$\begin{array}{r|rrrr} -4 & 1 & 8 & 8 & -32 \\ & & -4 & -16 & 32 \\ \hline & 1 & 4 & -8 & 0 \end{array}$$

Since $P(-4) = 0$, the number -4 is a root of $P(x)$. ■

Finding Factors of Polynomials

Observe in Example 2 that the synthetic division indicates that the polynomial $x^3 + 8x^2 + 8x - 32$ can be rewritten as $(x + 4) \cdot (x^2 + 4x - 8) + 0$, or simply $(x + 4)(x^2 + 4x - 8)$. We see that since -4 was a zero, or root, of $P(x)$, the binomial $x + 4$ is a factor of $P(x)$.

To generalize this result note that if we divide any polynomial $P(x)$ by $x - r$, we obtain a quotient and a remainder, related as follows:

$$P(x) = (x - r) \cdot Q(x) + P(r). \qquad \text{Here } P(r) \text{ represents the remainder.}$$

Then if $P(r) = 0$, we have

$$P(x) = (x - r) \cdot Q(x),$$

which is to say that if $P(r) = 0$, $x - r$ is a factor of $P(x)$.

If we recall from the principle of zero products (see Section 3.5) that when $x - r$ is a factor of $P(x)$, we have $P(r) = 0$, the following theorem is justified.

THE FACTOR THEOREM

For any polynomial $P(x)$, if $P(r) = 0$, then $x - r$ is a factor of $P(x)$; and if $x - r$ is a factor of $P(x)$, then $P(r) = 0$.

EXAMPLE 3 Determine whether $x - 3$ is a factor of $P(x) = x^3 - 2x^2 - 5x + 6$.

Solution We use either substitution or synthetic division to determine whether $P(3) = 0$. Using synthetic division, we have

$$\begin{array}{r|rrrr} 3 & 1 & -2 & -5 & 6 \\ & & 3 & 3 & -6. \\ \hline & 1 & 1 & -2 & 0 \end{array}$$

Thus since $P(3) = 0$ we see that $x - 3$ is a factor of $P(x)$. ■

EXAMPLE 4 Factor the polynomial $P(x) = x^3 - 2x^2 - 5x + 6$ and solve the equation $P(x) = 0$.

Solution The synthetic division that we used in Example 3 indicates that $P(x) = (x - 3)(x^2 + x - 2)$. By factoring the trinomial, we have $P(x) = (x - 3)(x + 2)(x - 1)$. The solutions of the equation $P(x) = 0$ are 3, -2, and 1. ■

EXAMPLE 5 Let $P(x) = x^3 + 2x^2 - 5x - 6$. Factor $P(x)$ and solve the equation $P(x) = 0$.

Solution We look for a linear factor of the form $x - r$. Let us try $x - 1$. We use synthetic division to see whether $P(1) = 0$.

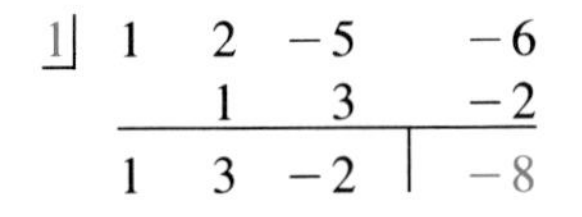

$$\begin{array}{r|rrrr} 1 & 1 & 2 & -5 & -6 \\ & & 1 & 3 & -2 \\ \hline & 1 & 3 & -2 & -8 \end{array}$$

Since $P(1) \neq 0$, we know that $x - 1$ is not a factor of $P(x)$. We try $x + 1$ or $x - (-1)$ in the form $x - r$.

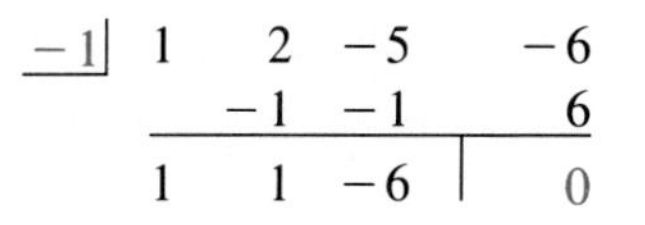

Since $P(-1) = 0$, we know that $x + 1$ is one factor and the quotient, $x^2 + x - 6$, is another. Thus

$$P(x) = (x + 1)(x^2 + x - 6).$$

The trinomial is easily factored, so we have

$$P(x) = (x + 1)(x + 3)(x - 2).$$

The solutions of $P(x) = 0$ are -1, -3, and 2. ■

Notice from Examples 4 and 5 that if $P(x)$ is factored so that

$$P(x) = (x - r) \cdot Q(x),$$

then the degree of $Q(x)$ is one less than the degree of $P(x)$. If $Q(x)$ is quadratic, we can either factor or use the quadratic formula to find the remaining roots. When the degree of $Q(x)$ is 3 or more, we again use synthetic division, this time to find a linear factor of $Q(x)$, and repeat this process until a quadratic factor is obtained. Then we either factor or use the quadratic formula.

EXAMPLE 6 Let $P(x) = x^4 - x^3 - 7x^2 + x + 6$. Factor $P(x)$ and solve the equation $P(x) = 0$.

Solution We look for a linear factor of the form $x - r$. Let us try $x - 1$. We use synthetic division to see whether $P(1) = 0$.

$$\begin{array}{r|rrrr|r} 1 & 1 & -1 & -7 & 1 & 6 \\ & & 1 & 0 & -7 & -6 \\ \hline & 1 & 0 & -7 & -6 & 0 \end{array}$$

Since $P(1) = 0$, we know that $x - 1$ is one factor and the quotient, $x^3 - 7x - 6$, is another. To factor the polynomial $Q(x) = x^3 - 7x - 6$, we again look for a linear factor of the form $x - r$. We try $x - 2$ and use synthetic division to see whether $Q(2) = 0$.

$$\begin{array}{r|rrr|r} 2 & 1 & 0 & -7 & -6 \\ & & 2 & 4 & -6 \\ \hline & 1 & 2 & -3 & -12 \end{array}$$

Since $Q(2) \neq 0$, we know that $x - 2$ is not a factor of $Q(x)$. We next try $x + 2$, using the form $x - (-2)$.

$$\begin{array}{r|rrr|r} -2 & 1 & 0 & -7 & -6 \\ & & -2 & 4 & 6 \\ \hline & 1 & -2 & -3 & 0 \end{array}$$

Since $Q(-2) = 0$, we know that $x + 2$ and $x^2 - 2x - 3$ are factors of $Q(x)$. Thus

$$
\begin{aligned}
P(x) &= (x-1)(x^3 - 7x - 6) && \text{From the first synthetic division}\\
&= (x-1)(x+2)(x^2 - 2x - 3) && \text{Factoring } Q(x)\\
&= (x-1)(x+2)(x+1)(x-3). && \text{Factoring } x^2 - 2x - 3
\end{aligned}
$$

We have now factored $P(x)$. The solutions of $P(x) = 0$ are $1, -2, -1$, and 3. ■

The search for a number r for which $P(r) = 0$ may seem like looking for a needle in a haystack. In Section 8.4 we will discover methods that will help us in our search. For the problems in this section, it will suffice to limit our search to all positive and negative factors of the constant term in $P(x)$.

EXERCISE SET 8.3

Use synthetic division to find the function values.

1. $P(x) = x^3 - 6x^2 + 11x - 6$; find $P(1)$, $P(-2)$, and $P(3)$.
2. $P(x) = x^3 + 7x^2 - 12x - 3$; find $P(-3)$, $P(-2)$, and $P(1)$.
3. $P(x) = x^5 - 2x^4 - 7x^3 + x^2 + 20$; find $P(10)$ and $P(-8)$.
4. $P(x) = 2x^5 - 5x^4 + 2x^3 - x + 15$; find $P(-10)$ and $P(7)$.
5. $P(x) = 2x^5 - 3x^4 + 2x^3 - x + 8$; find $P(20)$ and $P(-3)$.
6. $P(x) = x^5 - 10x^4 + 20x^3 - 5x - 100$; find $P(-10)$ and $P(5)$.
7. $P(x) = x^4 - 16$; find $P(2)$, $P(-2)$, and $P(3)$.
8. $P(x) = x^5 + 32$; find $P(2)$, $P(-2)$, and $P(3)$.

Using synthetic division, determine whether the numbers are roots of the polynomials.

9. $-3, 2$; $P(x) = 3x^3 + 5x^2 - 6x + 18$
10. $-4, 2$; $P(x) = 3x^3 + 11x^2 - 2x + 8$
11. $-2, 1, 5$; $P(x) = x^3 - 2x^2 - 13x - 10$
12. $2, 5, -3$; $P(x) = x^3 + 6x^2 - x - 30$
13. $-3, \frac{1}{2}$; $P(x) = x^3 - \frac{7}{2}x^2 + x - \frac{3}{2}$
14. $i, -i, -2$; $P(x) = x^3 + 2x^2 + x + 2$

Factor the polynomial $P(x)$. Then solve the equation $P(x) = 0$. To find r for which $P(r) = 0$, consider only positive and negative factors of the constant term in $P(x)$.

15. $P(x) = x^3 + 4x^2 + x - 6$
16. $P(x) = x^3 + 5x^2 - 2x - 24$
17. $P(x) = x^3 - 6x^2 + 3x + 10$
18. $P(x) = x^3 + 2x^2 - 13x + 10$
19. $P(x) = x^3 - x^2 - 14x + 24$
20. $P(x) = x^3 - 3x^2 - 10x + 24$
21. $P(x) = x^3 + 6x^2 - x - 30$
22. $P(x) = x^3 - 4x^2 - 9x + 36$
23. $P(x) = x^4 + 7x^3 + 11x^2 - 7x - 12$
24. $P(x) = x^4 + 6x^3 + x^2 - 24x - 20$
25. $P(x) = x^4 + 2x^3 - 13x^2 - 14x + 24$
26. $P(x) = x^4 + 10x^3 + 35x^2 + 50x + 24$
27. $P(x) = x^4 + 2x^3 - 3x^2 - 8x - 4$
28. $P(x) = x^4 + 4x^3 - 5x^2 - 36x - 36$
29. $P(x) = x^4 - x^3 - 19x^2 + 49x - 30$
30. $P(x) = x^4 + 11x^3 + 41x^2 + 61x + 30$

SKILL MAINTENANCE

31. Solve: $\dfrac{b}{a+b} = c$, for a.
32. Factor by grouping: $x^3 - 3x^2 + 2x - 6$.
33. Rewrite without fractional exponents: $x^{7/4}$.
34. Simplify: $(3 - 2i)^2$.

SYNTHESIS

Solve.

35. $\dfrac{6x^2}{x^2 + 11} + \dfrac{60}{x^3 - 7x^2 + 11x - 77} = \dfrac{1}{x - 7}$

36. $\dfrac{2x^2}{x^2 - 1} + \dfrac{4}{x + 3} = \dfrac{32}{x^3 + 3x^2 - x - 3}$

37. Find k so that $x + 2$ is a factor of $x^3 - kx^2 + 3x + 7k$.

38. ▦ Given that

$$f(x) = 2.13x^5 - 42.1x^3 + 17.5x^2 + 0.953x - 1.98,$$

find $f(3.21)$ (a) by synthetic division; (b) by substitution.

39. For what values of k will the remainder be the same when $x^2 + kx + 4$ is divided by $x - 1$ or $x + 1$?

40. When $x^2 - 3x + 2k$ is divided by $x + 2$, the remainder is 7. Find the value of k.

8.4 Rational Roots of Polynomial Equations

It is often quite difficult to find the roots of a polynomial. However, if all of a polynomial's coefficients are integers, there is a procedure that will yield all the rational roots. Although this procedure is largely trial and error, because of the following result (which is justified near the end of this section), only certain numbers need to be considered when searching for roots that are rational numbers.

THE RATIONAL ROOTS THEOREM

Let

$$P(x) = a_n x^n + a_{n-1} x^{n-1} + \cdots + a_1 x + a_0,$$

where all the coefficients are integers and n is a positive integer. Consider a rational number c/d that is written in reduced form. If c/d is a root of $P(x)$, then c is a factor of a_0 and d is a factor of a_n.

EXAMPLE 1 Let $P(x) = 3x^4 - 11x^3 + 10x - 4$. Find all rational roots of $P(x)$. If possible, find any other roots and factor $P(x)$ completely.

Solution By the rational roots theorem, if c/d is a root of $P(x)$, then c must be a factor of -4 and d must be a factor of 3. Thus the possibilities for c and d are

$$c: \quad 1, -1, 2, -2, 4, -4 \qquad d: \quad 1, -1, 3, -3.$$

Then the resulting possibilities for c/d are

$$\frac{c}{d}: \quad 1, -1, 2, -2, 4, -4, \frac{1}{3}, -\frac{1}{3}, \frac{2}{3}, -\frac{2}{3}, \frac{4}{3}, -\frac{4}{3}.$$

To find which are roots, we could use substitution, but synthetic division is usually more efficient. It is easiest to first consider the integers. Then we consider the fractions, if the integers do not produce all the roots.

We try 1.

$$\begin{array}{r|rrrrr} 1 & 3 & -11 & 0 & 10 & -4 \\ & & 3 & -8 & -8 & 2 \\ \hline & 3 & -8 & -8 & 2 & -2 \end{array}$$

We try -1.

$$\begin{array}{r|rrrrr} -1 & 3 & -11 & 0 & 10 & -4 \\ & & -3 & 14 & -14 & 4 \\ \hline & 3 & -14 & 14 & -4 & 0 \end{array}$$

Since $P(1) = -2$, 1 is not a root; but $P(-1) = 0$, so -1 is a root. Using the results of the synthetic division, we can express $P(x)$ as follows:

$$P(x) = (x + 1)(3x^3 - 14x^2 + 14x - 4).$$

We now use $3x^3 - 14x^2 + 14x - 4$ and check the other possible roots. We use synthetic division again to see whether -1 is a double root.

$$\begin{array}{r|rrrr} -1 & 3 & -14 & 14 & -4 \\ & & -3 & 17 & -31 \\ \hline & 3 & -17 & 31 & -35 \end{array}$$

It is not. The student can check that none of the remaining integers (2, -2, 4, or -4) are roots. Let's try $\frac{2}{3}$.

$$\begin{array}{r|rrrr} 2/3 & 3 & -14 & 14 & -4 \\ & & 2 & -8 & 4 \\ \hline & 3 & -12 & 6 & 0 \end{array}$$

Since $P\left(\frac{2}{3}\right) = 0$, $\frac{2}{3}$ is a root. Again, using the results of the synthetic division, we can express $P(x)$ as

$$P(x) = (x + 1)\left(x - \tfrac{2}{3}\right)(3x^2 - 12x + 6) = (x + 1)\left(x - \tfrac{2}{3}\right) \cdot 3 \cdot (x^2 - 4x + 2).$$

Since the factor $x^3 - 4x + 2$ is quadratic, we can use the quadratic formula to find that the other roots are $2 + \sqrt{2}$ and $2 - \sqrt{2}$. These are irrational numbers. Thus the rational roots are -1 and $\frac{2}{3}$. A complete factorization would be $P(x) = (x + 1)\left(x - \frac{2}{3}\right) \cdot 3 \cdot \left(x - (2 + \sqrt{2})\right)\left(x - (2 - \sqrt{2})\right)$ or $P(x) = (x + 1)(3x - 2) \cdot \left(x - 2 - \sqrt{2}\right)\left(x - 2 + \sqrt{2}\right)$. ■

Upper and Lower Bounds on Roots

Recall that when a polynomial $P(x)$ is divided by $x - a$, we obtain a quotient $Q(x)$ and a remainder R, related as follows:

$$P(x) = (x - a) \cdot Q(x) + R.$$

Let us consider the case in which a is positive, and R and all the coefficients of $Q(x)$ are non*negative*. Then $P(b)$ will be positive for all positive $b > a$:

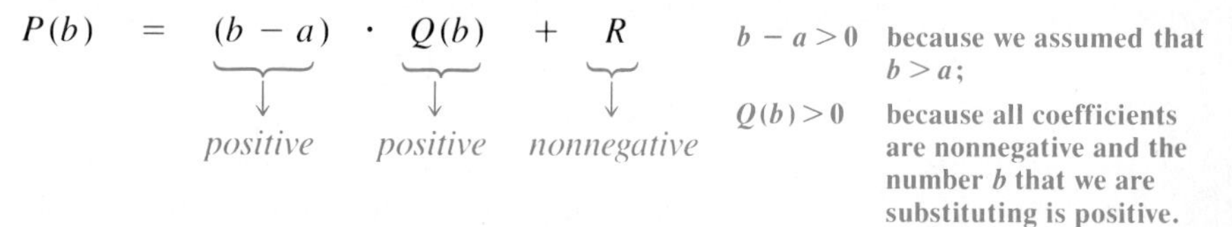

If R and all the coefficients of $Q(x)$ are non*positive*, $Q(b)$ will be negative and R will be nonpositive, thus guaranteeing that $P(b)$ will be negative for all positive $b > a$. In terms of synthetic division, this tells us the following:

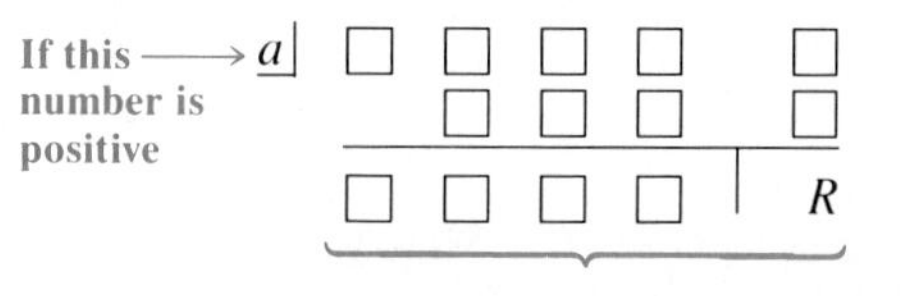

and the bottom row is either all nonnegative or all nonpositive

then there is no root greater than a. In other words, the number a is an **upper bound** to all the roots of $P(x)$. Of course, if the number R is 0, we know that a is actually a root as well as an upper bound. We state this result as a theorem.

THE UPPER BOUND THEOREM

If when a polynomial is divided by $x - a$, where a is positive, the remainder and coefficients of the quotient are either all nonnegative or all nonpositive, then a is an upper bound to the real roots of the polynomial (no root is greater than a).

A similar result for **lower bounds** follows. Its proof is outlined at the end of this section.

THE LOWER BOUND THEOREM

If a polynomial is divided by $x - a$, where a is negative, and the remainder and all coefficients of the quotient alternate sign (with zero considered a sign change), then a is a lower bound to the real roots of the polynomial (no root is less than a).

In terms of synthetic division, the lower bound theorem tells us the following:

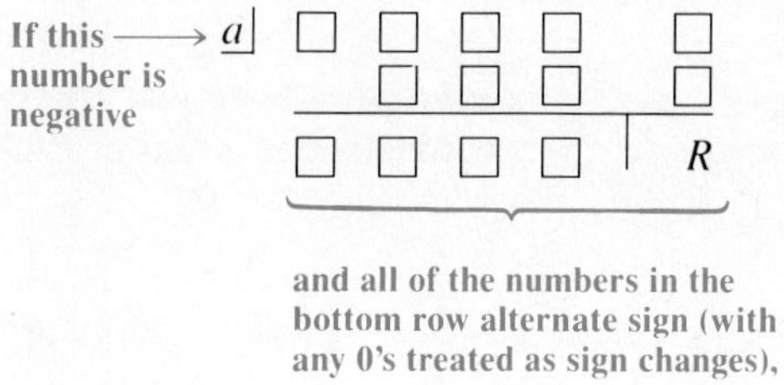

If this number is negative $\longrightarrow a$ | R

and all of the numbers in the bottom row alternate sign (with any 0's treated as sign changes),

then there is no root less than a.

These last two theorems suggest that we be alert for patterns in the bottom line of our synthetic division when searching for roots.

EXAMPLE 2 Let $P(x) = 2x^3 + 11x^2 + 10x - 8$. Find all rational roots of $P(x)$. If possible, find any other roots as well.

Solution We look for rational roots c/d where the possibilities for c and d are

$$c: \quad 1, -1, 2, -2, 4, -4, 8, -8 \qquad d: \quad 1, -1, 2, -2.$$

Thus the possibilities for c/d are

$$\frac{c}{d}: \quad 1, -1, 2, -2, 4, -4, 8, -8, \tfrac{1}{2}, -\tfrac{1}{2}.$$

We try 1.

$$\begin{array}{r|rrrr} 1 & 2 & 11 & 10 & -8 \\ & & 2 & 13 & 23 \\ \hline & 2 & 13 & 23 & 15 \end{array} \leftarrow \textbf{All nonnegative}$$

Since $P(1) = 15$, 1 is not a root. Because the bottom line of the synthetic division consists of nonnegatives, we conclude by the upper bound theorem that all real roots are less than 1. We try a value less than 1, -4.

$$\begin{array}{r|rrrr} -4 & 2 & 11 & 10 & -8 \\ & & -8 & -12 & 8 \\ \hline & 2 & 3 & -2 & 0 \end{array}$$

Since $P(-4) = 0$, -4 is a root and we can express $P(x)$ as follows:

$$P(x) = (x + 4)(2x^2 + 3x - 2)$$

or

$$P(x) = (x + 4)(2x - 1)(x + 2). \qquad \textbf{Factoring}$$

By using the principle of zero products and setting each factor equal to zero, we see that the roots of $P(x)$ are -4, $\frac{1}{2}$, and -2. ■

EXAMPLE 3 Let $P(x) = x^4 - 4x^3 - 3x^2 + 18x$. Find all rational roots of $P(x)$. If possible, find any other roots as well.

Solution Note that $P(x) = x(x^3 - 4x^2 - 3x + 18)$; so by the principle of zero products, 0 is a root of $P(x)$. Now look for roots of $x^3 - 4x^2 - 3x + 18$.

$$c: \quad 1, -1, 2, -2, 3, -3, 6, -6, 9, -9, 18, -18 \qquad d: \quad 1, -1$$

$$\frac{c}{d}: \quad 1, -1, 2, -2, 3, -3, 6, -6, 9, -9, 18, -18$$

We try 6.

$$\begin{array}{r|rrrr} 6 & 1 & -4 & -3 & 18 \\ & & 6 & 12 & 54 \\ \hline & 1 & 2 & 9 & 72 \end{array} \quad \leftarrow \textbf{All nonnegative}$$

Thus no roots are greater than 6. We try -3.

$$\begin{array}{r|rrrr} -3 & 1 & -4 & -3 & 18 \\ & & -3 & 21 & -54 \\ \hline & 1 & -7 & 18 & -36 \end{array} \quad \leftarrow \textbf{Signs alternate}$$

Using the lower bound theorem, we conclude that no root is less than -3. We try 3.

$$\begin{array}{r|rrrr} 3 & 1 & -4 & -3 & 18 \\ & & 3 & -3 & -18 \\ \hline & 1 & -1 & -6 & 0 \end{array}$$

This indicates that 3 is a root. We rewrite $P(x)$.

$$\begin{aligned} P(x) &= x(x^3 - 4x^2 - 3x + 18) \\ &= x(x - 3)(x^2 - x - 6) && \textbf{Using the last synthetic division} \\ &= x(x - 3)(x + 2)(x - 3) && \textbf{Factoring} \end{aligned}$$

By the principle of zero products, the solutions are 0, 3, and -2. ■

If, as in Example 3, a factor appears two or more times, we say that we have a **repeated root.** If the factor occurs twice, we say that we have a **double root,** or a root of **multiplicity two.** Roots that arise three times have a multiplicity of three and so on.

Notice that, in Example 1, $P(x)$ was a fourth-degree polynomial with four roots. In Example 2, $P(x)$ had degree three and three roots. Although the fourth-degree polynomial of Example 3 had but three roots, one of the roots had a multiplicity of two. This is generalized in the following result.

> Any polynomial of degree n has n roots, where roots of multiplicity m are counted m times.

Searching for rational roots can be a tedious process, even when the theorems of this section are enlisted for help. A search for *irrational* roots can be even more difficult, although for Exercises 39–50 we introduce *Descartes' rule of signs* as a way of determining how many positive real roots or negative real roots exist. We close this section with the two proofs mentioned earlier.

Proof (rational roots theorem): Suppose that c/d is a root of $P(x)$. It follows that

$$a_n\left(\frac{c}{d}\right)^n + a_{n-1}\left(\frac{c}{d}\right)^{n-1} + \cdots + a_1\left(\frac{c}{d}\right) + a_0 = 0. \tag{1}$$

We multiply by d^n and get the equations

$$a_nc^n + a_{n-1}c^{n-1}d + \cdots + a_1cd^{n-1} + a_0d^n = 0 \tag{2}$$

and
$$a_nc^n = (-a_{n-1}c^{n-1} - \cdots - a_1cd^{n-2} - a_0d^{n-1})d.$$

Thus d is a factor of a_nc^n. Now d has no factor in common with c, other than 1 or -1, because c/d was assumed to be in reduced form. Thus d has no factor in common with c^n. So d must be a factor of a_n.

In a similar way, we can show from Eq. (2) that

$$a_0d^n = (-a_nc^{n-1} - a_{n-1}c^{n-2}d - \cdots - a_1d^{n-1})c.$$

Thus c is a factor of a_0d^n. As ac is not a factor of d^n, it must be a factor of a_0. ■

Figure 1

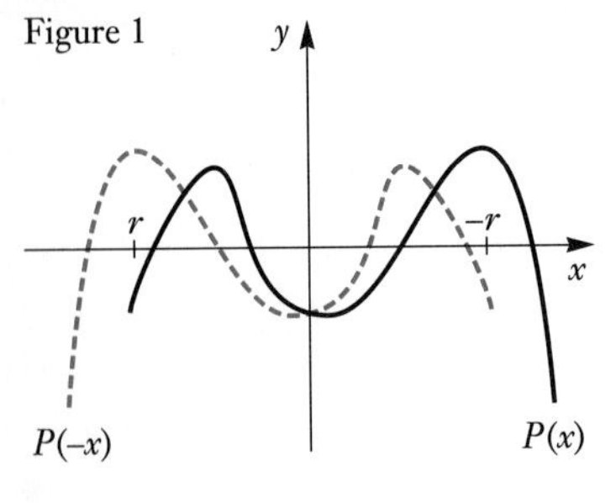

Proof (lower bound theorem): First observe that the graphs of $P(x)$ and $P(-x)$ are reflections of each other across the y-axis and that any *upper* bound for the roots of $P(-x)$, when negated, becomes a *lower* bound for the roots of $P(x)$ (see figure).

We will show that when the signs of the coefficients alternate signs as stated in the theorem, $-r$ is an upper bound for the roots of $P(-x)$.

Now suppose that r is negative and that

$$P(x) = (a_nx^n + a_{n-1}x^{n-1} + a_{n-2}x^{n-2} + \cdots + a_2x^2 + a_1x + a_0) \cdot (x - r) + P(r),$$

where $a_n, a_{n-1}, a_{n-2}, \ldots, a_2, a_1, a_0$, and $P(-r)$ alternate signs.

Four possibilities exist:

Case 1 If n is even and a_n is positive we have

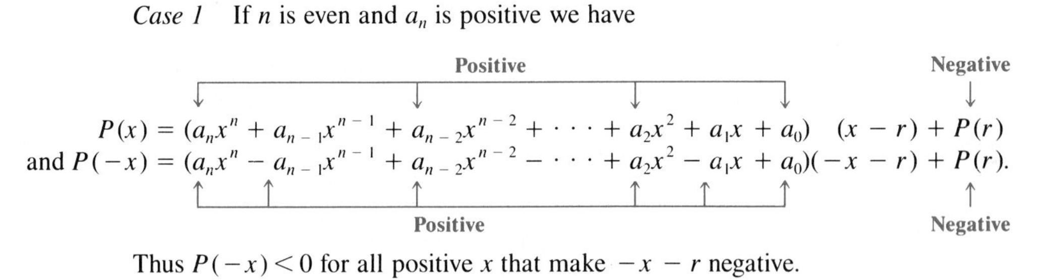

Thus $P(-x) < 0$ for all positive x that make $-x - r$ negative.

Case 2 If n is even and a_n is negative we have

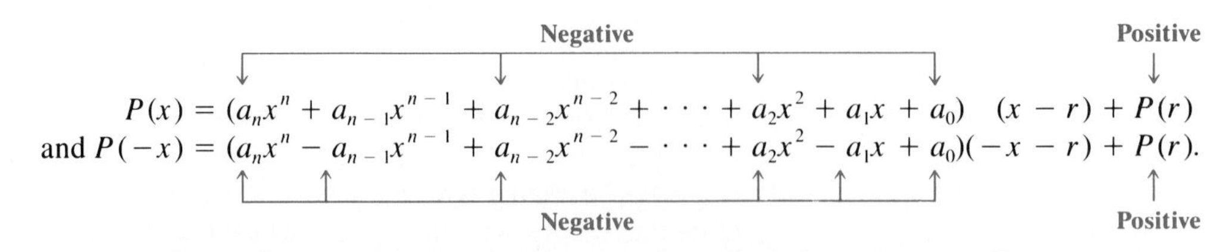

Thus, for all positive x such that $-x - r < 0$, we have $P(-x) > 0$.

Case 3 If n is odd and a_n is negative we have

$$\begin{aligned} P(x) &= (\overbrace{a_n x^n + a_{n-1}x^{n-1} + a_{n-2}x^{n-2} + \cdots + a_2x^2 + a_1x + a_0}^{\text{Positive}})\ (x - r) + \overbrace{P(r)}^{\text{Negative}} \\ \text{and } P(-x) &= (\underbrace{-a_n x^n + a_{n-1}x^{n-1} - a_{n-2}x^{n-2} + \cdots + a_2x^2 - a_1x + a_0}_{\text{Positive}})(-x - r) + \underbrace{P(r)}_{\text{Negative}}. \end{aligned}$$

Thus for all $x > 0$ such that $-x - r < 0$ we have $P(-x) < 0$.

Case 4 A similar examination of the signs of the coefficients of $P(-x)$ for n odd and $a_n > 0$ reveals that for all $x > 0$ such that $-x - r < 0$ we have $P(-x) > 0$.

In each case, $-x - r < 0$ when $x > -r$. Thus $-r$ is an upper bound for the roots of $P(-x)$. By symmetry, r is a lower bound for the roots of $P(x)$. ■

EXERCISE SET 8.4

Use the rational roots theorem to list all *possible* rational roots.

1. $x^5 - 3x^2 + 1$
2. $x^7 + 37x^5 - 6x^2 + 12$
3. $15x^6 + 47x^2 + 2$
4. $10x^{25} + 3x^{17} - 35x + 6$

Find the smallest positive integer that is guaranteed by the upper bound theorem to be an upper bound to the real roots.

5. $3x^4 - 15x^2 + 2x - 3$
6. $4x^4 - 14x^2 + 4x - 2$
7. $6x^3 - 17x^2 - 3x - 1$
8. $5x^3 - 15x^2 + 5x - 4$

Find the largest negative integer that is guaranteed by the lower bound theorem to be a lower bound to the real roots.

9. $3x^4 - 15x^3 + 2x - 3$
10. $4x^4 - 17x^3 + 3x - 2$
11. $6x^3 + 15x^2 + 3x - 1$
12. $6x^3 + 12x^2 + 5x - 3$

Find the rational roots, if they exist, of each polynomial or equation. If possible, find the other roots. Then write the equation or polynomial in factored form.

13. $x^3 + 3x^2 - 2x - 6$
14. $x^3 - x^2 - 3x + 3 = 0$
15. $x^3 - 3x + 2 = 0$
16. $x^3 - 3x + 4$
17. $x^3 - 5x^2 + 11x - 19$
18. $x^3 - 7x^2 - 2x + 23$
19. $5x^4 - 4x^3 + 19x^2 - 16x - 4 = 0$
20. $3x^4 - 4x^3 + x^2 + 6x - 2$
21. $x^4 - 3x^3 - 20x^2 - 24x - 8$
22. $x^4 + 5x^3 - 27x^2 + 31x - 10$
23. $x^3 - 4x^2 + 2x + 4 = 0$
24. $x^3 - 8x^2 + 17x - 4$
25. $x^3 + 8$
26. $x^3 - 8 = 0$
27. $\frac{1}{3}x^3 - \frac{1}{2}x^2 - \frac{1}{6}x + \frac{1}{6}$ (*Hint:* Factor out $\frac{1}{6}$.)
28. $\frac{2}{3}x^3 - \frac{1}{2}x^2 + \frac{2}{3}x - \frac{1}{2}$ (*Hint:* Factor out $\frac{1}{6}$.)

Find only the rational roots.

29. $x^4 + 32$
30. $x^6 + 8 = 0$
31. $x^3 - x^2 - 4x + 3 = 0$
32. $2x^3 + 3x^2 + 2x + 3$
33. $x^4 + 2x^3 + 2x^2 - 4x - 8 = 0$
34. $x^4 + 6x^3 + 17x^2 + 36x + 66 = 0$
35. $x^5 - 5x^4 + 5x^3 + 15x^2 - 36x + 20$
36. $x^5 - 3x^4 - 3x^3 + 9x^2 - 4x + 12$

SKILL MAINTENANCE

Simplify.

37. $(7 - 2i)(2 + 3i)$
38. i^{43}

SYNTHESIS

A useful rule for determining the number of positive or negative roots of a polynomial is *Descartes' rule of signs*. To use this rule on any polynomial $P(x) = a_n x^n + a_{n-1}x^{n-1} + \cdots + a_1 x + a_0$ with real coefficients, simply count the number of variations of sign as the coefficients are read from left to right (for example, $5x^3 - 7x^2 + 3x + 6 = 0$ has two sign changes—from 5 to -7 and from -7 to 3). The number of *positive* real roots is either the same as the number of sign changes or less than the number of sign changes by a positive even integer. The number of *negative* real roots is either the number of sign changes in $P(-x)$ or less than the number of sign changes in $P(-x)$ by a positive even integer. A root of multiplicity m must be counted m times.

What does Descartes' rule of signs tell you about the number of positive real roots?

39. $3x^5 - 2x^2 + x - 1$

40. $5x^6 - 3x^3 + x^2 - x$

41. $6x^7 + 2x^2 + 5x + 4 = 0$

42. $-3x^5 - 7x^3 - 4x - 5 = 0$

43. $3p^{18} + 2p^4 - 5p^2 + p + 3$

44. $5t^{12} - 7t^4 + 3t^2 + t + 1$

What does Descartes' rule of signs tell you about the number of negative real roots?

45. $3x^5 - 2x^2 + x - 1$

46. $5x^6 - 3x^3 + x^2 - x$

47. $6x^7 + 2x^2 + 5x + 4 = 0$

48. $-3x^5 - 7x^3 - 4x - 5 = 0$

49. $3p^{18} + 2p^3 - 5p^2 + p + 3$

50. $5t^{11} - 7t^4 + 3t^2 + t + 1$

51. An open box of volume 48 cm^3 can be made from a piece of tin 10 cm on a side by cutting a square from each corner and folding up the edges. What is the length of a side of the squares?

52. An open box of volume 500 cm^3 can be made from a piece of tin 20 cm on a side by cutting a square from each corner and folding up the edges. What is the length of a side of the squares?

53. Show that $\sqrt{5}$ is irrational by considering the equation $x^2 - 5 = 0$.

54. Generalize the result of Exercise 53 to find which positive integers have rational square roots.

8.5 Graphs of Polynomial Functions

Graphs of first-degree linear polynomial functions are lines; graphs of second-degree quadratic functions are parabolas. We now consider polynomials of higher degree. In general, graphing such functions is difficult and is best done using tools of calculus not studied in this text. Nevertheless, we will consider graphing in some detail, making use of what we know about roots. Were we to use a computer software graphing package or a graphing calculator, the graphs of polynomial functions could be generated quite easily.

The following facts will be useful:

> The domain of a polynomial function is the set of real numbers. The graph of a polynomial function is a continuous curve.

Because the graph of a polynomial function is continuous, we know that if, for some polynomial function f, $f(a)$ and $f(b)$ have opposite signs, then the graph of f must cross the x-axis somewhere between a and b.

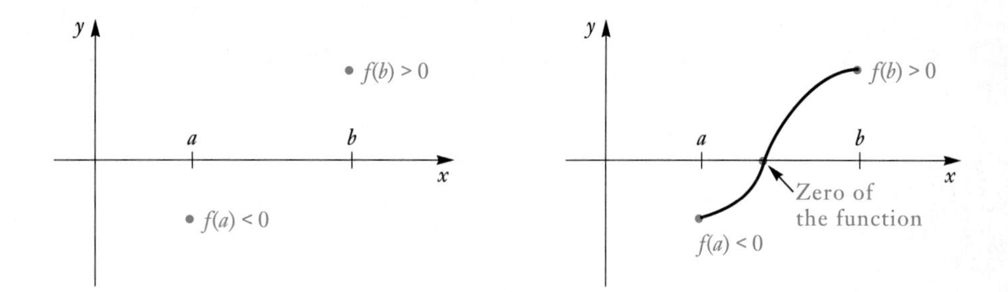

Another fact that follows from the preceding can be helpful in graphing. Suppose we have determined all the real-number zeros of a polynomial, which often is no small task. Then between successive zeros, the values of the polynomial are either all positive or all negative. If there was a sign change, then there would be an additional zero. Try drawing a continuous curve between successive zeros. You assume you have found all the zeros, so the curve cannot cross the x-axis. Thus values between zeros are either all positive or all negative. By using *test values* between successive zeros, we know the sign of the function on the entire interval.

EXAMPLE 1 Sketch a graph of the polynomial function f given by

$$f(x) = 2x^3 + x^2 - 8x - 4.$$

Solution We first try to factor the function in order to find the zeros. In this case we can use factoring by grouping:

$$\begin{aligned} f(x) &= 2x^3 + x^2 - 8x - 4 \\ &= x^2(2x + 1) - 4(2x + 1) \\ &= (2x + 1)(x^2 - 4) \qquad \textbf{2x + 1 is a common factor.} \\ &= (2x + 1)(x - 2)(x + 2). \end{aligned}$$

We see that the solutions of the equation $f(x) = 0$ are -2, -0.5, and 2. These are also the zeros of the polynomial function. The zeros divide the number line into four intervals: $(-\infty, -2)$, $(-2, -0.5)$, $(-0.5, 2)$, and $(2, \infty)$.

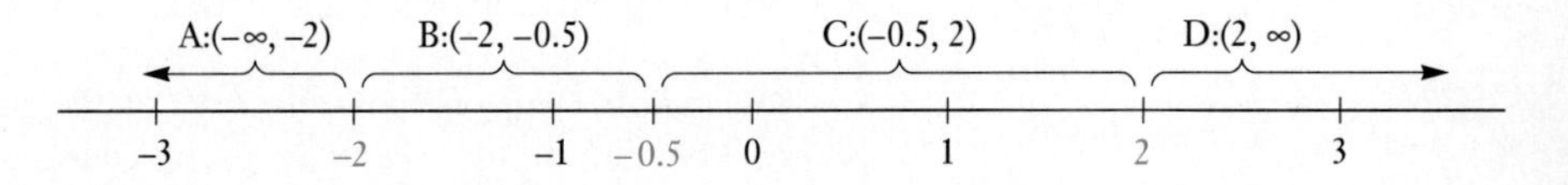

We try test numbers as follows in each interval. These also give us function values that can be used for graphing.

A: Test -3, $f(-3) = 2(-3)^3 + (-3)^2 - 8(-3) - 4 = -25$

B: Test -1, $f(-1) = 2(-1)^3 + (-1)^2 - 8(-1) - 4 = 3$

C: Test 1, $f(1) = 2(1)^3 + (1)^2 - 8(1) - 4 = -9$

D: Test 3, $f(3) = 2(3)^3 + (3)^2 - 8(3) - 4 = 35$

Since $f(-3) = -25$, the function values on the interval $(-\infty, -2)$ are all negative and the graph is below the x-axis on this interval. Since $f(-1) = 3$, the function values on the interval $(-2, -0.5)$ are all positive and the graph is above the x-axis on this interval. Since $f(1) = -9$, the function values on the interval $(-0.5, 2)$ are all negative and the graph is below the x-axis on this interval. Since $f(3) = 35$, the function values on the interval $(2, \infty)$ are all positive and the graph is above the x-axis on this interval. We summarize this information in the following table.

Interval	$(-\infty, -2)$	$(-2, -0.5)$	$(-0.5, 2)$	$(2, \infty)$
Test Value	$f(-3) = -25$	$f(-1) = 3$	$f(1) = -9$	$f(3) = 35$
Sign of $f(x)$	Negative	Positive	Negative	Positive
Location of Points on Graph	Below x-axis	Above x-axis	Below x-axis	Above x-axis

Using these results and calculating additional function values, we complete the graph as follows.

Note that as x gets very positive (large), $f(x)$ grows rapidly, and when x gets very negative, the value of $f(x)$ quickly becomes very negative. This reflects the relative importance of the leading term, $2x^3$.

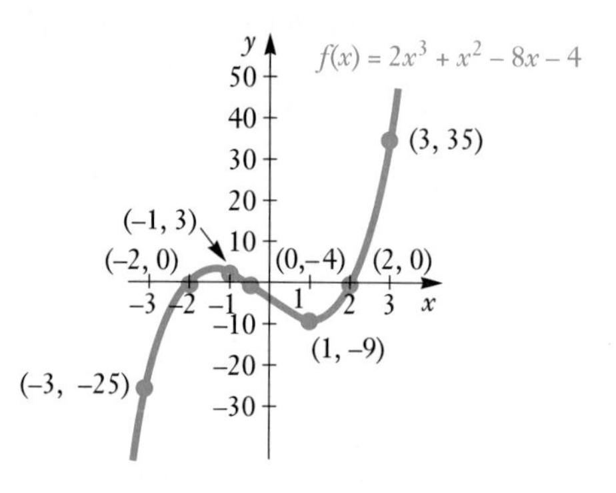

The points $\left(-\frac{4}{3}, \frac{100}{27}\right)$ and $(1, -9)$ of the preceding graph are called *turning points*. They are points where the graph changes from increasing to decreasing or from decreasing to increasing. In calculus you will learn skills that will allow you to find these points easily. Note that a polynomial function of degree n has at most $n - 1$ turning points and n zeros.

EXAMPLE 2 Sketch the graph of the polynomial function f given by

$$f(x) = x^4 - 4x^2 + 3.$$

Solution We first factor the function as follows:

$$\begin{aligned} f(x) &= x^4 - 4x^2 + 3 \qquad \text{This is in } \textit{quadratic form} \text{ (see Section 5.5).} \\ &= (x^2 - 3)(x^2 - 1) \\ &= \left(x + \sqrt{3}\right)\left(x - \sqrt{3}\right)(x + 1)(x - 1). \end{aligned}$$

The zeros of the function are $-\sqrt{3}$, -1, 1, and $\sqrt{3}$.

These zeros divide the real number line into the five open intervals given in the following table. We try a test value in each interval and determine the sign of the function for values of x in the interval. The results are summarized in the following table.

Interval	$\left(-\infty, -\sqrt{3}\right)$	$\left(-\sqrt{3}, -1\right)$	$(-1, 1)$	$\left(1, \sqrt{3}\right)$	$\left(\sqrt{3}, \infty\right)$
Test Value	$f(-2) = 3$	$f(-1.5) = -0.9375$	$f(0) = 3$	$f(1.5) = -0.9375$	$f(2) = 3$
Sign of $f(x)$	Positive	Negative	Positive	Negative	Positive
Location of Points on Graph	Above x-axis	Below x-axis	Above x-axis	Below x-axis	Above x-axis

It is important to observe that since the degree of $f(x)$ is even and the leading coefficient is positive, the value of $f(x)$ will grow as we move far to the left or far to the right on the x-axis. Also, since $f(x) = f(-x)$, the graph of f is symmetric with respect to the y-axis (f is an even function). Using this information and calculating some extra function values as needed, we sketch the graph.

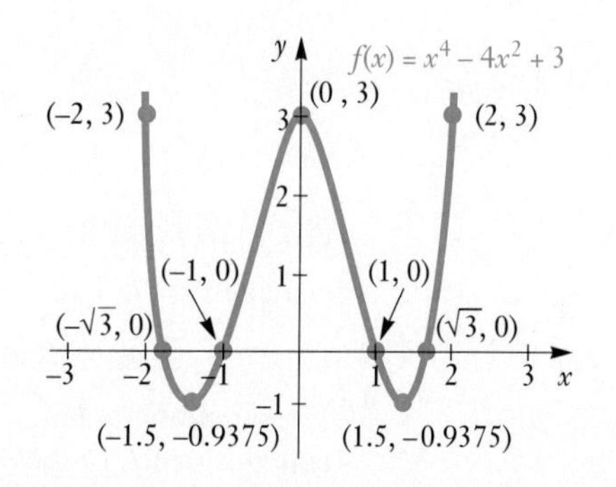

■

EXAMPLE 3 Sketch the graph of the polynomial function f given by

$$f(x) = 2x^4 - x^3 - 2x^2 - 11x + 6.$$

Solution

a) We note that, since the degree of this polynomial is even and the leading coefficient is positive, as we move far to the left or far to the right, function values will become very large.

b) Next make a table of values using synthetic division. We use values for x that might be rational roots and look for upper and lower bounds to the roots. In doing the synthetic division, we need only write the bottom row—this will save us some work. From the rational roots theorem, we find that the possible rational roots are $-1, 1, -2, 2, -3, 3, -6, 6, -\frac{1}{2}, \frac{1}{2}, -\frac{3}{2}$, and $\frac{3}{2}$.

x	2	-1	-2	-11	6	(x, y)	
1	2	1	-1	-12	-6	$(1, -6)$	These numbers are from synthetic division.
-1	2	-3	1	-12	18	$(-1, 18)$	Since signs alternate, -1 is a lower bound to the roots.
3	2	5	13	28	90	$(3, 90)$	Since all numbers are positive, 3 is an upper bound to the roots.
2	2	3	4	-3	0	$(2, 0)$	2 is a root.

c) Since 2 is a root, by the factor theorem we have

$$f(x) = (x - 2)Q(x),$$

where $Q(x) = 2x^3 + 3x^2 + 4x - 3$.

We now search for roots of $Q(x)$. By the rational roots theorem the possible rational roots of $Q(x)$ are $-1, 1, -3, 3, -\frac{1}{2}, \frac{1}{2}, -\frac{3}{2}$, and $\frac{3}{2}$. Since we saw that 1, 3, and -1 are not roots of $f(x)$, we need not consider them as possible roots of $Q(x)$. And since -1 is a lower bound to the roots, we need not consider -3 or $-\frac{3}{2}$. Thus the only possible rational roots worth considering are $-\frac{1}{2}, \frac{1}{2}$, and $\frac{3}{2}$.

We try $\frac{1}{2}$.

$$\begin{array}{r|rrrr} 1/2 & 2 & 3 & 4 & -3 \\ & & 1 & 2 & 3 \\ \hline & 2 & 4 & 6 & 0 \end{array}$$

Since $Q\left(\frac{1}{2}\right) = 0$, $\frac{1}{2}$ is a root. Again using the results of the synthetic division, we can express $f(x)$ as

$$f(x) = (x - 2)\left(x - \tfrac{1}{2}\right)(2x^2 + 4x + 6).$$

The quadratic formula reveals that the other two roots of $f(x)$ are complex. The real zeros of $f(x)$ are $\frac{1}{2}$ and 2.

d) The two real zeros divide the real number line into three open intervals. We try test values as in Examples 1 and 2, using some work from part (b).

Interval	$\left(-\infty, \frac{1}{2}\right)$	$\left(\frac{1}{2}, 2\right)$	$(2, \infty)$
Test Value	$f(0) = 6$	$f(1) = -6$	$f(3) = 90$
Sign of $f(x)$	Positive	Negative	Positive
Location of Points on Graph	Above x-axis	Below x-axis	Above x-axis

We use the information in the preceding tables, calculate some extra function values, if needed, and sketch the graph.

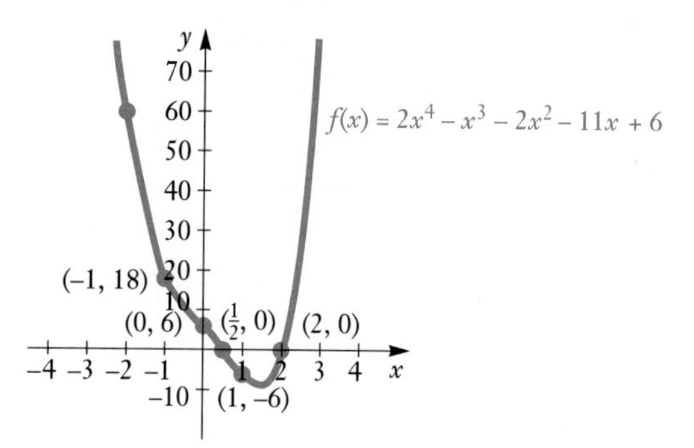

■

EXAMPLE 4 Sketch the graph of the polynomial function

$$f(x) = 2x^3 - 9x^2 + x + 3.$$

Solution

a) We note that, since the degree of this polynomial is odd and the leading coefficient is positive, as we move far to the right function values increase without bound and as we move far to the left function values decrease without bound.

b) We form the following table using synthetic division. The possible rational roots are -1, 1, -3, 3, $-\frac{1}{2}$, $\frac{1}{2}$, $-\frac{3}{2}$, and $\frac{3}{2}$.

x	2	-9	1	3	(x, y)	
-1	2	-11	12	-9	$(-1, -9)$	**-1 is a lower bound to the roots.**
1	2	-7	-6	-3	$(1, -3)$	
3	2	-3	-8	-21	$(3, -21)$	
$-\frac{1}{2}$	2	-10	6	0	$(-\frac{1}{2}, 0)$	**$-\frac{1}{2}$ is a root.**

c) By the factor theorem, we have

$$f(x) = \left(x + \tfrac{1}{2}\right)(2x^2 - 10x + 6)$$

or

$$f(x) = (2x + 1)(x^2 - 5x + 3).$$

The quadratic formula tells us that $(5 - \sqrt{13})/2$ and $(5 + \sqrt{13})/2$ are the other two roots.

d) We use a test value in each open interval, making use of the work in part (b).

Interval	$\left(-\infty, -\frac{1}{2}\right)$	$\left(-\frac{1}{2}, \frac{5-\sqrt{13}}{2}\right)$	$\left(\frac{5-\sqrt{13}}{2}, \frac{5+\sqrt{13}}{2}\right)$	$\left(\frac{5+\sqrt{13}}{2}, \infty\right)$
Test Value	$f(-1) = -9$	$f(0) = 3$	$f(3) = -21$	$f(5) = 33$
Sign of $f(x)$	Negative	Positive	Negative	Positive
Location of Points on Graph	Below x-axis	Above x-axis	Below x-axis	Above x-axis

Using the above information, we sketch the graph.

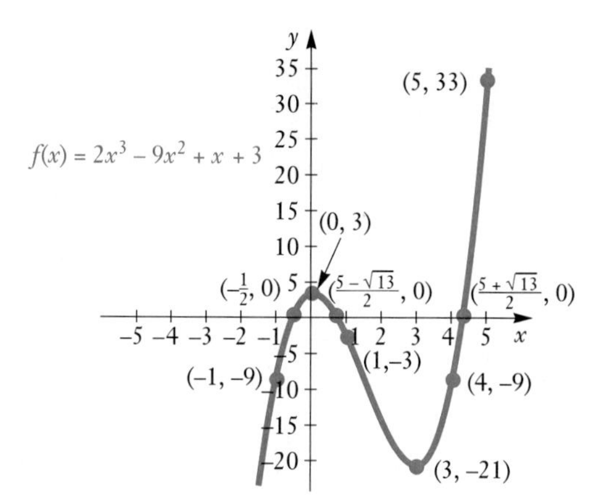

■

The following guidelines may assist you in graphing polynomial functions.

To graph polynomial functions:

1. Look at the degree of the polynomial and its leading coefficient. This gives a lot of information about the general shape of the graph.
2. Look for symmetries, as covered in Section 8.2.
3. Make a table of values, using synthetic division.
4. Find the y-intercept and as many x-intercepts as possible (the latter are roots of the polynomial). In doing this, recall the theorems about roots.
5. Plot the points, and connect them appropriately.

EXERCISE SET 8.5

Sketch a graph of each polynomial function after first factoring each polynomial.

1. $f(x) = x^4 - x^3$

2. $f(x) = x^3 - 4x$

3. $f(x) = 9x^2 - x^4$

4. $f(x) = 4x^2 - x^4$

5. $f(x) = x^3 + x^2 - 2x$

6. $f(x) = -x^3 - x^2 + 6x$

7. $f(x) = x^4 - 4x^2$
8. $f(x) = 25x - x^3$
9. $f(x) = x^4 - 9x^2 + 20$
10. $f(x) = x^4 - 3x^2 + 2$
11. $f(x) = x^3 - 3x^2 - 4x + 12$
12. $f(x) = x^3 - 2x^2 - 9x + 18$
13. $f(x) = -x^4 - 3x^3 - 3x^2$
14. $f(x) = -3x^4 - x^3 + 2x^2$

Sketch a graph of each polynomial function. Use the theorems about roots (see Examples 3 and 4).

15. $f(x) = x^3 + 2x^2 - 5x - 6$
16. $f(x) = x^3 - 3x^2 - 6x + 8$
17. $f(x) = x^4 - 6x^3 + 11x^2 - 6x$
18. $f(x) = x^4 + 2x^3 - 5x^2 - 6x$
19. $f(x) = x^3 + 4x^2 + x - 6$
20. $f(x) = x^3 + 2x^2 - 11x - 12$
21. $f(x) = 2x^3 - 7x^2 - 17x + 10$
22. $f(x) = 3x^3 + 8x^2 - 33x + 10$
23. $f(x) = -x^4 + x^3 + 4x^2 - 2x - 4$
24. $f(x) = x^5 - x^4 - 9x^3 + 3x^2 + 18x$
25. $f(x) = 2x^4 - 9x^3 + 3x^2 + 11x - 3$
26. $f(x) = x^4 + 3x^3 - 3x^2 - 12x - 4$

SKILL MAINTENANCE

27. Determine the nature of the solutions of the equation $x^2 - 8x + 15 = 0$.
28. Solve: $|x| \leq 5$.
29. Simplify: $\dfrac{x^3 - 8}{x^4 - 16}$.
30. Rewrite with fractional exponents: $\sqrt[3]{5ab}$.

SYNTHESIS

Computer–calculator exercises. There are many computer software packages and calculators that will graph polynomials and solve equations. Use a computer or a calculator to graph the following functions and approximate their roots.

31. $f(x) = x^3 - 9x^2 + 27x + 50$
32. $f(x) = x^3 - 3x + 1$
33. $f(x) = x^4 + 4x^3 - 36x^2 - 160x + 300$
34. $f(x) = x^6 + 4x^5 - 54x^4 - 160x^3 + 641x^2 + 828x - 1260$

8.6 Rational Functions

A **rational function** is a function that is a quotient of two polynomials, $p(x)/q(x)$, with $q(x)$ not the zero polynomial. Here are some examples:

$$f(x) = \frac{x^2 + 3x - 5}{x + 4}, \qquad f(x) = \frac{8}{x^2 + 1}, \qquad f(x) = \frac{6x^5 - 7x + 11}{4}.$$

The domain of a rational function consists of all inputs for which the divisor, $q(x)$, is not zero. Polynomial functions are themselves special kinds of rational functions, since $q(x)$ can be the polynomial 1. Here we are interested in graphing rational functions in which the denominator is not a constant. We begin with the simplest such functions.

Graphs

EXAMPLE 1 Graph: $f(x) = \dfrac{1}{x}$.

Solution We consider this function from a viewpoint that will help us in setting up the general scheme for graphing rational functions.

Note that the domain of this function consists of all real numbers except 0.

Since $f(-x) = 1/(-x) = -(1/x) = -f(x)$, the function is odd. Thus the graph is symmetric with respect to the origin.

Now we find some points, keeping in mind the symmetry of the graph.

x	1	2	3	4	5	$\frac{1}{2}$	$\frac{1}{3}$	$\frac{1}{4}$	$\frac{1}{5}$
$f(x)$	1	$\frac{1}{2}$	$\frac{1}{3}$	$\frac{1}{4}$	$\frac{1}{5}$	2	3	4	5

We plot these points. We use symmetry to get other points in the third quadrant.

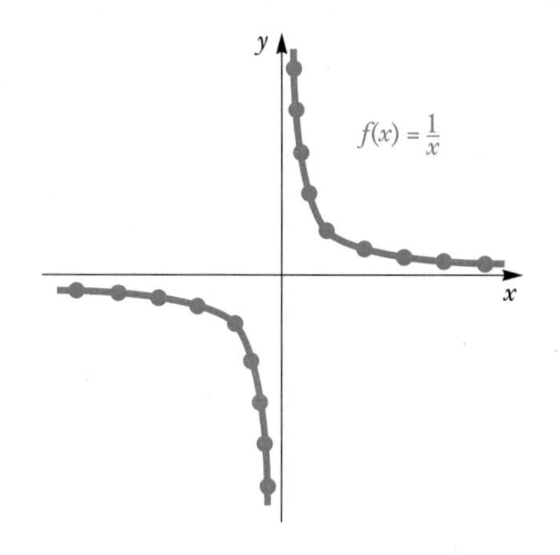

The points indicated by • are obtained from the table. Those marked • are obtained by reflection across the origin. ■

Asymptotes

Note that the curve in Example 1 does not touch either axis but comes very close. As $|x|$ becomes very large, the curve comes very near to the x-axis. In fact, we can find points as close to the x-axis as we please by choosing x large enough or x small enough. We say that the curve approaches the axes *asymptotically,* and we say that the x-axis is a **horizontal asymptote** and the y-axis a **vertical asymptote** to the curve.

Using the ideas of Section 7.4, we can easily graph certain variations of the preceding function.

EXAMPLE 2 Graph $g(x) = -\dfrac{1}{x}$.

Solution The graph of $g(x) = -1/x$ is a reflection of $f(x) = 1/x$ across the x-axis (or the y-axis; the result is the same). Both the x-axis and the y-axis are asymptotes to the curve. As a check, note that $g(1) = -1$.

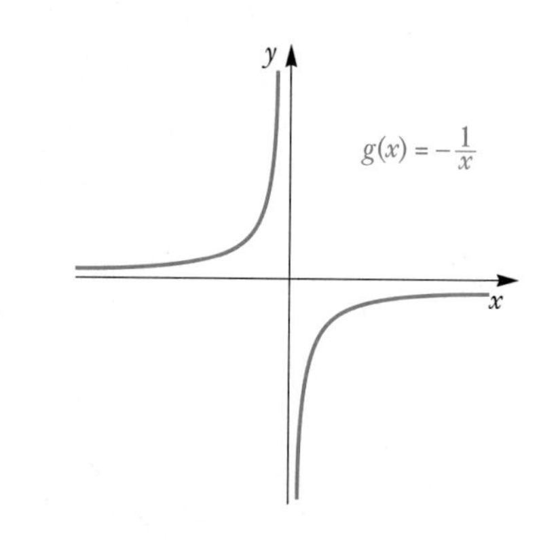

■

EXAMPLE 3 Graph $h(x) = \dfrac{1}{x - 2}$.

Solution The graph of $h(x) = 1/(x - 2)$ is a translation, or shift, of $f(x) = 1/x$ to the right 2 units. The x-axis, or $y = 0$, and the line $x = 2$ are asymptotes to the curve. As a check, note that $h(3) = 1$.

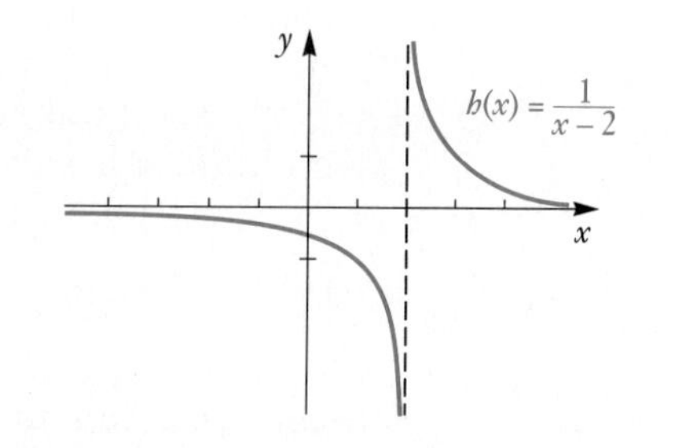

■

EXAMPLE 4 Graph $f(x) = \dfrac{1}{x^2}$.

Solution

a) Note that this function is defined for all x except 0. Note also that for all other values of x, $x^2 > 0$, so $1/x^2 > 0$. Thus, the entire graph is above the x-axis.

b) Since this function is defined for all x except 0, the line $x = 0$ is the vertical asymptote. As $|x|$ gets very close to 0, $f(x)$ becomes very large.

c) As $|x|$ gets very large, y approaches 0. Therefore the x-axis is the horizontal asymptote.

d) Since $f(x) = f(-x)$, this function is even. Thus it is symmetric with respect to the y-axis.

e) We compute some function values and, using those points and the preceding information, sketch the following graph. Points marked • are obtained from the table. Points marked • are obtained by reflection across the y-axis.

x	1	2	3	4	$\frac{1}{2}$	$\frac{1}{3}$
$f(x)$	1	$\frac{1}{4}$	$\frac{1}{9}$	$\frac{1}{16}$	4	9

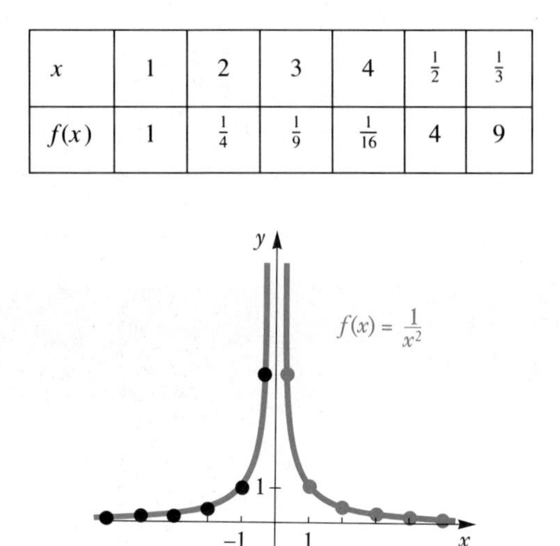

Occurrences of Asymptotes

It is important in graphing rational functions to determine where the asymptotes, if any, occur. Vertical asymptotes are easy to locate when a denominator is factored. The x-values that make a denominator 0 but *do not* make the numerator 0 are those of the vertical asymptotes.

EXAMPLE 5 Find the vertical asymptotes of the function $f(x) = \dfrac{3x - 2}{x(x - 5)(x + 3)}$.

Solution The vertical asymptotes are the lines $x = 0$, $x = 5$, and $x = -3$. ■

EXAMPLE 6 Determine the vertical asymptotes of the function $f(x) = \dfrac{x - 2}{x^3 - x}$.

Solution We factor the denominator:

$$x^3 - x = x(x + 1)(x - 1).$$

The vertical asymptotes are $x = 0$, $x = -1$, and $x = 1$. ■

Horizontal asymptotes occur when the degree of the numerator is the same as or less than that of the denominator. Let us first consider a function for which the degree of the numerator is less than that of the denominator:

$$f(x) = \frac{2x + 3}{x^3 - 2x^2 + 4}.$$

We multiply by $\frac{1/x^3}{1/x^3}$, to obtain

$$f(x) = \frac{\frac{2}{x^2} + \frac{3}{x^3}}{1 - \frac{2}{x} + \frac{4}{x^3}}.$$

We multiplied each term in the numerator and each term in the denominator by $1/x^3$.

Let us now consider what happens to the function values as $|x|$ becomes very large. Each expression with x in its denominator takes on smaller and smaller values, approaching 0. Thus the numerator of the complex fraction approaches 0 while the denominator approaches $1 - 0 + 0$; hence the complex fraction takes on values closer and closer to $0/1$, or 0. The x-axis, or $y = 0$, is therefore the horizontal asymptote. *Whenever the degree of a numerator is less than that of the denominator, the x-axis, or $y = 0$, will be an asymptote.*

Next, we consider a function for which the numerator and denominator have the same degree:

$$y = \frac{3x^2 + 2x - 4}{2x^2 - x + 1} = \frac{3x^2 + 2x - 4}{2x^2 - x + 1} \cdot \frac{1/x^2}{1/x^2}$$

Multiplying by 1, where $1 = \frac{1/x^2}{1/x^2}$

$$= \frac{3 + \frac{2}{x} - \frac{4}{x^2}}{2 - \frac{1}{x} + \frac{1}{x^2}}.$$

As $|x|$ gets very large, the numerator of the complex fraction approaches 3 and the denominator approaches 2. Therefore, the function gets very close to $\frac{3}{2}$ and the line $y = \frac{3}{2}$ is a horizontal asymptote. From this example, we can see that *when the degree of the numerator and the denominator are the same, the asymptote can be determined by dividing the leading coefficients of the two polynomials.*

EXAMPLE 7 Find the horizontal asymptote of the function $f(x) = \frac{5x^3 - x^2 + 7}{3x^5 + x - 10}$.

Solution Since the degree of the numerator is less than the degree of the denominator, the line $y = 0$ is the horizontal asymptote. ■

EXAMPLE 8 Find the horizontal asymptote of f, if $f(x) = \frac{-7x^4 - 10x^2 + 1}{11x^4 + x - 2}$.

Solution The numerator and the denominator have the same degree, so the line $y = -\frac{7}{11}$ is the horizontal asymptote. ■

There are asymptotes that are neither vertical nor horizontal. They are called **oblique,** and they occur when the degree of the numerator is greater than that of the denominator by 1. To find the asymptote we divide the numerator by the denominator. Consider

$$f(x) = \frac{2x^2 - 3x - 1}{x - 2}.$$

When we divide the numerator by the denominator, we obtain a quotient of $2x + 1$ and a remainder of 1. Thus

$$f(x) = 2x + 1 + \frac{1}{x - 2}.$$

Now we can see that when $|x|$ becomes very large, $1/(x-2)$ approaches 0, and the y-values thus approach $2x + 1$. This means that the graph comes closer and closer to the straight line $y = 2x + 1$.

EXAMPLE 9 Find and draw the asymptotes of the function

$$f(x) = \frac{2x^2 - 11x - 10}{x - 4}.$$

Solution

a) We first note that $x = 4$ is a vertical asymptote.
b) We divide, to obtain

$$f(x) = 2x - 3 - \frac{22}{x - 4}.$$

Thus the line $y = 2x - 3$ is an oblique asymptote.

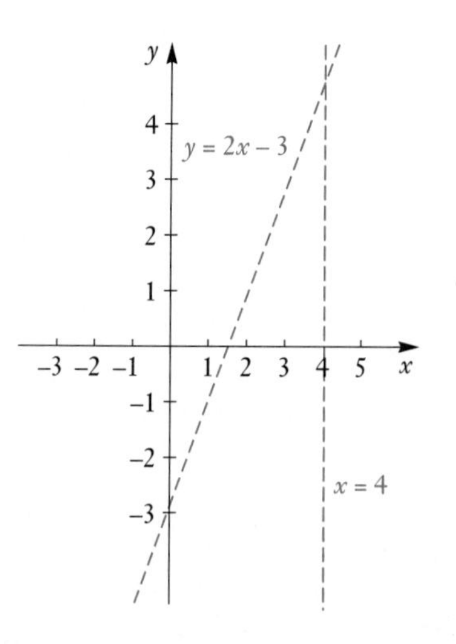

The following summarizes the conditions under which asymptotes occur.

Asymptotes of a rational function occur as follows.

1. *Vertical asymptotes.* Occur at any x-values that make the denominator zero but the numerator nonzero.
2. *The x-axis an asymptote.* Occurs when the degree of the denominator is greater than that of the numerator.
3. *Horizontal asymptote other than the x-axis.* Occurs when numerator and denominator have the same degree.
4. *Oblique asymptote.* Occurs when the degree of the numerator is greater than that of the denominator by 1.

Zeros or Roots

Zeros, or roots, of a rational function occur when the numerator is 0 but the denominator is not 0. The zeros of a function occur at points where the graph crosses the x-axis. Therefore, knowing the zeros helps in making a graph. If the numerator can be factored, the zeros are easy to determine.

EXAMPLE 10 Find the zeros of the rational function

$$f(x) = \frac{x^3 - x^2 - 6x}{x^2 - 3x + 2}.$$

Solution We factor the numerator and the denominator:

$$f(x) = \frac{x(x + 2)(x - 3)}{(x - 1)(x - 2)}.$$

The x-values making the numerator zero are 0, -2, and 3. Since none of these make the denominator zero, they are the zeros of the function. The graph will cross the x-axis at (0, 0), (-2, 0), and (3, 0). ■

The following is an outline of the procedure to be followed in graphing rational functions.

To graph rational functions:

a) Determine any symmetries.
b) Determine any horizontal or oblique asymptotes and sketch them.
c) Factor the denominator and the numerator.
  i) Determine any vertical asymptotes and sketch them.
  ii) Determine the zeros, if possible, and plot them.
d) Locate intervals where function values are positive and where they are negative.
e) Make a table of values and plot them.
f) Sketch the curve.

EXAMPLE 11 Graph $g(x) = \dfrac{1}{x^2 - 3}$.

Solution We follow the outline given in the preceding box.

a) The function is even, so the graph is symmetric about the y-axis.
b) Since the degree of the numerator is less than the degree of the denominator, the x-axis is a horizontal asymptote. There are no oblique asymptotes.
c) We can factor the denominator but not the numerator. We have

$$g(x) = \frac{1}{(x - \sqrt{3})(x + \sqrt{3})}.$$

The zeros of the denominator are $\pm\sqrt{3}$. Thus $x = \sqrt{3}$ and $x = -\sqrt{3}$ are vertical asymptotes of the graph of g. We draw these asymptotes with dashed lines. The function g has no zeros.
d) The vertical asymptotes divide the x-axis into intervals as follows. To find where the function is positive or negative, we use the procedure of Section 8.5. We try a test point in each interval.

Interval	$(-\infty, -\sqrt{3})$	$(-\sqrt{3}, \sqrt{3})$	$(\sqrt{3}, \infty)$
Test Value	$g(-4) = \frac{1}{13}$	$g(0) = -\frac{1}{3}$	$g(4) = \frac{1}{13}$
Sign of $g(x)$	Positive	Negative	Positive
Location of Points on Graphs	Above x-axis	Below x-axis	Above x-axis

e) We use the preceding information, compute some other functions values, and draw the graph.

x	$g(x)$, approx.
-5	0.05
-3	0.2
-2	1
$-\sqrt{2}$	-1
-1	-0.5
1	-0.5
$\sqrt{2}$	-1
2	1
3	0.2
5	0.05

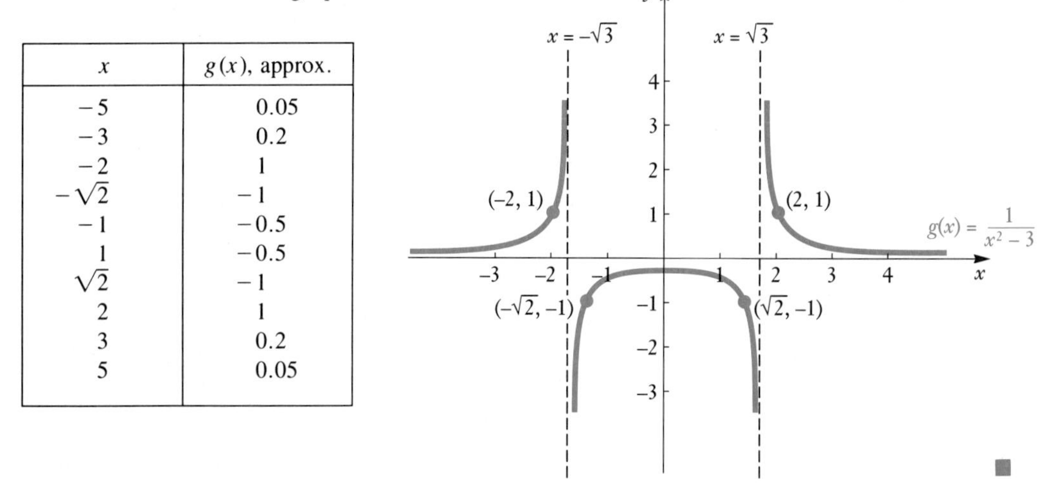

■

EXAMPLE 12 Graph $f(x) = \dfrac{x^3 - x^2 - 6x}{x^2 - 3x + 2}$.

Solution We follow the outline.

a) The function is neither even nor odd, and no symmetries are apparent.

b) The degree of the numerator is 1 more than the degree of the denominator, so we have an oblique asymptote. Dividing numerator by denominator will show that the line $y = x + 2$ is an oblique asymptote. Sketch the graph of $y = x + 2$ with a dashed line.

c) The numerator and denominator are easily factorable. We have

$$f(x) = \frac{x(x + 2)(x - 3)}{(x - 1)(x - 2)}.$$

The zeros of the denominator are 1 and 2. Thus the vertical asymptotes are $x = 1$ and $x = 2$. We sketch the vertical asymptotes, using dashed lines, and plot the zeros. The zeros of the function are 0, -2, and 3. We plot the points (0, 0), $(-2, 0)$, and (3, 0).

d) The zeros and vertical asymptotes divide the x-axis into intervals as follows. To find where the function is positive or negative, we try a test point in each interval.

Interval	$(-\infty, -2)$	$(-2, 0)$	$(0, 1)$	$(1, 2)$	$(2, 3)$	$(3, \infty)$
Test Value	$f(-3) = -0.9$	$f(-1) \approx 0.7$	$f(0.5) \approx -4.2$	$f(1.5) = 31.5$	$f(2.5) = -7.5$	$f(4) = 4$
Sign of $f(x)$	Negative	Positive	Negative	Positive	Negative	Positive
Location of Points on Graph	Below x-axis	Above x-axis	Below x-axis	Above x-axis	Below x-axis	Above x-axis

e) We next make a table of function values in addition to those found as test values and sketch the graph.

x	$f(x)$, approx.
-10	-7.9
-8	-5.9
-6	-3.9
-5	-2.9
5	5.8
6	7.2
8	9.5
10	11.7

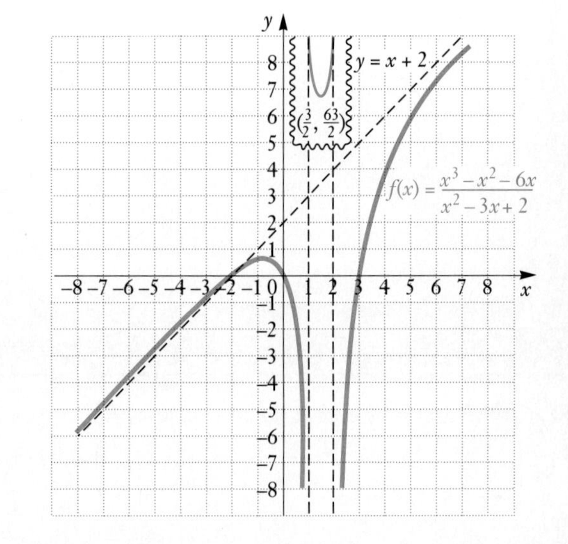

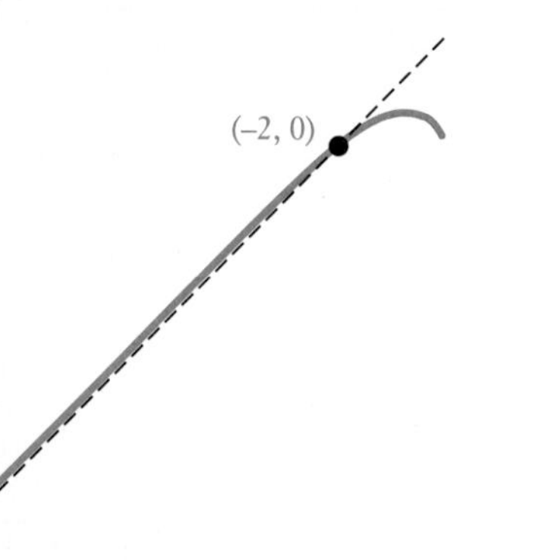

The lower left part of the graph is as shown in the margin. The curve crosses the oblique asymptote at $(-2, 0)$ and then, moving to the left, comes back close to the asymptote from above. Are there other points of intersection of the graph with the line $y = x + 2$? Setting $f(x) = x + 2$ and solving will show that the only solution is -2, so there are no other points of intersection. ■

It is sometimes written that graphs never cross their asymptotes. The curves in Examples 12 and 13 show that this is not always the case for horizontal or oblique asymptotes. But it is true that graphs of rational functions never cross *vertical* asymptotes.

EXAMPLE 13 Graph $f(x) = \dfrac{x^2 - 1}{x^2 + x - 6}$.

Solution Following the outline of steps, we obtain the following graph. Note that the graph crosses its horizontal asymptote and then approaches it from below as x gets very large. To determine where the horizontal asymptote $y = 1$ was crossed, we solved the equation $\dfrac{x^2 - 1}{x^2 + x - 6} = 1$.

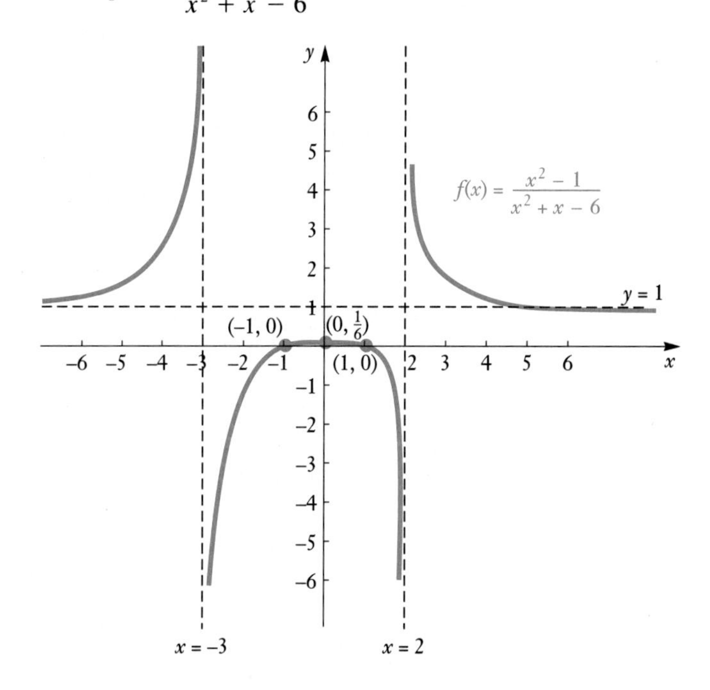

■

EXERCISE SET 8.6

Sketch the graph of each rational function. Be sure to list all the asymptotes.

1. $f(x) = \dfrac{1}{x - 3}$ **2.** $f(x) = \dfrac{1}{x - 5}$ **3.** $f(x) = \dfrac{-2}{x - 5}$ **4.** $f(x) = \dfrac{-3}{x - 3}$

5. $f(x) = \frac{2x + 1}{x}$

6. $f(x) = \frac{3x - 1}{x}$

7. $f(x) = \frac{1}{(x - 2)^2}$

8. $f(x) = \frac{-2}{(x - 3)^2}$

9. $f(x) = \frac{2}{x^2}$

10. $f(x) = \frac{1}{3x^2}$

11. $f(x) = \frac{1}{x^2 + 3}$

12. $f(x) = \frac{-1}{x^2 + 2}$

13. $f(x) = \frac{x - 1}{x + 2}$

14. $f(x) = \frac{x - 2}{x + 1}$

15. $f(x) = \frac{3x}{x^2 + 5x + 4}$

16. $f(x) = \frac{x + 3}{2x^2 - 5x - 3}$

17. $f(x) = \frac{x^2 - 4}{x - 1}$

18. $f(x) = \frac{x^2 - 9}{x + 1}$

19. $f(x) = \frac{x^2 + x - 2}{2x^2 + 1}$

20. $f(x) = \frac{x^2 - 2x - 3}{3x^2 + 2}$

21. $f(x) = \frac{x - 1}{x^2 - 2x - 3}$

22. $f(x) = \frac{x + 2}{x^2 + 2x - 15}$

23. $f(x) = \frac{x + 2}{(x - 1)^3}$

24. $f(x) = \frac{x - 3}{(x + 1)^3}$

25. $f(x) = \frac{(x - 3)^2}{x}$

26. $f(x) = \frac{(x + 2)^2}{x}$

27. $f(x) = \frac{x^3 + 2x^2 - 15x}{x^2 - 5x - 14}$

28. $f(x) = \frac{x^3 + 2x^2 - 3x}{x^2 - 25}$

29. $f(x) = \frac{5x^4}{x^4 + 1}$

30. $f(x) = \frac{x + 1}{x^2 + x - 6}$

31. $f(x) = \frac{x^2 - x - 2}{x + 2}$

32. $f(x) = \frac{x^2}{x^2 - x - 2}$

33. *Time is a function of speed.* The distance from Kansas City to Indianapolis is 500 miles. A car completes the trip in time t, in hours, at various speeds r, in miles per hour.

a) Express the time t required for the trip as a function of the speed r.

b) Sketch a graph of the function over the interval $(0, \infty)$.

34. *Electrical resistance.* The total resistance in a circuit with two resistors in parallel is given by

$$R = \frac{R_1 R_2}{R_1 + R_2},$$

where R_1 and R_2 are positive.

a) A circuit has a 10-ohm resistor and a variable resistor in parallel. Express the total resistance of the circuit as a function of the resistance of the variable resistor, r.

b) Graph the function.

c) What is the domain and range?

d) What number does the resistance approach as the resistance of the variable resistor gets larger and larger?

SKILL MAINTENANCE

35. Simplify: i^{75}.

36. Divide: $\frac{6x^2 - 2x}{2x^2 + 11x - 6} \div \frac{9x^2 - 1}{x^2 + 5x - 6}$.

37. If $f(x) = 2x + 3$, find $f(a + 5)$.

38. Simplify: $\dfrac{\frac{5}{x} - \frac{3}{y}}{\frac{3}{y} - \frac{2}{x}}$.

SYNTHESIS

Sketch a graph of each of the following functions.

39. $f(x) = \dfrac{x^3 + 4x^2 + x - 6}{x^2 - x - 2}$

40. $f(x) = \dfrac{2x^3 + x^2 - 8x - 4}{x^3 + x^2 - 9x - 9}$

Computer–calculator exercises. Sketch a graph of each of the following using a computer software graphing package or a graphing calculator.

41. $f(x) = \dfrac{x^4 + 3x^3 + 21x^2 - 50x + 80}{x^4 + 8x^3 - x^2 + 20x - 10}$

42. $f(x) = \dfrac{1}{x^{3.8} - 11} + 2$

43. $f(x) = \dfrac{(x^2 - 5)^4}{(x^3 + 7)^7}$

44. $f(x) = \dfrac{x^3 + 4}{x}$

45. The graph of a rational function follows. Find an equation for the function.

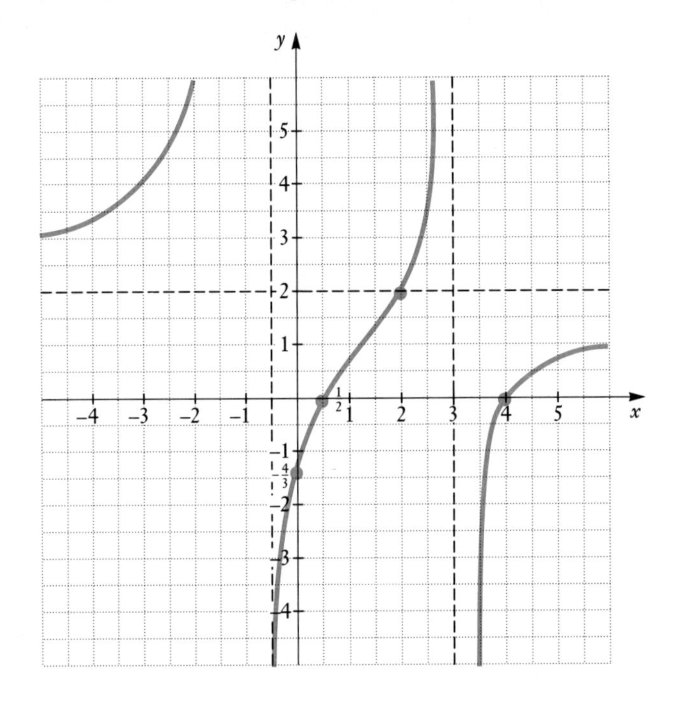

8.7 Polynomial and Rational Inequalities

Quadratic and Other Polynomial Inequalities

Inequalities such as the following are called *polynomial inequalities*:

$$x^3 - 4x^2 + x - 6 > 0, \quad x^2 + 3x - 10 < 0, \quad 5x^2 - 3x + 2 \geq 0, \quad 3x - 1 \leq 0.$$

Here we concentrate on inequalities of degree 2 or more. We shall consider three ways to solve such inequalities. The first two provide understanding, and the last yields the fastest method.

The first method for solving polynomial inequalities is to consider the graph of the related function in the plane.

EXAMPLE 1 Solve: $x^2 + 3x - 10 > 0$.

Solution Consider the function $f(x) = x^2 + 3x - 10$ and its graph. Its graph opens upward since the leading coefficient is positive.

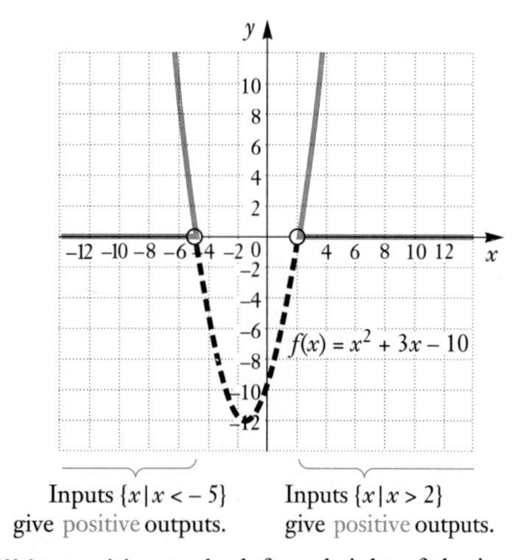

Function values will be positive to the left and right of the intercepts as shown. We find the intercepts by setting the polynomial equal to 0 and solving:

$$x^2 + 3x - 10 = 0$$

$$(x + 5)(x - 2) = 0$$

$$x + 5 = 0 \quad \text{or} \quad x - 2 = 0$$

$$x = -5 \quad \text{or} \quad x = 2.$$

Then the solution set of the inequality is

$$\{x \mid x < -5 \text{ or } x > 2\}, \quad \text{or} \quad (-\infty, -5) \cup (2, \infty).$$

We can solve any inequality by considering a graph of the related function and finding intercepts as in Example 1. In some cases we may need to use the quadratic formula to find the intercepts.

EXAMPLE 2 Solve $x^2 - 2x \leq 2$.

Solution We first find standard form with 0 on one side:

$$x^2 - 2x - 2 \leq 0.$$

Consider $f(x) = x^2 - 2x - 2$. Its graph opens upward. Function values will be nonpositive between and including its intercepts as shown. We find the intercepts by solving $f(x) = 0$. This time we will need the quadratic formula. The intercepts are $1 + \sqrt{3}$ and $1 - \sqrt{3}$.

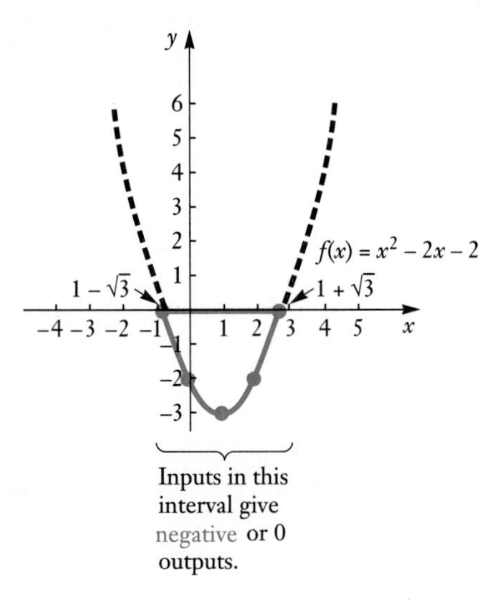

The solution set of the inequality is

$$\{x \mid 1 - \sqrt{3} \leq x \leq 1 + \sqrt{3}\}, \qquad \text{or} \qquad [1 - \sqrt{3}, 1 + \sqrt{3}]$$ ■

It should be pointed out that we need not actually draw graphs as in the preceding examples. Merely visualizing the graph will usually suffice.

Let us now consider another method of solving inequalities. This method works for any polynomial that we can factor into a product of first-degree polynomials.

EXAMPLE 3 Solve $x^2 + 2x - 3 > 0$.

Solution We factor the inequality, obtaining $(x + 3)(x - 1) > 0$. The solutions of $(x + 3)(x - 1) = 0$ are -3 and 1. They are not solutions of the inequality, but they divide the real number line in a natural way, as pictured here. The product $(x + 3)(x - 1)$ is positive or negative, for values other than -3 and 1, depending on the signs of the factors $x + 3$ and $x - 1$. We can determine this efficiently with a diagram as follows.

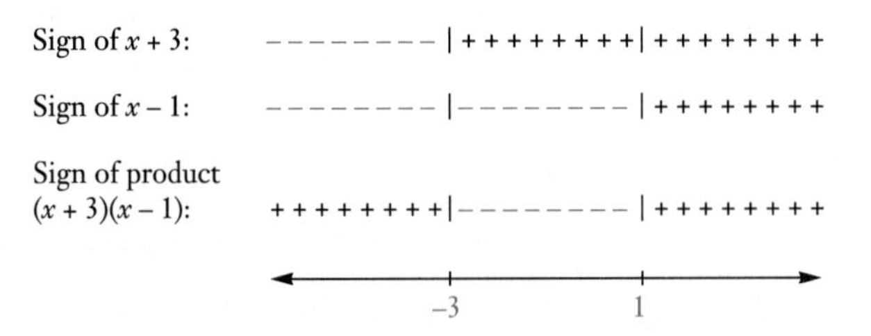

To set up the diagram, solve $x + 3 > 0$. We get $x > -3$. Thus $x + 3$ is positive for all numbers to the right of -3. We indicate this with $+$ signs. Accordingly, $x + 3$ is negative for all numbers to the left of -3. We indicate that with the $-$ signs.

Similarly, we solve $x - 1 > 0$ and get $x > 1$. Thus $x - 1$ is positive for all numbers to the right of 1 and negative for all numbers to the left of 1. We indicate this with the + and – signs.

For the product $(x + 3)(x - 1)$ to be positive, both factors must be positive or both must be negative. In the diagram, we see that this situation occurs when $x < -3$ and when $x > 1$. The solution set of the inequality is

$$\{x \mid x < -3 \quad or \quad x > 1\}, \qquad \text{or} \qquad (-\infty, -3) \cup (1, \infty).$$

EXAMPLE 4 Solve $4x(x + 1)(x - 1) < 0$.

Solution The solutions of $4x(x + 1)(x - 1) = 0$ are -1, 0, and 1. The product $4x(x + 1)(x - 1)$ is positive or negative, depending on the signs of the factors $4x$, $x + 1$, and $x - 1$. We determine this efficiently using a diagram.

Sign of $4x$:	− − − − − −	− − − − − −	+ + + + + +	+ + + + + +
Sign of $x + 1$:	− − − − − −	+ + + + + +	+ + + + + +	+ + + + + +
Sign of $x - 1$:	− − − − − −	− − − − − −	− − − − − −	+ + + + + +
Sign of product $4x(x + 1)(x - 1)$:	− − − − − −	+ + + + + +	− − − − − −	+ + + + + +
		−1	0	1

The product of three numbers is negative when it has an odd number of negative factors. We see from the diagram that the solution set is

$$\{x \mid x < -1 \quad or \quad 0 < x < 1\}, \qquad \text{or} \qquad (-\infty, -1) \cup (0, 1).$$

We now consider our final method for solving polynomial inequalities. The preceding method provides the understanding for this method. In Example 4 we see that the intercepts divide the number line into intervals. If a particular function has a positive output for one number in an interval, it will be positive for *all* the numbers in the interval. Thus we can merely make a test substitution in each interval to solve the inequality.

EXAMPLE 5 Solve: $x^2 + 3x - 10 < 0$.

Solution We set the polynomial equal to 0 and solve. The solutions of $x^2 + 3x - 10 = 0$ or $(x + 5)(x - 2) = 0$ are -5 and 2. We locate them on a number line as follows. Note that the numbers divide the number line into three intervals: A, B, and C.

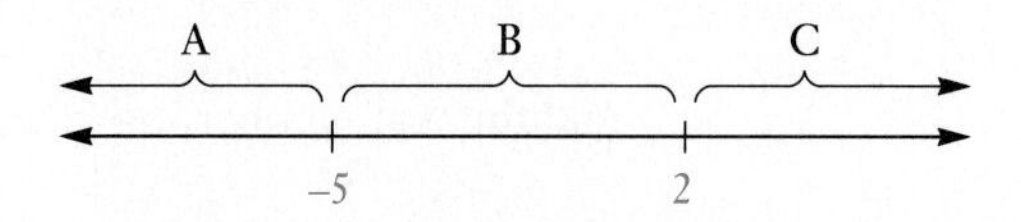

We pick a test number in interval A, say -7, and substitute -7 for x in the function $f(x) = x^2 + 3x - 10$:

$$f(-7) = (-7)^2 + 3(-7) - 10$$
$$= 49 - 21 - 10 = 18. \qquad f(-7) \text{ is positive.}$$

Since $f(-7) > 0$, the function will be positive for any number in interval A. Next we try a test number in interval B, say 1, and find $f(1)$:

$$f(1) = 1^2 + 3(1) - 10 = -6. \qquad f(1) \text{ is negative.}$$

Since $f(1) < 0$, the function will be negative for any number in interval B. Next we try a test number in interval C, say 4, and find $f(4)$:

$$f(4) = (4)^2 + 3(4) - 10$$
$$= 16 + 12 - 10 = 18. \qquad f(4) \text{ is positive.}$$

Since $f(4) > 0$, the function will be positive for any number in interval C. We are looking for numbers x for which $x^2 + 3x - 10 < 0$. Thus any number x in interval B is a solution. If the inequality would have been $\leq$ or $\geq$, we would also include the intercepts -5 and 2. The solution set is $\{x \mid -5 < x < 2\}$. ■

Let us review the last method.

To solve a polynomial inequality:

1. Get 0 on one side, set the polynomial on the other side equal to 0, and solve to find the intercepts.
2. Use the numbers found in step (1) to divide the number line into intervals.
3. Substitute a number from each interval into the related function. If the value is positive, then the function will be positive for all numbers in the interval. If the value is negative, then the function will be negative for all numbers in the interval.
4. Select the intervals for which the inequality is satisfied and write set-builder notation or interval notation for the solution set. Include the intercepts in the solution set if the inequality sign is $\leq$ or $\geq$.

EXAMPLE 6 Solve $x^3 + 2x^2 - 5x - 6 \geq 0$.

Solution Using the rational roots theorem, we find that -3, -1, and 2 are roots of the polynomial. Thus the above inequality is equivalent to

$$f(x) \geq 0,$$

where

$$f(x) = (x + 3)(x + 1)(x - 2),$$

and the real number line is divided into four intervals as follows.

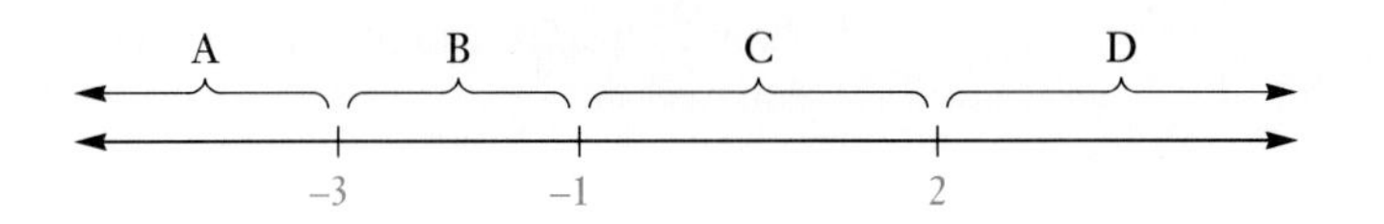

We try a test number in each interval and use the more convenient factored form to determine where positive function values occur:

A: Test -4, $f(-4) = (-4 + 3)(-4 + 1)(-4 - 2)$
$= \text{Neg.} \cdot \text{Neg.} \cdot \text{Neg.} = \text{Neg.}$

B: Test -2, $f(-2) = (-2 + 3)(-2 + 1)(-2 - 2)$
$= \text{Pos.} \cdot \text{Neg.} \cdot \text{Neg.} = \text{Pos.}$

C: Test 1, $f(1) = (1 + 3)(1 + 1)(1 - 2) = \text{Pos.} \cdot \text{Pos.} \cdot \text{Neg.} = \text{Neg.}$

D: Test 3, $f(3) = (3 + 3)(3 + 1)(3 - 2) = \text{Pos.} \cdot \text{Pos.} \cdot \text{Pos.} = \text{Pos.}$

Function values are positive in intervals B and D. Since the inequality symbol is $\geq$, we need to include the intercepts. The solution set is

$$\{x \mid -3 \leq x \leq -1 \quad or \quad 2 \leq x\}, \qquad \text{or} \qquad [-3, -1] \cup [2, \infty).$$

■

Rational Inequalities

We adapt the preceding method when an inequality involves rational expressions that are quotients of polynomials. We call these *rational inequalities.*

EXAMPLE 7 Solve: $\dfrac{x - 3}{x + 4} \geq 2.$

Solution We write the related equation by changing the $\geq$ symbol to an $=$ sign:

$$\frac{x - 3}{x + 4} = 2.$$

Then we solve this related equation. We multiply on both sides of the equation by the LCM, $x + 4$:

$$\begin{aligned}(x + 4) \cdot \frac{x - 3}{x + 4} &= (x + 4) \cdot 2 \\ x - 3 &= 2x + 8 \\ -11 &= x.\end{aligned}$$

In the case of rational inequalities, we also need to determine which replacements are not meaningful. These are those replacements that make the denominator 0. We set the denominator equal to 0 and solve:

$$\begin{aligned}x + 4 &= 0 \\ x &= -4.\end{aligned}$$

Now we use the numbers -11 and -4 to divide the number line into intervals as follows:

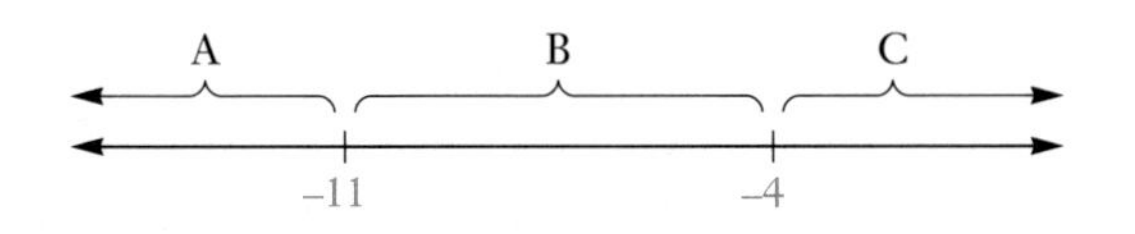

We try test numbers in each interval. We see if each satisfies the original inequality.

A: Test -15, $\quad \dfrac{-15-3}{-15+4} = \dfrac{-18}{-11} = \dfrac{18}{11} \not\geq 2$ $\quad$ **-15 *is not* a solution, so interval A is not part of the solution set.**

B: Test -8, $\quad \dfrac{-8-3}{-8+4} = \dfrac{-11}{-4} = \dfrac{11}{4} \geq 2$ $\quad$ **-8 *is* a solution, so interval B is part of the solution set.**

C: Test 1, $\quad \dfrac{1-3}{1+4} = \dfrac{-2}{5} = -\dfrac{2}{5} \not\geq 2$ $\quad$ **1 *is not* a solution, so interval C is not part of the solution set.**

The solution set includes the interval B. The number -11 is also included since the inequality symbol is $\geq$ and -11 is a solution of the related equation. The number -4 is not included since it is not a meaningful replacement. Thus the solution set of the original inequality is

$$\{x \mid -11 \leq x < -4\}, \quad \text{or} \quad [-11, -4).$$

There is an interesting visual interpretation of Example 7. If we graph the function $f(x) = \dfrac{x-3}{x+4}$, we see that the solutions of the inequality $\dfrac{x-3}{x+4} \geq 2$ may be found by inspection. We simply sketch the line $y = 2$ and locate all x-values for which $f(x) \geq 2$.

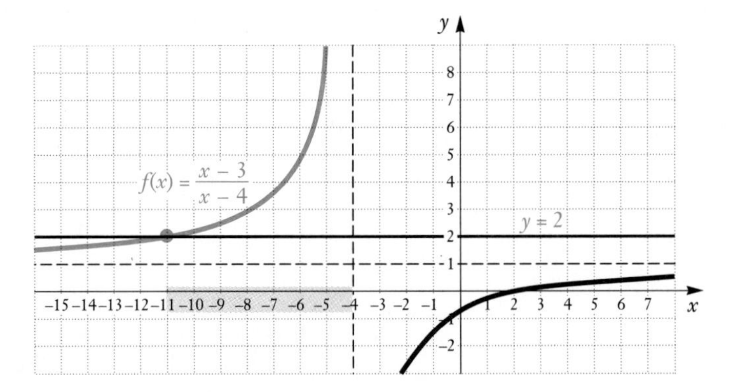

Because graphing rational functions can be very time-consuming, we normally just use test values.

To solve a rational inequality:

1. Change the inequality symbol to an equals sign and solve the related equation.
2. Find the replacements that are not meaningful.
3. Use the numbers found in steps (1) and (2) to divide the number line into intervals.
4. Substitute a number from each interval into the inequality. If the number is a solution, then the interval to which it belongs is part of the solution set.
5. Select the intervals for which the inequality is satisfied and write set-builder notation or interval notation for the solution set. If the inequality symbol is $\leq$ or $\geq$, then the solutions to step (1) should also be included in the solution set.

EXERCISE SET 8.7

Solve.

1. $(x+5)(x-3)>0$
2. $(x+4)(x-1)>0$
3. $(x-1)(x+2)\leq 0$
4. $(x+5)(x-3)\leq 0$
5. $x^2-x-2<0$
6. $x^2-3x-10<0$
7. $x^2\geq 1$
8. $x^2<25$
9. $9-x^2\leq 0$
10. $4-x^2\geq 0$
11. $x^2-2x+1\geq 0$
12. $x^2+6x+9<0$
13. $x^2+8<6x$
14. $x^2-12>4x$
15. $4x^2+7x<15$
16. $4x^2+7x\geq 15$
17. $2x^2+x>5$
18. $2x^2+x\leq 2$
19. $3x(x+2)(x-2)<0$
20. $5x(x+1)(x-1)>0$
21. $x^3-4x^2-7x+10\geq 0$
22. $x^3-3x^2-6x+8<0$
23. $x^3+4x^2+x-6<0$
24. $x^3-4x^2+x+6\leq 0$

Solve.

25. $\frac{1}{4-x}<0$
26. $\frac{-4}{2x+5}>0$
27. $3<\frac{1}{x}$
28. $\frac{1}{x}\leq 5$
29. $\frac{3x+2}{x-3}>0$
30. $\frac{5-2x}{4x+3}<0$
31. $\frac{x+2}{x}\leq 0$
32. $\frac{x}{x-3}\geq 0$
33. $\frac{x+1}{2x-3}\geq 1$
34. $\frac{x-1}{x-2}\geq 3$
35. $\frac{x+1}{x+2}\leq 3$
36. $\frac{x+1}{2x-3}\leq 1$
37. $\frac{x-6}{x}>1$
38. $\frac{x}{x+3}>-1$
39. $(x+1)(x-2)>(x+3)^2$
40. $(x-4)(x+3)>(x-1)^2$
41. $x^3-x^2>0$
42. $x^3-4x>0$
43. $x+\frac{4}{x}>4$
44. $\frac{1}{x^2}\leq\frac{1}{x^3}$
45. $\frac{1}{x^3}\leq\frac{1}{x^2}$
46. $x+\frac{1}{x}>2$
47. $\frac{2+x-x^2}{x^2+5x+6}<0$
48. $\frac{4}{x^2}-1>0$

SKILL MAINTENANCE

49. Solve: $|-3x+5|=7$.

50. Write a quadratic equation whose solutions are 3 and -5.

51. If $f(x) = \dfrac{2}{x+5}$ and $g(x) = 4 - x^2$, determine the domain of f/g.

52. Use the properties of exponents to simplify:

$$a^{2/5} \cdot a^{5/3}.$$

SYNTHESIS

Solve.

53. $x^4 - 2x^2 \leq 0$

54. $x^4 - 3x^2 > 0$

55. $\left|\dfrac{x+3}{x-4}\right| < 2$

56. $|x^2 - 5| = 5 - x^2$

57. $(7 - x)^{-2} < 0$

58. $(1 - x)^3 > 0$

59. $\left|1 + \dfrac{1}{x}\right| < 3$

60. $(x + 5)^{-2} > 0$

61. The base of a triangle is 4 cm greater than the height. Find the possible heights h such that the area of the triangle will be greater than 10 cm^2.

62. The length of a rectangle is 3 m greater than the width. Find the possible widths w such that the area of the rectangle will be greater than 15 m^2.

63. *Total profit.* A company determines that its total profit function is given by

$$P(x) = -3x^2 + 630x - 6000.$$

a) A company makes a profit for those nonnegative values of x for which $P(x) > 0$. Find the values of x for which the company makes a profit.

b) A company loses money for those nonnegative values of x for which $P(x) < 0$. Find the values of x for which the company loses money.

64. *Height of a thrown object.* The function

$$s(t) = -16t^2 + 32t + 1920$$

gives the height s of an object thrown from a cliff 1920 ft high, after time t, in seconds.

a) For what times is the height greater than 1920 ft?

b) For what times is the height less than 640 ft?

65. *Number of handshakes.* There are n people in a room. The number N of possible handshakes by the people is given by the function

$$N(n) = \frac{n(n-1)}{2}.$$

For what number of people n is $78 \leq N \leq 1225$?

66. *Number of diagonals.* A polygon with n sides has D diagonals, where D is given by the function

$$D(n) = \frac{n(n-3)}{2}.$$

For what number of sides n is $35 \leq D \leq 740$?

67. A company has the following total cost and total revenue functions to use in producing and selling x units of a certain product:

$$R(x) = 50x - x^2, \qquad C(x) = 5x + 350.$$

a) Find the break-even values.

b) Find the values of x that produce a profit.

c) Find the values of x that result in a loss.

68. A company has the following total cost and total revenue functions to use in producing and selling x units of a certain product:

$$R(x) = 80x - x^2, \qquad C(x) = 10x + 600.$$

a) Find the break-even values.

b) Find the values of x that produce a profit.

c) Find the values of x that result in a loss.

CHAPTER SUMMARY AND REVIEW 8

TERMS TO KNOW

Polynomial function
Synthetic division
Even function
Odd function
Symmetric with respect to an axis
Symmetric with respect to the origin
Root
Zero
Upper bound on roots
Lower bound on roots
Repeated root
Double root
Root of multiplicity two
Rational function
Asymptote
Horizontal asymptote
Vertical asymptote
Oblique asymptote
Polynomial inequality
Rational inequality

IMPORTANT PROPERTIES AND FORMULAS

THE REMAINDER THEOREM

The remainder obtained by dividing $P(x)$ by $x - r$ is $P(r)$.

THE FACTOR THEOREM

For any polynomial $P(x)$, if $P(r) = 0$, then $x - r$ is a factor of $P(x)$; and if $x - r$ is a factor of $P(x)$, then $P(r) = 0$.

THE RATIONAL ROOTS THEOREM

Let $P(x) = a_nx^n + a_{n-1}x^{n-1} + \cdots + a_1x + a_0$, where all the coefficients are integers and n is a positive integer. Consider a rational number c/d that is written in reduced form. If c/d is a root of $P(x)$, then c is a factor of a_0 and d is a factor of a_n.

THE UPPER BOUND THEOREM

If when a polynomial is divided by $x - a$, where a is positive, the remainder and coefficients of the quotient are either all nonnegative or all nonpositive, then a is an upper bound to the real roots of the polynomial. (No root is greater than a.)

THE LOWER BOUND THEOREM

If a polynomial is divided by $x - a$, where a is negative, and the remainder and all coefficients of the quotient alternate sign (with 0 considered a sign change), then a is a lower bound to the real roots of the polynomial. (No root is less than a.)

Any polynomial of degree n has n roots, where roots of multiplicity m are counted m times.

The domain of a polynomial function is the set of real numbers.

The graph of a polynomial function is a continuous curve.

To graph polynomial functions:

1. Look at the degree of the polynomial and its leading coefficient.
2. Look for symmetries.
3. Make a table of values using synthetic division.
4. Find the y-intercept and as many x-intercepts as possible.
5. Plot the points and connect them appropriately.

Asymptotes of a rational function occur as follows.

1. Vertical asymptotes occur at any x-values that make the denominator zero but the numerator nonzero.
2. The x-axis is an asymptote when the degree of the denominator is greater than that of the numerator.
3. Horizontal asymptotes other than the x-axis occur when the numerator and the denominator have the same degree.
4. Oblique asymptotes occur when the degree of the numerator is greater than that of the denominator by 1.

To graph rational functions:

a) Determine any symmetries.
b) Determine any horizontal or oblique asymptotes and sketch them.
c) Factor the denominator and the numerator.
  i) Determine any vertical asymptotes and sketch them.
  ii) Determine the zeros, if possible, and plot them.
d) Locate intervals where function values are positive and where they are negative.
e) Make a table of values and plot them.
f) Sketch the curve.

To solve a polynomial inequality:

1. Get 0 on one side, set the polynomial on the other side equal to 0, and solve to find the intercepts.
2. Use the numbers found in step (1) to divide the number line into intervals.
3. Substitute a number from each interval into the related function. If the value is positive, then the function will be positive for all numbers in the interval. If the value is negative, then the function will be negative for all numbers in the interval.
4. Select the intervals for which the inequality is satisfied, and write set-builder notation or interval notation for the solution set. Include the intercepts in the solution set if the inequality sign is $\leq$ or $\geq$.

To solve a rational inequality:

1. Change the inequality symbol to an equals sign and solve the related equation.
2. Find the replacements that are not meaningful.
3. Use the numbers found in steps (1) and (2) to divide the number line into intervals.
4. Substitute a number from each interval into the inequality. If the number is a solution, then the interval to which it belongs is part of the solution set.
5. Select the intervals for which the inequality is satisfied and write set-builder notation or interval notation for the solution set. If the inequality symbol is $\leq$ or $\geq$, then the solutions to step (1) should also be included in the solution set.

REVIEW EXERCISES

The review sections to be tested in addition to the material in this chapter are Sections 5.6, 6.2, 6.6, and 7.3.

Use synthetic division to find the quotient and the remainder.

1. $(x^3 + 3x^2 + 2x - 6) \div (x - 3)$

2. $(4x^3 + 6x^2 - 5) \div (x + 3)$

Determine whether the function is even, odd, both even and odd, or neither even nor odd.

3. $f(x) = 3x^2 - 2$

4. $f(x) = x + 3$

5. $f(x) = 3x^3$

6. $f(x) = x^5 - x^3$

Determine whether the graph is symmetric with respect to the x-axis, the y-axis, or the origin. If more than one type of symmetry exists or if none exists, state this.

7.

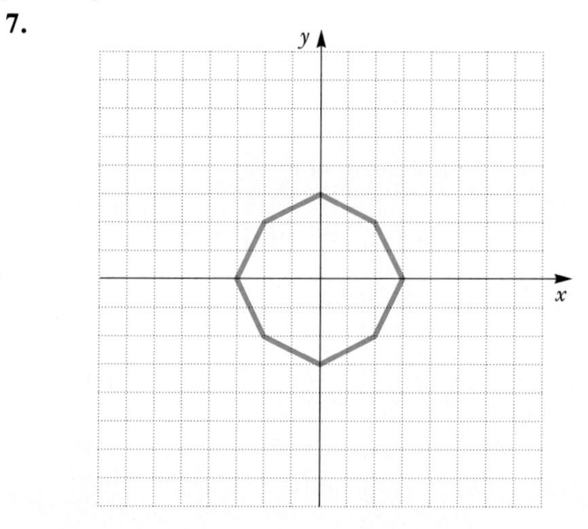

8.

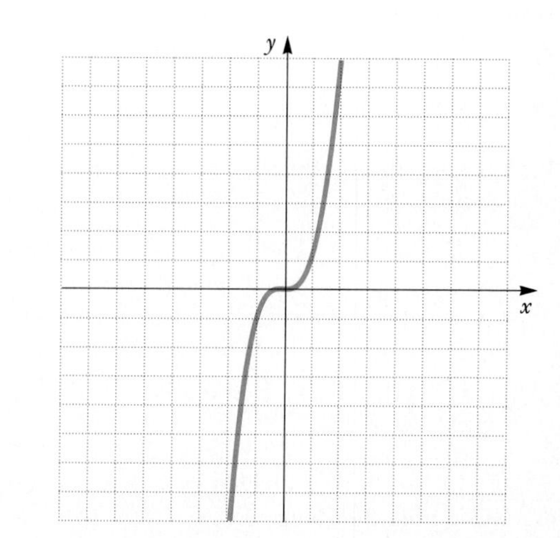

9. Use synthetic division to find $P(3)$:

$$P(x) = 2x^4 - 3x^3 + x^2 - 3x + 7.$$

10. Use synthetic division to determine whether -1 is a root of the polynomial $P(x) = 4x^4 + 4x^3 + 2x^2 + 7x + 5$.

11. Factor the polynomial $P(x)$. Then solve the equation $P(x) = 0$.

$$P(x) = x^3 + 7x^2 + 7x - 15$$

12. Find the smallest positive integer that is guaranteed by the upper bound theorem to be an upper bound to the roots of

$$2x^4 - 7x^2 + 2x - 1.$$

13. Find the largest negative integer that is guaranteed by the lower bound theorem to be a lower bound to the roots of

$$12x^3 + 24x^2 + 10x - 6.$$

14. Let $P(x) = x^4 - x^3 - 3x^2 - 9x - 108$. Find the rational roots of $P(x)$. If possible, find the other roots.

15. List all possible rational roots of $2x^6 - 12x^4 + 17x^2 + 12$.

16. Find the rational roots of $x^3 + x^2 - 3x - 6 = 0$.

Sketch the graph of the polynomial function.

17. $f(x) = x^3 + 3x^2 - 2x - 6$

18. $f(x) = x^3 + x^2 - 3x - 2$

Sketch the graph of the rational function.

19. $f(x) = \dfrac{x^2 + x - 6}{x^2 - x - 20}$

20. $f(x) = \dfrac{2x}{x^2 + 3x + 2}$

Solve.

21. $2x^2 - 3x - 2 > 0$

22. $(1 - x)(x + 4)(x - 2) \leq 0$

23. $\dfrac{x - 2}{x + 3} < 4$

24. $\dfrac{2}{x + 1} \geq x$

SKILL MAINTENANCE

25. Determine the nature of the solutions of

$$x^2 - 10x + 20.$$

26. Multiply: $(-1 + 3i)(-2 + 5i)$.

27. Find $f(a + 1) - f(a)$ if $f(x) = -2x + 5$.

28. Determine the domain of f/g if

$$f(x) = \frac{1}{x} \text{ and } g(x) = \frac{x - 1}{x + 1}.$$

SYNTHESIS

29. Find k such that $x + 3$ is a factor of $x^3 + kx^2 + kx - 15$.

30. The equation $x^2 - 8x + c = 0$ has a double root. Find it.

31. When $x^2 - 4x + 3k$ is divided by $x + 5$, the remainder is 33. Find the value of k.

32. Graph: $y = 1 - \dfrac{1}{x^2 + 4}$.

THINKING IT THROUGH

1. Explain how the fact that a function is even or odd can be used in graphing the function.

2. Explain how the knowledge of the degree of a polynomial function can be used in graphing the function.

3. Explain why the graph of a rational function may cross a horizontal or oblique asymptote, but will never cross a vertical asymptote.

4. Explain why, in solving polynomial or rational inequalities, a test substitution will suffice to determine the sign of the function for each interval.

CHAPTER TEST 8

Use synthetic division to find the quotient and the remainder.

1. $(x^3 + 5x^2 + 4x - 7) \div (x - 4)$

2. $(2x^3 + 3x^2 - 14x - 13) \div (x + 3)$

Determine whether the function is even, odd, or neither. Then list any symmetries that exist.

3. 4. 5.

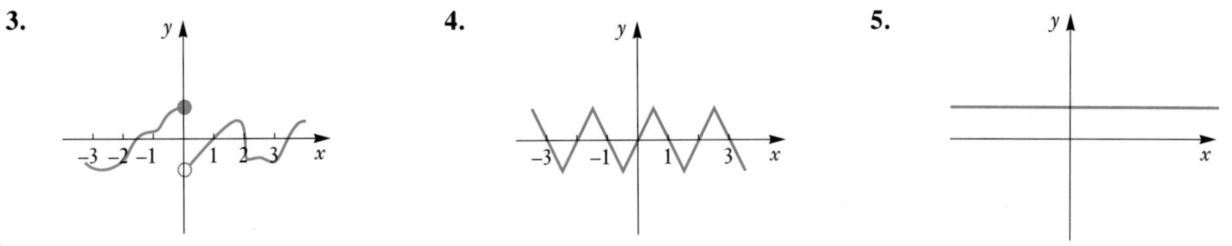

6. Use synthetic division to find $P(-5)$:

$$P(x) = 2x^4 - 7x^3 + 8x^2 - 10.$$

7. Factor the polynomial $P(x)$. Then solve the equation $P(x) = 0$.

$$P(x) = x^3 - 4x^2 + x + 6$$

8. List all possible rational roots of

$$3x^5 - 4x^3 - 2x + 6.$$

9. Find the smallest positive integer that is guaranteed by the upper bound theorem to be an upper bound to the real roots of $3x^4 - 2x^3 - 15x - 10$.

10. Let $P(x) = 2x^4 + 9x^3 + 5x^2 - 14x - 12$. Find the rational roots, if they exist, of $P(x)$. If possible, find the other roots. Then write the polynomial in factored form.

Sketch the graph of the polynomial function.

11. $f(x) = x^4 - 5x^2$

12. $f(x) = x^4 - 3x^3 + 2x^2$

13. Sketch the graph: $f(x) = \dfrac{x - 2}{x^2 - 2x - 15}$.

Solve.

14. $x^2 - 8x + 12 \geq 0$

15. $8x^2 + 10x - 3 < 0$

16. $\dfrac{x - 4}{2x + 3} \leq 1$

SKILL MAINTENANCE

17. Find $f(x + 2)$ if $f(x) = \dfrac{x - 1}{x - 2}$.

18. Find $(f + g)(-3)$ if $f(x) = x^2 + 5$ and $g(x) = 12 - 5x$.

19. Write a quadratic equation having the solutions -1 and $\frac{1}{3}$.

20. Simplify: $3i^{38} - 2i^{25}$.

SYNTHESIS

21. Solve: $\left|2 + \frac{1}{x}\right| < 6$.

22. Graph: $f(x) = \left|2 - \frac{1}{3x}\right|$.

23. Solve: $x^4 - 2x^3 + 3x^2 - 2x + 2 = 0$.

CHAPTER 9

Exponential and Logarithmic Functions

FEATURE PROBLEM

The population of the world passed 5.0 billion in 1987. The exponential growth rate was 2.8% per year. Find an exponential growth function for world population growth.

THE MATHEMATICS

The exponential growth function that fits the data is

$$P(t) = 5e^{0.028t},$$

where P is the population, in billions, t years after 1987.

The functions that we consider in this chapter are interesting not only from a purely intellectual point of view, but also for their rich applications to many fields. We will look at such applications as compound interest and population growth, to name just two.

The basis of the theory concerns exponents. We define some functions having variable exponents (*exponential functions*); the rest follows from those functions and their properties.

The review sections to be tested in addition to the material in this chapter are Sections 4.3, 5.5, 6.7, and 7.6.

9.1 Exponential Functions

The following graph shows the price of a first-class postage stamp and the year in which that price was instituted.

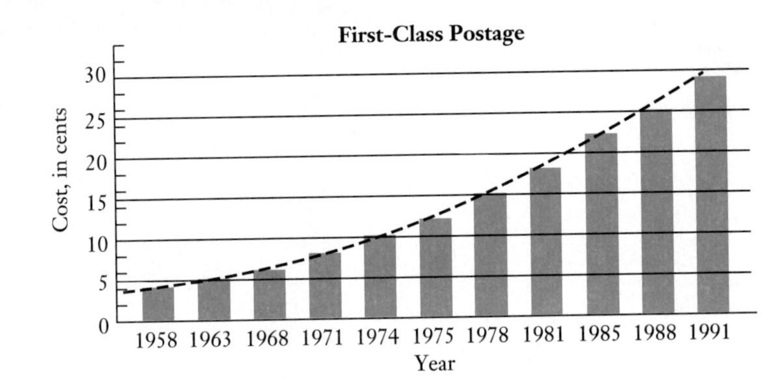

A curve drawn along the graph would approximate the graph of an *exponential function*. We now consider such graphs. We will also study graphs and properties of *logarithmic functions,* which are closely related to the exponential functions. You will see that these new functions appear in a variety of applications.

Graphing Exponential Functions

In Chapter 5 we gave meaning to exponential expressions with rational number exponents, such as

$$5^{1/4}, \quad 3^{-3/4}, \quad 7^{2.34}, \quad 8^{1.73}.$$

For example, $5^{1.73}$, or $5^{173/100}$, represents the 100th root of 5 raised to the 173rd power. We now give meaning to expressions with irrational exponents, such as

$$5^{\sqrt{3}}, \quad 7^{\pi}, \quad 9^{-\sqrt{2}}.$$

Consider $5^{\sqrt{3}}$. Let us think of rational numbers r close to $\sqrt{3}$ and look at 5^r. As r gets closer to $\sqrt{3}$, 5^r gets closer to some real number.

r closes in on $\sqrt{3}$.	5^r closes in on some real number p.
r	5^r
$1 < \sqrt{3} < 2$	$5 = 5^1 < p < 5^2 = 25$
$1.7 < \sqrt{3} < 1.8$	$15.426 = 5^{1.7} < p < 5^{1.8} = 18.120$
$1.73 < \sqrt{3} < 1.74$	$16.189 = 5^{1.73} < p < 5^{1.74} = 16.452$
$1.732 < \sqrt{3} < 1.733$	$16.241 = 5^{1.732} < p < 5^{1.733} = 16.267$

As r closes in on $\sqrt{3}$, 5^r closes in on some real number p. We define $5^{\sqrt{3}}$ to be the number p. To seven decimal places,

$$5^{\sqrt{3}} \approx 16.2424508.$$

Any positive irrational exponent can be defined in a similar way. Negative irrational exponents are then defined in the same way as negative integer exponents. Thus the expression a^x has meaning for *any* real number x. The general laws of exponents still hold, but we will not prove that here. We now define exponential functions.

> The function $f(x) = a^x$, where a is a positive constant, $a \neq 1$, is called the *exponential function*, base a.

We require the base a to be positive to avoid the possibility of taking even roots of negative numbers, such as the square root of -1, which is not a real number. The restriction $a \neq 1$ is made to avoid consideration of the constant function $f(x) = 1^x$, or $f(x) = 1$.

The following are examples of exponential functions:

$$f(x) = 2^x, \qquad f(x) = \left(\tfrac{1}{2}\right)^x, \qquad f(x) = (0.4)^x.$$

Note that, in contrast to polynomial functions like $f(x) = x^2$ and $f(x) = x^3$, the variable in an exponential function is in the *exponent.* Let us consider graphs of exponential functions.

EXAMPLE 1 Graph the exponential function $y = 2^x$.

Solution We compute some function values, thinking of y as $f(x)$, and list the results in a table.

$$f(0) = 2^0 = 1$$
$$f(1) = 2^1 = 2$$
$$f(2) = 2^2 = 4$$
$$f(3) = 2^3 = 8$$
$$f(-1) = 2^{-1} = \frac{1}{2^1} = \frac{1}{2}$$
$$f(-2) = 2^{-2} = \frac{1}{2^2} = \frac{1}{4}$$
$$f(-3) = 2^{-3} = \frac{1}{2^3} = \frac{1}{8}$$

x	y, or $f(x)$
0	1
1	2
2	4
3	8
-1	$\frac{1}{2}$
-2	$\frac{1}{4}$
-3	$\frac{1}{8}$

Next, we plot these points and connect them with a smooth curve.

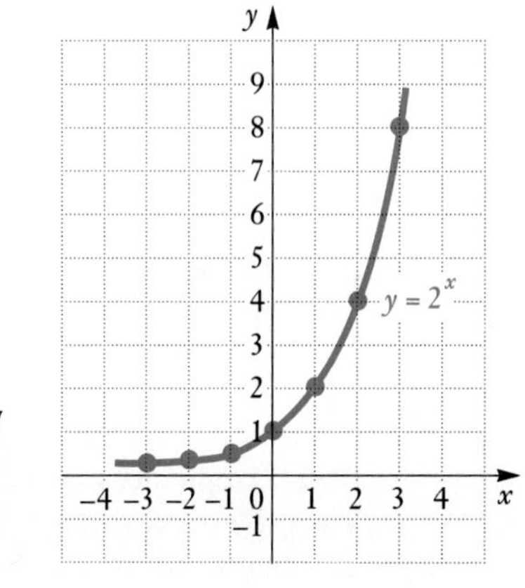

In graphing, be sure to plot enough points to determine how steeply the curve rises.

The curve comes very close to the x-axis but does not touch or cross it.

■

Note that as x increases, the function values increase indefinitely. As x decreases, the function values decrease, getting very close to 0 but never quite reaching it. Thus the x-axis is a horizontal asymptote.

EXAMPLE 2 Graph the exponential function $y = \left(\frac{1}{2}\right)^x$.

Solution We compute some function values, thinking of y as $f(x)$, and list the results in a table. Before we do this, note that

$$y = f(x) = \left(\tfrac{1}{2}\right)^x = (2^{-1})^x = 2^{-x}.$$

Then we have

$$f(0) = 2^{-0} = 1$$
$$f(1) = 2^{-1} = \frac{1}{2^1} = \frac{1}{2}$$
$$f(2) = 2^{-2} = \frac{1}{2^2} = \frac{1}{4}$$
$$f(3) = 2^{-3} = \frac{1}{2^3} = \frac{1}{8}$$
$$f(-1) = 2^{-(-1)} = 2^1 = 2$$
$$f(-2) = 2^{-(-2)} = 2^2 = 4$$
$$f(-3) = 2^{-(-3)} = 2^3 = 8.$$

x	y, or $f(x)$
0	1
1	$\frac{1}{2}$
2	$\frac{1}{4}$
3	$\frac{1}{8}$
−1	2
−2	4
−3	8

We plot these points and draw the curve. Note that this graph is a reflection, or mirror image, of the graph in Example 1 across the y-axis.

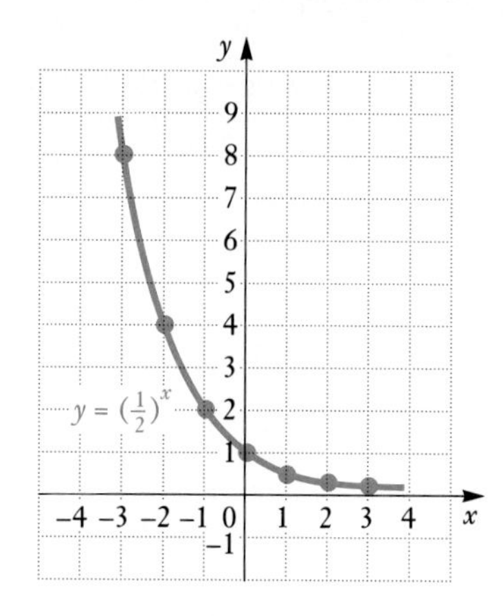

The preceding examples illustrate exponential functions with various bases. Let us list some of their characteristics.

A. When $a > 1$, the function $f(x) = a^x$ increases from left to right. The greater the value of a, the steeper the curve. (See the figure on the left, below.)

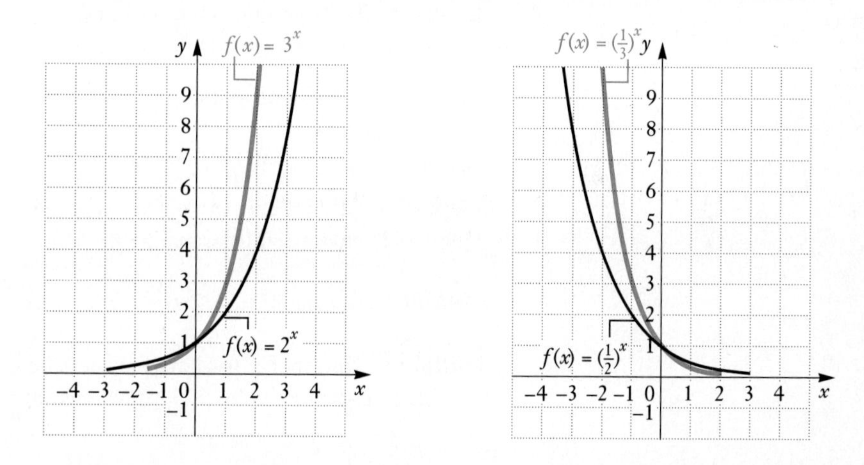

B. When $0 < a < 1$, the function $f(x) = a^x$ decreases from left to right. As a approaches 1, the curve becomes less steep. (See the figure on the right, above.)

Note that all such functions $f(x) = a^x$ go through the point (0, 1). That is, the y-intercept is (0, 1).

EXAMPLE 3 Graph: $y = 2^{x-2}$.

Solution We construct a table of values. Then we plot the points and connect them with a smooth curve. Be sure to note that $x - 2$ is the *exponent.* Think of y as $f(x)$.

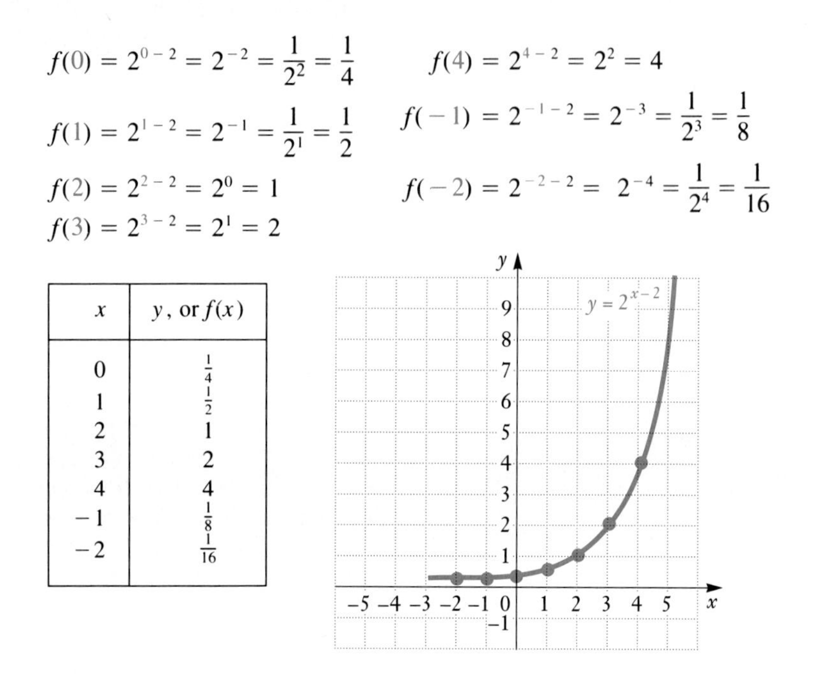

$$f(0) = 2^{0-2} = 2^{-2} = \frac{1}{2^2} = \frac{1}{4} \qquad f(4) = 2^{4-2} = 2^2 = 4$$

$$f(1) = 2^{1-2} = 2^{-1} = \frac{1}{2^1} = \frac{1}{2} \qquad f(-1) = 2^{-1-2} = 2^{-3} = \frac{1}{2^3} = \frac{1}{8}$$

$$f(2) = 2^{2-2} = 2^0 = 1 \qquad f(-2) = 2^{-2-2} = 2^{-4} = \frac{1}{2^4} = \frac{1}{16}$$

$$f(3) = 2^{3-2} = 2^1 = 2$$

x	y, or $f(x)$
0	$\frac{1}{4}$
1	$\frac{1}{2}$
2	1
3	2
4	4
-1	$\frac{1}{8}$
-2	$\frac{1}{16}$

The graph is the same as the graph of $y = 2^x$, shifted two units to the right. ■

Equations with x and y Interchanged

It will be helpful in later work if we are able to graph an equation in which the x and y in $y = 2^x$ have been interchanged.

EXAMPLE 4 Graph: $x = 2^y$.

Solution Note that x is alone on one side of the equation. We can find ordered pairs that are solutions by picking values for y and then computing values for x.

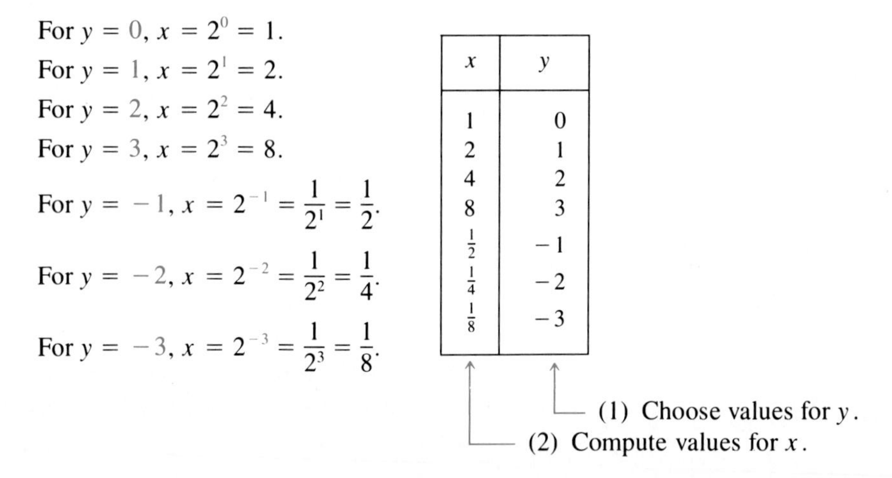

For $y = 0$, $x = 2^0 = 1$.

For $y = 1$, $x = 2^1 = 2$.

For $y = 2$, $x = 2^2 = 4$.

For $y = 3$, $x = 2^3 = 8$.

For $y = -1$, $x = 2^{-1} = \frac{1}{2^1} = \frac{1}{2}$.

For $y = -2$, $x = 2^{-2} = \frac{1}{2^2} = \frac{1}{4}$.

For $y = -3$, $x = 2^{-3} = \frac{1}{2^3} = \frac{1}{8}$.

x	y
1	0
2	1
4	2
8	3
$\frac{1}{2}$	-1
$\frac{1}{4}$	-2
$\frac{1}{8}$	-3

(1) Choose values for y.
(2) Compute values for x.

We plot the points and connect them with a smooth curve. Note that the curve does not touch or cross the y-axis.

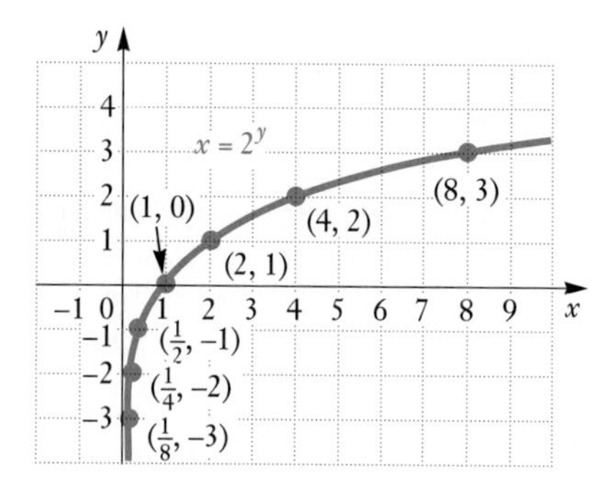

Note too that this curve looks just like the graph of $y = 2^x$, except that it is reflected across the line $y = x$, as shown here.

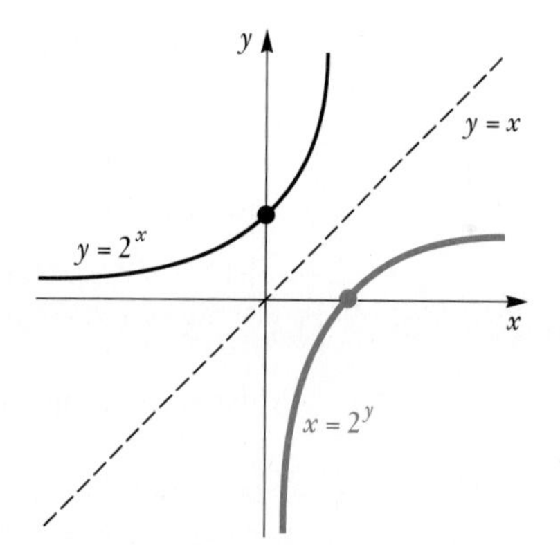

Applications of Exponential Functions

EXAMPLE 5 *Interest compounded annually.* The amount of money A that a principal P will be worth after t years at interest rate i, compounded annually, is given by the formula

$$A = P(1 + i)^t.$$

Suppose that \$100,000 is invested at 8% interest, compounded annually.

a) Find a function for the amount in the account after t years.
b) Find the amount of money in the account at $t = 0$, $t = 4$, $t = 8$, and $t = 10$.
c) Graph the function.

Solution

a) If $P = \$100{,}000$ and $i = 8\% = 0.08$, we can substitute these values and form the following function:

$$A(t) = \$100{,}000(1 + 0.08)^t$$
$$= \$100{,}000(1.08)^t.$$

b) To find the function values, a calculator with a power key might be helpful.

$$A(0) = \$100{,}000(1.08)^0 = \$100{,}000(1) = \$100{,}000$$
$$A(4) = \$100{,}000(1.08)^4 = \$100{,}000(1.36048896) \approx \$136{,}048.90$$
$$A(8) = \$100{,}000(1.08)^8 \approx \$100{,}000(1.85093021) \approx \$185{,}093.02$$
$$A(10) = \$100{,}000(1.08)^{10} \approx \$100{,}000(2.158924997) \approx \$215{,}892.50$$

c) We use the function values computed in part (b), and others if we wish, and draw the graph as follows. Note that the axes are scaled differently because of the large numbers.

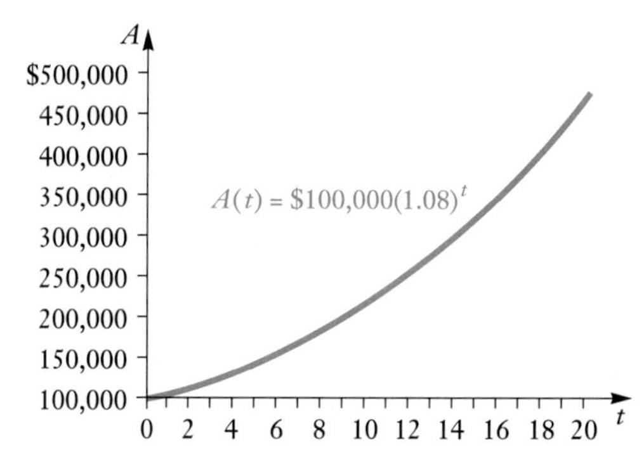

EXERCISE SET 9.1

Graph.

1. $y = f(x) = 2^x$
2. $y = f(x) = 3^x$
3. $y = 5^x$
4. $y = 6^x$
5. $y = 2^{x+1}$
6. $y = 2^{x-1}$
7. $y = 3^{x-2}$
8. $y = 3^{x+2}$
9. $y = 2^x - 3$
10. $y = 2^x + 1$
11. $y = 5^{x+3}$
12. $y = 6^{x-4}$
13. $y = \left(\frac{1}{2}\right)^x$
14. $y = \left(\frac{1}{3}\right)^x$
15. $y = \left(\frac{1}{5}\right)^x$
16. $y = \left(\frac{1}{4}\right)^x$
17. $y = 2^{2x-1}$
18. $y = 3^{4-x}$
19. $y = 2^{x-1} - 3$
20. $y = 2^{x+3} - 4$

Graph.

21. $x = 2^y$
22. $x = 6^y$
23. $x = \left(\frac{1}{2}\right)^y$
24. $x = \left(\frac{1}{3}\right)^y$
25. $x = 5^y$
26. $x = 3^y$
27. $x = \left(\frac{2}{3}\right)^y$
28. $x = \left(\frac{4}{3}\right)^y$

Graph both equations using the same set of axes.

29. $y = 2^x,\ x = 2^y$
30. $y = 3^x,\ x = 3^y$
31. $y = \left(\frac{1}{2}\right)^x,\ x = \left(\frac{1}{2}\right)^y$
32. $y = \left(\frac{1}{4}\right)^x,\ x = \left(\frac{1}{4}\right)^y$

Solve.

33. *Interest compounded annually.* Suppose that \$50,000 is invested at 9% interest, compounded annually.

a) Find a function for the amount in the account after t years.

b) Find the amount of money in the account at $t = 0$, $t = 4$, $t = 8$, and $t = 10$.

c) Graph the function.

34. *Turkey consumption.* The amount of turkey consumed by each person in this country is increasing exponentially. Assuming $t = 0$ corresponds to 1937, the amount of turkey, in pounds per person, consumed t years after 1937 is given by the exponential function

$$N(t) = 2.3(3)^{0.033t}.$$

a) How much turkey was consumed per person in 1940? in 1950? in 1980? in 1988?

b) How much will be consumed per person in 2000?

c) Graph the function.

35. *Recycling aluminum cans.* It is known that $\frac{1}{4}$ of all aluminum cans distributed will be recycled each year. A beverage company distributes 250,000 cans. The number still in use after time t, in years, is given by the exponential function

$$N(t) = 250{,}000\left(\tfrac{1}{4}\right)^t.$$

a) How many cans are still in use after 0 years? 1 year? 4 years? 10 years?

b) Graph the function.

36. *Salvage value.* An office machine is purchased for \$5200. Its value each year is about 80% of the value of the preceding year. Its value, in dollars, after t years is given by the exponential function

$$V(t) = 5200(0.8)^t.$$

a) Find the value of the machine after 0 years, 1 year, 2 years, 5 years, and 10 years.

b) Graph the function.

37. *Compact discs.* The number of compact discs purchased each year is increasing exponentially. The number N, in millions, purchased is given by the exponential function

$$N(t) = 7.5(6)^{0.5t},$$

where t is the number of years after 1985.

a) Find the number of compact discs sold in 1985, 1986, 1988, 1990, 1995, and 2000.

b) Graph the function.

38. *Growth of bacteria.* The bacteria *Escherichi coli* are commonly found in the bladder of human beings. Suppose that 3000 of the bacteria are present at time $t = 0$. Then t minutes later, the number of bacteria present will be

$$N(t) = 3000(2)^{t/20}.$$

a) How many bacteria will be present after 10 minutes? 20 minutes? 30 minutes? 40 minutes? 60 minutes?

b) Graph the function.

SKILL MAINTENANCE

39. Multiply and simplify: $x^{-5} \cdot x^3$.

40. Simplify: $(x^{-3})^4$.

41. Divide and simplify: $\frac{x^{-3}}{x^4}$.

42. Simplify: 5^0.

SYNTHESIS

43. Approximate each of the following to six decimal places.

a) 2^3 **b)** $2^{3.1}$ **c)** $2^{3.14}$

d) $2^{3.141}$ **e)** $2^{3.1415}$ **f)** $2^{3.14159}$

Without using a calculator, determine which number is larger.

44. 5^{π} or π^5 **45.** $\pi^{1.3}$ or $\pi^{2.4}$ **46.** $\sqrt{8^3}$ or $8^{\sqrt{3}}$

Graph each of the following. You will find a calculator with a power key most helpful.

47. $f(x) = (2.3)^x$ **48.** $f(x) = (3.8)^x$ **49.** $g(x) = (0.125)^x$ **50.** $g(x) = (0.9)^x$

Graph.

51. $y = 2^x + 2^{-x}$ **52.** $y = \left(\frac{1}{2}\right)^x - 1$ **53.** $y = 3^x + 3^{-x}$

54. $y = 2^{-(x-1)^2}$ **55.** $y = |2^{x^2} - 1|$ **56.** $y = |2^x - 2|$

Graph both equations using the same set of axes.

57. $y = 3^{-(x-1)}$, $x = 3^{-(y-1)}$ **58.** $y = 1^x$, $x = 1^y$

59. *Typing speed.* Jim is studying typing one semester in college. After Jim has studied for t hours, his speed, in words per minute, is given by the exponential function

$$S(t) = 200[1 - (0.99)^t].$$

a) How fast can Jim type after he has studied for 10 hours? 20 hours? 40 hours? 85 hours?
b) Graph the function.

9.2 Composite and Inverse Functions

Composite Functions

In the real world functions frequently occur in which some quantity depends on a variable that, in turn, depends on another variable. For instance, the number of employees hired by a firm may depend on the firm's profits, which may in turn depend on the number of items the firm produces. Functions like this are called **composite functions.**

There exists a function g that gives a correspondence between women's shoe sizes in the United States and those in Italy. It is given by $g(x) = 2(x + 12)$, where x is a shoe size in the United States and $g(x)$ is a shoe size in Italy. For example, a shoe size of 4 in the United States corresponds to a shoe size of $g(4) = 2(4 + 12)$, or 32, in Italy.

There is also a function that gives a correspondence between women's shoe sizes in Italy and those in Britain. The function is given by $f(x) = \frac{1}{2}x - 14$, where x is a shoe size in Italy and $f(x)$ is the corresponding shoe size in Britain. For example, a shoe size of 32 in Italy corresponds to a shoe size in Britain of $f(32) = \frac{1}{2}(32) - 14$, or 2.

It seems reasonable to assume that a shoe size of 4 in the United States corresponds to a shoe size of 2 in Britain and that there is a function h that describes this correspondence. Can we find a formula for h? If we look at the following tables, we might guess that such a formula is $h(x) = x - 2$, and that is indeed correct. But, for more complicated formulas, we would need to do some algebra.

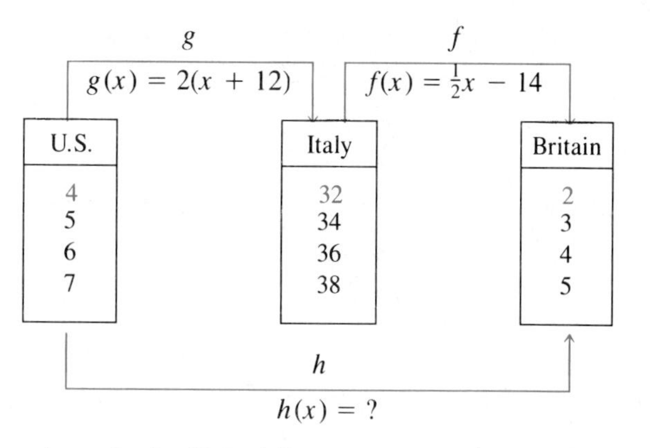

A shoe size x in the United States corresponds to a shoe size $g(x)$ in Italy, where

$$g(x) = 2(x + 12).$$

Now $2(x + 12)$ is a shoe size in Italy. If we replace x in $f(x)$ by $g(x)$, or $2(x + 12)$, we can find the shoe size in Britain that corresponds to a shoe size x in the United States:

$$\begin{aligned} f(g(x)) &= \tfrac{1}{2}[2(x + 12)] - 14 = \tfrac{1}{2}[2x + 24] - 14 \\ &= x + 12 - 14 = x - 2. \end{aligned}$$

This gives a formula for h: $h(x) = x - 2$. Thus a shoe size of 4 in the United States corresponds to a shoe size of $h(4) = 4 - 2$, or 2, in Britain. The function h is called the *composition* of f and g and is denoted $f \circ g$.

> The *composite function* $f \circ g$, the *composition* of f and g, is defined as
>
> $$f \circ g(x) = f(g(x)).$$

We can visualize the composition of functions as follows.

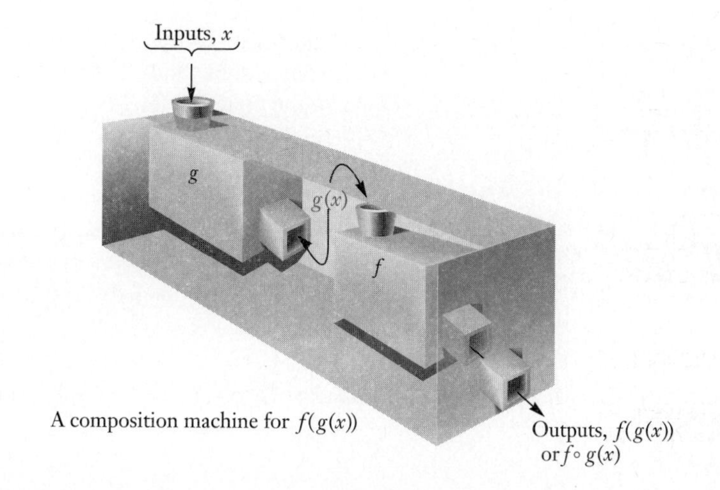

A composition machine for $f(g(x))$

To find $f \circ g(x)$, we substitute $g(x)$ for x in $f(x)$.

EXAMPLE 1 Given $f(x) = 3x$ and $g(x) = 1 + x^2$:

a) Find $f \circ g(5)$ and $g \circ f(5)$.
b) Find $f \circ g(x)$ and $g \circ f(x)$.

Solution Consider each function separately:

$$f(x) = 3x \qquad \text{This function multiplies each input by 3.}$$

and

$$g(x) = 1 + x^2. \qquad \text{This function adds 1 to the square of each input.}$$

a) To find $f \circ g(5)$, we first find $g(5)$ by substituting in the formula for g: Square 5 and add 1, to get 26. Then we substitute the result, 26, into the formula for f: Multiply 26 by 3.

$$\begin{aligned} f \circ g(5) &= f(g(5)) = f(1 + 5^2) \\ &= f(26) = 3(26) = 78 \end{aligned}$$

To find $g \circ f(5)$, we first find $f(5)$ by substituting into the formula for f: Multiply 5 by 3, to get 15. Then we substitute the result, 15, into the formula for g: Square 15 and add 1.

$$\begin{aligned} g \circ f(5) &= g(f(5)) = g(3 \cdot 5) \\ &= g(15) = 1 + 15^2 = 1 + 225 = 226 \end{aligned}$$

b) We see that $f \circ g$ first does what g does (adds 1 to the square) and then does what f does (multiplies by 3). We find $f \circ g(x)$ by substituting $g(x)$ for x in the equation for $f(x)$:

$$\begin{aligned} f \circ g(x) &= f(g(x)) = f(1 + x^2) && \text{Substituting } 1 + x^2 \text{ for } g(x) \\ &= 3(1 + x^2) = 3 + 3x^2. && \textit{These} \text{ parentheses indicate multiplication.} \end{aligned}$$

Next, observe that $g \circ f$ first does what f does (multiplies by 3) and then does what g does (adds 1 to the square). We find $g \circ f(x)$ by substituting $f(x)$ for x in the equation for $g(x)$:

$$\begin{aligned} g \circ f(x) &= g(f(x)) = g(3x) && \text{Substituting } 3x \text{ for } f(x) \\ &= 1 + (3x)^2 \\ &= 1 + 9x^2. \end{aligned}$$

■

Note in Example 1 that $f \circ g(5) \neq g \circ f(5)$ and that, in general, $f \circ g(x) \neq g \circ f(x)$.

EXAMPLE 2 Given $f(x) = \sqrt{x}$ and $g(x) = x - 1$, find $f \circ g(x)$ and $g \circ f(x)$.

Solution

$$f \circ g(x) = f(g(x)) = f(x - 1) = \sqrt{x - 1}$$
$$g \circ f(x) = g(f(x)) = g(\sqrt{x}) = \sqrt{x} - 1$$ ■

It is important to be able to recognize how a function can be expressed as a composition.

EXAMPLE 3 Find $f(x)$ and $g(x)$ such that when $h(x) = (7x + 3)^2$, we have $h(x) = f \circ g(x)$.

Solution This is $7x + 3$ to the 2nd power. Two functions that can be used for the composition are $f(x) = x^2$ and $g(x) = 7x + 3$. We can check by forming the composition:

$$h(x) = f \circ g(x) = f(g(x)) = f(7x + 3) = (7x + 3)^2.$$

This is the most "obvious" answer to the question. There can be other less obvious answers. For example, if

$$g(x) = 7x + 4 \qquad \text{and} \qquad f(x) = (x - 1)^2,$$

then

$$\begin{aligned} f \circ g(x) = f(g(x)) &= f(7x + 4) \\ &= (7x + 4 - 1)^2 = (7x + 3)^2. \end{aligned}$$ ■

Inverse and One-to-One Functions

Let us consider the following two functions.

COST OF A 60-SECOND COMMERCIAL DURING THE SUPER BOWL

Domain (set of inputs)		Range (set of outputs)
1967	→	$80,000
1970	→	$200,000
1977	→	$324,000
1981	→	$550,000
1983	→	$800,000
1985	→	$1,100,000
1988	→	$1,350,000

U.S. SENATORS

Domain (set of inputs)		Range (set of outputs)
Mitchell	→	Maine
Cohen	→	Maine
Mack	→	Florida
Graham	→	Florida
Leahy	→	Vermont
Jeffords	→	Vermont
D'Amato	→	New York
Moynihan	→	New York

Suppose we reverse the arrows. We would then obtain what is called the **inverse correspondence.** Are these new correspondences functions?

COST OF A 60-SECOND COMMERCIAL DURING THE SUPER BOWL	
Range (set of outputs)	**Domain** (set of inputs)
1967 ←	\$80,000
1970 ←	\$200,000
1977 ←	\$324,000
1981 ←	\$550,000
1983 ←	\$800,000
1985 ←	\$1,100,000
1988 ←	\$1,350,000

U.S. SENATORS	
Range (set of outputs)	**Domain** (set of inputs)
Mitchell ←	Maine
Cohen ←	Maine
Mack ←	Florida
Graham ←	Florida
Leahy ←	Vermont
Jeffords ←	Vermont
D'Amato ←	New York
Moynihan ←	New York

We see that the inverse of the first correspondence is a function, but the inverse of the second correspondence is not a function.

Recall that for each input a function provides exactly one output. However, nothing in our definition of function precludes having the same output for two or more different inputs. That is, it is possible with a function for different inputs to correspond to the same output in the range. When this possibility is *excluded,* the inverse is also a function.

In the Super Bowl function, different inputs have different outputs. It is an example of what is called a **one-to-one** function. In the U.S. Senator function, the input *Cohen* has the output *Maine* and the input *Mitchell* also has the output *Maine*. It is not a one-to-one function.

> A function f is *one-to-one* if different inputs have different outputs. That is, if for any $a \neq b$, we have $f(a) \neq f(b)$, the function f is one-to-one. If a function is one-to-one, then its inverse correspondence is also a function.

How can we tell graphically whether a function is one-to-one and thus has an inverse that is a function?

EXAMPLE 4 Determine whether $f(x) = 2^x$ is one-to-one and thus has an inverse that is a function.

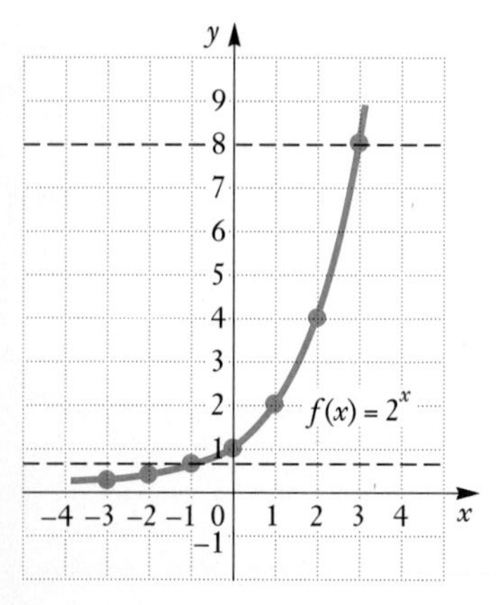

Solution The graph of $f(x) = 2^x$ is shown here. We constructed this graph in Example 1 of Section 9.1. Try to find a horizontal line that crosses the graph more than once. There is no such line. Also, choose two different inputs, say -1 and 3. Now $f(-1) = 2^{-1} = \frac{1}{2}$ and $f(3) = 2^3 = 8$. The outputs are different. This will be the case for *any* pair of inputs that are different.

Thus the function is one-to-one and its inverse is a function. ■

THE HORIZONTAL LINE TEST

A function is one-to-one and has an inverse that is a function if there is no horizontal line that crosses the graph more than once.

In Section 6.2 we learned that the graph of a function must pass the vertical line test. We see here that for a function to have an inverse that is a function, it must pass the horizontal line test as well.

EXAMPLE 5 Determine whether the function $f(x) = x^2$ is one-to-one and has an inverse that is also a function.

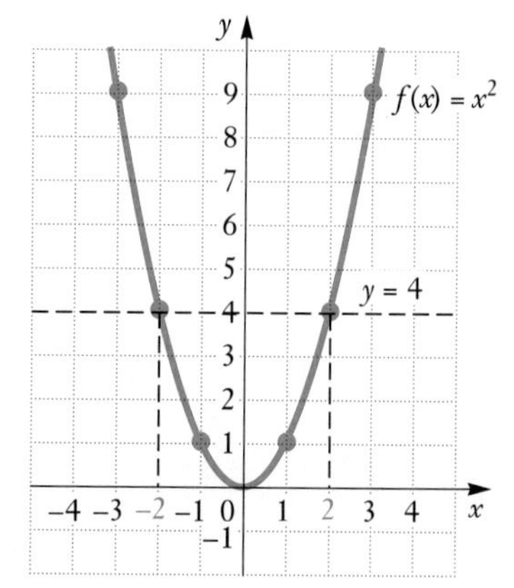

Solution The graph of $f(x) = x^2$ is shown here. Many horizontal lines cross the graph more than once, in particular the line $y = 4$. Note that where the line crosses, the first coordinates are -2 and 2. Although these are different inputs, they have the same output. That is, $-2 \neq 2$, but

$$f(-2) = (-2)^2 = 4 = 2^2 = f(2).$$

Thus the function is not one-to-one and does not have an inverse that is a function. ■

Consider the Super Bowl function again. Note that if we start with the input 1985, we get $1,100,000 as an output. Then consider the inverse function. If we start with the input $1,100,000, we get 1985 as the output. Thus the inverse takes us back where we started. This does not happen with the U.S. Senator function. If we start with Cohen as an input, we get Maine as an output. If we go to the inverse and start with Maine, we have two possibilities for outputs, Cohen and Mitchell. The U.S. Senator function is not one-to-one and its inverse is not a function.

Finding Formulas for Inverses

If the inverse of a function f is also a function, it can be named f^{-1} (read "f-inverse").

The -1 in f^{-1} is *not* an exponent!

Suppose a function is described by a formula. If it has an inverse that is a function, how do we find a formula for its inverse? For any equation in two variables, if we interchange the variables, we obtain an equation of the inverse correspondence. If it is a function, we proceed, as in the next example, to find a formula for f^{-1}.

EXAMPLE 6 Given $f(x) = x + 1$:

a) Determine whether the function is one-to-one.

b) If it is one-to-one, find a formula for $f^{-1}(x)$.

Solution

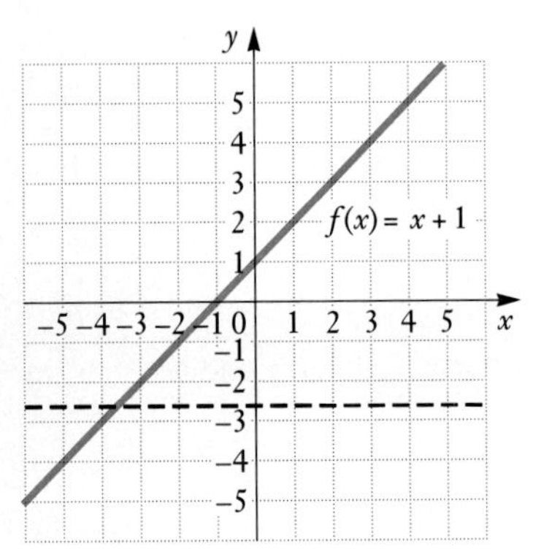

a) The graph of $f(x) = x + 1$ is shown here. It passes the horizontal line test, so it is one-to-one. Thus its inverse is a function.

b) To find a formula, we first think of this function as $y = x + 1$. Generally, we are given x-values, or inputs, and asked to find y-values, or outputs. Suppose we start with $x = 5$. Then $y = 5 + 1 = 6$. If we are given the output 6 and asked for the input from which it came, we could set $y = 6$ in the equation and solve for x:

$$6 = x + 1$$
$$5 = x.$$

Suppose the output is 13 and we want the input from which *it* came. We set $y = 13$ and solve for x:

$$13 = x + 1$$
$$12 = x.$$

In general, if the output is y and we want to get the input from which y came, we solve for x:

$$y = x + 1$$
$$y - 1 = x.$$

The variables that describe a function are called "dummy" variables, or blanks. We could express the inverse function $y - 1 = x$ as

$$(\ ) - 1 = [\ ],$$

where the output, [ ], depends on the input, ( ).

The inputs of the inverse are what we substitute into the function to get outputs, and x is traditionally used for that variable. The outputs are represented by y. Thus the function $(\ ) - 1 = [\ ]$ can be expressed as

$$x - 1 = y.$$

We have essentially interchanged the original variables and solved for y.

Or, using $f^{-1}(x)$ for y, we finally obtain a formula for the inverse:

$$f^{-1}(x) = x - 1.$$

■

Note in Example 6 that f pairs any x with $x + 1$ (this function adds 1 to each number in its domain). Its inverse, f^{-1}, pairs any number x with $x - 1$ (this function subtracts 1 from each member of its domain). Thus the function and its inverse reverse each other.

To Find a Formula for the Inverse of a Function f

If a function is one-to-one, it has an inverse that is a function. A formula for the inverse can be found as follows:

1. Replace $f(x)$ by y.
2. Solve the equation for x.
3. Interchange x and y.
4. Replace y by $f^{-1}(x)$.

EXAMPLE 7 Given $g(x) = 2x - 3$:

a) Determine whether the function is one-to-one.
b) If it is one-to-one, find a formula for $g^{-1}(x)$.

Solution

a) The function $g(x) = 2x - 3$ is linear. Any linear function that is not constant will pass the horizontal line test. Thus g is one-to-one.

b) We replace $g(x)$ by y and solve for x.

$$y = 2x - 3$$
$$y + 3 = 2x$$
$$\frac{y + 3}{2} = x.$$

Next we interchange x and y:

$$\frac{x + 3}{2} = y.$$

Then we replace y by $g^{-1}(x)$:

$$g^{-1}(x) = \frac{x + 3}{2}.$$

■

Let us consider inverses of functions in terms of a function machine. Suppose that the function g programmed into a machine has an inverse that is also a function. Suppose then that the function machine has a reverse switch. When the switch is thrown, the machine is then programmed to perform the inverse function g^{-1}. Inputs then enter at the opposite end, and the entire process is reversed.

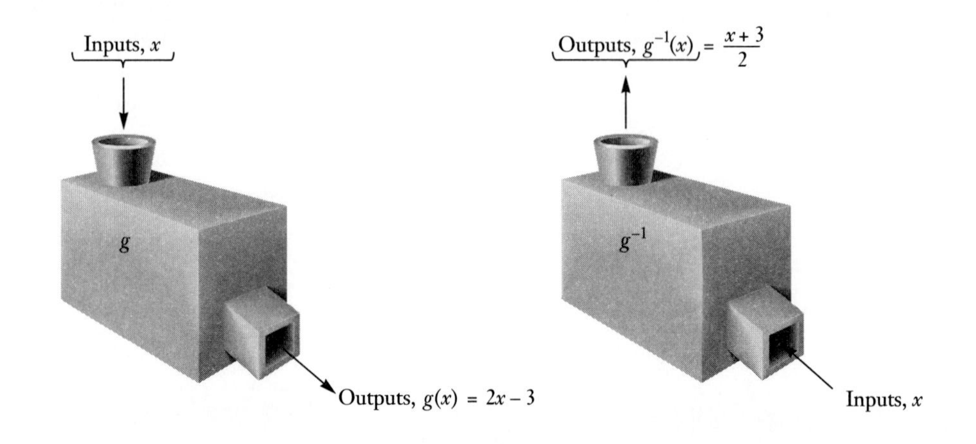

For example, examine the function $g(x) = 2x - 3$ and its inverse $g^{-1}(x) = (x + 3)/2$, from Example 7. Consider the input 5. Now

$$g(5) = 2(5) - 3 = 10 - 3 = 7.$$

The output is 7. Now we use 7 for the input in the inverse:

$$\begin{aligned} g^{-1}(7) &= \frac{7 + 3}{2} \\ &= \frac{10}{2} = 5. \end{aligned}$$

The function g takes 5 to 7. The inverse function g^{-1} takes the number 7 back to 5.

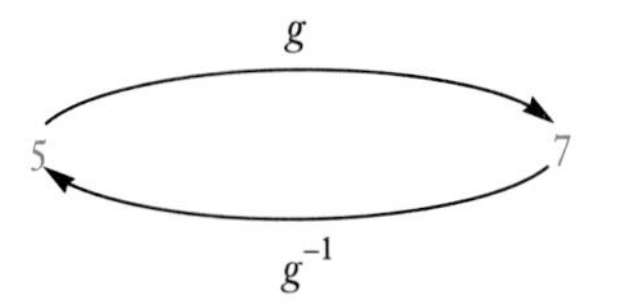

Graphing Functions and Their Inverses

How do the graphs of a function and its inverse compare?

EXAMPLE 8 Graph $g(x) = 2x - 3$ and $g^{-1}(x) = (x + 3)/2$, using the same set of axes. Then compare.

Solution The graph of each function follows. Note that the graph of g^{-1} can be obtained by reflecting the graph of g across the line $y = x$. That is, if you graph $g(x) = 2x - 3$ and $y = x$ and fold the paper along the line $y = x$, the graph of $g^{-1}(x) = (x + 3)/2$ will be "flipped" across the line.

When we interchange y and x in finding a formula for the inverse, we are, in effect, flipping the graph of $g(x) = 2x - 3$ over the line $y = x$. For example, when the y-intercept of the graph of g, $(0, -3)$, is reversed, we get the x-intercept of the graph of g^{-1}, $(-3, 0)$.

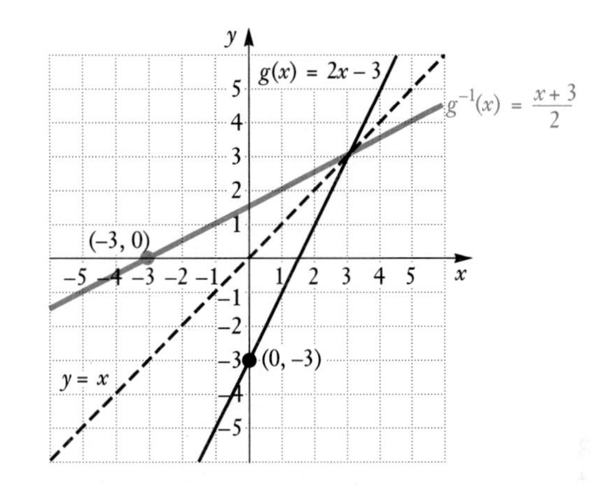

■

The graph of f^{-1} is a reflection of the graph of f across the line $y = x$.

EXAMPLE 9 Consider $f(x) = x^3 + 2$.

a) Determine whether the function is one-to-one.
b) If it is one-to-one, find a formula for its inverse.
c) Graph the inverse.

Solution

a) The graph of $f(x) = x^3 + 2$ follows. It passes the horizontal line test and thus has an inverse.

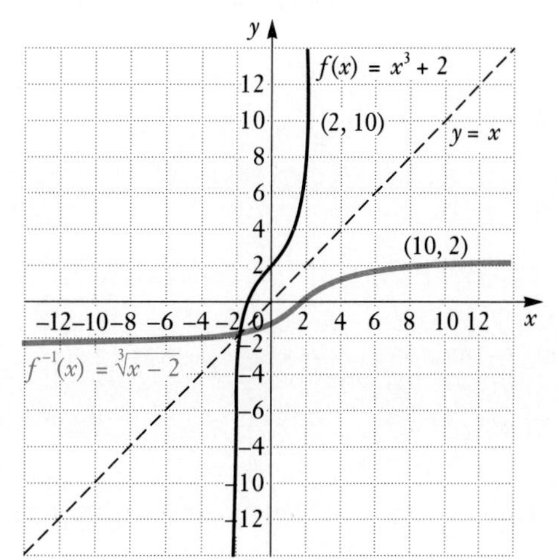

b) We replace $f(x)$ by y and solve for x:

$$y = x^3 + 2$$
$$y - 2 = x^3$$
$$\sqrt[3]{y - 2} = x.$$

Since any number has only one cube root, we can solve for x.

Next, we interchange x and y:

$$\sqrt[3]{x - 2} = y.$$

We replace y by $f^{-1}(x)$:

$$f^{-1}(x) = \sqrt[3]{x - 2}.$$

c) To find the graph, flip the graph of $f(x) = x^3 + 2$ over the line $y = x$, as we did on page 409. It can also be found by substituting to find function values. The graph is shown using the same set of axes. ■

Inverse Functions and Composition

> If a function f is one-to-one, then f^{-1} is the unique function such that
>
> **a)** $f^{-1} \circ f(x) = x$ and
> **b)** $f \circ f^{-1}(x) = x$.

The first condition asserts that if we start with an input x for the function f and find its output, then the inverse will take the output back to x. The second condition asserts that if we start with an input x for the function f^{-1}, the inverse takes it to its output, and if we apply the original function to that output, we will get back to x.

EXAMPLE 10 Let $f(x) = 2x + 1$. Show that $f^{-1}(x) = (x - 1)/2$.

Solution We find $f^{-1} \circ f(x)$ and $f \circ f^{-1}(x)$ and check to see that each is x.

a) $f^{-1} \circ f(x) = f^{-1}(f(x)) = f^{-1}(2x + 1) = \dfrac{(2x + 1) - 1}{2} = \dfrac{2x}{2} = x$

b) $f \circ f^{-1}(x) = f(f^{-1}(x)) = f\left(\dfrac{x - 1}{2}\right) = 2\left(\dfrac{x - 1}{2}\right) + 1 = x - 1 + 1 = x$ ■

EXERCISE SET 9.2

Find $f \circ g(x)$ and $g \circ f(x)$.

1. $f(x) = 3x^2 + 2$, $g(x) = 2x - 1$

2. $f(x) = 4x + 3$, $g(x) = 2x^2 - 5$

3. $f(x) = 4x^2 - 1$, $g(x) = 2/x$

4. $f(x) = 3/x$, $g(x) = 2x^2 + 3$

5. $f(x) = x^2 + 1$, $g(x) = x^2 - 1$

6. $f(x) = 1/x^2$, $g(x) = x + 2$

Find $f(x)$ and $g(x)$ such that $h(x) = f \circ g(x)$. Answers may vary.

7. $h(x) = (5 - 3x)^2$

8. $h(x) = 4(3x - 1)^2 + 9$

9. $h(x) = (3x^2 - 7)^5$
10. $h(x) = \sqrt{5x + 2}$
11. $h(x) = \dfrac{1}{x - 1}$
12. $h(x) = \dfrac{3}{x} + 4$
13. $h(x) = \dfrac{1}{\sqrt{7x + 2}}$
14. $h(x) = \sqrt{x - 7} - 3$
15. $h(x) = \dfrac{x^3 + 1}{x^3 - 1}$
16. $h(x) = \left(\sqrt{x} + 5\right)^4$

Determine whether each function is one-to-one.

17. $f(x) = 3x - 4$
18. $f(x) = 5 - 2x$
19. $f(x) = x^2 - 3$
20. $f(x) = 1 - x^2$
21. $g(x) = 3^x$
22. $g(x) = \left(\frac{1}{2}\right)^x$
23. $g(x) = |x|$
24. $h(x) = |x| - 1$
25. $f(x) = |x + 3|$
26. $f(x) = |x - 2|$
27. $g(x) = \dfrac{-2}{x}$
28. $h(x) = \dfrac{1}{x}$

Given the function, (a) determine if it is one-to-one; (b) if it is one-to-one, find a formula for the inverse.

29. $f(x) = x + 2$
30. $f(x) = x + 7$
31. $f(x) = 5 - x$
32. $f(x) = 9 - x$
33. $g(x) = x - 5$
34. $g(x) = x - 8$
35. $f(x) = 3x$
36. $f(x) = 4x$
37. $g(x) = 3x + 2$
38. $g(x) = 4x + 7$
39. $h(x) = \dfrac{4}{x + 3}$
40. $h(x) = \dfrac{1}{x - 8}$
41. $f(x) = \dfrac{1}{x}$
42. $f(x) = \dfrac{3}{x}$
43. $f(x) = \dfrac{2x + 1}{3}$
44. $f(x) = \dfrac{3x + 2}{5}$
45. $g(x) = \dfrac{x - 3}{x + 4}$
46. $g(x) = \dfrac{2x - 1}{5x + 3}$
47. $f(x) = x^3 - 1$
48. $f(x) = x^3 + 5$
49. $g(x) = (x - 2)^3$
50. $g(x) = (x + 7)^3$
51. $f(x) = \sqrt[3]{x}$
52. $f(x) = \sqrt[3]{x - 4}$
53. $f(x) = 2x^2 + 3, \quad x \geq 0$
54. $f(x) = 3x^2 - 2, \quad x \geq 0$

Graph the function and its inverse using the same set of axes.

55. $f(x) = \frac{1}{2}x - 3$
56. $g(x) = x + 4$
57. $f(x) = x^3$
58. $f(x) = x^3 - 1$
59. $f(x) = 2^x$
60. $f(x) = 3^x$
61. $g(x) = \left(\frac{1}{2}\right)^x$
62. $f(x) = \left(\frac{2}{3}\right)^x$
63. $f(x) = 3 - x^2, \quad x \geq 0$
64. $f(x) = x^2 - 1, \quad x \leq 0$
65. Let $f(x) = \frac{4}{5}x$. Show that
$$f^{-1}(x) = \tfrac{5}{4}x.$$
66. Let $f(x) = (x + 7)/3$. Show that
$$f^{-1}(x) = 3x - 7.$$
67. Let $f(x) = (1 - x)/x$. Show that
$$f^{-1}(x) = \frac{1}{x + 1}.$$
68. Let $f(x) = x^3 - 5$. Show that
$$f^{-1}(x) = \sqrt[3]{x + 5}.$$

69. *Women's dress sizes in the United States and France.* A size-6 dress in the United States is size 38 in France. A function that will convert dress sizes in the United States to those in France is

$$f(x) = x + 32.$$

a) Find the dress sizes in France that correspond to sizes 8, 10, 14, and 18 in the United States.

b) Determine whether this function has an inverse that is a function. If so, find a formula for the inverse.

c) Use the inverse function to find dress sizes in the United States that correspond to sizes 40, 42, 46, and 50 in France.

70. *Women's dress sizes in the United States and Italy.* A size-6 dress in the United States is size 36 in Italy. A function that will convert dress sizes in the United States to those in Italy is

$$f(x) = 2(x + 12).$$

a) Find the dress sizes in Italy that correspond to sizes 8, 10, 14, and 18 in the United States.

b) Determine whether this function has an inverse that is a function. If so, find a formula for the inverse.

c) Use the inverse function to find dress sizes in the United States that correspond to sizes 40, 44, 52, and 60 in Italy.

SKILL MAINTENANCE

71. Find an equation of variation if y varies directly as x and $y = 7.2$ when $x = 0.8$.

72. Find an equation of variation if y varies inversely as x and $y = 3.5$ when $x = 6.1$.

SYNTHESIS

73. Does the constant function $f(x) = 4$ have an inverse that is a function? If so, find a formula. If not, explain why.

74. An organization determines that the cost per person of chartering a bus is given by the function

$$C(x) = \frac{100 + 5x}{x},$$

where x = the number of people in the group and $C(x)$ is in dollars. Determine $C^{-1}(x)$ and explain how this inverse function could be used.

9.3 Logarithmic Functions

In this section we consider a type of function called a *logarithm function,* or *logarithmic function*. Such functions have many applications to problem solving.

Graphs of Logarithmic Functions

Consider the exponential function $f(x) = 2^x$. Does this function have an inverse that is a function? We see from the graph that this function is one-to-one and does have an inverse f^{-1} that is a function.

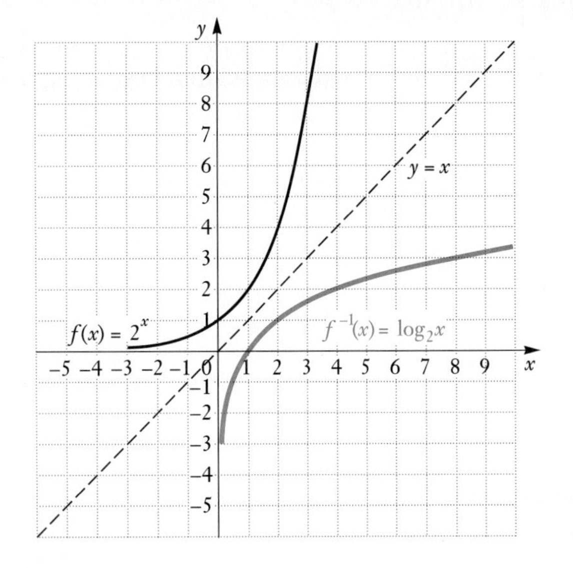

Consider the input 3 and the original function

$$f(x) = 2^x.$$

Then

$$f(3) = 2^3 = 8.$$

This also tells us that $f^{-1}(8) = 3$. If we did not know this, then to find $f^{-1}(8)$, we would be looking for a number t such that

$$8 = 2^t.$$

Now we can probably think this out, but suppose that we wanted to find $f^{-1}(13)$. The input for the inverse would be the number 13. Then we would be looking for t such that

$$13 = 2^t$$

We know that the inverse exists, but we do not have a way to name $f^{-1}(13)$ as yet. Mathematicians have invented a name for the inverse of $f(x) = 2^x$. It is the *logarithmic function, base* 2, denoted

$$f^{-1}(x) = \log_2 x.$$

We read "$\log_2 x$" as "the logarithm, base 2, of x." This means

$\log_2 8$ *is the power to which we raise 2 to get 8.*

Thus, $\log_2 8 = 3$. Similarly,

$\log_2 13$ *is the power to which we raise 2 to get 13.*

We have no simpler way to write this for now.

For any exponential function $f(x) = a^x$, the inverse is called a **logarithmic function, base *a*.** The graph of the inverse can, of course, be obtained by reflecting the graph of $f(x) = a^x$ across the line $y = x$. It will be helpful to remember that the inverse of $f(x) = a^x$ is given by $f^{-1}(x) = \log_a x$. Normally, we use a number a that is greater than 1 for the logarithm base.

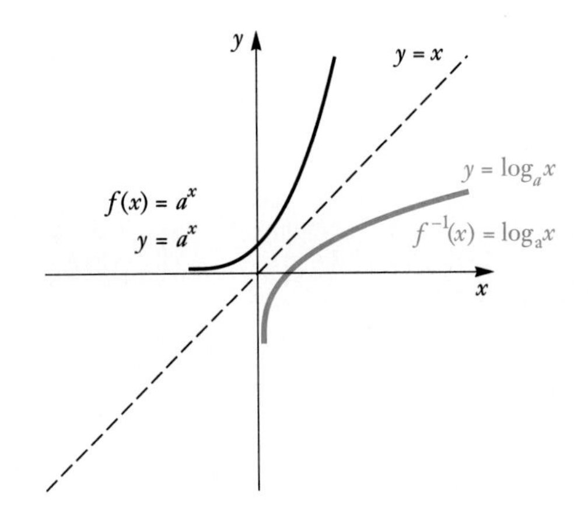

We define $y = \log_a x$ as that number y such that $a^y = x$, where $x > 0$ and a is a positive constant other than 1.

It is helpful in dealing with logarithmic functions to remember that the logarithm of a number is an *exponent.* It is the exponent y in a^y. You might also think to yourself, "The logarithm, base a, of a number x is the power to which a must be raised in order to get x."

A logarithm is an exponent.

The following is a comparison of exponential and logarithmic functions.

Exponential Function	Logarithmic Function
$y = a^x$ $f(x) = a^x$ $a > 0, a \neq 1$ x is any real number. Inputs can be any real numbers. $y > 0$ Outputs are positive.	$x = a^y$ $f(x) = \log_a x$ $a > 0, a \neq 1$ y is any real number. Outputs can be any real numbers. $x > 0$ Inputs are positive.

Why do we exclude 1 from being a logarithmic base? If we did include it, we would be considering $x = 1^y = 1$. The graph of this equation is a vertical line and is not a function.

EXAMPLE 1 Graph: $y = f(x) = \log_5 x$.

Solution The equation $y = \log_5 x$ is equivalent to $5^y = x$. We can find ordered pairs that are solutions by picking values for y and computing the x-values.

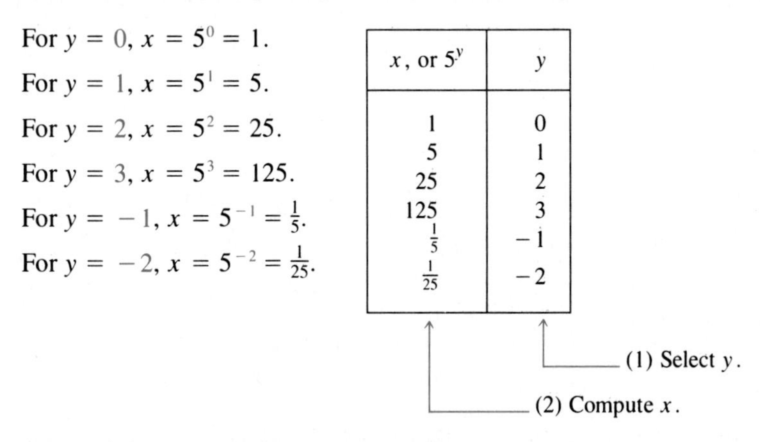

For $y = 0$, $x = 5^0 = 1$.

For $y = 1$, $x = 5^1 = 5$.

For $y = 2$, $x = 5^2 = 25$.

For $y = 3$, $x = 5^3 = 125$.

For $y = -1$, $x = 5^{-1} = \frac{1}{5}$.

For $y = -2$, $x = 5^{-2} = \frac{1}{25}$.

x, or 5^y	y
1	0
5	1
25	2
125	3
$\frac{1}{5}$	-1
$\frac{1}{25}$	-2

(1) Select y.

(2) Compute x.

We plot the set of ordered pairs and connect the points with a smooth curve. The graph of $y = 5^x$ has been shown only for reference.

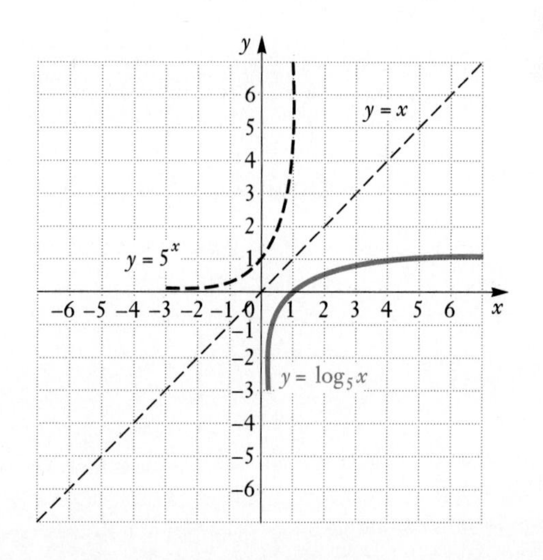

■

Converting from Exponential Equations to Logarithmic Equations

We use the definition of logarithms to convert from *exponential equations* to *logarithmic equations.*

$$y = \log_a x \quad \text{is equivalent to} \quad a^y = x.$$

Be sure to memorize this relationship! It is probably the most important definition in the chapter. Often this definition will be a justification for a proof or a procedure that we are considering.

EXAMPLES Convert to logarithmic equations.

2. $8 = 2^x \Leftrightarrow x = \log_2 8$ — The exponent is the logarithm. The double arrow indicates equivalence. The base remains the same.

3. $y^{-1} = 4 \Leftrightarrow -1 = \log_y 4$

4. $a^b = c \Leftrightarrow b = \log_a c$ ■

Converting from Logarithmic Equations to Exponential Equations

We also use the definition of logarithms to convert from logarithmic equations to exponential equations.

EXAMPLES Convert to exponential equations.

5. $y = \log_3 5 \Leftrightarrow 3^y = 5$ — The logarithm is the exponent. The base remains the same.

6. $-2 = \log_a 7 \Leftrightarrow a^{-2} = 7$

7. $a = \log_b d \Leftrightarrow b^a = d$ ■

Solving Certain Logarithmic Equations

Certain equations involving logarithms can be solved by first converting to exponential equations.

EXAMPLE 8 Solve: $\log_2 x = -3$.

Solution

$$\log_2 x = -3$$

$$2^{-3} = x \quad \text{Converting to an exponential equation}$$

$$\tfrac{1}{8} = x \quad \text{Computing } 2^{-3}$$

Check: For $\frac{1}{8}$ to be the solution, $\log_2 \frac{1}{8}$ should equal -3. Since $2^{-3} = \frac{1}{8}$, we know that $\frac{1}{8}$ checks and is the solution. ■

EXAMPLE 9 Solve: $\log_x 16 = 2$.

Solution

$$\log_x 16 = 2$$

$$x^2 = 16 \qquad \textbf{Converting to an exponential equation}$$

$$x = 4 \qquad \text{or} \qquad x = -4 \qquad \textbf{Principle of positive and negative roots}$$

Check: $\log_4 16 = 2$ because $4^2 = 16$. Thus 4 is a solution. Because all logarithm bases must be positive, $\log_{-4} 16$ is not defined. Therefore, -4 is not a solution. Logarithm bases must be positive because logarithms are defined in terms of exponential functions that are defined only for positive bases. ■

Solving an equation like $\log_b a = x$ amounts to finding the logarithm, base b, of the number a. We have done this in Example 1 when graphing $y = \log_5 x$. To think of finding logarithms as solving equations may help in some cases.

EXAMPLE 10 Find $\log_{10} 1000$.

Solution

METHOD 1. Let $\log_{10} 1000 = x$. Then

$$10^x = 1000 \qquad \textbf{Converting to an exponential equation}$$

$$10^x = 10^3 \qquad \textbf{Writing 1000 as a power of 10}$$

$$x = 3. \qquad \textbf{The exponents must be the same.}$$

Therefore, $\log_{10} 1000 = 3$.

METHOD 2. Think of the meaning of $\log_{10} 1000$. It is the exponent to which we raise 10 to get 1000. That exponent is 3. Therefore, $\log_{10} 1000 = 3$. ■

EXAMPLE 11 Find $\log_{10} 0.01$.

Solution

METHOD 1. Let $\log_{10} 0.01 = x$. Then

$$10^x = 0.01 \qquad \textbf{Converting to an exponential equation}$$

$$10^x = \tfrac{1}{100}$$

$$10^x = 10^{-2} \qquad \textbf{Writing } \tfrac{1}{100} \textbf{ as a power of 10}$$

$$x = -2.$$

Therefore, $\log_{10} 0.01 = -2$.

METHOD 2. $\log_{10} 0.01$ is the exponent to which we raise 10 to get 0.01. If we note that $0.01 = 1/100 = 1/10^2$, it follows that the exponent is -2. Therefore, $\log_{10} 0.01 = -2$. ■

Here are some other examples. Think out mentally how they can be found and compare them.

$\log_{10} 100 = 2$	$\log_{64} 64 = 1$
$\log_{10} 10 = 1$	$\log_8 64 = 2$
$\log_{10} 1 = 0$	$\log_4 64 = 3$
$\log_{10} 0.1 = -1$	$\log_2 64 = 6$

EXAMPLE 12 Find $\log_5 1$.

Solution

METHOD 1. Let $\log_5 1 = x$. Then

$$5^x = 1 \quad \text{Converting to an exponential equation}$$
$$5^x = 5^0 \quad \text{Writing 1 as a power of 5}$$
$$x = 0.$$

Therefore, $\log_5 1 = 0$.

METHOD 2. $\log_5 1$ is the exponent to which we raise 5 to get 1. That exponent is 0. Therefore, $\log_5 1 = 0$. ■

Example 12 illustrates an important property of logarithms.

For any base a,

$$\log_a 1 = 0.$$

The logarithm, base a, of 1 is always 0.

The proof follows from the fact that $a^0 = 1$ is equivalent to the logarithmic equation $\log_a 1 = 0$.

Another property follows similarly. We know that $a^1 = a$ for any real number a. In particular, it holds for any positive number a. This is equivalent to the logarithmic equation $\log_a a = 1$.

For any base a,

$$\log_a a = 1.$$

EXERCISE SET 9.3

Graph.

1. $y = \log_2 x$
2. $y = \log_{10} x$
3. $y = \log_6 x$
4. $y = \log_3 x$
5. $f(x) = \log_4 x$
6. $f(x) = \log_5 x$
7. $f(x) = \log_{1/2} x$
8. $f(x) = \log_{2.5} x$

Graph both functions using the same set of axes.

9. $f(x) = 3^x, \quad f^{-1}(x) = \log_3 x$
10. $f(x) = 4^x, \quad f^{-1}(x) = \log_4 x$

Convert to logarithmic equations.

11. $10^3 = 1000$
12. $10^2 = 100$
13. $5^{-3} = \frac{1}{125}$
14. $4^{-5} = \frac{1}{1024}$
15. $8^{1/3} = 2$
16. $16^{1/4} = 2$
17. $10^{0.3010} = 2$
18. $10^{0.4771} = 3$
19. $e^2 = t$
20. $p^k = 3$
21. $Q^t = x$
22. $p^m = V$
23. $e^2 = 7.3891$
24. $e^3 = 20.0855$
25. $e^{-2} = 0.1353$
26. $e^{-4} = 0.0183$

Convert to exponential equations.

27. $t = \log_3 8$
28. $h = \log_7 10$
29. $\log_5 25 = 2$
30. $\log_6 6 = 1$
31. $\log_{10} 0.1 = -1$
32. $\log_{10} 0.01 = -2$
33. $\log_{10} 7 = 0.845$
34. $\log_{10} 3 = 0.4771$
35. $\log_e 20 = 2.9957$
36. $\log_e 10 = 2.3026$
37. $\log_t Q = k$
38. $\log_m P = a$
39. $\log_e 0.25 = -1.3863$
40. $\log_e 0.989 = -0.0111$
41. $\log_r T = -x$
42. $\log_c M = -w$

Solve.

43. $\log_3 x = 2$
44. $\log_4 x = 3$
45. $\log_x 36 = 2$
46. $\log_x 64 = 3$
47. $\log_2 x = -1$
48. $\log_3 x = -2$
49. $\log_8 x = \frac{1}{3}$
50. $\log_{32} x = \frac{1}{5}$

Find each of the following.

51. $\log_{10} 100$
52. $\log_{10} 100{,}000$
53. $\log_{10} 0.1$
54. $\log_{10} 0.001$
55. $\log_{10} 1$
56. $\log_{10} 10$
57. $\log_5 625$
58. $\log_2 64$
59. $\log_5 \frac{1}{25}$
60. $\log_2 \frac{1}{16}$
61. $\log_3 1$
62. $\log_4 4$
63. $\log_e e$
64. $\log_e 1$
65. $\log_{27} 9$
66. $\log_8 2$
67. $\log_e e^3$
68. $\log_e e^{-4}$
69. $\log_{10} 10^t$
70. $\log_3 3^p$

SKILL MAINTENANCE

Simplify.

71. $\dfrac{\dfrac{3}{x} - \dfrac{2}{xy}}{\dfrac{2}{x^2} + \dfrac{1}{xy}}$

72. $\dfrac{\dfrac{4 + x}{x^2 + 2x + 1}}{\dfrac{3}{x + 1} - \dfrac{2}{x + 2}}$

Rename.

73. 8^{-4}
74. $x^{4/5}$
75. $t^{-2/3}$
76. 5^1

SYNTHESIS

77. Graph both equations using the same set of axes.

$$y = \left(\tfrac{3}{2}\right)^x, \qquad y = \log_{3/2} x$$

Graph.

78. $y = \log_2 (x - 1)$

79. $y = \log_3 |x + 1|$

Solve.

80. $|\log_3 x| = 3$

81. $\log_{125} x = \frac{2}{3}$

82. $\log_\pi \pi^4 = x$

83. $\log_{\sqrt{5}} x = -3$

84. $\log_b b^{2x^2} = x$

85. $\log_4 (3x - 2) = 2$

86. $\log_8 (2x + 1) = -1$

87. $\log_x \sqrt[5]{36} = \frac{1}{10}$

88. $\log_{10} (x^2 + 21x) = 2$

Simplify.

89. $\log_{1/4} \frac{1}{64}$

90. $\log_{81} 3^{\log_3 81}$

91. $\log_{10} (\log_4(\log_3 81))$

92. $\log_2 (\log_2(\log_4 256))$

93. $\log_{\sqrt{3}} \frac{1}{81}$

94. $\log_{1/5} 25$

9.4 Properties of Logarithmic Functions

Logarithmic functions are important in many applications and in more advanced mathematics. We now establish some basic properties that are useful in manipulating expressions involving logarithms. As the proofs of these properties reveal, the properties of logarithms are related to the properties of exponents.

Logarithms of Products

One of the interesting properties special to logarithmic functions is the following.

> **PROPERTY 1**
>
> For any positive numbers M and N,
>
> $$\log_a MN = \log_a M + \log_a N.$$
>
> (The logarithm of a product is the sum of the logarithms of the factors. The number a can be any logarithm base.)

EXAMPLE 1 Express as a sum of logarithms: $\log_2 (4 \cdot 16)$.

Solution

$$\log_2 (4 \cdot 16) = \log_2 4 + \log_2 16 \qquad \text{By Property 1}$$

As a check, note that

$$\log_2 (4 \cdot 16) = \log_2 64 = 6$$

and that

$$\log_2 4 + \log_2 16 = 2 + 4 = 6.$$

■

EXAMPLE 2 Express as a single logarithm: $\log_{10} 0.01 + \log_{10} 1000$.

Solution

$$\log_{10} 0.01 + \log_{10} 1000 = \log_{10} (0.01 \times 1000) \quad \text{By Property 1}$$
$$= \log_{10} 10$$

■

A Proof of Property 1. Let $\log_a M = x$ and $\log_a N = y$. Converting to exponential equations, we have $a^x = M$ and $a^y = N$.

Now we multiply the latter two equations, to obtain

$$MN = a^x \cdot a^y = a^{x+y}.$$

Converting back to a logarithmic equation, we get

$$\log_a MN = x + y.$$

Remembering what x and y represent, we get

$$\log_a MN = \log_a M + \log_a N,$$

as desired. ■

Logarithms of Powers

The second basic property is as follows.

PROPERTY 2

For any positive number M and any real number p,

$$\log_a M^p = p \cdot \log_a M.$$

(The logarithm of a power of M is the exponent times the logarithm of M. The number a can be any logarithm base.)

EXAMPLES Express as a product.

3. $\log_a 9^{-5} = -5 \log_a 9$ **By Property 2**

4. $\log_a \sqrt[4]{5} = \log_a 5^{1/4}$ **Writing exponential notation**

$= \frac{1}{4} \log_a 5$ **By Property 2**

■

A Proof of Property 2. Let $x = \log_a M$. We then convert to an exponential equation, to get $a^x = M$. Raising both sides to the pth power, we obtain

$$(a^x)^p = M^p, \qquad \text{or} \qquad a^{xp} = M^p.$$

Converting back to a logarithmic equation, we get

$$\log_a M^p = xp.$$

But $x = \log_a M$, so substituting, we have

$$\log_a M^p = (\log_a M)p = p \cdot \log_a M,$$

as desired. ■

Logarithms of Quotients

Here is the third basic property.

PROPERTY 3

For any positive numbers M and N,

$$\log_a \frac{M}{N} = \log_a M - \log_a N.$$

(The logarithm of a quotient is the logarithm of the dividend minus the logarithm of the divisor. The number a can be any logarithm base.)

EXAMPLE 5 Express as a difference of logarithms: $\log_t (6/U)$.

Solution

$$\log_t \frac{6}{U} = \log_t 6 - \log_t U \qquad \text{By Property 3}$$ ■

EXAMPLE 6 Express as a single logarithm: $\log_b 17 - \log_b 27$.

Solution

$$\log_b 17 - \log_b 27 = \log_b \frac{17}{27} \qquad \text{By Property 3}$$ ■

A Proof of Property 3. Our proof makes use of Property 1 and Property 2:

$$\begin{aligned} \log_a \frac{M}{N} &= \log_a MN^{-1} \\ &= \log_a M + \log_a N^{-1} && \text{By Property 1} \\ &= \log_a M + (-1)\log_a N && \text{By Property 2} \\ &= \log_a M - \log_a N. \end{aligned}$$ ■

Using the Properties Together

EXAMPLES Express in terms of logarithms of x, y, z, m, and n.

7. $\log_a \frac{x^2y^3}{z^4} = \log_a (x^2y^3) - \log_a z^4$ Using Property 3

$= \log_a x^2 + \log_a y^3 - \log_a z^4$ Using Property 1

$= 2 \log_a x + 3 \log_a y - 4 \log_a z$ Using Property 2

8. $\log_a \sqrt[4]{\frac{xy}{z^3}} = \log_a \left(\frac{xy}{z^3}\right)^{1/4}$ Writing exponential notation

$= \frac{1}{4} \cdot \log_a \frac{xy}{z^3}$ Using Property 2

$= \frac{1}{4}(\log_a xy - \log_a z^3)$ Using Property 3. Parentheses are important.

$= \frac{1}{4}(\log_a x + \log_a y - 3 \log_a z)$ Using Properties 1 and 2

9. $\log_b \frac{xy}{m^3n^4} = \log_b xy - \log_b m^3n^4$ Using Property 3

$= (\log_b x + \log_b y) - (\log_b m^3 + \log_b n^4)$ Using Property 1

$= \log_b x + \log_b y - \log_b m^3 - \log_b n^4$ Removing parentheses

$= \log_b x + \log_b y - 3 \log_b m - 4 \log_b n$ Using Property 2 ■

EXAMPLES Express as a single logarithm.

10. $\frac{1}{2} \log_a x - 7 \log_a y + \log_a z$

$= \log_a \sqrt{x} - \log_a y^7 + \log_a z$ Using Property 2: recall that $x^{\frac{1}{2}} = \sqrt{x}$.

$= \log_a \frac{\sqrt{x}}{y^7} + \log_a z$ Using Property 3

$= \log_a \frac{z\sqrt{x}}{y^7}$ Using Property 1

11. $\log_a \frac{b}{\sqrt{x}} + \log_a \sqrt{bx} = \log_a \frac{b}{\sqrt{x}} \sqrt{bx}$ Using Property 1

$= \log_a b\sqrt{b}$ Removing a factor of 1: $\sqrt{x}/\sqrt{x}$

$= \log_a b^{3/2}$, or $\frac{3}{2} \log_a b$ Since $b\sqrt{b} = b^1 \cdot b^{\frac{1}{2}}$ ■

EXAMPLES Suppose that, for some choice of a,

$$\log_a 2 = 0.301 \quad \text{and} \quad \log_a 3 = 0.477.$$

Find each of the following.

12. $\log_a 6$ $\log_a 6 = \log_a(2 \cdot 3) = \log_a 2 + \log_a 3$ Property 1

$= 0.301 + 0.477$

$= 0.778$

13. $\log_a \frac{2}{3}$ $\log_a \frac{2}{3} = \log_a 2 - \log_a 3$ Property 3

$= 0.301 - 0.477$

$= -0.176$

14. $\log_a 81$ $\quad \log_a 81 = \log_a 3^4 = 4 \log_a 3$ **Property 2**

$= 4(0.477)$

$= 1.908$

15. $\log_a \frac{1}{3}$ $\quad \log_a \frac{1}{3} = \log_a 1 - \log_a 3$ **Property 3**

$= 0 - 0.477$ **Recall that $\log_a 1 = 0$ because $a^0 = 1$**

$= -0.477$

16. $\log_a 2a$ $\quad \log_a 2a = \log_a 2 + \log_a a$ **Property 1**

$= 0.301 + 1$

$= 1.301$

17. $\log_a 5$ $\quad$ *No way to find using these properties.*
($\log_a 5 \neq \log_a 2 + \log_a 3$) ■

The Logarithm of the Base to a Power

The final property that we will consider is as follows.

PROPERTY 4

For any base a,

$$\log_a a^k = k.$$

The logarithm, base a, of a to a power is the power.

A Proof of Property 4. The proof involves Property 2 as well as the fact that $\log_a a = 1$:

$$\log_a a^k = k(\log_a a) \quad \text{Using Property 2}$$
$$= k \cdot 1 \quad \text{Using } \log_a a = 1$$
$$= k.$$
■

If you forget Property 4, you can apply Property 2 and the fact that $\log_a a = 1$. However, if you remember the definition of logarithm, it should be easy to see that k is the power to which you raise a in order to get a^k.

EXAMPLES Simplify.

18. $\log_3 3^7 = 7$ $\quad$ **7 is the power to which you raise 3 in order to get 3^7.**

19. $\log_{10} 10^{5.6} = 5.6$

20. $\log_e e^{-t} = -t$ ■

EXERCISE SET 9.4

Express as a sum of logarithms.

1. $\log_2 (32 \cdot 8)$

2. $\log_3 (27 \cdot 81)$

3. $\log_4 (64 \cdot 16)$

4. $\log_5 (25 \cdot 125)$

5. $\log_c Bx$

6. $\log_t 5Y$

Express as a single logarithm.

7. $\log_a 6 + \log_a 70$
8. $\log_b 65 + \log_b 2$
9. $\log_c K + \log_c y$
10. $\log_t H + \log_t M$

Express as a product.

11. $\log_a x^3$
12. $\log_b t^5$
13. $\log_c y^6$
14. $\log_{10} y^7$
15. $\log_b C^{-3}$
16. $\log_c M^{-5}$

Express as a difference of logarithms.

17. $\log_a \frac{67}{5}$
18. $\log_t \frac{T}{7}$
19. $\log_b \frac{3}{4}$
20. $\log_a \frac{y}{x}$

Express as a single logarithm.

21. $\log_a 15 - \log_a 7$
22. $\log_b 42 - \log_b 7$

Express in terms of logarithms of w, x, y, and z.

23. $\log_a x^2y^3z$
24. $\log_a xy^4z^3$
25. $\log_b \frac{xy^2}{z^3}$
26. $\log_b \frac{x^2y^5}{w^4z^7}$
27. $\log_c \sqrt[3]{\frac{x^4}{y^3z^2}}$
28. $\log_a \sqrt{\frac{x^6}{y^5z^8}}$
29. $\log_a \sqrt[4]{\frac{x^8y^{12}}{a^3z^5}}$
30. $\log_a \sqrt[3]{\frac{x^6y^3}{a^2z^7}}$

Express as a single logarithm and simplify, if possible.

31. $\frac{2}{3} \log_a x - \frac{1}{2} \log_a y$
32. $\frac{1}{2} \log_a x + 3 \log_a y - 2 \log_a x$
33. $\log_a 2x + 3(\log_a x - \log_a y)$
34. $\log_a x^2 - 2 \log_a \sqrt{x}$
35. $\log_a \frac{a}{\sqrt{x}} - \log_a \sqrt{ax}$
36. $\log_a (x^2 - 4) - \log_a (x - 2)$

Given $\log_b 3 = 1.099$ and $\log_b 5 = 1.609$, find each of the following.

37. $\log_b 15$
38. $\log_b \frac{3}{5}$
39. $\log_b \frac{5}{3}$
40. $\log_b \frac{1}{3}$
41. $\log_b \frac{1}{5}$
42. $\log_b \sqrt{b}$
43. $\log_b \sqrt{b^3}$
44. $\log_b 3b$
45. $\log_b 5b$
46. $\log_b 9$
47. $\log_b 25$
48. $\log_b 75$

Simplify.

49. $\log_t t^9$
50. $\log_p p^4$
51. $\log_e e^m$
52. $\log_Q Q^{-2}$

Solve for x.

53. $\log_3 3^4 = x$
54. $\log_5 5^7 = x$
55. $\log_e e^x = -7$
56. $\log_a a^x = 2.7$

SKILL MAINTENANCE

Solve.

57. $\sqrt{x + 7} = 5$
58. $\sqrt[3]{5x - 4} = 4$
59. $t^4 - 8t^2 + 12 = 0$
60. $a^{-2} + a^{-1} - 6 = 0$

SYNTHESIS

Express as a single logarithm and simplify, if possible.

61. $\log_a (x^8 - y^8) - \log_a (x^2 + y^2)$
62. $\log_a (x + y) + \log_a (x^2 - xy + y^2)$

Express as a sum or difference of logarithms.

63. $\log_a \sqrt{1 - s^2}$

64. $\log_a \dfrac{c - d}{\sqrt{c^2 - d^2}}$

65. If $\log_a x = 2$, $\log_a y = 3$, and $\log_a z = 4$, what is

$\log_a \dfrac{\sqrt[3]{x^2 z}}{\sqrt[3]{y^2 z^{-2}}}$?

Solve.

66. $(x + 3) \cdot \log_a a^x = x$

67. $\log_a 5x = \log_a 5 + \log_a x$

68. If $\log_a x = 2$, what is $\log_a (1/x)$?

69. If $\log_a x = 2$, what is $\log_{1/a} x$?

Determine whether each of the following is true. Assume a, x, P, and $Q > 0$.

70. $\dfrac{\log_a P}{\log_a Q} = \log_a \dfrac{P}{Q}$

71. $\dfrac{\log_a P}{\log_a Q} = \log_a P - \log_a Q$

72. $\log_a 3x = \log_a 3 + \log_a x$

73. $\log_a 3x = 3 \log_a x$

74. $\log_a (P + Q) = \log_a P + \log_a Q$

75. $\log_a x^2 = 2 \log_a x$

Prove the following.

76. $\log_a \left(\dfrac{1}{x}\right) = -\log_a x, \quad a > 0, x > 0$

77. $\log_a \left(\dfrac{x + \sqrt{x^2 - 3}}{3}\right) = -\log_a \left(x - \sqrt{x^2 - 3}\right),$
$a > 0, x \geq \sqrt{3}$

9.5 Calculators, Natural Logarithms, and Changing Bases

Any positive number different from 1 can be used as the base of a logarithmic function. However, some numbers are easier to use than others, and there are logarithm bases that fit into certain applications more naturally than others. Base-10 logarithms, called **common logarithms,** are useful because they are the same base as our "commonly" used decimal system for naming numbers.

Before calculators became so widely available, common logarithms were extensively used in calculations. In fact, that is why logarithms were invented. Another logarithm base widely used today is, interestingly enough, an irrational number, named e. We will consider e and natural logarithms later in this section. We first consider common logarithms.

Common Logarithms on a Calculator

Before the invention of calculators, tables were developed for finding common logarithms. Here we find common logarithms using calculators.

The abbreviation *log,* with no base written, is used for base-10 logarithms, or common logarithms. Thus

$$\log 17 \quad \text{means} \quad \log_{10} 17.$$

On scientific calculators, the key for common logarithms is usually marked [log]. To find the common logarithm of a number, we enter that number and press the [log] key. If your calculator works differently, press the [log] key first or consult an owner's manual.

EXAMPLE 1 Find log 53,128.

Solution We enter 53,128 and then press the [log] key. We find that

$$\log 53{,}128 \approx 4.7253. \quad \text{Rounded to four decimal places}$$

EXAMPLE 2 Find log 0.000128.

Solution We enter 0.000128 and then press the [log] key. We find that

$$\log 0.000128 \approx -3.8928. \quad \text{Rounded to four decimal places}$$

The inverse of a logarithmic function is, of course, an exponential function. The inverse of finding a logarithm is called finding an *antilogarithm.* To find an antilogarithm, we *exponentiate*:

$$\text{for} \quad f(x) = \log x, \quad f^{-1}(x) = \text{antilog } x = 10^x.$$

Generally, there is no key on a calculator marked "antilog." It is up to you to know that to find the inverse, or antilogarithm, you must use the [10^x] key, if there is one. If there is no such key, then you must raise 10 to the x power using a key marked [x^y]. Often the [log] key serves as the [10^x] key after a "shift" key is pushed.

EXAMPLE 3 Find antilog 2.1792.

Solution We enter 2.1792 and then press the [10^x] key. We find that

$$\text{antilog } 2.1792 = 10^{2.1792} \approx 151.078.$$

EXAMPLE 4 Find antilog (-4.678834).

Solution

$$\text{antilog } (-4.678834) = 10^{-4.678834} = 0.00002095$$

The Base e and Natural Logarithms on a Calculator

When interest is computed n times a year, the compound interest formula is

$$A = P\left(1 + \frac{i}{n}\right)^{nt},$$

where A is the amount that an initial investment P will be worth after t years at interest rate i. Suppose that \$1 is an initial investment at 100% interest for 1 year (no bank would pay this). The preceding formula becomes a function A defined in terms of the number of compounding periods n:

$$A(n) = \left(1 + \frac{1}{n}\right)^n.$$

Let us find some function values. We round to six decimal places. We use a calculator with a power key $\boxed{x^y}$.

n	$A(n) = \left(1 + \frac{1}{n}\right)^n$
1 (compounded annually)	\$2.00
2 (compounded semiannually)	\$2.25
3	\$2.370370
4 (compounded quarterly)	\$2.441406
5	\$2.488320
100	\$2.704814
365 (compounded daily)	\$2.714567
8760 (compounded hourly)	\$2.718127

The numbers in this table get closer and closer to a very important number in mathematics, called e. Being irrational, its decimal representation does not terminate or repeat.

$$e \approx 2.7182818284\ldots$$

Logarithms to the base e are called **natural logarithms.**

The abbreviation "ln" is generally used with natural logarithms. Thus

$$\ln 53 \quad \text{means} \quad \log_e 53.$$

The calculator key $\boxed{\ln}$ is used for natural logarithms.

EXAMPLE 5 Find ln 4568.

Solution We enter 4568 and then press the $\boxed{\ln}$ key. We find that

$$\ln 4568 \approx 8.4268. \quad \text{Rounded to four decimal places}$$ ■

To find the antilogarithm, base e, we use the $\boxed{e^x}$ key, if there is one. If not, we use a power key $\boxed{x^y}$ and an approximation for e, say, 2.71828. Often the $\boxed{\ln}$ key serves as the $\boxed{e^x}$ key after a "shift" key is pressed.

EXAMPLE 6 Find antilog$_e$ 2.1792.

Solution The problem gives the exponent. We enter 2.1792 and then press the $\boxed{e^x}$ key. We find that

$$\begin{aligned}\text{antilog}_e\, 2.1792 &= e^{2.1792}\\ &\approx 8.8392.\end{aligned}$$ ■

Changing Logarithmic Bases

Most calculators give the values of both common logarithms and natural logarithms. To find a logarithm with some other base, we can use the following conversion formula.

CHANGE-OF-BASE FORMULA

For any logarithmic bases a and b, and any positive number M,

$$\log_b M = \frac{\log_a M}{\log_a b}.$$

Proof. Let $x = \log_b M$. Then, writing an equivalent exponential equation, we have $b^x = M$. Next we take the logarithmic base a on both sides. This gives us

$$\log_a b^x = \log_a M.$$

By Property 2,

$$x \log_a b = \log_a M,$$

and solving for x, we obtain

$$x = \frac{\log_a M}{\log_a b}.$$

But $x = \log_b M$, so we have

$$\log_b M = \frac{\log_a M}{\log_a b},$$

which is the change-of-base formula. ■

EXAMPLE 7 Find $\log_5 8$ using common logarithms.

Solution Let $a = 10$, $b = 5$, and $M = 8$. Then substitute into the change-of-base formula:

$$\log_5 8 = \frac{\log_{10} 8}{\log_{10} 5} \quad \text{Substituting}$$

$$\approx \frac{0.9031}{0.6990} \quad \text{When using your calculator, you need not round before dividing.}$$

$$\approx 1.2920.$$

To check, we use a calculator with an $\boxed{x^y}$ key to verify that

$$5^{1.2920} \approx 8.$$ ■

We can also use base e for a conversion.

EXAMPLE 8 Find $\log_4 31$ using natural logarithms.

Solution Substituting e for a, 4 for b, and 31 for M, we have

$$\log_4 31 = \frac{\log_e 31}{\log_e 4} \quad \text{Using the change-of-base formula}$$
$$= \frac{\ln 31}{\ln 4}$$
$$\approx \frac{3.4340}{1.3863}$$
$$\approx 2.4771.$$

Graphs of Exponential and Logarithmic Functions, Base *e*

EXAMPLE 9 Graph $f(x) = e^x$ and $f(x) = e^{-x}$.

Solution We use a calculator with an $\boxed{e^x}$ key to find approximate values of e^x and e^{-x}. Using these values, we can draw the graphs of the functions.

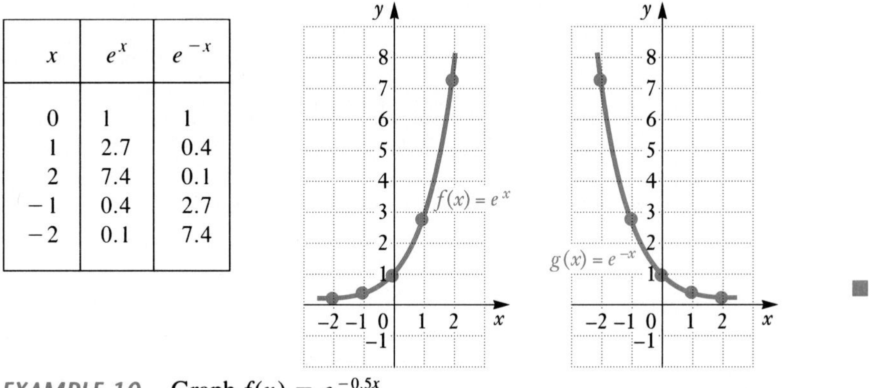

x	e^x	e^{-x}
0	1	1
1	2.7	0.4
2	7.4	0.1
−1	0.4	2.7
−2	0.1	7.4

EXAMPLE 10 Graph $f(x) = e^{-0.5x}$.

Solution We find some solutions with a calculator, plot them, and then draw the graph. For example, $f(2) = e^{-0.5(2)} = e^{-1} \approx 0.4$.

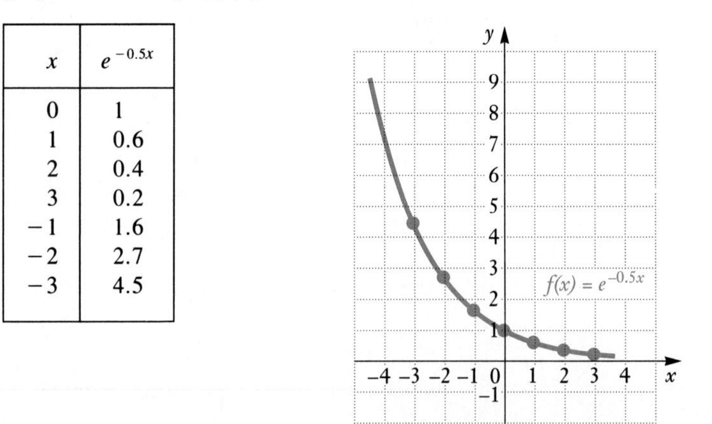

x	$e^{-0.5x}$
0	1
1	0.6
2	0.4
3	0.2
−1	1.6
−2	2.7
−3	4.5

EXAMPLE 11 Graph $g(x) = \ln x$.

Solution We find some solutions with a calculator and then draw the graph. As expected, the graph is a reflection across the line $y = x$ of the graph of $y = e^x$.

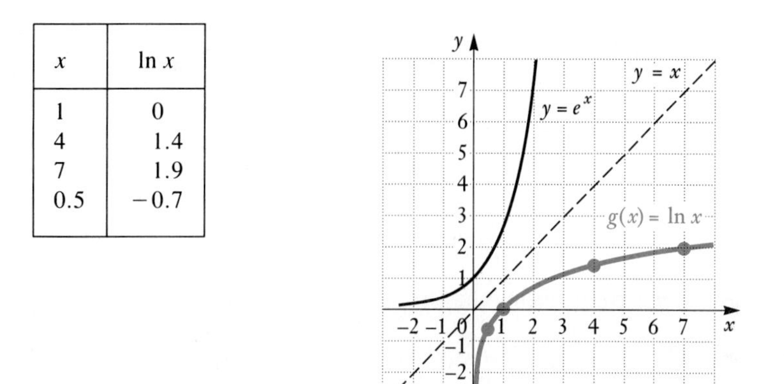

x	$\ln x$
1	0
4	1.4
7	1.9
0.5	−0.7

■

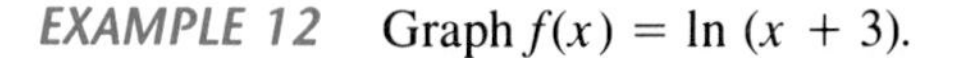

EXAMPLE 12 Graph $f(x) = \ln(x + 3)$.

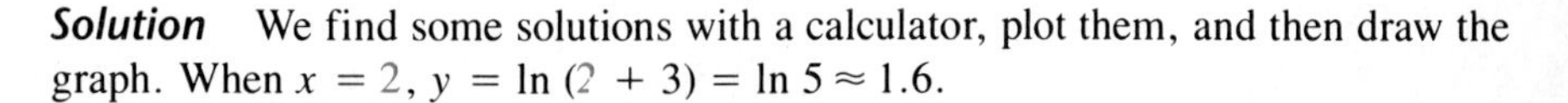

Solution We find some solutions with a calculator, plot them, and then draw the graph. When $x = 2$, $y = \ln(2 + 3) = \ln 5 \approx 1.6$.

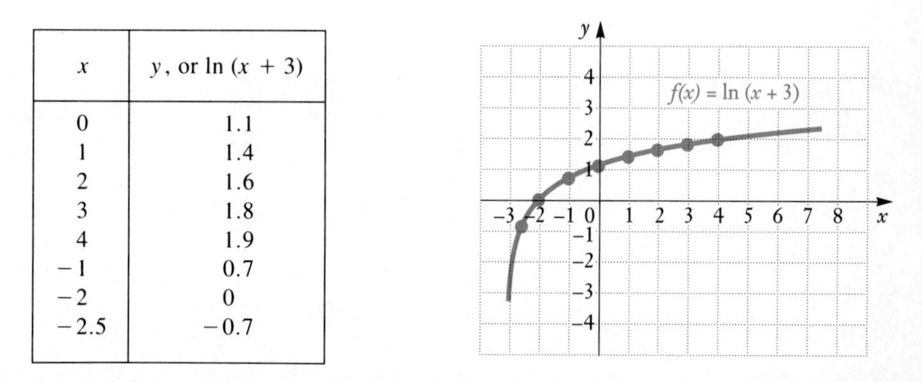

x	y, or $\ln(x + 3)$
0	1.1
1	1.4
2	1.6
3	1.8
4	1.9
−1	0.7
−2	0
−2.5	−0.7

Note that the graph of $y = \ln(x + 3)$ is a horizontal translation (three units to the left) of the graph of $y = \ln x$. ■

EXERCISE SET 9.5

Use a calculator to find the following common logarithms and antilogarithms.

1. log 2
2. log 5
3. log 6.34
4. log 5.02
5. log 45
6. log 74
7. log 437
8. log 295
9. log 13,400
10. log 93,100
11. log 0.052
12. log 0.387

13. antilog 3
14. antilog 5
15. antilog 2.7
16. antilog 14.8
17. antilog 0.477133
18. antilog 0.06532
19. antilog (−0.5465)
20. antilog (−0.3404)
21. $10^{-2.9523}$
22. $10^{4.8982}$

Find the following logarithms and antilogarithms, base e, using a calculator.

23. ln 2
24. ln 3
25. ln 62
26. ln 30
27. ln 4365
28. ln 901.2
29. ln 0.0062
30. ln 0.00073
31. antilog_e 3.6052
32. antilog_e 4.9312
33. antilog_e (−6.0751)
34. antilog_e (−2.3001)
35. antilog_e 0.00567
36. antilog_e 0.01111
37. antilog_e 34
38. antilog_e 56
39. $e^{2.0325}$
40. $e^{-1.3783}$

Find the following logarithms using the change-of-base formula.

41. $\log_6 100$
42. $\log_3 18$
43. $\log_2 10$
44. $\log_7 50$
45. $\log_{200} 30$
46. $\log_{100} 30$
47. $\log_{0.5} 5$
48. $\log_{0.1} 3$
49. $\log_2 0.2$
50. $\log_2 0.08$
51. $\log_\pi 58$
52. $\log_\pi 200$

Graph.

53. $f(x) = e^x$
54. $f(x) = e^{3x}$
55. $f(x) = e^{-3x}$
56. $f(x) = e^{-x}$
57. $f(x) = e^{x-1}$
58. $f(x) = e^{x+2}$
59. $f(x) = e^{-x} - 3$
60. $f(x) = e^x + 3$
61. $f(x) = 5e^{0.2x}$
62. $f(x) = 8e^{0.6x}$
63. $f(x) = 20e^{-0.5x}$
64. $f(x) = 10e^{-0.4x}$
65. $f(x) = \ln(x + 4)$
66. $f(x) = \ln(x + 1)$
67. $f(x) = 2 - \ln x$
68. $f(x) = 3 - \ln x$
69. $f(x) = 3 \ln x$
70. $f(x) = 2 \ln x$
71. $f(x) = \ln(x - 2)$
72. $f(x) = \ln(x - 3)$
73. $f(x) = \frac{1}{2} \ln x$
74. $f(x) = 0.8 \ln x$
75. $f(x) = \ln x - 3$
76. $f(x) = \ln x + 2$
77. $f(x) = 1 - e^{-x}$
78. $f(x) = 1 - e^{-0.2x}$

SKILL MAINTENANCE

Solve for x.

79. $4x^2 - 25 = 0$
80. $5x^2 - 7x = 0$

Solve.

81. $x^{1/2} - 6x^{1/4} + 8 = 0$
82. $2y - 7\sqrt{y} + 3 = 0$

SYNTHESIS

83. Find a formula for converting common logarithms to natural logarithms.
84. Find a formula for converting natural logarithms to common logarithms.

Solve for x.

85. $\log 374x = 4.2931$
86. $\log 95x^2 = 3.0177$
87. $\log 692 + \log x = \log 3450$
88. $\dfrac{4.31}{\ln x} = \dfrac{28}{3.01}$

9.6 Exponential and Logarithmic Equations and Problem Solving

Solving Exponential Equations

Equations with variables in exponents, such as $5^x = 12$ and $2^{7x} = 64$, are called **exponential equations.** Sometimes, as is the case with $2^{7x} = 64$, we can write each side as a power of the same number:

$$2^{7x} = 2^6.$$

Then the exponents are the same and we can set them equal and solve:

$$7x = 6$$
$$x = \tfrac{6}{7}.$$

We use the following property.

> For any $a > 0$, $a \neq 1$,
>
> $a^x = a^y$ is equivalent to $x = y$.

This follows from the fact that $f(x) = a^x$ is a one-to-one function, in which case two outputs a^x and a^y are equal only if their inputs are equal.

EXAMPLE 1 Solve: $2^{3x-5} = 16$.

Solution Note that $16 = 2^4$. Thus we can write each side as a power of the same number:

$$2^{3x-5} = 2^4.$$

Since the base is the same, 2, the exponents must be the same. Thus,

$$3x - 5 = 4$$
$$3x = 9$$
$$x = 3.$$

The check is left for the student. The solution is 3. ■

When it does not seem possible to write each side as a power of the same base, we can take the common or natural logarithm on each side and then use Property 2.

EXAMPLE 2 Solve: $5^x = 12$.

Solution

$$\begin{aligned}
5^x &= 12 \\
\log 5^x &= \log 12 && \text{Taking the common logarithm on both sides} \\
x \log 5 &= \log 12 && \text{Property 2} \\
x &= \frac{\log 12}{\log 5} && \text{Solving for } x
\end{aligned}$$

Caution! This is *not* $\log 12 - \log 5$.

This is an exact answer. We cannot simplify further, but we can approximate using a calculator or a table:

$$x = \frac{\log 12}{\log 5} \approx \frac{1.0792}{0.6990} \approx 1.544.$$

You can check this answer by finding $5^{1.544}$ using the $\boxed{x^y}$ key on a calculator. ■

If we prefer, we can take the logarithm with e as the base. This will often ease our work.

EXAMPLE 3 Solve: $e^{0.06t} = 1500$.

Solution We take the natural logarithm on both sides:

$$\begin{aligned}
\ln e^{0.06t} &= \ln 1500 && \text{Taking the natural logarithm on both sides} \\
0.06t &= \ln 1500 && \text{Here we use Property 4: } \log_a a^k = k. \\
0.06t &\approx 7.3132 && \text{Using a calculator} \\
t &\approx 121.89.
\end{aligned}$$

■

Solving Logarithmic Equations

Equations containing logarithmic expressions are called **logarithmic equations.** We solved some such equations in Section 9.3. We did so by converting to an equivalent exponential equation.

EXAMPLE 4 Solve: $\log_2 x = 3$.

Solution We obtain an equivalent exponential equation:

$$\begin{aligned}
x &= 2^3 \\
x &= 8.
\end{aligned}$$

The solution is 8. ■

> To solve logarithmic equations, first try to obtain a single logarithmic expression on one side and then write an equivalent exponential equation.

EXAMPLE 5 Solve: $\log_4 (8x - 6) = 3$.

Solution We already have a single logarithmic expression, so we write an equivalent exponential equation:

$$8x - 6 = 4^3 \quad \text{Writing an equivalent exponential equation}$$
$$8x - 6 = 64$$
$$8x = 70$$
$$x = \frac{70}{8}, \quad \text{or } \frac{35}{4}.$$

The check is left for the student. The solution is $\frac{35}{4}$. ■

EXAMPLE 6 Solve: $\log x + \log (x - 3) = 1$.

Solution We have common logarithms here. It will help to write in the base, 10.

$$\log_{10} x + \log_{10} (x - 3) = 1$$
$$\log_{10} [x(x - 3)] = 1 \quad \text{Using Property 1 to obtain a single logarithm}$$
$$x(x - 3) = 10^1 \quad \text{Writing an equivalent exponential equation}$$
$$x^2 - 3x = 10$$
$$x^2 - 3x - 10 = 0$$
$$(x + 2)(x - 5) = 0 \quad \text{Factoring}$$
$$x + 2 = 0 \quad \text{or} \quad x - 5 = 0 \quad \text{Principle of zero products}$$
$$x = -2 \quad \text{or} \quad x = 5$$

Check: For -2:

$\log x + \log (x - 3) = 1$	
$\log (-2) + \log (-2 - 3)$	1

The number -2 *does not check* because negative numbers do not have logarithms.

For 5:

$\log x + \log (x - 3) = 1$	
$\log 5 + \log (5 - 3)$	1
$\log 5 + \log 2$	
$\log 10$	
1	TRUE

The solution is 5. ■

Applications

There are many important applications involving logarithmic and exponential functions.

EXAMPLE 7 *Chemistry: pH of substances.* In chemistry the pH of a substance is defined as follows:

$$\text{pH} = -\log [\text{H}^+],$$

where $[\text{H}^+]$ is the hydrogen ion concentration in moles per liter.

a) The hydrogen ion concentration of pineapple juice is 1.6×10^{-4} moles per liter. Find the pH.
b) The pH of a common hair rinse is 2.9. Find the hydrogen ion concentration.

Solution

a) To find the pH of pineapple juice, we substitute 1.6×10^{-4} for $[\text{H}^+]$ in the formula for pH:

$$\begin{aligned}\text{pH} = -\log [\text{H}^+] &= -\log (1.6 \times 10^{-4}) = -[\log 1.6 + \log 10^{-4}] \\ &= -[\log 1.6 + (-4)] \\ &\approx -[0.2041 + (-4)] \\ &= -(-3.7959) \approx 3.8.\end{aligned}$$

The pH of pineapple juice is about 3.8.

b) To find the hydrogen ion concentration of the hair rinse, we substitute 2.9 for pH in the formula and solve for $[\text{H}^+]$:

$$2.9 = -\log [\text{H}^+]$$

$$-2.9 = \log [\text{H}^+] \qquad \text{Multiplying by } -1 \text{ on both sides}$$

$$10^{-2.9} = [\text{H}^+] \qquad \text{Writing an equivalent exponential equation}$$

$$0.0013 \approx [\text{H}^+]$$

$$[\text{H}^+] \approx 1.3 \times 10^{-3} \text{ moles per liter.} \qquad \text{Writing scientific notation: a number between 1 and 10, times a power of 10}$$

■

EXAMPLE 8 *Earthquake magnitude.* The magnitude R (measured on the Richter scale) of an earthquake of intensity I is defined as

$$R = \log \frac{I}{I_0},$$

where I_0 is a minimum intensity used for comparison. We can think of I_0 as a threshold intensity, which is the weakest earthquake that can be recorded on a seismograph. When one earthquake is 10 times as intense as another, its magnitude on the Richter scale is 1 higher. If an earthquake is 100 times as intense as another, its magnitude on the Richter scale is 2 higher, and so on. Thus an earthquake whose magnitude is 7 on the Richter scale is 10 times as intense as an earthquake whose magnitude is 6. The San Francisco (Loma Prieta) earthquake of 1989 had an intensity of $10^{7.2} I_0$. What was its magnitude on the Richter scale?

Solution We substitute into the formula:

$$R = \log\frac{10^{7.2} I_0}{I_0} = \log 10^{7.2} = 7.2.$$

■

EXAMPLE 9 *Interest compounded annually.* Suppose that \$100,000 is invested at 8% interest, compounded annually. In t years, it will grow to the amount A given by the function

$$A(t) = \$100{,}000(1.08)^t.$$

(See Example 5 in Section 9.1.)

a) How long will it take until there is \$500,000 in the account?

b) Let T = the amount of time it takes for the \$100,000 to double itself. Find the *doubling time, T*.

Solution

a) We set $A(t) = \$500{,}000$ and solve for t:

$$500{,}000 = 100{,}000(1.08)^t$$

$$\frac{500{,}000}{100{,}000} = (1.08)^t$$

$$5 = (1.08)^t$$

$$\log 5 = \log(1.08)^t \quad \text{Taking the common logarithm on both sides}$$

$$\log 5 = t \log 1.08 \quad \text{Property 2}$$

$$\frac{\log 5}{\log 1.08} = t.$$

We simplify further by using a calculator or a table and approximating:

$$t = \frac{\log 5}{\log 1.08} \approx \frac{0.69897}{0.03342} \approx 20.9.$$

It will take about 20.9 years for the \$100,000 to grow to \$500,000.

b) To find the doubling time T, we set $A(t) = \$200{,}000$ and $t = T$ and solve for T:

$$200{,}000 = 100{,}000(1.08)^T$$

$$2 = (1.08)^T \quad \text{Dividing by 100,000 on both sides}$$

$$\log 2 = \log(1.08)^T \quad \text{Taking the common logarithm on both sides}$$

$$\log 2 = T \log 1.08 \quad \text{Property 2}$$

$$T = \frac{\log 2}{\log 1.08} \approx \frac{0.30103}{0.03342} \approx 9.0.$$

The doubling time is about 9 years.

■

Exponential Growth

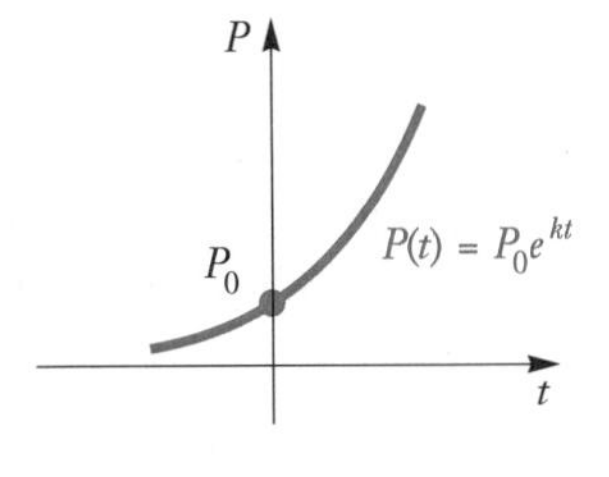

The equation

$$P(t) = P_0e^{kt}$$

is an effective model of many kinds of growth, ranging from growing populations to investments that are increasing in value. In this equation P_0 is the population at time 0, P is the population at time t, and k is a positive constant that depends on the situation. The constant k is often called the **exponential growth rate.** You should regard the exponential growth rate as a population's rate of growth at any *instant* in time. Since the population is continually growing, the percent of total growth after one year will exceed the exponential growth rate.

EXAMPLE 10 *Growth of the United States.* In 1985 the population of the United States was 234 million and the exponential growth rate was 0.8% per year.

a) Find the exponential growth function.
b) What would you expect the population to be in the year 2000?

Solution

a) In 1985, at $t = 0$, the population was 234 million. We substitute 234 for P_0 and 0.8%, or 0.008, for k to obtain the exponential growth function:

$$P(t) = 234e^{0.008t}.$$

b) In 2000 we have $t = 15$. That is, 15 years have passed. To find the population in 2000, we substitute 15 for t:

$$\begin{aligned} P(15) &= 234e^{0.008(15)} && \textbf{Substituting 15 for } t \\ &= 234e^{0.12} \\ &\approx 234(1.1275) && \textbf{Finding } e^{0.12} \textbf{ using a calculator} \\ &\approx 263.8 \text{ million.} \end{aligned}$$

The population of the United States in 2000 will be about 263.8 million. ■

EXAMPLE 11 *Business: Interest compounded continuously.* Suppose an amount of money P_0 is invested in a savings account at interest rate k, compounded continuously. That is, suppose that interest is computed every "instant" and added to the amount in the account. The balance $P(t)$, after t years, is given by the exponential function

$$P(t) = P_0e^{kt}.$$

a) Suppose that \$2000 is invested and grows to \$2983.65 in 5 years. Find the exponential growth function.
b) After what amount of time will the \$2000 double itself?

Solution

a) At $t = 0$, $P(0) = \$2000$. Thus the exponential growth equation is

$$P(t) = 2000e^{kt}, \qquad \text{where } k \text{ must still be determined.}$$

We know that at $t = 5$, $P(5) = \$2983.65$. We substitute and solve for k:

$$2983.65 = 2000e^{k(5)} = 2000e^{5k}$$

$$\frac{2983.65}{2000} = e^{5k} \qquad \text{Dividing on both sides by 2000}$$

$$1.491825 = e^{5k}$$

$$\ln 1.491825 = \ln e^{5k} \qquad \text{Taking the natural logarithm on both sides}$$

$$0.4 \approx 5k \qquad \text{Finding } \ln 1.491825 \text{ on a calculator and simplifying } \ln e^{5k}$$

$$\frac{0.4}{5} \approx 0.08 \approx k.$$

The interest rate is about 0.08, or 8%, compounded continuously. Note that since interest is being compounded continuously, the interest earned each year is more than 8%. The exponential growth function is

$$P(t) = 2000e^{0.08t}.$$

b) To find the doubling time T, we set $P(T) = \$4000$ and solve for T:

$$4000 = 2000e^{0.08T}$$

$$2 = e^{0.08T}$$

$$\ln 2 = \ln e^{0.08T} \qquad \text{Taking the natural logarithm on both sides}$$

$$\ln 2 = 0.08T$$

$$\frac{\ln 2}{0.08} = T \qquad \text{Dividing}$$

$$\frac{0.693147}{0.08} \approx T$$

$$8.7 \approx T.$$

Thus the original investment of \$2000 will double in about 8.7 years. ■

Comparing Examples 9 and 11, we see that for any specified interest rate, continuous compounding gives the highest yield and the shortest doubling time.

Exponential Decay

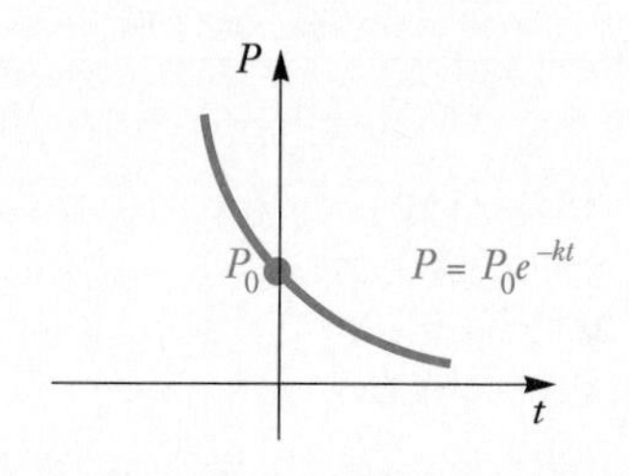

The function

$$P(t) = P_0e^{-kt}, \qquad k > 0,$$

is an effective model of the decline, or decay, of a population or quantity. An example is the decay of a radioactive substance. Here P_0 is the amount of the substance at time $t = 0$, P is the amount of the substance remaining at time t, and k is a positive constant that depends on the situation. The constant k is called the **decay rate.** The **half-life** of a substance is the amount of time necessary for half of the substance to decay.

EXAMPLE 12 *Carbon dating.* The radioactive element carbon-14 has a half-life of 5570 years. The percentage of carbon-14 present in the remains of animal bones can be used to determine age. How old is an animal bone that has lost 40% of its carbon-14?

Solution We first find k. To do this, we use the concept of half-life. When $t = 5570$ (the half-life), P will be half of P_0. Then

$$0.5P_0 = P_0e^{-k(5570)}, \qquad \text{or} \qquad 0.5 = e^{-5570k}.$$

We take the natural logarithm on both sides:

$$\ln 0.5 = \ln e^{-5570k} = -5570k.$$

Then

$$k = \frac{\ln 0.5}{-5570} \approx \frac{-0.6931}{-5570} \approx 0.00012.$$

Now we have the function

$$P(t) = P_0e^{-0.00012t}.$$

(*Note:* This equation can be used for any subsequent carbon-dating problem.) If an animal bone has lost 40% of its carbon-14 from an initial amount P_0, then $60\%(P_0)$ is the amount present. To find the age t of the bone, we solve the following equation for t:

$$\begin{aligned} 0.6\,P_0 &= P_0e^{-0.00012t} && \text{We want to find } t \text{ for which } P(t) = 0.6\,P_0. \\ 0.6 &= e^{-0.00012t} \\ \ln 0.6 &= \ln e^{-0.00012t} \\ -0.5108 &\approx -0.00012t \\ t &\approx \frac{-0.5108}{-0.00012} \\ t &\approx 4257. \end{aligned}$$

The animal bone is about 4257 years old. ■

EXERCISE SET 9.6

Solve.

1. $2^x = 8$
2. $3^x = 81$
3. $4^x = 256$
4. $5^x = 125$
5. $2^{2x} = 32$
6. $4^{3x} = 64$
7. $3^{5x} = 27$
8. $5^{7x} = 625$
9. ▦ $2^x = 9$
10. ▦ $2^x = 30$
11. ▦ $2^x = 10$
12. ▦ $2^x = 33$

13. $5^{4x-7} = 125$
14. $4^{3x+5} = 16$
15. $3^{x^2} \cdot 3^{4x} = \frac{1}{27}$
16. $3^{5x} \cdot 9^{x^2} = 27$
17. 🖩 $4^x = 7$
18. 🖩 $8^x = 10$
19. 🖩 $e^t = 100$
20. 🖩 $e^t = 1000$
21. 🖩 $e^{-t} = 0.1$
22. 🖩 $e^{-t} = 0.01$
23. 🖩 $e^{-0.02t} = 0.06$
24. 🖩 $e^{0.07t} = 2$
25. 🖩 $2^x = 3^{x-1}$
26. 🖩 $3^{x+2} = 5^{x-1}$
27. 🖩 $(2.8)^x = 41$
28. 🖩 $(3.4)^x = 80$
29. 🖩 $20 - (1.7)^x = 0$
30. 🖩 $125 - (4.5)^y = 0$

Solve.

31. $\log_3 x = 3$
32. $\log_5 x = 4$
33. $\log_2 x = -3$
34. $\log_4 x = \frac{1}{2}$
35. $\log x = 1$
36. $\log x = 3$
37. $\log x = -2$
38. $\log x = -3$
39. $\ln x = 2$
40. $\ln x = 1$
41. $\ln x = -1$
42. $\ln x = -3$
43. $\log_5 (2x - 7) = 3$
44. $\log_2 (7 - 6x) = 5$
45. $\log x + \log (x - 9) = 1$
46. $\log x + \log (x + 9) = 1$
47. $\log x - \log (x + 3) = -1$
48. $\log (x + 9) - \log x = 1$
49. $\log_2 (x + 1) + \log_2 (x - 1) = 3$
50. $\log_4 (x + 3) - \log_4 (x - 5) = 2$
51. $\log_4 (x + 6) - \log_4 x = 2$
52. $\log_2 x + \log_2 (x - 2) = 3$
53. $\log_4 (x + 3) + \log_4 (x - 3) = 2$
54. $\log_5 (x + 4) + \log_5 (x - 4) = 2$

🖩 Solve.

55. *Interest compounded annually.* Suppose that \$50,000 is invested at 6% interest, compounded annually. After t years it grows to the amount A given by the function

$$A(t) = \$50{,}000(1.06)^t.$$

a) After what amount of time will there be \$450,000 in the account?
b) Find the doubling time.

56. *Turkey consumption.* The amount of turkey consumed by each person in one year in the United States is increasing exponentially. Assume that $t = 0$ corresponds to 1937. The amount of turkey, in pounds per person, consumed t years after 1937 is given by the function

$$N(t) = 2.3(3)^{0.033t}.$$

a) After what amount of time will the consumption rate be 20 lb of turkey per year?
b) What is the doubling time on the consumption of turkey?

57. *Recycling aluminum cans.* It is known that one fourth of all aluminum cans distributed will be recycled each year. A beverage company distributes 250,000 cans. The number still in use after t years is given by the function

$$N(t) = 250{,}000\left(\tfrac{1}{4}\right)^t.$$

a) After how many years will 60,000 cans still be in use?
b) After what amount of time will only 10 cans still be in use?

58. *Salvage value.* An office machine is purchased for \$5200. Its value each year is about 80% of the preceding year. Its value in dollars after t years is given by the exponential function

$$V(t) = \$5200(0.8)^t.$$

a) After what amount of time will the salvage value be \$1200?
b) After what amount of time will the salvage value be half the original value?

59. *Interest compounded continuously.* Suppose that P_0 is invested in a savings account where interest is compounded continuously at 9% per year. That is, the balance $P(t)$ after time t, in years, is

$$P(t) = P_0 e^{kt}.$$

a) Find the exponential growth function for $P(t)$ in terms of P_0 and 0.09.
b) Suppose that $1000 is invested. What is the balance after 1 year? after 2 years?
c) When will an investment of $1000 double itself?

60. *Interest compounded continuously.* Suppose that P_0 is invested in a savings account where interest is compounded continuously at 10% per year. That is, the balance $P(t)$ after time t, in years, is

$$P(t) = P_0 e^{kt}.$$

a) Find the exponential function for $P(t)$ in terms of P_0 and 0.10.
b) Suppose that $20,000 is invested. What is the balance after 1 year? after 2 years?
c) When will an investment of $20,000 double itself?

61. *Population growth.* The exponential growth rate of the population of Europe west of the USSR is 1% per year. What is the doubling time?

62. *Population growth.* The exponential growth rate of the population of Mexico is 3.5% per year (one of the highest in the world). What is the doubling time?

63. *Compact discs.* The number of compact discs purchased each year is increasing exponentially. The number N, in millions, purchased is given by

$$N(t) = 7.5(6)^{0.5t},$$

where $t = 0$ corresponds to 1985, $t = 1$ corresponds to 1986, and so on, t being the number of years after 1985.

a) After what amount of time will one billion compact discs be sold in a year?
b) What is the doubling time on the sale of compact discs?

64. *Growth of bacteria.* The bacteria *Escherichi coli* are commonly found in the bladder of human beings. Suppose that 3000 of the bacteria are present at time $t = 0$. Then t minutes later the number of bacteria present will be

$$N(t) = 3000(2)^{t/20}.$$

a) After what amount of time will there be 60,000 bacteria?
b) If 100,000,000 bacteria accumulate, a bladder infection can occur. What amount of time would have to pass for a possible bladder infection to occur?
c) What is the doubling time?

65. *Psychology: Forgetting.* Students in an English class took a final exam. They took equivalent forms of the exam in monthly intervals thereafter. The average score $S(t)$, in percent, after t months was found to be given by

$$S(t) = 68 - 20 \log (t + 1), \quad t \geq 0.$$

a) What was the average score when they initially took the test, $t = 0$?
b) What was the average score after 4 months? 24 months?
c) Graph the function.
d) After what time t was the average score 50?

66. *Business: Advertising.* A model for advertising response is given by

$$N(a) = 2000 + 500 \log a, \quad a \geq 1,$$

where $N(a)$ = the number of units sold and a = the amount spent on advertising, in thousands of dollars.

a) How many units were sold after spending $1000 ($a = 1$) on advertising?
b) How many units were sold after spending $8000?
c) Graph the function.
d) How much would have to be spent to sell 5000 units?

Consider the pH formula of Example 7 for Exercises 67–74.

Find the pH, given the hydrogen ion concentration.

67. A common brand of mouthwash:

$$[H^+] = 6.3 \times 10^{-7} \text{ moles per liter}$$

68. A common brand of insect repellent:

$$[H^+] = 4.0 \times 10^{-8} \text{ moles per liter}$$

69. Eggs:

$$[H^+] = 1.6 \times 10^{-8} \text{ moles per liter}$$

70. Tomatoes:

$$[H^+] = 6.3 \times 10^{-5} \text{ moles per liter}$$

Find the hydrogen ion concentration of each substance, given the pH.

71. Tap water: pH = 7

72. Rainwater: pH = 5.4

73. Orange juice: pH = 3.2

74. Wine: pH = 4.8

75. The San Francisco earthquake of 1906 had an intensity of $10^{8.25}$ times I_0. What was its magnitude on the Richter scale?

76. In 1986 there was an earthquake near Cleveland, Ohio. It had an intensity of 10^5 times I_0. What was its magnitude on the Richter scale?

77. *Oil demand.* The exponential growth rate of the demand for oil in the United States is 10% per year. When will the demand be double that of 1990?

78. *Coal demand.* The exponential growth rate of the demand for coal in the world is 4% per year. When will the demand be double that of 1990?

79. *Population growth.* The population of Dallas, Texas, was 844,401 in 1970. In 1982 it was 943,848. Assuming that growth was exponential:

a) Find the value k ($P_0 = 844{,}401$). Write the function.

b) Estimate the population of Dallas in 1996.

80. *Population growth.* The population of Memphis, Tennessee, was 623,988 in 1970. In 1980 it was 646,174. Assuming that growth was exponential:

a) Find the value of k ($P_0 = 623{,}988$). Write the function.

b) Estimate the population of Memphis in 2000.

81. An ivory tusk has lost 20% of its carbon-14. How old is the tusk?

82. A piece of wood has lost 10% of its carbon-14. How old is the piece of wood?

83. The decay rate of iodine-131 is 9.6% per day. What is its half-life?

84. The decay rate of krypton-85 is 6.3% per year. What is its half-life?

85. The half-life of polonium is 3 minutes. What is its decay rate?

86. The half-life of lead is 22 years. What is its decay rate?

87. *Value of a Van Gogh painting.* The Van Gogh painting *Irises,* shown on the right, sold for $84,000 in 1947 and was sold again for $53,900,000 in 1987. Assume that the growth in the value V of the painting is exponential.

a) Find the exponential growth rate k and determine the exponential growth function, assuming $V_0 = 84{,}000$.

b) Estimate the value of the painting in 1997.

c) What is the doubling time for the value of the painting?

d) How long after 1947 will the value of the painting be $1 billion?

Van Gogh's *Irises,* a 28-by-32-inch oil on canvas.

88. *Value of a baseball card.* The collecting of baseball cards and other memorabilia has become a profitable hobby. The following card contains a photograph of Dale Murphy in his rookie season of 1977. The value of that card in 1983 was $17.50. Its value in 1987 was $48.00, an increase that occurred because Murphy has turned out to be such an outstanding player. Assume that the value of the card has grown exponentially.

a) Find the value k and determine the exponential growth function if $V_0 = \$17.50$.

b) Estimate the value of the card in 1995 and in 2000.

c) What is the doubling time for the value of the card?

d) In what year will the value of the card be $2000?

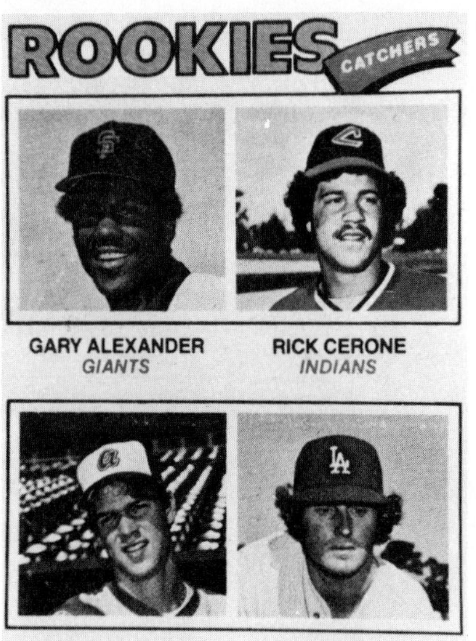

SKILL MAINTENANCE

Simplify.

89. $\dfrac{\dfrac{x-5}{x+3}}{\dfrac{x}{x-3}+\dfrac{2}{x+3}}$

90. $\dfrac{\dfrac{3}{a}+\dfrac{5}{b}}{\dfrac{2}{a^2}-\dfrac{4}{b^2}}$

91. What is the minimum product of two numbers whose difference is 8? What are the numbers?

92. The area of a cube varies directly as the square of the length of a diagonal of a side. If a cube has an area of 24 cm² when the length of a diagonal is $2\sqrt{2}$ cm, how long is a diagonal when the area of a cube is 48 cm²?

SYNTHESIS

Solve.

93. $8^x = 16^{3x+9}$

94. $27^x = 81^{2x-3}$

95. $\log_6 (\log_2 x) = 0$

96. $\log_x (\log_3 27) = 3$

97. $\log \sqrt[3]{x} = \sqrt{\log x}$

98. $\log \sqrt[4]{x} = \sqrt{\log x}$

99. $\log_5 \sqrt{x^2+1} = 1$

100. $\log \sqrt[3]{x^2} + \log \sqrt[3]{x^4} = \log 2^{-3}$

101. $\log_5 \sqrt{x^2-9} = 1$

102. $x \log \frac{1}{8} = \log 8$

103. $\log (\log x) = 5$

104. $2^{x^2+4x} = \frac{1}{8}$

105. $\log x^2 = (\log x)^2$

106. $\log_5 |x| = 4$

107. $\log x^{\log x} = 25$

108. $\log \sqrt{2x} = \sqrt{\log 2x}$

109. $\log_a a^{x^2+4x} = 21$

110. $(\log_a x)^{-1} = \log_a x$

111. *Typing speed.* Kristin is studying typing one semester in college. After she has studied for t hours, her speed, in words per minute, is given by

$$S(t) = 200[1 - (0.99)^t].$$

a) When will Kristin's speed be 100 words per minute?

b) After the course has been completed, Kristin's speed is 150 words per minute. How many hours did she study?

112. Examine the restriction on t in Exercise 65.

a) What upper limit might be placed on t?

b) In practice, would this upper limit ever be enforced? Why or why not?

CHAPTER SUMMARY AND REVIEW 9

TERMS TO KNOW

Exponential function	Logarithmic function	Doubling time
Composite functions	Exponential equation	Exponential growth rate
One-to-one function	Logarithmic equation	Decay rate
Inverse function	Common logarithm	Half-life
Horizontal line test	Natural logarithm	

IMPORTANT PROPERTIES AND FORMULAS

Exponential function:	$f(x) = a^x$
Interest compounded annually:	$A = P(1 + i)^t$
Composition of f and g:	$f \circ g(x) = f(g(x))$

TO FIND A FORMULA FOR THE INVERSE OF A FUNCTION

If a function is one-to-one, it has an inverse that is a function. A formula for the inverse can be found as follows:

1. Replace $f(x)$ by y.
2. Solve the equation for x.
3. Interchange x and y.
4. Replace y by $f^{-1}(x)$.

Definition of logarithms: $y = \log_a x$ is that number y such that $x = a^y$, where $x > 0$ and a is a positive constant other than 1.

$$y = \log_a x \text{ is equivalent to } x = a^y$$

Properties of logarithms:

$$\log_a MN = \log_a M + \log_a N, \quad \log_a \frac{M}{N} = \log_a M - \log_a N,$$

$$\log_a M^p = p \cdot \log_a M, \quad \log_a 1 = 0,$$

$$\log_a a = 1, \quad \log_a a^k = k,$$

$$\log M = \log_{10} M, \quad e \approx 2.7182818284\ldots,$$

$$\ln M = \log_e M, \quad \log_b M = \frac{\log_a M}{\log_a b}$$

For any $a > 0$, $a \neq 1$, $a^x = a^y$ is equivalent to $x = y$.

$\text{pH} = -\log\,[\text{H}^+]$

Earthquake magnitude: $R = \log \frac{I}{I_0}$

Growth: $P(t) = P_0 e^{kt}$

Interest compounded continuously: $P(t) = P_0 e^{kt}$, where P_0 is the principal invested for t years at interest rate k

Decay: $P(t) = P_0 e^{-kt}$

Carbon dating: $P(t) = P_0 e^{-0.00012t}$

REVIEW EXERCISES

The review exercises to be tested in addition to the material in this chapter are Sections 4.3, 5.5, 6.7, and 7.6.

Graph.

1. $f(x) = 3^{x-2}$

2. $x = 3^{y-2}$

3. $y = \log_3 x$

4. $f(x) = \log_{1/2} x$

5. Find $f \circ g(x)$ and $g \circ f(x)$ if $f(x) = x^2$ and $g(x) = 3x - 5$.

6. Determine whether $f(x) = 4 - x^2$ is one-to-one.

Find a formula for the inverse.

7. $f(x) = x + 2$

8. $g(x) = \dfrac{2x - 3}{7}$

9. $f(x) = \dfrac{2}{x + 5}$

10. $g(x) = 8x^3$

Convert to an exponential equation.

11. $\log_4 16 = x$

12. $\log_{10} 2 = 0.3010$

13. $\log_{1/2} 8 = -3$

14. $\log_{16} 8 = \frac{3}{4}$

Convert to a logarithmic equation.

15. $10^4 = 10{,}000$

16. $25^{1/2} = 5$

17. $7^{-2} = \dfrac{1}{49}$

18. $(2.718)^3 = 20.1$

Express in terms of logarithms of x, y, and z.

19. $\log_a x^4 y^2 z^3$

20. $\log_a \dfrac{xy}{z^2}$

21. $\log \sqrt[4]{\dfrac{z^2}{x^3 y}}$

22. $\log_q \left(\dfrac{x^2 y}{z^4}\right)^{1/3}$

Express as a single logarithm.

23. $\log_a 8 + \log_a 15$

24. $\log_a 72 - \log_a 12$

25. $\frac{1}{2} \log a - \log b - 2 \log c$

26. $\frac{1}{3}[\log_a x - 2 \log_a y]$

Simplify.

27. $\log_m m$

28. $\log_m 1$

29. $\log_m m^{17}$

30. $\log_m m^{-7}$

Given $\log_a 2 = 1.8301$ and $\log_a 7 = 5.0999$, find each of the following.

31. $\log_a 14$

32. $\log_a \frac{2}{7}$

33. $\log_a 28$

34. $\log_a 3.5$

35. $\log_a \sqrt{7}$

36. $\log_a \frac{1}{4}$

Find each of the following using a calculator.

37. $\log 0.00627$

38. $\log 72{,}800{,}000$

39. antilog 4.4742

40. antilog (-1.4425)

41. antilog 2.3294

42. $\log 0.004937$

43. $\log 394{,}900$

44. antilog (-6.7889)

45. $\ln 23{,}912.2$

46. $\ln 0.06774$

47. $\text{antilog}_e\,(-10.56)$

48. $\text{antilog}_e\,45$

Find each of the following logarithms using the change-of-base formula and a calculator.

49. $\log_5 2$

50. $\log_{12} 70$

Solve.

51. $\log_3 x = -2$

52. $\log_x 32 = 5$

53. $\log x = -4$

54. $\ln x = 2$

55. $4^{2x-5} = 16$

56. ▦ $4^x = 8.3$

57. $\log_4 16 = x$

58. $\log (x^2 - 9) - \log (x - 3) = 1$

59. $\log_4 x + \log_4 (x - 6) = 2$

60. $\log x + \log (x - 15) = 2$

61. $\log_3 (x - 4) = 3 - \log_3 (x + 4)$

62. ▦ *Forgetting.* In a business class, students were tested at the end of the course on a final exam. They were tested again after 6 months. The forgetting formula was determined to be

$$S(t) = 62 - 18 \log (t + 1),$$

where t is the time, in months, after taking the first test.

a) Determine the average score when they first took the test (when $t = 0$).

b) What was the average score after 6 months?

c) After what time was the average score 34?

63. ▦ *Cost of a prime rib dinner.* The average cost C of a prime rib dinner was \$4.65 in 1962. In 1986 it was \$15.81. Assume that the growth followed the exponential growth function.

a) Find k and write the exponential growth function.

b) How much will a prime rib dinner cost in 2002?

c) When will the average cost of a prime rib dinner be \$20?

d) What is the doubling time?

64. ▦ The population of a city doubled in 16 years. What was the exponential growth rate?

65. ▦ How long will it take \$7600 to double itself if it is invested at 8.4%, compounded continuously?

66. ▦ How old is a skeleton that has lost 34% of its carbon-14?

67. ▦ What is the pH of a substance whose hydrogen ion concentration is 2.3×10^{-7} moles per liter?

68. An earthquake has an intensity of $10^{8.3}$ times I_0. What is its amplitude on the Richter scale?

SKILL MAINTENANCE

69. Solve $aT^2 + bT = Q$ for T.

70. Solve: $x^4 + 16 = 17x^2$.

71. Simplify: $\dfrac{\dfrac{1}{ab} - \dfrac{2}{bc}}{\dfrac{2}{ac} + \dfrac{3}{ab}}$.

72. Find an equation of variation in which y varies inversely as x, and $y = 72$ when $x = \frac{1}{8}$.

SYNTHESIS

Solve.

73. $\log_2(\log_x 9^4) = 3$

74. $3^{x^2} = \dfrac{1}{81(81^x)}$

THINKING IT THROUGH

1. Compare the following properties of exponents to the properties of logarithms:

$$a^n a^m = a^{n+m}, \qquad \frac{a^n}{a^m} = a^{n-m}, \qquad (a^n)^m = a^{nm}.$$

2. Describe the types of equations we are able to solve after studying this chapter that are not like those we studied earlier.

3. Find the logarithms, base 2, of 2, 4, 8, 64, $\frac{1}{2}$, $\frac{1}{16}$, and 1. Discuss your results.

CHAPTER TEST 9

Graph.

1. $f(x) = 2^{x+3}$

2. $f(x) = \log_7 x$

3. Find $f \circ g(x)$ and $g \circ f(x)$ if $f(x) = x + x^2$ and $g(x) = 5x - 2$.

4. Determine whether $f(x) = 2 - |x|$ is one-to-one.

Find a formula for the inverse.

5. $f(x) = 4x - 3$

6. $g(x) = \dfrac{x-2}{4}$

7. $f(x) = \dfrac{x+1}{x-2}$

Convert to a logarithmic equation.

8. $4^{-3} = x$

9. $256^{1/2} = 16$

Convert to an exponential equation.

10. $\log_4 16 = 2$

11. $m = \log_7 49$

12. Express in terms of logarithms of a, b, and c:

$$\log \frac{a^3 b^{1/2}}{c^2}$$

13. Express as a single logarithm:

$$\tfrac{1}{3}\log_a x - 3\log_a y + 2\log_a z.$$

Simplify.

14. $\log_t t^{23}$

15. $\log_p p$

16. $\log_c 1$

Given that $\log_a 2 = 0.301$, $\log_a 6 = 0.778$, and $\log_a 7 = 0.845$, find each of the following.

17. $\log_a \frac{2}{7}$

18. $\log_a \sqrt{24}$

19. $\log_a 21$

Find each of the following using a calculator.

20. $\log 0.0123$

21. antilog 5.6484

22. antilog (-7.2614)

23. log 12,340

24. $\ln 0.01234$

25. $\text{antilog}_e\,(5.6774)$

26. Find $\log_{18} 31$ using the change-of-base formula and a calculator.

Solve.

27. $\log_x 25 = 2$

28. $\log_4 x = \frac{1}{2}$

29. $\log x = 4$

30. $5^{4-3x} = 125$

31. ▦ $7^x = 1.2$

32. ▦ $\ln x = \frac{1}{4}$

33. $\log (x^2 - 1) - \log (x - 1) = 1$

34. ▦ *Walking speed.* The average walking speed R of people living in a city of population P, in thousands, is given by

$$R = 0.37 \ln P + 0.05,$$

where R is in feet per second.

a) The population of Akron, Ohio, is 660,000. Find the average walking speed.

b) A city has an average walking speed of 2.6 ft/sec. Find the population.

35. ▦ *Population of the USSR.* The population of the USSR was 209 million in 1959, and the exponential growth rate was 1% per year.

a) Write an exponential function describing the population of the USSR.

b) What will the population be in 1996? in 2010?

c) When was the population 250 million?

d) What is the doubling time?

36. ▦ The population of a city doubled in 20 years. What was the exponential growth rate?

37. ▦ How long will it take an investment to double itself if it is invested at 7.6%, compounded continuously?

38. ▦ How old is an animal bone that has lost 43% of its carbon-14?

39. An earthquake has an intensity of $10^{4.7}$ times I_0. What is its amplitude on the Richter scale?

40. The hydrogen ion concentration of water is 1.0×10^{-7}. What is the pH?

SKILL MAINTENANCE

41. Solve: $y - 9\sqrt{y} + 8 = 0$.

42. Solve $S = at^2 - bt$ for t.

43. Find an equation of variation in which y varies directly as x, and $y = \frac{1}{3}$ when $x = \frac{1}{2}$.

44. Simplify:

$$\frac{\dfrac{1}{x^2 - 4}}{\dfrac{1}{x + 2} + \dfrac{1}{x - 2}}.$$

SYNTHESIS

45. Solve: $\log_5 |2x - 7| = 4$.

46. If $\log_a x = 2$, $\log_a y = 3$, and $\log_a z = 4$, find

$$\log_a \frac{\sqrt[3]{x^2 z}}{\sqrt[3]{y^2 z^{-1}}}.$$

CUMULATIVE REVIEW 1-9

Evaluate.

1. $3rs + 2st$ for $r = 2$, $s = 3$, and $t = 5$

2. $|-x|$ for $x = -2\frac{1}{4}$

3. $(2 - y)^2$ for $y = -2$

4. $-x^3 + 2x^2 + 3x - 5$ for $x = -1$

5. $\dfrac{t^3 - 1}{t^2 + t + 1}$ for $t = -2$

6. x^0 for $x = -|2|$

Perform the indicated operations and simplify, if possible.

7. $(2x^3 + 3x) + (2x^2 - 4x - 7) + (-x^3 + 3)$

8. $(3s + 2t)^2$

9. $(4x + 5y)(4x - 5y)$

10. $\dfrac{18x^6y}{6x^3y}$

11. $\dfrac{x^2 + 3x + 2}{x^2 - 2x + 1} \cdot \dfrac{2x^2 - 2}{x^2 + 4x + 4}$

12. $\dfrac{x^2 + 3x + 2}{x^2 - 2x + 1} \div \dfrac{2x^2 - 2}{x^2 + 4x + 4}$

13. $\dfrac{2}{2x + 1} - \dfrac{1}{1 - 2x} - \dfrac{4x}{4x^2 - 1}$

14. $(6 + 2i)(6 - 2i)$

15. $\dfrac{2 + i}{2 - i}$

Perform the indicated operations and simplify, if possible. Assume that all radicands represent positive numbers.

16. $\sqrt[4]{80x^8} \cdot \sqrt[4]{25x^6}$

17. $\dfrac{\sqrt[3]{81x^5y^2}}{\sqrt[3]{3x^2y}}$

18. $(\sqrt{x} + \sqrt{2y})(\sqrt{x} + \sqrt{3y})$

19. $4\sqrt{8x} - 2\sqrt{50x} + \sqrt{18x}$

Factor.

20. $x^3 - 216$

21. $8x^6 - 162$

22. $x^2 + 35 - 12x$

23. $4x^2 - 12x + 9$

24. Rewrite without negative exponents: $\left(-\frac{1}{2}\right)^{-3}$.

25. Use fractional exponents to write a single radical expression: $\sqrt[3]{x} \cdot \sqrt{2x}$.

Find each of the following if $f(x) = x^2 + x + 1$ and $g(x) = x + 1$.

26. $f(x + 1)$

27. $(f + g)(3)$

28. $f \circ g(x)$

29. $g^{-1}(x)$

Graph.

30. $5x = 15 + 3y$

31. $y = 2x^2 - 4x - 1$

32. $y = \log_3 x$

33. $y = 3^x$

34. $5x = 25$

35. $f(x) = x^4 - 5x^2 + 6$

36. Graph: $f(x) = 2(x + 3)^2 + 1$.
- **a)** Label the vertex.
- **b)** Draw the line of symmetry.
- **c)** Find the maximum or minimum value.

37. Find the slope of the line given by $2y - x = 7$.

38. Find an equation of the line having slope $-\frac{2}{3}$ and containing $(-3, 1)$.

39. Use synthetic division to find the quotient and remainder: $(2x^4 - x^2 + x + 5) \div (x - 2)$.

40. Determine whether the function is even, odd, or neither even nor odd: $f(x) = 2x^3 + 5x - 1$.

41. Factor the polynomial $P(x)$. Then solve the equation $P(x) = 0$.

$$P(x) = x^3 - 3x - 2$$

42. List all possible rational roots of $7x^8 + 6x^5 - 1$.

43. Convert to a logarithmic equation: $81^{1/4} = 3$.

Find each of the following.

44. $\log_7 1$

45. $\log 10{,}000$

46. $\log_{16} 4$

47. ▦ $\ln 3.1$

48. $\ln e^{15}$

49. ▦ $\log 37.5$

50. Express in terms of logarithms of x, y, and z:

$$\log\left(\frac{x^2z^3}{y}\right).$$

51. Express as a single logarithm and simplify, if possible:

$$3 \log x - \tfrac{1}{2} \log y - 2 \log z.$$

Solve.

52. $8(2x - 3) = 6 - 4(2 - 3x)$

53. $(5x - 2)(4x + 20) = 0$

54. $\sqrt{x - 1} = \sqrt{x + 4} - 1$

55. $x - 8\sqrt{x} + 15 = 0$

56. $|2x - 3| \geq 9$

57. $|2x + 5| < 3$

58. $x(x - 3) = 10$

59. $2x^2 + 5 = x$

60. $\frac{7}{x^2 - 5x} - \frac{2}{x - 5} = \frac{4}{x}$

61. $\frac{8}{x + 1} + \frac{11}{x^2 - x + 1} = \frac{24}{x^3 + 1}$

62. $3x^2 + 75 = 0$

63. $2x^2 - 2x - 3 = 0$

64. $x^2 - 10 > 3x$

65. $\frac{x + 1}{x - 1} \leq 0$

66. $\log_8 x = 1$

67. $\log_x 49 = 2$

68. $9^x = 27$

69. $\log x - \log (x - 8) = 1$

70. 🖩 $3^{5x} = 7$

71. Twenty-four plus five times a number is eight times the number. Find the number.

72. Phil can build a shed from a lumber kit in 10 hr. Jenny can build the same shed in 12 hr. How long would it take Phil and Jenny, working together, to build the shed?

73. A boat can move at a speed of 5 km/h in still water. The boat travels 42 km downstream in the same time that it takes to travel 12 km upstream. What is the speed of the stream?

74. What is the minimum product of two numbers whose difference is 14? What are the numbers that yield this product?

75. *Forgetting.* Students in a biology class took a final exam. A forgetting formula for the average exam grade was determined to be

$$S(t) = 78 - 15 \log (t + 1),$$

where t is the number of months after the final was taken.

a) The average score when the students first took the test is when $t = 0$. Find the students' average score on the final exam.

🖩 b) What would the average score be on a retest after 4 months?

76. *Population growth.* The population of Europe west of Russia was 430 million in 1961, and the exponential growth rate was 1% per year.

a) Write an exponential equation describing the growth of the population of Europe.

🖩 b) Predict what the population will be in 2001; in 2010.

77. y varies directly as the square of x and inversely as z, and $y = 2$ when $x = 5$ and $z = 100$. What is y when $x = 3$ and $z = 4$?

SYNTHESIS

Solve.

78. $\frac{5}{3x - 3} + \frac{10}{3x + 6} = \frac{5x}{x^2 + x - 2}$

79. $\log \sqrt{2x} = \sqrt{\log 2x}$

80. $4 \leq |3 - x| \leq 6$

CHAPTER 10 Conic Sections

FEATURE PROBLEM

In Washington, D.C., there is a large grassy area south of the White House known as the *Ellipse.* It is actually an ellipse with a maximum length of 1048 ft and a maximum width of 898 ft. Assuming a coordinate system is superimposed on the area in such a way that the center is at the origin and the maximum length and maximum width are on the x- and y-axes of the coordinate system, find an equation of the ellipse.

THE MATHEMATICS

The equation has the general form

$$\frac{x^2}{a^2} + \frac{y^2}{b^2} = 1,$$

where a and b can be found from the given information. This is an equation of an *ellipse.*

The ellipse described in the Feature Problem is a curve known as a *conic section,* meaning that the curve is formed as a cross section of a cone. In this chapter we will study equations whose graphs are conic sections. We have already studied two other conic sections, *lines* and *parabolas,* in some detail in Chapters 6 and 7. There are many applications involving conics, and we will consider some of them in this chapter. The review sections to be tested in addition to the material in this chapter are Sections 4.5, 5.2, 7.2, and 7.5.

10.1 The Distance Formula and Circles

The Distance Formula

For our work with conics in this and subsequent sections, we now develop a formula for finding the distance between two points whose coordinates are known.

Suppose that two points are on a horizontal line, thus having the same second coordinate. We can find the distance between them by subtracting their first coordinates. This difference may be negative, depending on the order in which we subtract. So to make sure we get a positive number, we take the absolute value of this difference. The distance between two points on a horizontal line (x_1, y) and (x_2, y) is thus $|x_2 - x_1|$. Similarly, the distance between two points on a vertical line (x, y_1) and (x, y_2) is $|y_2 - y_1|$.

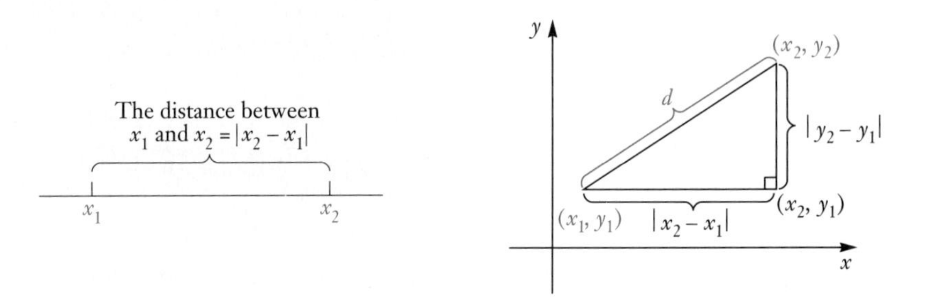

Now consider any two points (x_1, y_1) and (x_2, y_2) not on a horizontal or vertical line. These points are vertices of a right triangle, as shown. The other vertex is (x_2, y_1). The legs of this triangle have the lengths $|x_2 - x_1|$ and $|y_2 - y_1|$. Now by the Pythagorean theorem, we obtain a relation between the length of the hypotenuse d and the lengths of the legs:

$$d^2 = |x_2 - x_1|^2 + |y_2 - y_1|^2.$$

We may now dispense with the absolute value signs because squares of numbers are never negative. Thus we have

$$d^2 = (x_2 - x_1)^2 + (y_2 - y_1)^2.$$

By taking the square root, we obtain the distance between two points:

> The distance between any two points (x_1, y_1) and (x_2, y_2) is given by
> $$d = \sqrt{(x_2 - x_1)^2 + (y_2 - y_1)^2}$$

EXAMPLE 1 Find the distance between $(5, -1)$ and $(-4, 6)$.

Solution We substitute into the distance formula:

$$d = \sqrt{(-4-5)^2 + [6-(-1)]^2} \qquad \textbf{Substituting}$$
$$d = \sqrt{(-9)^2 + 7^2}$$
$$d = \sqrt{130}$$
$$d \approx 11.402. \qquad \textbf{Using a calculator}$$

■

We can use the distance formula to determine whether three points are vertices of a right triangle.

EXAMPLE 2 Determine whether the points $A(-2, 2)$, $B(4, -3)$ and $C(-2, -3)$ are vertices of a right triangle.

Solution First we find the squares of the distances between the points:

$$d_1^2 = [4-(-2)]^2 + (-3-2)^2$$
$$= (6)^2 + (-5)^2 = 61, \qquad \textbf{The distance between } A \textbf{ and } B \textbf{ is } \sqrt{61}.$$
$$d_2^2 = [-2-(-2)]^2 + (-3-2)^2$$
$$= (0)^2 + (-5)^2 = 25, \qquad \textbf{The distance between } A \textbf{ and } C \textbf{ is } \sqrt{25}\textbf{, or 5.}$$
$$d_3^2 = (-2-4)^2 + [-3-(-3)]^2$$
$$= (-6)^2 + (0)^2 = 36. \qquad \textbf{The distance between } B \textbf{ and } C \textbf{ is } \sqrt{36}\textbf{, or 6.}$$

Since $d_2^2 + d_3^2 = d_1^2$, the points *are* vertices of a right triangle. ■

Midpoints of Segments

The distance formula can be used to verify or derive a formula for finding the coordinates of the *midpoint* of a segment when the coordinates of the endpoints are known. We will not derive the formula but simply state it. The proof will be left to the exercises.

If the endpoints of a segment are (x_1, y_1) and (x_2, y_2), then the coordinates of the midpoint are

$$\left(\frac{x_1 + x_2}{2}, \frac{y_1 + y_2}{2}\right).$$

(We obtain the coordinates of the midpoint by averaging the coordinates of the endpoints.)

EXAMPLE 3 Find the midpoint of the segment with endpoints $(-2, 3)$ and $(4, -6)$.

Solution Using the midpoint formula, we obtain

$$\left(\frac{-2 + 4}{2}, \frac{3 + (-6)}{2}\right), \quad \text{or} \quad \left(\frac{2}{2}, \frac{-3}{2}\right), \quad \text{or} \quad \left(1, -\frac{3}{2}\right).$$ ■

Circle

Circles

When a plane intersects a cone perpendicular to the axis of the cone, as shown, a circle is formed.

A **circle** is a set of points in a plane that are a fixed distance r, called the **radius,** from a fixed point (h, k), called the **center.** If a point (x, y) is on the circle, then by the definition of a circle and the distance formula, it must follow that

$$r = \sqrt{(x - h)^2 + (y - k)^2},$$

or

$$r^2 = (x - h)^2 + (y - k)^2.$$

The equation, in standard form, of a circle with center (h, k) and radius r is

$$(x - h)^2 + (y - k)^2 = r^2.$$

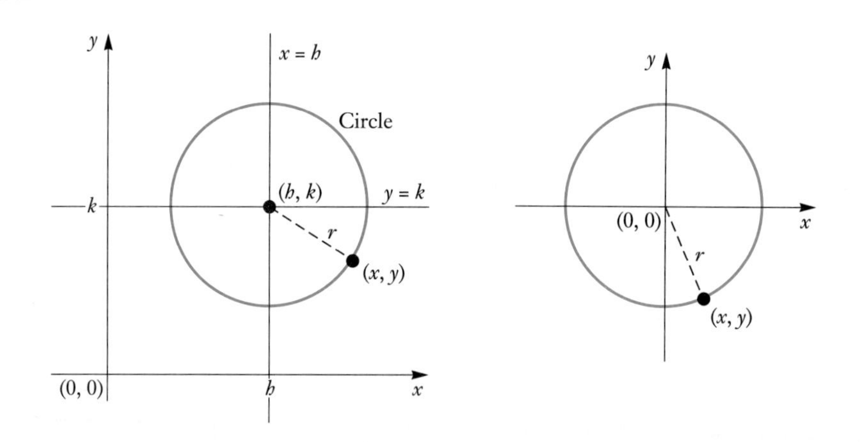

EXAMPLE 4 Find an equation of the circle having center $(4, -5)$ and radius 6.

Solution Using the standard form, we obtain

$$(x - 4)^2 + [y - (-5)]^2 = 6^2,$$

or

$$(x - 4)^2 + (y + 5)^2 = 36.$$

■

EXAMPLE 5 Find the center and the radius of $(x - 2)^2 + (y + 3)^2 = 16$. Then graph the circle.

Solution We can first write standard form: $(x - 2)^2 + (y - (-3))^2 = 4^2$. Then the center is $(2, -3)$ and the radius is 4. Now the graph is easy to draw, as shown, using a compass.

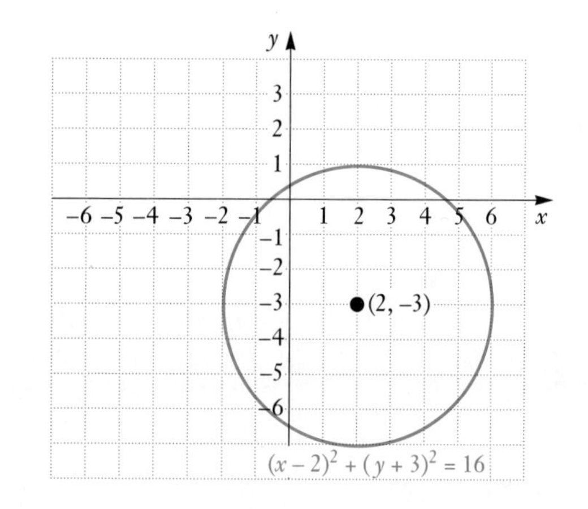

■

Completing the square allows us to find the standard form for the equation of a circle.

EXAMPLE 6 Find the center and the radius of the circle

$$x^2 + y^2 + 8x - 2y + 15 = 0.$$

Solution We complete the square twice to get standard form:

$$(x^2 + 8x + \quad) + (y^2 - 2y + \quad) = -15.$$

We take half the coefficient of the x-term and square it, obtaining 16. We add $16 - 16$ in the first parentheses. Similarly, we add $1 - 1$ in the second parentheses:

$$(x^2 + 8x + 16 - 16) + (y^2 - 2y + 1 - 1) = -15.$$

Next we do some rearranging and factoring:

$$(x^2 + 8x + 16) + (y^2 - 2y + 1) - 16 - 1 = -15$$
$$(x + 4)^2 + (y - 1)^2 = 2.$$

This is equivalent to standard form: $(x - (-4))^2 + (y - 1)^2 = 2$

The center is $(-4, 1)$, and the radius is $\sqrt{2}$. ■

EXAMPLE 7 Find an equation of a circle with center $(-2, -3)$ that passes through the point $(-5, 1)$.

Solution Since $(-2, -3)$ is the center, we have

$$(x + 2)^2 + (y + 3)^2 = r^2.$$

To determine the radius r, we find the distance between the center, $(-2, -3)$ and a point on the circle $(-5, 1)$:

$$\begin{aligned} r &= \sqrt{(x_1 - h)^2 + (y_1 - k)^2} \\ &= \sqrt{(-5 - (-2)) + (1 - (-3))^2} \\ &= \sqrt{9 + 16} \\ &= 5. \end{aligned}$$

Thus $(x + 2)^2 + (y + 3)^2 = 25$ is an equation of the circle. ■

EXERCISE SET 10.1

Find the distance between the pair of points. Where appropriate, use Table 1 in the Appendix or a calculator to find an approximation.

1. $(-3, -2)$ and $(1, 1)$
2. $(5, 9)$ and $(-1, 6)$
3. $(0, -7)$ and $(3, -4)$
4. $(2, 2)$ and $(-2, -2)$
5. $(9, 5)$ and $(6, 1)$
6. $(1, 10)$ and $(7, 2)$
7. $(5, 6)$ and $(5, -2)$
8. $(5, 6)$ and $(0, 6)$
9. $(8.6, -3.4)$ and $(-9.2, -3.4)$
10. $(5.9, 2)$ and $(3.7, -3.7)$
11. $(-1, 3k)$ and $(6, 2k)$
12. $(a, -3)$ and $(2a, 5)$
13. $(0, \sqrt{7})$ and $(\sqrt{6}, 0)$
14. $(\sqrt{d}, -\sqrt{3c})$ and $(\sqrt{d}, \sqrt{3c})$
15. $(6m, -7n)$ and $(-2m, n)$
16. $(\frac{5}{7}, \frac{1}{14})$ and $(\frac{1}{7}, \frac{11}{14})$
17. $(-3\sqrt{3}, 1 - \sqrt{6})$ and $(\sqrt{3}, 1 + \sqrt{6})$
18. $(5.989, 2.001)$ and $(3.712, -7.784)$

Determine whether the points are vertices of a right triangle.

19. $(9, 6)$, $(-1, 2)$, and $(1, -3)$
20. $(-5, -8)$, $(1, 6)$, and $(5, -4)$
21. $(-5, 1)$, $(-1, -2)$, and $(4, 10)$
22. $(-5, \sqrt{2})$, $(-3, 2 + \sqrt{2})$, and $(-1, \sqrt{2})$

Find the midpoint of the segment with the following endpoints.

23. $(-3, 6)$ and $(2, -8)$
24. $(6, 7)$ and $(7, -9)$
25. $(8, 5)$ and $(-1, 2)$
26. $(-1, 2)$ and $(1, -3)$
27. $(-8, -5)$ and $(6, -1)$
28. $(8, -2)$ and $(-3, 4)$

29. $(-3.4, 8.1)$ and $(2.9, -8.7)$

30. $(4.1, 6.9)$ and $(5.2, -6.9)$

31. $\left(\frac{1}{6}, -\frac{3}{4}\right)$ and $\left(-\frac{1}{3}, \frac{5}{6}\right)$

32. $\left(-\frac{4}{5}, -\frac{2}{3}\right)$ and $\left(\frac{1}{8}, \frac{3}{4}\right)$

33. $\left(\sqrt{2}, -1\right)$ and $\left(\sqrt{3}, 4\right)$

34. $\left(9, 2\sqrt{3}\right)$ and $\left(-4, 5\sqrt{3}\right)$

35. $(-a, b)$ and (a, b)

36. $\left(2 - \sqrt{3}, 5\sqrt{2}\right)$ and $\left(2 + \sqrt{3}, 3\sqrt{2}\right)$

Find the center and the radius of the circle. Then graph the circle.

37. $x^2 + y^2 = 36$

38. $x^2 + y^2 = 25$

39. $(x + 1)^2 + (y + 3)^2 = 4$

40. $(x - 2)^2 + (y + 3)^2 = 1$

41. $(x - 8)^2 + (y + 3)^2 = 40$

42. $(x + 5)^2 + (y - 1)^2 = 75$

43. $x^2 + y^2 = 2$

44. $x^2 + y^2 = 3$

45. $(x - 5)^2 + y^2 = \frac{1}{4}$

46. $x^2 + (y - 1)^2 = \frac{1}{25}$

47. $x^2 + y^2 + 8x - 6y - 15 = 0$

48. $x^2 + y^2 + 6x - 4y - 15 = 0$

49. $x^2 + y^2 - 8x + 2y + 13 = 0$

50. $x^2 + y^2 + 6x + 4y + 12 = 0$

51. $x^2 + y^2 - 4x = 0$

52. $x^2 + y^2 + 6x = 0$

53. $x^2 + y^2 + 10y - 75 = 0$

54. $x^2 + y^2 - 8x - 84 = 0$

55. $x^2 + y^2 + 7x - 3y - 10 = 0$

56. $x^2 + y^2 - 21x - 33y + 17 = 0$

57. $4x^2 + 4y^2 = 1$

58. $25x^2 + 25y^2 = 1$

Find an equation of the circle satisfying the given conditions.

59. Center $(0, 0)$, radius 7

60. Center $(0, 0)$, radius 4

61. Center $(-2, 7)$, radius $\sqrt{5}$

62. Center $(5, 6)$, radius $2\sqrt{3}$

63. Center $(-4, 3)$, radius $4\sqrt{3}$

64. Center $(-2, 7)$, radius $2\sqrt{5}$

65. Center $(-7, -2)$, radius $5\sqrt{2}$

66. Center $(-5, -8)$, radius $3\sqrt{2}$

67. Center $(-8, t)$, radius 1.3

68. Center $(2.7, k)$, radius $2p$

69. Center $(0, 0)$, passing through $(-3, 4)$

70. Center $(3, -2)$, passing through $(11, -2)$

71. Center $(-4, 1)$, passing through $(-2, 5)$

72. Center $(-3, -3)$, passing through $(1.8, 2.6)$

SKILL MAINTENANCE

73. A rectangle 10 in. long and 6 in. wide is bordered by a strip of uniform width. If the perimeter of the larger rectangle is twice that of the smaller rectangle, what is the width of the border?

74. One airplane flies 60 mph faster than another. To fly a certain distance, the faster plane takes 4 hr and the slower plane takes 4 hr and 24 min. What is the distance?

Complete the square and graph.

75. $f(x) = 2x^2 - 10x + 7$

76. $f(x) = -3x^2 + 24x - 50$

SYNTHESIS

77. Find the point on the y-axis that is equidistant from $(2, 10)$ and $(6, 2)$.

78. Find the point on the x-axis that is equidistant from $(-1, 3)$ and $(-8, -4)$.

Three points lie on the same line if the sum of two of the distances between the points is the other distance. Determine whether the following points lie on the same line.

79. $A(-4, 13)$, $B(0, 1)$, and $C(5, -14)$

80. $A(1, 7)$, $B(-3, -1)$, and $C(4, 12)$

81. Prove the midpoint formula by showing that

i) the distance from (x_1, y_1) to $\left(\frac{x_1 + x_2}{2}, \frac{y_1 + y_2}{2}\right)$ equals the distance from (x_2, y_2) to $\left(\frac{x_1 + x_2}{2}, \frac{y_1 + y_2}{2}\right)$; and

ii) the points (x_1, y_1), $\left(\frac{x_1 + x_2}{2}, \frac{y_1 + y_2}{2}\right)$, and (x_2, y_2) lie on the same line.

Find an equation of a circle satisfying the given conditions.

82. Center $(3, -5)$ and tangent (touching at one point) to the y-axis

83. Center $(-7, -4)$ and tangent to the x-axis

84. The endpoints of a diameter are $(7, 3)$ and $(-1, -3)$.

85. Center $(-3, 5)$ with a circumference of 8π units

86. A ferris wheel has a radius of 24.3 ft. Assuming that the center is 30.6 ft off the ground and that the origin is below the center as in the following figure, find an equation of the circle.

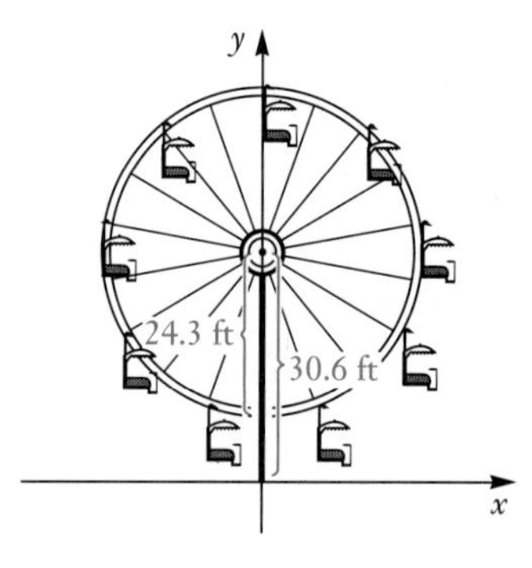

10.2 Conic Sections: Parabolas

When a cone is cut by a plane that is parallel to a side, as shown in the figure, the conic section formed is a **parabola.** The graphs of quadratic functions (see Chapter 7) are parabolas.

Many antennas have a parabolic cross section.

Parabola in three dimensions:

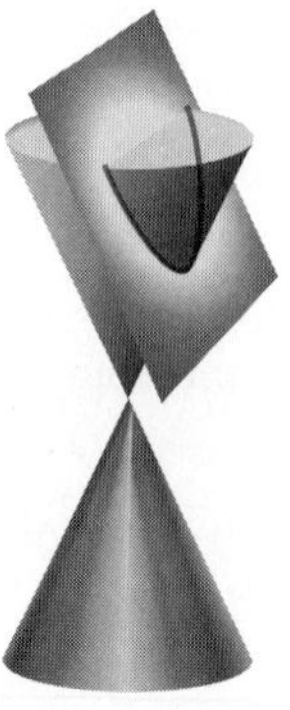

Parabola in a plane:

Parabolas have many applications in electricity, mechanics, and optics. The cross section of a satellite dish is a parabola. Cables that support bridges that are straight lines are shaped like parabolas. (Free-hanging cables have a different shape, called a "catenary.")

PARABOLAS

Parabolas have equations as follows:

$y = ax^2 + bx + c,$ (Line of symmetry parallel to y-axis)

$x = ay^2 + by + c.$ (Line of symmetry parallel to x-axis)

We devoted considerable study to the graphing of parabolas in Chapter 7. Thus our graphing here is somewhat a review. The chief difference is that all parabolas in Chapter 7 opened upward or downward. Here we also consider horizontal parabolas, which open to the right or to the left.

EXAMPLE 1 Graph: $y = x^2 - 4x + 8$.

Solution

a) To graph $y = x^2 - 4x + 8$, we first find the vertex. We can do so in two ways. One way is to complete the square, as follows:

$$\begin{aligned} y &= x^2 - 4x + 8 \\ &= (x^2 - 4x + 4) + 8 - 4 && \text{Adding and subtracting 4} \\ &= (x - 2)^2 + 4. && \text{Factoring and simplifying} \end{aligned}$$

The vertex is $(2, 4)$.

By factoring and completing the square, it can be shown (in Exercise 31 of this section) that *the x-coordinate of the vertex of the parabola given by $y = ax^2 + bx + c$ is found using the formula $x = -b/(2a)$*. Thus, instead of completing the square as we did here, we could have evaluated the formula:

$$\begin{aligned} x &= -\frac{b}{2a} \\ &= -\frac{-4}{2(1)} = 2. \end{aligned}$$

To find the y-coordinate of the vertex, we substitute 2 for x:

$$y = x^2 - 4x + 8 = 2^2 - 4(2) + 8 = 4.$$

The vertex is $(2, 4)$.

b) We choose some x-values on both sides of the vertex and graph the parabola. Since the coefficient of x^2 is 1, which is positive, we know that the graph opens upward. Be sure to find y when $x = 0$. This gives us the y-intercept.

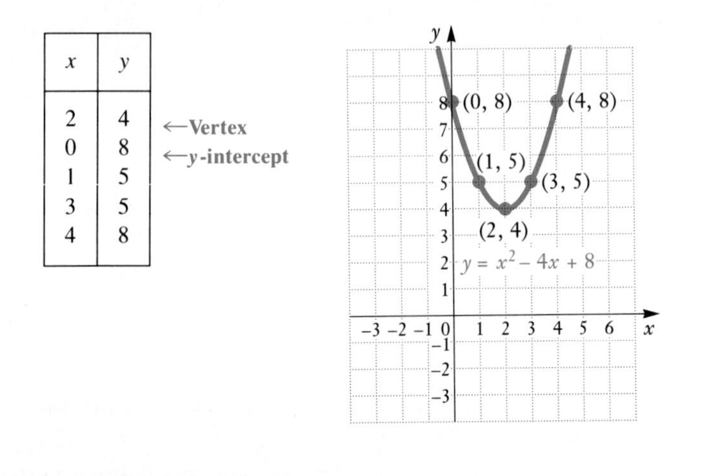

x	y	
2	4	←Vertex
0	8	←y-intercept
1	5	
3	5	
4	8	

> To graph an equation in the form $y = ax^2 + bx + c$:
>
> **a)** Find the vertex (h, k) either by completing the square to find an equivalent equation $y = a(x - h)^2 + k$, or by using $-b/(2a)$ to find the x-coordinate and substituting to find the y-coordinate.
>
> **b)** Choose other values for x on both sides of the vertex, and compute the corresponding y-values.
>
> **c)** The graph opens upward for $a > 0$ and downward for $a < 0$.

EXAMPLE 2 Graph: $x = y^2 - 4y + 8$.

Solution This equation is like that in Example 1 except that x and y are interchanged. The vertex is (4, 2) instead of (2, 4). To find ordered pairs, we first choose values for y on each side of the vertex. Then we compute values for x. A table is shown, together with the graph. Note that in this table the x- and y-values of the table in Example 1 are interchanged.

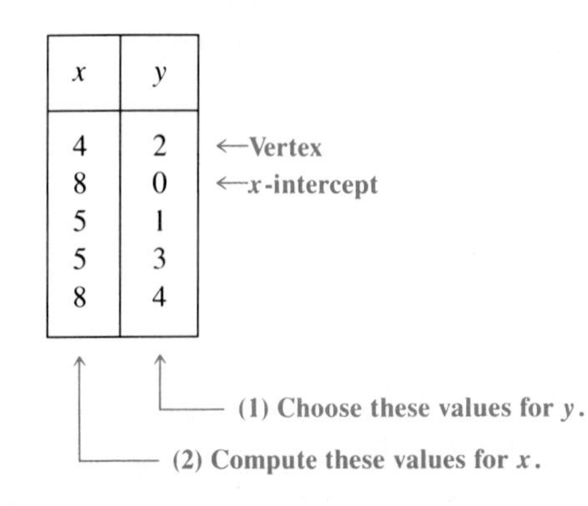

x	y	
4	2	←Vertex
8	0	←x-intercept
5	1	
5	3	
8	4	

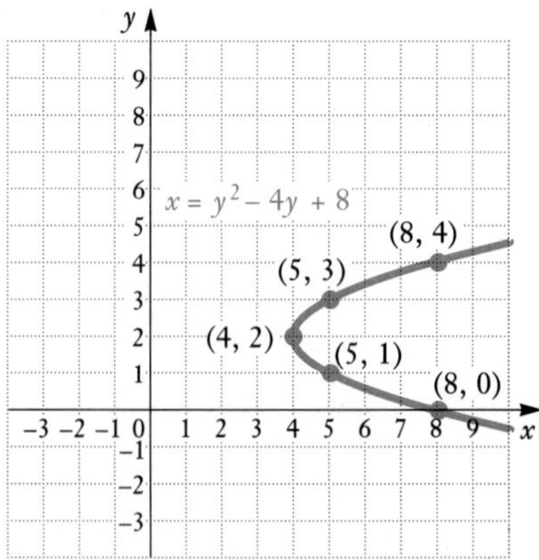

> To graph an equation in the form $x = ay^2 + by + c$:
>
> **a)** Find the vertex (h, k) either by completing the square to find an equivalent equation $x = a(y - k)^2 + h$ or by using $-b/(2a)$ to find the y-coordinate and substituting to find the x-coordinate.
>
> **b)** Choose other values for y that are above and below the vertex, and compute the corresponding x-values.
>
> **c)** The graph opens to the right if $a > 0$ and to the left if $a < 0$.

EXAMPLE 3 Graph: $x = -2y^2 + 10y - 7$.

Solution

a) We first find the vertex. We can do this by completing the square, as follows, or we can use $-b/(2a)$.

$$\begin{aligned} x &= -2y^2 + 10y - 7 \\ &= -2(y^2 - 5y) - 7 \\ &= -2\left(y^2 - 5y + \tfrac{25}{4}\right) - 7 - (-2)\tfrac{25}{4} \qquad \textbf{Adding and subtracting } (-2)\tfrac{25}{4} \\ &= -2\left(y - \tfrac{5}{2}\right)^2 + \tfrac{11}{2}. \qquad \textbf{Factoring and simplifying} \end{aligned}$$

The vertex is $\left(\frac{11}{2}, \frac{5}{2}\right)$.

For practice we also find the vertex by first computing the second coordinate, $y = -b/(2a)$, and then substituting to find the first coordinate:

$$y = -\frac{b}{2a} = -\frac{10}{2(-2)} = \frac{5}{2}$$

$$x = -2y^2 + 10y - 7 = -2\left(\tfrac{5}{2}\right)^2 + 10\left(\tfrac{5}{2}\right) - 7 = \tfrac{11}{2}$$

b) To find ordered pairs, we first choose values for y. Then we compute values for x. A table is given, together with the graph. The graph opens to the left because the coefficient of y^2, -2, is negative.

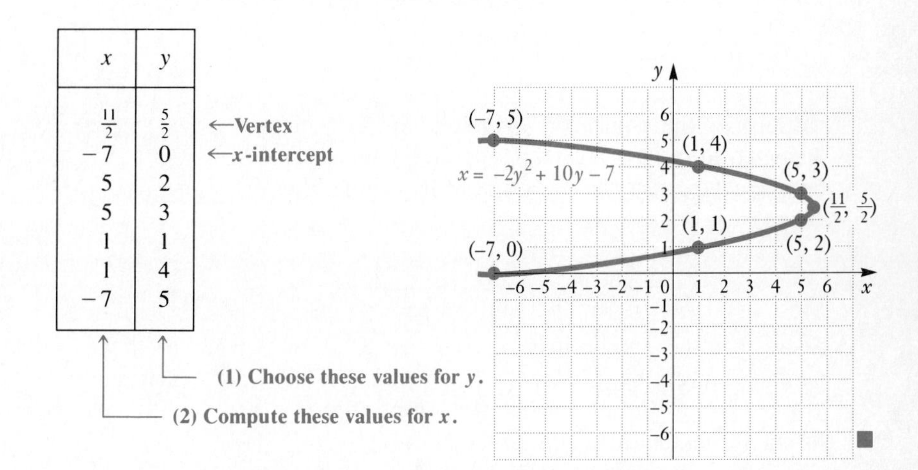

x	y	
$\frac{11}{2}$	$\frac{5}{2}$	← Vertex
-7	0	← x-intercept
5	2	
5	3	
1	1	
1	4	
-7	5	

(1) Choose these values for y.

(2) Compute these values for x.

■

EXERCISE SET 10.2

Graph.

1. $y = x^2$
2. $x = y^2$
3. $x = y^2 + 4y + 1$
4. $y = x^2 - 2x + 3$
5. $y = -x^2 + 4x - 5$
6. $x = 4 - 3y - y^2$
7. $x = y^2 + 1$
8. $x = 2y^2$
9. $x = -1 \cdot y^2$
10. $x = y^2 - 1$
11. $x = -y^2 + 2y$
12. $x = y^2 + y - 6$
13. $x = 8 - y - y^2$
14. $y = x^2 + 2x + 1$
15. $y = x^2 - 2x + 1$
16. $y = -\frac{1}{2}x^2$
17. $x = -y^2 + 2y + 3$
18. $x = -y^2 - 2y + 3$
19. $x = -2y^2 - 4y + 1$
20. $x = 2y^2 + 4y - 1$

SKILL MAINTENANCE

21. Simplify: $\sqrt[3]{\dfrac{8a^{11}}{27b^6}}$.

Solve.

22. $-2x^2 - 4x + 1 = 0$
23. $x^2 + 2 = 0$
24. $2x^2 + 5x - 7 = 0$

SYNTHESIS

25. Graph the equation $x = y^2 - y - 6$. Use the graph to approximate the solutions of each of the following equations.
    a) $y^2 - y - 6 = 2$ (*Hint:* Graph $x = 2$ on the same set of axes as the graph of $x = y^2 - y - 6$.)
    b) $y^2 - y - 6 = -3$

Using the same set of axes, graph the pair of equations. Try to discover a way to obtain one graph from the other without computing points for the second graph. (*Hint:* Review Section 9.2.)

26. $y = x^2, \quad x = y^2$
27. $y = -2x^2 + 3, \quad x = -2y^2 + 3$
28. $y = x^2 + 2x - 3, \quad x = y^2 + 2y - 3$
29. $y = 4 - x^2, \quad x = 4 - y^2$
30. *Horsepower of an engine.* The horsepower of a certain kind of engine is given by the formula

$$H = \frac{D^2N}{2.5},$$

where N is the number of cylinders and D is the diameter, in inches, of each piston. Graph this equation, assuming that $N = 6$ (a 6-cylinder engine). Let D run from 2.5 to 8.

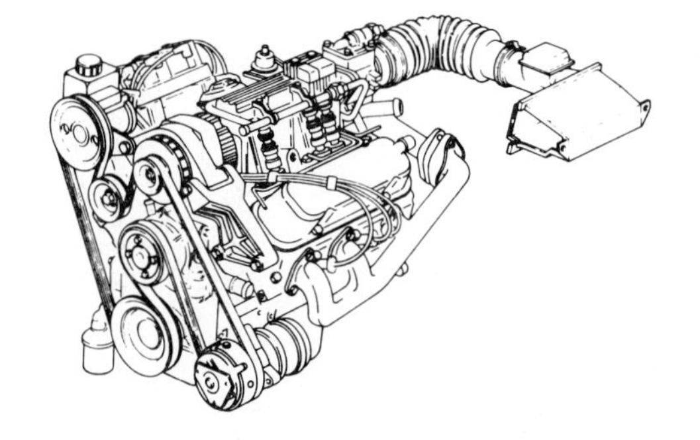

31. Show that the x-coordinate of the vertex of the parabola given by $y = ax^2 + bx + c$ $(a \neq 0)$ is $x = -b/(2a)$. (*Hint:* Complete the square after factoring a from the first two terms.)
32. From your work in Exercise 31, determine a formula that can be used to find the y-coordinate of the vertex of the parabola given by $y = ax^2 + bx + c$.

10.3 Conic Sections: Ellipses

Ellipses

When a cone is cut at an angle, as shown, the conic section formed is an **ellipse.** You can draw an ellipse by sticking two tacks in a piece of cardboard. Then tie a string to the tacks, place a pencil as shown, and draw.

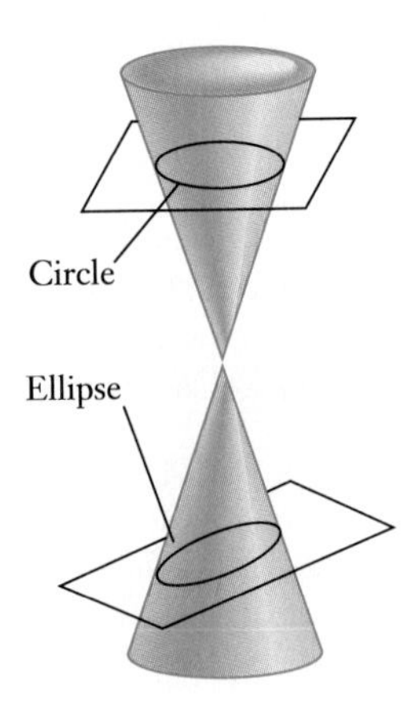

An Ellipse in a Plane

The formal mathematical definition is related to this method of drawing the ellipse. The ellipse is defined as the set of all points in a plane such that the *sum* of the distances from two fixed points F_1 and F_2 (called *foci*) is constant. In the figure shown, the tacks are at the foci.

An *ellipse* is the set of all points P in a plane such that the sum of the distances from P to two fixed points F_1 and F_2 in the plane is constant. The points F_1 and F_2 are called the *foci* (singular, *focus*) of the ellipse.

Ellipses have equations as follows. The proof will be left to the exercises.

EQUATION OF AN ELLIPSE

An ellipse with its center at the origin has an equation, in standard form,

$$\frac{x^2}{a^2} + \frac{y^2}{b^2} = 1, \qquad a, b > 0.$$

Ellipses with centers other than the origin are discussed in Exercises 23–26.

In graphing ellipses it helps to first find the intercepts. If we replace x by 0, we can find the y-intercepts:

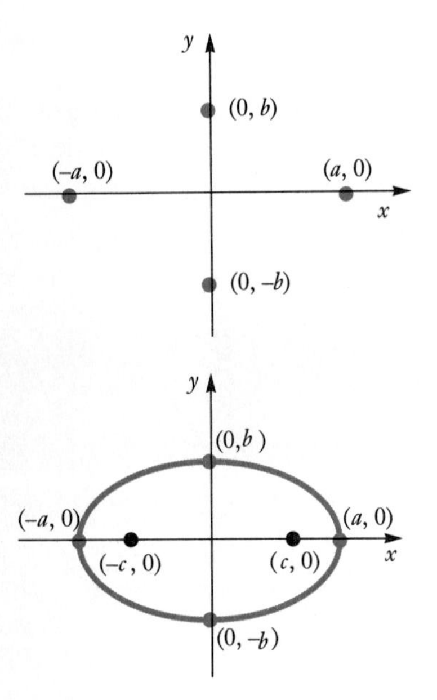

$$\frac{0^2}{a^2} + \frac{y^2}{b^2} = 1$$

$$\frac{y^2}{b^2} = 1$$

$$y^2 = b^2$$

$$y = \pm b.$$

Thus the y-intercepts are $(0, b)$ and $(0, -b)$. Similarly, the x-intercepts are $(a, 0)$ and $(-a, 0)$. If $a^2 > b^2$, then the foci are on the x-axis and have coordinates $(-c, 0)$ and $(c, 0)$, where $c^2 = a^2 - b^2$. If $b^2 > a^2$, then the foci are on the y-axis and have coordinates $(0, -c)$ and $(0, c)$, where $c^2 = b^2 - a^2$. (If $a^2 = b^2$, then the graph is a circle.)

For the ellipse

$$\frac{x^2}{a^2} + \frac{y^2}{b^2} = 1,$$

the x-intercepts are $(-a, 0)$ and $(a, 0)$, and the y-intercepts are $(0, -b)$ and $(0, b)$.

$$\frac{x^2}{a^2} + \frac{y^2}{b^2} = 1, \quad a^2 > b^2 \qquad\qquad \frac{x^2}{a^2} + \frac{y^2}{b^2} = 1, \quad b^2 > a^2$$

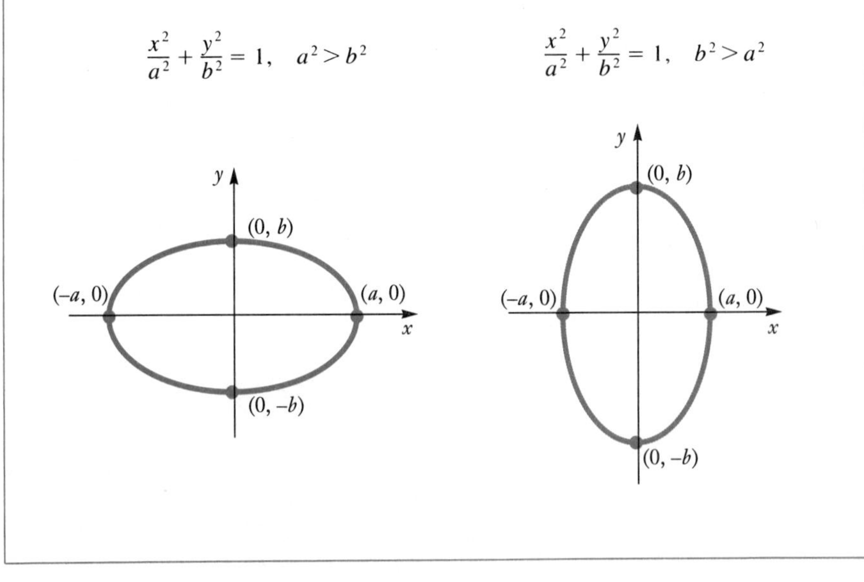

Plotting these points and filling in an oval-shaped curve, we get a graph of the ellipse. If a more precise graph is desired, we can plot more points.

EXAMPLE 1 Graph the ellipse

$$\frac{x^2}{4} + \frac{y^2}{9} = 1.$$

Solution Note that

$$\frac{x^2}{4} + \frac{y^2}{9} = \frac{x^2}{2^2} + \frac{y^2}{3^2}.$$

Thus the x-intercepts are $(-2, 0)$ and $(2, 0)$, and the y-intercepts are $(0, 3)$ and $(0, -3)$. We plot these points and connect them with an oval-shaped curve. To be more accurate, we might find some other points on the curve.

We let $x = 1$ and solve for y:

$$\frac{1^2}{4} + \frac{y^2}{9} = 1$$

$$36\left(\frac{1}{4} + \frac{y^2}{9}\right) = 36 \cdot 1$$

$$36 \cdot \frac{1}{4} + 36 \cdot \frac{y^2}{9} = 36$$

$$9 + 4y^2 = 36$$

$$4y^2 = 27$$

$$y^2 = \frac{27}{4}$$

$$y = \pm\sqrt{\frac{27}{4}}$$

$$y \approx \pm 2.6.$$

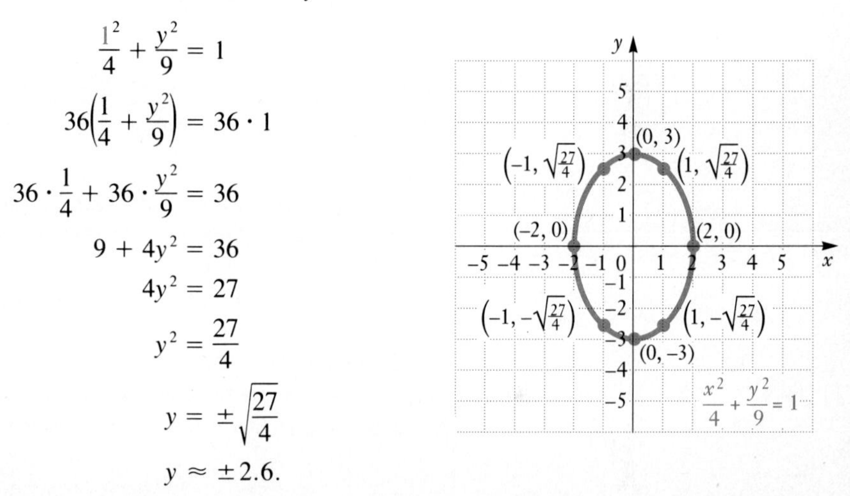

Thus $(1, 2.6)$ and $(1, -2.6)$ can also be plotted and used to draw the graph. Similarly, the points $(-1, 2.6)$ and $(-1, -2.6)$ can also be computed and plotted. ■

EXAMPLE 2 Graph: $4x^2 + 25y^2 = 100$.

Solution This equation is not in standard form. We get it in standard form by multiplying on both sides by $\frac{1}{100}$:

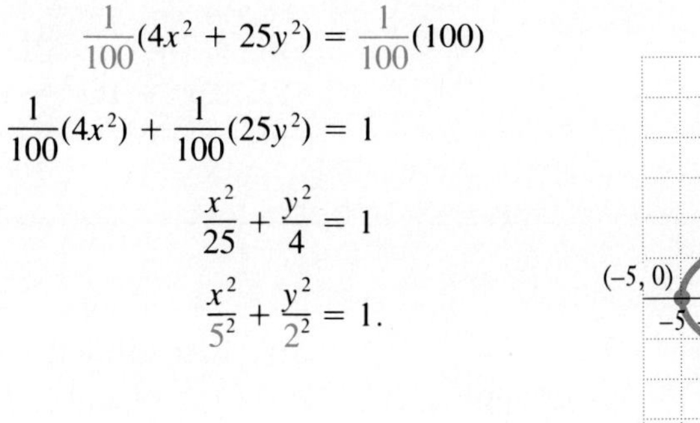

$$\frac{1}{100}(4x^2 + 25y^2) = \frac{1}{100}(100)$$

$$\frac{1}{100}(4x^2) + \frac{1}{100}(25y^2) = 1$$

$$\frac{x^2}{25} + \frac{y^2}{4} = 1$$

$$\frac{x^2}{5^2} + \frac{y^2}{2^2} = 1.$$

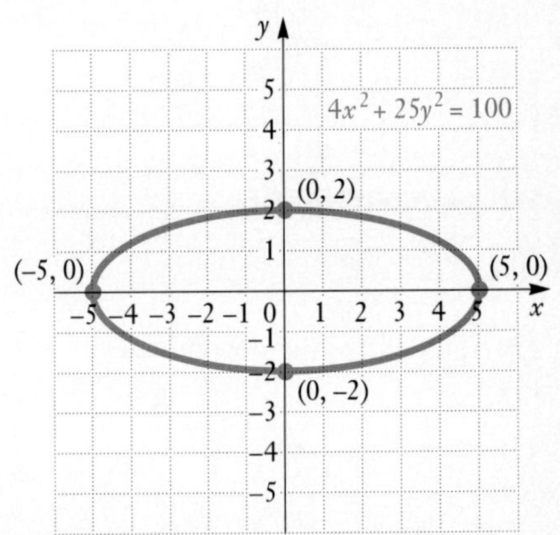

The x-intercepts are $(-5, 0)$ and $(5, 0)$. The y-intercepts are $(0, -2)$ and $(0, 2)$. We plot the intercepts and connect them with an oval-shaped curve. Other points can also be computed and plotted. ■

Applications

Ellipses have many applications. For example, earth satellites travel in elliptical orbits. The planets travel around the sun in elliptical orbits with the sun at one focus.

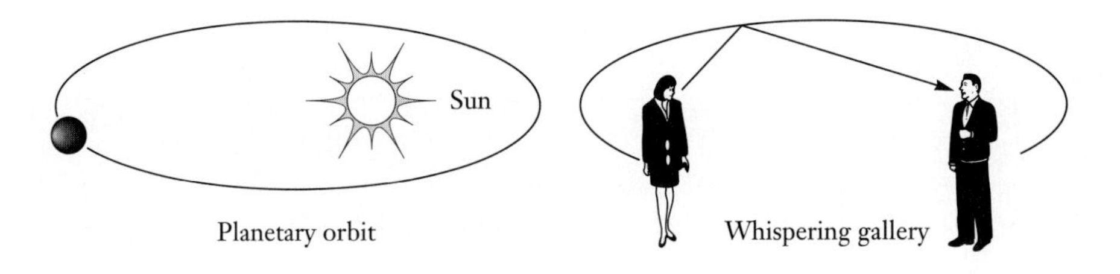

An interesting application found in some museums is the *whispering gallery,* which is elliptical. Persons with their heads at the foci can whisper and hear each other clearly, while persons at other positions cannot hear them. This happens when sound waves emanating at one focus are reflected to the other focus, being concentrated there.

A dentist often uses a reflector light. So that the light does not hit you in the eyes, it is covered and reflected. The reflection is directed toward your mouth. One focus is at the light source and the other is at your mouth.

EXERCISE SET 10.3

Graph the ellipse.

1. $\frac{x^2}{4} + \frac{y^2}{1} = 1$

2. $\frac{x^2}{1} + \frac{y^2}{4} = 1$

3. $\frac{x^2}{16} + \frac{y^2}{25} = 1$

4. $\frac{x^2}{9} + \frac{y^2}{25} = 1$

5. $4x^2 + 9y^2 = 36$

6. $9x^2 + 4y^2 = 36$

7. $16x^2 + 9y^2 = 144$

8. $9x^2 + 16y^2 = 144$

9. $2x^2 + 3y^2 = 6$

10. $5x^2 + 7y^2 = 35$

11. $4x^2 + 9y^2 = 1$

12. $25x^2 + 16y^2 = 1$

13. $5x^2 + 12y^2 = 60$

14. $8x^2 + 3y^2 = 24$

SKILL MAINTENANCE

Given $\frac{3 - \sqrt{a}}{2 + \sqrt{a}}$:

15. Rationalize the denominator.

16. Rationalize the numerator.

Solve.

17. $(1 + x)(1 - x) = 3x^2 - 4$

18. $(2 + x)^2 = 5x^2 - 3x + 2(7x - 8)$

SYNTHESIS

19. The maximum distance of the planet Mars from the sun is 2.48×10^8 miles. The minimum distance is 3.46×10^7 miles. The sun is at one focus of the elliptical orbit. Find the distance from the sun to the other focus.

20. An eccentric person builds a pool table in the shape of an ellipse with a hole at one focus and a tiny dot at the other. Guests are amazed at how many bank shots the owner of the pool table makes. Explain.

21. The President's oval office is an ellipse 31 ft wide and 38 ft long. Show in a sketch precisely where the President and an adviser could sit to best use the room's acoustics.

22. Find an equation of an ellipse that has x-intercepts $(-9, 0)$ and $(9, 0)$ and y-intercepts $(0, -11)$ and $(0, 11)$.

The equation (standard form) of an ellipse centered at the origin is

$$\frac{x^2}{a^2} + \frac{y^2}{b^2} = 1.$$

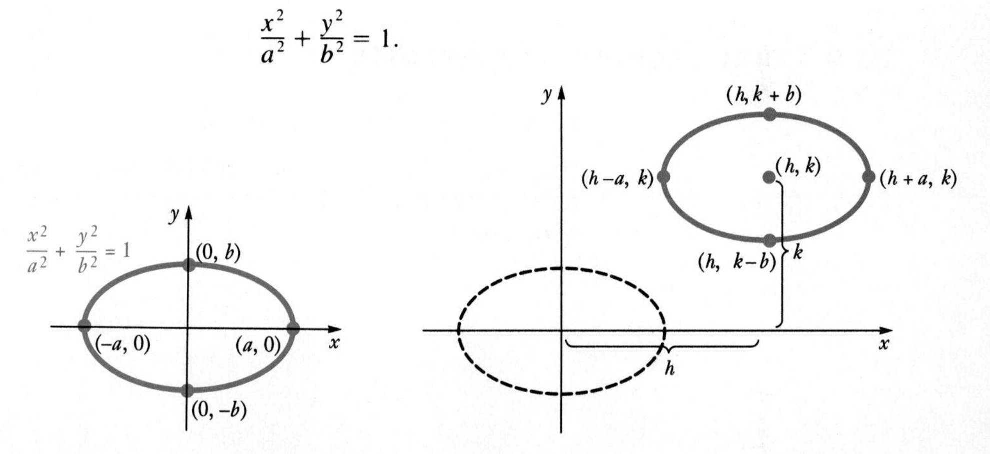

The intercepts, or *vertices,* are $(a, 0)$, $(-a, 0)$, $(0, b)$, and $(0, -b)$. The equation (standard form) of an ellipse with center (h, k) is

$$\frac{(x-h)^2}{a^2} + \frac{(y-k)^2}{b^2} = 1.$$

The vertices are $(h + a, k)$, $(h - a, k)$, $(h, k + b)$, and $(h, k - b)$.

For each ellipse, complete the square if necessary and write standard form. Find the center and the vertices. Then graph the ellipse.

23. $4(x - 1)^2 + 9(y + 2)^2 = 36$

24. $16x^2 + y^2 + 96x - 8y + 144 = 0$

25. $4x^2 + 25y^2 - 8x + 50y = 71$

26. $25x^2 + 9y^2 + 150x - 36y + 260 = 0$

Find the equation of the ellipse with the given vertices.

27. $(-8, 0)$, $(8, 0)$, $(0, -2)$, $(0, 2)$

28. $(0, 2)$, $(5, -1)$, $(10, 2)$, $(5, 5)$

29. Find an equation of the ellipse that contains the point $(1, 2\sqrt{3})$ and two of whose vertices are $(2, 0)$ and $(-2, 0)$.

30. Describe the graph of

$$\frac{(x-h)^2}{a^2} + \frac{(y-k)^2}{b^2} = 1$$

when $a^2 = b^2$.

31. Let $(-c, 0)$ and $(c, 0)$ be the foci of an ellipse centered at the origin. Any point $P(x, y)$ is on the ellipse if the sum of the distances from the foci to P is some constant. Use $2a$ to represent this constant.

a) Show that an equation for the ellipse is given by

$$\frac{x^2}{a^2} + \frac{y^2}{a^2 - c^2} = 1.$$

b) Substitute b^2 for $a^2 - c^2$ to get standard form.

10.4 Conic Sections: Hyperbolas

Hyperbolas (Standard Form)

When a cone is cut as shown on the left, the conic section formed is a parabola. If both parts of the cone are cut by a plane, as shown, the conic section is called a **hyperbola.**

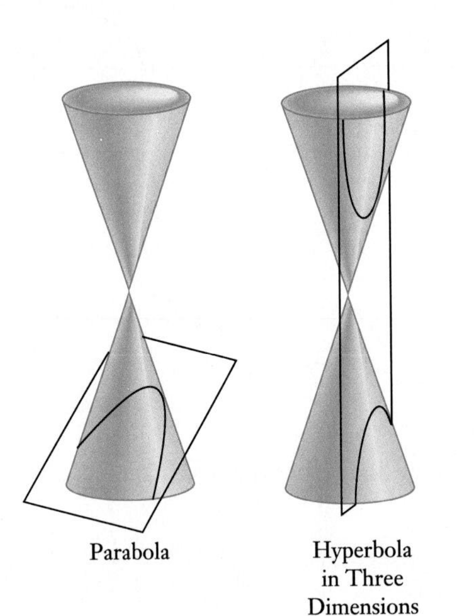

Parabola

Hyperbola in Three Dimensions

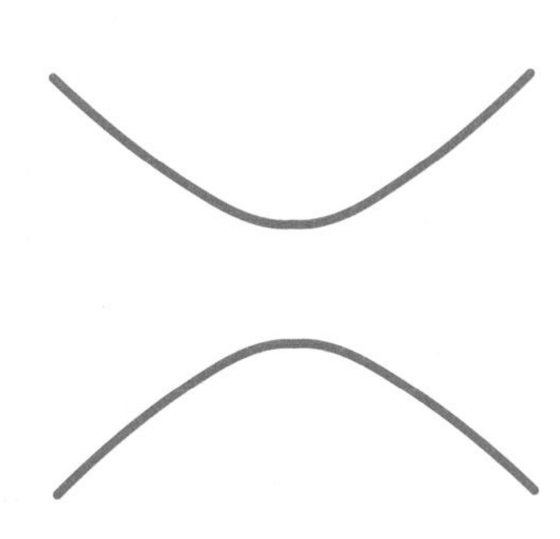

Hyperbola in a Plane

A hyperbola looks somewhat like a pair of parabolas, but the shapes are actually different. A hyperbola has two vertices, and the line through the vertices is known as an *axis.* The point halfway between the vertices is called the *center.* Here we consider only hyperbolas that are centered at the origin.

HYPERBOLAS

Hyperbolas with their centers at the origin have equations as follows:

$$\frac{x^2}{a^2} - \frac{y^2}{b^2} = 1, \qquad \text{(Axis horizontal)}$$

$$\frac{y^2}{b^2} - \frac{x^2}{a^2} = 1. \qquad \text{(Axis vertical)}$$

Note carefully that these equations have a 1 on the right.

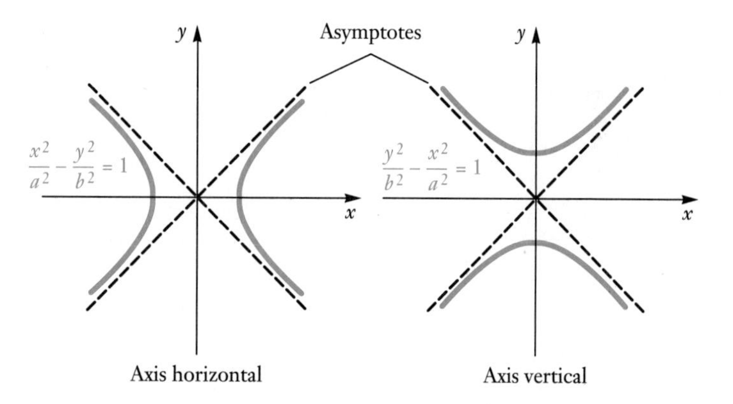

Hyperbolas with horizontal or vertical axes and centers *not* at the origin are discussed in the exercise set.

To graph a hyperbola, it helps to begin by graphing the lines called *asymptotes*.

ASYMPTOTES OF A HYPERBOLA

For hyperbolas with equations as given in the preceding box, the asymptotes are the lines

$$y = \frac{b}{a}x \qquad \text{and} \qquad y = -\frac{b}{a}x.$$

As a hyperbola gets farther away from the origin, it gets closer and closer to its asymptotes. The larger $|x|$ gets, the closer the graph gets to an asymptote. The asymptotes act to "constrain" the graph of a hyperbola. Parabolas *do not* have asymptotes. There are no lines to constrain parabolas.

The next thing to do is to find the vertices. Then it is easy to sketch the curve.

EXAMPLE 1 Graph: $\dfrac{x^2}{4} - \dfrac{y^2}{9} = 1.$

Solution

a) Note that

$$\frac{x^2}{4} - \frac{y^2}{9} = \frac{x^2}{2^2} - \frac{y^2}{3^2},$$

so $a = 2$ and $b = 3$. The asymptotes are thus

$$y = \frac{3}{2}x \qquad \text{and} \qquad y = -\frac{3}{2}x.$$

We sketch them.

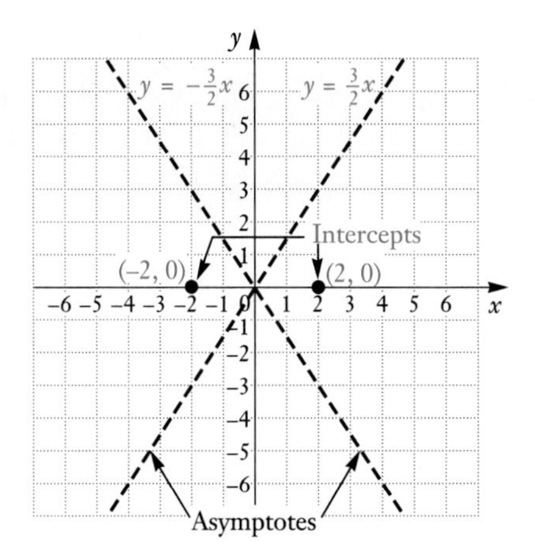

b) Next we find the intercepts, or *vertices*. When we replace y with 0 and solve for x, we see that $x^2/2^2 = 1$, so $x = \pm 2$. The intercepts are (2, 0) and $(-2, 0)$. There are intercepts on only one axis. If we replace x with 0, we see that $y^2/9 = -1$, and this equation has no real number solutions.

c) We plot the intercepts and sketch the graph. Through each intercept, we draw a smooth curve that approaches the asymptotes closely.

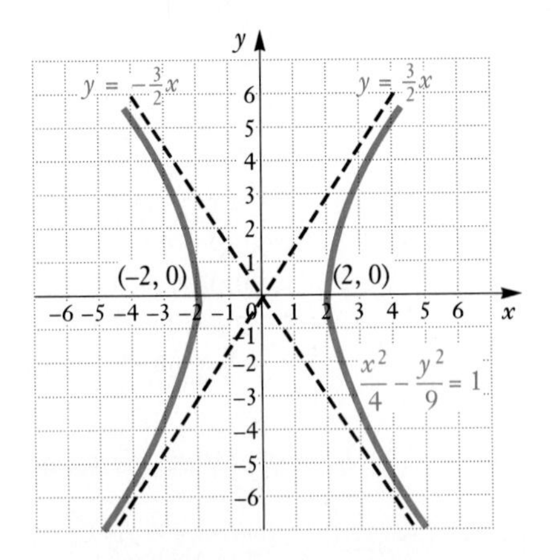

■

EXAMPLE 2 Graph: $\frac{y^2}{36} - \frac{x^2}{4} = 1$.

Solution

a) Note that

$$\frac{y^2}{36} - \frac{x^2}{4} = \frac{y^2}{6^2} - \frac{x^2}{2^2} = 1.$$

The intercept distance is the number in color in the term without the minus sign. There is a y in that term, so the intercepts are on the y-axis. The asymptotes are $y = \frac{6}{2}x$ and $y = -\frac{6}{2}x$, or $y = 3x$ and $y = -3x$.

b) Using the numbers 6 and 2, we can quickly sketch a rectangle to use as a guide. Thinking of ± 2 as x-coordinates and ± 6 as y-coordinates, we form all possible ordered pairs: $(2, 6)$, $(2, -6)$, $(-2, 6)$, and $(-2, -6)$. We plot these pairs and lightly sketch a rectangle through them. The asymptotes pass through the corners (see Figure A).

c) The intercepts are $(0, 6)$ and $(0, -6)$. We now draw curves through the intercepts toward the asymptotes, as shown in Figure B.

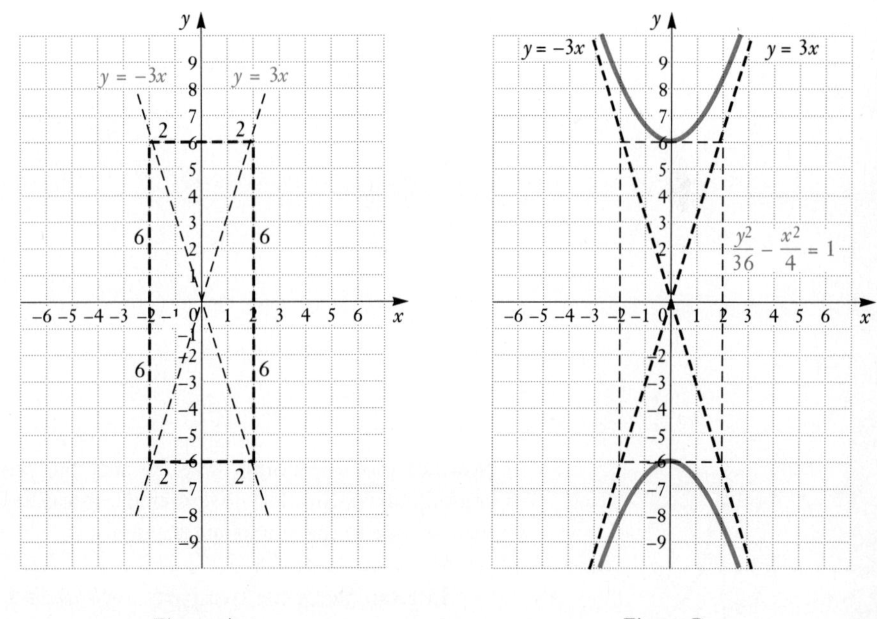

Figure A Figure B

■

Hyperbolas (Nonstandard Form)

The equations that we have just seen for hyperbolas are the standard ones, but there are other hyperbolas. We consider some of them.

> Hyperbolas having the x- and y-axes as asymptotes have equations as follows:
>
> $$xy = c, \qquad \text{where } c \text{ is a nonzero constant.}$$

EXAMPLE 3 Graph: $xy = -8$.

Solution We first solve for y:

$$y = -\frac{8}{x}.$$

Next, we find some solutions, keeping the results in a table. Note that we cannot use 0 for x and that for large values of $|x|$, y will be close to 0.

x	y
2	−4
−2	4
4	−2
−4	2
1	−8
−1	8
8	−1
−8	1

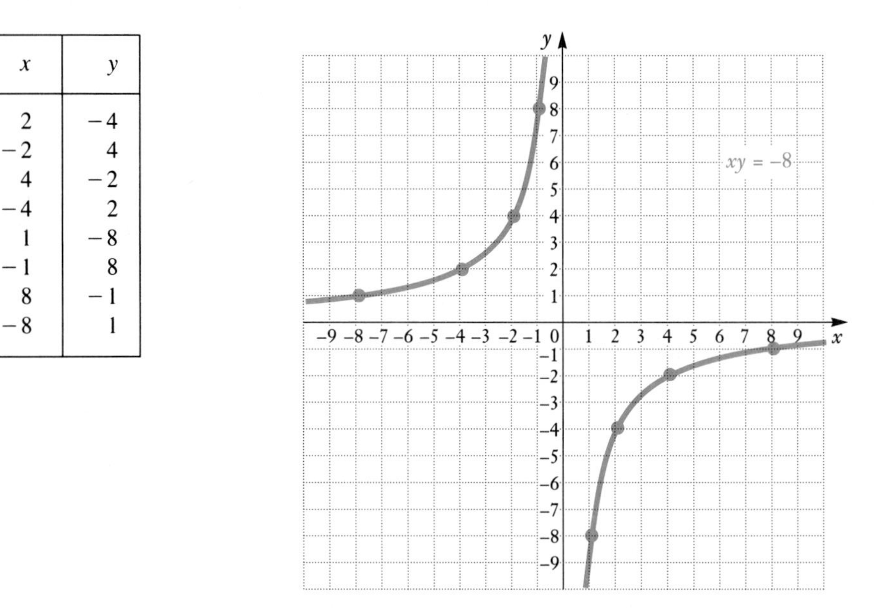

Now we plot the points. We connect the points in the second quadrant with a smooth curve. Similarly, we connect those in the fourth quadrant. Remember that the x-axis and the y-axis are asymptotes. ■

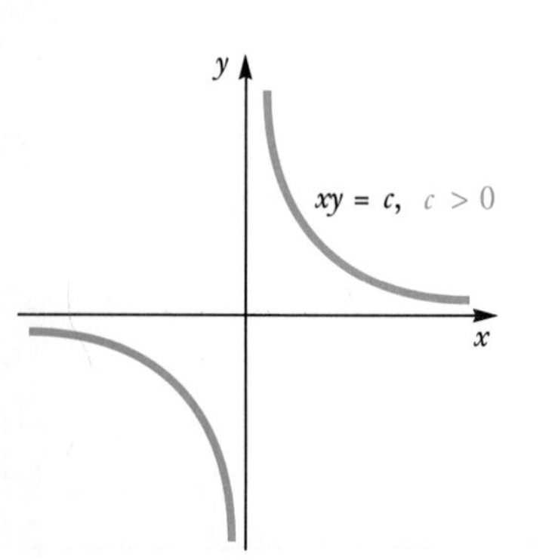

In Example 3, the two parts, or *branches,* of the hyperbola are in the second and fourth quadrants. If the constant c is positive, the two branches will be in the first and third quadrants, as shown in the graph to the left.

Hyperbolas also have many applications. A jet breaking the sound barrier creates a sonic boom whose wave front has the shape of a cone. The cone intersects the ground in one branch of a hyperbola. Some comets travel in hyperbolic orbits and a cross section of an amphitheater may be hyperbolic in shape.

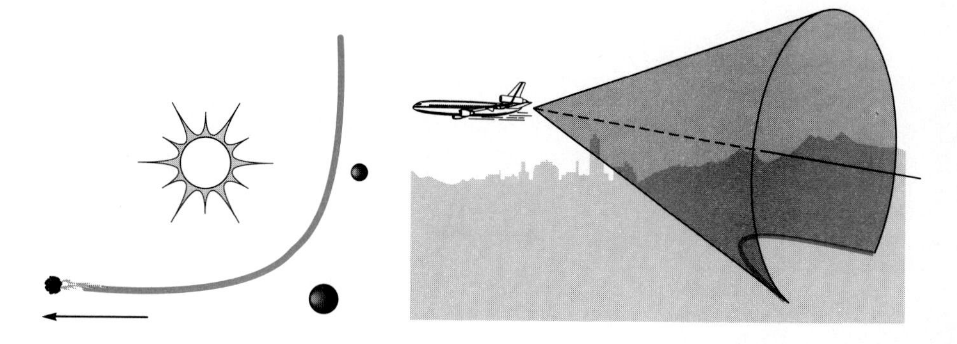

Classifying Graphs of Equations

The following is a summary of the equations of conic sections and their graphs.

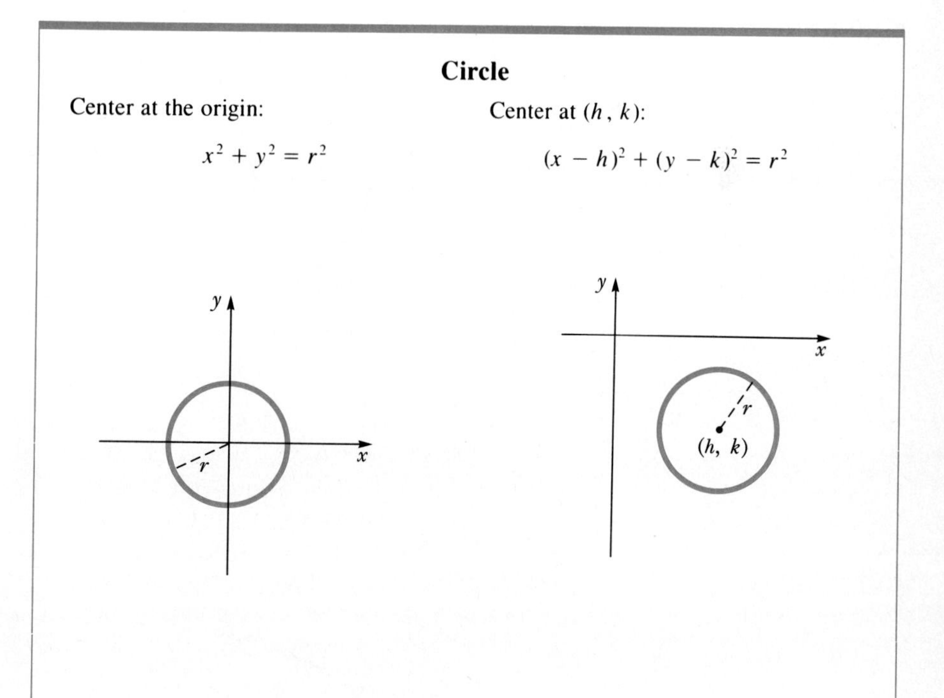

Circle

Center at the origin:

$$x^2 + y^2 = r^2$$

Center at (h, k):

$$(x - h)^2 + (y - k)^2 = r^2$$

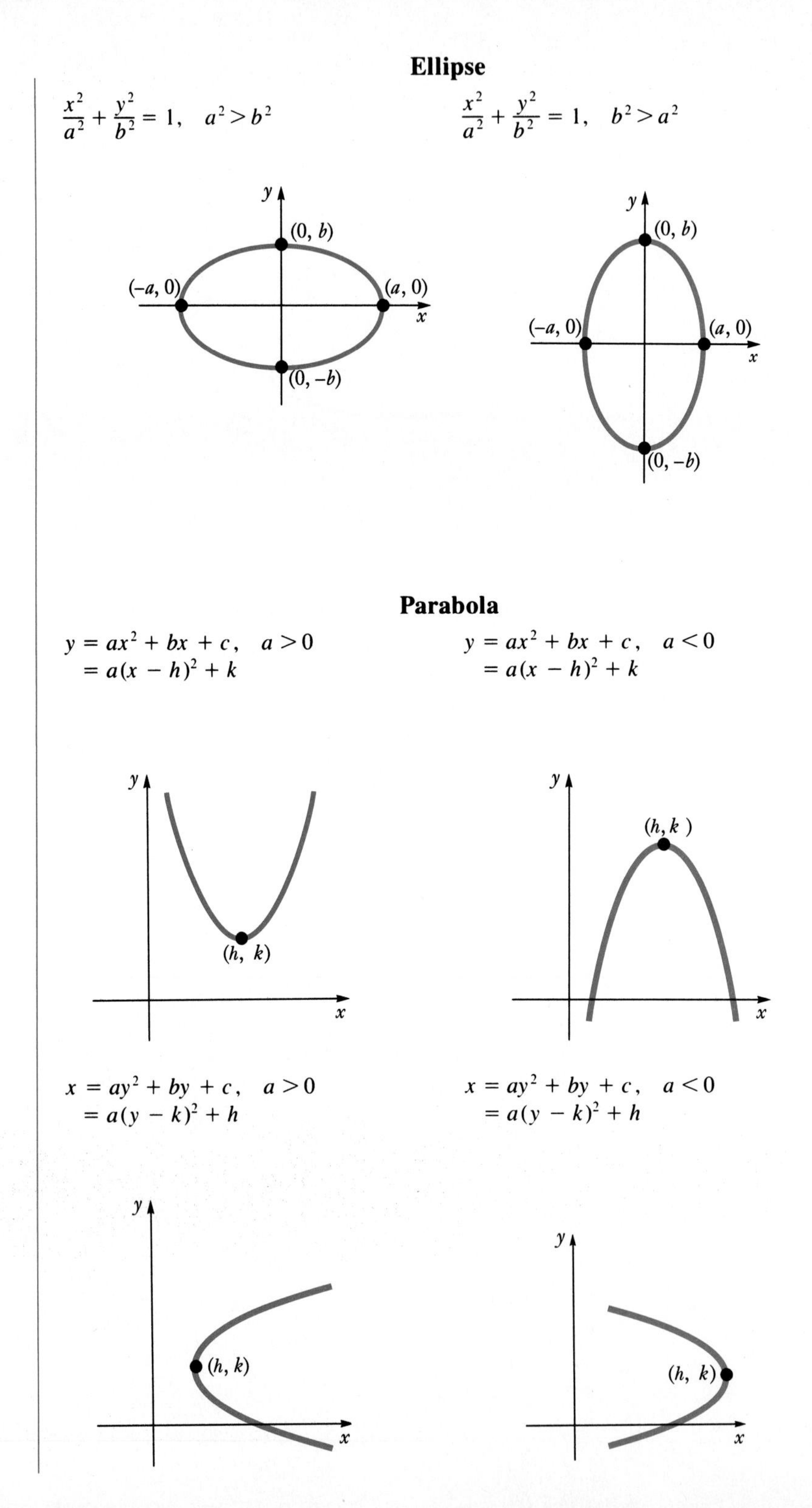

Ellipse
$\frac{x^2}{a^2} + \frac{y^2}{b^2} = 1, \quad a^2 > b^2$
$\frac{x^2}{a^2} + \frac{y^2}{b^2} = 1, \quad b^2 > a^2$
y
x
(0, b)
(−a, 0)
(a, 0)
(0, −b)
y
x
(0, b)
(−a, 0)
(a, 0)
(0, −b)
Parabola
$y = ax^2 + bx + c, \quad a > 0$
$= a(x - h)^2 + k$
$y = ax^2 + bx + c, \quad a < 0$
$= a(x - h)^2 + k$
y
x
(h, k)
y
x
(h, k)
$x = ay^2 + by + c, \quad a > 0$
$= a(y - k)^2 + h$
$x = ay^2 + by + c, \quad a < 0$
$= a(y - k)^2 + h$
y
x
(h, k)
y
x
(h, k)

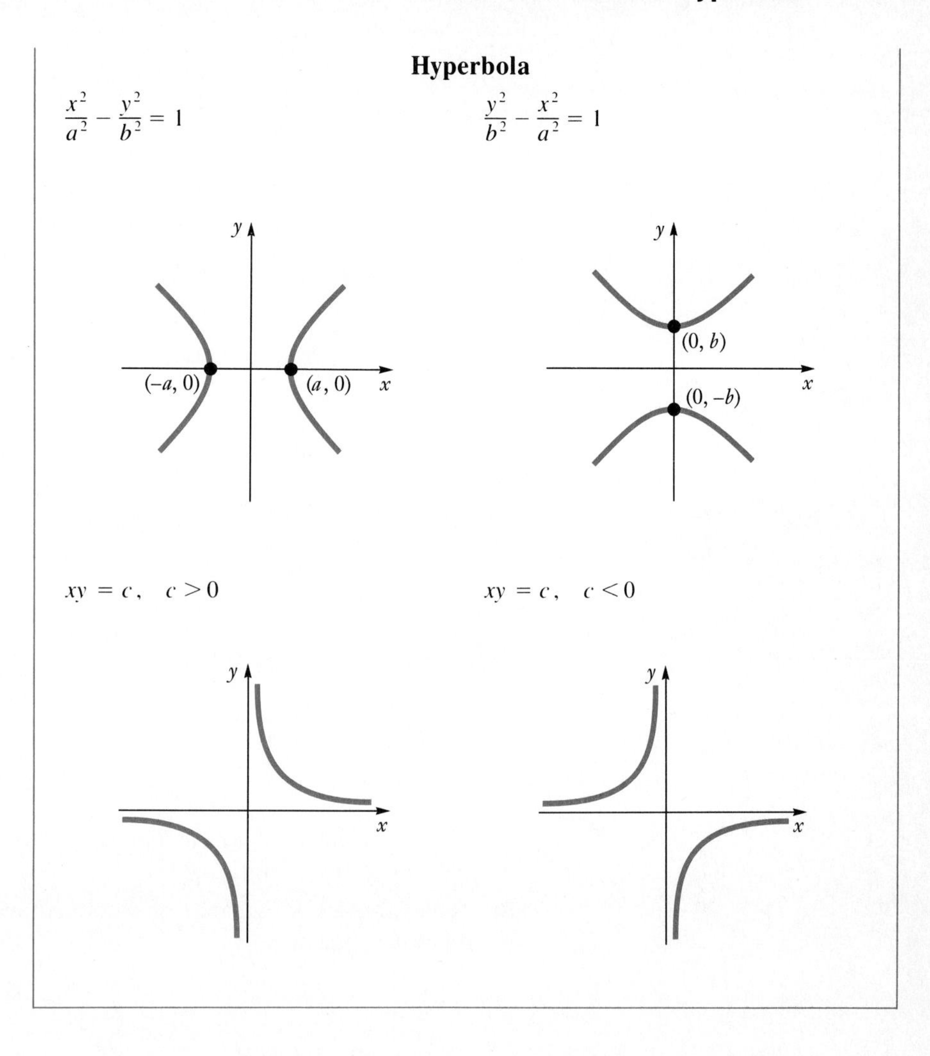

Suppose we encounter an equation that is not in one of the preceding forms. Sometimes we can find an equivalent equation that does fit one of the forms and then classify it as a circle, an ellipse, a parabola, or a hyperbola.

EXAMPLE 4 Classify the equation $5x^2 = 20 - 5y^2$ as a circle, an ellipse, a parabola, or a hyperbola.

Solution Suppose we get the terms with variables on one side by adding $5y^2$:

$$5x^2 + 5y^2 = 20.$$

The fact that x and y are *both* squared tells us that we do not have a parabola. The fact that the squared terms are *added* tells us that we do not have a hyperbola. Do we have a circle? To find out, we need to get $x^2 + y^2$ by itself. We can do that by factoring the 5 out of both terms on the left and then multiplying by $\frac{1}{5}$ on both sides:

$$5(x^2 + y^2) = 20$$
$$x^2 + y^2 = 4$$
$$x^2 + y^2 = 2^2.$$

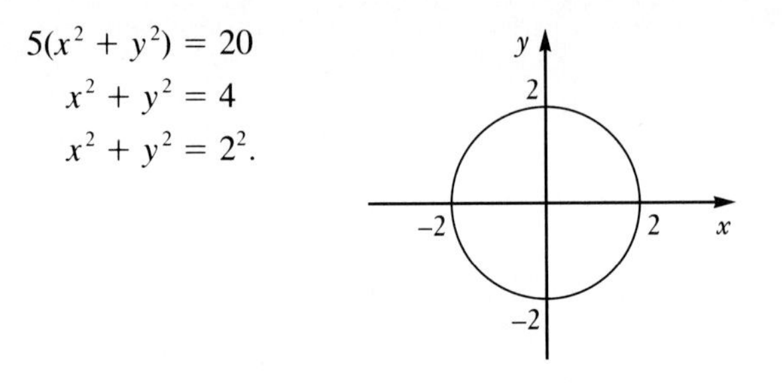

The graph is a circle with center at the origin and radius 2. ■

EXAMPLE 5 Classify the graph of each equation as a circle, an ellipse, a parabola, or a hyperbola.

a) $x + 3 + 8y = y^2$
b) $x^2 = y^2 + 4$
c) $x^2 = 16 - 4y^2$

Solution

a) Since only one of the variables is squared, we can find the following equivalent equation:

$$x = y^2 - 8y - 3.$$

Thus the graph is a horizontal parabola that opens to the right since the coefficient of y^2, 1, is positive.

b) Both variables are squared, so we do not have a parabola. We can obtain the equivalent equation

$$\frac{x^2}{2^2} - \frac{y^2}{2^2} = 1.$$

The minus sign in this form tells us that the graph of the equation is a hyperbola.

c) Both variables are squared, so the graph is not a parabola. We obtain the following equivalent equation:

$$x^2 + 4y^2 = 16.$$

If the coefficients of the terms were the same, we might have the graph of a circle as in Example 4, but they are not. We have a plus sign between the squared terms, so the graph is not a hyperbola. We divide on both sides and obtain the equivalent equation

$$\frac{x^2}{16} + \frac{y^2}{4} = 1$$

whose graph is an ellipse. ■

EXERCISE SET 10.4

Graph.

1. $\frac{x^2}{16} - \frac{y^2}{16} = 1$

2. $\frac{y^2}{9} - \frac{x^2}{9} = 1$

3. $\frac{y^2}{16} - \frac{x^2}{9} = 1$

4. $\frac{x^2}{9} - \frac{y^2}{4} = 1$

5. $\frac{x^2}{25} - \frac{y^2}{36} = 1$

6. $\frac{y^2}{9} - \frac{x^2}{25} = 1$

7. $x^2 - y^2 = 4$

8. $y^2 - x^2 = 25$

9. $4y^2 - 9x^2 = 36$

10. $25x^2 - 16y^2 = 400$

Graph.

11. $xy = 6$

12. $xy = -4$

13. $xy = -9$

14. $xy = 3$

15. $xy = -1$

16. $xy = -2$

17. $xy = 2$

18. $xy = 1$

Classify the graph of the equation as a circle, an ellipse, a parabola, or a hyperbola.

19. $x^2 + y^2 - 10x + 8y - 40 = 0$

20. $y + 1 = 2x^2$

21. $9x^2 - 4y^2 - 36 = 0$

22. $1 - 3y = 2y^2 - x$

23. $4x^2 + 25y^2 - 100 = 0$

24. $y^2 + x^2 = 7$

25. $x^2 + y^2 = 2x + 4y + 4$

26. $2y + 13 + x^2 = 8x - y^2$

$= 64 - y$

28. $y = \frac{\quad}{x}$

29. $x - \frac{3}{y} = 0$

30. $x - 4 = y^2 + 5$

31. $y + 6x = x^2 + 6$

32. $x^2 = 16 + y^2$

33. $9y^2 = 36 + 4x^2$

34. $3x^2 + 5y^2 + x^2 = y^2 + 49$

SKILL MAINTENANCE

35. Simplify: $\sqrt[3]{125t^{15}}$.

36. Solve: $2x^2 + 10 = 0$.

37. Rationalize the denominator:

$$\frac{4\sqrt{2} - 5\sqrt{3}}{6\sqrt{3} - 8\sqrt{2}}.$$

38. An airplane travels 500 mi at a certain speed. A larger plane travels 1620 mi at a speed that is 320 mph faster, but takes 1 hr longer. Find the speed of each plane.

SYNTHESIS

Find an equation of a hyperbola satisfying the given conditions.

39. Having intercepts $(0, 8)$ and $(0, -8)$ and asymptotes $y = 4x$ and $y = -4x$

40. Having intercepts $(8, 0)$ and $(-8, 0)$ and asymptotes $y = 4x$ and $y = -4x$

Hyperbolas with center at (h, k) have standard equations as follows, where the asymptotes are

$$y - k = \pm\frac{b}{a}(x - h).$$

The vertices are labeled in the figures.

$$\frac{(x-h)^2}{a^2} - \frac{(y-k)^2}{b^2} = 1 \qquad \frac{(y-k)^2}{b^2} - \frac{(x-h)^2}{a^2} = 1$$

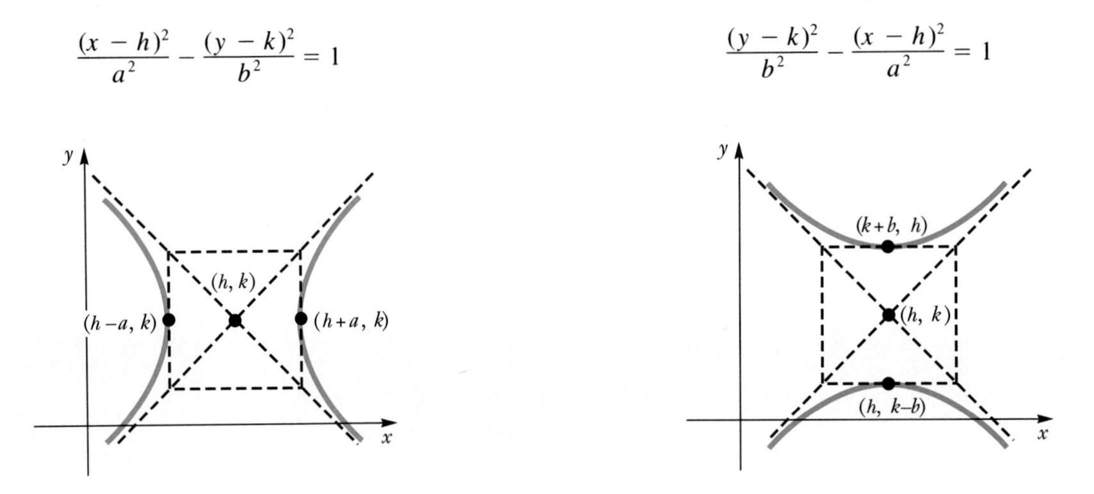

For each of the following equations of hyperbolas, complete the square, if necessary, and write in standard form. Find the center, the vertices, and the asymptotes. Then graph the hyperbola.

41. $\frac{(x-2)^2}{9} - \frac{(y+1)^2}{16} = 1$

42. $4x^2 - y^2 + 24x + 4y + 28 = 0$

43. $4y^2 - 25x^2 - 8y - 100x - 196 = 0$

44. $x^2 - y^2 - 2y - 4x = 6$

CHAPTER SUMMARY AND REVIEW 10

TERMS TO KNOW

Conic section
Circle
Center
Standard form
Parabola
Ellipse
Foci
Hyperbola
Asymptotes
Vertices

IMPORTANT PROPERTIES AND FORMULAS

THE DISTANCE FORMULA

The distance between any two points (x_1, y_1) and (x_2, y_2) is given by

$$d = \sqrt{(x_2 - x_1)^2 + (y_2 - y_1)^2}.$$

THE MIDPOINT FORMULA

If the endpoints of a segment are (x_1, y_1) and (x_2, y_2), then the coordinates of the midpoint are

$$\left(\frac{x_1 + x_2}{2}, \frac{y_1 + y_2}{2}\right).$$

CIRCLE

Center at (h, k): $(x - h)^2 + (y - k)^2 = r^2$

PARABOLA

$y = ax^2 + bx + c,$	Vertex at (h, k)
$= a(x - h)^2 + k;$	Opens up for $a > 0$, down for $a < 0$.
$x = ay^2 + by + c,$	Vertex at (h, k)
$= a(y - k)^2 + h$	Opens right for $a > 0$, left for $a < 0$.

ELLIPSE

Center at the origin

Foci on x-axis: $\frac{x^2}{a^2} + \frac{y^2}{b^2} = 1, \quad a^2 > b^2$ Foci on y-axis: $\frac{x^2}{a^2} + \frac{y^2}{b^2} = 1, \quad b^2 > a^2$

HYPERBOLA

$$\frac{x^2}{a^2} - \frac{y^2}{b^2} = 1, \qquad \frac{y^2}{b^2} - \frac{x^2}{a^2} = 1, \qquad xy = c$$

(See also the summary of graphs at the end of Section 10.4.)

REVIEW EXERCISES

The review sections to be tested in addition to the material in this chapter are Sections 4.5, 5.2, 7.2, and 7.5.

Find the distance between the given points.

1. $(2, 6)$ and $(6, 6)$

2. $(-1, 1)$ and $(-5, 4)$

3. $(4, 7)$ and $(-3, -2)$

4. $(2, 3a)$ and $(-1, a)$

Determine whether the points are vertices of a right triangle.

5. $(3, 2)$, $(-1, 5)$, and $(0, 4)$

6. $(5, 2)$, $(-1, 5)$, and $(4, 0)$

Find the midpoint of the segment with the given endpoints.

7. $(1, 6)$ and $(7, 6)$

8. $(-1, 1)$ and $(-5, 4)$

9. $(4, 7)$ and $(-3, -2)$

10. $(2, 3a)$ and $(-1, a)$

Find the center and the radius.

11. $(x + 2)^2 + (y - 3)^2 = 2$

12. $(x - 5)^2 + y^2 = 49$

13. $x^2 + y^2 - 6x - 2y + 1 = 0$

14. $x^2 + y^2 + 8x - 6y - 10 = 0$

15. Find an equation of the circle with center $(-4, 3)$ and radius $4\sqrt{3}$.

16. Find an equation of the circle with center $(7, -2)$ and radius $2\sqrt{5}$.

17. Find an equation of the circle having its center at (3, 4) and passing through the origin.

Classify the equation as a circle, an ellipse, a parabola, or a hyperbola. Then graph.

18. $4x^2 + 4y^2 = 100$

19. $9x^2 + 2y^2 = 18$

20. $y = -x^2 + 2x - 3$

21. $\frac{y^2}{9} - \frac{x^2}{4} = 1$

22. $xy = 9$

23. $x = y^2 + 2y - 2$

24. $xy = -3$

25. $x^2 + y^2 + 6x - 8y - 39 = 0$

SKILL MAINTENANCE

26. Simplify: $\frac{\sqrt[3]{81a^8b^{10}}}{\sqrt[3]{3a^6b}}$.

27. Solve: $x^2 + 2x + 5 = 0$.

28. Graph: $y = x^2 + 2x + 5$.

29. The speed of a moving sidewalk at an airport is 5 ft/sec. A person can walk 55 ft forward on the moving sidewalk in the same time it takes to walk 5 ft in the opposite direction. At what rate would the person walk on a nonmoving sidewalk?

SYNTHESIS

30. Find the point on the x-axis that is equidistant from $(-3, 4)$ and $(5, 6)$.

31. Find an equation of the circle having a diameter with endpoints $(-3, 5)$ and $(7, 3)$.

32. Find an equation of the ellipse with the following vertices: $(-7, 0)$, $(7, 0)$, $(0, -3)$, and $(0, 3)$.

THINKING IT THROUGH

1. Which, if any, of the conic sections studied in this chapter are functions?

2. Compare the number of x- and y-intercepts each type of conic section may have.

3. Explain the differences between the graphs of a parabola and a hyperbola.

CHAPTER TEST 10

Find the distance between the given points.

1. $(4, -1)$ and $(-5, 8)$

2. $(3, -a)$ and $(-3, a)$

Find the midpoint of the segment with the given endpoints.

3. $(4, -1)$ and $(-5, 8)$

4. $(3, -a)$ and $(-3, a)$

Find the center and the radius of the circle.

5. $(x + 2)^2 + (y - 3)^2 = 64$

6. $x^2 + y^2 + 4x - 6y + 4 = 0$

7. Find an equation of the circle with center $(2, -5)$ and radius $3\sqrt{2}$.

Classify the equation as a circle, an ellipse, a parabola, or a hyperbola. Then graph.

8. $y = x^2 - 4x - 1$

9. $x^2 + y^2 + 2x + 6y + 6 = 0$

10. $\frac{x^2}{9} - \frac{y^2}{4} = 1$

11. $16x^2 + 4y^2 = 64$

12. $xy = -5$

13. $x = -y^2 + 4y$

SKILL MAINTENANCE

14. Solve: $x^2 + 2x = 5$.

15. Simplify: $\sqrt{\frac{48x^2y}{3xy}}$.

16. A boat travels 6 mi upstream in the same time it takes to travel 30 mi downstream. The speed of the stream is 4 mph. Find the speed of the boat in still water.

17. Find the x-intercepts: $f(x) = 3x^2 - x - 5$.

SYNTHESIS

18. Find an equation of the ellipse with the following vertices: $(1, 3)$, $(6, 6)$, $(11, 3)$, and $(6, 0)$.

19. Find the point on the y-axis that is equidistant from $(-3, -5)$ and $(4, -7)$.

CHAPTER 11

Systems of Equations and Inequalities

FEATURE PROBLEM

A chemist has one solution of acid and water that is 25% acid and a second that is 65% acid. How many liters of each should be mixed together to get 8 L of a solution that is 40% acid?

THE MATHEMATICS

We let x = the amount of the 25% solution and y = the amount of the 65% solution and translate the information to a *system of equations*:

$$x + y = 8,$$

$$0.25x + 0.65y = 0.4(8).$$

To solve the problem, we solve the system for x and y.

The most difficult part of solving problems in algebra is almost always translating the problem situation to mathematical language. Once you get an equation, the rest is usually straightforward. In this chapter we study *systems of equations* and how to solve them using graphing, substitution, and elimination. One of the great advantages of using a system of equations is that many problem situations then become easier to translate to mathematical language.

Systems of equations have extensive application to many fields, such as psychology, sociology, business, education, engineering, and science. Systems of inequalities are also useful in a branch of mathematics called *linear programming.* This chapter includes a brief introduction to linear programming as well as a study of *matrices,* which can also be used to solve systems of equations.

The review sections to be tested in addition to the material in this chapter are Sections 6.3, 8.1, 8.3, and 9.6.

11.1 Systems of Equations in Two Variables

Identifying Solutions

Recall that a **conjunction** of sentences is formed by joining them with the word *and*. Here is an example:

$$x + y = 11 \quad \textit{and} \quad 3x - y = 5.$$

A **solution** of a sentence with two variables, such as $x + y = 11$, is an ordered pair. Some pairs in the solution set of $x + y = 11$ are

$$(5, 6), \quad (12, -1), \quad (4, 7), \quad (8, 3).$$

Some pairs in the solution set of $3x - y = 5$ are

$$(0, -5), \quad (4, 7), \quad (-2, -11), \quad (9, 22).$$

The solution set of the sentence

$$x + y = 11 \quad \textit{and} \quad 3x - y = 5$$

consists of all pairs that make *both* sentences true. That is, it is the *intersection* of the solution sets. Note that (4, 7) is a solution of the conjunction; in fact, it is the only solution.

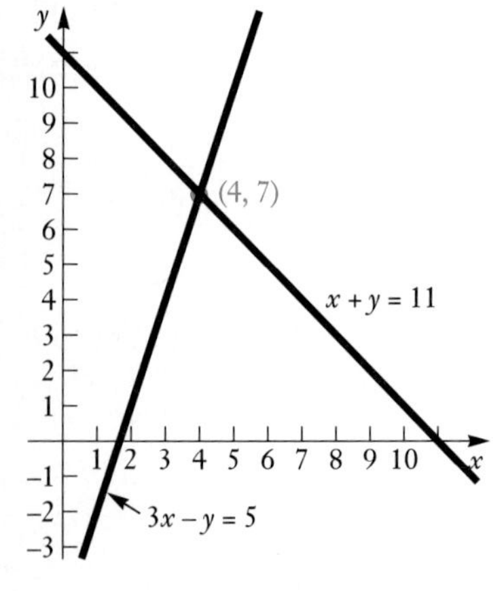

Solving Systems of Equations Graphically

One way to find solutions of a conjunction of equations is by trial and error. Another way is to graph the equations and look for points of intersection. For example, the graph on the left shows the solution sets of $x + y = 11$ and $3x - y = 5$. Their intersection is the single ordered pair (4, 7).

We often refer to a conjunction of equations as a **system of equations.** We usually omit the word *and* and very often write one equation under the other.

The graph of each equation in a system of two linear equations in two variables is a line.

Given the graphs of two lines, the following can happen:

a) The lines have no point in common—they are parallel. The system has no solution. (See graph (a).)
b) The lines have exactly one point in common. The system has exactly one solution. (See graph (b).)
c) The lines are the same—they have infinitely many points in common. The system has infinitely many solutions. (See graph (c).)

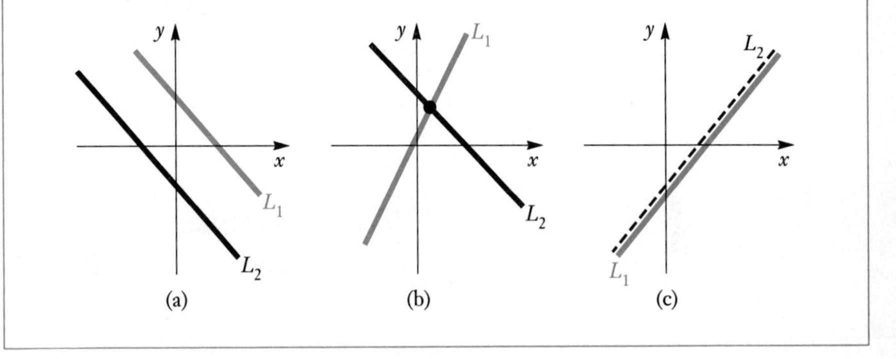

We will consider cases (a) and (c) in Section 11.3. We now consider two algebraic methods for solving systems of linear equations that have one solution. Both methods yield precise answers and avoid the difficulty of searching a graph for a point of intersection.

The Substitution Method

To use the substitution method, we solve one equation for one of the variables. Then we substitute in the other equation and solve.

EXAMPLE 1 Solve the system

$$x + y = 11, \quad (1)$$
$$3x - y = 5. \quad (2)$$

Solution First we solve Eq. (1) for y. (We could just as well solve for x.)

$$y = 11 - x$$

Then we substitute $11 - x$ for y in Eq. (2). This gives an equation in one variable, which we know how to solve:

$$3x - (11 - x) = 5$$
$$x = 4. \qquad \text{Solving for } x\text{. The student should be able to write the missing steps.}$$

Now we substitute 4 for x in either Eq. (1) or (2) and solve for y. Let us use Eq. (1):

$$4 + y = 11 \qquad \text{Solving for } y$$
$$y = 7.$$

The solution is (4, 7). We list the coordinates of the solution in alphabetical order, 4 for x and 7 for y.

Check:

$x + y = 11$		
$4 + 7$	11	
11		TRUE

$3x - y = 5$		
$3 \cdot 4 - 7$	5	
$12 - 7$		
5		TRUE

■

Gauss–Jordan Elimination with Equations

The next algebraic method to consider for solving systems of equations is called **Gauss–Jordan elimination,** or simply, **elimination.** It is an adaptation of methods developed by two German mathematicians, Carl Friedrich Gauss (1777–1855) and Wilhelm Jordan (1842–1899).

The **elimination method** makes use of the addition and multiplication principles for solving equations. Gauss–Jordan elimination is a special **algorithm,** or step-by-step procedure, that can be programmed on a computer. Our goal is to "transform" a system into an "equivalent system" for which the solution is obvious. Two systems are equivalent if they have exactly the same solutions.

Suppose we want to solve the system

$$\begin{aligned} 3x - 4y &= -1, \\ -3x + 2y &= 0. \end{aligned}$$

Our goal is to carry out certain procedures to obtain an equivalent system of the type

$$Ax + By = C, \qquad (1)$$
$$Dy = E. \qquad (2)$$

When we obtain such a system, we can easily solve for the variables by multiplying on each side of Eq. (2) by the reciprocal of the coefficient of y. After we have solved for y, we substitute into Eq. (1) to find x.

Now considering the system given above, we add the left-hand sides, obtaining $-2y$, and then add the right-hand sides, obtaining -1. When we do this, we often say that we "added the two equations." In this way, we eliminate the x-term in the second equation to obtain a system of equations equivalent to (having the same solutions as) the original:

$$\begin{aligned} 3x - 4y &= -1, \\ -3x + 2y &= 0. \end{aligned}$$

$$3x - 4y = -1 \qquad (1)$$
$$-2y = -1 \qquad (2)$$

$$\begin{aligned} 3x - 4y &= -1 \\ -3x + 2y &= 0 \\ \hline -2y &= -1 \end{aligned} \quad \text{Adding}$$

The system is now in the form

$$Ax + By = C,$$
$$Dy = E.$$

We solve for y and then substitute into the first equation to find x:

$$-2y = -1 \qquad \longrightarrow 3x - 4\left(\tfrac{1}{2}\right) = -1$$
$$y = \tfrac{1}{2}; \qquad 3x - 2 = -1$$
$$3x = 1$$
$$x = \tfrac{1}{3}.$$

This kind of substitution is often called *back-substitution*.

We now know that the solution of the original system is $\left(\frac{1}{3}, \frac{1}{2}\right)$ because the solutions of the equivalent system, $y = \frac{1}{2}$ *and* $x = \frac{1}{3}$, are obvious. On this and subsequent problems, the student can carry out a check by substituting into the original system as we did in Example 1.

EXAMPLE 2 Solve: $5x + 3y = 7,$
$3x - 5y = -23.$

Solution We first multiply the second equation by 5 to make the x-coefficient a multiple of 5:

$$5x + 3y = 7,$$
$$15x - 25y = -115.$$

Now we multiply the first equation by -3 and add it to the second equation. This eliminates the x-term:

$$5x + 3y = 7, \qquad -15x - 9y = -21 \quad \textbf{Multiplying by } -3$$
$$-34y = -136. \leftarrow \qquad 15x - 25y = -115$$
$$-34y = -136 \quad \textbf{Adding}$$

Next, we solve the second equation for y. Then we substitute the result into the first equation to find x:

$$-34y = -136 \qquad \longrightarrow 5x + 3(4) = 7$$
$$y = 4; \qquad 5x + 12 = 7$$
$$5x = -5$$
$$x = -1.$$

The solution is $(-1, 4)$. ■

We can take some preliminary steps to simplify the elimination method. One is to first write the equations in the form $Ax + By = C$. Another is to interchange two equations before beginning. For example, if we have the system

$$5x + y = -2,$$
$$x + 7y = 3,$$

we may prefer to write the second equation first:

$$\begin{aligned} x + 7y &= 3, \\ 5x + y &= -2. \end{aligned}$$

This procedure accomplishes two things. One is that the x-coefficient is 1 in the first equation. When we have found y later, this will make it easier to solve for x. The other is that this makes the x-coefficient in the second equation a multiple of the first.

Something else that might be done is to multiply one or more equations by a power of 10 before beginning in order to eliminate decimal points. For example, if we have the system

$$\begin{aligned} -0.3x + 0.5y &= 0.3, \\ 0.01x - 0.4y &= 1.2, \end{aligned}$$

we can multiply the first equation by 10 and the second by 100 to clear decimals and transform the system to

$$\begin{aligned} -3x + 5y &= 3, \\ x - 40y &= 120. \end{aligned}$$

We might also multiply in order to clear equations of fractions.

We transform the original system to an equivalent system of equations using any of the following operations, or transformations.

Transformations Producing Equivalent Systems

Each of the following will produce an equivalent system of equations:

a) Interchanging any two equations.
b) Multiplying each number or term of an equation by the same nonzero number.
c) Multiplying each number or term of one equation by the same nonzero number and adding the result to another equation.

EXAMPLE 3 Solve:

$$\begin{aligned} 5x + y &= -2, \\ x + 7y &= 3. \end{aligned}$$

Solution We first interchange the equations so that the x-coefficient of the second equation will be a multiple of the first:

$$\begin{aligned} x + 7y &= 3, \\ 5x + y &= -2. \end{aligned}$$

Next, we multiply the first equation by -5 and add the result to the second equation. This eliminates the x-term in the second equation:

$$x + 7y = 3,$$
$$-34y = -17.$$

$$\begin{aligned} -5x - 35y &= -15 \\ 5x + \quad y &= -2 \\ \hline -34y &= -17 \end{aligned} \quad \text{Adding}$$

Now we solve the second equation for y. Then we substitute the result in the first equation to find x.

$$-34y = -17 \qquad\qquad x + 7\left(\tfrac{1}{2}\right) = 3$$
$$y = \tfrac{1}{2} \qquad\qquad x + \tfrac{7}{2} = 3$$
$$x = 3 - \tfrac{7}{2}, \quad \text{or} \quad -\tfrac{1}{2}$$

The solution is $\left(-\frac{1}{2}, \frac{1}{2}\right)$. ■

Problem Solving

As you have already noticed, in the five-step problem-solving process, the most difficult and time-consuming part is the translation of a problem situation to mathematical language. In many cases this task becomes easier if we translate to more than one equation in more than one variable.

EXAMPLE 4 An airplane flies the 3000-mi distance from Los Angeles to New York, with a tailwind, in 5 hr. On the return trip, against the wind, it makes the trip in 6 hr. Find the speed of the plane and the speed of the wind.

Solution

1. Familiarize. We first make a drawing.

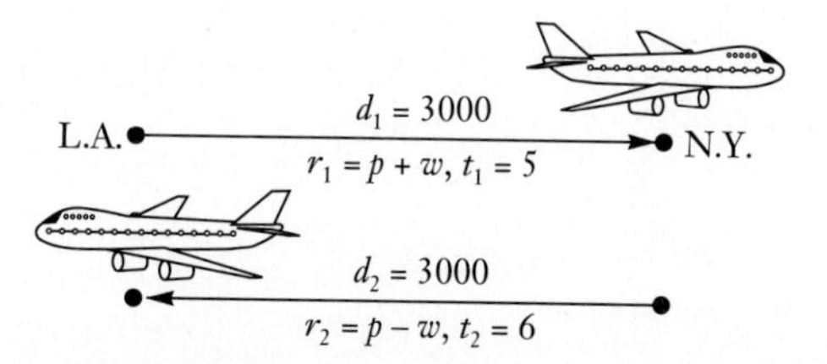

We let p represent the speed of the plane in still air and w represent the speed of the wind. Recall that distance, speed, and time are related by the motion formula, $d = rt$. Sometimes we may need to use other formulas that we can derive from this one, namely, $r = d/t$ and $t = d/r$.

From the figure, we see that the distances are the same. When the plane flies east with the wind, its speed (or rate) is $p + w$. When it flies west against the wind, its speed is $p - w$. We list the information in a table, the columns of the table coming from the formula $d = rt$.

	Distance	Speed	Time
East (with the wind)	3000	$p + w$	5
West (against the wind)	3000	$p - w$	6

2. **Translate.** Using $d = rt$ in each row of the table, we get an equation. Thus we have a system of equations:

$$3000 = (p + w)5 = 5p + 5w,$$
$$3000 = (p - w)6 = 6p - 6w.$$

3. **Carry out.** We solve the system

$$5p + 5w = 3000,$$
$$6p - 6w = 3000.$$

There is a common factor in each equation. We multiply by $\frac{1}{5}$ in the first equation and $\frac{1}{6}$ in the second equation to eliminate the common factors. Then we have

$$p + w = 600,$$
$$p - w = 500.$$

Next, we multiply the first equation by -1 and add it to the second equation. This eliminates the p-term:*

$$p + w = 600,$$
$$-2w = -100.$$

$$\begin{aligned} -p - w &= -600 \\ p - w &= 500 \\ \hline -2w &= -100 \quad \text{Adding} \end{aligned}$$

Then we solve the second equation for w and substitute into the first equation to find p:

$$-2w = -100 \qquad p + 50 = 600$$
$$w = 50; \qquad p = 550.$$

4. **Check.** We leave the check to the student.

5. **State.** The solution of the system of equations is (550, 50). That is, the speed of the plane is 550 mph and the speed of the wind is 50 mph. ■

EXAMPLE 5 Wine A is 5% alcohol and wine B is 15% alcohol. How many liters of each should be mixed to get a 10-L mixture that is 12% alcohol?

---

*We could also solve by adding the first equation, $p + w = 600$ to the second equation, $p - w = 500$, thereby eliminating the w-term. The approach used emphasizes the Gauss–Jordan algorithm.

Solution

1. **Familiarize.** We organize the information in a table.

	Amount of Solution	Percentage of Alcohol	Amount of Alcohol in Solution
A	x liters	5%	$5\%x$
B	y liters	15%	$15\%y$
Mixture	10 liters	12%	0.12×10, or 1.2 liters

Note that we have used x for the number of liters of A and y for the number of liters of B. To get the amount of alcohol, we multiply by the percentages.

2. **Translate.** If we add x and y in the first column, we get 10, and this gives us one equation:

$$x + y = 10. \qquad \text{We must finish with a 10-L mixture.}$$

If we add the amounts of alcohol in the third column, we get 1.2, and this gives us another equation:

$$5\%x + 15\%y = 1.2. \qquad \text{The alcohol in the mix comes from wine A and wine B.}$$

After changing percents to decimals and then clearing of decimals, we have this system:

$$x + y = 10,$$
$$5x + 15y = 120.$$

3. **Carry out.** Then solve the system. We leave this to the student. The solution is (3, 7). That is, 3 L of wine A and 7 L of wine B are possibilities for a solution to the original problem.

4. **Check.** We add the amounts of wine: 3 L + 7 L = 10 L. Thus the amount of wine checks. Next we check the amount of alcohol:

$$5\%(3) + 15\%(7) = 0.15 + 1.05, \quad \text{or} \quad 1.2 \text{ L}.$$

Thus the amount of alcohol checks.

5. **State.** Using 3 L of wine A and 7 L of wine B will yield a 10-L mixture that is 12% alcohol. ■

EXERCISE SET 11.1

1. Determine whether $\left(\frac{1}{2}, 1\right)$ is a solution of the system

$$3x + y = \tfrac{5}{2},$$
$$2x - y = \tfrac{1}{4}.$$

2. Determine whether $\left(-2, \frac{1}{4}\right)$ is a solution of the system

$$x + 4y = -1,$$
$$2x + 8y = -2.$$

Solve graphically. Be sure to check.

3. $x + y = 2,$ $3x + y = 0$

4. $x + y = 1,$ $3x + y = 7$

5. $y + 1 = 2x,$ $y - 1 = 2x$

6. $y + 1 = 2x,$ $3y = 6x - 3$

7. $x + y = 4,$ $x - y = 2$

8. $x - y = 3,$ $x + y = 5$

9. $2x - y = 4,$ $5x - y = 13$

10. $3x + y = 5,$ $x - 2y = 4$

11. $4x - y = 9,$ $x - 3y = 16$

12. $2y = 6 - x,$ $3x - 2y = 6$

Solve using the substitution method.

13. $3x + 5y = 3,$ $x = 8 - 4y$

14. $2x - 3y = 13,$ $y = 5 - 4x$

15. $9x - 2y = 3,$ $3x - 6 = y$

16. $x = 3y - 3,$ $x + 2y = 9$

17. $5m + n = 8$ $3m - 4n = 14$

18. $4x + y = 1,$ $x - 2y = 16$

19. $4x + 12y = 4,$ $-5x + y = 11$

20. $-3b + a = 7,$ $5a + 6b = 14$

21. $3x - y = 1,$ $2x + 2y = 2$

22. $5p + 7q = 1,$ $4p - 2q = 16$

23. $3x - y = 7,$ $2x + 2y = 5$

24. $5x + 3y = 4,$ $x - 4y = 3$

Solve using the elimination method.

25. $x + 3y = 7,$ $-x + 4y = 7$

26. $x + y = 9,$ $2x - y = -3$

27. $2x + y = 6,$ $x - y = 3$

28. $x - 2y = 6,$ $-x + 3y = -4$

29. $9x + 3y = -3,$ $2x - 3y = -8$

30. $6x - 3y = 18,$ $6x + 3y = -12$

31. $5x + 3y = 19,$ $2x - 5y = 11$

32. $3x + 2y = 3,$ $9x - 8y = -2$

33. $5r - 3s = 24,$ $3r + 5s = 28$

34. $5x - 7y = -16,$ $2x + 8y = 26$

35. $0.3x - 0.2y = 4,$ $0.2x + 0.3y = 1$

36. $0.7x - 0.3y = 0.5,$ $-0.4x + 0.7y = 1.3$

37. $\frac{1}{2}x + \frac{1}{3}y = 4,$ $\frac{1}{4}x + \frac{1}{3}y = 3$

38. $\frac{2}{3}x + \frac{1}{7}y = -11,$ $\frac{1}{7}x - \frac{1}{3}y = -10$

Problem Solving

39. Find two numbers whose sum is -10 and whose difference is 1.

40. Find two numbers whose sum is -1 and whose difference is 10.

41. A boat travels 46 km downstream in 2 hr. It travels 51 km upstream in 3 hr. Find the speed of the boat and the speed of the stream.

42. An airplane travels 3000 km with a tailwind in 3 hr. It travels 3000 km with a headwind in 4 hr. Find the speed of the plane and the speed of the wind.

43. Antifreeze A is 18% alcohol and antifreeze B is 10% alcohol. How many liters of each should be mixed to get 20 L of a mixture that is 15% alcohol?

44. Beer A is 6% alcohol and beer B is 2% alcohol. How many liters of each should be mixed to get 50 L of a mixture that is 3.2% alcohol?

45. Two cars leave town traveling in opposite directions. One travels at a speed of 80 km/h and the other at 96 km/h. In how many hours will they be 528 km apart?

46. A train leaves a station and travels north at a speed of 75 km/h. Two hours later, a second train leaves on a parallel track and travels north at 125 km/h. How far from the station will they meet?

47. Two planes travel toward each other from cities that are 780 km apart at speeds of 190 and 200 km/h. They started at the same time. In how many hours will they meet?

48. Two motorcycles travel toward each other from Chicago and Indianapolis, which are about 350 km apart, at speeds of 110 and 90 km/h. They started at the same time. In how many hours will they meet?

49. One week, a business sold 40 scarves. White ones cost \$4.95 and printed ones cost \$7.95. In all, \$282 worth of scarves were sold. How many of each kind were sold?

50. One day, a store sold 30 sweatshirts. White ones cost \$9.95 and yellow ones cost \$10.50. In all, \$310.60 worth of sweatshirts were sold. How many of each color were sold?

51. Paula is 12 years older than her brother Bob. Four years from now, Bob will be $\frac{2}{3}$ as old as Paula. How old are they now?

52. Carlos is 8 years older than his sister Maria. Four years ago, Maria was $\frac{2}{3}$ as old as Carlos. How old are they now?

53. The perimeter of a lot is 190 m. The width is one fourth the length. Find the dimensions.

54. The perimeter of a rectangular field is 628 m. The width of the field is 6 m less than the length. Find the dimensions.

55. The perimeter of a rectangle is 384 m. The length is 82 m greater than the width. Find the length and the width.

56. The perimeter of a rectangle is 86 cm. The length is 19 cm greater than the width. Find the area.

57. *Business.* Two investments are made that total \$15,000. For a certain year, these investments yield \$1432 in simple interest. Part of the \$15,000 is invested at 9% and part at 10%. Find the amount invested at each rate.

58. *Business.* For a certain year, \$3900 is received in interest from two investments. A certain amount is invested at 5%, and \$10,000 more than this is invested at 6%. Find the amount invested at each rate. (*Hint:* Express each equation in standard form $Ax + By = C$.)

SKILL MAINTENANCE

59. Graph: $y = -\frac{2}{3}x + 1$.

60. Determine an equation for a line with slope $\frac{7}{8}$ and y-intercept $(0, -3)$.

61. Use synthetic division to find the quotient and the remainder:

$$(x^3 - 2x^2 + 2x - 5) \div (x + 2).$$

62. Use synthetic division to find $P(3)$:

$$P(x) = x^5 - 4x^4 + 2x^2 - 5x + 1.$$

SYNTHESIS

Solve.

63. $\dfrac{x+y}{4} - \dfrac{x-y}{3} = 1,$
$\dfrac{x-y}{2} + \dfrac{x+y}{4} = -9$

64. $\dfrac{x+y}{2} - \dfrac{y-x}{3} = 0,$
$\dfrac{x+y}{3} - \dfrac{x+y}{4} = 0$

Each of the following is a system of equations that is *not* linear. But each is *linear in form,* in that an appropriate substitution, say u for $1/x$ and v for $1/y$, yields a linear system. Solve for the new variable and then solve for the original variable.

65. $\dfrac{1}{x} - \dfrac{3}{y} = 2,$
$\dfrac{6}{x} + \dfrac{5}{y} = -34$

66. $2\sqrt[3]{x} + \sqrt{y} = 0,$
$5\sqrt[3]{x} + 2\sqrt{y} = -5$

67. $3|x| + 5|y| = 30,$
$5|x| + 3|y| = 34$

68. $15x^2 + 2y^3 = 6,$
$25x^2 - 2y^3 = -6$

69. Nancy jogs and walks to the university each day. She averages 4 km/h walking and 8 km/h jogging. The distance from home to the university is 6 km, and she makes the trip in 1 hr. How far does she jog in a trip?

70. James and Joan are mathematics professors. They have a total of 46 years of teaching experience. Two years ago James had taught 2.5 times as many years as Joan. How long has each taught?

71. A limited edition of a book published by a historical society was offered for sale to its membership. The cost was one book for $12 or two books for $20. The society sold 880 books, and the total amount of money taken in was $9840. How many members ordered two books?

72. The ten's digit of a two-digit positive number is 2 more than three times the unit's digit. If the digits are interchanged, the new number is 13 less than half the given number. Find the given integer. (*Hint:* Let x = the ten's-place digit and y = the unit's-place digit; then $10x + y$ is the number.)

73. An automobile radiator contains 16 L of antifreeze and water. This mixture is 30% antifreeze. How much of this mixture should be drained and replaced with pure antifreeze so that there will be 50% antifreeze?

74. You are in line at a ticket window. There are two more people ahead of you in line than there are behind you. In the entire line, there are three times as many people as there are behind you. How many people are ahead of you in the line?

Business: Supply and demand. Find the equilibrium point (x_E, p_E) for the following supply and demand functions. (The equilibrium point is the point at which consumer demand equals producer's supply.)

75. Demand: $x + 43p = 800$
Supply: $x - 16p = 210$

76. Demand: $x = 8800 - 30p$
Supply: $x = 7000 + 15p$

77. Demand: $x = 760 - 13p$
Supply: $x = 430 + 2p$

78. Demand: $x + 60p = 2000$
Supply: $x - 94p = 460$

79. A student, out hiking for the weekend, is standing on a railroad bridge, as shown in the figure. A train is approaching from the direction shown by the arrow. If the student runs at a speed of 10 mph toward the train, she will reach point P on the bridge at the same instant that the train does. If she runs to point Q at the other end of the bridge at a speed of 10 mph, she will reach point Q also at the same instant that the train does. How fast, in miles per hour, is the train traveling?

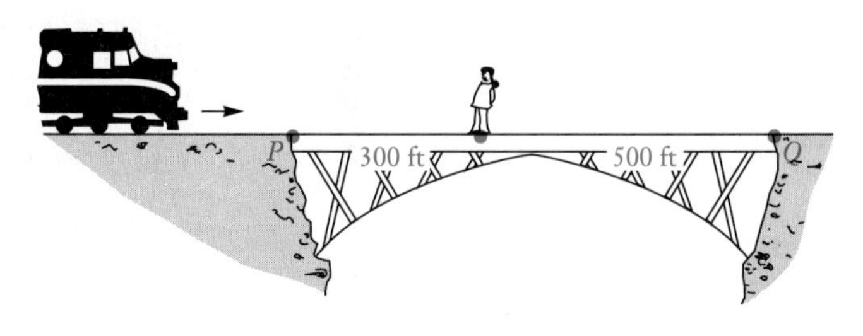

11.2 Systems of Equations in Three Variables

Identifying Solutions

A **linear equation in three variables** is an equation equivalent to one of the type $Ax + By + Cz = D$. We now solve systems of these linear equations in three variables.

A **solution** of a system of equations in three variables is an ordered triple that makes all three equations true.

EXAMPLE 1 Determine whether $(2, -1, 0)$ is a solution of the system

$$\begin{aligned} 4x + 2y + 5z &= 6, \\ 2x - y + z &= 5, \\ x + 2y - z &= 2. \end{aligned}$$

Solution We substitute $(2, -1, 0)$ into each of the three equations:

$$\begin{array}{r|l} \multicolumn{2}{c}{4x + 2y + 5z = 6} \\ \hline 4(2) + 2(-1) + 5(0) & 6 \\ 8 - 2 + 0 & \\ 6 & \quad \text{TRUE} \end{array} \qquad \begin{array}{r|l} \multicolumn{2}{c}{2x - y + z = 5} \\ \hline 2(2) - (-1) + 0 & 5 \\ 4 + 1 + 0 & \\ 5 & \quad \text{TRUE} \end{array}$$

$$\begin{array}{r|l} \multicolumn{2}{c}{x + 2y - z = 2} \\ \hline 2 + 2(-1) - 0 & 2 \\ 2 - 2 - 0 & \\ 0 & \quad \text{FALSE} \end{array}$$

Since $(2, -1, 0)$ is a solution of two of the equations but not *all* of the equations, it is not a solution of the system. ■

Solving Systems of Equations in Three or More Variables

Graphical methods of solving linear equations in three variables are unsatisfactory because a three-dimensional coordinate system is required. The substitution method becomes cumbersome for most systems of more than two equations. Therefore, we will use the elimination method. It is essentially the same as for systems of two equations.

Our goal is to transform the original system to an equivalent one of the form

$$\begin{aligned} Ax + By + Cz &= D, \\ Ey + Fz &= G, \\ Hz &= K. \end{aligned}$$

Then we solve the third equation for z and back-substitute to find the other variables.

EXAMPLE 2 Solve

$$\begin{aligned} 2x - 4y + 6z &= 22, && \textbf{(P1)} \\ 4x + 2y - 3z &= 4, && \textbf{(P2)} \\ 3x + 3y - z &= 4, && \textbf{(P3)} \end{aligned}$$

where (P1), (P2), and (P3) indicate the equation that is in the first, second, and third position, respectively. We will maintain this positional order throughout the solution and refer to the equations by their positional number.

Solution We begin by multiplying (P3) by 2, to make each x-coefficient a multiple of the first.* Then we have the following:

$$\begin{aligned} 2x - 4y + 6z &= 22, && \textbf{(P1)} \\ 4x + 2y - 3z &= 4, && \textbf{(P2)} \\ 6x + 6y - 2z &= 8. && \textbf{(P3)} \end{aligned}$$

*By proceeding in this manner, we avoid fractions. The method will work when fractions are allowed but is more difficult.

Next, we multiply (P1) by -2 and add it to (P2). We also multiply (P1) by -3 and add it to (P3). Then we have the following:

$$\begin{array}{r} -4x + 8y - 12z = -44 \\ 4x + 2y - 3z = 4 \\ \hline 10y - 15z = -40 \end{array} \qquad \begin{aligned} 2x - 4y + 6z &= 22, && \text{(P1)} \\ 10y - 15z &= -40, && \text{(P2)} \\ 18y - 20z &= -58. && \text{(P3)} \end{aligned} \qquad \begin{array}{r} -6x + 12y - 18z = -66 \\ 6x + 6y - 2z = 8 \\ \hline 18y - 20z = -58 \end{array}$$

Now we multiply (P3) by -5 to make the y-coefficient a multiple of the y-coefficient in (P2):

$$\begin{aligned} 2x - 4y + 6z &= 22, && \text{(P1)} \\ 10y - 15z &= -40, && \text{(P2)} \\ -90y + 100z &= 290. && \text{(P3)} \end{aligned}$$

Next, we multiply (P2) by 9 and add it to (P3):

$$\begin{array}{r} 90y - 135z = -360 \\ -90y + 100z = 290 \\ \hline -35z = -70 \end{array} \qquad \begin{aligned} 2x - 4y + 6z &= 22, && \text{(P1)} \\ 10y - 15z &= -40, && \text{(P2)} \\ -35z &= -70. && \text{(P3)} \end{aligned}$$

Now we solve (P3) for z:

$$\begin{aligned} -35z &= -70 \\ z &= 2. \end{aligned}$$

Next, we back-substitute 2 for z in (P2) and solve for y:

$$\begin{aligned} 10y - 15(2) &= -40 \\ 10y - 30 &= -40 \\ 10y &= -10 \\ y &= -1. \end{aligned}$$

Finally, we back-substitute -1 for y and 2 for z in (P1) and solve for x:

$$\begin{aligned} 2x - 4(-1) + 6(2) &= 22 \\ 2x + 4 + 12 &= 22 \\ 2x + 16 &= 22 \\ 2x &= 6 \\ x &= 3. \end{aligned}$$

The solution is $(3, -1, 2)$. To be sure that computational errors have not been made, one can check by substituting 3 for x, -1 for y, and 2 for z in all three original equations. If all are true, then the triple is a solution. ■

Although the solution of a system of three linear equations in three variables is difficult to find graphically, it is of interest to "see" what a solution might be. The graph of a linear equation in three variables is a plane. Thus the solution set of such a system is the intersection of three planes. Some possibilities are shown in the following figures.

Problem Solving

Solving systems of three or more equations is important in many applications. Systems of equations arise very often in the use of statistics, for example, in such fields as the social sciences. They also occur in problems of business, science, and engineering.

EXAMPLE 3 A triangle's largest angle is 70° greater than its smallest angle. The largest angle is twice as large as the remaining angle. Find the measure of each angle.

Solution

1. **Familiarize.** The first thing to do with a problem like this is to make a drawing, or a sketch.

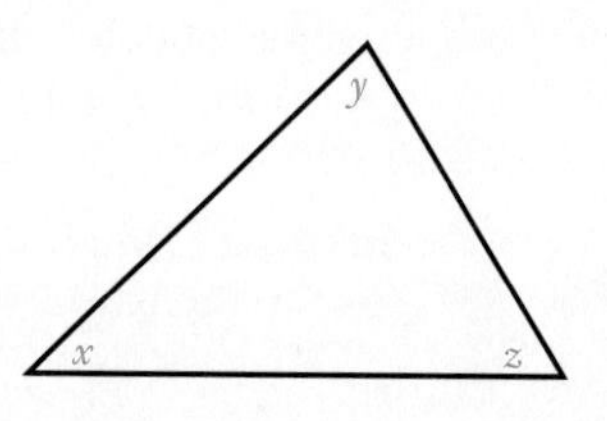

We don't know the size of any angle, so we have used x, y, and z for the measures of the angles. Recall that the measures of the angles of a triangle add up to 180°.

2. **Translate.** The geometric fact about triangles gives us one equation:

$$x + y + z = 180.$$

There are two statements in the problem that we can translate almost directly.

The largest angle	is	70°	greater than	the smallest angle.
z	$=$	70	$+$	x

The largest angle	is	twice as large as the remaining angle.
z	$=$	$2y$

We now have a system of three equations:

$$\begin{aligned} x + y + z &= 180, \\ x + 70 &= z, \\ 2y &= z; \end{aligned} \qquad \text{or} \qquad \begin{aligned} x + y + z &= 180, \\ x - z &= -70, \\ 2y - z &= 0. \end{aligned}$$

3. **Carry out.** We solve the system. The details are left to the student, but the solution is (30, 50, 100).

4. **Check.** The sum of the numbers is 180, so that checks. The largest angle measures 100° and the smallest measures 30°. The largest angle is thus 70° greater than the smallest. The remaining angle measures 50°. The largest angle measures 100°, so it is twice as large. We do have an answer to the problem.

5. **State.** The measures of the angles of the triangle are 30°, 50°, and 100°. ■

Mathematical Models and Problem Solving

In a situation in which a quadratic function will serve as a mathematical model, we may wish to find an equation, or formula, for the function. Recall that for a linear model, we can find an equation if we know two data points. For a quadratic function, we need three data points.

EXAMPLE 4 In a certain situation, it is believed that a quadratic function will be a good model. Find an equation of the function, given the data points $(1, -4)$, $(-1, -6)$, and $(2, -9)$.

Solution We want to find a quadratic function of the form

$$f(x) = ax^2 + bx + c$$

containing the three given points, that is, a function for which the equation will be true when we substitute any of the ordered pairs of numbers into it. When we substitute, we get for $(1, -4)$: $\quad -4 = a \cdot 1^2 + b \cdot 1 + c;$

for $(-1, -6)$: $\quad -6 = a(-1)^2 + b(-1) + c;$

for $(2, -9)$: $\quad -9 = a \cdot 2^2 + b \cdot 2 + c.$

We now have a system of equations in the three unknowns a, b, and c:

$$\begin{aligned} a + b + c &= -4, \\ a - b + c &= -6, \\ 4a + 2b + c &= -9. \end{aligned}$$

We solve this system of equations, obtaining $(-2, 1, -3)$. Thus the function we are looking for is

$$f(x) = -2x^2 + x - 3.$$

■

EXAMPLE 5 *The cost of operating an automobile at various speeds.* Under certain conditions, it is found that the cost of operating an automobile as a function of speed is approximated by a quadratic function. Use the following data to find an equation of the function. Then use the equation to determine the cost of operating the automobile at 80 mph.

Speed, in Miles per Hour	Operating Cost per Mile, in Cents
10	22
20	20
50	20

Solution We use the three data points to obtain a, b, and c in the equation $f(x) = ax^2 + bx + c$:

$$\begin{aligned} 22 &= 100a + 10b + c, && \text{Using 22 for } f(x) \text{ and 10 for } x \\ 20 &= 400a + 20b + c, && \text{Using 20 for } f(x) \text{ and 20 for } x \\ 20 &= 2500a + 50b + c. && \text{Using 20 for } f(x) \text{ and 50 for } x \end{aligned}$$

We solve this system of equations, obtaining $(0.005, -0.35, 25)$. Thus,

$$f(x) = 0.005x^2 - 0.35x + 25.$$

To find the cost of operating at 80 mph, we find $f(80)$:

$$f(80) = 0.005(80)^2 - 0.35(80) + 25 = 29¢.$$

A graph of the cost function of Example 5 is shown.

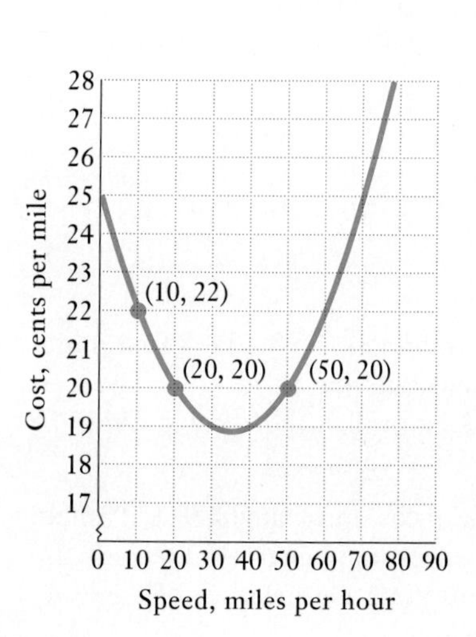

It should be noted that this cost function can give approximate results only within a certain interval. For example, $f(0) = 25$, meaning that it costs 25 cents per mile to stand still. This, of course, is absurd in the sense of mileage, although one does incur costs in owning a car whether one drives it or not. ■

EXERCISE SET 11.2

1. Determine whether $(1, -2, 3)$ is a solution of the system

$$\begin{aligned} x + y + z &= 2, \\ x - 2y - z &= 2, \\ 3x + 2y + z &= 2. \end{aligned}$$

2. Determine whether $(2, -1, -2)$ is a solution of the system

$$\begin{aligned} x + y - 2z &= 5, \\ 2x - y - z &= 7, \\ -x - 2y + 3z &= 6. \end{aligned}$$

Solve.

3. $x + y + z = 6,$ $2x - y + 3z = 9,$ $-x + 2y + 2z = 9$

4. $2x - y + z = 10,$ $4x + 2y - 3z = 10,$ $x - 3y + 2z = 8$

5. $2x - y - 3z = -1,$ $2x - y + z = -9,$ $x + 2y - 4z = 17$

6. $x - y + z = 6,$ $2x + 3y + 2z = 2,$ $3x + 5y + 4z = 4$

7. $2x - 3y + z = 5,$ $x + 3y + 8z = 22,$ $3x - y + 2z = 12$

8. $6x - 4y + 5z = 31,$ $5x + 2y + 2z = 13,$ $x + y + z = 2$

9. $3a - 2b + 7c = 13,$ $a + 8b - 6c = -47,$ $7a - 9b - 9c = -3$

10. $x + y + z = 0,$ $2x + 3y + 2z = -3,$ $-x + 2y - 3z = -1$

11. $2x + 3y + z = 17,$ $x - 3y + 2z = -8,$ $5x - 2y + 3z = 5$

12. $2x + y - 3z = -4,$ $4x - 2y + z = 9,$ $3x + 5y - 2z = 5$

13. $2x + y + z = -2,$ $2x - y + 3z = 6,$ $3x - 5y + 4z = 7$

14. $2x + y + 2z = 11,$ $3x + 2y + 2z = 8,$ $x + 4y + 3z = 0$

15. $4a + 9b = 8,$ $8a + 6c = -1,$ $6b + 6c = -1$

16. $3p + 2r = 11,$ $q - 7r = 4,$ $p - 6q = 1$

17. $x + y + z = 57,$ $-2x + y = 3,$ $x - z = 6$

18. $x + y + z = 105,$ $10y - z = 11,$ $2x - 3y = 7$

19. $2a - 3b = 2,$ $7a + 4c = \frac{3}{4},$ $2c - 3b = 1$

20. $a - 3c = 6,$ $b + 2c = 2,$ $7a - 3b - 5c = 14$

Problem Solving

21. The sum of three numbers is 26. Twice the first minus the second is 2 less than the third. The third is the second minus three times the first. Find the numbers.

22. The sum of three numbers is 5. The first number minus the second plus the third is 1. The first minus the third is 3 more than the second. Find the numbers.

23. In triangle ABC the measure of angle B is three times the measure of angle A. The measure of angle C is 30° greater than the measure of angle A. Find the angle measures.

24. In triangle ABC the measure of angle B is 2° more than three times the measure of angle A. The measure of angle C is 8° more than the measure of angle A. Find the angle measures.

25. A farmer picked strawberries on three days. She picked a total of 87 quarts. On Tuesday she picked 15 quarts more than on Monday. On Wednesday she picked 3 quarts fewer than on Tuesday. How many quarts did she pick each day?

26. Gina sells magazines part time. On Thursday, Friday, and Saturday, she sold \$66 worth. On Thursday she sold \$3 more than on Friday. On Saturday she sold \$6 more than on Thursday. How much did she take in each day?

27. Sawmills A, B, and C can produce 7400 board-feet of lumber per day. Mills A and B together can produce 4700 board-feet, while B and C together can produce 5200 board-feet. How many board-feet can each mill produce by itself?

28. A factory has three polishing machines, A, B, and C. When all three are working, 5700 lenses can be polished in one week. When only A and B are working, 3400 lenses can be polished in one week. When only B and C are working, 4200 lenses can be polished in one week. How many lenses can be polished in a week by each machine?

29. Three welders, A, B, and C, can weld 37 linear feet per hour when working together. If A and B together can weld 22 linear feet per hour, and A and C together can weld 25 linear feet per hour, how many linear feet per hour can each weld alone?

30. When three pumps, A, B, and C, are running together, they can pump 3700 gallons per hour. When only A and B are running, 2200 gallons per hour can be pumped. When only A and C are running, 2400 gallons per hour can be pumped. What is the pumping capacity of each pump?

31. *Nutrition.* A dietician in a hospital prepares meals under the guidance of a physician. Suppose that for a particular patient a physician prescribes a meal to have 800 calories, 55 g of protein, and 220 mg of vitamin C. The dietician prepares the meal using steak (each 3-oz serving contains 300 cal, 20 g of protein, and no vitamin C), baked potatoes (one baked potato contains 100 cal, 5 g of protein, and 20 mg of vitamin C), and broccoli (one 156-g serving contains 50 cal, 5 g of protein, and 100 mg of vitamin C). How many servings of each food are needed to satisfy the physician's requirements? (*Hint:* Let s = the number of servings of steak, p = the number of baked potatoes, and b = the number of servings of broccoli. Find an equation for the total number of calories, the total amount of protein, and the total amount of vitamin C.)

32. Repeat Exercise 31 but replace the broccoli with asparagus, for which one 180-g serving contains 50 calories, 5 g of protein, and 44 mg of vitamin C. Which meal would you prefer eating?

33. *Curve fitting.* Find numbers a, b, and c such that a quadratic function $ax^2 + bx + c$ fits the data points $(1, 4)$, $(-1, -2)$, and $(2, 13)$. Write the equation for the function.

34. *Curve fitting.* Find numbers a, b, and c such that a quadratic function $ax^2 + bx + c$ fits the data points $(1, 4)$, $(-1, 6)$, and $(-2, 16)$. Write the equation for the function.

35. *Predicting earnings.* A business earns \$38 in the first week, \$66 in the second week, and \$86 in the third week. The manager graphs the points (1, 38), (2, 66), and (3, 86) and finds that a quadratic function might fit the data.

a) Find a quadratic function that fits the data.
b) Using the model, predict the earnings for the fourth week.

36. *Predicting earnings.* A business earns \$1000 in its first month, \$2000 in the second month, and \$8000 in the third month. The manager plots the points (1, 1000), (2, 2000), and (3, 8000) and finds that a quadratic function might fit the data.

a) Find a quadratic function that fits the data.
b) Using the model, predict the earnings for the fourth month.

37. *Biomedical: Death rate as a function of sleep.* (This problem is based on a study by Dr. Harold J. Morowitz.)

Average Number of Hours of Sleep, x	Death Rate per Year per 100,000 Males, y
5	1121
7	626
9	967

a) Use the given data points to find a quadratic function $f(x) = ax^2 + bx + c$ that fits the data.
b) Use the model to find the death rate of males who sleep 4 hr, 6 hr, and 10 hr.

38. *Shoe size and life expectancy.* In a recent study published in *Orthopedic Quarterly,* a team of Swedish orthopedists hypothesized a correlation between shoe size and life expectancy that closely fits a quadratic function. Data are given in the following table.

Shoe Size (men)	Life Expectancy, in Years
8	72
11	82
14	69

a) Find a quadratic function that fits the data.
b) What is the life expectancy of a man with a shoe size of 10?
c) A man's life expectancy is 79. What is his shoe size?

SKILL MAINTENANCE

39. Factor $P(x)$ and then solve $P(x) = 0$ for $P(x) = x^3 + x^2 - x - 1$.

40. Factor: $8x^3 - 27$.

Solve.

41. $2^{3x} = 64$

42. $\log_7 x = 3$

SYNTHESIS

Solve. *Hint for Exercises* 43 *and* 44: Let u represent $1/x$, v represent $1/y$, and w represent $1/z$ and then solve for u, v, and w before finding x, y, and z.

43. $\frac{2}{x} + \frac{2}{y} - \frac{3}{z} = 3,$
$\frac{1}{x} - \frac{2}{y} - \frac{3}{z} = 9,$
$\frac{7}{x} - \frac{2}{y} + \frac{9}{z} = -39$

44. $\frac{2}{x} - \frac{1}{y} - \frac{3}{z} = -1,$
$\frac{2}{x} - \frac{1}{y} + \frac{1}{z} = -9,$
$\frac{1}{x} + \frac{2}{y} - \frac{4}{z} = 17$

45. Pipes A, B, and C are connected to the same tank. When all three pipes are running, they can fill the tank in 3 hr. When pipes A and C are running, they can fill the tank in 4 hr. When pipes A and B are running, they can fill the tank in 8 hr. How long would it take each, running alone, to fill the tank?

46. When A, B, and C work together, they can do a job in 2 hr. When B and C work together, they can do the job in 4 hr. When A and B work together, they can do the job in $\frac{12}{5}$ hr. How long would it take each, working alone, to do the job?

47. A theater had 100 people in attendance. The audience consisted of men, women, and children. The ticket prices were $10 for men, $3 for women, and 50 cents for children. The total amount of money taken in was $100. How many men, women, and children were in attendance? Does there seem to be some information missing? Do some careful reasoning.

48. Art, Bob, Carl, Denny, and Fred are on the same bowling team. They are all being truthful in the following comments regarding the last game they bowled.

Art: My score was a prime number. Fred finished third.

Bob: None of us bowled a score over 200.

Carl: Art beat me by exactly 23 pins. Denny's score was divisible by 10.

Denny: The sum of our five scores was exactly 885 pins. Bob's score was divisible by 8.

Fred: Art beat Bob by fewer than 10 pins. Denny beat Bob by exactly 14 pins.

Determine the score of each bowler in the game.

11.3 Special Cases

In Sections 11.1 and 11.2, each system had *exactly one* solution. Here we consider special cases where systems have no solution or infinitely many solutions.

Consistent and Inconsistent Systems

We say that a system of equations is **consistent** if it has at least one solution. We say that a system of equations is **inconsistent** if it has no solution.

Let us consider a system that does not have a solution and see what happens when we apply the elimination method.

EXAMPLE 1 Solve using the elimination method. Classify the system as consistent or inconsistent.

$$x - 3y = 1,$$
$$-2x + 6y = 5$$

Solution Let us first look at what happens graphically. We graph each equation and find where they intersect. It turns out that the lines are parallel and have no point of intersection.

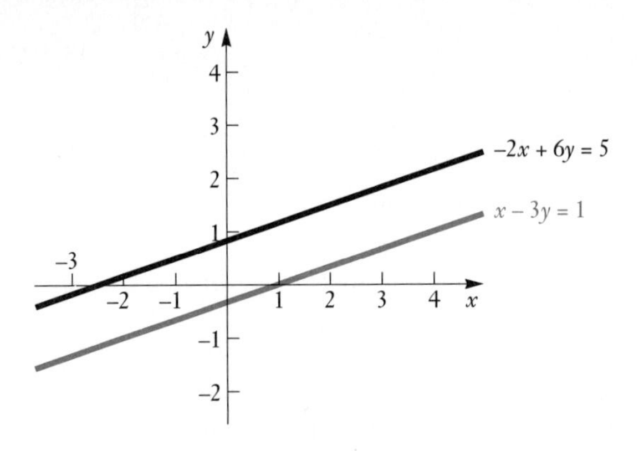

We see graphically that the system has no solution.

Let us see what happens if we apply the elimination method. We multiply the first equation by 2 and add the result to the second equation. This gives us

$$\begin{array}{r} 2x - 6y = 2 \\ -2x + 6y = 5 \\ \hline 0 = 7 \end{array} \quad \longrightarrow \quad \begin{array}{r} x - 3y = 1, \\ 0 = 7. \end{array}$$

The second equation says that $0 \cdot x + 0 \cdot y = 7$. There are no numbers x and y for which this is true.

Whenever we obtain a statement such as $0 = 7$, which is obviously false, we will know that the system we are trying to solve has no solutions. It is *inconsistent.* The solution set is the empty set, $\emptyset$. ■

Dependent and Independent Systems

Any system of equations in which *removing* one or more equations results in an equivalent system is said to be **dependent.** In other words, if there exists a system of fewer equations with the same solution(s), then the original system is dependent. Otherwise, the system is **independent.** Since, in Example 1, neither equation could be removed without altering the solution, that system was *independent.*

EXAMPLE 2 Solve the system, using the elimination method. Classify it as consistent or inconsistent, dependent or independent.

$$\begin{aligned} 2x + 3y &= 6, \\ 4x + 6y &= 12 \end{aligned}$$

Solution Let us look at what happens graphically. We graph each equation.

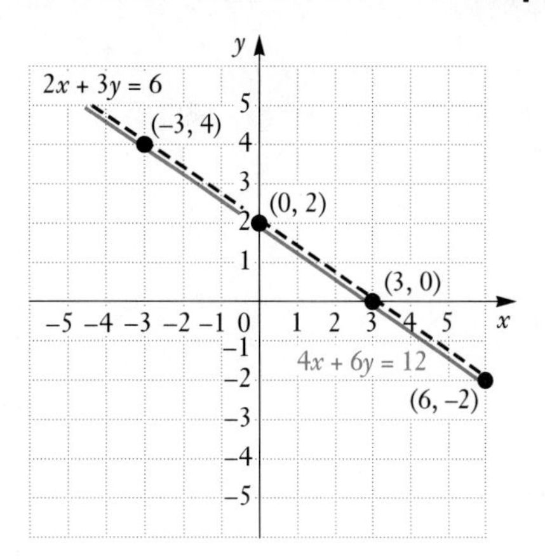

Some of the ordered pairs that solve the first equation are

$$(0, 2), \qquad (3, 0), \qquad (6, -2), \qquad \text{and} \qquad (-3, 4).$$

Some ordered pairs that solve the second equation are

$$(0, 2), \qquad (3, 0), \qquad (6, -2), \qquad \text{and} \qquad (-3, 4).$$

Indeed, the solution sets of the two equations are the same:

$$\{(x, y)|2x + 3y = 6\} = \{(x, y)|4x + 6y = 12\}.$$

If we remove one of the equations from the system, we still get the same solution set. Thus,

$$\begin{aligned} 2x + 3y &= 6, \\ 4x + 6y &= 12 \end{aligned} \qquad \text{is equivalent to} \qquad 2x + 3y = 6.$$

Thus the system is *dependent*. It is also *consistent* because it has a solution.

What happens when we apply the elimination method? We multiply the first equation by -2 and add. This gives us

$$\begin{aligned} -4x - 6y &= -12 \\ 4x + 6y &= 12 \\ \hline 0 &= 0 \end{aligned} \quad \longrightarrow \quad \begin{aligned} 2x + 3y &= 6, \\ 0 &= 0. \end{aligned}$$

The equation $0 = 0$ is equivalent to $0x + 0y = 0$, which is true for any values of x and y. Thus it is true for any pair of numbers x and y that constitute a solution of the system. Therefore the equation $0 = 0$ contributes nothing to the system and can be ignored. We then analyze any remaining equations to see whether the system they form has a solution. In this case we know that the equation $2x + 3y = 6$ has infinitely many solutions, so the system is *consistent* and *dependent*. ■

EXAMPLE 3 Solve using the elimination method. Classify the system as consistent or inconsistent, dependent or independent.

$$\begin{aligned} x - 3y &= 1, \\ x + y &= 3, \\ 5x - 7y &= 9 \end{aligned}$$

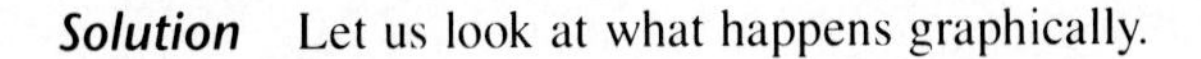

Solution Let us look at what happens graphically.

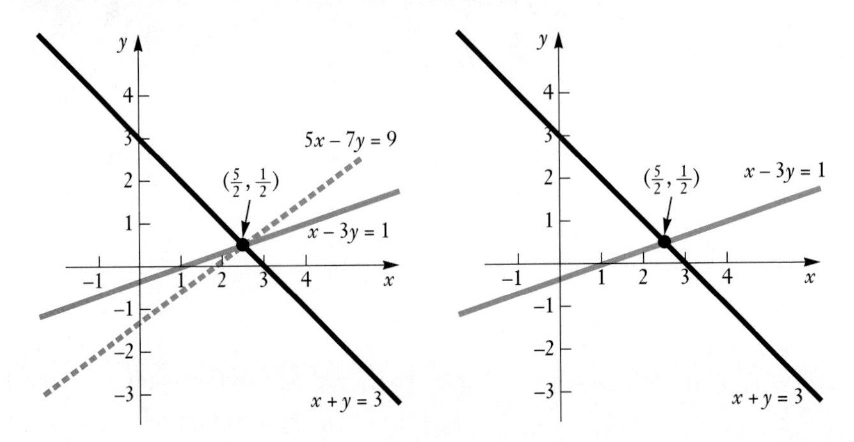

The solution set of the original system is $\{(\frac{5}{2}, \frac{1}{2})\}$ so we know the system is *consistent.* If we ignore one equation, say $5x - 7y = 9$, we obtain the graphs shown on the right. These still intersect at exactly one point, $(\frac{5}{2}, \frac{1}{2})$. Thus,

$$\begin{aligned} x - 3y &= 1, \\ x + y &= 3, \\ 5x - 7y &= 9 \end{aligned} \quad \text{is equivalent to} \quad \begin{aligned} x - 3y &= 1, \\ x + y &= 3, \end{aligned}$$

which means that the system is *dependent.*

Suppose we were to apply Gauss–Jordan elimination to the original system. We would start with

$$\begin{aligned} x - 3y &= 1, \\ x + y &= 3, \\ 5x - 7y &= 9. \end{aligned}$$

We'd then multiply the first equation by -1 and add the result to the second equation. We'd also multiply the first equation by -5 and add the result to the third equation:

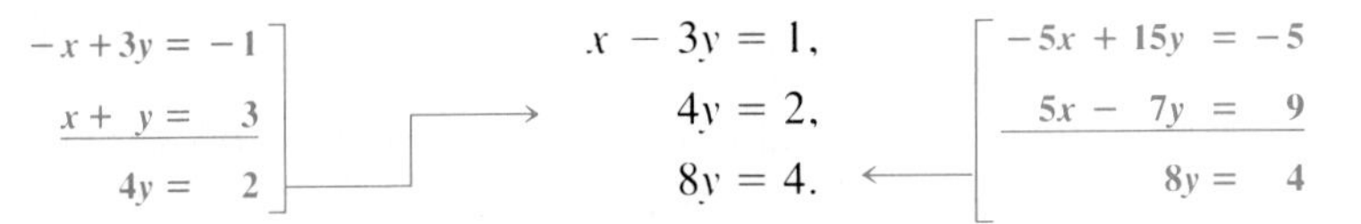

After multiplying the second equation by -2 we would then add the result to the third equation:

$$\begin{aligned} -8y &= -4 \\ 8y &= 4 \\ \hline 0 &= 0 \end{aligned} \quad \longrightarrow \quad \begin{aligned} x - 3y &= 1, \\ 4y &= 2, \\ 0 &= 0. \end{aligned} \quad \text{True for all } x \text{ and } y$$

The equation $0 = 0$ is true for all values of x and y. This is how we could tell algebraically that the system is *dependent.* To determine consistency, we would consider the first two equations, find that the solution is $x = \frac{5}{2}$ and $y = \frac{1}{2}$, and conclude that the system is *consistent.* ■

In solving a system, how do we know that it is dependent?

> If, at some stage, we find that two of the equations are identical, then we know that the system is *dependent.* If we obtain an obviously true statement, such as $0 = 0$, then we know that the system is *dependent.* We cannot know whether such a system is consistent or inconsistent without further analysis.

A dependent and consistent system may have an infinite number of solutions. In such a case for systems of two variables, we can describe the solutions by expressing one variable in terms of the other.

EXAMPLE 4 Solve:

$$2x + 3y = 1,$$
$$4x + 6y = 2.$$

Solution We multiply the first equation by -2 and add. This gives us

$$\begin{aligned} -4x - 6y &= -2 \\ 4x + 6y &= 2 \\ \hline 0 &= 0 \end{aligned} \qquad \longrightarrow \qquad \begin{aligned} 2x + 3y &= 1, \\ 0 &= 0. \end{aligned}$$

Now we know that the system is dependent. The last equation contributes nothing, so we consider only the first one. Let us solve for x. We obtain

$$x = \frac{1 - 3y}{2}.$$

We can now describe the ordered pairs in the solution set, in terms of y only, as follows:

$$\left(\frac{1 - 3y}{2}, y\right).$$

Any value that we choose for y gives us a value for x, and thus an ordered pair in the solution set. Some of these solutions are

$$(-4, 3), \quad \left(-\frac{5}{2}, 2\right), \quad \left(\frac{7}{2}, -2\right).$$

■

When a system of three or more equations is dependent and consistent, we can describe its solutions by expressing one or more of the variables in terms of the others.

EXAMPLE 5 Solve:

$$\begin{aligned} x + 2y + 3z &= 4, &\quad (P1)\\ 2x - y + z &= 3, &\quad (P2)\\ 3x + y + 4z &= 7, &\quad (P3)\end{aligned}$$

where (P1), (P2), and (P3) indicate the equations in the first, second, and third positions, respectively.

Solution Each x-coefficient is a multiple of the first. Thus we can begin solving by multiplying (P1) by -2 and adding it to (P2). We also multiply (P1) by -3 and add it to (P3):

$$\begin{aligned} -2x - 4y - 6z &= -8\\ 2x - y + z &= 3\\ \hline -5y - 5z &= -5\end{aligned}$$

$$\begin{aligned} x + 2y + 3z &= 4, &\quad (P1)\\ -5y - 5z &= -5, &\quad (P2)\\ -5y - 5z &= -5. &\quad (P3)\end{aligned}$$

$$\begin{aligned} -3x - 6y - 9z &= -12\\ 3x + y + 4z &= 7\\ \hline -5y - 5z &= -5\end{aligned}$$

Now (P2) and (P3) are identical. We no longer have a system of three equations, but a system of two. Thus we know that the system is dependent. If we were to multiply (P2) by -1 and add it to (P3), we would obtain $0 = 0$. We proceed by multiplying (P2) by $-\frac{1}{5}$ since the coefficients have the common factor -5:

$$\begin{aligned} x + 2y + 3z &= 4, &\quad (P1)\\ y + z &= 1. &\quad (P2)\end{aligned}$$

We will express two of the variables in terms of the other one. Let us choose z. Then we solve (P2) for y:

$$y = 1 - z.$$

We substitute this value of y in (P1), obtaining

$$x + 2(1 - z) + 3z = 4.$$

Simplifying and solving for x, we get

$$x = 2 - z.$$

The solutions, then, are all of the form

$$(2 - z, 1 - z, z).$$

We obtain the solutions by choosing various values of z. If we let $z = 0$, we obtain the triple $(2, 1, 0)$. If we let $z = 5$, we obtain $(-3, -4, 5)$, and so on. ■

Homogeneous Equations

When all the terms of a polynomial have the same degree, we say that the polynomial is **homogeneous.** Here are some examples:

$$3x^2 + 5y^2, \qquad 4x + 5y - 2z, \qquad 17x^3 - 4y^3 + 57z^3.$$

An equation formed by a homogeneous polynomial set equal to 0 is called a **homogeneous equation.** Let us now consider a system of homogeneous linear (first degree) equations.

EXAMPLE 6 Solve:

$$\begin{aligned} 4x - 3y + z &= 0, \\ 2x \quad\quad - 3z &= 0, \\ -8x + 6y - 2z &= 0. \end{aligned}$$

Solution Any homogeneous system like this always has a solution (it can never be inconsistent) because (0, 0, 0) is a solution. This is called the **trivial solution.** There may or may not be other solutions. To find out, we proceed as in the case of nonhomogeneous equations.

First, we interchange the first two equations so that all the x-coefficients are multiples of the first:

$$\begin{aligned} 2x \quad\quad - 3z &= 0, && \text{(P1)} \\ 4x - 3y + z &= 0, && \text{(P2)} \\ -8x + 6y - 2z &= 0. && \text{(P3)} \end{aligned}$$

Now we multiply (P1) by -2 and add it to (P2). We also multiply (P1) by 4 and add it to (P3):

$$\begin{aligned} -4x \quad\quad + 6z &= 0 \\ 4x - 3y + z &= 0 \\ \hline -3y + 7z &= 0 \end{aligned} \qquad \begin{aligned} 2x \quad\quad - 3z &= 0, && \text{(P1)} \\ -3y + 7z &= 0, && \text{(P2)} \\ 6y - 14z &= 0. && \text{(P3)} \end{aligned} \qquad \begin{aligned} 8x \quad\quad - 12z &= 0 \\ -8x + 6y - 2z &= 0 \\ \hline 6y - 14z &= 0 \end{aligned}$$

Next, we multiply (P2) by 2 and add it to (P3):

$$\begin{aligned} -6y + 14z &= 0 \\ 6y - 14z &= 0 \\ \hline 0 &= 0 \end{aligned} \qquad \begin{aligned} 2x \quad\quad - 3z &= 0, && \text{(P1)} \\ -3y + 7z &= 0, && \text{(P2)} \\ 0 &= 0. && \text{(P3)} \end{aligned}$$

Now we know that the system is dependent; hence it has an infinite set of solutions. Next, we solve (P2) for y:

$$y = \tfrac{7}{3}z.$$

Typically, we would substitute $\frac{7}{3}z$ for y in (P1), but because the y-term is missing, we need only solve for x:

$$x = \tfrac{3}{2}z.$$

We can now describe the members of the solution set as follows:

$$\left(\tfrac{3}{2}z, \tfrac{7}{3}z, z\right).$$

Some of the ordered pairs in the solution set are

$$\left(\tfrac{3}{2}, \tfrac{7}{3}, 1\right), \qquad \left(3, \tfrac{14}{3}, 2\right), \qquad \left(-\tfrac{3}{2}, -\tfrac{7}{3}, -1\right).$$ ■

EXERCISE SET 11.3

Solve. If a system has more than one solution, list three of them.

1. $9x - 3y = 15,$
$6x - 2y = 10$

2. $2s - 3t = 9,$
$4s - 6t = 9$

3. $5c + 2d = 24,$
$30c + 12d = 10$

4. $3x + 2y = 18,$
$9x + 6y = 5$

Solve.

5. $3x + 2y = 5,$
$4y = 10 - 6x$

6. $5x + 2 = 7y,$
$-14y + 4 = -10x$

7. $12y - 8x = 6,$
$4x + 3 = 6y$

8. $16x - 12y = 10,$
$6y + 5 = 8x$

Solve. If a system has more than one solution, list three of them.

9. $x + 2y - z = -8,$
$2x - y + z = 4,$
$8x + y + z = 2$

10. $x + 2y - z = 4,$
$4x - 3y + z = 8,$
$5x - y = 12$

11. $2x + y - 3z = 1,$
$x - 4y + z = 6,$
$4x + 16y + 4z = 24$

12. $4x + 12y + 16z = 4,$
$3x + 4y + 5z = 3,$
$x + 8y + 11z = 1$

13. $2x + y - 3z = 0,$
$x - 4y + z = 0,$
$4x - 16y + 4z = 0$

14. $4x + 12y + 16z = 0,$
$3x + 4y + 5z = 0,$
$x + 8y + 11z = 0$

15. $x + y - z = -3,$
$x + 2y + 2z = -1$

16. $x + y + 13z = 0,$
$x - y - 6z = 0$

17. $2x + y + z = 0,$
$x + y - z = 0,$
$x + 2y + 2z = 0$

18. $5x + 4y + z = 0,$
$10x + 8y - z = 0,$
$x - y - z = 0$

19. Classify each of the systems in the odd-numbered exercises 1–17 as consistent or inconsistent, dependent or independent.

20. Classify each of the systems in the even-numbered exercises 2–18 as consistent or inconsistent, dependent or independent.

SKILL MAINTENANCE

21. Determine the slope and the y-intercept of the graph of $3x + 2y = 8$.

22. Use synthetic division to find the quotient and the remainder:

$$(3x^3 - 7x^2 + 5x + 1) \div (x - 2).$$

23. Solve: $\ln (x - 3) = 2$.

24. Simplify: $(3 - i)^2$.

SYNTHESIS

Solve.

25. ▦ $4.026x - 1.448y = 18.32,$
$0.724y = -9.16 + 2.013x$

26. ▦ $0.0284y = 1.052 - 8.114x,$
$0.0142y + 4.057x = 0.526$

27. a) Solve:

$$\begin{aligned} w + x + y + z &= 4, \\ w + x + y + z &= 3, \\ w + x + y + z &= 3. \end{aligned}$$

b) Classify the system as consistent or inconsistent.
c) Classify the system as dependent or independent.

28. a) Solve:

$$\begin{aligned} w - 8x + 3y + 2z &= 0, \\ -w + 5x + y - z &= 0, \\ -w + 2x + 5y &= 0, \\ 3x - 4y - z &= 0. \end{aligned}$$

b) Classify the system as consistent or inconsistent.
c) Classify the system as dependent or independent.

Determine the constant k such that each system is dependent.

29. $6x - 9y = -3,$
$-4x + 6y = k$

30. $8x - 16y = 20,$
$10x - 20y = k$

31. An 18-hole golf course has par-3 holes, par-4 holes, and par-5 holes. A golfer who shoots par on every hole has a total of 72. There are the same number of par-3 holes as there are par-5 holes. There is at least one of each type of hole. How many of each type of hole are there on the golf course?

32. A two-digit number is such that the number is equal to four times the sum of the digits. Find the number.

11.4 Elimination Using Matrices

In solving systems of equations, we perform computations with the constants. The variables play no important role until the end. We can simplify writing by omitting the variables. For example, the system

$$\begin{aligned} 3x + 4y &= 5, \\ x - 2y &= 1 \end{aligned}$$

simplifies to

$$\begin{matrix} 3 & 4 & 5 \\ 1 & -2 & 1 \end{matrix}$$

if we leave off the variables and omit the operation and equals signs.

In this example we have written a rectangular array of numbers. Such an array is called a **matrix** (plural, **matrices**). We ordinarily write brackets around matrices. The following are matrices:

$$\begin{bmatrix} 4 & 1 & 3 & 5 \\ 1 & 0 & 1 & 2 \\ 6 & 3 & -2 & 0 \end{bmatrix}, \quad \begin{bmatrix} 6 & 2 & 1 & 4 & 7 \\ 1 & 2 & 1 & 3 & 1 \\ 4 & 0 & -2 & 0 & -3 \end{bmatrix}, \quad \begin{bmatrix} 1 & 2 \\ 145 & 0 \\ -7 & 9 \\ 8 & 1 \\ 0 & 0 \end{bmatrix}.$$

The **rows** of a matrix are horizontal, and the **columns** are vertical.

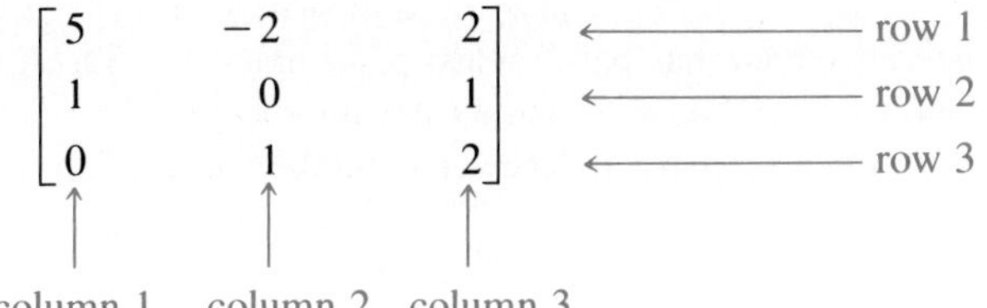

Let us now use matrices to solve systems of linear equations.

EXAMPLE 1 Solve:

$$\begin{aligned} 2x - y + 4z &= -3, \\ x \qquad - 4z &= 5, \\ 6x - y + 2z &= 10. \end{aligned}$$

Solution We first write a matrix, using only the constants. Where there are missing terms, we must write 0's. Note that we have included a vertical dashed line to separate the coefficients from the constants at the end of each equation.

$$\left[\begin{array}{rrr|r} 2 & -1 & 4 & -3 \\ 1 & 0 & -4 & 5 \\ 6 & -1 & 2 & 10 \end{array}\right]$$

The individual numbers are called *elements*.

We do exactly the same calculations using the matrix that we would do if we wrote the entire equations. The first step, if possible, is to interchange the rows so that each number in the first column below the first number is a multiple of that number. We do this by interchanging rows 1 and 2:

$$\left[\begin{array}{rrr|r} 1 & 0 & -4 & 5 \\ 2 & -1 & 4 & -3 \\ 6 & -1 & 2 & 10 \end{array}\right].$$

This corresponds to interchanging the first two equations.

Next, we multiply the first row by -2 and add it to the second row:

$$\left[\begin{array}{ccc|c} 1 & 0 & -4 & 5 \\ 0 & -1 & 12 & -13 \\ 6 & -1 & 2 & 10 \end{array}\right].$$

This corresponds to multiplying new equation (P1) by -2 and adding it to new equation (P2).*

Now we multiply the first row by -6 and add it to the third row:

$$\left[\begin{array}{ccc|c} 1 & 0 & -4 & 5 \\ 0 & -1 & 12 & -13 \\ 0 & -1 & 26 & -20 \end{array}\right].$$

This corresponds to multiplying equation (P1) by -6 and adding it to equation (P3).

Next we multiply row 2 by -1 and add it to the third row:

$$\left[\begin{array}{ccc|c} 1 & 0 & -4 & 5 \\ 0 & -1 & 12 & -13 \\ 0 & 0 & 14 & -7 \end{array}\right].$$

This corresponds to multiplying equation (P2) by -1 and adding it to equation (P3).

If we now put the variables back, we have

$$\begin{aligned} x \qquad - 4z &= 5, \\ -y + 12z &= -13, \\ 14z &= -7. \end{aligned}$$

Now we proceed as before. We solve (P3) for z and get $z = -\frac{1}{2}$. Next we back-substitute $-\frac{1}{2}$ for z in (P2) and solve for y: $-y + 12\left(-\frac{1}{2}\right) = -13$, so $y = 7$. Since there is no y-term in (P1), we need only substitute $-\frac{1}{2}$ for z in (P1) and solve for x: $x - 4\left(-\frac{1}{2}\right) = 5$, so $x = 3$. The solution is $\left(3, 7, -\frac{1}{2}\right)$. ■

Note that in the preceding example our goal was to get the matrix in the form

$$\left[\begin{array}{ccc|c} a & b & c & d \\ 0 & e & f & g \\ 0 & 0 & h & k \end{array}\right],$$

where there are just 0's below the **main diagonal,** formed by a, e, and h. Then we put the variables back and complete the solution.

All the operations used in the preceding example correspond to operations with the equations and produce equivalent systems of equations. We call the matrices **row-equivalent** and the operations that produce them **row-equivalent operations.**

*Recall that (P1), (P2), and (P3) indicate the equations that are in the first, second, and third position, respectively.

Row-Equivalent Operations

Each of the following row-equivalent operations produces an equivalent matrix:

a) Interchanging any two rows of a matrix.
b) Multiplying each element of a row by the same nonzero number.
c) Multiplying each element of a row by a nonzero number and adding the result to another row.

The best overall method for solving systems of equations is by row-equivalent matrices; even computers are programmed to use them.

EXERCISE SET 11.4

Solve using matrices.

1. $4x + 2y = 11,$
$3x - y = 2$

2. $3x - 3y = 11,$
$9x - 2y = 5$

3. $x + 2y - 3z = 9,$
$2x - y + 2z = -8,$
$3x - y - 4z = 3$

4. $x - y + 2z = 0,$
$x - 2y + 3z = -1,$
$2x - 2y + z = -3$

5. $5x - 3y = -2,$
$4x + 2y = 5$

6. $3x + 4y = 7,$
$-5x + 2y = 10$

7. $4x - y - 3z = 1,$
$8x + y - z = 5,$
$2x + y + 2z = 5$

8. $3x + 2y + 2z = 3,$
$x + 2y - z = 5,$
$2x - 4y + z = 0$

9. $p + q + r = 1,$
$p + 2q + 3r = 4,$
$4p + 5q + 6r = 7$

10. $m + n + t = 9,$
$m - n - t = -15,$
$m + n + t = 3$

11. $-2w + 2x + 2y - 2z = -10,$
$w + x + y + z = -5,$
$3w + x - y + 4z = -2,$
$w + 3x - 2y + 2z = -6$

12. $-w + 2x - 3y + z = -8,$
$-w + x + y - z = -4,$
$w + x + y + z = 22,$
$-w + x - y - z = -14$

SKILL MAINTENANCE

13. Graph: $y = \frac{4}{7}x + 1$.

14. Use synthetic division to find the quotient and the remainder:

$$(2x^3 - 11x^2 + 17x - 1) \div (x - 3).$$

Solve.

15. $3^{1-5x} = 81$

16. $\log_5 (x + 2) = 3$

SYNTHESIS

17. A collection of 34 coins consists of dimes and nickels. The total value is $1.90. How many dimes and how many nickels are there?

18. A collection of 43 coins consists of dimes and quarters. The total value is $7.60. How many dimes and how many quarters are there?

19. A collection of 22 coins consists of nickels, dimes, and quarters. The total value is $2.90. There are 6 more nickels than dimes. How many of each type of coin are there?

20. A collection of 18 coins consists of nickels, dimes, and quarters. The total value is $2.55. There are 2 more quarters than dimes. How many of each type of coin are there?

21. A tobacco dealer has two kinds of tobacco. One is worth \$4.05 per pound and the other is worth \$2.70 per pound. The dealer wants to blend the two tobaccos to get a 15-lb mixture worth \$3.15 per pound. How much of each kind of tobacco should be used?

22. A grocer mixes candy worth \$0.80 per pound with nuts worth \$0.70 per pound to get a 20-lb mixture worth \$0.77 per pound. How many pounds of candy and how many pounds of nuts are used?

23. One year, \$8950 was received in simple interest from two investments. A certain amount was invested at $12\frac{1}{2}\%$, and \$10,000 more than this was invested at 13%. Find the amount of principal invested at each rate.

24. Solve:

$$ax + by = c,$$
$$dx + ey = f.$$

Solve.

25. $4.83x + 9.06y = -39.42,$
$-1.35x + 6.67y = -33.99$

26. $3.11x - 2.04y = -24.39,$
$7.73x + 5.19y = -35.48$

27. $3.55x - 1.35y + 1.03z = 9.16,$
$-2.14x + 4.12y + 3.61z = -4.50,$
$5.48x - 2.44y - 5.86z = 0.813$

28. $4.12x - 1.35y - 18.2z = 601.3,$
$-3.41x + 68.9y + 38.7z = 1777,$
$0.955x - 0.813y - 6.53z = 160.2$

11.5 Determinants and Cramer's Rule

Determinants of 2 × 2 Matrices

If a matrix has the same number of rows and columns, it is called a **square matrix.** With every square matrix is associated a number called its **determinant,** defined as follows for 2 × 2 matrices.

The determinant of the matrix $\begin{bmatrix} a & c \\ b & d \end{bmatrix}$ is denoted $\begin{vmatrix} a & c \\ b & d \end{vmatrix}$ and is defined as follows:

$$\begin{vmatrix} a & c \\ b & d \end{vmatrix} = ad - bc.$$

EXAMPLE 1 Evaluate: $\begin{vmatrix} \sqrt{2} & -3 \\ -4 & -\sqrt{2} \end{vmatrix}$.

Solution $\begin{vmatrix} \sqrt{2} & -3 \\ -4 & -\sqrt{2} \end{vmatrix}$ The arrows indicate the products involved.

$$= \sqrt{2}(-\sqrt{2}) - (-4)(-3) = -2 - 12 = -14$$

■

Cramer's Rule: 2 × 2 Systems

Determinants have many uses. One of these is in solving systems of linear equations where the number of variables is the same as the number of equations and the constants are not all 0. Let us consider a system of two equations:

$$a_1x + b_1y = c_1,$$
$$a_2x + b_2y = c_2.$$

Using the methods of the preceding sections, we can solve. We obtain

$$x = \frac{c_1b_2 - c_2b_1}{a_1b_2 - a_2b_1}, \qquad y = \frac{a_1c_2 - a_2c_1}{a_1b_2 - a_2b_1}.$$

The numerators and denominators of the expressions for x and y can be written as determinants.

CRAMER'S RULE: 2 × 2 SYSTEMS

The solution of the system

$$a_1x + b_1y = c_1,$$
$$a_2x + b_2y = c_2,$$

if it is unique, is given by

$$x = \frac{\begin{vmatrix} c_1 & b_1 \\ c_2 & b_2 \end{vmatrix}}{\begin{vmatrix} a_1 & b_1 \\ a_2 & b_2 \end{vmatrix}}, \qquad y = \frac{\begin{vmatrix} a_1 & c_1 \\ a_2 & c_2 \end{vmatrix}}{\begin{vmatrix} a_1 & b_1 \\ a_2 & b_2 \end{vmatrix}}.$$

The equations above make sense only if the determinant in the denominator is not 0. If the denominator *is* 0, then one of two things happens.

1. If the denominator is 0 and the other two determinants in the numerators are also 0, then the system of equations is dependent.
2. If the denominator is 0 and at least one of the other determinants in the numerators is not 0, then the system is inconsistent.

To use Cramer's rule, we compute the three determinants and compute x and y as shown above. Note that the denominator in both cases contains the coefficients of x and y, in the same position as in the original equations. For x, the numerator is obtained by replacing the x-coefficients (the a's) by the c's. For y, the numerator is obtained by replacing the y-coefficients (the b's) by the c's.

EXAMPLE 2 Solve using Cramer's rule:

$$2x + 5y = 7,$$
$$5x - 2y = -3.$$

Solution We have

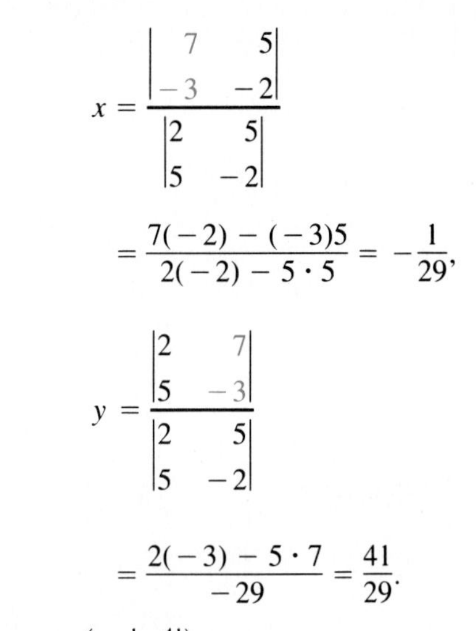

$$x = \frac{\begin{vmatrix} 7 & 5 \\ -3 & -2 \end{vmatrix}}{\begin{vmatrix} 2 & 5 \\ 5 & -2 \end{vmatrix}} = \frac{7(-2) - (-3)5}{2(-2) - 5 \cdot 5} = -\frac{1}{29},$$

$$y = \frac{\begin{vmatrix} 2 & 7 \\ 5 & -3 \end{vmatrix}}{\begin{vmatrix} 2 & 5 \\ 5 & -2 \end{vmatrix}} = \frac{2(-3) - 5 \cdot 7}{-29} = \frac{41}{29}.$$

The solution is $\left(-\frac{1}{29}, \frac{41}{29}\right)$. ■

Cramer's Rule: 3 × 3 Systems

A similar method has been developed for solving systems of three equations in three unknowns. Before stating the rule, though, we must develop some terminology.

> The *determinant* of a three-by-three matrix is defined as follows:
>
> $$\begin{vmatrix} a_1 & b_1 & c_1 \\ a_2 & b_2 & c_2 \\ a_3 & b_3 & c_3 \end{vmatrix} = a_1 \cdot \begin{vmatrix} b_2 & c_2 \\ b_3 & c_3 \end{vmatrix} - a_2 \cdot \begin{vmatrix} b_1 & c_1 \\ b_3 & c_3 \end{vmatrix} + a_3 \cdot \begin{vmatrix} b_1 & c_1 \\ b_2 & c_2 \end{vmatrix}.$$

The two-by-two determinants on the right can be obtained from the three-by-three determinant by crossing out the row and the column in which the a-coefficients occur.

EXAMPLE 3 Evaluate:

$$\begin{vmatrix} -1 & 0 & 1 \\ -5 & 1 & -1 \\ 4 & 8 & 1 \end{vmatrix} = -1 \cdot \begin{vmatrix} 1 & -1 \\ 8 & 1 \end{vmatrix} - (-5) \cdot \begin{vmatrix} 0 & 1 \\ 8 & 1 \end{vmatrix} + 4 \cdot \begin{vmatrix} 0 & 1 \\ 1 & -1 \end{vmatrix}$$

$$= -1(1 + 8) + 5(0 - 8) + 4(0 - 1)$$

$$= -9 - 40 - 4 = -53.$$ ■

CRAMER'S RULE: 3 × 3 SYSTEMS

The solution of the system

$$a_1x + b_1y + c_1z = d_1,$$
$$a_2x + b_2y + c_2z = d_2,$$
$$a_3x + b_3y + c_3z = d_3$$

is found by considering the following determinants:

$$D = \begin{vmatrix} a_1 & b_1 & c_1 \\ a_2 & b_2 & c_2 \\ a_3 & b_3 & c_3 \end{vmatrix}, \quad D_x = \begin{vmatrix} d_1 & b_1 & c_1 \\ d_2 & b_2 & c_2 \\ d_3 & b_3 & c_3 \end{vmatrix},$$

$$D_y = \begin{vmatrix} a_1 & d_1 & c_1 \\ a_2 & d_2 & c_2 \\ a_3 & d_3 & c_3 \end{vmatrix}, \quad D_z = \begin{vmatrix} a_1 & b_1 & d_1 \\ a_2 & b_2 & d_2 \\ a_3 & b_3 & d_3 \end{vmatrix}.$$

The solution, if it is unique, is given by

$$x = \frac{D_x}{D}, \qquad y = \frac{D_y}{D}, \qquad z = \frac{D_z}{D}.$$

Note that we obtain the determinant D_x in the numerator for x from D by replacing the x-coefficients by d_1, d_2, and d_3. A similar thing happens with D_y and D_z. When $D = 0$, Cramer's rule cannot be used. If $D = 0$ and D_x, D_y, and D_z are 0, the system is dependent. If $D = 0$ and at least one of D_x, D_y, or D_z is not 0, then the system is inconsistent.

EXAMPLE 4 Solve using Cramer's rule:

$$x - 3y + 7z = 13,$$
$$x + y + z = 1,$$
$$x - 2y + 3z = 4.$$

Solution We have

$$D = \begin{vmatrix} 1 & -3 & 7 \\ 1 & 1 & 1 \\ 1 & -2 & 3 \end{vmatrix} = -10, \qquad D_x = \begin{vmatrix} 13 & -3 & 7 \\ 1 & 1 & 1 \\ 4 & -2 & 3 \end{vmatrix} = 20,$$

$$D_y = \begin{vmatrix} 1 & 13 & 7 \\ 1 & 1 & 1 \\ 1 & 4 & 3 \end{vmatrix} = -6, \qquad D_z = \begin{vmatrix} 1 & -3 & 13 \\ 1 & 1 & 1 \\ 1 & -2 & 4 \end{vmatrix} = -24.$$

Then

$$x = \frac{D_x}{D} = \frac{20}{-10} = -2,$$

$$y = \frac{D_y}{D} = \frac{-6}{-10} = \frac{3}{5},$$

$$z = \frac{D_z}{D} = \frac{-24}{-10} = \frac{12}{5}.$$

The solution is $\left(-2, \frac{3}{5}, \frac{12}{5}\right)$. In practice, it is not necessary to evaluate D_z. When we have found values for x and y, we can substitute them into one of the equations and find z. ■

EXERCISE SET 11.5

Evaluate.

1. $\begin{vmatrix} -2 & -\sqrt{5} \\ -\sqrt{5} & 3 \end{vmatrix}$

2. $\begin{vmatrix} \sqrt{5} & -3 \\ 4 & 2 \end{vmatrix}$

3. $\begin{vmatrix} x & 4 \\ x & x^2 \end{vmatrix}$

4. $\begin{vmatrix} y^2 & -2 \\ y & 3 \end{vmatrix}$

5. $\begin{vmatrix} 3 & 1 & 2 \\ -2 & 3 & 1 \\ 3 & 4 & -6 \end{vmatrix}$

6. $\begin{vmatrix} 3 & -2 & 1 \\ 2 & 4 & 3 \\ -1 & 5 & 1 \end{vmatrix}$

7. $\begin{vmatrix} x & 0 & -1 \\ 2 & x & x^2 \\ -3 & x & 1 \end{vmatrix}$

8. $\begin{vmatrix} x & 1 & -1 \\ x^2 & x & x \\ 0 & x & 1 \end{vmatrix}$

Solve using Cramer's rule.

9. $-2x + 4y = 3,$
 $3x - 7y = 1$

10. $5x - 4y = -3,$
 $7x + 2y = 6$

11. $\sqrt{3}x + \pi y = -5,$
 $\pi x - \sqrt{3}y = 4$

12. $\pi x - \sqrt{5}y = 2,$
 $\sqrt{5}x + \pi y = -3$

13. $3x + 2y - z = 4,$
 $3x - 2y + z = 5,$
 $4x - 5y - z = -1$

14. $3x - y + 2z = 1,$
 $x - y + 2z = 3,$
 $-2x + 3y + z = 1$

15. $6y + 6z = -1,$
 $8x + 6z = -1,$
 $4x + 9y = 8$

16. $3x + 5y = 2,$
 $2x - 3z = 7,$
 $4y + 2z = -1$

SKILL MAINTENANCE

17. Use synthetic division to find $f(3)$:
 $f(x) = 2x^5 - 3x^4 + 5x^2 - 7x + 9.$

18. Factor: $125x^3 - 27$.

19. Solve: $\log(x^2 - 9x) = 1$.

20. Determine the slope and the y-intercept of the graph of $3x - 5y = 2$.

SYNTHESIS

Solve.

21. $\begin{vmatrix} x & 5 \\ -4 & x \end{vmatrix} = 24$

22. $\begin{vmatrix} y & 2 \\ 3 & y \end{vmatrix} = y$

23. $\begin{vmatrix} x & -3 \\ -1 & x \end{vmatrix} \geq 0$

24. $\begin{vmatrix} y & -5 \\ -2 & y \end{vmatrix} < 0$

25. $\begin{vmatrix} x+3 & 4 \\ x-3 & 5 \end{vmatrix} = -7$

26. $\begin{vmatrix} m+3 & -3 \\ m+3 & -4 \end{vmatrix} = 3m - 5$

27. $\begin{vmatrix} 2 & x & 1 \\ 1 & 2 & -1 \\ 3 & 4 & -2 \end{vmatrix} = -6$

28. $\begin{vmatrix} x & 2 & x \\ 3 & -1 & 1 \\ 1 & -2 & 2 \end{vmatrix} = -10$

Rewrite the expression using determinants. Answers may vary.

29. $2L + 2W$

30. $\pi r + \pi h$

31. $a^2 + b^2$

32. $\frac{1}{2}h(a + b)$

11.6 Systems of Inequalities and Linear Programming

A **graph** of an inequality is a drawing that represents its solutions. An inequality in one variable can be graphed on a number line. An inequality in two variables can be graphed on a coordinate plane.

Solutions of Inequalities in Two Variables

The solutions of inequalities in two variables are ordered pairs.

EXAMPLE 1 Determine whether $(-3, 2)$ is a solution of $5x - 4y \leq 13$.

Solution We use alphabetical order of variables. We replace x by -3 and y by 2.

$5x - 4y \leq 13$	
$5(-3) - 4 \cdot 2$	13
$-15 - 8$	
-23	TRUE

Since $-23 \leq 13$ is true, $(-3, 2)$ is a solution. ■

EXAMPLE 2 Determine whether $(6, -7)$ is a solution of $5x - 4y \leq 13$.

Solution We use alphabetical order of variables. We replace x by 6 and y by -7.

$5x - 4y \leq 13$	
$5(6) - 4(-7)$	13
$30 + 28$	
58	FALSE

Since $58 \leq 13$ is false, $(6, -7)$ is not a solution. ■

Graphs of Linear Inequalities in Two Variables

A **linear inequality** is an inequality that is equivalent to

$$Ax + By < C, \quad Ax + By \leq C, \quad Ax + By > C, \quad \text{or} \quad Ax + By \geq C.$$

That is, there is a first-degree polynomial on one side and a constant on the other. To graph linear inequalities, we use what we already know about graphing linear equations.

EXAMPLE 3 Graph: $y < x$.

Solution We first graph the line $y = x$ for comparison. Every solution of $y = x$ is an ordered pair like (4, 4). The first and second coordinates are the same. The graph of $y = x$ is shown at the left.

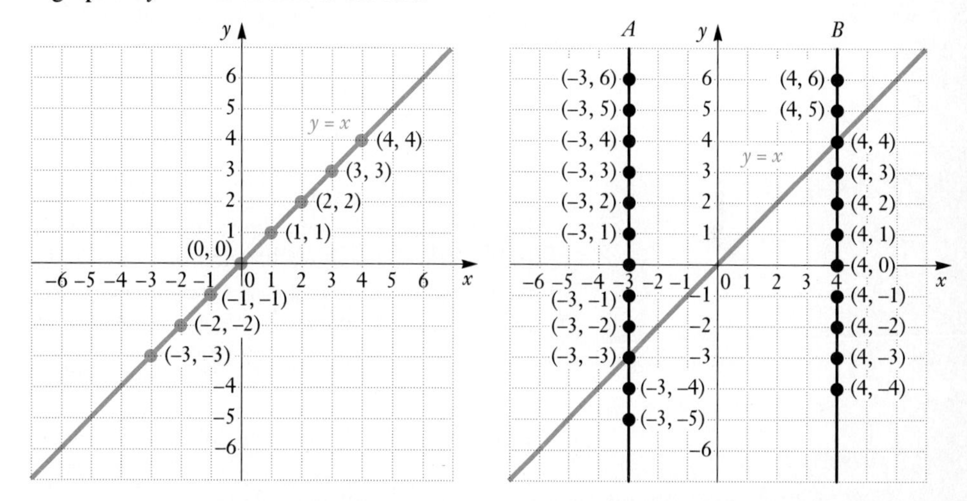

Now look at the graph to the right. We consider a vertical line A and ordered pairs on it. For all points above $y = x$, the second coordinate is greater than the first, $y > x$. For all points below the line, $y < x$. The same thing happens for vertical line B and indeed for any vertical line. If we could draw all such bottom parts of the vertical lines, we would obtain the solutions of $y < x$. We see this on the following graph at the left. We generally use color shading to indicate these solutions, as shown on the graph at the right. The half-plane below $y = x$ represents the graph of $y < x$. Points on $y = x$ are not in the graph, so we draw it dashed.

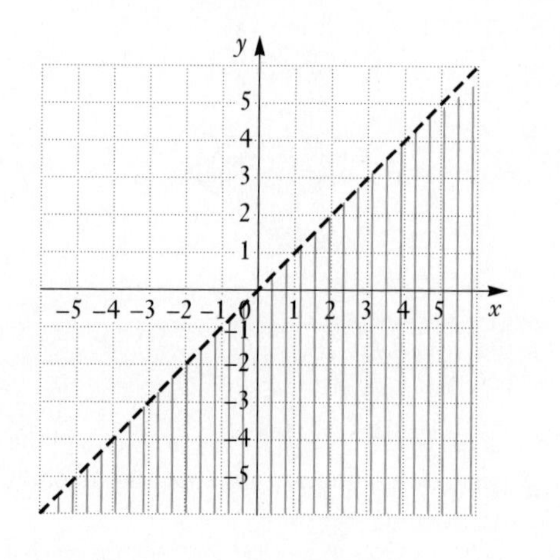

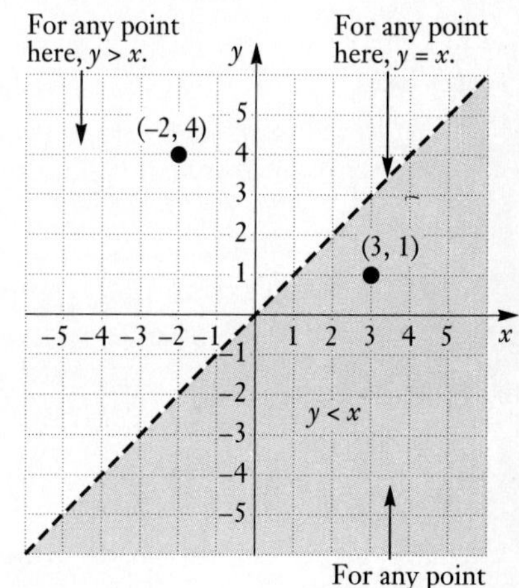

Had we not known which half-plane to shade, we could have determined this by considering a test point on either side of the line. We consider the pair (3, 1) as a test point and substitute to see if it is a solution:

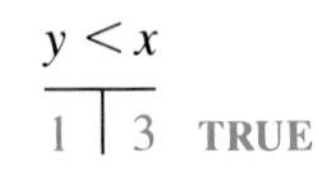

We see that (3, 1) is a solution, and in fact any point on the same side of $y = x$ as (3, 1) is a solution. The points in that half-plane are solutions of $y > x$. ■

EXAMPLE 4 Graph: $3y - 2x \geq 1$.

Solution First we sketch the line $3y - 2x = 1$. Because points on the line $3y - 2x = 1$ are also in the graph of $3y - 2x \geq 1$, we draw the line solid. This indicates that all points on the line are solutions. The rest of the solutions are either in the half-plane above the line or in the half-plane below the line. To determine which, we select a point that is not on the line and determine whether it is a solution of $3y - 2x \geq 1$. We try $(-2, 4)$ as a test point:

$3y - 2x$	≥ 1
$3(4) - 2(-2)$	1
$12 + 4$	
16	TRUE

We see that $16 \geq 1$ is true. Thus $(-2, 4)$ is a solution. All of the points in the half-plane containing $(-2, 4)$ are solutions, and the points in the opposite half-plane are not solutions. We shade the half-plane and obtain the graph as follows:

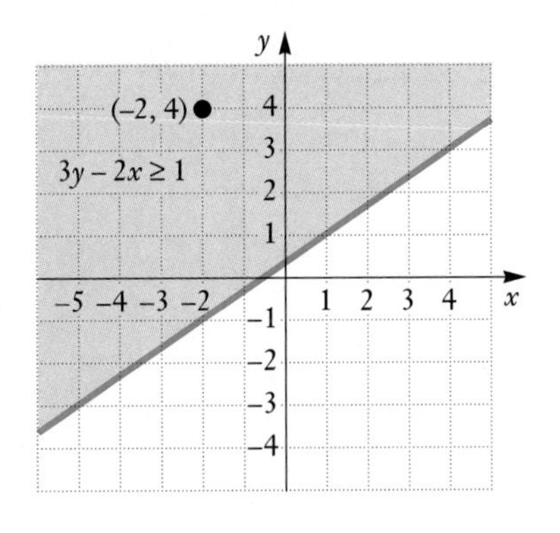

■

Graphs of linear inequalities are half-planes, sometimes including the line along the edge. The equation for that line is called a **related equation.** We graph linear inequalities as follows:

To graph a linear inequality in two variables:

1. Replace the inequality symbol with an equals sign and graph this related equation. If the inequality symbol is $<$ or $>$, draw the line dashed. If the inequality symbol is $\leq$ or $\geq$, draw the line solid.
2. The graph consists of a half-plane, either above or below or left or right of the line, and, if the line is solid, the line as well. To determine which half-plane to shade, pick a point not on the line as a test point. Substitute to find whether that point is a solution of the inequality. If so, shade the half-plane containing that point. If not, shade the opposite half-plane.

EXAMPLE 5 Graph: $6x - 2y > 12$.

Solution We first graph the line $6x - 2y = 12$. The intercepts are $(0, -6)$ and $(2, 0)$. The point $(3, 3)$ is also on the graph. This line forms the boundary of the solutions of the inequality. In this case, points on the line are not solutions of the inequality, so we draw a dashed line. To determine which half-plane to shade, we consider a test point not on the line. We try $(0, 0)$ and substitute:

$6x - 2y$	> 12
$6(0) - 2(0)$	12
$0 + 0$	
0	FALSE

Since this inequality is false, the point $(0, 0)$ is not a solution; no point in the half-plane containing $(0, 0)$ is a solution. Thus the points in the opposite half-plane are solutions. The graph is shown below.

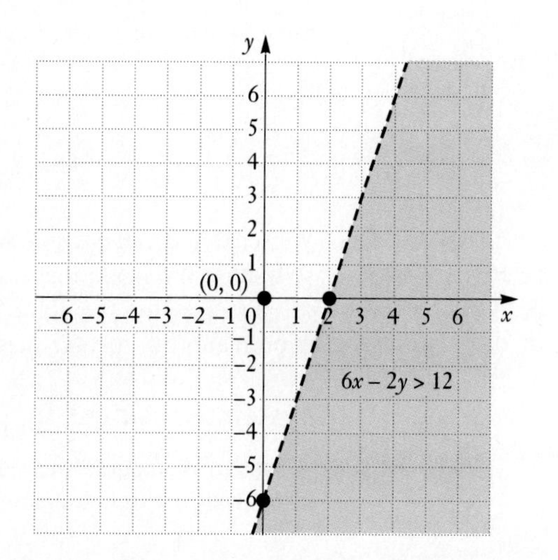

■

The point (0, 0) is the easiest to use as a test point unless the line goes through the origin. If the graph of the related equation passes through the origin, we must test some other point not on the line. The point (1, 1) is often another convenient point to try.

EXAMPLE 6 Graph $x > -3$ on a plane.

Solution There is a missing variable in this inequality. If we graph the inequality on a number line, we see that its graph is as follows:

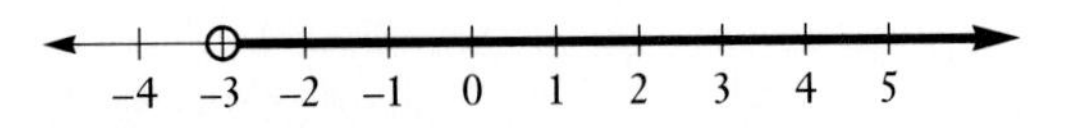

But we can also write this inequality as $x + 0y > -3$ and consider graphing it on the plane. We use the same technique that we have used with the other examples. We first graph the related equation $x = -3$ in the plane. We draw the graph with a dashed line.

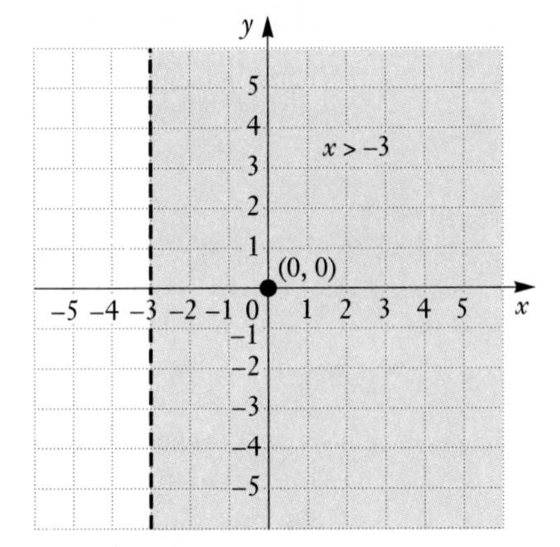

The rest of the graph is a half-plane to the right or left of the line $x = -3$. To determine which, we consider a test point (0, 0):

$x + 0y > -3$	
$0 + 0(0)$	-3
0	TRUE

We see that (0, 0) is a solution, so all the pairs in the half-plane containing (0, 0) are solutions. We shade that half-plane.

We see that the solutions of $x > -3$ are all those ordered pairs whose first coordinates are greater than -3. ■

Systems of Linear Inequalities

A system of inequalities generally has a very large solution set. The most useful thing to do in problem situations involving such inequalities is to graph the solution set. To graph a system of equations, we graph the individual equations and then find the intersection of the individual graphs. For a system of inequalities, we do the same thing; that is, we graph them and look for the intersection.

EXAMPLE 7 Graph the solutions of the system of inequalities

$$y - x \leq 4,$$
$$x + y \leq 4.$$

Solution First we graph the inequality $y - x \leq 4$. We graph $y - x = 4$ using a solid line. We consider (0, 0) as a test point and find that it is a solution, so we shade all points on that side of the line using color shading. The arrows at the ends of the line also indicate the half-plane that contains the solutions.

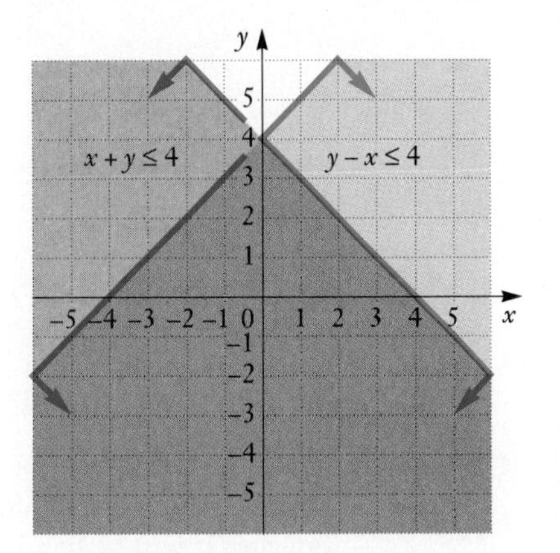

Next, we graph $x + y \leq 4$. We graph $x + y = 4$ using a solid line and consider (0, 0) as a test point. Again, since (0, 0) is a solution, we shade that side of the line using gray shading this time. The solution set of the system consists of the region that is shaded both color and gray and parts of the lines $y - x = 4$ and $x + y = 4$. ■

EXAMPLE 8 Graph: $-1 < y \leq 2$.

Solution This is a conjunction of two inequalities

$$-1 < y \quad \textit{and} \quad y \leq 2.$$

It will be true for any y that is both greater than -1 and less than or equal to 2. Here we show the graph of $y > -1$ in color and the graph of $y \leq 2$ in gray. The arrows indicate which half-plane is to be shaded.

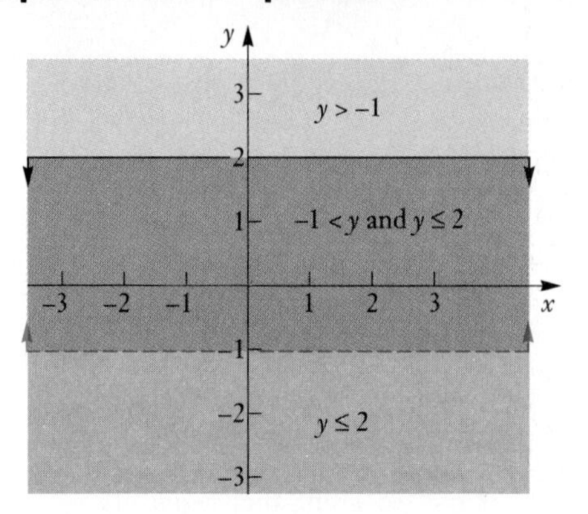

Since our inequality is a conjunction, the solution is the region that has been shaded twice.

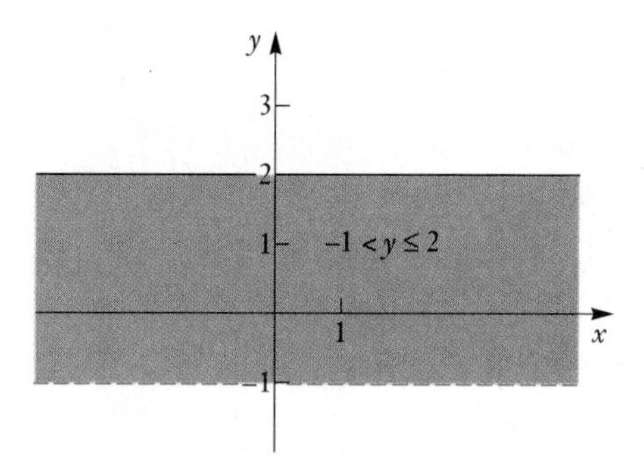

A system of inequalities may have a graph that consists of a polygon and its interior. It will be helpful later in this section if we are able to find the vertices, or corners, of such a graph.

EXAMPLE 9 Graph the system of inequalities. Find the coordinates of any vertices formed.

$$\begin{aligned} 2x + y &\geq 2, \\ 4x + 3y &\leq 12, \\ \tfrac{1}{2} \leq x &\leq 2, \\ y &\geq 0 \end{aligned}$$

Solution The separate graphs are shown on the left and the graph of the intersection, which is the graph of the system, is shown on the right.

We find the vertex $\left(\frac{1}{2}, 1\right)$ by solving the system

$$\begin{aligned} 2x + y &= 2, \\ x &= \tfrac{1}{2}. \end{aligned}$$

We find the vertex (1, 0) by solving the system

$$2x + y = 2,$$
$$y = 0.$$

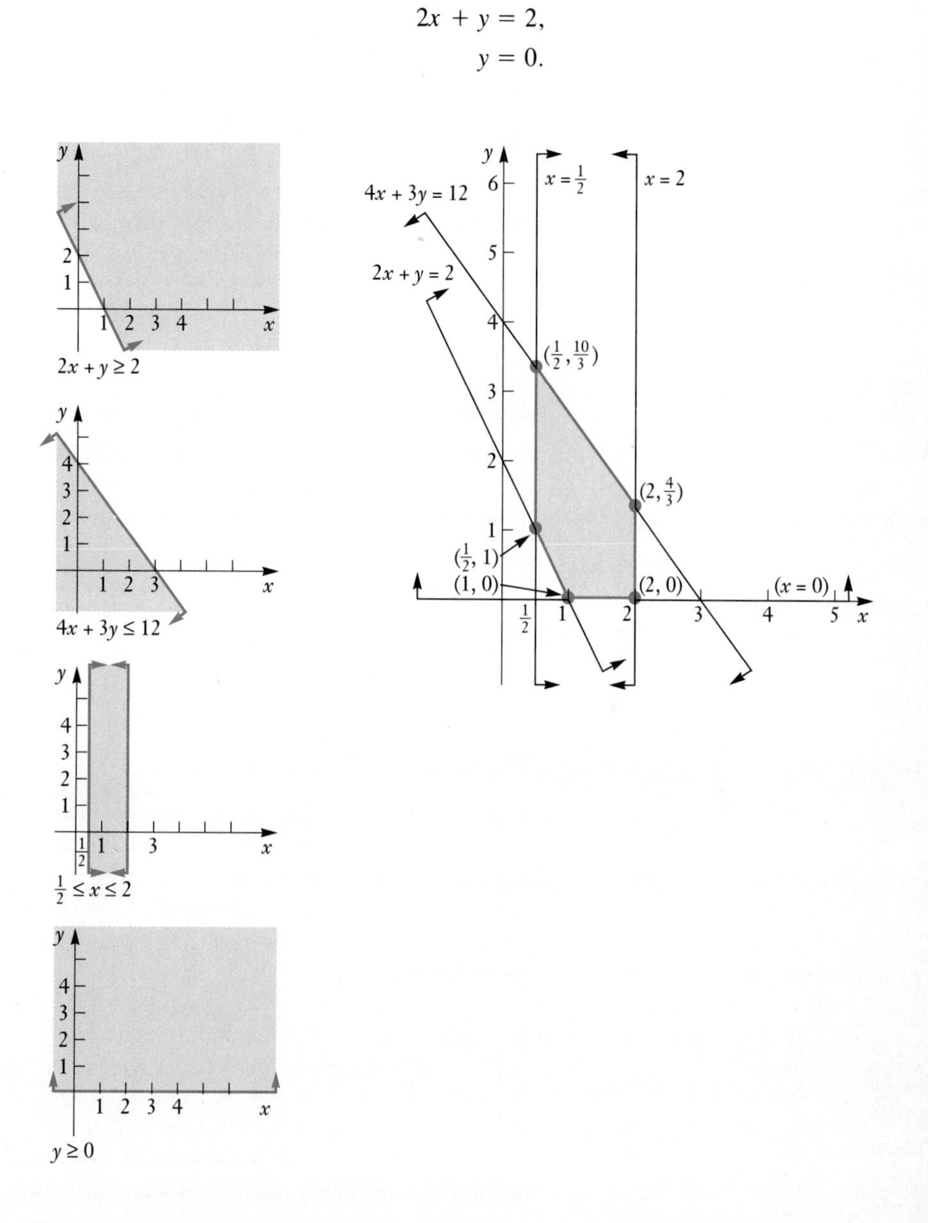

We find the vertex (2, 0) from the system

$$x = 2,$$
$$y = 0.$$

The vertices $\left(2, \frac{4}{3}\right)$ and $\left(\frac{1}{2}, \frac{10}{3}\right)$ were found by solving, respectively, the systems

$$\begin{aligned} x &= 2, \\ 4x + 3y &= 12; \end{aligned} \quad \text{and} \quad \begin{aligned} x &= \tfrac{1}{2}, \\ 4x + 3y &= 12. \end{aligned}$$

■

An Application: Linear Programming

There are many problems in real life for which we want to find a greatest value (a maximum) or a least value (a minimum). For example, if you are in business, you would like to know how to make the *most* profit. Or you might like to know how to make your expenses the *least* possible. Many such problems can be solved using systems of inequalities and a branch of mathematics known as **linear programming.** The following result is of primary importance to linear programming.

Suppose a linear function of two variables $F = ax + by + c$ is defined over a system of linear inequalities. Then maximum or minimum values of the function will occur at certain vertices. To find the maximum or minimum:

1. Graph the system of inequalities and find the vertices.
2. Compute the function values at the vertices. The largest and smallest of those values are the maximum and minimum of the function.

This theorem was proved during World War II. Linear programming was developed then to deal with the complicated process of shipping men and supplies to Europe. Let us consider an example.

EXAMPLE 10 You are taking a test in which items of type A are worth 10 points and items of type B are worth 15 points. It takes 3 minutes to complete each item of type A and 6 minutes to complete each item of type B. The total time allowed is 60 minutes, and you are not allowed to answer more than 16 questions. Assuming that all of your answers are correct, how many items of each type should you answer to get the best score?

Solution Let x = the number of items of type A and y = the number of items of type B. The total score T is a linear function of the two variables x and y:

$$T = 10x + 15y.$$

This function has a domain that is a set of ordered pairs of numbers (x, y). This domain is determined by the following inequalities, also called *constraints:*

Total number of questions allowed, not more than 16	$x + y \leq 16,$
Time, not more than 60 min	$3x + 6y \leq 60,$
Numbers of items answered will not be negative	$x \geq 0,$
	$y \geq 0.$

We now graph the domain of the function T by graphing this system of inequalities. We will determine the vertices, if any are formed.

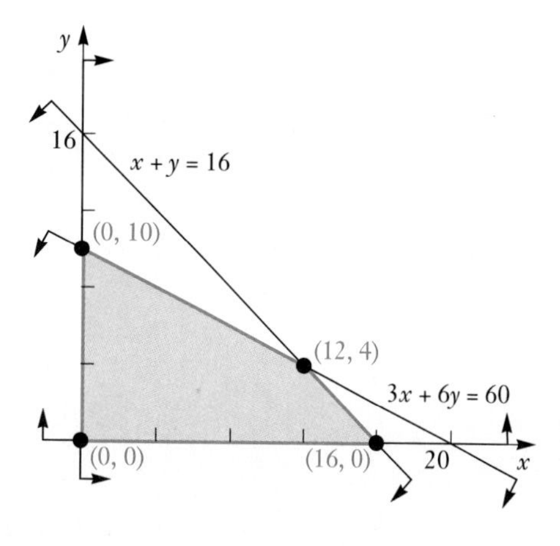

We know that the maximum and minimum values occur at certain vertices of the polygon. All we need do to find these values is to substitute the coordinates of the vertices in $T = 10x + 15y$.

Vertices (x, y)	Score $T = 10x + 15y$	
$(0, 0)$	$T = 10(0) + 15(0) = 0$	←Minimum
$(16, 0)$	$T = 10(16) + 15(0) = 160$	
$(12, 4)$	$T = 10(12) + 15(4) = 180$	←Maximum
$(0, 10)$	$T = 10(0) + 15(10) = 150$	

From the table, we see that the minimum value is 0 and the maximum is 180. To get this maximum, you must answer 12 items of type A and 4 items of type B. ■

EXAMPLE 11 Find the maximum and minimum values of $F = 9x + 40y$, subject to the constraints

$$y - x \geq 1,$$
$$y - x \leq 3,$$
$$2 \leq x \leq 5.$$

Solution We graph the system of inequalities, determine the vertices, and find the function values for those ordered pairs.

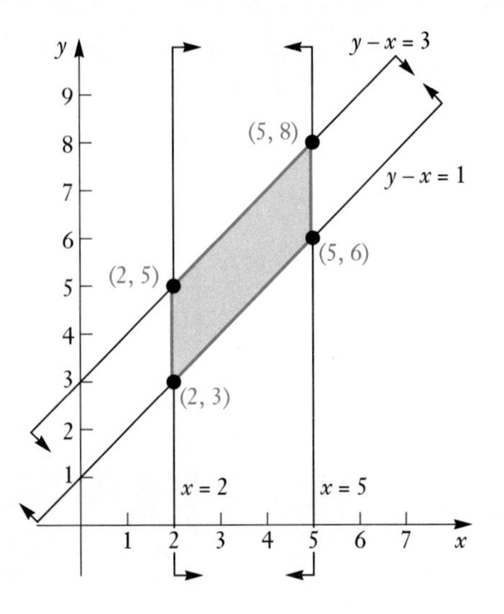

Vertices (x, y)	Value of Function $F = 9x + 40y$	
(2, 3)	$F = 9(2) + 40(3) = 138$	←Minimum
(2, 5)	$F = 9(2) + 40(5) = 218$	
(5, 6)	$F = 9(5) + 40(6) = 285$	
(5, 8)	$F = 9(5) + 40(8) = 365$	←Maximum

The maximum value of F is 365 when $x = 5$ and $y = 8$. The minimum value of F is 138 when $x = 2$ and $y = 3$. ■

EXAMPLE 12 A company manufactures motorcycles and bicycles. To stay in business, it must produce at least 10 motorcycles each month, but it does not have the facilities to produce more than 60 motorcycles. It also does not have the facilities to produce more than 120 bicycles. The total production of motorcycles and bicycles cannot exceed 160. The profit on a motorcycle is \$134 and on a bicycle, \$20. Find the number of each that should be manufactured in order to maximize profit.

Solution Let x = the number of motorcycles to be produced and y = the number of bicycles to be produced. The profit P is given by

$$P = \$134x + \$20y,$$

subject to the constraints

$10 \le x \le 60,$ **From 10 to 60 motorcycles will be produced.**

$0 \le y \le 120,$ **No more than 120 bicycles can be made.**

$x + y \le 160.$ **Total production cannot exceed 160.**

We graph the system of inequalities, determine the vertices, and find the function values for those ordered pairs.

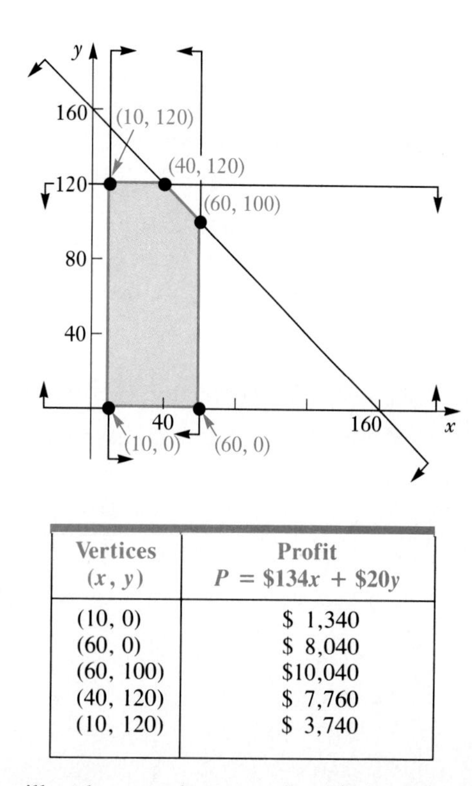

Vertices (x, y)	Profit $P = \$134x + \$20y$
(10, 0)	\$ 1,340
(60, 0)	\$ 8,040
(60, 100)	\$10,040
(40, 120)	\$ 7,760
(10, 120)	\$ 3,740

Thus the company will make a maximum profit of \$10,040 by producing 60 motorcycles and 100 bicycles. ■

EXERCISE SET 11.6

Determine whether the ordered pair is a solution of the inequality.

1. $(-4, 2);\ 2x + y > -5$

2. $(3, -6);\ 4x + 2y \le 0$

3. $(8, 14);\ 2y - 3x \ge 5$

4. $(7, 20);\ 3x - y < -1$

Graph the inequality on a plane, using graph paper.

5. $y > 2x$

6. $2y < x$

7. $y + x \ge 0$

8. $y + x < 0$

9. $y > x - 3$

10. $y \le x + 4$

11. $x + y < 4$

12. $x - y \ge 5$

13. $3x - 2y \le 6$

14. $2x - 5y < 10$

15. $3y + 2x \ge 6$

16. $2y + x \le 4$

17. $3x - 2 \le 5x + y$

18. $2x - 6y \ge 8 + 2y$

19. $x < -4$

20. $y \ge 5$

21. $y > -3$

22. $x \le 5$

23. $-4 < y < -1$

24. $-1 < y < 4$

25. $-4 \le x \le 4$

26. $-3 \le x \le 3$

27. $y \ge |x|$

28. $y \le |x|$

Graph the system of inequalities. Find the coordinates of any vertices formed and shade the solution set.

29. $y \le x,$
$y \ge 3 - x$

30. $y \ge x,$
$y \le x - 5$

31. $y \ge x,$
$y \le x - 4$

32. $y \ge x,$
$y \le 2 - x$

33. $y \geq -3,$
$x \geq 1$

34. $y \leq -2,$
$x \geq 2$

35. $x \leq 3,$
$y \geq 2 - 3x$

36. $x \geq -2,$
$y \leq 3 - 2x$

37. $x + y \leq 1,$
$x - y \leq 2$

38. $y + 3x \geq 0,$
$y + 3x \leq 2$

39. $2y - x \leq 2,$
$y + 3x \geq -1$

40. $y \leq 2x + 1,$
$y \geq -2x + 1,$
$x \leq 2$

41. $x - y \leq 2,$
$x + 2y \geq 8,$
$y \leq 4$

42. $x + 2y \leq 12,$
$2x + y \leq 12,$
$x \geq 0,$
$y \geq 0$

43. $4y - 3x \geq -12,$
$4y + 3x \geq -36,$
$y \leq 0,$
$x \leq 0$

44. $8x + 5y \leq 40,$
$x + 2y \leq 8,$
$x \geq 0,$
$y \geq 0$

45. $3x + 4y \geq 12,$
$5x + 6y \leq 30,$
$1 \leq x \leq 3$

46. $y - x \geq 1,$
$y - x \leq 3,$
$2 \leq x \leq 5$

Find the maximum and minimum values of the function and the values of x and y for which they occur.

47. $P = 17x - 3y + 60$, subject to

$$6x + 8y \leq 48,\quad 0 \leq y \leq 4,\quad 0 \leq x \leq 7.$$

48. $Q = 28x - 4y + 72$, subject to

$$5x + 4y \geq 20,\quad 0 \leq y \leq 4,\quad 0 \leq x \leq 3.$$

49. $F = 5x + 36y$, subject to

$$5x + 3y \leq 34,\quad 3x + 5y \leq 30,\quad x \geq 0,\quad y \geq 0.$$

50. $G = 16x + 14y$, subject to

$$3x + 2y \leq 12,\quad 7x + 5y \leq 29,\quad x \geq 0,\quad y \geq 0.$$

51. The Hockeypuck Biscuit Factory makes two types of biscuits, Biscuit Jumbos and Mitimite Biscuits. The oven can cook at most 200 biscuits per day. Jumbos each require 2 oz of flour, Mitimites require 1 oz of flour, and there is at most 300 oz of flour available. The income from Jumbos is $0.10 and from Mitimites is $0.08. How many of each type of biscuit should be made in order to maximize income? What is the maximum income?

52. A student owns a car and a moped. The student has at most 12 gal of gasoline to be used between the car and the moped. The car's tank holds at most 10 gal and the moped's 3 gal. The mileage for the car is 20 mpg and for the moped is 100 mpg. How many gallons of gasoline should each vehicle use if the student wants to travel as far as possible? What is the maximum number of miles?

53. You are about to take a test that contains questions of type A worth 10 points and questions of type B worth 25 points. You must do at least 3 questions of type A, but time restricts doing more than 12. You must do at least 4 questions of type B, but time restricts doing more than 15. You can do no more than 20 questions in total. How many of each type of question must you do to maximize your score? What is this maximum score?

54. You are about to take a test that contains questions of type A worth 4 points and questions of type B worth 7 points. You must do at least 5 questions of type A, but time restricts doing more than 10. You must do at least 3 questions of type B, but time restricts doing more than 10. You can do no more than 18 questions in total. How many of each type of question must you do to maximize your score? What is this maximum score?

55. A lumber company can convert logs into either lumber or plywood. In a given week, the mill can turn out 400 units of production, of which 100 units of lumber and 150 units of plywood are required by regular customers. The profit on a unit of lumber is \$20 and on a unit of plywood is \$30. How many units of each type should the mill produce in order to maximize the profit?

56. A farm consists of 240 acres of cropland. The farmer wishes to plant this acreage in corn or oats. Profit per acre in corn production is \$40 and in oats, \$30. An additional restriction is that the total number of hours of labor during the production period is 320. Each acre of land in corn production uses 2 hours of labor during the production period, while production of oats requires 1 hour per acre. Find how the land should be divided between corn and oats in order to give maximum profit.

57. A woman is planning to invest up to \$40,000 in corporate or municipal bonds, or both. The least she is allowed to invest in corporate bonds is \$6000, and she does not want to invest more than \$22,000 in corporate bonds. She also does not want to invest more than \$30,000 in municipal bonds. The interest on corporate bonds is 8% and on municipal bonds is $7\frac{1}{2}$%. This is simple interest for one year. How much should she invest in each type of bond to maximize her income? What is the maximum income?

58. A man is planning to invest up to \$22,000 in bank X or bank Y, or both. He wants to invest at least \$2000 but no more than \$14,000 in bank X. Bank Y does not insure more than a \$15,000 investment, so he will invest no more than that in bank Y. The interest in bank X is 6% and in bank Y, $6\frac{1}{2}$%. This is simple interest for one year. How much should he invest in each bank to maximize his income? What is the maximum income?

59. A pipe tobacco company has 3000 lb of English tobacco, 2000 lb of Virginia tobacco, and 500 lb of Latakia tobacco. To make one batch of Smello tobacco, it takes 12 lb of English tobacco and 4 lb of Latakia. To make one batch of Roppo tobacco, it takes 8 lb of English and 8 lb of Virginia tobacco. The profit is \$10.56 per batch for Smello and \$6.40 for Roppo. How many batches of each kind of tobacco should be made to yield maximum profit? What is the maximum profit? (*Hint:* Organize the information in a table.)

60. It takes a tailoring firm 2 hr of cutting and 4 hr of sewing to make a knit suit. To make a worsted suit, it takes 4 hr of cutting and 2 hr of sewing. At most 20 hr per day are available for cutting and at most 16 hr per day are available for sewing. The profit on a knit suit is \$34 and on a worsted suit is \$31. How many of each kind of suit should be made to maximize profit?

SKILL MAINTENANCE

61. Find an equation for a line with slope $-\frac{2}{9}$ and y-intercept $(0, 4)$.

Solve.

62. $e^{2x-5} = 20$

63. $\log_3 (2x - 8) = 4$

64. $3x^2 + 25x - 9 = 0$

SYNTHESIS

Graph each system.

65. $y \geq x^2 - 2,$
$y \leq 2 - x^2$

66. $y < x + 1,$
$y \geq x^2$

67. *Hockey wins and losses.* A hockey team figures that it needs at least 60 points for the season to make the playoffs. A win w is worth 2 points and a tie t is worth 1 point. Find a system of inequalities that describes the situation. Graph the system.

68. *Elevators.* Many elevators have a capacity of 1 metric ton (1000 kg). An elevator contains c children, each weighing 35 kg, and a adults, each weighing 75 kg. Find a system of inequalities that asserts that the elevator is overloaded. Graph the system.

69. *Allocation of resources in a manufacturing process.* A furniture manufacturer produces chairs and sofas. The chairs require 20 ft of wood, 1 lb of foam rubber, and 2 sq yd of material. The sofas require 100 ft of wood, 50 lb of foam rubber, and 20 sq yd of material. The manufacturer has in stock 1900 ft of wood, 500 lb of foam rubber, and 240 sq yd of material. The chairs can be sold for \$20 each and the sofas for \$300 each. How many of each should be produced to maximize income?

11.7 Nonlinear Systems of Equations

When we studied systems of linear equations, we solved them both graphically and algebraically. Here we study systems in which at least one equation is of second degree. We will use graphical and then algebraic methods of solving.

Systems Involving One Nonlinear Equation

We consider a system involving an equation of a circle and an equation of a line. Let's think about the possible ways in which a circle and a line can intersect. The three possibilities are shown in the figure. For L_1 there is no point of intersection; hence the system of equations has no real solution. For L_2 there is one point of intersection, hence one real solution. For L_3 there are two points of intersection, hence two real solutions.

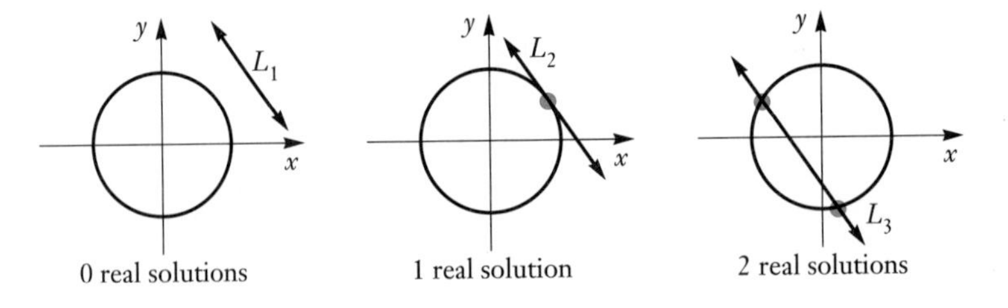

EXAMPLE 1 Solve this system graphically:

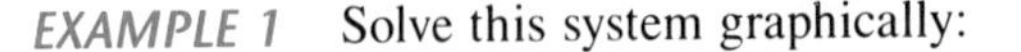

$$x^2 + y^2 = 25,$$
$$3x - 4y = 0.$$

Solution We graph the two equations, using the same axes. The points of intersection have coordinates that must satisfy both equations. The solutions seem to be (4, 3) and (−4, −3). We check.

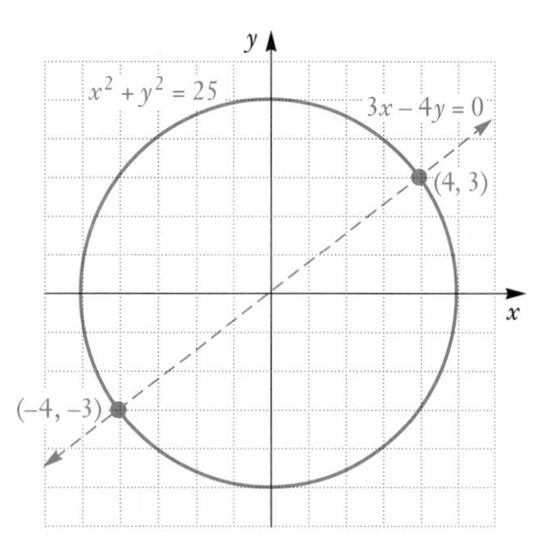

Check: For (−4, −3). You should do the check for (4, 3).

$3x - 4y = 0$	
$3(-4) - 4(-3)$	0
$-12 + 12$	
0	TRUE

$x^2 + y^2 = 25$	
$(-4)^2 + (-3)^2$	25
$16 + 9$	
25	TRUE

■

Remember that we used both *Gauss–Jordan elimination* and *substitution* to solve systems of linear equations. In solving systems where one equation is of first degree and one is of second degree, it is preferable to use the *substitution* method.

EXAMPLE 2 Solve the system of Example 1 algebraically:

$$x^2 + y^2 = 25, \quad (1)$$

$$3x - 4y = 0. \quad (2)$$

Solution First solve the linear equation (2) for x:

$$x = \frac{4}{3}y.$$

Then substitute $\frac{4}{3}y$ for x in Eq. (1) and solve for y:

$$\left(\frac{4}{3}y\right)^2 + y^2 = 25$$

$$\frac{16}{9}y^2 + y^2 = 25$$

$$\frac{16}{9}y^2 + \frac{9}{9}y^2 = 25$$

$$\frac{25}{9}y^2 = 25$$

$$\frac{9}{25} \cdot \frac{25}{9}y^2 = \frac{9}{25} \cdot 25$$

$$y^2 = 9$$

$$y = \pm 3.$$

Now substitute these numbers into the linear equation $x = \frac{4}{3}y$ and solve for x:

$$\text{For } y = 3, \qquad x = \frac{4}{3}(3), \qquad \text{or} \qquad 4;$$

$$\text{For } y = -3, \qquad x = \frac{4}{3}(-3), \qquad \text{or} \qquad -4.$$

The pairs $(4, 3)$ and $(-4, -3)$ check, hence are solutions. ■

EXAMPLE 3 Solve the system

$$y + 3 = 2x, \qquad (1)$$

$$x^2 + 2xy = -1. \qquad (2)$$

Solution First solve the linear equation (1) for y:

$$y = 2x - 3.$$

Then substitute $2x - 3$ for y in Eq. (2) and solve for x:

$$x^2 + 2x(2x - 3) = -1$$

$$x^2 + 4x^2 - 6x = -1$$

$$5x^2 - 6x + 1 = 0$$

$$(5x - 1)(x - 1) = 0$$

$$5x - 1 = 0 \quad \text{or} \quad x - 1 = 0$$

$$x = \frac{1}{5} \quad \text{or} \quad x = 1.$$

Now substitute these numbers into the linear equation $y = 2x - 3$ and solve for y:

$$\text{For } x = \frac{1}{5}, \qquad y = 2\left(\frac{1}{5}\right) - 3, \qquad \text{or} \qquad -\frac{13}{5};$$

$$\text{For } x = 1, \qquad y = 2(1) - 3, \qquad \text{or} \qquad -1.$$

The pairs $\left(\frac{1}{5}, -\frac{13}{5}\right)$ and $(1, -1)$ check and are solutions. ■

Systems of Second-Degree Equations

We now consider systems of two second-degree equations. The following figures show some ways in which a circle and a hyperbola can intersect.

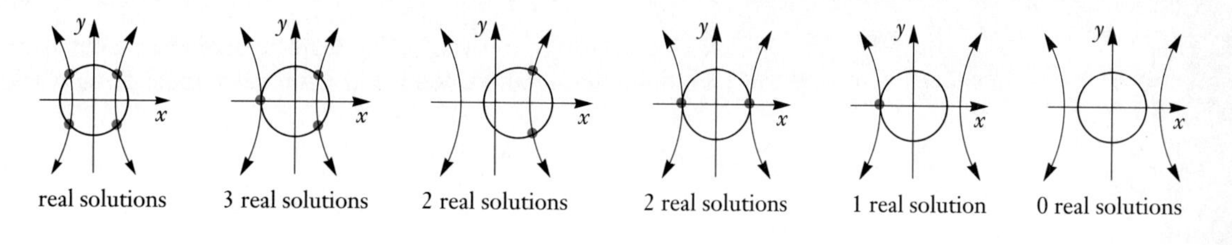

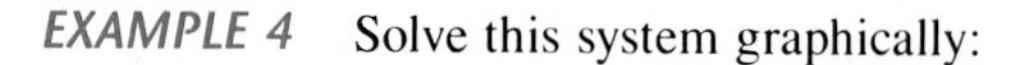

EXAMPLE 4 Solve this system graphically:

$$x^2 + y^2 = 25,$$
$$\frac{x^2}{25} - \frac{y^2}{25} = 1.$$

Solution We graph the two equations using the same axes. The points of intersection have coordinates that must satisfy both equations; the solutions seem to be $(5, 0)$ and $(-5, 0)$.

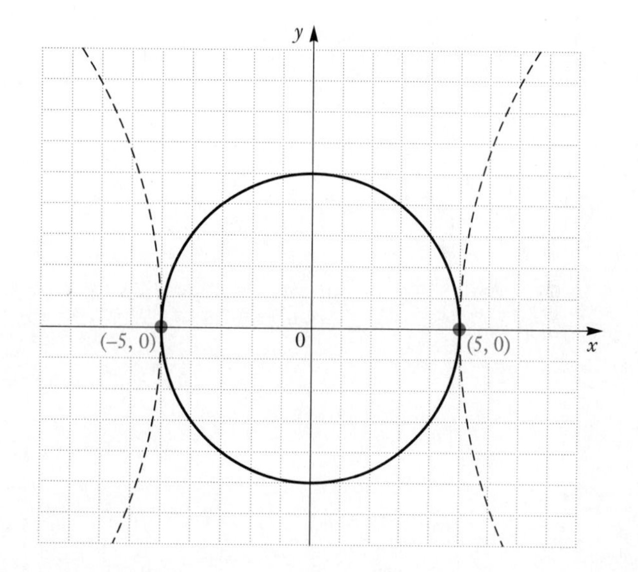

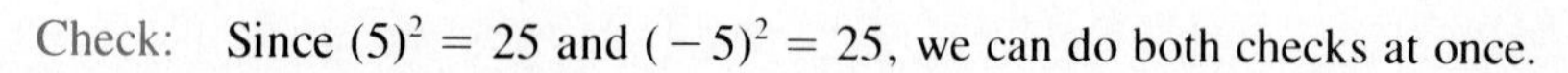

Check: Since $(5)^2 = 25$ and $(-5)^2 = 25$, we can do both checks at once.

$$\begin{array}{c|c} \frac{x^2}{25} - \frac{y^2}{25} = 1 & \\ \hline \frac{(\pm 5)^2}{25} - \frac{0^2}{25} & 1 \\ \frac{25}{25} - 0 & \\ 1 & \text{TRUE} \end{array} \qquad \begin{array}{c|c} x^2 + y^2 = 25 & \\ \hline (\pm 5)^2 + 0^2 & 25 \\ 25 + 0 & \\ 25 & \text{TRUE} \end{array}$$

The solutions are $(-5, 0)$ and $(5, 0)$. ■

To solve systems of two second-degree equations, we can use either substitution or an elimination method. We begin with an example using elimination.

EXAMPLE 5 Solve this system:

$$2x^2 + 5y^2 = 22, \quad (1)$$
$$3x^2 - y^2 = -1. \quad (2)$$

Solution Here we multiply the second equation by 5 and add the equations:

$$\begin{aligned} 2x^2 + 5y^2 &= 22 \\ 15x^2 - 5y^2 &= -5 \quad \text{Multiplying by 5} \\ \hline 17x^2 \qquad &= 17 \quad \text{Adding} \\ x^2 &= 1 \\ x &= \pm 1. \end{aligned}$$

If $x = 1$, $x^2 = 1$, and if $x = -1$, $x^2 = 1$, so substituting 1 or -1 for x in Eq. (2), we have

$$\begin{aligned} 3 \cdot (\pm 1)^2 - y^2 &= -1 \\ 3 - y^2 &= -1 \\ -y^2 &= -4 \\ y^2 &= 4 \\ y &= \pm 2. \end{aligned}$$

Thus, if $x = 1$, $y = 2$ or $y = -2$; and if $x = -1$, $y = 2$ or $y = -2$. The possible solutions are $(1, 2)$, $(1, -2)$, $(-1, 2)$, and $(-1, -2)$.

Check: Since $(2)^2 = 4$, $(-2)^2 = 4$, $(1)^2 = 1$, and $(-1)^2 = 1$, we can check all four pairs at once.

$2x^2 + 5y^2 = 22$			$3x^2 - y^2 = -1$		
$2(\pm 1)^2 + 5(\pm 2)^2$	22		$3(\pm 1)^2 - (\pm 2)^2$	-1	
$2 + 20$			$3 - 4$		
22		TRUE	-1		TRUE

The solutions are $(1, 2)$, $(1, -2)$, $(-1, 2)$, and $(-1, -2)$. ■

When a product of variables is in one equation and the other is of the form $Ax^2 + By^2 = C$, we often solve for one of the variables in the equation with the product and then use substitution.

EXAMPLE 6 Solve the system

$$x^2 + 4y^2 = 20, \quad (1)$$
$$xy = 4. \quad (2)$$

Solution Here we use the substitution method. First solve Eq. (2) for y:

$$y = \frac{4}{x}.$$

Then substitute $4/x$ for y in Eq. (1) and solve for x:

$$x^2 + 4\left(\frac{4}{x}\right)^2 = 20$$

$$x^2 + \frac{64}{x^2} = 20$$

$$x^4 + 64 = 20x^2 \qquad \text{Multiplying by } x^2$$

$$x^4 - 20x^2 + 64 = 0$$

$$u^2 - 20u + 64 = 0 \qquad \text{Letting } u = x^2$$

$$(u - 16)(u - 4) = 0 \qquad \text{Factoring}$$

$$u = 16 \quad \text{or} \quad u = 4 \qquad \text{Principle of zero products}$$

Then $x^2 = 16$ or $x^2 = 4$, so $x = \pm 4$ or $x = \pm 2$. Since $y = 4/x$, if $x = 4$, $y = 1$; if $x = -4$, $y = -1$; if $x = 2$, $y = 2$; if $x = -2$, $y = -2$. The solutions are $(4, 1)$, $(-4, -1)$, $(2, 2)$, and $(-2, -2)$. ■

EXERCISE SET 11.7

Solve. Where possible, sketch the graphs to confirm the solutions.

1. $x^2 + y^2 = 25,$ $y - x = 1$

2. $x^2 + y^2 = 100,$ $y - x = 2$

3. $4x^2 + 9y^2 = 36,$ $3y + 2x = 6$

4. $9x^2 + 4y^2 = 36,$ $3x + 2y = 6$

5. $y^2 = x + 3,$ $2y = x + 4$

6. $y = x^2,$ $3x = y + 2$

7. $x^2 - xy + 3y^2 = 27,$ $x - y = 2$

8. $2y^2 + xy + x^2 = 7,$ $x - 2y = 5$

9. $x^2 + 4y^2 = 25,$ $x + 2y = 7$

10. $y^2 - x^2 = 16,$ $2x - y = 1$

11. $x^2 - xy + 3y^2 = 5,$ $x - y = 2$

12. $m^2 + 3n^2 = 10,$ $m - n = 2$

13. $3x + y = 7,$ $4x^2 + 5y = 24$

14. $2y^2 + xy = 5,$ $4y + x = 7$

15. $a + b = 7,$ $ab = 4$

16. $p + q = -6,$ $pq = -7$

17. $2a + b = 1,$ $b = 4 - a^2$

18. $4x^2 + 9y^2 = 36,$ $x + 3y = 3$

19. $a^2 + b^2 = 89,$ $a - b = 3$

20. $xy = 4,$ $x + y = 5$

Solve.

21. $x^2 + y^2 = 25,$ $y^2 = x + 5$

22. $y = x^2,$ $x = y^2$

23. $x^2 + y^2 = 9,$ $x^2 - y^2 = 9$

24. $y^2 - 4x^2 = 4,$ $4x^2 + y^2 = 4$

25. $x^2 + y^2 = 25,$ $xy = 12$

26. $x^2 - y^2 = 16,$ $x + y^2 = 4$

27. $x^2 + y^2 = 4,$ $16x^2 + 9y^2 = 144$

28. $x^2 + y^2 = 25,$ $25x^2 + 16y^2 = 400$

29. $x^2 + y^2 = 16,$ $y^2 - 2x^2 = 10$

30. $x^2 + y^2 = 14,$ $x^2 - y^2 = 4$

31. $x^2 + y^2 = 5,$ $xy = 2$

32. $x^2 + y^2 = 20$ $xy = 8$

33. $x^2 + y^2 = 13,$ $xy = 6$

34. $x^2 + 4y^2 = 20,$ $xy = 4$

35. $3xy + x^2 = 34$ $2xy - 3x^2 = 8$

36. $2xy + 3y^2 = 7,$ $3xy - 2y^2 = 4$

37. $xy - y^2 = 2,$ $2xy - 3y^2 = 0$

38. $4a^2 - 25b^2 = 0,$ $2a^2 - 10b^2 = 3b + 4$

39. $x^2 - y = 5,$ $x^2 + y^2 = 25$

40. $ab - b^2 = -4,$ $ab - 2b^2 = -6$

SKILL MAINTENANCE

41. Find an equation for a linear function whose graph has slope -2 and y-intercept $(0, -3)$.

42. Use synthetic division to find the quotient and the remainder:

$$(-3x^3 + 2x^2 - 11x + 8) \div (x - 4).$$

43. Factor: $x^3 - 2x^2 - 5x + 6$.

44. Solve: $e^{3x-5} = 17$.

SYNTHESIS

When the solution of a nonlinear system of equations involves complex numbers, the graphs of the equations will not intersect. Solve the systems in Exercises 45–48 and sketch the graphs to confirm the solutions.

45. $x^2 + y^2 = 5,$
$x - y = 8$

46. $4x^2 + 9y^2 = 36,$
$y - x = 8$

47. $x^2 + y^2 = 25,$
$9x^2 + 4y^2 = 36$

48. $x^2 + y^2 = 1,$
$9x^2 - 16y^2 = 144$

49. The square of a certain number exceeds twice the square of another number by $\frac{1}{8}$. The sum of their squares is $\frac{5}{16}$. Find the numbers.

50. Find an equation of the circle that passes through $(4, -20)$, $(10, -2)$, and $(-4, -4)$.

51. Four squares with sides 5 in. long are cut from the corners of a rectangular metal sheet that has an area of 340 in^2. The edges are bent up to form an open box with a volume of 350 in^3. Find the dimensions of the box.

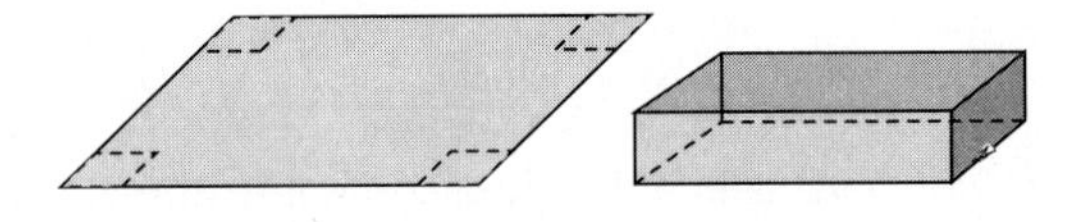

Solve for x and y.

52. $x^2 + xy = a,$
$y^2 + xy = b$

53. $x^2 - y^2 = a^2 - b^2,$
$x + y = a - b$

Solve.

54. $p^2 + q^2 = 13,$
$\frac{1}{pq} = -\frac{1}{6}$

55. $a + b = \frac{5}{6},$
$\frac{a}{b} + \frac{b}{a} = \frac{13}{6}$

CHAPTER SUMMARY AND REVIEW 11

TERMS TO KNOW

Conjunction
Solution
System of equations
Substitution method
Gauss–Jordan elimination
Elimination method
Algorithm
Back-substitution

Inconsistent
Dependent
Independent
Homogeneous
Homogeneous equation
Trivial solution
Matrix
Matrices

Main diagonal
Row-equivalent matrices
Row-equivalent operations
Square matrix
Determinant
Cramer's rule
Graph of an inequality
Linear inequality

Linear equation in three variables
Consistent
Rows
Columns
Elements
Related equation
Linear programming

IMPORTANT PROPERTIES AND FORMULAS

Given the graphs of two lines, the following can happen:

a) The lines have no point in common—they are parallel. The system has no solution.
b) The lines have exactly one point in common. The system has exactly one solution.
c) The lines are the same—they have infinitely many points in common. The system has infinitely many solutions.

TRANSFORMATIONS PRODUCING EQUIVALENT SYSTEMS

Each of the following will produce an equivalent system of equations:

a) Interchanging any two equations.
b) Multiplying each number or term of an equation by the same nonzero number.
c) Multiplying each number or term of one equation by the same nonzero number and adding the result to another equation.

While solving a system using Gauss–Jordan elimination,

a) obtaining an obviously false statement, such as $0 = 7$, means that the system is inconsistent; and
b) obtaining an obviously true statement, such as $0 = 0$, means that the system is dependent.

The determinant of a 2×2 matrix:

$$\begin{vmatrix} a & c \\ b & d \end{vmatrix} = ad - bc.$$

The determinant of a 3×3 matrix:

$$\begin{vmatrix} a_1 & b_1 & c_1 \\ a_2 & b_2 & c_2 \\ a_3 & b_3 & c_3 \end{vmatrix} = a_1 \cdot \begin{vmatrix} b_2 & c_2 \\ b_3 & c_3 \end{vmatrix} - a_2 \cdot \begin{vmatrix} b_1 & c_1 \\ b_3 & c_3 \end{vmatrix} + a_3 \cdot \begin{vmatrix} b_1 & c_1 \\ b_2 & c_2 \end{vmatrix}.$$

CRAMER'S RULE: 2 × 2 SYSTEMS

The solution of the system

$$a_1x + b_1y = c_1,$$
$$a_2x + b_2y = c_2,$$

if it is unique, is given by

$$x = \frac{\begin{vmatrix} c_1 & b_1 \\ c_2 & b_2 \end{vmatrix}}{\begin{vmatrix} a_1 & b_1 \\ a_2 & b_2 \end{vmatrix}}, \qquad y = \frac{\begin{vmatrix} a_1 & c_1 \\ a_2 & c_2 \end{vmatrix}}{\begin{vmatrix} a_1 & b_1 \\ a_2 & b_2 \end{vmatrix}}.$$

CRAMER'S RULE: 3 × 3 SYSTEMS

The solution of the system

$$\begin{aligned} a_1x + b_1y + c_1z &= d_1, \\ a_2x + b_2y + c_2z &= d_2, \\ a_3x + b_3y + c_3z &= d_3 \end{aligned}$$

is found by considering the following determinants:

$$D = \begin{vmatrix} a_1 & b_1 & c_1 \\ a_2 & b_2 & c_2 \\ a_3 & b_3 & c_3 \end{vmatrix}, \qquad D_x = \begin{vmatrix} d_1 & b_1 & c_1 \\ d_2 & b_2 & c_2 \\ d_3 & b_3 & c_3 \end{vmatrix},$$

$$D_y = \begin{vmatrix} a_1 & d_1 & c_1 \\ a_2 & d_2 & c_2 \\ a_3 & d_3 & c_3 \end{vmatrix}, \qquad D_z = \begin{vmatrix} a_1 & b_1 & d_1 \\ a_2 & b_2 & d_2 \\ a_3 & b_3 & d_3 \end{vmatrix}.$$

The solution, if it is unique, is given by

$$x = \frac{D_x}{D}, \qquad y = \frac{D_y}{D}, \qquad z = \frac{D_z}{D}.$$

To graph a linear inequality in two variables:

1. Replace the inequality symbol with an equals sign and graph this related equation. If the inequality symbol is $<$ or $>$, draw the line dashed. If the inequality symbol is $\leq$ or $\geq$, draw the line solid.
2. The graph consists of a half-plane, either above or below or left or right of the line, and, if the line is solid, the line as well. To determine which half-plane to shade, pick a point not on the line as a test point. Substitute to find whether that point is a solution of the inequality. If so, shade the half-plane containing that point. If not, shade the opposite half-plane.

Suppose a linear function of two variables $F = ax + by + c$ is defined over a system of linear inequalities. Then maximum or minimum values of the function will occur at certain vertices. To find the maximum or minimum:

1. Graph the system of inequalities and find the vertices.
2. Compute the function values at the vertices. The largest and smallest of those values are the maximum and minimum of the function.

REVIEW EXERCISES

The review sections to be tested in addition to the material in this chapter are Sections 6.3, 8.1, 8.3, and 9.6.

Solve graphically. Be sure to check.

1. $4x - y = 10,$
$2x + 3y = 12$

2. $y = 3x + 7,$
$3x + 2y = -4$

Solve using the substitution method.

3. $7x - 4y = 6,$
$y - 3x = -2$

4. $y = x + 2,$
$y - x = 8$

5. $9x - 6y = 2,$
$x = 4y + 5$

Solve using the elimination method.

6. $8x - 2y = 10,$
$-4y - 3x = -17$

7. $4x - 7y = 18,$
$9x + 14y = 40$

8. $3x - 5y = -4,$
$5x - 3y = 4$

Solve.

9. Jimmy has \$20.00 to spend at the store. He can spend all the money for two record albums and a poster, or he can buy one record album and two posters and have \$1.00 left over. What is the price of a record album? What is the price of a poster?

10. A train leaves town at noon traveling north at a speed of 44 mph. One hour later, another train, going 55 mph, travels north on a parallel track. How many hours will the second train travel before it overtakes the first train?

11. A beaker of alcohol contains a solution that is 30% alcohol. In another beaker is a solution that is 50% alcohol. How much of each should be mixed to obtain 40 L of a solution that is 45% alcohol?

12. A family invested \$5000, part at 10% and the remainder at 10.5%. The annual income from both investments is \$517. What is the amount invested at each rate?

Solve. If a system has more than one solution, list three of them.

13. $x + 2y + z = 10,$
$2x - y + z = 8,$
$3x + y + 4z = 2$

14. $3x + 2y + z = 3,$
$6x - 4y - 2z = -34,$
$-x + 3y - 3z = 14$

15. $2x - 5y - 2z = -4,$
$7x + 2y - 5z = -6,$
$-2x + 3y + 2z = 4$

16. $-5x + 5y = -6,$
$2x - 2y = 4$

17. $3x + y = 2,$
$x + 3y + z = 0,$
$x + z = 2$

18. $3x + 4y = 6,$
$1.5x - 3 = -2y$

19. $x + y + 2z = 1,$
$x - y + z = 1,$
$x + 2y + z = 2$

Solve.

20. In triangle *ABC* the measure of angle *A* is four times the measure of angle *C*, and the measure of angle *B* is 45° more than the measure of angle *C*. What are the measures of the angles of the triangle?

21. Find the three-digit number in which the sum of the digits is 11, the ten's digit is three less than the sum of the hundred's and one's digits, and the one's digit is five less than the hundred's digit.

22. Lynn has \$194 in her purse, consisting of \$20, \$5, and \$1 bills. The number of \$1 bills is one less than the total number of \$20 and \$5 bills. If she has 39 bills in her purse, how many of each denomination does she have?

23. Find numbers a, b, and c such that a quadratic function $ax^2 + bx + c$ fits the data points (1, 4), (2, 9) and (−2, 13). Write the equation for the function.

24. Classify each of the systems in Exercises 13–19 as consistent or inconsistent.

25. Classify each of the systems in Exercises 13–19 as dependent or independent.

Solve using matrices. Show your work.

26. $3x + 4y = -13,$
$5x + 6y = 8$

27. $3x - y + z = -1,$
$2x + 3y + z = 4,$
$5x + 4y + 2z = 5$

Evaluate.

28. $\begin{vmatrix} -2 & 4 \\ -3 & 5 \end{vmatrix}$

29. $\begin{vmatrix} 2 & 3 & 0 \\ 1 & 4 & -2 \\ 2 & -1 & 5 \end{vmatrix}$

Solve using Cramer's rule. Show your work.

30. $2x + 3y = 6,$
$x - 4y = 14$

31. $2x + y + z = -2,$
$2x - y + 3z = 6,$
$3x - 5y + 4z = 7$

Graph the system. Find the coordinates of any vertices formed.

32. $y \geq -3,$
$x \geq 2$

33. $x + 3y > -1,$
$x + 3y < 4$

34. $x - 3y \leq 3,$
$x + 3y \geq 9,$
$y \leq 6$

35. $2x + y \geq 9,$
$4x + 3y \geq 23,$
$x + 3y \geq 8,$
$x \geq 0,$
$y \geq 0$

36. Maximize and minimize $T = 6x + 10y$, subject to

$$x + y \leq 10,$$
$$5x + 10y \geq 50,$$
$$x \geq 2,$$
$$y \geq 0.$$

37. You are about to take a test that contains questions of type A worth 7 points and questions of type B worth 12 points. The total number of questions worked must be at least 8. If you know that type A questions take 10 min each to complete and type B questions take 8 min each and that the maximum time for the test is 80 min, how many of each type of question must you do in order to maximize your score? What is this maximum score?

Solve.

38. $x^2 - y^2 = 33,$
$x + y = 11$

39. $x^2 - 2x + 2y^2 = 8,$
$2x + y = 6$

40. $x^2 - y = 3,$
$2x - y = 3$

41. $x^2 + y^2 = 25,$
$x^2 - y^2 = 7$

42. $x^2 - y^2 = 3,$
$y = x^2 - 3$

43. $x^2 + y^2 = 18,$
$2x + y = 3$

44. $x^2 + y^2 = 100,$
$2x^2 - 3y^2 = -120$

45. $x^2 + 2y^2 = 12,$
$xy = 4$

SKILL MAINTENANCE

46. Determine the slope and the y-intercept of the line $y = \frac{1}{3}x - 2$.

47. Solve: $2^{2x+5} = \frac{1}{4}$.

48. Use synthetic division to find the quotient and the remainder: $(5x^4 + 4x^3 + 3x + 2) \div (x + 2)$.

49. Use synthetic division to find $P(-10)$ when $P(x) = 3x^5 + 2x^4 - x^3 + 1$.

SYNTHESIS

Solve.

50. $5^{x+y} = 25,$
$2^{2x-y} = 64$

51. $\frac{2}{3x} + \frac{4}{5y} = 8,$
$\frac{5}{4x} - \frac{3}{2y} = -6$

52. $x^2 + y^2 = 2,$
$y - x = 5$

53. The sum of two numbers is 36, and the product is 4. Find the sum of the reciprocals of the numbers.

THINKING IT THROUGH

1. Briefly compare the strengths and weaknesses of the graphical, substitution, and elimination methods as applied to the solution of systems of two linear equations in two variables.

2. List a system of equations with no solution. (Answers may vary.)

3. List a system of equations with infinitely many solutions. (Answers may vary.)

4. Describe the kinds of problems that are more readily solved using systems of equations.

CHAPTER TEST 11

Solve. If a system has more than one solution, list three of them.

1. $x + 3y = -8,$
$4x - 3y = 23$

2. $4y + 2x = 18,$
$3x + 6y = 26$

3. $2x + 4y = -6,$
$y = 3x - 9$

4. $4x - 6y = 3,$
$6x - 4y = -3$

5. $6x + 2y - 4z = 15,$
$-3x - 4y + 2z = -6,$
$4x - 6y + 3z = 8$

6. $-3x + y - 2z = 8,$
$-x + 2y - z = 5,$
$2x + y + z = -3$

7. $2x + 2y = 0,$
$4x + 4z = 4,$
$2x + y + z = 2$

8. $3x + 3z = 0,$
$2x + 2y = 2,$
$3y + 3z = 3$

9. Classify each of the systems in Questions 1–8 as consistent or inconsistent.

10. Classify each of the systems in Questions 1–8 as dependent or independent.

11. A plane flew for 5 hr with a 20-km/h tailwind and returned in 7 hr against the same wind. Find the speed of the plane in still air.

12. Mixture A is 34% salt and the rest water. Mixture B is 61% salt and the rest water. How many pounds of each mixture would be needed to obtain 120 lb of a mixture that is 50% salt?

13. The graph of $f(x) = ax^2 + bx + c$ contains the points $(-2, 3)$, $(1, 1)$, and $(0, 3)$. Find a, b, and c and write the equation for the function.

14. A student has a total of 225 on three tests. The sum of the scores on the first and second tests exceeds the third score by 61. The first score exceeds the second by 6. Find the three scores.

Solve using matrices.

15. $7x - 8y = 10,$
$9x + 5y = -2$

16. $x + 3y - 3z = 12,$
$3x - y + 4z = 0,$
$-x + 2y - z = 1$

Evaluate.

17. $\begin{vmatrix} -3 & 1 \\ -4 & \frac{2}{3} \end{vmatrix}$

18. $\begin{vmatrix} 1 & 1 & -2 \\ 0 & 2 & -6 \\ 4 & 0 & 3 \end{vmatrix}$

Solve using Cramer's rule.

19. $7x - 3y = 31,$
$4x + 2y = 14$

20. $x - 2y + 7z = 11,$
$2x + y - 3z = -5,$
$6x + z = 1$

Graph. Find the coordinates of any vertices.

21. $x + y \geq 3,$
$x - y \geq 5$

22. $2y - x \geq -7,$
$2y + 3x \leq 15,$
$y \leq 0,$
$x \leq 0$

23. Find the maximum and minimum values of $F = 134x + 20y$, subject to

$$x + y \leq 160,$$
$$10 \leq x \leq 60,$$
$$0 \leq y \leq 120.$$

24. You are about to take a test that contains questions of type A worth 5 points and type B worth 12 points. You must complete the test in 72 min. Type A questions take 4 min each to complete, and type B questions take 8 min. The total number of problems worked must not exceed 12. If you are told that you must work at least 2 questions of type B, how many of each type of question must you do in order to maximize your score? What is this maximum score?

Solve.

25. $\frac{x^2}{16} + \frac{y^2}{9} = 1,$
$3x + 4y = 12$

26. $x^2 + y^2 = 16,$
$\frac{x^2}{16} - \frac{y^2}{9} = 1$

27. $x^2 + y^2 = 100,$
$2y + x = 20$

28. $x^2 + y^2 = 13,$
$xy = 6$

SKILL MAINTENANCE

29. Use synthetic division to find the quotient and the remainder: $(x^5 + x^4 - x^3 + x^2) \div (x - 1)$.

30. Use synthetic division to determine whether 2 and -3 are roots of $P(x) = 2x^3 - 9x^2 + 7x + 6$.

31. Graph: $f(x) = -2$.

32. Solve: $\log (x - 7) = 1$.

SYNTHESIS

33. Find an equation of the circle that passes through the points $(-3, 8)$, $(4, 1)$, and $(-3, -4)$.

34. Two squares are such that the sum of their areas is $8\ m^2$ and the difference of their areas is $2\ m^2$. Find the length of a side of each square.

35. Solve:

$$\begin{aligned} 3x^2 - 4y^2 + z^2 &= -2, \\ 5x^2 + y^2 - 2z^2 &= 1, \\ 7x^2 + 3y^2 + 2z^2 &= 19. \end{aligned}$$

CHAPTER 12

Sequences, Series, and Combinatorics

FEATURE PROBLEM

The state of Michigan runs a 6-out-of-44-number lotto twice a week that pays at least $1.5 million. You purchase a card for $1 and pick any 6 numbers from 1 to 44. If your 6 numbers match those that the state draws, you win. How many 6-number combinations are there for the drawing?

THE MATHEMATICS

The number of combinations is

$$
\begin{aligned}
{}_{44}C_6 &= \binom{44}{6} \\
&= \frac{44 \cdot 43 \cdot 42 \cdot 41 \cdot 40 \cdot 39}{6 \cdot 5 \cdot 4 \cdot 3 \cdot 2 \cdot 1} \\
&= 7{,}059{,}052.
\end{aligned}
$$

The first part of this chapter is devoted to *sequences* and *series*. The idea of a sequence is a familiar one. For example, when a baseball coach writes a batting order, a sequence is being formed. When the members of a sequence are numbers, we can think of adding them. Such a sum is called a *series*.

In the last part of this chapter we begin a brief study of *probability*. The theory of probability has important applications to business, medicine, sociology, psychology, science, and to games of chance.

The review sections to be tested in addition to the material in this chapter are Sections 9.3, 9.4, 10.1, and 11.1.

12.1 Sequences and Series

In this section we discuss sets of numbers, considered in order, and their sums.

Sequences

Suppose that \$1000 is invested at 8%, compounded annually. The amounts to which the account will grow after 1 year, 2 years, 3 years, 4 years, and so on, are as follows:

$$\begin{array}{cccc} ① & ② & ③ & ④ \\ \downarrow & \downarrow & \downarrow & \downarrow \\ \$1080.00, & \$1166.40, & \$1259.71, & \$1360.49, \ldots . \end{array}$$

Note that we can think of this as a function that pairs 1 with the number \$1080.00, 2 with the number \$1166.40, 3 with the number \$1259.71, 4 with the number \$1360.49, and so on. A sequence is thus a *function*, where the domain is a set of consecutive positive integers.

If we keep computing the amounts in the account forever, we obtain an **infinite sequence:**

$$\$1080.00, \$1166.40, \$1259.71, \$1360.49, \$1469.33, \$1586.87, \ldots .$$

The three dots at the end indicate that the sequence goes on without stopping. If we stop after a certain number of years, we obtain a **finite sequence:**

$$\$1080.00, \$1166.40, \$1259.71, \$1360.49.$$

SEQUENCES

An *infinite sequence* is a function having for its domain the set of positive integers: $\{1, 2, 3, 4, 5, \ldots\}$.

A *finite sequence* is a function having for its domain a set of positive integers $\{1, 2, 3, 4, 5, \ldots, n\}$, for some positive integer n.

As another example, consider the sequence given by

$$a(n) = 2^n, \quad \text{or} \quad a_n = 2^n.$$

The notation a_n means the same as $a(n)$ but is more commonly used with sequences. Some of the function values (also known as *terms* of the sequence) are as follows:

$$a_1 = 2^1 = 2,$$
$$a_2 = 2^2 = 4,$$
$$a_3 = 2^3 = 8,$$
$$a_6 = 2^6 = 64.$$

The first term of the sequence is a_1, the fifth term is a_5, and the nth term, or **general term,** is a_n. This sequence can also be denoted in the following ways:

2, 4, 8, . . . ; or
2, 4, 8, . . . , 2^n, ... 	The 2^n emphasizes that the nth term of this sequence is found by raising 2 to the nth power.

EXAMPLE 1 Find the first 4 terms and the 57th term of the sequence whose general term is given by $a_n = (-1)^n/(n+1)$.

Solution

$$a_1 = \frac{(-1)^1}{1+1} = -\frac{1}{2}$$
$$a_2 = \frac{(-1)^2}{2+1} = \frac{1}{3}$$
$$a_3 = \frac{(-1)^3}{3+1} = -\frac{1}{4}$$
$$a_4 = \frac{(-1)^4}{4+1} = \frac{1}{5}$$
$$a_{57} = \frac{(-1)^{57}}{57+1} = -\frac{1}{58}.$$

■

Note that the expression $(-1)^n$ causes the signs of the terms to alternate between positive and negative, depending on whether n is even or odd.

Finding the General Term

When a sequence is described merely by naming the first few terms, we do not know for sure what the general term is, but we are expected to make a guess by looking for a pattern.

EXAMPLES For each sequence, make a guess at the general term.

2. 1, 4, 9, 16, 25, . . .
These are squares of numbers, so the general term may be n^2.

3. $\sqrt{1}, \sqrt{2}, \sqrt{3}, \sqrt{4}, \ldots$
These are square roots of numbers, so the general term may be $\sqrt{n}$.

4. $-1, 2, -4, 8, -16, \ldots$
   These are powers of 2 with alternating signs, so the general term may be $(-1)^n[2^{n-1}]$.

5. $2, 4, 8, \ldots$
   If we see the pattern of powers of 2, we will see 16 as the next term and guess 2^n for the general term. Then the sequence could be written with more terms as

$$2, 4, 8, 16, 32, 64, 128, \ldots.$$

   If we see that we can get the second term by adding 2, the third term by adding 4, and the next term by adding 6, and so on, we will see 14 as the next term. A general term for the sequence is then $n^2 - n + 2$, and the sequence can be written with more terms as

$$2, 4, 8, 14, 22, 32, 44, 58, \ldots.$$

■

Example 5 illustrates that with few given terms, the possibility of not being certain about the nth term is greater.

Sums and Series

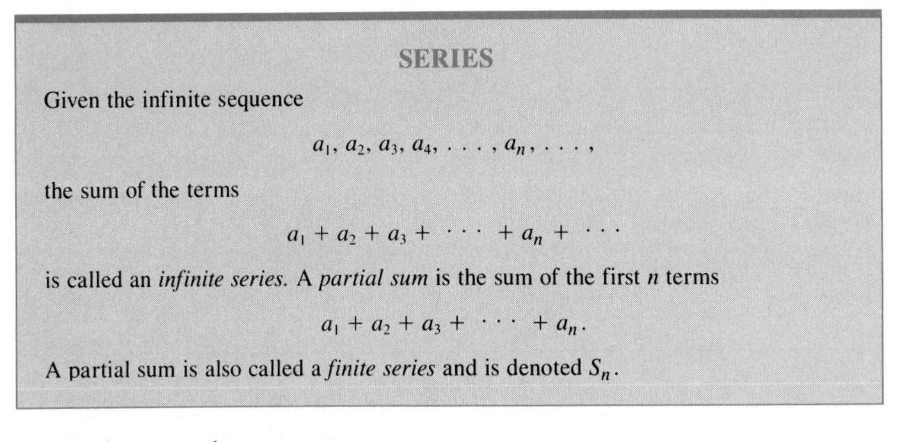

SERIES

Given the infinite sequence

$$a_1, a_2, a_3, a_4, \ldots, a_n, \ldots,$$

the sum of the terms

$$a_1 + a_2 + a_3 + \cdots + a_n + \cdots$$

is called an *infinite series.* A *partial sum* is the sum of the first n terms

$$a_1 + a_2 + a_3 + \cdots + a_n.$$

A partial sum is also called a *finite series* and is denoted S_n.

For instance, the sequence

$$3, 5, 7, 9, \ldots, 2n + 1$$

has the following partial sums:

$S_1 = 3,$ **This is the first term of the given sequence.**

$S_2 = 3 + 5 = 8,$ **This is the sum of the first two terms.**

$S_3 = 3 + 5 + 7 = 15,$ **The sum of the first three terms**

$S_4 = 3 + 5 + 7 + 9 = 24.$ **The sum of the first four terms**

EXAMPLE 6 For the sequence $-2, 4, -6, 8, -10, 12, -14$, (a) find S_3; (b) find S_5.

Solution

a) $S_3 = -2 + 4 + (-6) = -4$
b) $S_5 = -2 + 4 + (-6) + 8 + (-10) = -6$ ■

Sigma Notation

The Greek letter Σ (sigma) can be used to simplify notation when a series has a formula for the general term.

The sum of the first four terms of the sequence 3, 5, 7, 9, . . . , $2k + 1$ can be named as follows, using what is called *sigma notation*, or *summation notation*:

$$\sum_{k=1}^{4} (2k + 1).$$

This is read "the sum as k goes from 1 to 4 of $(2k + 1)$." The letter k is called the *index of summation*. Sometimes the index of summation starts at a number other than 1.

EXAMPLES Find and evaluate each sum.

7. $\sum_{k=1}^{5} k^2 = 1^2 + 2^2 + 3^2 + 4^2 + 5^2 = 1 + 4 + 9 + 16 + 25 = 55$

8. $\sum_{k=1}^{4} (-1)^k(2k) = (-1)^1(2 \cdot 1) + (-1)^2(2 \cdot 2) + (-1)^3(2 \cdot 3) + (-1)^4(2 \cdot 4)$
$= -2 + 4 - 6 + 8 = 4$

9. $\sum_{k=0}^{3} (2^k + 5) = (2^0 + 5) + (2^1 + 5) + (2^2 + 5) + (2^3 + 5)$
$= 6 + 7 + 9 + 13 = 35$ ■

EXAMPLES Write sigma notation for each sum.

10. $1 + 4 + 9 + 16 + 25$
This is a sum of squares, $1^2 + 2^2 + 3^2 + 4^2 + 5^2$, so the general term is k^2. Sigma notation is

$$\sum_{k=1}^{5} k^2.$$

11. $-1 + 3 - 5 + 7$
These are odd integers with alternating signs. Therefore the general term is $(-1)^k(2k - 1)$, beginning with $k = 1$. Note that $2k - 1$ is a formula for finding the kth odd natural number. Sigma notation is

$$\sum_{k=1}^{4} (-1)^k(2k - 1).$$

12. $3 + 9 + 27 + 81 + \cdots$

This is a sum of powers of 3, and it is also an infinite series. We use the symbol ∞ to represent infinity and name the infinite series using sigma notation as follows:

$$\sum_{k=1}^{\infty} 3^k.$$

EXERCISE SET 12.1

In each of the following, the nth term of a sequence is given. In each case find the first 4 terms; the 10th term, a_{10}; and the 15th term, a_{15}.

1. $a_n = 3n + 1$

2. $a_n = 3n - 1$

3. $a_n = \dfrac{n}{n+1}$

4. $a_n = n^2 + 1$

5. $a_n = n^2 - 2n$

6. $a_n = \dfrac{n^2 - 1}{n^2 + 1}$

7. $a_n = n + \dfrac{1}{n}$

8. $a_n = \left(-\dfrac{1}{2}\right)^{n-1}$

9. $a_n = (-1)^n n^2$

10. $a_n = (-1)^n(n + 3)$

11. $a_n = (-1)^{n+1}(3n - 5)$

12. $a_n = (-1)^n(n^3 - 1)$

13. $a_n = \dfrac{n+2}{n+5}$

14. $a_n = \dfrac{2n - 1}{3n - 4}$

Find the indicated term of the sequence.

15. $a_n = 4n - 7$; a_8

16. $a_n = 5n + 11$; a_9

17. $a_n = (3n + 4)(2n - 5)$; a_7

18. $a_n = (3n + 2)^2$; a_6

19. $a_n = (-1)^{n-1}(3.4n - 17.3)$; a_{12}

20. $a_n = (-2)^{n-2}(45.68 - 1.2n)$; a_{23}

21. $a_n = 5n^2(4n - 100)$; a_{11}

22. $a_n = 4n^2(11n + 31)$; a_{22}

23. $a_n = \left(1 + \dfrac{1}{n}\right)^2$; a_{20}

24. $a_n = \left(1 - \dfrac{1}{n}\right)^3$; a_{15}

25. $a_n = \log 10^n$; a_{43}

26. $a_n = \ln e^n$; a_{67}

27. $a_n = 1 + \dfrac{1}{n^2}$; a_{38}

28. $a_n = 2 - \dfrac{1000}{n}$; a_{100}

For each sequence find the general term, or nth term, a_n, or a rule for finding a_n. Answers may vary.

29. $1, 3, 5, 7, 9, \ldots$

30. $3, 9, 27, 81, 243, \ldots$

31. $-2, 6, -18, 54, \ldots$

32. $-2, 3, 8, 13, 18, \ldots$

33. $\frac{2}{3}, \frac{3}{4}, \frac{4}{5}, \frac{5}{6}, \frac{6}{7}, \ldots$

34. $\sqrt{2}, \sqrt{4}, \sqrt{6}, \sqrt{8}, \sqrt{10}, \ldots$

35. $\sqrt{3}, 3, 3\sqrt{3}, 9, 9\sqrt{3}, \ldots$

36. $1 \cdot 2, 2 \cdot 3, 3 \cdot 4, 4 \cdot 5, \ldots$

37. $-1, -4, -7, -10, -13, \ldots$

38. $\log 1, \log 10, \log 100, \log 1000, \ldots$

For each sequence find the indicated partial sum.

39. $1, 2, 3, 4, 5, 6, 7, \ldots; S_7$

40. $1, -3, 5, -7, 9, -11, \ldots; S_8$

41. $2, 4, 6, 8, \ldots; S_5$

42. $1, \frac{1}{4}, \frac{1}{9}, \frac{1}{16}, \frac{1}{25}, \ldots; S_5$

Rename and evaluate the sum.

43. $\sum_{k=1}^{5} \frac{1}{2k}$

44. $\sum_{k=1}^{6} \frac{1}{2k+1}$

45. $\sum_{k=0}^{5} 2^k$

46. $\sum_{k=4}^{7} \sqrt{2k-1}$

47. $\sum_{k=7}^{10} \log k$

48. $\sum_{k=0}^{4} \pi k$

49. $\sum_{k=1}^{8} \frac{k}{k+1}$

50. $\sum_{k=1}^{4} \frac{k-2}{k+3}$

51. $\sum_{k=1}^{5} (-1)^k$

52. $\sum_{k=1}^{5} (-1)^{k+1}$

53. $\sum_{k=1}^{8} (-1)^{k+1} 3^k$

54. $\sum_{k=1}^{7} (-1)^k 4^{k+1}$

55. $\sum_{k=1}^{6} \frac{2}{k^2+1}$

56. $\sum_{k=1}^{10} k(k+1)$

57. $\sum_{k=0}^{5} (k^2 - 2k + 3)$

58. $\sum_{k=0}^{5} (k^2 - 3k + 4)$

59. $\sum_{k=1}^{10} \frac{1}{k(k+1)}$

60. $\sum_{k=1}^{10} \frac{2^k}{2^k+1}$

Write sigma notation.

61. $\frac{1}{2} + \frac{2}{3} + \frac{3}{4} + \frac{4}{5} + \frac{5}{6} + \frac{6}{7}$

62. $3 + 6 + 9 + 12 + 15$

63. $-2 + 4 - 8 + 16 - 32 + 64$

64. $\frac{1}{1^2} + \frac{1}{2^2} + \frac{1}{3^2} + \frac{1}{4^2} + \frac{1}{5^2}$

65. $4 - 9 + 16 - 25 + \cdots + (-1)^n n^2$

66. $9 - 16 + 25 + \cdots + (-1)^{n+1} n^2$

67. $5 + 10 + 15 + 20 + 25 + \cdots$

68. $7 + 14 + 21 + 28 + 35 + \cdots$

69. $\frac{1}{1 \cdot 2} + \frac{1}{2 \cdot 3} + \frac{1}{3 \cdot 4} + \frac{1}{4 \cdot 5} + \cdots$

70. $\frac{1}{1 \cdot 2^2} + \frac{1}{2 \cdot 3^2} + \frac{1}{3 \cdot 4^2} + \frac{1}{4 \cdot 5^2} + \cdots$

SKILL MAINTENANCE

Simplify.

71. $\log_3 3$

72. $\log_3 1$

73. $\log_3 3^7$

74. $\log_c c$

SYNTHESIS

Find the first five terms of the sequence, and then find S_5.

75. $a_n = \frac{1}{2^n} \log 1000^n$

76. $a_n = i^n, i = \sqrt{-1}$

77. $a_n = \ln (1 \cdot 2 \cdot 3 \cdots n)$

78. **a)** Find the first few terms of the sequence $a_n = n^2 - n + 41$.

b) What pattern do you observe?

c) Find the 41st term. Does the pattern you found in part (b) still hold?

Find decimal notation, rounded to six decimal places, for the first six terms of the sequence.

79. ▦ $a_n = \left(1 + \frac{1}{n}\right)^n$

80. ▦ $a_n = \sqrt{n+1} - \sqrt{n}$

81. Find a formula for S_n, given that $a_n = \frac{1}{n} \cdot \frac{1}{n+1}$.

Some sequences are given by a *recursive definition*. The value of the first term, a_1, is given, and then we are told how to find each subsequent term from the term preceding it in the sequence. Find the first six terms of each of the following recursively defined sequences.

82. $a_1 = 1, a_{n+1} = 3a_n - 2$

83. $a_1 = 0, a_{n+1} = a_n^2 + 4$

84. A single cell of bacterium divides into two every 15 min. Suppose the same rate of division is maintained for 4 hr. Give a sequence that lists the number of cells after successive 15-min periods.

85. The value of an office machine is \$5200. Its scrap value each year is 75% of its value the year before. Give a sequence that lists the scrap value of the machine at the start of each year for a 10-year period.

86. A student gets \$4.20 for working in a warehouse for a publishing company. Each year the student gets a \$0.15 hourly raise. Give a sequence that lists the hourly salary of the student over a 10-year period.

12.2 Arithmetic Sequences and Series

In this section, we concentrate on a particular kind of sequence called an arithmetic sequence. If we start with a particular first term and then add the same number successively, we obtain an arithmetic sequence. We will also study arithmetic series.

Arithmetic Sequences

Consider this sequence:

$$2, 5, 8, 11, 14, 17, \ldots .$$

Note that adding 3 to any term produces the next term. In other words, the difference between any term and the preceding one is 3. This is an example of an **arithmetic** (pronounced ăr′ĭth-mĕt′-ĭk) **sequence.**

ARITHMETIC SEQUENCE

A sequence is *arithmetic* if there exists a number d, called the *common difference*, such that $a_n = a_{n-1} + d$, for any $n \geq 2$.

Arithmetic sequences are also called *arithmetic progressions.*

EXAMPLES The following are arithmetic sequences. Identify the first term, a_1, and the common difference, d.

	Sequence	*First Term, a_1*	*Common Difference, d*
1.	$4, 9, 14, 19, 24, \ldots$	4	5
2.	$34, 27, 20, 13, 6, -1, -8, \ldots$	34	-7
3.	$2, 2\frac{1}{2}, 3, 3\frac{1}{2}, 4, 4\frac{1}{2}, \ldots$	2	$\frac{1}{2}$

To obtain d for the sequence in Example 1, we choose any term beyond the first, say the second term, 9, and subtract the preceding term from it: $9 - 4 = 5$. Then we can check by adding 5 to each term to see if we obtain the next:

$$a_1 = 4,$$
$$a_2 = 4 + 5 = 9,$$
$$a_3 = 9 + 5 = 14,$$
$$a_4 = 14 + 5 = 19, \quad \text{and so on.}$$

We now find a formula for the general, or nth, term of any arithmetic sequence. Let us denote the common difference by d and write out the first few terms:

$$a_1,$$
$$a_2 = a_1 + d,$$
$$a_3 = a_2 + d = (a_1 + d) + d = a_1 + 2d, \quad \text{Substituting for } a_2$$
$$a_4 = a_3 + d = (a_1 + 2d) + d = a_1 + 3d. \quad \text{Substituting for } a_3$$

Note that the coefficient of d in each case is 1 less than the number of the term, n.

Generalizing, we obtain the following formula.

Formula 1

The nth term of an arithmetic sequence is given by

$$a_n = a_1 + (n - 1)d, \quad \text{for any } n \geq 1.$$

EXAMPLE 4 Find the 14th term of the arithmetic sequence 4, 7, 10, 13, ...

Solution First we note that $a_1 = 4$, $d = 3$, and $n = 14$. Then using Formula 1, we obtain

$$a_n = a_1 + (n - 1)d$$
$$a_{14} = 4 + (14 - 1) \cdot 3 = 4 + 13 \cdot 3 = 4 + 39 = 43.$$

The 14th term is 43. ■

EXAMPLE 5 In the sequence in Example 4, which term is 301? That is, what is n if $a_n = 301$?

Solution We substitute into Formula 1 and solve for n:

$$\begin{aligned} a_n &= a_1 + (n - 1)d \\ 301 &= 4 + (n - 1) \cdot 3 \\ 301 &= 4 + 3n - 3 \\ 300 &= 3n \\ 100 &= n. \end{aligned}$$

The term 301 is the 100th term of the sequence. ■

Given two terms and their places in an arithmetic sequence, we can construct the sequence.

EXAMPLE 6 The third term of an arithmetic sequence is 8, and the sixteenth term is 47. Find a_1 and d and construct the sequence.

Solution We know that $a_3 = 8$ and $a_{16} = 47$. Thus we would have to add d thirteen times to get from 8 to 47. That is,

$$8 + 13d = 47.$$

Solving, we obtain

$$\begin{aligned} 13d &= 39 \\ d &= 3. \end{aligned}$$

Since $a_3 = 8$, we subtract d twice to get to a_1. Thus,

$$a_1 = 8 - 2 \cdot 3 = 2.$$

The sequence is 2, 5, 8, 11, ... Note that we could have subtracted d from a_{16} fifteen times. ■

In general, d should be subtracted $(n - 1)$ times from a_n in order to find a_1.

Sum of the First *n* Terms of an Arithmetic Sequence

Suppose we add the first four terms of the sequence

$$3, 5, 7, 9, 11, \ldots.$$

We get what is called an **arithmetic series:**

$$3 + 5 + 7 + 9, \qquad \text{or} \qquad 24.$$

The sum of the first n terms of a sequence is denoted S_n. Thus for the preceding sequence, $S_4 = 24$. We want to find a formula for S_n when the sequence is arithmetic. We can denote an arithmetic sequence as follows.

This is the next-to-last term. If you add d to this term, the result is a_n.

$$a_1, (a_1 + d), (a_1 + 2d), \ldots, \underbrace{(a_n - 2d)}, \overbrace{(a_n - d)}, a_n$$

This term is two terms back from the last. If you add d to this term, you get the next-to-last term, $a_n - d$.

Then S_n is given by

$$S_n = a_1 + (a_1 + d) + (a_1 + 2d) + \cdots + (a_n - 2d) + (a_n - d) + a_n. \qquad (1)$$

Rearranging the order of addition, we have

$$S_n = a_n + (a_n - d) + (a_n - 2d) + \cdots + (a_1 + 2d) + (a_1 + d) + a_1. \qquad (2)$$

Then if we add corresponding terms of each side of Eqs. (1) and (2), we get

$$\begin{aligned} 2S_n = {} & [a_1 + a_n] + [(a_1 + d) + (a_n - d)] \\ & + [(a_1 + 2d) + (a_n - 2d)] + \cdots + [(a_n - 2d) + (a_1 + 2d)] \\ & + [(a_n - d) + (a_1 + d)] + [a_n + a_1]. \end{aligned}$$

This simplifies to

$$2S_n = (a_1 + a_n) + (a_1 + a_n) + (a_1 + a_n) + \cdots + (a_1 + a_n).$$

Since there are n binomials $(a_1 + a_n)$ being added, it follows that

$$2S_n = n(a_1 + a_n),$$

from which we obtain the following formula.

Formula 2

The sum of the first n terms of an arithmetic sequence is given by

$$S_n = \frac{n}{2}(a_1 + a_n).$$

EXAMPLE 7 Find the sum of the first 100 natural numbers.

Solution The sum is

$$1 + 2 + 3 + \cdots + 99 + 100.$$

This is the sum of the first 100 terms of the arithmetic sequence for which

$$a_1 = 1, \qquad a_n = 100, \qquad \text{and} \qquad n = 100.$$

Then substituting in the formula

$$S_n = \frac{n}{2}(a_1 + a_n),$$

we get

$$S_{100} = \tfrac{100}{2}(1 + 100) = 50(101) = 5050.$$ ■

The preceding formula is useful when we know a_1 and a_n, the first and last terms, but it often happens that a_n is not known. Thus we need a formula in terms of a_1, n, and d. We substitute the expression for a_n, given in Formula 1, $a_n = a_1 + (n - 1)d$, into Formula 2:

$$S_n = \frac{n}{2}(a_1 + [a_1 + (n - 1)d]).$$

This gives us the following.

Formula 3

The sum of the first n terms of an arithmetic sequence is given by

$$S_n = \frac{n}{2}[2a_1 + (n - 1)d].$$

EXAMPLE 8 Find the sum of the first 15 terms of the arithmetic sequence 4, 7, 10, 13, ...

Solution Note that

$$a_1 = 4, \qquad d = 3, \qquad \text{and} \qquad n = 15.$$

Here we use Formula 3 since we do not know a_n:

$$S_n = \frac{n}{2}[2a_1 + (n - 1)d].$$

We get

$$S_{15} = \tfrac{15}{2}[2 \cdot 4 + (15 - 1)3] = \tfrac{15}{2}[8 + 14 \cdot 3] = \tfrac{15}{2}[8 + 42]$$
$$= \tfrac{15}{2}[50] = 375.$$ ■

Problem Solving

For some problem-solving situations, the translation may involve sequences or series. We look at some examples.

EXAMPLE 9 Chris takes a job, starting with an hourly wage of \$14.25, and is promised a raise of 15¢ per hour every 2 months for 5 years. At the end of 5 years, what will be Chris's hourly wage?

Solution

1. **Familiarize.** One thing to do is write down the hourly wage for several two-month time periods.

Beginning:	14.25,
After two months:	14.40,
After four months:	14.55,
and so on.	

What appears is a *sequence of numbers*: 14.25, 14.40, 14.55, ... Is it an arithmetic sequence? Yes, because we add 0.15 each time to get the next term.

We ask ourselves what we know about arithmetic sequences. We recall, or look up, the pertinent formula(s). There are three:

$$a_n = a_1 + (n - 1)d, \qquad S_n = \frac{n}{2}(a_1 + a_n), \qquad S_n = \frac{n}{2}[2a_1 + (n - 1)d].$$

In this case, we are not looking for a *sum*, so it is probably the first formula that will give us our answer. We want to know the last term in a sequence. We will need to know a_1, n, and d. From our list above, we see that

$$a_1 = 14.25 \qquad \text{and} \qquad d = 0.15.$$

What is n? That is, how many terms are in the sequence? Each year there are 6 raises, since Chris gets a raise every 2 months. There are 5 years, so the total number of raises will be $5 \cdot 6$, or 30. There will be 31 terms: the original wage and 30 increased rates.

2. **Translate.** We want to find the 31st term of an arithmetic sequence, with $a_1 = 14.25$, $d = 0.15$, and $n = 31$. We substitute into the formula:

$$a_{31} = 14.25 + (31 - 1) \cdot 0.15.$$

3. **Carry out.** We calculate to obtain $18.75.

4. **Check.** We can check the calculations. We can also calculate in a slightly different way for another check. For example, at the end of a year, there will be 6 raises, for a total raise of $0.90. At the end of 5 years, the total raise will be 5 × $0.90, or $4.50. If we add that to the original wage of $14.25, we obtain $18.75. The answer checks.

5. **State.** At the end of 5 years, Chris's hourly wage will be $18.75. ■

Example 9 is one in which the calculations or the translation could be done in a number of ways. It is possible to solve the problem in several different, and very good, manners. Here is a point to remember: There is often a variety of ways in which a problem can be solved. You should use the one that is best or easiest for you. In this chapter, however, we will concentrate on the use of sequences and series and their related formulas in problem solving.

EXAMPLE 10 A stack of telephone poles has 30 poles in the bottom row. There are 29 poles in the second row, 28 in the next row, and so on. How many poles are in the stack?

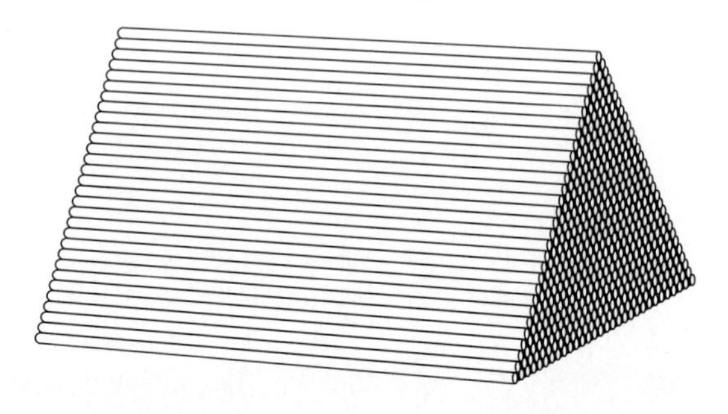

Solution

1. **Familiarize.** A picture will help in this case. The following figure shows the ends of the poles and the way in which they stack. There are 30 poles on the bottom, and we see that there will be one fewer in each succeeding row. How many rows will there be?

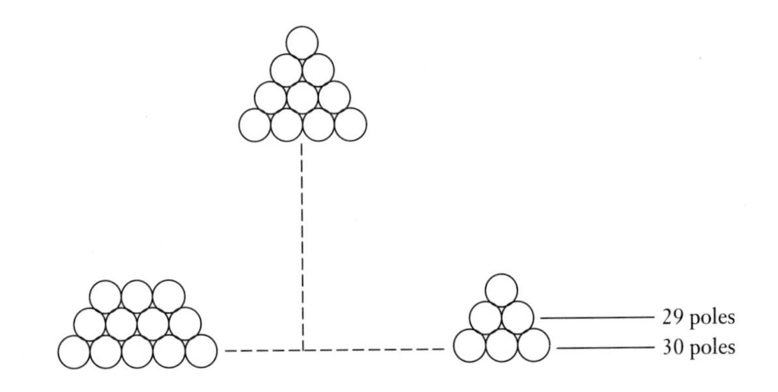

We go from 30 poles in a row, down to one pole in the top row, so there must be 30 rows.

We want the sum

$$30 + 29 + 28 + \cdots + 1.$$

Thus we want the sum of an arithmetic sequence. We recall, or look up if necessary, the formula for the sum of an arithmetic sequence:

$$S_n = \frac{n}{2}(a_1 + a_n).$$

2. **Translate.** We want to find the sum of an arithmetic sequence, where $a_1 = 30$ and, since there are 30 terms, $n = 30$. There is just one pole on top, so $a_{30} = 1$.

Substituting into the formula, we have

$$S_{30} = \frac{30}{2}(30 + 1).$$

3. **Carry out.** We calculate, and find that $S_{30} = 465$.
4. **Check.** In this case, we can check the calculations by doing them again. A longer, harder way would be to do the entire addition:

$$30 + 29 + 28 + \cdots + 1.$$

5. **State.** The answer is that there are 465 poles in the stack. ■

EXERCISE SET 12.2

Find the first term and the common difference.

1. 2, 7, 12, 17, . . .

2. 1.06, 1.12, 1.18, 1.24, . . .

3. 7, 3, -1, -5, . . .

4. -9, -6, -3, 0, . . .

5. $\frac{3}{2}$, $\frac{9}{4}$, 3, $\frac{15}{4}$, . . .

6. $\frac{3}{5}$, $\frac{1}{10}$, $-\frac{2}{5}$, . . .

7. \$2.12, \$2.24, \$2.36, \$2.48, . . .

8. \$214, \$211, \$208, \$205, . . .

9. Find the 12th term of the arithmetic sequence 2, 6, 10, ...

10. Find the 11th term of the arithmetic sequence 0.07, 0.12, 0.17, ...

11. Find the 17th term of the arithmetic sequence 7, 4, 1, ...

12. Find the 14th term of the arithmetic sequence 3, $\frac{7}{3}$, $\frac{5}{3}$, ...

13. Find the 13th term of the arithmetic sequence \$1200, \$964.32, \$728.64, ...

14. Find the 10th term of the arithmetic sequence \$2345.78, \$2967.54, \$3589.30, ...

15. In the sequence of Exercise 9, what term is 106?

16. In the sequence of Exercise 10, what term is 1.67?

17. In the sequence of Exercise 11, what term is -296?

18. In the sequence of Exercise 12, what term is -27?

19. Find a_{17} when $a_1 = 5$ and $d = 6$.

20. Find a_{20} when $a_1 = 14$ and $d = -3$.

21. Find a_1 when $d = 4$ and $a_8 = 33$.

22. Find a_1 when $d = 8$ and $a_{11} = 26$.

23. Find n when $a_1 = 5$, $d = -3$, and $a_n = -76$.

24. Find n when $a_1 = 25$, $d = -14$, and $a_n = -507$.

25. In an arithmetic sequence, $a_{17} = -40$ and $a_{28} = -73$. Find a_1 and d. Write the first 5 terms of the sequence.

26. In an arithmetic sequence $a_{17} = \frac{25}{3}$ and $a_{32} = \frac{95}{6}$. Find a_1 and d. Write the first 5 terms of the sequence.

27. Find the sum of the first 20 terms of the series $5 + 8 + 11 + 14 + \cdots$.

28. Find the sum of the first 14 terms of the series $11 + 7 + 3 + \cdots$.

29. Find the sum of the first 300 natural numbers.

30. Find the sum of the first 400 natural numbers.

31. Find the sum of the even numbers from 2 to 100, inclusive.

32. Find the sum of the odd numbers from 1 to 99, inclusive.

33. Find the sum of the multiples of 7 from 7 to 98, inclusive.

34. Find the sum of all multiples of 4 that are between 14 and 523.

35. If an arithmetic series has $a_1 = 2$, $d = 5$, and $n = 20$, what is S_n?

36. If an arithmetic series has $a_1 = 7$, $d = -3$, and $n = 32$, what is S_n?

Problem Solving

37. A gardener is making a triangular planting, with 35 plants in the front row, 31 in the second row, 27 in the third row, and so on. If the pattern is consistent, how many plants will there be in the last row? How many plants are there altogether?

38. A formation of a marching band has 14 marchers in the front row, 16 in the second row, 18 in the third row, and so on, for 25 rows. How many marchers are in the last row? How many marchers are there altogether?

39. How many poles will be in a pile of telephone poles if there are 50 in the first layer, 49 in the second, and so on, until there is one in the last layer?

40. If 10¢ is saved on October 1, 20¢ on October 2, 30¢ on October 3, and so on, how much is saved during October? (October has 31 days.)

41. A family saves money in an arithmetic sequence: \$600 the first year, \$700 the second, and so on, for 20 years. How much do they save in all (disregarding interest)?

42. A student saves \$30 on August 1, \$50 on August 2, \$70 on August 3, and so on. How much would be saved in August? (August has 31 days.)

43. Theaters are often built with more seats per row as the rows move toward the back. Suppose the main floor of a theater has 28 seats in the first row, 32 in the second, 36 in the third, and so on, for 50 rows. How many seats are on the main floor?

44. A person sets up an investment such that it will return \$5000 the first year, \$6125 the second year, \$7250 the third year, and so on, for 25 years. How much in all is received from the investment?

SKILL MAINTENANCE

Convert to an exponential equation.

45. $\log_a P = k$

46. $\ln t = a$

Find an equation of the circle satisfying the given conditions.

47. Center $(0, 0)$, radius 9

48. Center $(-2, 5)$, radius $3\sqrt{2}$

SYNTHESIS

49. Find a formula for the sum of the first n natural numbers:

$$1 + 2 + 3 + \cdots + n.$$

50. Find a formula for the sum of the first n consecutive odd numbers starting with 1:

$$1 + 3 + 5 + \cdots + (2n - 1).$$

51. Find three numbers in an arithmetic sequence such that the sum of the first and third is 10 and the product of the first and second is 15.

52. Find the first term and the common difference for the arithmetic sequence where $a_2 = 4p - 3q$ and $a_4 = 10p + q$.

53. ▦ Find the first 10 terms of the arithmetic sequence for which $a_1 = \$8760$ and $d = -\$798.23$.

54. ▦ Find the sum of the first 10 terms of the sequence given in Exercise 53.

55. Prove that if p, m, and q form an arithmetic sequence (as listed), then

$$m = \frac{p + q}{2}.$$

56. *Business: Straight-line depreciation.* A company buys an office machine for \$5200 on January 1 of a given year. The machine is expected to last for 8 years, at the end of which time its *trade-in*, or *salvage, value* will be \$1100. If the company figures the decline in value to be the same each year, then the *book values*, or *salvage values*, after t years, $0 \le t \le 8$, form an arithmetic sequence given by

$$a_t = C - t\left(\frac{C - S}{N}\right).$$

where C = the original cost of the item (\$5200), N = the years of expected life (8), and S = the salvage value (\$1100).

a) Find the formula for a_t for the straight-line depreciation of the office machine.

b) Find the salvage value after 0 years, 1 year, 2 years, 3 years, 4 years, 7 years, and 8 years.

12.3 Geometric Sequences and Series

For arithmetic sequences we added a certain number to each term to get the next term. With the kind of sequence we consider now, we *multiply* each term by a certain number to get the next term. These are *geometric sequences.* We also consider geometric series.

Geometric Sequences

Consider the sequence

$$2, 6, 18, 54, 162, \ldots.$$

If we multiply each term by 3, we obtain the next term. Sequences in which each term can be multiplied by a certain number in order to get the next term are called **geometric.** We usually call this number r. We refer to it as the *common ratio* because we can get r by dividing any term by the preceding term.

GEOMETRIC SEQUENCE

A sequence is *geometric* if there exists a number r, called the *common ratio*, such that

$$\frac{a_{n+1}}{a_n} = r, \quad \text{or} \quad a_{n+1} = a_n \cdot r \quad \text{for any } n \ge 1.$$

A geometric sequence is also called a *geometric progression*.

EXAMPLES The following are geometric sequences. Identify the common ratio.

Sequence	*Common Ratio*	
1. 3, 6, 12, 24, 48, 96, . . .	2	$\frac{6}{3} = 2, \frac{12}{6} = 2$, and so on
2. 3, −6, 12, −24, 48, −96, . . .	−2	$\frac{-6}{3} = -2, \frac{12}{-6} = -2$, and so on
3. \$5200, \$3900, \$2925, \$2193.75, . . .	0.75	$\frac{\$3900}{\$5200} = 0.75$, $\frac{\$2925}{\$3900} = 0.75$
4. \$1000, \$1080, \$1166.40, . . .	1.08	$\frac{\$1080}{\$1000} = 1.08$
5. $1, \frac{1}{2}, \frac{1}{4}, \frac{1}{8}, \ldots$	$\frac{1}{2}$	$\frac{\frac{1}{2}}{1} = \frac{1}{2}, \frac{\frac{1}{4}}{\frac{1}{2}} = \frac{1}{2}$

■

We now find a formula for the general, or nth, term of any geometric sequence. Let a_1 be the first term and let r be the common ratio. We write out the first few terms as follows:

$$a_1,$$
$$a_2 = a_1 r,$$
$$a_3 = a_2 r = (a_1 r)r = a_1 r^2, \quad \text{Substituting for } a_2$$
$$a_4 = a_3 r = (a_1 r^2)r = a_1 r^3. \quad \text{Substituting for } a_3$$

Note that the exponent is 1 less than the number of the term.

Generalizing, we obtain the following.

Formula 4

The nth term of a geometric sequence is given by

$$a_n = a_1 r^{n-1}, \quad \text{for any } n \geq 1.$$

EXAMPLE 6 Find the 7th term of the geometric sequence 4, 20, 100, ...

Solution First we note that

$$a_1 = 4 \quad \text{and} \quad n = 7.$$

To find the common ratio, we can divide any term by its predecessor, provided it has one. Since the second term is 20 and the first is 4, we get

$$r = \frac{20}{4}, \quad \text{or} \quad 5.$$

Then, using the formula

$$a_n = a_1 r^{n-1},$$

we have

$$a_7 = 4 \cdot 5^{7-1} = 4 \cdot 5^6 = 4 \cdot 15{,}625 = 62{,}500.$$

EXAMPLE 7 Find the 10th term of the geometric sequence

$$64, -32, 16, -8, \ldots .$$

Solution First we note that

$$a_1 = 64, \qquad n = 10, \qquad \text{and} \qquad r = \frac{-32}{64}, \quad \text{or} \ -\frac{1}{2}.$$

Then using the formula

$$a_n = a_1 r^{n-1},$$

we have

$$a_{10} = 64 \cdot \left(-\frac{1}{2}\right)^{10-1} = 64 \cdot \left(-\frac{1}{2}\right)^9 = 2^6 \cdot \left(-\frac{1}{2^9}\right) = -\frac{1}{2^3} = -\frac{1}{8}.$$

Sum of the First n Terms of a Geometric Sequence

We want to find a formula for the sum S_n of the first n terms of a geometric sequence

$$a_1, a_1 r, a_1 r^2, a_1 r^3, \ldots, a_1 r^{n-1}, \ldots .$$

The **geometric series** is given by

$$S_n = a_1 + a_1 r + a_1 r^2 + \cdots + a_1 r^{n-2} + a_1 r^{n-1}. \tag{1}$$

We want to develop a formula that allows us to find this sum without a great deal of adding. If we multiply on both sides of Eq. (1) by r, we have

$$rS_n = a_1 r + a_1 r^2 + a_1 r^3 + \cdots + a_1 r^{n-1} + a_1 r^n. \tag{2}$$

When we multiply on both sides of Eq. (1) by -1, we get

$$-S_n = -a_1 - a_1 r - a_1 r^2 - \cdots - a_1 r^{n-2} - a_1 r^{n-1}. \tag{3}$$

Then, when we add corresponding sides of Eqs. (2) and (3), the highlighted terms are opposites of each other and sum to 0. Thus we get

$$rS_n - S_n = a_1 r^n - a_1,$$

or

$$(r - 1)S_n = a_1(r^n - 1), \qquad \text{Factoring}$$

from which we get the following formula.

Formula 5

The sum of the first n terms of a geometric sequence is given by

$$S_n = \frac{a_1(r^n - 1)}{r - 1}, \qquad \text{for any } r \neq 1.$$

EXAMPLE 8 Find the sum of the first 7 terms of the geometric sequence 3, 15, 75, 375, ...

Solution First we note that

$$a_1 = 3, \qquad n = 7, \qquad \text{and} \qquad r = \frac{15}{3}, \qquad \text{or } 5.$$

Then, using the formula

$$S_n = \frac{a_1(r^n - 1)}{r - 1},$$

we have

$$S_7 = \frac{3(5^7 - 1)}{5 - 1} = \frac{3(78{,}125 - 1)}{4} = \frac{3(78{,}124)}{4} = 58{,}593.$$

Infinite Geometric Series

Suppose we consider the sum of the terms of an infinite geometric sequence, such as 2, 4, 8, 16, 32, ... We get what is called an **infinite geometric series:**

$$2 + 4 + 8 + 16 + 32 + \cdots.$$

Here, as n grows larger and larger, the sum of the first n terms, S_n, becomes larger and larger without bound. There are also infinite series that get closer and closer to some specific number. Here is an example:

$$\frac{1}{2} + \frac{1}{4} + \frac{1}{8} + \frac{1}{16} + \cdots + \frac{1}{2^n} + \cdots.$$

Let's consider S_n for some values of n:

$$\begin{aligned}
S_1 &= \tfrac{1}{2} &&= \tfrac{1}{2} = 0.5,\\
S_2 &= \tfrac{1}{2} + \tfrac{1}{4} &&= \tfrac{3}{4} = 0.75,\\
S_3 &= \tfrac{1}{2} + \tfrac{1}{4} + \tfrac{1}{8} &&= \tfrac{7}{8} = 0.875,\\
S_4 &= \tfrac{1}{2} + \tfrac{1}{4} + \tfrac{1}{8} + \tfrac{1}{16} &&= \tfrac{15}{16} = 0.9375,\\
S_5 &= \tfrac{1}{2} + \tfrac{1}{4} + \tfrac{1}{8} + \tfrac{1}{16} + \tfrac{1}{32} &&= \tfrac{31}{32} = 0.96875.
\end{aligned}$$

We can describe S_n as follows:

$$S_n = 1 - \frac{1}{2^n}.$$

Note that the value of S_n is less than 1 for any value of n, but as n gets larger and larger, the values of S_n get closer and closer to 1. We say that 1 is the *limit* of S_n and that 1 is the *sum* of this *infinite geometric series.* The sum of an infinite series, if it exists, is denoted S_∞. It can be shown (but we will not do it here) that the sum of the terms of an infinite geometric series exists if and only if $|r| < 1$ (that is, the absolute value of the common ratio is less than 1).

We want to find a formula for the sum of an infinite geometric series. We first consider the sum of the first n terms:

$$S_n = \frac{a_1(r^n - 1)}{r - 1} = \frac{a_1 - a_1 r^n}{1 - r}.$$ Using the distributive law and multiplying by $\frac{-1}{-1}$

For $|r| < 1$, it follows that values of r^n get closer and closer to 0 as n gets larger. (Choose a number between -1 and 1 and check this by finding larger and larger powers on your calculator.) As r^n gets closer and closer to 0, so does $a_1 r^n$. Thus S_n gets closer and closer to $a_1/(1 - r)$.

Formula 6

When $|r| < 1$, the sum of an infinite geometric series is given by

$$S_\infty = \frac{a_1}{1 - r}.$$

EXAMPLE 9 Determine whether this infinite geometric series has a sum. If so, find it.

$$1 + 3 + 9 + 27 + \cdots$$

Solution $|r| = |3| = 3$, and since $|r| \nless 1$, the series does *not* have a sum. ■

EXAMPLE 10 Determine whether this infinite geometric series has a sum. If so, find it.

$$1 - \tfrac{1}{2} + \tfrac{1}{4} - \tfrac{1}{8} + \tfrac{1}{16} - \cdots$$

Solution

a) $|r| = \left|-\tfrac{1}{2}\right| = \tfrac{1}{2}$, and since $|r| < 1$, the series does have a sum.
b) The sum is given by

$$S_\infty = \frac{1}{1 - \left(-\frac{1}{2}\right)} = \frac{1}{\frac{3}{2}} = \frac{2}{3}.$$ ■

EXAMPLE 11 Find fractional notation for 0.63636363 . . . .

Solution We can express this as

$$0.63 + 0.0063 + 0.000063 + \cdots.$$

This is an infinite geometric series, where $a_1 = 0.63$ and $r = 0.01$. Since $|r| < 1$, this series has a sum:

$$S_\infty = \frac{a_1}{1 - r} = \frac{0.63}{1 - 0.01} = \frac{0.63}{0.99} = \frac{63}{99}.$$

Thus fractional notation for 0.63636363 . . . is $\frac{63}{99}$, or $\frac{7}{11}$. ■

Problem Solving

For some problem-solving situations, the translation may involve geometric sequences or series.

EXAMPLE 12 Suppose someone offered you a job for the month of September (30 days) under the following conditions. You will be paid \$0.01 for the first day, \$0.02 for the second, \$0.04 for the third, and so on, doubling your previous day's salary each day. How much would you earn? (Would you take the job? Make a guess before reading further.)

Solution

1. **Familiarize.** You earn \$0.01 the first day, \$0.01(2) the second day, \$0.01(2)(2) the third day, and so on. The amounts form a geometric sequence with $a_1 = \$0.01$, $r = 2$, and $n = 30$.
2. **Translate.** The amount earned is the geometric series

$$\$0.01 + \$0.01(2) + \$0.01(2^2) + \$0.01(2^3) + \cdots + \$0.01(2^{29}),$$

where

$$a_1 = \$0.01, \qquad n = 30, \qquad \text{and} \qquad r = 2.$$

3. **Carry out.** Then, using the formula

$$S_n = \frac{a_1(r^n - 1)}{r - 1},$$

we have

$$\begin{aligned} S_{30} &= \frac{\$0.01(2^{30} - 1)}{2 - 1} \\ &\approx \$0.01(1{,}073{,}700{,}000 - 1) && \text{Use a calculator to approximate } 2^{30}. \\ &\approx \$0.01(1{,}073{,}700{,}000) && 1{,}073{,}700{,}000 - 1 \approx 1{,}073{,}700{,}000 \\ &= \$10{,}737{,}000. \end{aligned}$$

4. **Check.** We check by repeating the calculations. That will be left to the student.
5. **State.** Since the salary is more than \$10 million a month, most people would take the job! ■

EXAMPLE 13 A student loan is in the amount of \$600. Interest is to be 12%, compounded annually, and the entire amount is to be paid after 10 years. How much is to be paid back?

Solution

1. **Familiarize.** Suppose we let P represent any principal amount. At the end of one year, the amount owed will be $P + 0.12P$, or $1.12P$. That amount will be the principal for the second year. The amount owed at the end of the second year will be 1.12(New principal) = $1.12(1.12P)$, or $(1.12)^2P$. Thus the amount owed at the beginning of successive years is as follows:

$$\overset{①}{\downarrow}\ P,\quad \overset{②}{\downarrow}\ 1.12P,\quad \overset{③}{\downarrow}\ 1.12^2P,\quad \overset{④}{\downarrow}\ 1.12^3P,\quad \text{and so on.}$$

   We have a geometric sequence. The amount owed at the beginning of the 11th year will be the amount owed at the end of the 10th year.

2. **Translate.** We have a geometric sequence with $a_1 = 600$, $r = 1.12$, and $n = 11$. The appropriate formula is

$$a_n = a_1r^{n-1}.$$

3. **Carry out.** We substitute and calculate:

$$\begin{aligned} a_{11} &= 600(1.12)^{11-1} \\ &= 600(1.12)^{10} \\ &\approx 600(3.1058482) && \textbf{Use a calculator to approximate } 1.12^{10}. \\ &\approx 1863.51. && \textbf{Rounded to the nearest hundredth} \end{aligned}$$

4. **Check.** We repeat the calculations. We might also estimate an answer to see whether the result is reasonable.

5. **State.** The answer is that \$1863.51 is to be paid back at the end of 10 years. ■

EXAMPLE 14 A super-rebound ball rebounds to 60% of the height from which it has fallen or is dropped. The ball is dropped from a height of 20 ft and begins to bounce.

a) After bouncing and rebounding 9 times, how far has it traveled upward (the total rebound distance)?

b) Approximately how far will the ball have traveled upward (bounced) before it comes to rest?

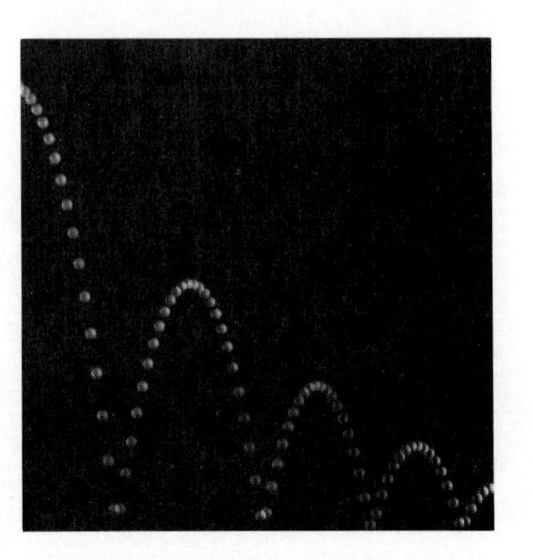

Solution

1. Familiarize. Let's do some calculations and look for a pattern.

First fall:	20 ft
First rebound:	0.6×20, or 12 ft
Second fall:	12 ft, or 0.6×20
Second rebound:	0.6×12, or $0.6(0.6 \times 20)$, which is 7.2 ft
Third fall:	7.2 ft, or $0.6(0.6 \times 20)$
Third rebound:	0.6×7.2, or $0.6(0.6(0.6 \times 20))$, which is 4.32 ft

The rebound distances form a geometric sequence:

$$\underset{\textcircled{1}}{0.6 \times 20}, \quad \underset{\textcircled{2}}{0.6^2 \times 20}, \quad \underset{\textcircled{3}}{0.6^3 \times 20}, \quad \underset{\textcircled{4}}{0.6^4 \times 20}, \ldots,$$

or

$$12, \quad 0.6 \times 12, \quad 0.6^2 \times 12, \quad 0.6^3 \times 12, \ldots.$$

2. Translate.

a) The total rebound distance after 9 bounces is the sum of a geometric sequence. The first term is 12 and the common ratio is 0.6. There will be 9 terms, so we can use Formula 5:

$$S_n = \frac{a_1(r^n - 1)}{r - 1}.$$

b) Theoretically, the ball will never stop bouncing. Actually, it will eventually stop. We can approximate the total distance bounced by considering an infinite number of bounces. We use Formula 6:

$$S_\infty = \frac{a_1}{1 - r}.$$

3. **Carry out.**

 a) We substitute into the formula and calculate:

$$S_9 = \frac{12[(0.6)^9 - 1]}{0.6 - 1}$$
$$\approx 29.70. \qquad \text{Using a calculator}$$

 b) We substitute and calculate:

$$S_\infty = \frac{12}{1 - 0.6}$$
$$= 30.$$

4. **Check.** We can do the calculations again.

5. **State.**

 a) In 9 bounces, the ball will have traveled upward a total distance of about 29.70 ft.

 b) The ball will travel upward about 30 ft before it comes to rest. ■

EXERCISE SET 12.3

For each geometric sequence, find the common ratio.

1. 2, 4, 8, 16, . . .

2. $12, \ -4, \ \frac{4}{3}, \ -\frac{4}{9}, \ldots$

3. $1, \ -1, \ 1, \ -1, \ldots$

4. $-5, \ -0.5, \ -0.05, \ -0.005, \ldots$

5. $\frac{1}{2}, \ -\frac{1}{4}, \ \frac{1}{8}, \ -\frac{1}{16}, \ldots$

6. $\frac{2}{3}, \ -\frac{4}{3}, \ \frac{8}{3}, \ -\frac{16}{3}, \ldots$

7. $75, \ 15, \ 3, \ \frac{3}{5}, \ldots$

8. 6.275, 0.6275, 0.06275, . . .

9. $\frac{1}{x}, \ \frac{1}{x^2}, \ \frac{1}{x^3}, \ldots$

10. $5, \ \frac{5m}{2}, \ \frac{5m^2}{4}, \ \frac{5m^3}{8}, \ldots$

11. \$780, \$858, \$943.80, \$1038.18, . . .

12. \$5600, \$5320, \$5054, \$4801.30, . . .

For each geometric sequence, find the indicated term.

13. 2, 4, 8, 16, . . . ; the 6th term

14. $2, \ -10, \ 50, \ -250, \ldots$; the 9th term

15. $2, \ 2\sqrt{3}, \ 6, \ldots$; the 9th term

16. $1, \ -1, \ 1, \ -1, \ldots$; the 57th term

17. ▦ $\frac{8}{243}, \ \frac{8}{81}, \ \frac{8}{27}, \ldots$; the 10th term

18. ▦ $\frac{7}{625}, \ \frac{-7}{125}, \ \frac{7}{25}, \ldots$; the 13th term

19. ▦ \$1000, \$1080, \$1166.40, . . . ; the 12th term

20. ▦ \$1000, \$1070, \$1144.90, . . . ; the 11th term

For each geometric sequence, find the nth, or general, term.

21. 1, 3, 9, . . .

22. 25, 5, 1, . . .

23. $1, \ -1, \ 1, \ -1, \ldots$

24. 2, 4, 8, . . .

25. $\frac{1}{x}, \ \frac{1}{x^2}, \ \frac{1}{x^3}, \ldots$

26. $5, \ \frac{5m}{2}, \ \frac{5m^2}{4}, \ldots$

For Exercises 27–34 use Formula 5 to find the indicated sum.

27. The sum of the first 7 terms of the geometric series

$$6 + 12 + 24 + \cdots$$

28. The sum of the first 6 terms of the geometric series

$$16 - 8 + 4 - \cdots$$

29. The sum of the first 7 terms of the geometric series

$$\tfrac{1}{18} - \tfrac{1}{6} + \tfrac{1}{2} - \cdots$$

30. The sum of the first 5 terms of the geometric series

$$6 + 0.6 + 0.06 + \cdots$$

31. The sum of the first 8 terms of the series

$$1 + x + x^2 + x^3 + \cdots$$

32. The sum of the first 10 terms of the series

$$1 + x^2 + x^4 + x^6 + \cdots$$

33. ▦ The sum of the first 16 terms of the geometric sequence

$$\$200, \$200(1.06), \$200(1.06)^2, \ldots$$

34. ▦ The sum of the first 23 terms of the geometric sequence

$$\$1000, \$1000(1.08), \$1000(1.08)^2, \ldots$$

Determine whether the infinite geometric series has a sum. If so, find it.

35. $4 + 2 + 1 + \cdots$

36. $7 + 3 + \frac{9}{7} + \cdots$

37. $25 + 20 + 16 + \cdots$

38. $12 + 9 + \frac{27}{4} + \cdots$

39. $100 - 10 + 1 - \frac{1}{10} + \cdots$

40. $-6 + 18 - 54 + 162 - \cdots$

41. $8 + 40 + 200 + \cdots$

42. $-6 + 3 - \frac{3}{2} + \frac{3}{4} - \cdots$

43. $0.3 + 0.03 + 0.003 + \cdots$

44. $0.37 + 0.0037 + 0.000037 + \cdots$

45. $\$500(1.02)^{-1} + \$500(1.02)^{-2} + \$500(1.02)^{-3} + \cdots$

46. $\$1000(1.08)^{-1} + \$1000(1.08)^{-2} + \$1000(1.08)^{-3} + \cdots$

Find fractional notation for the infinite sum. (These are geometric series.)

47. 0.4444 . . .

48. 9.999999 . . .

49. 0.55555 . . .

50. 0.66666 . . .

51. 0.15151515 . . .

52. 0.12121212 . . .

Solve. Use a calculator as needed for evaluating formulas.

53. A ping-pong ball is dropped from a height of 16 ft and always rebounds one fourth of the distance fallen. How high does it rebound the 6th time?

54. Approximate the total of the rebound heights of the ball in Exercise 53.

55. Gaintown has a current population of 100,000, and the population is increasing by 3% each year. What will the population be in 15 years?

56. How long will it take for the population of Gaintown to double? (See Exercise 55.)

57. A student borrows $1200. The loan is to be repaid in 13 years at 12% interest, compounded annually. How much will be repaid at the end of 13 years?

58. A piece of paper is 0.01 in. thick. It is folded repeatedly in such a way that its thickness is doubled each time for 20 times. How thick is the result?

59. A superball dropped from the top of the Washington Monument (556 ft high) always rebounds three fourths of the distance fallen. How far (up and down) will the ball have traveled when it hits the ground for the 6th time?

60. Approximate the total distance that the ball of Exercise 59 will have traveled when it comes to rest.

61. Suppose someone offered you a job for the month of February (28 days) under the following conditions. You will be paid $0.01 the 1st day, $0.02 the 2nd, $0.04 the 3rd, and so on, doubling your previous day's salary each day. How much would you earn?

62. *An annuity.* A person decides to save money in a savings account for retirement. At the beginning of each year, $1000 is invested at 11%, compounded annually. How much will be in the retirement fund at the end of 40 years?

SKILL MAINTENANCE

Express in terms of logarithms of x, y, and z.

63. $\log_a x^2y^3z^5$

64. $\log_a \dfrac{x^2y^3}{z^5}$

Solve each system.

65. $5x - 2y = -3,$
$2x + 5y = -24$

66. $x - 2y + 3z = 4,$
$2x - y + z = -1,$
$4x + y + z = 1$

SYNTHESIS

67. Find the sum of the first n terms of

$$1 + x + x^2 + \cdots .$$

68. Find the sum of the first n terms of

$$x^2 - x^3 + x^4 - x^5 + \cdots .$$

69. The sides of a square are each 16 cm long. A second square is inscribed by joining the midpoints of the sides, successively. In the second square we repeat the process, inscribing a third square. If this process is continued indefinitely, what is the sum of all of the areas of all the squares? (*Hint:* Use an infinite geometric series.)

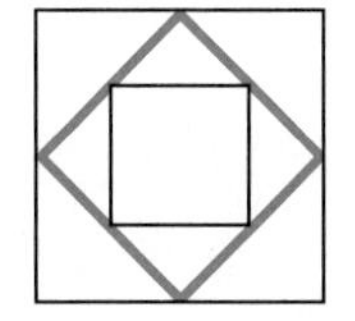

70. The infinite series

$$S_\infty = 2 + \frac{1}{2} + \frac{1}{2\cdot 3} + \frac{1}{2\cdot 3\cdot 4} + \frac{1}{2\cdot 3\cdot 4\cdot 5} + \frac{1}{2\cdot 3\cdot 4\cdot 5\cdot 6} + \cdots$$

is not geometric, but it does have a sum. Find values of S_1, S_2, S_3, S_4, S_5, and S_6. Make a conjecture about the value of S_∞.

12.4 Combinatorics: Permutations and Combinations

To study probability, it is first necessary to study the theory of counting, called **combinatorics.** Such a study concerns itself with determining the number of ways in which a set can be arranged or combined, certain objects can be chosen, or a succession of events can occur.

EXAMPLE 1 How many 3-letter code symbols can be formed with the letters A, B, C *without* repetition?

Solution Examples of such symbols are ABC, CBA, ACB, and so on. Consider placing the letters in these frames.

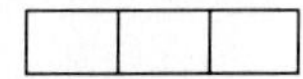

We can select any of the 3 letters for the first letter in the symbol. Once this letter has been selected, the second can be selected from the 2 remaining letters. The third letter is already determined, since only 1 possibility is left. The possibilities can be arrived at with a **tree diagram.**

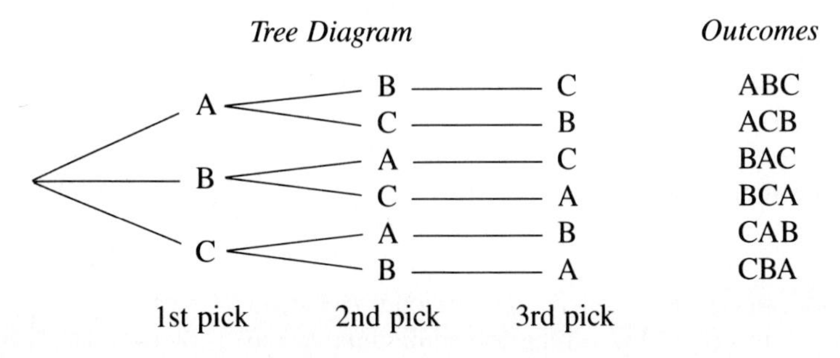

There are $3 \cdot 2 \cdot 1$, or 6, possibilities. The set of all of them is as follows:

$$\{ABC, ACB, BAC, BCA, CAB, CBA\}.$$

Suppose we perform an experiment such as selecting letters (as in the preceding example), flipping a coin, or drawing a card. The results are called **outcomes.** An **event** is a set of outcomes. The following principle concerns events that occur together, or are combined.

FUNDAMENTAL COUNTING PRINCIPLE

Given a combined action, or event, in which the first action can be performed in n_1 ways, the second action can be performed in n_2 ways, and so on, the total number of ways in which the combined action can be performed is the product

$$n_1 \cdot n_2 \cdot n_3 \cdot \cdots \cdot n_k.$$

EXAMPLE 2 How many 3-letter code symbols can be formed with the letters A, B, and C *with* repetition?

Solution There are 3 choices for the first letter and, since we allow repetition, 3 choices for the second and 3 for the third. Thus by the fundamental counting principle, there are $3 \cdot 3 \cdot 3$, or 27, choices.

Permutations

We now turn our attention to the part of combinatorics that deals with the study of *permutations.* The study of permutations involves *order* and *arrangement.*

A *permutation* of a set of n objects is an ordered arrangement of all n objects.

Consider, for example, a set of 4 objects:

$$\{A, B, C, D\}.$$

To find the number of ordered arrangements of the set, we select a first letter: There are 4 choices. Then we select a second letter: There are 3 choices. Then we select a third letter: There are 2 choices. Finally, there is 1 choice for the last selection. Thus by the fundamental counting principle, there are $4 \cdot 3 \cdot 2 \cdot 1$, or 24, permutations of a set of 4 objects.

We can find a formula for the total number of permutations of all objects in a set of n objects. We have n choices for the first selection, $n - 1$ for the second, $n - 2$ for the third, and so on. For the nth selection, there is only 1 choice.

Permutations of n Objects

The total number of permutations of a set of n objects, denoted ${}_nP_n$, is given by

$${}_nP_n = n(n-1)(n-2)\cdots(3)(2)(1).$$

EXAMPLE 3 Find (a) ${}_4P_4$ and (b) ${}_7P_7$.

Solution

a) ${}_4P_4 = 4 \cdot 3 \cdot 2 \cdot 1 = 24$
b) ${}_7P_7 = 7 \cdot 6 \cdot 5 \cdot 4 \cdot 3 \cdot 2 \cdot 1 = 5040$ ■

EXAMPLE 4 In how many different ways can 9 different letters be placed in 9 mailboxes, one letter to a box?

Solution

$${}_9P_9 = 9 \cdot 8 \cdot 7 \cdot 6 \cdot 5 \cdot 4 \cdot 3 \cdot 2 \cdot 1 = 362{,}880$$ ■

Factorial Notation

Products of successive natural numbers, such as $7 \cdot 6 \cdot 5 \cdot 4 \cdot 3 \cdot 2 \cdot 1$ and $9 \cdot 8 \cdot 7 \cdot 6 \cdot 5 \cdot 4 \cdot 3 \cdot 2 \cdot 1$, are used so often that it is convenient to adopt a notation for them.

For the product $7 \cdot 6 \cdot 5 \cdot 4 \cdot 3 \cdot 2 \cdot 1$, we write 7!, read "7-factorial."

For any natural number n,

$$n! = n(n-1)(n-2)\cdots(3)(2)(1).$$

Here are some examples.

$$
\begin{aligned}
7! &= 7 \cdot 6 \cdot 5 \cdot 4 \cdot 3 \cdot 2 \cdot 1 = 5040 \\
6! &= 6 \cdot 5 \cdot 4 \cdot 3 \cdot 2 \cdot 1 = 720 \\
5! &= 5 \cdot 4 \cdot 3 \cdot 2 \cdot 1 = 120 \\
4! &= 4 \cdot 3 \cdot 2 \cdot 1 = 24 \\
3! &= 3 \cdot 2 \cdot 1 = 6 \\
2! &= 2 \cdot 1 = 2 \\
1! &= 1 = 1
\end{aligned}
$$

We also define 0! to be 1. We do this so that certain formulas and theorems can be stated concisely and with a consistent pattern.

We can now simplify the permutation formula as follows:

$$_nP_n = n!$$

Note that $8! = 8 \cdot 7!$. We can see this as follows. By definition of factorial notation,

$$
\begin{aligned}
8! &= 8 \cdot 7 \cdot 6 \cdot 5 \cdot 4 \cdot 3 \cdot 2 \cdot 1 \\
&= 8 \cdot (7 \cdot 6 \cdot 5 \cdot 4 \cdot 3 \cdot 2 \cdot 1) \\
&= 8 \cdot 7!.
\end{aligned}
$$

Generalizing, we get the following.

For any natural number n, $n! = n(n-1)!$.

By using this result repeatedly, we can further manipulate factorial notation.

EXAMPLE 5 Rewrite 7! with a factor of 5!.

Solution

$$7! = 7 \cdot 6 \cdot 5!$$

■

Permutations of *n* Objects Taken *r* at a Time

Consider a set of 6 objects, say {A, B, C, D, E, F}. How many ordered arrangements are there having 3 members without repetition? We can select the first object in 6 ways. There are then 5 choices for the second and then 4 choices for the third. By the fundamental counting principle, there are then $6 \cdot 5 \cdot 4$ ways to construct the ordered arrangement. In other words, there are $6 \cdot 5 \cdot 4$ permutations of a set of 6 objects taken 3 at a time. Note that if we multiply by 1 we have

$$6 \cdot 5 \cdot 4 = \frac{6 \cdot 5 \cdot 4 \cdot 3 \cdot 2 \cdot 1}{3 \cdot 2 \cdot 1}, \quad \text{or} \quad \frac{6!}{3!}.$$

A *permutation* of a set of n objects taken r at a time is an ordered arrangement of r objects taken from the set.

Consider a set of n objects and the selecting of an ordered arrangement of r objects. The first object can be selected in n ways. The second can be selected in $n - 1$ ways, and so on. The rth can be selected in $n - (r - 1)$ ways. By the fundamental counting principle, the total number of permutations is

$$n(n - 1)(n - 2) \cdots [n - (r - 1)].$$

We now multiply by 1:

$$n(n - 1)(n - 2) \cdots [n - (r - 1)] \frac{(n - r)!}{(n - r)!}$$
$$= \frac{n(n - 1)(n - 2)(n - 3) \cdots [n - (r - 1)](n - r)!}{(n - r)!}.$$

The numerator is now the product of all natural numbers from n to 1, hence is $n!$. Thus the total number of permutations is

$$\frac{n!}{(n - r)!}.$$

This gives us the following result.

PERMUTATIONS OF n OBJECTS TAKEN r AT A TIME

The number of permutations of a set of n objects taken r at a time, denoted ${}_nP_r$, is given by

$${}_nP_r = n(n - 1)(n - 2) \cdots [n - (r - 1)]$$
$$= \frac{n!}{(n - r)!}.$$

Formula (1) is most useful in application, but formula (2) will be important in a later development.

EXAMPLE 6 Compute ${}_6P_4$ using both of the above formulas.

Solution Using formula (1), we have

$${}_6P_4 = 6 \cdot 5 \cdot 4 \cdot 3$$ Note that the 6 in ${}_6P_4$ shows where to start and

$$= 360.$$ the 4 in ${}_6P_4$ shows how many factors there are.

Using formula (2) of Theorem 2, we have

$$_6P_4 = \frac{6!}{(6-4)!} = \frac{6!}{2!} = \frac{6 \cdot 5 \cdot 4 \cdot 3 \cdot 2 \cdot 1}{2 \cdot 1} = 6 \cdot 5 \cdot 4 \cdot 3 = 360.$$

■

EXAMPLE 7 In how many ways can the letters of the set {A, B, C, D, E, F, G} be arranged without repetition to form code words of (a) 5 letters? (b) 2 letters?

Solution

a) $_7P_5 = 7 \cdot 6 \cdot 5 \cdot 4 \cdot 3 = 2520$
b) $_7P_2 = 7 \cdot 6 = 42$ ■

EXAMPLE 8 A baseball manager arranges the batting order as follows: The 4 infielders will bat first, then the other 5 players will follow. How many different batting orders are possible?

Solution The infielders can bat in 4! different ways; the rest in 5! different ways. Then by the fundamental counting principle, we have $_4P_4 \cdot {_5P_5} = 4! \cdot 5!$, or 2880, possible batting orders. ■

Combinations

If you play cards, you know that in most situations the *order* in which you hold cards *is not important*! It is just the contents of the hand, or set, of cards. We may sometimes make selections from a set *without regard to order.* Such selections are called **combinations.**

A → B → C	C A B
Permutation:	*Combination:*
Order considered!	Order *not* considered!

EXAMPLE 9 Find all the combinations of taking 3 elements from the set of 5 elements {A, B, C, D, E}. How many are there?

Solution The combinations are

{A, B, C}, {A, B, D}, {A, B, E}, {A, C, D}, {A, C, E},
{A, D, E}, {B, C, D}, {B, C, E}, {B, D, E}, {C, D, E}.

There are 10 combinations of 5 objects taken 3 at a time. ■

When we find all the combinations of 5 objects taken 3 at a time, we are finding all the 3-element subsets. When we are naming a set, the order of the listing is *not* considered. Thus,

{A, C, B} names the same set as {A, B, C}.

A *combination* of r objects chosen from a set of n objects is a subset of the set of n objects.

Because the elements of a subset may be listed in any order, it is important to remember that *when thinking of combinations we do not think about order.*

EXAMPLE 10 Find all the subsets of the set {A, B, C}. Identify these as combinations. How many subsets are there in all?

Solution

a) The empty set has 0 elements in it. It is denoted $\emptyset$. The empty set is a subset of every set. In this case, it is the combination of 3 objects taken 0 at a time. There is 1 such combination, $\emptyset$.

b) The following are all the one-element subsets of {A, B, C}:

{A}, {B}, {C}.

These are the combinations of 3 objects taken 1 at a time. There are 3 such combinations.

c) The following are all the two-element subsets of {A, B, C}:

{A, B}, {A, C}, {B, C}.

These are the combinations of 3 objects taken 2 at a time. These are 3 such combinations.

d) The following are all the three-element subsets of {A, B, C}:

{A, B, C}.

These are the combinations of 3 objects taken 3 at a time. There is only 1 such combination. A set is always a subset of itself.

The total number of subsets is $1 + 3 + 3 + 1$, or 8. ■

We want to develop a formula for computing the number of combinations of n objects taken r at a time without actually listing the combinations, or subsets.

The notation for the number of combinations taken r at a time from a set of n objects is denoted ${}_nC_r$.

We call ${}_nC_r$ **combination notation.** In Example 10 we see that

$${}_3C_0 = 1, \qquad {}_3C_1 = 3, \qquad {}_3C_2 = 3, \qquad \text{and} \qquad {}_3C_3 = 1.$$

We can derive some general results here. First, it is always true that ${}_nC_n = 1$ because a set with n objects has only 1 subset with n objects, the set itself. Second, ${}_nC_1 = n$ because a set with n objects has n subsets with 1 element each. Finally, ${}_nC_0 = 1$ because a set with n objects has only one subset with 0 elements, namely, the empty set $\emptyset$.

We want to derive a general formula for ${}_nC_r$, for any $r \leq n$. Let us return to Example 9 and compare the number of combinations with the number of permutations.

Combinations		*Permutations*					
{A, B, C}	⟶	ABC	BCA	CAB	CBA	BAC	ACB
{A, B, D}	⟶	ABD	BDA	DAB	DBA	BAD	ADB
{A, B, E}	⟶	ABE	BEA	EAB	EBA	BAE	AEB
{A, C, D}	⟶	ACD	CDA	DAC	DCA	CAD	ADC
{A, C, E}	⟶	ACE	CEA	EAC	ECA	CAE	AEC
{A, D, E}	⟶	ADE	DEA	EAD	EDA	DAE	AED
{B, C, D}	⟶	BCD	CDB	DBC	DCB	CBD	BDC
{B, C, E}	⟶	BCE	CEB	EBC	ECB	CBE	BEC
{B, D, E}	⟶	BDE	DEB	EBD	EDB	DBE	BED
{C, D, E}	⟶	CDE	DEC	ECD	EDC	DCE	CED

Note that each combination of 3 objects, say {A, C, E}, yields 3!, or 6, permutations, as shown. It follows that

$$3! \cdot {}_5C_3 = 60 = {}_5P_3 = 5 \cdot 4 \cdot 3,$$

so

$${}_5C_3 = \frac{{}_5P_3}{3!} = \frac{5 \cdot 4 \cdot 3}{3 \cdot 2 \cdot 1} = 10. \qquad \text{Dividing by 3!}$$

In general, the number of combinations of n objects taken r at a time, ${}_nC_r$, times the number of permutations of these r objects, $r!$, must equal the number of permutations of n objects taken r at a time:

$$r! \cdot {}_nC_r = {}_nP_r$$

$${}_nC_r = \frac{{}_nP_r}{r!} = \frac{1}{r!} \cdot {}_nP_r = \frac{1}{r!} \cdot \frac{n!}{(n-r)!} = \frac{n!}{r!(n-r)!}. \qquad \text{Dividing by } r!$$

This now gives us two formulas for computing ${}_nC_r$.

COMBINATIONS OF n OBJECTS TAKEN r AT A TIME

The total number of combinations of n objects taken r at a time, denoted ${}_nC_r$, is given by

$$ {}_nC_r = \frac{n!}{r!(n-r)!}, \tag{1} $$

or

$$ {}_nC_r = \frac{{}_nP_r}{r!} = \frac{n(n-1)(n-2)\cdots[n-(r-1)]}{r!}. \tag{2} $$

There is another kind of notation that is also used for ${}_nC_r$. It is called **binomial coefficient notation.** The reason for such terminology will be seen later.

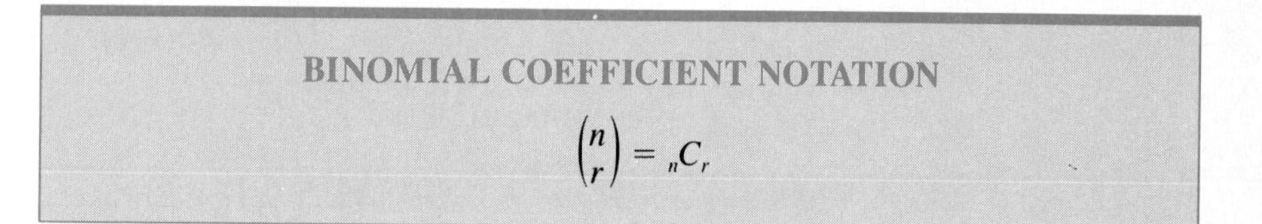

BINOMIAL COEFFICIENT NOTATION

$$ \binom{n}{r} = {}_nC_r $$

You should be able to use either notation and either formula.

EXAMPLE 11 Evaluate $\binom{7}{5}$, using both of the above formulas.

Solution

a) By formula (1),

$$ \binom{7}{5} = \frac{7!}{5!2!} = \frac{7\cdot6\cdot5\cdot4\cdot3\cdot2\cdot1}{5\cdot4\cdot3\cdot2\cdot1\cdot2\cdot1} = \frac{7\cdot6\cdot5\cdot4\cdot3}{5\cdot4\cdot3\cdot2\cdot1} = \frac{7\cdot6}{2\cdot1} = 21. $$

b) By formula (2),

The 7 tells where to start.

$$ \binom{7}{5} = \frac{7\cdot6\cdot5\cdot4\cdot3}{5\cdot4\cdot3\cdot2\cdot1} = \frac{7\cdot6}{2\cdot1} = 21. $$

The 5 tells us how many factors there are in both numerator and denominator and where to start the denominator. ■

> CAUTION!
>
> $$\binom{n}{r} \text{ does not mean } n \div r, \text{ or } \frac{n}{r}.$$

The method in Example 11(b), using formula (2), is easiest to carry out, but in some situations formula (1) does become useful.

EXAMPLE 12 Evaluate $\binom{n}{0}$ and $\binom{n}{2}$.

Solution We use formula (1) for the first expression and formula (2) for the second. Then

$$\binom{n}{0} = \frac{n!}{0!(n-0)!} = \frac{n!}{1 \cdot n!} = 1,$$

using formula (1), and

$$\binom{n}{2} = \frac{n(n-1)}{2!} = \frac{n(n-1)}{2}, \qquad \text{or} \qquad \frac{n^2 - n}{2},$$

using formula (2). ■

Note that

$$\binom{7}{2} = \frac{7 \cdot 6}{2 \cdot 1} = 21,$$

so that from Example 11,

$$\binom{7}{5} = \binom{7}{2}.$$

This says that the number of 5-element subsets of a set of 7 objects is the same as the number of 2-element subsets of a set of 7 objects. When 5 elements are chosen from a set, we simultaneously choose *not* to include 2 elements. To see this, consider such a set:

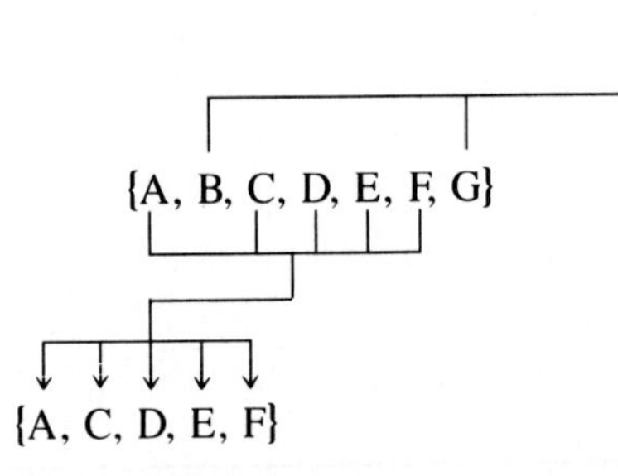

Whenever we form a subset with 5 elements, we leave behind a subset with 2 elements, and vice versa.

Thus the numbers of each type of subset are the same. In general:

> For any whole numbers r and n, $r \leq n$,
>
> $$\binom{n}{r} = \binom{n}{n-r} \quad \text{and} \quad {}_nC_r = {}_nC_{n-r}.$$
>
> The number of subsets of size r of a set with n objects is the same as the number of subsets of size $n - r$. The number of combinations of n objects taken r at a time is the same as the number of combinations of n objects taken $n - r$ at a time.

This observation provides an alternative way to compute. For example, instead of computing ${}_{52}C_{48}$, it is a lot easier to compute ${}_{52}C_4$.

We now solve problems involving combinations.

EXAMPLE 13 *Michigan lotto.* The state of Michigan runs a 6-out-of-44-number lotto twice a week that pays at least \$1.5 million. You purchase a card for \$1 and pick any 6 numbers from 1 to 44. How many possible 6-number combinations are there for drawing?

Solution

No order is implied here. You pick any 6 numbers from 1 to 44. Thus the number of combinations is

$${}_{44}C_6 = \binom{44}{6} = \frac{44 \cdot 43 \cdot 42 \cdot 41 \cdot 40 \cdot 39}{6 \cdot 5 \cdot 4 \cdot 3 \cdot 2 \cdot 1} = 7{,}059{,}052.$$

■

EXAMPLE 14 How many committees can be formed from a group of 5 governors and 7 senators if each committee contains 3 governors and 4 senators?

Solution The 3 governors can be selected in ${}_5C_3$ ways and the 4 senators can be selected in ${}_7C_4$ ways. If we use the fundamental counting principle, it follows that the number of possible committees is

$${}_5C_3 \cdot {}_7C_4 = 10 \cdot 35 = 350.$$

■

EXERCISE SET 12.4

Evaluate.

1. ${}_4P_3$
2. ${}_7P_5$
3. ${}_{10}P_7$
4. ${}_{10}P_3$
5. How many 5-digit numbers can be named using the digits 5, 6, 7, 8, and 9 without repetition? with repetition?
6. How many 4-digit numbers can be named using the digits 2, 3, 4, and 5 without repetition? with repetition?
7. In how many ways can 5 students be arranged in a straight line?
8. In how many ways can 7 athletes be arranged in a straight line?

9. How many 7-digit phone numbers can be formed with the digits 0, 1, 2, 3, 4, 5, 6, 7, 8, and 9, assuming that no digit is used more than once and the first digit is not 0?

10. A program is planned to have 5 rock numbers and 4 speeches. In how many ways can this be done if a rock number and a speech are to alternate and the rock numbers come first?

11. A penny, nickel, dime, quarter, and half-dollar are arranged in a straight line.

a) Considering just the coins, in how many ways can they be lined up?
b) Considering the coins and heads and tails, in how many ways can they be lined up?

12. A penny, nickel, dime, and quarter are arranged in a straight line.

a) Considering just the coins, in how many ways can they be lined up?
b) Considering the coins and heads and tails, in how many ways can they be lined up?

13. 🖩 Compute ${}_{52}P_4$.

14. 🖩 Compute ${}_{50}P_5$.

15. 🖩 A state forms its license plates by first listing a number that corresponds to the county in which the car owner lives (the names of the counties are alphabetized and the number is its location in that order). Then the plate lists a letter of the alphabet, and this is followed by a number from 1 to 9999. How many such plates are possible if there are 80 counties?

16. How many code symbols can be formed using 4 out of 5 letters of A, B, C, D, E if the letters

a) are not repeated?
b) can be repeated?
c) are not repeated but must begin with D?
d) are not repeated but must end with DE?

17. *Zip Codes.* A Zip Code in Dallas, Texas, is 75247. A Zip Code in Cambridge, Massachusetts, is 02142.

a) How many Zip Codes are possible if any of the digits 0 to 9 can be used?
b) If each post office has its own Zip Code, how many possible post offices can there be?

18. *Zip Codes.* Zip Codes are sometimes given using a 9-digit number like 75247-5456, where the last 4 digits represent a post office box number.

a) How many 9-digit Zip Codes are possible?
b) There are 243 million people in the United States. If each person has a Zip Code and there are enough post office boxes, are there enough Zip Codes?

19. *Social security numbers.* A social security number is a 9-digit number like 293-36-0391.

a) How many social security numbers can there be?
b) There are 243 million people in the United States. Can each person have a social security number?

20. 🖩 How "long" is 15!? You own 15 different books and decide to actually make up all possible arrangements of the books on a shelf. About how long, in years, would it take if you can make one arrangement per second?

Evaluate.

21. ${}_{13}C_2$

22. ${}_{9}C_6$

23. $\binom{13}{11}$

24. $\binom{9}{3}$

25. $\binom{7}{1}$

26. $\binom{8}{8}$

27. $\frac{{}_5P_3}{3!}$

28. $\frac{{}_{10}P_5}{5!}$

29. $\binom{6}{0}$

30. $\binom{6}{3}$

31. ${}_{12}C_{11}$

32. ${}_{12}C_{10}$

33. $\binom{m}{2}$

34. $\binom{m}{m}$

In each of the following exercises, give an expression for the answer in terms of permutation notation, combination notation, factorial notation, or other products. Then evaluate.

35. There are 23 students in a fraternity. How many sets of 4 officers can be selected?

36. How many basketball games can be played in a 9-team league if each team plays all other teams once? twice?

37. On a test a student is to select 6 out of 10 questions. In how many ways can the student do this?

38. On a test, a student is to select 7 out of 11 questions. In how many ways can the student do this?

39. How many lines are determined by 8 points, no 3 of which are collinear? How many triangles are determined by the same points?

40. How many lines are determined by 7 points, no 3 of which are collinear? How many triangles are determined by the same points?

41. Of the first 10 questions on a test, a student must answer 7. On the second 5 questions, the student must answer 3. In how many ways can this be done?

42. Of the first 8 questions on a test, a student must answer 6. On the second 4 questions, the student must answer 3. In how many ways can this be done?

43. Suppose the Senate of the United States consists of 58 Democrats and 42 Republicans. How many committees made up of 6 Democrats and 4 Republicans can be formed? You need not simplify the expression.

44. Suppose the Senate of the United States consists of 63 Republicans and 37 Democrats. How many committees made up of 8 Republicans and 12 Democrats can be formed? You need not simplify the expression.

45. How many 5-card poker hands consisting of 3 aces and 2 cards that are not aces are possible with a 52-card deck? (See Section 12.6 for a description of a 52-card deck.)

46. How many 5-card poker hands consisting of 2 kings and 3 cards that are not kings are possible with a 52-card deck?

47. Bresler's Ice Cream, a national firm, sells ice cream in 33 flavors.

a) How many 3-dip cones are possible if order of flavors is to be considered and no flavor is repeated?
b) How many 3-dip cones are possible if order is to be considered and a flavor can be repeated?
c) How many 3-dip cones are possible if order is not considered and no flavor is repeated?

48. Baskin-Robbins Ice Cream, a national firm, sells ice cream in 31 flavors.

a) How many 2-dip cones are possible if order of flavors is to be considered and no flavor is repeated?
b) How many 2-dip cones are possible if order is to be considered and a flavor can be repeated?
c) How many 2-dip cones are possible if order is not considered and no flavor is repeated?

SKILL MAINTENANCE

49. Find the distance between the points $(3, -5)$ and $(-4, 2)$.

50. Find the midpoint of the segment with endpoints $(-4, 5)$ and $(2, -8)$.

SYNTHESIS

Solve for n.

51. ${}_nP_5 = 7 \cdot {}_nP_4$

52. ${}_nP_4 = 8 \cdot {}_{n-1}P_3$

53. ${}_nP_5 = 9 \cdot {}_{n-1}P_4$

54. ${}_nP_4 = 8 \cdot {}_nP_3$

55. In a single-elimination sports tournament consisting of n teams, a team is eliminated when it loses one game. How many games are required to complete the tournament?

56. In a double-elimination softball tournament consisting of n teams, a team is eliminated when it loses two games. At most, how many games are required to complete the tournament?

57. ▦ How many 5-card poker hands are possible with a 52-card deck?

58. ▦ How many 13-card bridge hands are possible with a 52-card deck?

59. There are 8 points on a circle. How many triangles can be inscribed with these points as vertices?

60. There are n points on a circle. How many quadrilaterals can be inscribed with these points as vertices?

61. A set of 5 parallel lines crosses another set of 8 parallel lines at angles that are not right angles. How many parallelograms are formed?

62. Prove: For any natural numbers n and $r \leq n$,

$$\binom{n}{r-1} + \binom{n}{r} = \binom{n+1}{r}.$$

12.5 The Binomial Theorem

Binomial Expansions Using Pascal's Triangle

Consider the following expanded powers of $(a + b)^n$, where $a + b$ is any binomial and n is a whole number. Look for patterns.

$$\begin{aligned}
(a+b)^0 &= 1\\
(a+b)^1 &= a+b\\
(a+b)^2 &= a^2+2ab+b^2\\
(a+b)^3 &= a^3+3a^2b+3ab^2+b^3\\
(a+b)^4 &= a^4+4a^3b+6a^2b^2+4ab^3+b^4\\
(a+b)^5 &= a^5+5a^4b+10a^3b^2+10a^2b^3+5ab^4+b^5
\end{aligned}$$

Each expansion is a polynomial. There are some patterns to be noted in the expansions.

1. In each term the sum of the exponents is n.
2. The exponents of a start with n and decrease to 0. The last term has no factor of a. The first term has no factor of b. The exponents of b start in the second term with 1 and increase to n, or we can think of them starting in the first term with 0 and increasing to n.
3. There is one more term than the power n. That is, there are $n + 1$ terms in the expansion of $(a + b)^n$.
4. Now we consider the coefficients. The first and last coefficients are 1, and the coefficients have a symmetry to them. They start at 1 and increase through certain values about "half"-way and then decrease through these same values back to 1. Let us explore this further.

   Suppose we wanted to find an expansion of $(a + b)^8$. If the patterns we have noticed were to continue, then we know that there are 9 terms in the expansion, which would be in the following form:

   $$a^8 + c_1a^7b + c_2a^6b^2 + c_3a^5b^3 + c_4a^4b^4 + c_3a^3b^5 + c_2a^2b^6 + c_1ab^7 + b^8.$$

How can we determine these coefficients? We can answer this question in two different ways. Your instructor may direct you regarding which to learn. The first method seems to be the easiest, but is not always. It involves writing down the coefficients in a triangular array as follows. We form what is known as **Pascal's triangle:**

$(a + b)^0$:						1					
$(a + b)^1$:					1		1				
$(a + b)^2$:				1		2		1			
$(a + b)^3$:			1		3		3		1		
$(a + b)^4$:		1		4		6		4		1	
$(a + b)^5$:	1		5		10		10		5		1

There are many patterns in the triangle. Find as many as you can.

Perhaps you discovered a way to write the next row of numbers, given the numbers in the row above it. There are always 1's on the outside. Each remaining number is found by adding the two numbers above:

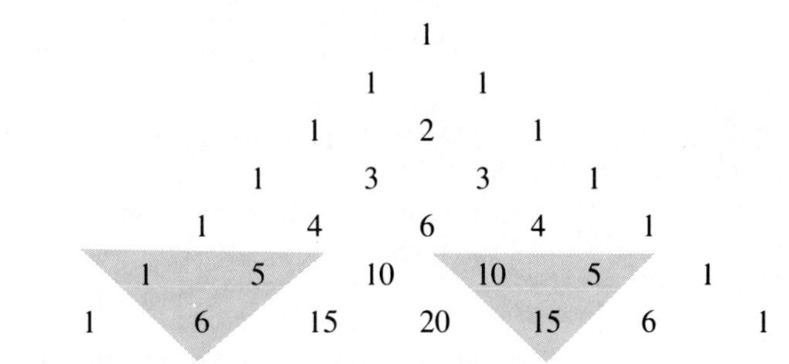

We see that in the last row

the 1st number is 1;

the 2nd number is 1 + 5, or 6;

the 3rd number is 5 + 10, or 15;

the 4th number is 10 + 10, or 20;

the 5th number is 10 + 5, or 15; and

the 6th number is 5 + 1, or 6.

Thus the expansion of $(a + b)^6$ is

$$(a + b)^6 = a^6 + 6a^5b + 15a^4b^2 + 20a^3b^3 + 15a^2b^4 + 6ab^5 + b^6.$$

To find the expansion for $(a + b)^8$, we complete two more rows of Pascal's triangle:

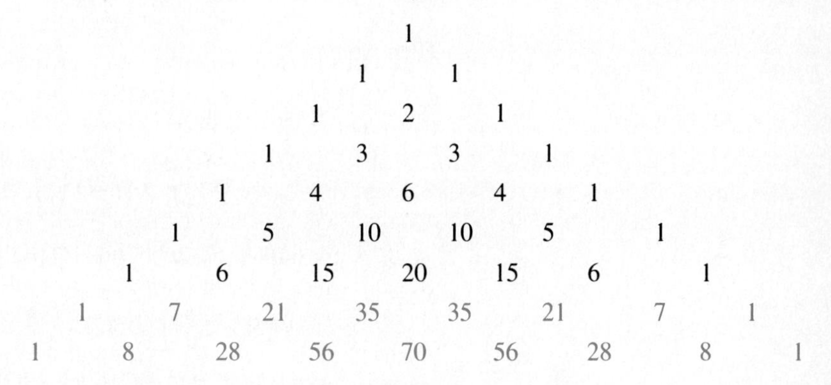

Thus the expansion of $(a+b)^8$ is

$$(a+b)^8 = a^8 + 8a^7b + 28a^6b^2 + 56a^5b^3 + 70a^4b^4 + 56a^3b^5 + 28a^2b^6 + 8ab^7 + b^8.$$

We can generalize our results as follows:

> **THE BINOMIAL THEOREM (FORM 1)**
>
> For any binomial $a+b$ and any natural number n,
>
> $$(a+b)^n = c_0a^nb^0 + c_1a^{n-1}b^1 + c_2a^{n-2}b^2 + \cdots + c_{n-1}a^1b^{n-1} + c_na^0b^n,$$
>
> where the numbers $c_0, c_1, c_2, \ldots, c_n$ are from the $(n+1)$st row of Pascal's triangle.

EXAMPLE 1 Expand: $(u-v)^5$.

Solution Note that $a = u$, $b = -v$, and $n = 5$. We use the 6th row of Pascal's triangle:

$$1 \quad 5 \quad 10 \quad 10 \quad 5 \quad 1$$

Then we have

$$\begin{aligned}(u-v)^5 &= 1(u)^5 + 5(u)^4(-v)^1 + 10(u)^3(-v)^2 \\ &\quad + 10(u)^2(-v)^3 + 5(u)(-v)^4 + 1(-v)^5 \\ &= u^5 - 5u^4v + 10u^3v^2 - 10u^2v^3 + 5uv^4 - v^5.\end{aligned}$$

Note that the signs of the terms alternate between + and −. When the power of $-v$ is odd, the sign is −. ■

EXAMPLE 2 Expand: $\left(2t + \frac{3}{t}\right)^6$.

Solution Note that $a = 2t$, $b = 3/t$, and $n = 6$. We use the 7th row of Pascal's triangle:

$$1 \quad 6 \quad 15 \quad 20 \quad 15 \quad 6 \quad 1.$$

Then we have

$$\begin{aligned}\left(2t+\frac{3}{t}\right)^6 &= (2t)^6 + 6(2t)^5\left(\frac{3}{t}\right)^1 + 15(2t)^4\left(\frac{3}{t}\right)^2 + 20(2t)^3\left(\frac{3}{t}\right)^3 \\ &\quad + 15(2t)^2\left(\frac{3}{t}\right)^4 + 6(2t)^1\left(\frac{3}{t}\right)^5 + \left(\frac{3}{t}\right)^6 \\ &= 64t^6 + 6(32t^5)\left(\frac{3}{t}\right) + 15(16t^4)\left(\frac{9}{t^2}\right) + 20(8t^3)\left(\frac{27}{t^3}\right) \\ &\quad + 15(4t^2)\left(\frac{81}{t^4}\right) + 6(2t)\left(\frac{243}{t^5}\right) + \frac{729}{t^6} \\ &= 64t^6 + 576t^4 + 2160t^2 + 4320 + 4860t^{-2} + 2916t^{-4} + 729t^{-6}.\end{aligned}$$ ■

Binomial Expansion Using Factorial Notation

The disadvantage in using Pascal's triangle is that one must compute all the preceding rows in the table to obtain the row needed for the expansion. The following method avoids this difficulty. It will also enable us to find a specific term—say the 8th term—without computing all the other terms in the expansion. This method is useful in such courses as finite mathematics and calculus and uses, for obvious reasons, the **binomial coefficient notation** $\binom{n}{r}$ developed in Section 12.4.

We can restate the binomial theorem as follows.

THE BINOMIAL THEOREM (FORM 2)

For any binomial $a + b$ and any natural number n,

$$(a+b)^n = \binom{n}{0}a^n + \binom{n}{1}a^{n-1}b + \binom{n}{2}a^{n-2}b^2 + \cdots + \binom{n}{n}b^n.$$

As justification of the binomial theorem, note that

$$(a+b)^n = \underbrace{(a+b)\cdot(a+b)\cdot(a+b)\cdot\cdots\cdot(a+b)}_{n \text{ of these}}$$

and that in carrying out the multiplication we must select either an a or b from each of the n binomials. Thus there is only one way to form a^n and that is by selecting only the a's in the binomials and avoiding the b's. The coefficient of the $a^{n-1}b$ term must be n, or $\binom{n}{1}$, since there are n ways in which to choose 1 b and $n - 1$ a's:

$$\left.\begin{array}{l}(a+b)\cdot(a+b)\cdot(a+b)\cdot\cdots\cdot(a+b)\\(a+b)\cdot(a+b)\cdot(a+b)\cdot\cdots\cdot(a+b)\\(a+b)\cdot(a+b)\cdot(a+b)\cdot\cdots\cdot(a+b)\\\cdot\\\cdot\\\cdot\\(a+b)\cdot(a+b)\cdot(a+b)\cdot\cdots\cdot(a+b).\end{array}\right\}\binom{n}{1}\text{ ways to form } a^{n-1}b$$

Similarly, the expansion of $(a + b)^n$ will contain $\binom{n}{2}$ terms that are $a^{n-2}b^2$ since there are $\binom{n}{2}$ ways of choosing 2 b's (and simultaneously $n - 2$ a's) from the n binomials. Generalizing, we see that there are $\binom{n}{r}$ ways of choosing r b's (and thus $n - r$ a's), so the coefficient of $a^{n-r}b^r$ is $\binom{n}{r}$.

EXAMPLE 3 Expand: $(3x + y)^4$.

Solution We use the binomial theorem (Form 2) with $a = 3x$, $b = y$, and $n = 4$:

$$\begin{aligned}(3x+y)^4 &= \binom{4}{0}(3x)^4 + \binom{4}{1}(3x)^3(y) + \binom{4}{2}(3x)^2(y)^2 + \binom{4}{3}(3x)(y)^3 + \binom{4}{4}(y)^4\\&= \frac{4!}{0!4!}3^4x^4 + \frac{4!}{1!3!}3^3x^3y + \frac{4!}{2!2!}3^2x^2y^2 + \frac{4!}{3!1!}3xy^3 + \frac{4!}{4!0!}y^4\\&= 81x^4 + 108x^3y + 54x^2y^2 + 12xy^3 + y^4.\end{aligned}$$

■

EXAMPLE 4 Expand: $(x^2 - 2y)^5$.

Solution Note that $a = x^2$, $b = -2y$, and $n = 5$. Then, using the binomial theorem, we have

$$\begin{aligned}(x^2 - 2y)^5 &= \binom{5}{0}(x^2)^5 + \binom{5}{1}(x^2)^4(-2y) + \binom{5}{2}(x^2)^3(-2y)^2 \\ &\quad + \binom{5}{3}(x^2)^2(-2y)^3 + \binom{5}{4}x^2(-2y)^4 + \binom{5}{5}(-2y)^5 \\ &= \frac{5!}{0!5!}x^{10} + \frac{5!}{1!4!}x^8(-2y) + \frac{5!}{2!3!}x^6(-2y)^2 + \frac{5!}{3!2!}x^4(-2y)^3 \\ &\quad + \frac{5!}{4!1!}x^2(-2y)^4 + \frac{5!}{5!0!}(-2y)^5 \\ &= x^{10} - 10x^8y + 40x^6y^2 - 80x^4y^3 + 80x^2y^4 - 32y^5.\end{aligned}$$

■

Note that in the binomial theorem (Form 2), $\binom{n}{0}a^nb^0$ gives us the first term, $\binom{n}{1}a^{n-1}b^1$ gives us the second term, $\binom{n}{2}a^{n-2}b^2$ gives us the third term, and so on. This can be generalized to give a method for finding a specific term without writing the entire expansion.

The $(r + 1)$st term of $(a + b)^n$ is

$$\binom{n}{r}a^{n-r}b^r.$$

EXAMPLE 5 Find the 5th term in the expansion of $(2x - 5y)^6$.

Solution First, we note that $5 = 4 + 1$. Thus $r = 4$, $a = 2x$, $b = -5y$, and $n = 6$. Then the 5th term of the expansion is

$$\binom{6}{4}(2x)^{6-4}(-5y)^4, \quad \text{or} \quad \frac{6!}{4!2!}(2x)^2(-5y)^4, \quad \text{or} \quad 37{,}500x^2y^4.$$

■

EXAMPLE 6 Find the 8th term in the expansion of $(3x - 2)^{10}$.

Solution First, we note that $8 = 7 + 1$. Thus, $r = 7$, $a = 3x$, $b = -2$, and $n = 10$. Then the 8th term of the expansion is

$$\binom{10}{7}(3x)^{10-7}(-2)^7, \quad \text{or} \quad \frac{10!}{7!3!}(3x)^3(-128), \quad \text{or} \quad -414{,}720x^3.$$

■

It is because of the binomial theorem that $\binom{n}{r}$ is called a *binomial coefficient.* It should now be apparent why 0! is defined to be 1. In the binomial expansion, we want $\binom{n}{0}$ to equal 1 and we also want the definition

$$\binom{n}{r} = \frac{n!}{r!(n-r)!}$$

to hold for all whole numbers n and r. Thus we must have

$$\binom{n}{0} = \frac{n!}{0!(n-0)!} = \frac{n!}{0!n!} = 1.$$

This will be satisfied if 0! is defined to be 1.

EXERCISE SET 12.5

Expand. Use both of the methods shown in this section.

1. $(m + n)^5$
2. $(a - b)^4$
3. $(x - y)^6$
4. $(p + q)^7$
5. $(x^2 - 3y)^5$
6. $(3c - d)^7$
7. $(3c - d)^6$
8. $(t^{-2} + 2)^6$
9. $(x - y)^3$
10. $(x - y)^5$
11. $\left(\frac{1}{x} + y\right)^7$
12. $(2s - 3t^2)^3$
13. $\left(a - \frac{2}{a}\right)^9$
14. $\left(2x + \frac{1}{x}\right)^9$
15. $(a^2 + b^3)^5$
16. $(x^3 + 2)^6$
17. $(\sqrt{3} - t)^4$
18. $(\sqrt{5} + t)^6$
19. $(x^{-2} + x^2)^4$
20. $\left(\frac{1}{\sqrt{x}} - \sqrt{x}\right)^6$

Find the indicated term of the binomial expression.

21. 3rd, $(a + b)^6$
22. 6th, $(x + y)^7$
23. 12th, $(a - 2)^{14}$
24. 11th, $(x - 3)^{12}$
25. 5th, $(2x^3 - \sqrt{y})^8$
26. 4th, $\left(\frac{1}{b^2} + \frac{b}{3}\right)^7$
27. Middle, $(2u - 3v^2)^{10}$
28. Middle two, $(\sqrt{x} + \sqrt{3})^5$

SKILL MAINTENANCE

29. Graph: $f(x) = \log_7 x$.
30. Express as a single logarithm:

$$\log_a 17 - \log_a 8.$$

31. Find the center and the radius of the circle. Then graph the circle.

$$x^2 + y^2 - 10x + 2y + 22 = 0$$

32. Solve using the substitution method:

$$4x - 5y = 7$$
$$x + y = 6.$$

SYNTHESIS

Expand.

33. $(\sqrt{2} - i)^4$, where $i^2 = -1$
34. $(1 + i)^6$, where $i^2 = -1$
35. ▦ At one point in a past season, Darryl Strawberry had a batting average of 0.313. Suppose he came to bat 5 times in a game. The probability of his getting exactly 3 hits is the 3rd term of the binomial expansion of $(0.313 + 0.687)^5$. Find that term and use your calculator to estimate the probability.
36. ▦ The probability that a woman will be either widowed or divorced is 85%. Suppose 8 women are interviewed. The probability that exactly 5 of them will be either widowed or divorced in their lifetime is the 6th term of the binomial expansion of $(0.15 + 0.85)^8$. Find that term and use your calculator to estimate the probability.

37. ▦ In reference to Exercise 35, the probability that Strawberry will get at most 3 hits is found by adding the last 4 terms of the binomial expansion of $(0.313 + 0.687)^5$. Find these terms and use your calculator to estimate the probability.

38. ▦ In reference to Exercise 36, the probability that at least 6 of them will be widowed or divorced is found by adding the last 3 terms of the binomial expansion of $(0.15 + 0.85)^8$. Find these terms and use your calculator to estimate the probability.

39. Find the term of

$$\left(\frac{3x^2}{2} - \frac{1}{3x}\right)^{12}$$

that does not contain x.

40. Find the middle term of $(x^2 - 6y^{3/2})^8$.

12.6 Probability

We say that when a coin is tossed, the chances that it will fall heads are 1 out of 2, or the **probability** that it will fall heads is $\frac{1}{2}$. Of course this does not mean that if a coin is tossed ten times, it will necessarily fall heads exactly five times. If the coin is tossed a great number of times, however, it will fall heads very nearly half of them.

Experimental and Theoretical Probability

If we toss a coin a great number of times, say 1000, and count the number of heads, we can determine the probability of getting a head. If there are 503 heads, we would calculate the probability of getting a head to be

$$\frac{503}{1000}, \qquad \text{or} \qquad 0.503.$$

This is an **experimental** determination of probability. Such a determination of probability is quite common. Here, for example, are some probabilities that have been determined *experimentally*:

1. If you kiss someone who has a cold, the probability of your catching a cold is 0.07.
2. A person just released from prison has an 80% probability of returning.

If we consider a coin and *reason* that it is just as likely to fall heads as tails, we would calculate the probability to be $\frac{1}{2}$. This is a **theoretical** determination of probability. Here, for example, are some probabilities that have been determined *theoretically:*

1. If there are 30 people in a room, the probability that two of them have the same birthday (excluding year of birth) is 0.706.
2. If a deck of 52 playing cards is thoroughly shuffled and a card is selected, the probability that the card is a jack is $\frac{1}{13}$, or about 0.077.

Experimental Probabilities

We first consider experimental determination of probability. The basic principle we use in computing such probabilities is as follows.

Principle P (Experimental)

An experiment is performed in which n observations are made. If a situation E, or event, occurs m times out of the n observations, then we say that the *experimental probability* of that event is given by

$$P(E) = \frac{m}{n}.$$

EXAMPLE 1 *Sociological survey.* An actual experiment was conducted to determine the number of people who are left-handed, right-handed, or both. The results are shown in the graph.

a) Determine the probability that a person is left-handed.
b) Determine the probability that a person is ambidextrous (uses both hands equally well).

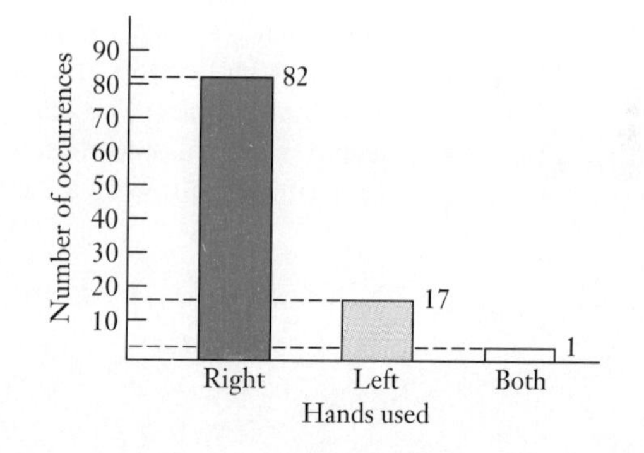

Solution

a) The number of people who were right-handed was 82, the number who were left-handed was 17, and there was 1 person who was ambidextrous. The total number of observations was 82 + 17 + 1, or 100. Thus the probability that a person is left-handed is P, where

$$P = \frac{17}{100}.$$

b) The probability that a person is ambidextrous is P, where

$$P = \frac{1}{100}.$$

■

EXAMPLE 2 *TV ratings.* The major television networks and others such as cable TV are always concerned about the percentages of homes that have TVs and are watching their programs. Because contacting every home in the country is too costly and unmanageable, a sample, or portion, of the homes are contacted. This is done by an electronic device attached to the TVs of about 1400 homes across the country. Viewing information is then fed into a computer. The following are the results of a recent survey.

Network	CBS	ABC	NBC	Other or not watching
Number of Homes Watching	258	231	206	705

What is the probability that a home was tuned to CBS during the time period considered?

Solution The probability that a home was tuned to CBS is P, where

$$P = \frac{258}{1400} \approx 0.184 = 18.4\%.$$

Theoretical Probabilities

We need some terminology before we can continue. Suppose we perform an experiment such as flipping a coin, throwing a dart, drawing a card from a deck, or checking an item off an assembly line for quality. The results of an experiment are called **outcomes.** The set of all possible outcomes is called the **sample space.** An **event** is a set of outcomes, that is, a subset of the sample space. For example, for the experiment "throwing a dart," suppose the dartboard is as shown.

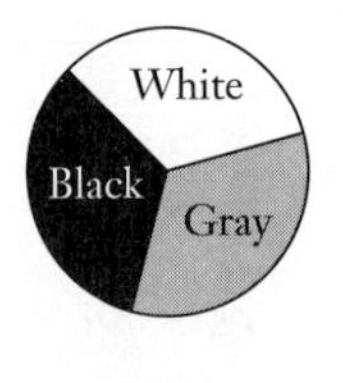

Then one event is

{black}, (the outcome is "hitting black")

which is a subset of the sample space

{black, white, gray}, (sample space)

assuming that the dart must hit the target somewhere.

We denote the probability that an event E occurs as $P(E)$. For example, "getting a head" may be denoted by H. Then $P(H)$ represents the probability of getting a head. When all the outcomes of an experiment have the same probability of occurring, we say that they are *equally likely.* To see the distinction between events that are equally likely and those that are not, consider the dartboards shown here.

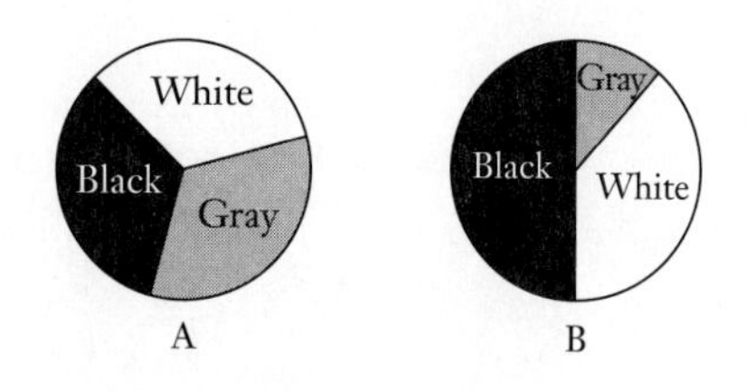

For dartboard A, the events hitting *black, white,* and *gray* are equally likely, but for board B they are not. When a sample space is comprised of equally likely events, it is easy to calculate certain probabilities.

Principle P (Theoretical)

If an event E can occur m ways out of n possible equally likely outcomes of a sample space S, then the *theoretical probability* of that event is given by

$$P(E) = \frac{m}{n}.$$

A die (pl., dice) is a cube, with six faces, each containing a number of dots from 1 to 6.

EXAMPLE 3 What is the probability of rolling a 3 on a die?

Solution On a fair die, there are 6 equally likely outcomes and there is 1 way to get a 3. By Principle P, $P(3) = \frac{1}{6}$. ■

EXAMPLE 4 What is the probability of rolling an even number on a die?

Solution The event is getting an *even* number. It can occur in 3 ways (getting 2, 4, or 6). The number of equally likely outcomes is 6. By Principle P, $P(\text{even}) = \frac{3}{6}$, or $\frac{1}{2}$. ■

EXAMPLE 5 Suppose we select, without looking, one marble from a bag containing 3 red marbles and 4 green marbles. What is the probability of selecting a red marble?

Solution There are 7 equally likely ways of selecting any marble, and since the number of ways of getting a red marble is 3,

$$P(\text{selecting a red marble}) = \tfrac{3}{7}. \qquad \blacksquare$$

We now use a number of examples related to a standard deck of 52 playing cards. Such a deck is made up as shown in the following figure.

A DECK OF 52 CARDS

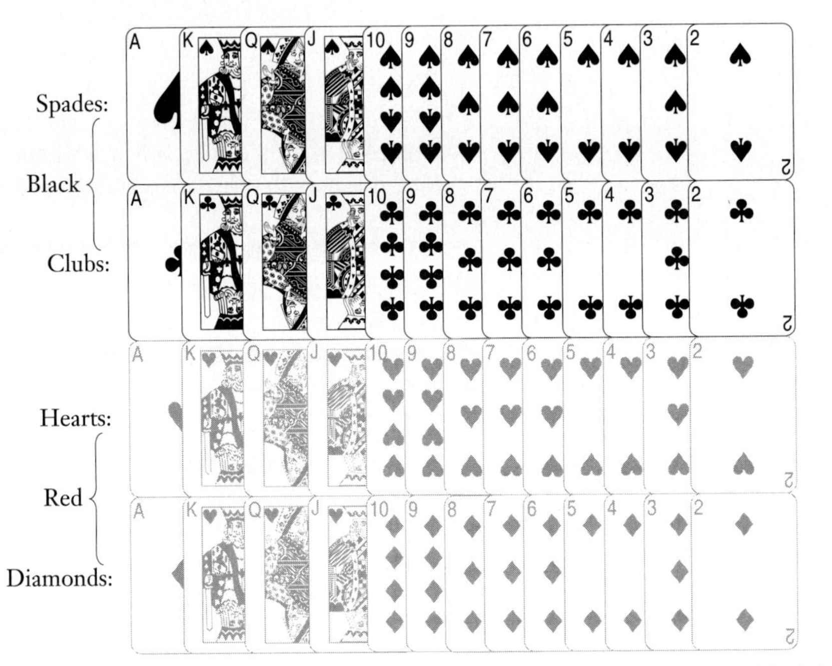

EXAMPLE 6 What is the probability of drawing an ace from a well-shuffled deck of 52 cards?

Solution Since there are 52 outcomes (cards in the deck) and they are equally likely (from a well-shuffled deck) and there are 4 ways to obtain an ace, by Principle P we have

$$P(\text{drawing an ace}) = \frac{4}{52}, \quad \text{or} \quad \frac{1}{13}. \qquad \blacksquare$$

The following are some results that follow from Principle P.

> If an event E cannot occur, then $P(E) = 0$.

For example, in coin tossing, the event that a coin would land on its edge has probability 0.

If an event E is certain to occur (that is, every trial is a success), then $P(E) = 1$.

For example, in coin tossing, the event that a coin falls either heads or tails has probability 1.

In general:

The probability that an event E will occur is a number from 0 to 1:

$$0 \leq P(E) \leq 1.$$

In the following examples we use the combinations that we studied in Section 12.4 to calculate theoretical probabilities.

EXAMPLE 7 Suppose 2 cards are drawn from a well-shuffled deck of 52 cards. What is the probability that both of them are spades?

Solution The number of ways n of drawing 2 cards from a deck of 52 is ${}_{52}C_2$. Now 13 of the 52 cards are spades, so the number of ways m of drawing 2 *spades* is ${}_{13}C_2$. Thus,

$$P(\text{getting 2 spades}) = \frac{m}{n} = \frac{{}_{13}C_2}{{}_{52}C_2} = \frac{78}{1326} = \frac{1}{17}.$$

■

EXAMPLE 8 Suppose 2 people are selected at random from a group that consists of 6 men and 4 women. What is the probability that both of them are women?

Solution The number of ways of selecting 2 people from a group of 10 is ${}_{10}C_2$. The number of ways of selecting 2 women from a group of 4 is ${}_4C_2$. Thus the probability of selecting 2 women from the group of 10 is P, where

$$P = \frac{{}_4C_2}{{}_{10}C_2} = \frac{6}{45} = \frac{2}{15}.$$

■

EXAMPLE 9 Suppose 3 people are selected at random from a group that consists of 6 men and 4 women. What is the probability that 1 man and 2 women are selected?

Solution The number of ways of selecting 3 people from a group of 10 is ${}_{10}C_3$. One man can be selected in ${}_6C_1$ ways, and 2 women can be selected in ${}_4C_2$ ways. By the fundamental counting principle, the number of ways of selecting 1 man and 2 women is ${}_6C_1 \cdot {}_4C_2$. Thus the probability is

$$P = \frac{{}_6C_1 \cdot {}_4C_2}{{}_{10}C_3}, \quad \text{or} \quad \frac{3}{10}.$$

■

EXAMPLE 10 What is the probability of getting a total of 8 on a roll of a pair of dice? (Assume that the dice are different, say one blue and one black.)

Solution On each die, there are 6 possible outcomes. The outcomes are paired so there are $6 \cdot 6$, or 36, possible ways in which the two can fall.

Blue die						
6	(1, 6)	(2, 6)	(3, 6)	(4, 6)	(5, 6)	(6, 6)
5	(1, 5)	(2, 5)	(3, 5)	(4, 5)	(5, 5)	(6, 5)
4	(1, 4)	(2, 4)	(3, 4)	(4, 4)	(5, 4)	(6, 4)
3	(1, 3)	(2, 3)	(3, 3)	(4, 3)	(5, 3)	(6, 3)
2	(1, 2)	(2, 2)	(3, 2)	(4, 2)	(5, 2)	(6, 2)
1	(1, 1)	(2, 1)	(3, 1)	(4, 1)	(5, 1)	(6, 1)
	1	2	3	4	5	6

The pairs that total 8 are as shown. Thus there are 5 possible ways of getting a total of 8, so the probability is $\frac{5}{36}$. ■

Origin and Use of Probability

A desire to calculate odds in games of chance gave rise to the theory of probability. Today the theory of probability and its closely related field, mathematical statistics, have many applications, most of them not related to games of chance. Opinion polls, with such uses as predicting elections, are a familiar example. Quality control, in which a prediction about the percentage of faulty items manufactured is made, is an important application, among many, in business. Still other applications are in the areas of genetics, medicine, and the kinetic theory of gases.

EXERCISE SET 12.6

1. In an actual survey, 100 people were polled to determine the probability of a person wearing either glasses or contact lenses. Of those polled, 57 wore either glasses or contacts. What is the probability that a person wears either glasses or contacts? What is the probability that a person wears neither?

2. In another survey 100 people were polled and asked to select a number from 1 to 5. The results are shown in the following table.

Number of Choices	1	2	3	4	5
Number of People Who Selected That Number	18	24	23	23	12

What is the probability that the number selected is 1? 2? 3? 4? 5? What general conclusion might a psychologist make from this experiment?

Linguistics. An experiment was conducted to determine the relative occurrence of various letters of the English alphabet. A paragraph from a newspaper, one from a textbook, and one from a magazine were considered. In all, there were 1044 letters. The number of occurrences of each letter of the alphabet is listed in the following table.

Letter	A	B	C	D	E	F	G	H	I	J	K	L	M
Number of Occurrences	78	22	33	33	140	24	22	63	60	2	9	35	30

Letter	N	O	P	Q	R	S	T	U	V	W	X	Y	Z
Number of Occurrences	74	74	27	4	67	67	95	31	10	22	8	13	1

Round answers to Exercises 3–6 to three decimal places.

3. What is the probability of the occurrence of the letter A? E? I? O? U?

4. What is the probability of a vowel occurring?

5. What is the probability of a consonant occurring?

6. Which letter has the least probability of occurring? What is the probability of this letter not occurring?

Suppose we draw a card from a well-shuffled deck of 52 cards.

7. How many equally likely outcomes are there?

8. What is the probability of drawing a queen?

9. What is the probability of drawing a heart?

10. What is the probability of drawing a club?

11. What is the probability of drawing a 4?

12. What is the probability of drawing a red card?

13. What is the probability of drawing a black card?

14. What is the probability of drawing an ace or a two?

15. What is the probability of drawing a 9 or a king?

Suppose we select, without looking, one marble from a bag containing 4 red marbles and 10 green marbles.

16. What is the probability of selecting a red marble?

17. What is the probability of selecting a green marble?

18. What is the probability of selecting a purple marble?

19. What is the probability of selecting a white marble?

Suppose 4 cards are drawn from a well-shuffled deck of 52 cards.

20. What is the probability that all 4 are spades?

21. What is the probability that all 4 are hearts?

22. If 4 marbles are drawn at random all at once from a bag containing 8 white marbles and 6 black marbles, what is the probability that 2 will be white and 2 will be black?

23. From a group of 8 men and 7 women, a committee of 4 is chosen. What is the probability that 2 men and 2 women will be chosen?

24. What is the probability of getting a total of 6 on a roll of a pair of dice?

25. What is the probability of getting a total of 3 on a roll of a pair of dice?

26. What is the probability of getting snake eyes (a total of 2) on a roll of a pair of dice?

27. What is the probability of getting box-cars (a total of 12) on a roll of a pair of dice?

28. From a bag containing 5 nickels, 8 dimes, and 7 quarters, 5 coins are drawn at random, all at once. What is the probability of getting 2 nickels, 2 dimes, and 1 quarter?

29. From a bag containing 6 nickels, 10 dimes, and 4 quarters, 6 coins are drawn at random, all at once. What is the probability of getting 3 nickels, 2 dimes, and 1 quarter?

Roulette. A roulette wheel contains slots numbered 00, 0, 1, 2, 3, . . . , 35, 36. Eighteen of the slots numbered 1 through 36 are colored red and eighteen are colored black. The 00 and 0 slots are uncolored. The wheel is spun, and a ball is rolled around the rim until it falls into a slot. What is the probability that the ball falls in:

30. a black slot?

31. a red slot?

32. a red or black slot?

33. the 00 slot?

34. the 0 slot?

35. either the 00 or 0 slot? (Here the house always wins.)

36. an odd-numbered slot?

SKILL MAINTENANCE

37. Convert to an exponential equation:

$$7 = \log_c 15.$$

38. Express as a single logarithm:

$$\log_a 7 + \log_a 5.$$

39. Find an equation of a circle with center $(-5, 3)$ and radius 12.

40. Find two numbers whose sum is -12 and whose difference is 6.

SYNTHESIS

Five-card poker hands and probabilities. In part (a) of each problem, give a reasoned expression as well as the answer. Read all the problems before beginning.

41. ▦ How many 5-card poker hands can be dealt from a standard 52-card deck?

42. ▦ A *royal flush* consists of a 5-card hand with A-K-Q-J-10 of the same suit.

a) How many royal flushes are there?
b) What is the probability of getting a royal flush?

43. ▦ A *straight flush* consists of 5 cards in sequence in the same suit, but excludes royal flushes. An ace can be used low, before a two.

a) How many straight flushes are there?
b) What is the probability of getting a straight flush?

44. ▦ *Four of a kind* is a 5-card hand in which 4 of the cards are of the same denomination, such as J-J-J-J-6, 7-7-7-7-A, or 2-2-2-2-5.

a) How many are there?
b) What is the probability of getting four of a kind?

45. ▦ A *full house* consists of a pair and three of a kind, such as Q-Q-Q-4-4.

a) How many are there?
b) What is the probability of getting a full house?

46. ▦ A *pair* is a 5-card hand in which just 2 of the cards are of the same denomination, such as Q-Q-8-A-3.

a) How many are there?
b) What is the probability of getting a pair?

47. ▦ *Three of a kind* is a 5-card hand in which exactly 3 of the cards are of the same denomination and the other 2 are *not* of the same denomination, such as Q-Q-Q-10-7.

a) How many are there?
b) What is the probability of getting three of a kind?

48. ▦ A *flush* is a 5-card hand in which all the cards are of the same suit, but not all in sequence (not a straight flush or royal flush).

a) How many are there?
b) What is the probability of getting a flush?

49. ▦ *Two pairs* is a hand like Q-Q-3-3-A.

a) How many are there?
b) What is the probability of getting two pairs?

50. ▦ A *straight* is any 5 cards in sequence, but not of the same suit—for example, 4 of spades, 5 of spades, 6 of diamonds, 7 of hearts, and 8 of clubs.

a) How many are there?
b) What is the probability of getting a straight?

CHAPTER SUMMARY AND REVIEW 12

TERMS TO KNOW

Sequence
Infinite sequence
Finite sequence
Series
Common difference
Geometric sequence
Common ratio
Infinite geometric series
Fundamental counting principle
Permutation
Factorial notation

Infinite series
Partial sum
Finite series
Sigma notation
Arithmetic sequence
Limit or sum of an infinite geometric series
Outcome
Event
Combination
Binomial theorem
Pascal's triangle
Probability

IMPORTANT PROPERTIES AND FORMULAS

Arithmetic sequence: $a_n = a_{n-1} + d$

nth term of an arithmetic sequence: $a_n = a_1 + (n-1)d$

Sum of the first n terms of an arithmetic sequence: $S_n = \frac{n}{2}(a_1 + a_n)$,

$$S_n = \frac{n}{2}[2a_1 + (n-1)d]$$

Geometric sequence: $a_{n+1} = a_n r$

nth term of a geometric sequence: $a_n = a_1 r^{n-1}$

Sum of the first n terms of a geometric sequence: $S_n = \frac{a_1(r^n - 1)}{r - 1}$

Sum of an infinite geometric sequence: $S_\infty = \frac{a_1}{1 - r}, |r| < 1$

FUNDAMENTAL COUNTING PRINCIPLE

Given a combined action, or event, in which the first action can be performed in n_1 ways, the second action can be performed in n_2 ways, and so on, then the total number of ways the combined action can be performed is the product

$$n_1 \cdot n_2 \cdot n_3 \cdot \cdots \cdot n_k.$$

PERMUTATIONS OF n OBJECTS TAKEN r AT A TIME

The number of permutations of a set of n objects taken r at a time, denoted ${}_nP_r$, is given by

$$ {}_nP_r = n(n-1)(n-2) \cdot \cdots \cdot [n - (r-1)] \quad (1)$$

$$ = \frac{n!}{(n-r)!}. \quad (2)$$

COMBINATIONS OF n OBJECTS TAKEN r AT A TIME

The total number of combinations of n objects taken r at a time, denoted ${}_nC_r$ is given by

$$ {}_nC_r = \frac{n!}{r!(n-r)!}, \quad (1)$$

or,

$$ {}_nC_r = \frac{{}_nP_r}{r!} = \frac{n(n-1)(n-2)\ldots[n-(r-1)]}{r!} \quad (2)$$

Binomial coefficient notation: $\binom{n}{r} = {}_nC_r$

Binomial theorem: $(a+b)^n = \binom{n}{0}a^n + \binom{n}{1}a^{n-1}b + \binom{n}{2}a^{n-2}b^2 + \cdots + \binom{n}{n}b^n$

If an event E cannot occur, then $P(E) = 0$.
If an event E is certain to occur, then $P(E) = 1$.
The probability that an event E will occur is a number from 0 to 1:

$$0 \le P(E) \le 1.$$

REVIEW EXERCISES

Review sections to be tested in addition to the material in this chapter are Sections 9.3, 9.4, 10.1, and 11.1.

1. Find the 10th term in the arithmetic sequence $\frac{3}{4}, \frac{13}{12}, \frac{17}{12}, \ldots$.

2. Find the 6th term in the arithmetic sequence $a - b, a, a + b, \ldots$.

3. Find the sum of the first 18 terms of the arithmetic sequence 4, 7, 10, ...

4. Find the sum of the first 30 positive integers.

5. The first term of an arithmetic sequence is 5. The 17th term is 53. Find the 3rd term.

6. The common difference in an arithmetic sequence is 3. The 10th term is 23. Find the first term.

7. For a geometric sequence, $a_1 = -2$, $r = 2$, and $a_n = -64$. Find n and S_n.

8. For a geometric sequence, $r = \frac{1}{2}$, $n = 5$, and $S_n = \frac{31}{2}$. Find a_1 and a_n.

9. Determine whether this geometric sequence has a sum:

$$25, 27.5, 30.25, 33.275, \ldots.$$

10. Determine whether this geometric sequence has a sum:

$$0.27, 0.0027, 0.000027, \ldots.$$

11. Find this infinite sum. The series is geometric.

$$\frac{1}{2} - \frac{1}{6} + \frac{1}{18} - \cdots$$

12. Find fractional notation for $2.\overline{13}$.

13. An auditorium has 31 seats in the first row, 33 seats in the second row, 35 seats in the third row, and so on, for 18 rows. How many seats are there in the 17th row?

14. A golf ball is dropped from a height of 30 ft to the pavement, and the rebound is one fourth of the distance that it drops. If, after each descent, it continues to rebound one fourth of the distance dropped, what is the total distance that the ball has traveled when it reaches the pavement on its 10th descent?

15. You receive 10¢ on the first day of the year, 12¢ on the 2nd day, 14¢ on the 3rd day, and so on. How much will you receive on the 365th day? What is the sum of all these 365 gifts?

16. The present population of a city is 30,000. Its population is supposed to double every 10 yr. What will its population be at the end of 80 yr?

17. The sides of a square are each 16 in. long. A second square is inscribed by joining the midpoints of the sides, successively. In the second square, we repeat the process, inscribing a third square. If this process is continued indefinitely, what is the sum of the perimeters of all of the squares? (*Hint:* Use an infinite geometric series.)

18. A pendulum is moving back and forth in such a way that it traverses an arc 10 cm in length, and thereafter arcs are $\frac{4}{7}$ the length of the previous arc. What is the sum of the arc lengths that the pendulum traverses?

19. Write Σ notation for this sequence:

$$0 + 3 + 8 + 15 + 24 + 35 + 48.$$

20. In how many different ways can 6 books be arranged on a shelf?

21. If 9 different signal flags are available, how many different displays are possible using 4 flags in a row?

22. The winner of a contest can choose any 8 of 15 prizes. How many different selections can be made?

23. The Greek alphabet contains 24 letters. How many fraternity or sorority names can be formed using 3 different letters?

24. A manufacturer of houses has one floor plan but achieves variety by having 3 different colored roofs, 4 different ways of attaching the garage, and 3 different types of entrance. Find the number of different houses that can be produced.

25. Find the 12th term of $(a + x)^{18}$. Do not multiply out the factorials.

26. Find the 4th term of $(a + x)^{12}$.

Expand.

27. $(m + n)^7$

28. $(x^2 + 3y)^4$

29. $(5i + 1)^6$, where $i^2 = -1$

30. Before an election, a poll was conducted to see which candidate was favored. Three people were running for a particular office. During the polling, 86 favored A, 97 favored B, and 23 favored C. Assuming that the poll is a valid indicator of the election, what is the probability that a person will vote for A? for B? for C?

31. What is the probability of rolling a 10 on a roll of a pair of dice? on a roll of one die?

32. From a deck of 52 cards, 1 card is drawn. What is the probability that it is a club?

33. From a deck of 52 cards, 3 are drawn at random without replacement. What is the probability that 2 are aces and 1 is a king?

SKILL MAINTENANCE

34. Find $\log_7 \dfrac{1}{49}$.

35. Solve: $\log_3 3^5 = x$.

36. Find the midpoint of the segment with endpoints $(7, -4)$ and $(-3, 5)$.

37. Solve using the elimination method:

$$7x - 2y = 9,$$
$$2x + 3y = 5.$$

SYNTHESIS

38. The zeros of this polynomial form an arithmetic sequence. Find them.

$$x^4 - 4x^3 - 4x^2 + 16x$$

39. Write the first 3 terms of the infinite geometric sequence with $S_\infty = \frac{3}{11}$ and $r = 0.01$.

40. Write the first 3 terms of the infinite geometric sequence with $r = -\frac{1}{3}$ and $S_\infty = \frac{3}{8}$.

41. Simplify:

$$\sum_{r=0}^{10} (-1)^r \binom{10}{r} (\log x)^{10-r} (\log y)^r.$$

Solve for n.

42. $\binom{n}{n-1} = 36$

43. $26 \cdot \binom{n}{1} = \binom{n}{3}$

THINKING IT THROUGH

1. Explain the relationship between a sequence and a series.

2. Compare the kinds of problems for which permutations would be used with those problems for which combinations would be used.

3. Explain the use of permutations and combinations in probability.

4. Give some examples of situations in which experimental probability would be more practical to use than theoretical probability.

CHAPTER TEST 12

1. Find the 18th term of the arithmetic sequence $\frac{1}{4}$, 1, $\frac{7}{4}$, $\frac{5}{2}$, ...

2. The 2nd term of an arithmetic sequence is 9, and the 9th term is 37. Find the common difference.

3. Which term of the arithmetic sequence 1, $\frac{3}{2}$, 2, $\frac{5}{2}$, . . . is $\frac{31}{2}$?

4. Find the 8th term of the geometric sequence

$$0.2,\ 0.6,\ 1.8,\ \ldots.$$

5. Find the sum

$$\sum_{k=1}^{5} \left(\frac{1}{2}\right)^{k+1}.$$

6. Which of the following infinite geometric sequences have sums?

a) 2, 0.2, 0.02, 0.002, . . .
b) 3, −6, 12, −24, 48, . . .
c) $\frac{1}{20}$, $\frac{1}{10}$, $\frac{1}{5}$, $\frac{2}{5}$, . . .

7. Find the sum of the infinite geometric sequence

$$25,\ -5,\ 1,\ -\tfrac{1}{5},\ \ldots.$$

8. A student made deposits in a savings account as follows: \$20.50 the first month, \$26 the second month, \$31.50 the third month, and so on, for 2 years. What was the sum of the deposits?

9. A publishing company prints only $\frac{3}{5}$ as many books with each new printing of a book. If 100,000 copies of a book are printed originally, how many will be printed in the 5th printing?

10. Find fractional notation for $0.12\overline{88}$.

11. Write $\sum$ notation for this sequence.

$$0 + 2 + 6 + 12 + 20 + 30 + 42$$

12. In how many different ways can 6 people be seated in a row?

13. On a test, a student must answer 4 out of 7 questions. In how many ways can this be done?

14. From a group of 20 seniors and 14 juniors, how many committees consisting of 3 seniors and 2 juniors are possible?

15. Ben and Jerry's Homemade Inc., an international firm, sells ice cream in 20 flavors. How many 3-dip cones are possible if order of flavors is to be considered and no flavor is repeated?

16. Expand $(3a + 2b)^5$.

17. Find the 3rd term of $(2a + b)^7$.

18. Expand $(x - \sqrt{2})^5$.

19. What is the probability of getting a total of 7 on a roll of a pair of dice?

20. From a deck of 52 cards, 1 card is drawn. What is the probability of drawing a 3 or a queen?

21. If 3 marbles are drawn at random all at once from a bag containing 5 green marbles, 7 red marbles, and 4 white marbles, what is the probability that 2 will be green and 1 will be white?

SKILL MAINTENANCE

22. Convert to a logarithmic equation:

$$5^4 = 625.$$

23. Express as a single logarithm:

$$\log_b 5 - \log_b 7.$$

24. Find two numbers whose sum is -15 and whose difference is 7.

25. Find the distance between the points $(-4, 1)$ and $(2, -5)$.

SYNTHESIS

26. Find 4 numbers in an arithmetic sequence such that twice the second minus the fourth is 1 and the sum of the first and third is 14.

27. How many diagonals does a dodecagon (12-sided polygon) have?

28. Solve for n:

$$\binom{n}{6} = 3 \cdot \binom{n-1}{5}.$$

29. Solve for a:

$$\sum_{r=0}^{5} 9^{5-r} a^r = 0.$$

CUMULATIVE REVIEW 1-12

Simplify.

1. $(-9x^2y^3)(5x^4y^{-7})$

2. $|-3.5 + 9.8|$

3. $2y - [3 - 4(5 - 2y) - 3y]$

4. $(10 \cdot 8 - 9 \cdot 7)^2 - 54 \div 9 - 3$

5. Evaluate $\dfrac{ab - ac}{bc}$ when $a = -2$, $b = 3$, and $c = -4$.

Perform the indicated operations and simplify.

6. $(5a^2 - 3ab - 7b^2) - (2a^2 + 5ab + 8b^2)$

7. $(-3x^2 + 4x^3 - 5x - 1) + (9x^3 - 4x^2 + 7 - x)$

8. $(2a - 1)(3a + 5)$

9. $(3a^2 - 5y)^2$

10. $\dfrac{1}{x - 2} - \dfrac{4}{x^2 - 4} + \dfrac{3}{x + 2}$

11. $\dfrac{x^2 - 6x + 8}{3x + 9} \cdot \dfrac{x + 3}{x^2 - 4}$

12. $\dfrac{3x + 3y}{5x - 5y} \div \dfrac{3x^2 + 3y^2}{5x^3 - 5y^3}$

13. $\dfrac{x - \dfrac{a^2}{x}}{1 + \dfrac{a}{x}}$

Factor.

14. $4x^2 - 12x + 9$

15. $27a^3 - 8$

16. $a^3 + 3a^2 - ab^2 - 3b^2$

17. $15y^4 + 33y^2 - 36$

18. For the function described by

$$f(x) = 3x^2 - 4x,$$

find $f(-2)$.

19. Divide:

$$(7x^4 - 5x^3 + x^2 - 4) \div (x - 2).$$

Solve.

20. $9(x - 1) - 3(x - 2) = 1$

21. $x^2 - 2x = 48$

22. $\frac{6}{x} + \frac{6}{x + 2} = \frac{5}{2}$

23. $\frac{7x}{x - 3} - \frac{21}{x} + 11 = \frac{63}{x^2 - 3x}$

24. $5x + 3y = 2,$
$3x + 5y = -2$

25. $x + y - z = 0,$
$3x + y + z = 6,$
$x - y + 2z = 5$

26. $\sqrt{x - 5} = 5 - \sqrt{x}$

27. $x^4 - 29x^2 + 100 = 0$

28. $x^2 + y^2 = 8,$
$x^2 - y^2 = 2$

29. 🖩 $5^x = 8$

30. $\log(x^2 - 25) - \log(x + 5) = 3$

31. $\log_4 x = -2$

32. $7^{2x+3} = 49$

33. $|2x - 1| \leq 5$

34. $7x^2 + 14 = 0$

35. $x^2 + 4x = 3$

36. $|2y + 3| > 7$

37. $2x^2 + x < 1$

38. $\frac{x - 1}{x + 2} \geq 0$

Solve.

39. You are taking a test in which items of type A are worth 10 points and items of type B are worth 12 points. It takes 4 minutes to complete each item of type A and 5 minutes to complete each item of type B. The total time allowed is 60 minutes, and you are not allowed to answer more than 14 questions. Assuming that all of your answers are correct, how many items of each type should you answer to get the best score?

40. A telephone company charges \$0.40 for the first minute and \$0.25 for every other minute of a long-distance call placed before 5 P.M. The rates after 5 P.M. drop to \$0.30 for the first minute and \$0.20 for every other minute of the call. A certain call placed before 5 P.M. costs \$4.20. How much would a call of the same duration placed after 5 P.M. cost?

41. Find three consecutive integers whose sum is 198.

42. A pentagon with all five sides congruent has a perimeter equal to that of an octagon with all eight sides congruent. One side of the pentagon is two less than three times one side of the octagon. What is the perimeter of each figure?

43. A chemist has two solutions of ammonia and water. Solution A is 6% ammonia. Solution B is 2% ammonia. How many liters of each solution are needed to obtain 80 L of a solution that is 3.2% ammonia?

44. An airplane can fly 190 mi with the wind in the same time it takes to fly 160 mi against the wind. The speed of the wind is 30 mph. How fast can the plane fly in still air?

45. Person A can do a certain job in 21 min. Person B can do the same job in 14 min. How long would it take to do the job if the two worked together?

46. The centripetal force F of an object moving in a circle varies directly as the square of the velocity v and inversely as the radius r of the circle. If $F = 8$ when $v = 1$ and $r = 10$, what is F when $v = 2$ and $r = 16$?

47. A farmer wants to fence in a rectangular area next to a river. (Note that no fence will be needed along the river.) What is the area of the largest region that can be fenced in with 100 ft of fencing?

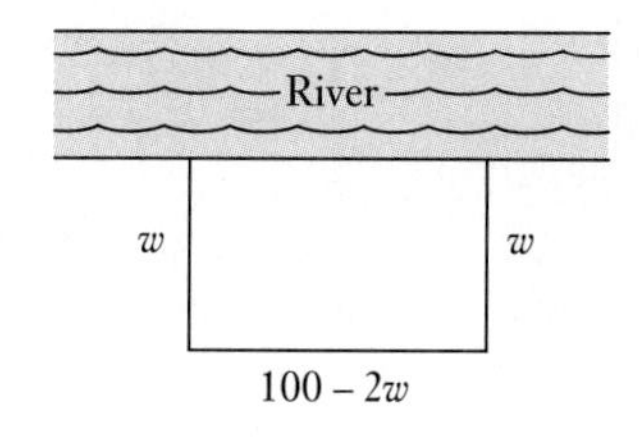

Graph.

48. $3x - y = 6$

49. $x^2 + 4x + y^2 - 2y + 3 = 0$

50. $\dfrac{x^2}{25} + \dfrac{y^2}{4} = 1$

51. $y = \log_2 x$

52. $f(x) = \dfrac{x - 1}{x^2 - x - 2}$

53. $2x - 3y < -6$

54. Graph: $f(x) = -2(x - 3)^2 + 1$.

a) Label the vertex.
b) Draw the line of symmetry.
c) Find the maximum or minimum value.

55. Solve $V = P - Prt$ for r.

56. Find an equation of the line containing the points $(1, 5)$ and $(-1, 3)$.

57. Find an equation of the line containing the point $(-1, 4)$ and perpendicular to the line whose equation is $3x - y = 6$.

Evaluate.

58. $\begin{vmatrix} -5 & -7 \\ 4 & 6 \end{vmatrix}$

59. $\begin{vmatrix} 2 & -1 & 1 \\ 1 & 2 & 0 \\ 3 & -1 & 1 \end{vmatrix}$

60. List all possible rational roots of $2x^3 - x^2 - 5x - 2$.

61. Factor: $P(x) = 2x^3 - x^2 - 5x - 2$.

62. Add and simplify: $2\sqrt{12} + 4\sqrt{27}$.

63. Multiply and simplify: $\sqrt{8x}\,\sqrt{8x^3y}$.

64. Simplify: $(25x^{4/3}y^{1/2})^{3/2}$.

65. Divide and simplify:

$$\frac{\sqrt[3]{15x}}{\sqrt[3]{3y^2}}.$$

66. Rationalize the denominator:

$$\frac{1 - \sqrt{x}}{1 + \sqrt{x}}.$$

67. Write a single radical expression:

$$\frac{\sqrt[3]{(x + 1)^5}}{\sqrt{(x + 1)^3}}.$$

68. Multiply these complex numbers:

$$(3 + 2i)(4 - 7i).$$

69. Write a quadratic equation whose solutions are $5\sqrt{2}$ and $-5\sqrt{2}$.

70. Find the center and the radius of the circle

$$x^2 + y^2 - 4x + 6y - 23 = 0.$$

71. Express as a single logarithm:

$$\tfrac{2}{3}\log_a x - \tfrac{1}{2}\log_a y + 5\log_a z.$$

72. Convert to an exponential equation:

$$\log_a c = 5.$$

Find each of the following using a calculator.

73. log 5677.2

74. antilog (−4.8904)

75. ln 5677.2

76. antilog_e (−4.8904)

Population growth. The Virgin Islands has one of the highest exponential growth rates in the world, 9.6%. In 1970 the population was 75,150.

77. Write an exponential equation describing the growth of the population of the Virgin Islands.

78. Predict the population for 1995.

79. Find the 14th term of the arithmetic sequence 3, 9, 15, ...

80. Find the 21st term of the arithmetic sequence 19, 12, 5, ...

81. Find the sum of the first 25 terms of the arithmetic series $-1 + 2 + 5 + \cdots$.

82. Find the general term of the geometric sequence 16, 4, 1, ...

83. Find the 15th term of the geometric sequence 0.0025, 0.005, 0.01, ...

84. Find the sum of the first 9 terms of the geometric series $x + 1.5x + 2.25x + \cdots$.

85. On Mark's 9th birthday, his grandmother opened a savings account for him with $100. The account draws 6% interest compounded annually. If Mark neither deposits nor withdraws any money from the bank, how much will be in the account on his 18th birthday?

Evaluate.

86. ${}_{10}P_4$

87. $\binom{10}{4}$

88. What is the 6th term in the binomial expansion of $(2x + y)^9$?

89. What is the probability of drawing a heart from a well-shuffled deck of 52 cards?

SYNTHESIS

Solve.

90. $\frac{9}{x} - \frac{9}{x + 12} = \frac{108}{x^2 + 12x}$

91. $\log_2 (\log_3 x) = 2$

92. y varies directly as the cube of x and x is multiplied by 0.5. What is the effect on y?

93. Divide these complex numbers:

$$\frac{2\sqrt{6} + 4\sqrt{5}i}{2\sqrt{6} - 4\sqrt{5}i}.$$

94. Diaphantos, a famous mathematician, spent $\frac{1}{6}$ of his life as a child, $\frac{1}{12}$ as a young man, and $\frac{1}{7}$ as a bachelor. Five years after he was married, he had a son who died 4 years before his father at half his father's final age. How long did Diaphantos live?

Tables

TABLE 1 Powers, Roots and Reciprocals

n	n^2	n^3	$\sqrt{n}$	$\sqrt[3]{n}$	$\sqrt{10n}$	$\frac{1}{n}$	n	n^2	n^3	$\sqrt{n}$	$\sqrt[3]{n}$	$\sqrt{10n}$	$\frac{1}{n}$
1	1	1	1.000	1.000	3.162	1.0000	**51**	2,601	132,651	7.141	3.708	22.583	.0196
2	4	8	1.414	1.260	4.472	.5000	**52**	2,704	140,608	7.211	3.733	22.804	.0192
3	9	27	1.732	1.442	5.477	.3333	**53**	2,809	148,877	7.280	3.756	23.022	.0189
4	16	64	2.000	1.587	6.325	.2500	**54**	2,916	157,464	7.348	3.780	23.238	.0185
5	25	125	2.236	1.710	7.071	.2000	**55**	3,025	166,375	7.416	3.803	23.452	.0182
6	36	216	2.449	1.817	7.746	.1667	**56**	3,136	175,616	7.483	3.826	23.664	.0179
7	49	343	2.646	1.913	8.367	.1429	**57**	3,249	185,193	7.550	3.849	23.875	.0175
8	64	512	2.828	2.000	8.944	.1250	**58**	3,364	195,112	7.616	3.871	24.083	.0172
9	81	729	3.000	2.080	9.487	.1111	**59**	3,481	205,379	7.681	3.893	24.290	.0169
10	100	1,000	3.162	2.154	10.000	.1000	**60**	3,600	216,000	7.746	3.915	24.495	.0167
11	121	1,331	3.317	2.224	10.488	.0909	**61**	3,721	226,981	7.810	3.936	24.698	.0164
12	144	1,728	3.464	2.289	10.954	.0833	**62**	3,844	238,328	7.874	3.958	24.900	.0161
13	169	2,197	3.606	2.351	11.402	.0769	**63**	3,969	250,047	7.937	3.979	25.100	.0159
14	196	2,744	3.742	2.410	11.832	.0714	**64**	4,096	262,144	8.000	4.000	25.298	.0156
15	225	3,375	3.873	2.466	12.247	.0667	**65**	4,225	274,625	8.062	4.021	25.495	.0154
16	256	4,096	4.000	2.520	12.648	.0625	**66**	4,356	287,496	8.124	4.041	25.690	.0152
17	289	4,913	4.123	2.571	13.038	.0588	**67**	4,489	300,763	8.185	4.062	25.884	.0149
18	324	5,832	4.243	2.621	13.416	.0556	**68**	4,624	314,432	8.246	4.082	26.077	.0147
19	361	6,859	4.359	2.668	13.784	.0526	**69**	4,761	328,509	8.307	4.102	26.268	.0145
20	400	8,000	4.472	2.714	14.142	.0500	**70**	4,900	343,000	8.367	4.121	26.458	.0143
21	441	9,261	4.583	2.759	14.491	.0476	**71**	5,041	357,911	8.426	4.141	26.646	.0141
22	484	10,648	4.690	2.802	14.832	.0455	**72**	5,184	373,248	8.485	4.160	26.833	.0139
23	529	12,167	4.796	2.844	15.166	.0435	**73**	5,329	389,017	8.544	4.179	27.019	.0137
24	576	13,824	4.899	2.884	15.492	.0417	**74**	5,476	405,224	8.602	4.198	27.203	.0135
25	625	15,625	5.000	2.924	15.811	.0400	**75**	5,625	421,875	8.660	4.217	27.386	.0133
26	676	17,576	5.099	2.962	16.125	.0385	**76**	5,776	438,976	8.718	4.236	27.568	.0132
27	729	19,683	5.196	3.000	16.432	.0370	**77**	5,929	456,533	8.775	4.254	27.749	.0130
28	784	21,952	5.292	3.037	16.733	.0357	**78**	6,084	474,552	8.832	4.273	27.928	.0128
29	841	24,389	5.385	3.072	17.029	.0345	**79**	6,241	493,039	8.888	4.291	28.107	.0127
30	900	27,000	5.477	3.107	17.321	.0333	**80**	6,400	512,000	8.944	4.309	28.284	.0125
31	961	29,791	5.568	3.141	17.607	.0323	**81**	6,561	531,441	9.000	4.327	28.460	.0123
32	1,024	32,768	5.657	3.175	17.889	.0312	**82**	6,724	551,368	9.055	4.344	28.636	.0122
33	1,089	35,937	5.745	3.208	18.166	.0303	**83**	6,889	571,787	9.110	4.362	28.810	.0120
34	1,156	39,304	5.831	3.240	18.439	.0294	**84**	7,056	592,704	9.165	4.380	28.983	.0119
35	1,225	42,875	5.916	3.271	18.708	.0286	**85**	7,225	614,125	9.220	4.397	29.155	.0118
36	1,296	46,656	6.000	3.302	18.974	.0278	**86**	7,396	636,056	9.274	4.414	29.326	.0116
37	1,369	50,653	6.083	3.332	19.235	.0270	**87**	7,569	658,503	9.327	4.431	29.496	.0115
38	1,444	54,872	6.164	3.362	19.494	.0263	**88**	7,744	681,472	9.381	4.448	29.665	.0114
39	1,521	59,319	6.245	3.391	19.748	.0256	**89**	7,921	704,969	9.434	4.465	29.833	.0112
40	1,600	64,000	6.325	3.420	20.000	.0250	**90**	8,100	729,000	9.487	4.481	30.000	.0111
41	1,681	68,921	6.403	3.448	20.248	.0244	**91**	8,281	753,571	9.539	4.498	30.166	.0110
42	1,764	74,088	6.481	3.476	20.494	.0238	**92**	8,464	778,688	9.592	4.514	30.332	.0109
43	1,849	79,507	6.557	3.503	20.736	.0233	**93**	8,649	804,357	9.644	4.531	30.496	.0108
44	1,936	85,184	6.633	3.530	20.976	.0227	**94**	8,836	830,584	9.695	4.547	30.659	.0106
45	2,025	91,125	6.708	3.557	21.213	.0222	**95**	9,025	857,375	9.747	4.563	30.822	.0105
46	2,116	97,336	6.782	3.583	21.448	.0217	**96**	9,216	884,736	9.798	4.579	30.984	.0104
47	2,209	103,823	6.856	3.609	21.679	.0213	**97**	9,409	912,673	9.849	4.595	31.145	.0103
48	2,304	110,592	6.928	3.634	21.909	.0208	**98**	9,604	941,192	9.899	4.610	31.305	.0102
49	2,401	117,649	7.000	3.659	22.136	.0204	**99**	9,801	970,299	9.950	4.626	31.464	.0101
50	2,500	125,000	7.071	3.684	22.361	.0200	**100**	10,000	1,000,000	10.000	4.642	31.623	.0100

TABLE 2 Common Logarithms

x	0	1	2	3	4	5	6	7	8	9
1.0	.0000	.0043	.0086	.0128	.0170	.0212	.0253	.0294	.0334	.0374
1.1	.0414	.0453	.0492	.0531	.0569	.0607	.0645	.0682	.0719	.0755
1.2	.0792	.0828	.0864	.0899	.0934	.0969	.1004	.1038	.1072	.1106
1.3	.1139	.1173	.1206	.1239	.1271	.1303	.1335	.1367	.1399	.1430
1.4	.1461	.1492	.1523	.1553	.1584	.1614	.1644	.1673	.1703	.1732
1.5	.1761	.1790	.1818	.1847	.1875	.1903	.1931	.1959	.1987	.2014
1.6	.2041	.2068	.2095	.2122	.2148	.2175	.2201	.2227	.2253	.2279
1.7	.2304	.2330	.2355	.2380	.2405	.2430	.2455	.2480	.2504	.2529
1.8	.2553	.2577	.2601	.2625	.2648	.2672	.2695	.2718	.2742	.2765
1.9	.2788	.2810	.2833	.2856	.2878	.2900	.2923	.2945	.2967	.2989
2.0	.3010	.3032	.3054	.3075	.3096	.3118	.3139	.3160	.3181	.3201
2.1	.3222	.3243	.3263	.3284	.3304	.3324	.3345	.3365	.3385	.3404
2.2	.3424	.3444	.3464	.3483	.3502	.3522	.3541	.3560	.3579	.3598
2.3	.3617	.3636	.3655	.3674	.3692	.3711	.3729	.3747	.3766	.3784
2.4	.3802	.3820	.3838	.3856	.3874	.3892	.3909	.3927	.3945	.3962
2.5	.3979	.3997	.4014	.4031	.4048	.4065	.4082	.4099	.4116	.4133
2.6	.4150	.4166	.4183	.4200	.4216	.4232	.4249	.4265	.4281	.4298
2.7	.4314	.4330	.4346	.4362	.4378	.4393	.4409	.4425	.4440	.4456
2.8	.4472	.4487	.4502	.4518	.4533	.4548	.4564	.4579	.4594	.4609
2.9	.4624	.4639	.4654	.4669	.4683	.4698	.4713	.4728	.4742	.4757
3.0	.4771	.4786	.4800	.4814	.4829	.4843	.4857	.4871	.4886	.4900
3.1	.4914	.4928	.4942	.4955	.4969	.4983	.4997	.5011	.5024	.5038
3.2	.5051	.5065	.5079	.5092	.5105	.5119	.5132	.5145	.5159	.5172
3.3	.5185	.5198	.5211	.5224	.5237	.5250	.5263	.5276	.5289	.5307
3.4	.5315	.5328	.5340	.5353	.5366	.5378	.5391	.5403	.5416	.5428
3.5	.5441	.5453	.5465	.5478	.5490	.5502	.5514	.5527	.5539	.5551
3.6	.5563	.5575	.5587	.5599	.5611	.5623	.5635	.5647	.5658	.5670
3.7	.5682	.5694	.5705	.5717	.5729	.5740	.5752	.5763	.5775	.5786
3.8	.5798	.5809	.5821	.5832	.5843	.5855	.5866	.5877	.5888	.5899
3.9	.5911	.5922	.5933	.5944	.5955	.5966	.5977	.5988	.5999	.6010
4.0	.6021	.6031	.6042	.6053	.6064	.6075	.6085	.6096	.6107	.6117
4.1	.6128	.6138	.6149	.6160	.6170	.6180	.6191	.6201	.6212	.6222
4.2	.6232	.6243	.6253	.6263	.6274	.6284	.6294	.6304	.6314	.6325
4.3	.6335	.6345	.6355	.6365	.6375	.6385	.6395	.6405	.6415	.6425
4.4	.6435	.6444	.6454	.6464	.6474	.6484	.6493	.6503	.6513	.6522
4.5	.6532	.6542	.6551	.6561	.6571	.6580	.6590	.6599	.6609	.6618
4.6	.6628	.6637	.6646	.6656	.6665	.6675	.6684	.6693	.6702	.6712
4.7	.6721	.6730	.6739	.6749	.6758	.6767	.6776	.6785	.6794	.6803
4.8	.6812	.6821	.6830	.6839	.6848	.6857	.6866	.6875	.6884	.6893
4.9	.6902	.6911	.6920	.6928	.6937	.6946	.6955	.6964	.6972	.6981
5.0	.6990	.6998	.7007	.7016	.7024	.7033	.7042	.7050	.7059	.7067
5.1	.7076	.7084	.7093	.7101	.7110	.7118	.7126	.7135	.7143	.7152
5.2	.7160	.7168	.7177	.7185	.7193	.7202	.7210	.7218	.7226	.7235
5.3	.7243	.7251	.7259	.7267	.7275	.7284	.7292	.7300	.7308	.7316
5.4	.7324	.7332	.7340	.7348	.7356	.7364	.7372	.7380	.7388	.7396
x	**0**	**1**	**2**	**3**	**4**	**5**	**6**	**7**	**8**	**9**

TABLE 2 (continued)

x	0	1	2	3	4	5	6	7	8	9
5.5	.7404	.7412	.7419	.7427	.7435	.7443	.7451	.7459	.7466	.7474
5.6	.7482	.7490	.7497	.7505	.7513	.7520	.7528	.7536	.7543	.7551
5.7	.7559	.7566	.7574	.7582	.7589	.7597	.7604	.7612	.7619	.7627
5.8	.7634	.7642	.7649	.7657	.7664	.7672	.7679	.7686	.7694	.7701
5.9	.7709	.7716	.7723	.7731	.7738	.7745	.7752	.7760	.7767	.7774
6.0	.7782	.7789	.7796	.7803	.7810	.7818	.7825	.7832	.7839	.7846
6.1	.7853	.7860	.7868	.7875	.7882	.7889	.7896	.7903	.7910	.7917
6.2	.7924	.7931	.7938	.7945	.7952	.7959	.7966	.7973	.7980	.7987
6.3	.7993	.8000	.8007	.8014	.8021	.8028	.8035	.8041	.8048	.8055
6.4	.8062	.8069	.8075	.8082	.8089	.8096	.8102	.8109	.8116	.8122
6.5	.8129	.8136	.8142	.8149	.8156	.8162	.8169	.8176	.8182	.8189
6.6	.8195	.8202	.8209	.8215	.8222	.8228	.8235	.8241	.8248	.8254
6.7	.8261	.8267	.8274	.8280	.8287	.8293	.8299	.8306	.8312	.8319
6.8	.8325	.8331	.8338	.8344	.8351	.8357	.8363	.8370	.8376	.8382
6.9	.8388	.8395	.8401	.8407	.8414	.8420	.8426	.8432	.8439	.8445
7.0	.8451	.8457	.8463	.8470	.8476	.8482	.8488	.8494	.8500	.8506
7.1	.8513	.8519	.8525	.8531	.8537	.8543	.8549	.8555	.8561	.8567
7.2	.8573	.8579	.8585	.8591	.8597	.8603	.8609	.8615	.8621	.8627
7.3	.8633	.8639	.8645	.8651	.8657	.8663	.8669	.8675	.8681	.8686
7.4	.8692	.8698	.8704	.8710	.8716	.8722	.8727	.8733	.8739	.8745
7.5	.8751	.8756	.8762	.8768	.8774	.8779	.8785	.8791	.8797	.8802
7.6	.8808	.8814	.8820	.8825	.8831	.8837	.8842	.8848	.8854	.8859
7.7	.8865	.8871	.8876	.8882	.8887	.8893	.8899	.8904	.8910	.8915
7.8	.8921	.8927	.8932	.8938	.8943	.8949	.8954	.8960	.8965	.8971
7.9	.8976	.8982	.8987	.8993	.8998	.9004	.9009	.9015	.9020	.9025
8.0	.9031	.9036	.9042	.9047	.9053	.9058	.9063	.9069	.9074	.9079
8.1	.9085	.9090	.9096	.9101	.9106	.9112	.9117	.9122	.9128	.9133
8.2	.9138	.9143	.9149	.9154	.9159	.9165	.9170	.9175	.9180	.9186
8.3	.9191	.9196	.9201	.9206	.9212	.9217	.9222	.9227	.9232	.9238
8.4	.9243	.9248	.9253	.9258	.9263	.9269	.9274	.9279	.9284	.9289
8.5	.9294	.9299	.9304	.9309	.9315	.9320	.9325	.9330	.9335	.9340
8.6	.9345	.9350	.9355	.9360	.9365	.9370	.9375	.9380	.9385	.9390
8.7	.9395	.9400	.9405	.9410	.9415	.9420	.9425	.9430	.9435	.9440
8.8	.9445	.9450	.9455	.9460	.9465	.9469	.9474	.9479	.9484	.9489
8.9	.9494	.9499	.9504	.9509	.9513	.9518	.9523	.9528	.9533	.9538
9.0	.9542	.9547	.9552	.9557	.9562	.9566	.9571	.9576	.9581	.9586
9.1	.9590	.9595	.9600	.9605	.9609	.9614	.9619	.9624	.9628	.9633
9.2	.9638	.9643	.9647	.9652	.9657	.9661	.9666	.9671	.9675	.9680
9.3	.9685	.9689	.9694	.9699	.9703	.9708	.9713	.9717	.9722	.9727
9.4	.9731	.9736	.9741	.9745	.9750	.9754	.9759	.9763	.9768	.9773
9.5	.9777	.9782	.9786	.9791	.9795	.9800	.9805	.9809	.9814	.9818
9.6	.9823	.9827	.9832	.9836	.9841	.9845	.9850	.9854	.9859	.9863
9.7	.9868	.9872	.9877	.9881	.9886	.9890	.9894	.9899	.9903	.9908
9.8	.9912	.9917	.9921	.9926	.9930	.9934	.9939	.9943	.9948	.9952
9.9	.9956	.9961	.9965	.9969	.9974	.9978	.9983	.9987	.9991	.9996
x	0	1	2	3	4	5	6	7	8	9

TABLE 3 Geometric Formulas

Plane Geometry:

Rectangle
Area: $A = lw$
Perimeter: $P = 2l + 2w$

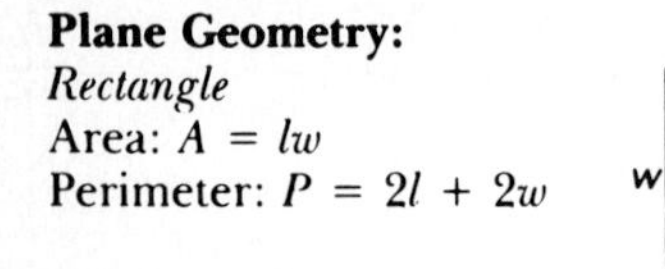

Square
Area: $A = s^2$
Perimeter: $P = 4s$

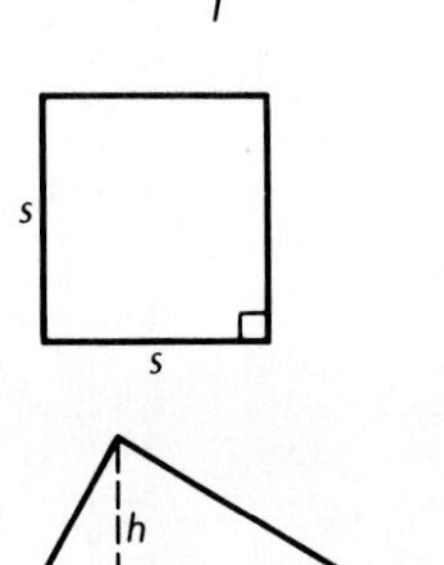

Triangle
Area: $A = \frac{1}{2}bh$

Sum of Angle Measures:
$A + B + C = 180°$

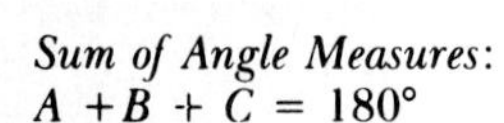

Right Triangle
Pythagorean Theorem (Equation):
$a^2 + b^2 = c^2$

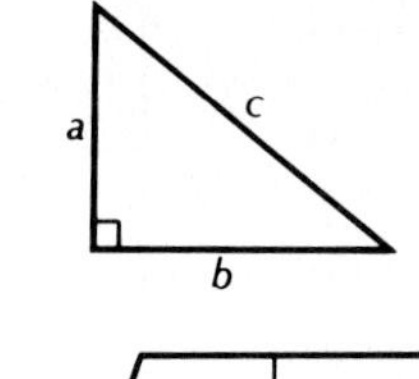

Parallelogram
Area: $A = bh$

Trapezoid
Area: $A = \frac{1}{2}h(a + b)$

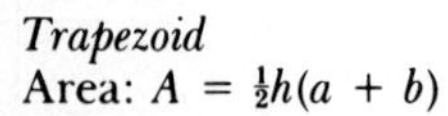

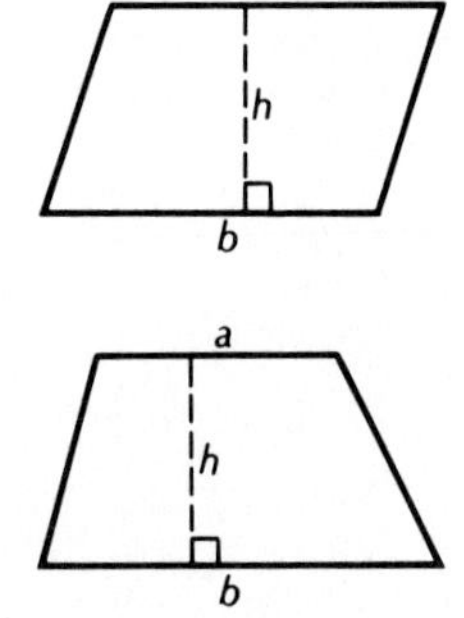

Circle
Area: $A = \pi r^2$
Circumference:
$C = \pi D = 2\pi r$
($\frac{22}{7}$ and 3.14 are different approximations for π)

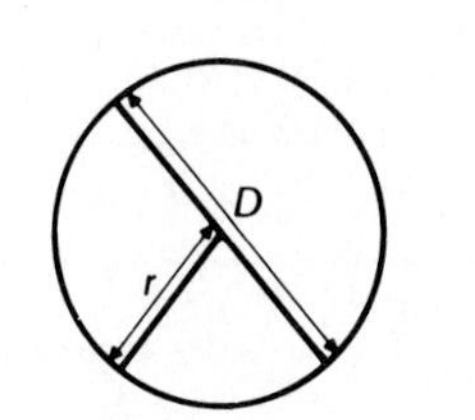

Solid Geometry:

Rectangular Solid
Volume: $V = lwh$

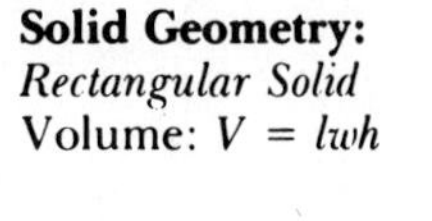

Cube
Volume: $V = s^3$

Right Circular Cylinder
Volume: $V = \pi r^2 h$
Lateral Surface Area:
$L = 2\pi rh$
Total Surface Area:
$S = 2\pi rh + 2\pi r^2$

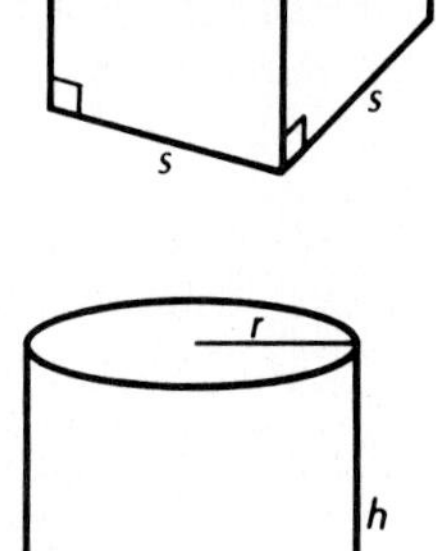

Right Circular Cone
Volume: $V = \frac{1}{3}\pi r^2 h$
Lateral Surface Area:
$L = \pi rs$
Total Surface Area:
$S = \pi r^2 + \pi rs$
Slant Height:
$s = \sqrt{r^2 + h^2}$

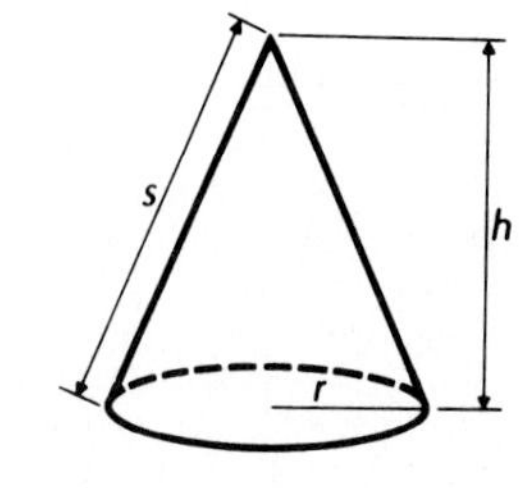

Sphere
Volume: $V = \frac{4}{3}\pi r^3$
Surface Area: $S = 4\pi r^2$

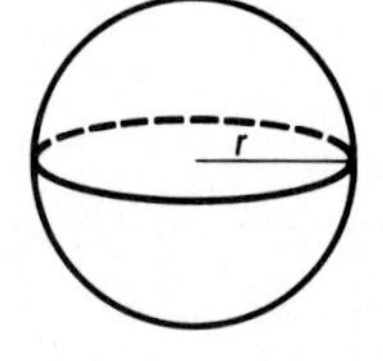

Answers

CHAPTER 1

Exercise Set 1.1, pp. 8–9

1. $n - 4$ **3.** $2x$ **5.** $0.32n$ **7.** $\frac{1}{2}x + 7$ **9.** $0.19t - 4$ **11.** $x - y + 5$ **13.** $xy - 4$ **15.** $0.35t + 1$ **17.** 13 **19.** 42 **21.** 7 **23.** 4 **25.** 25 **27.** 4 **29.** 3 *is* a solution. **31.** 7 *is not* a solution. **33.** 6 *is* a solution. **35.** 5 *is* a solution. **37.** 0.4 *is not* a solution. **39.** $\frac{17}{3}$ *is* a solution. **41.** $\{a, e, i, o, u\}$ or $\{a, e, i, o, u, y\}$ **43.** $\{2, 4, 6, 8, \ldots\}$ **45.** $\{5, 10, 15, 20, \ldots\}$ **47.** $\{x \mid x$ is an odd number between 10 and 30$\}$ **49.** $\{x \mid x$ is a whole number less than 5$\}$ **51.** $\{n \mid n$ is a multiple of 5 between 7 and 79$\}$ **53.** True **55.** True **57.** True **59.** True **61.** True **63.** True **65.** True **67.** False **69.** True **71.** $3(x + y)$ **73.** $\frac{n - m}{n + m}$ **75.** $\{0\}$ **77.** $\{x \mid x$ is a real number$\}$

79.

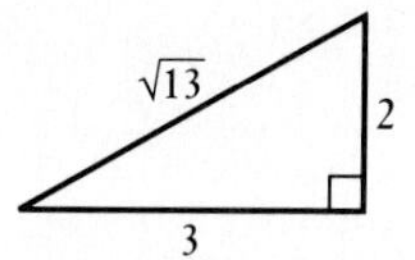

Exercise Set 1.2, pp. 17–18

1. 7 **3.** 9 **5.** 6.2 **7.** 0 **9.** $1\frac{7}{8}$ **11.** 4.21 **13.** -9 is less than or equal to -1; true **15.** -7 is greater than 1; false **17.** 3 is greater than or equal to -5; true **19.** -9 is less than -4; true **21.** -4 is greater than or equal to -4; true **23.** -5 is less than -5; false **25.** 17 **27.** -11 **29.** -3.2 **31.** $-\frac{11}{35}$ **33.** -8.5 **35.** $\frac{5}{9}$ **37.** -4.5 **39.** 0 **41.** -6.4 **43.** -7.29 **45.** 4.8 **47.** 0 **49.** $6\frac{1}{3}$ **51.** -7 **53.** 2.7 **55.** -1.79 **57.** 0 **59.** 0.03 **61.** 2 **63.** -5 **65.** 4 **67.** -17 **69.** -3.1 **71.** $-\frac{11}{10}$ **73.** 2.9 **75.** 7.9 **77.** -28 **79.** 24 **81.** -21 **83.** -7.2 **85.** 0 **87.** 5.44 **89.** 5 **91.** -5 **93.** -73 **95.** 0 **97.** 7 **99.** $\frac{1}{5}$ **101.** $-\frac{1}{9}$ **103.** $\frac{3}{2}$ **105.** $-\frac{11}{3}$ **107.** $\frac{5}{6}$ **109.** $-\frac{6}{5}$ **111.** $\frac{8}{27}$ **113.** $-\frac{3}{8}$ **115.** $-\frac{7}{3}$ **117.** $\frac{15x}{35}$ **119.** $-\frac{12a}{32}$ **121.** $10x$ **123.** $8a$ **125.** $0.7x - 5$ **127.** -6 **129.** 9 **131.** -4

Exercise Set 1.3, pp. 22–23

1. $2 \cdot 2 \cdot 2 \cdot 2$ **3.** $(1.4)(1.4)(1.4)(1.4)(1.4)$ **5.** $(-3)(-3)$ **7.** $-3x$ **9.** $(5ab)(5ab)(5ab)$ **11.** $10pq$ **13.** 10^6 **15.** x^7 **17.** $(3y)^3$ **19.** $(-2a)^3$ **21.** 23 **23.** 27 **25.** $-\frac{6}{11}$ **27.** $-\frac{65}{7}$ **29.** 117 **31.** 28 **33.** -5 **35.** $20\frac{2}{3}$ **37.** 66 **39.** 343 **41.** 60 **43.** -1075 **45.** $\frac{4}{5}$ **47.** -1 **49.** 40 **51.** 512; 32 **53.** -64; -4 **55.** Assuming algebraic logic, $\boxed{9}\ \boxed{+}\ \boxed{3}\ \boxed{\times}\ \boxed{4}\ \boxed{+}\ \boxed{1}\ \boxed{=}\ \boxed{\div}\ \boxed{2}\ \boxed{=}$ **57.** Assuming algebraic logic, $\boxed{8}\ \boxed{-}\ \boxed{2}\ \boxed{\times}\ \boxed{3}\ \boxed{+}\ \boxed{4}\ \boxed{\div}\ \boxed{1}\ \boxed{5}\ \boxed{=}$ **59.** Assuming algebraic logic, $\boxed{9}\ \boxed{-}\ \boxed{2}\ \boxed{x^y}\ \boxed{8}\ \boxed{+}\ \boxed{4}\ \boxed{=}\ \boxed{\div}\ \boxed{3}\ \boxed{=}$ **61.** Assuming algebraic logic, $\boxed{7}\ \boxed{-}\ \boxed{2}\ \boxed{\div}\ \boxed{3}\ \boxed{=}\ \boxed{x^y}\ \boxed{4}\ \boxed{-}\ \boxed{5}\ \boxed{\div}\ \boxed{8}\ \boxed{=}$ **63.** $p + q$, or $q + p$ **65.** Any number except 0 **67.** Any number except 2 **69.** $x^2 + 7$ **71.** $(x + 7)^2$ **73.** $\frac{x + 3}{(x + 3)^2}$ **75.** $2(x - 1)$

Exercise Set 1.4, pp. 28–29

1. $8 + y$ **3.** nm **5.** $9 + yx$, $yx + 9$, $xy + 9$ **7.** $ba + c$, $c + ba$, $c + ab$ **9.** $y^2 + x$ **11.** $t + xt^2$, $t + t^2x$, $t^2x + t$ **13.** $(2 \cdot 8) \cdot x$ **15.** $(x + 2y) + 5$ **17.** $\left(\frac{1}{2} + 3a\right) + b$ **19.** $\frac{3}{4}(xy)$ **21.** $3a + 3$ **23.** $4x - 4y$ **25.** $-10a - 15b$ **27.** $2ab - 2ac + 2ad$ **29.** $2\pi rh + 2\pi r$ **31.** $\frac{1}{2}ha + \frac{1}{2}hb$ **33.** $8(x + y)$ **35.** $9(p - 1)$ **37.** $7(x - 3)$ **39.** $x(y + 1)$ **41.** $2(x - y + z)$ **43.** $3(x + 2y - 1)$ **45.** $9a$ **47.** $-3b$ **49.** $15y$ **51.** $11a$ **53.** $-8t$ **55.** $10x$ **57.** $8x - 8y$ **59.** $2c + 10d$ **61.** $22x + 18$ **63.** $2x - 33y$ **65.** $-a - 5$ **67.** $m + 1$ **69.** $5d - 12$ **71.** $-7x + 14$ **73.** $-9x + 21$ **75.** $44a - 22$ **77.** -190 **79.** $-12y - 145$ **81.** $\frac{89}{48}$ **83.** 64 **85.** 8; $\frac{1}{8}$; no **87.** 16; 1; no **89.** **(a)** $17(x + 2)$, $17 \cdot 10 + 34 = 170 + 34 = 204$, $17(10 + 2) = 17 \cdot 12 = 204$; **(b)** Yes; distributive law

Review Exercises: Chapter 1, pp. 31–33

1. [1.1] $0.36n - 5$ **2.** [1.1] 18 **3.** [1.1] Yes **4.** [1.1] No **5.** [1.1] $\{4, 8, 12, \ldots\}$ **6.** [1.1] $\{n \mid n$ is a multiple of 3 between 17 and 49$\}$ **7.** [1.1] True **8.** [1.1] False **9.** [1.1]

True **10.** [1.1] False **11.** [1.2] 3.7 **12.** [1.2] -5 is less than or equal to -3, a true statement since -5 is to the left of -3. **13.** [1.2] -1.5 **14.** [1.2] $-3\frac{2}{3}$ **15.** [1.2] 2.1 **16.** [1.2] -4 **17.** [1.2] 16 **18.** [1.2] -8 **19.** [1.2] $\frac{7}{16}$ **20.** [1.2] $\frac{4}{9}$ **21.** [1.2] $-30x/18$ **22.** [1.2] $9a$ **23.** [1.3] $(-3ab)(-3ab)$ **24.** [1.3] $(4b)^5$ **25.** [1.3] $\frac{1}{8}$ **26.** [1.3] 6 **27.** [1.3] -130 **28.** [1.3] $-\frac{1}{4}$ **29.** [1.3] 1296, 162 **30.** [1.3] [5] [+] [3] [x^y] [5] [−] [12] [=] [÷] [6] [=] **31.** [1.4] $c^2b + a$, $a + c^2b$, or $bc^2 + a$ **32.** [1.4] $(5 + 3x) + 9$ **33.** [1.4] $14x - 7x^2$ **34.** [1.4] $5(x - 3y + 1)$ **35.** [1.4] $16a - 8b$ **36.** [1.4] $3x + 10$ **37.** [1.4] $19b + 23$ **38.** [1.1] $\{0\}$ **39.** [1.3] $x^2 + (x + 1)^2$ **40.** [1.4] 4, 4. No, for division to be associative we must have $(a \div b) \div c = a \div (b \div c)$ for *any* choice of a, b, and c.

Test: Chapter 1, p. 33

1. [1.1] $mn + 7$ **2.** [1.1] 39 **3.** [1.1] Yes **4.** [1.1] $\{x|x$ is an odd number between 8 and $16\} = \{9, 11, 13, 15\}$ **5.** [1.1] False **6.** [1.1] True **7.** [1.2] 4.9 **8.** [1.2] $2\frac{1}{3}$ **9.** [1.2] $\frac{3}{14}$ **10.** [1.2] -1.4 **11.** [1.2] 12.9 **12.** [1.2] $-\frac{11}{14}$ **13.** [1.2] $-6x$ **14.** [1.3] $(-4x)(-4x)(-4x)$ **15.** [1.3] $-\frac{4}{13}$ **16.** [1.3] -42 **17.** [1.3] $\frac{7}{4}$ **18.** [1.3] [5] [−] [3] [÷] [8] [=] [x^y] [3] [−] [2] [÷] [7] [=] **19.** [1.4] $-6x + 2xy$ **20.** [1.4] $3a(b + 2)$ **21.** [1.4] $-8x + 20$ **22.** [1.4] $-7b - 42 = -7(b + 6)$ **23.** [1.2] 7 **24.** [1.4] $x^2 + 2x - 4$

CHAPTER 2

Exercise Set 2.1, pp. 40–41

1. Equivalent **3.** Equivalent **5.** Equivalent **7.** Not equivalent **9.** Not equivalent **11.** 14.6 **13.** 8 **15.** 18 **17.** 5 **19.** 24 **21.** 7 **23.** 8 **25.** 21 **27.** 2 **29.** 2 **31.** $\frac{18}{5}$ **33.** 0 **35.** $\frac{4}{5}$ **37.** $\frac{4}{3}$ **39.** $\frac{37}{5}$ **41.** 13 **43.** 2 **45.** 2 **47.** 7 **49.** 5 **51.** $-\frac{51}{31}$ **53.** 5 **55.** 2 **57.** $\frac{3}{5}$ **59.** $-\frac{1}{2}$ **61.** -1 **63.** No solution **65.** All real numbers **67.** No solution **69.** All real numbers **71.** $\{1, 2, 3, 4, 5, 6, 7, 8, 9\}$; $\{x|x$ is a positive integer less than $10\}$ **73.** -2.2 **75.** -1.7 **77.** 54 **79.** 4 **81.** 0.00705 **83.** -4.1762 **85.** 8 **87.** -2 **89.** $\frac{224}{29}$

Exercise Set 2.2, pp. 48–49

1. Let x and $x + 9$ be the numbers; $x + (x + 9) = 81$ **3.** Let t be the swimmer's time; $(5 - 3.2)t = 2.7$ **5.** Let t be the boat's time; $(12 + 3)t = 35$ **7.** Let x, $x + 1$, and $x + 2$ be the angle measures; $x + (x + 1) + (x + 2) = 180$ **9.** Let x be the shorter length; $x + (x + 4) = 12$ **11.** Let x be the measure of the second angle; $3x + x + (2x - 12) = 180$ **13.** Let n be the first odd number; $n + 2(n + 2) + 3(n + 4) = 70$ **15.** Let s be the length of a side in the smaller square; $4s + 4(s + 2) = 100$ **17.** Let x be the first number; $x + (3x - 6) + \left[\frac{2}{3}(3x - 6) + 2\right] = 172$ **19.** Let x be the score on the next test; $\frac{93 + 89 + 72 + 80 + 96 + x}{6} = 88$ **21.** 4.1 **23.** 50.3 **25.** 320 **27.** 63 **29.** 165 and 330 **31.** 1.5 hr **33.** 14 and 16 **35.** A 6-m piece and a 4-m piece **37.** 96°, 32°, 52° **39.** $6a - 3ab$ **41.** $5(x - 2)$ **43.** 90 in^2 **45.** 1,200,000

Exercise Set 2.3, pp. 54–55

1. $w = \frac{A}{l}$ **3.** $I = \frac{W}{E}$ **5.** $r = \frac{d}{t}$ **7.** $l = \frac{V}{wh}$ **9.** $m = \frac{E}{c^2}$ **11.** $l = \frac{P - 2w}{2}$, or $\frac{P}{2} - w$ **13.** $a^2 = c^2 - b^2$ **15.** $r^2 = \frac{A}{\pi}$ **17.** $h = \frac{2}{11}W + 40$ **19.** $r^3 = \frac{3V}{4\pi}$ **21.** $h = \frac{2A}{b_1 + b_2}$ **23.** $m = \frac{rF}{v^2}$ **25.** $n = \frac{q_1 + q_2 + q_3}{A}$ **27.** $t = \frac{d_2 - d_1}{v}$ **29.** $d_1 = d_2 - vt$ **31.** $m = \frac{r}{1 + np}$ **33.** $a = \frac{y}{b - c^2}$ **35.** 12 cm **37.** \$1571.43 **39.** 9 ft **41.** 7 years **43.** 461.25 g **45.** 72% **47.** -58.44 **49.** $a = \frac{2s - 2v_i t}{t^2}$ **51.** $V_1 = \frac{P_2V_2T_1}{P_1T_2}$ **53.** $c = \frac{a}{x} - b$ **55.** $\frac{a}{c} = \frac{b}{d}$ **57.** $b = \frac{c}{a + d}$ **59.** 3 mos **61.** Take the square root on both sides. **63.** 7.4 cm

Exercise Set 2.4, pp. 62–64

1. No, no, no, yes **3.** No, yes, yes, no

5. $\{x|x > 4\}$, or $(4, \infty)$

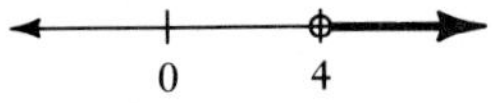

0 4

7. $\{t|t \le 6\}$, or $(-\infty, 6]$

0 6

9. $\{y|y < -3\}$, or $(-\infty, -3)$

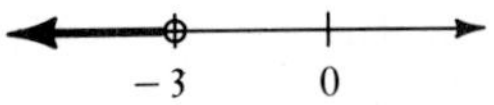

−3 0

11. $\{x|x \ge -6\}$, or $[-6, \infty)$

−6 0

13. $\{x|x > -5\}$, or $(-5, \infty)$

−5 0

15. $\{y|y < 6\}$, or $(-\infty, 6)$

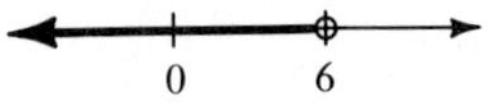

0 6

17. $\{a|a \le -21\}$, or $(-\infty, -21]$

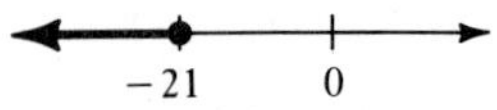

−21 0

19. $\{t|t \ge -5\}$, or $[-5, \infty)$

−5 0

21. $\{y|y > -6)$, or $(-6, \infty)$

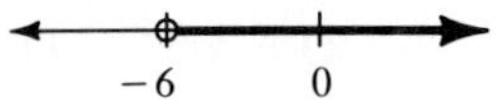

−6 0

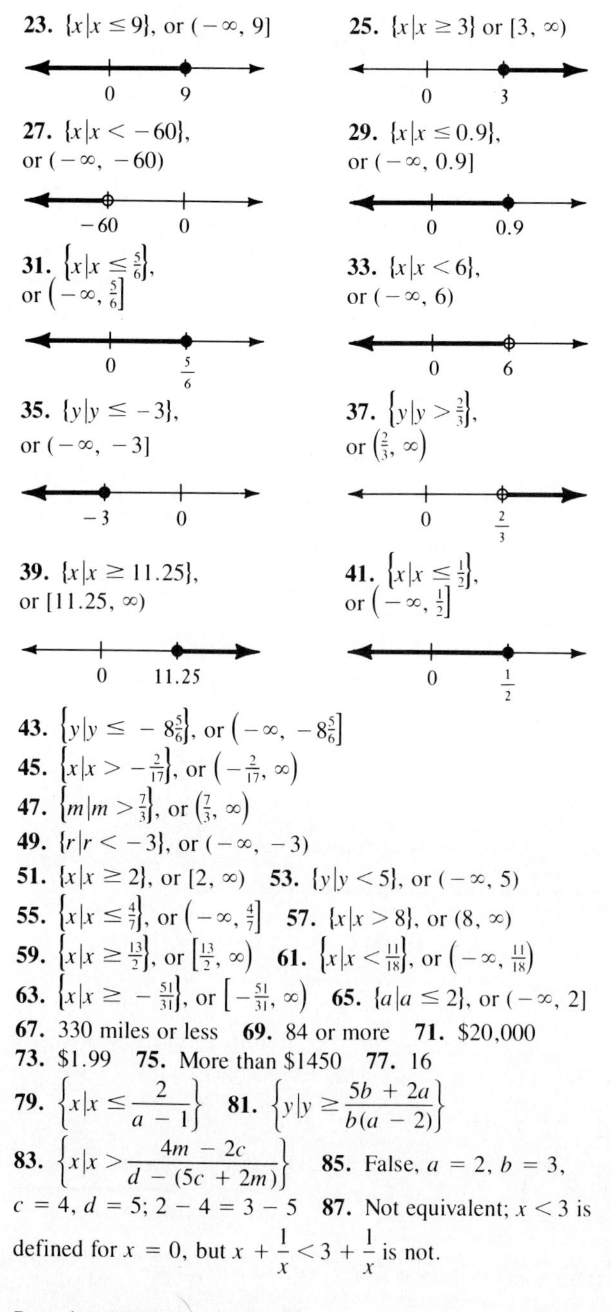

23. $\{x|x \leq 9\}$, or $(-\infty, 9]$ **25.** $\{x|x \geq 3\}$ or $[3, \infty)$

27. $\{x|x < -60\}$, or $(-\infty, -60)$ **29.** $\{x|x \leq 0.9\}$, or $(-\infty, 0.9]$

31. $\left\{x|x \leq \frac{5}{6}\right\}$, or $\left(-\infty, \frac{5}{6}\right]$ **33.** $\{x|x < 6\}$, or $(-\infty, 6)$

35. $\{y|y \leq -3\}$, or $(-\infty, -3]$ **37.** $\left\{y|y > \frac{2}{3}\right\}$, or $\left(\frac{2}{3}, \infty\right)$

39. $\{x|x \geq 11.25\}$, or $[11.25, \infty)$ **41.** $\left\{x|x \leq \frac{1}{2}\right\}$, or $\left(-\infty, \frac{1}{2}\right]$

43. $\left\{y|y \leq -8\frac{5}{6}\right\}$, or $\left(-\infty, -8\frac{5}{6}\right]$
45. $\left\{x|x > -\frac{2}{17}\right\}$, or $\left(-\frac{2}{17}, \infty\right)$
47. $\left\{m|m > \frac{7}{3}\right\}$, or $\left(\frac{7}{3}, \infty\right)$
49. $\{r|r < -3\}$, or $(-\infty, -3)$
51. $\{x|x \geq 2\}$, or $[2, \infty)$ **53.** $\{y|y < 5\}$, or $(-\infty, 5)$
55. $\left\{x|x \leq \frac{4}{7}\right\}$, or $\left(-\infty, \frac{4}{7}\right]$ **57.** $\{x|x > 8\}$, or $(8, \infty)$
59. $\left\{x|x \geq \frac{13}{2}\right\}$, or $\left[\frac{13}{2}, \infty\right)$ **61.** $\left\{x|x < \frac{11}{18}\right\}$, or $\left(-\infty, \frac{11}{18}\right)$
63. $\left\{x|x \geq -\frac{51}{31}\right\}$, or $\left[-\frac{51}{31}, \infty\right)$ **65.** $\{a|a \leq 2\}$, or $(-\infty, 2]$
67. 330 miles or less **69.** 84 or more **71.** \$20,000
73. \$1.99 **75.** More than \$1450 **77.** 16
79. $\left\{x|x \leq \frac{2}{a-1}\right\}$ **81.** $\left\{y|y \geq \frac{5b+2a}{b(a-2)}\right\}$
83. $\left\{x|x > \frac{4m-2c}{d-(5c+2m)}\right\}$ **85.** False, $a = 2$, $b = 3$, $c = 4$, $d = 5$; $2 - 4 = 3 - 5$ **87.** Not equivalent; $x < 3$ is defined for $x = 0$, but $x + \frac{1}{x} < 3 + \frac{1}{x}$ is not.

Exercise Set 2.5, pp. 69–71

1. $\{6, 8\}$ **3.** $\emptyset$ **5.** $\{1, 2, 3, 4\}$
7. **9.**

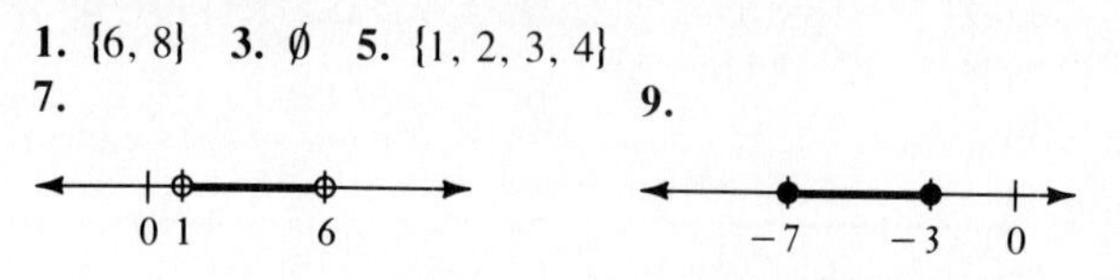

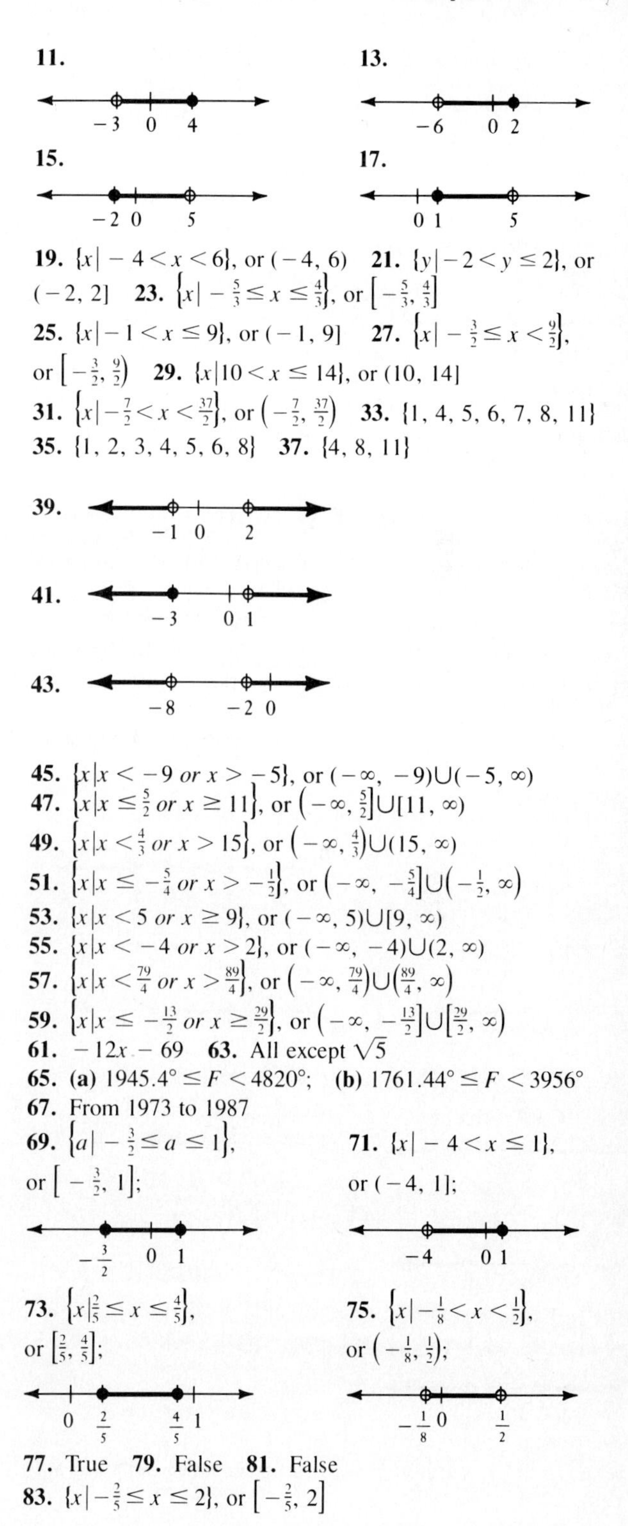

11. **13.** **15.** **17.**

19. $\{x|-4 < x < 6\}$, or $(-4, 6)$ **21.** $\{y|-2 < y \leq 2\}$, or $(-2, 2]$ **23.** $\left\{x|-\frac{5}{3} \leq x \leq \frac{4}{3}\right\}$, or $\left[-\frac{5}{3}, \frac{4}{3}\right]$
25. $\{x|-1 < x \leq 9\}$, or $(-1, 9]$ **27.** $\left\{x|-\frac{3}{2} \leq x < \frac{9}{2}\right\}$, or $\left[-\frac{3}{2}, \frac{9}{2}\right)$ **29.** $\{x|10 < x \leq 14\}$, or $(10, 14]$
31. $\left\{x|-\frac{7}{2} < x < \frac{37}{2}\right\}$, or $\left(-\frac{7}{2}, \frac{37}{2}\right)$ **33.** $\{1, 4, 5, 6, 7, 8, 11\}$
35. $\{1, 2, 3, 4, 5, 6, 8\}$ **37.** $\{4, 8, 11\}$

39. **41.** **43.**

45. $\{x|x < -9 \text{ or } x > -5\}$, or $(-\infty, -9)\cup(-5, \infty)$
47. $\left\{x|x \leq \frac{5}{2} \text{ or } x \geq 11\right\}$, or $\left(-\infty, \frac{5}{2}\right]\cup[11, \infty)$
49. $\left\{x|x < \frac{4}{3} \text{ or } x > 15\right\}$, or $\left(-\infty, \frac{4}{3}\right)\cup(15, \infty)$
51. $\left\{x|x \leq -\frac{5}{4} \text{ or } x > -\frac{1}{2}\right\}$, or $\left(-\infty, -\frac{5}{4}\right]\cup\left(-\frac{1}{2}, \infty\right)$
53. $\{x|x < 5 \text{ or } x \geq 9\}$, or $(-\infty, 5)\cup[9, \infty)$
55. $\{x|x < -4 \text{ or } x > 2\}$, or $(-\infty, -4)\cup(2, \infty)$
57. $\left\{x|x < \frac{79}{4} \text{ or } x > \frac{89}{4}\right\}$, or $\left(-\infty, \frac{79}{4}\right)\cup\left(\frac{89}{4}, \infty\right)$
59. $\left\{x|x \leq -\frac{13}{2} \text{ or } x \geq \frac{29}{2}\right\}$, or $\left(-\infty, -\frac{13}{2}\right]\cup\left[\frac{29}{2}, \infty\right)$
61. $-12x - 69$ **63.** All except $\sqrt{5}$
65. (a) $1945.4° \leq F < 4820°$; **(b)** $1761.44° \leq F < 3956°$
67. From 1973 to 1987
69. $\left\{a|-\frac{3}{2} \leq a \leq 1\right\}$, or $\left[-\frac{3}{2}, 1\right]$; **71.** $\{x|-4 < x \leq 1\}$, or $(-4, 1]$;
73. $\left\{x|\frac{2}{5} \leq x \leq \frac{4}{5}\right\}$, or $\left[\frac{2}{5}, \frac{4}{5}\right]$; **75.** $\left\{x|-\frac{1}{8} < x < \frac{1}{2}\right\}$, or $\left(-\frac{1}{8}, \frac{1}{2}\right)$;
77. True **79.** False **81.** False
83. $\left\{x|-\frac{2}{5} \leq x \leq 2\right\}$, or $\left[-\frac{2}{5}, 2\right]$

Exercise Set 2.6, pp. 77–79

1. $3|x|$ **3.** $9x^2$ **5.** $4x^2$ **7.** $8|y|$ **9.** $\frac{4}{|x|}$ **11.** $\frac{x^2}{|y|}$
13. 34 **15.** 11 **17.** 33
19. $\{-3, 3\}$; **21.** $\emptyset$
23. $\{0\}$; **25.** $\{-5.5, 5.5\}$;
27. $\{-9, 15\}$; **29.** $\left\{-\frac{1}{2}, \frac{7}{2}\right\}$;
31. $\left\{-\frac{3}{2}, \frac{17}{2}\right\}$ **33.** $\left\{-\frac{5}{4}, \frac{23}{4}\right\}$ **35.** $\{-11, 11\}$
37. $\{-8, 8\}$ **39.** $\left\{-\frac{11}{5}, \frac{11}{5}\right\}$ **41.** $\{-7, 8\}$ **43.** $\{-12, 2\}$
45. $\{-1, 2\}$ **47.** $\emptyset$
49. $\{x \mid -3 < x < 3\}$, or $(-3, 3)$;
51. $\{x \mid x \leq -2 \text{ or } x \geq 2\}$, or $(-\infty, -2] \cup [2, \infty)$;
53. $\{t \mid t \leq -5.5 \text{ or } t \geq 5.5\}$, or $(-\infty, -5.5] \cup [5.5, \infty)$;
55. $\{x \mid 2 < x < 4\}$, or $(2, 4)$;
57. $\{x \mid -7 \leq x \leq 3\}$, or $[-7, 3]$;
59. $\{x \mid x < 2 \text{ or } x > 4\}$, or $(-\infty, 2) \cup (4, \infty)$;
61. $\left\{x \mid -\frac{1}{2} \leq x \leq \frac{7}{2}\right\}$, or $\left[-\frac{1}{2}, \frac{7}{2}\right]$;
63. $\left\{y \mid y < -\frac{3}{2} \text{ or } y > \frac{17}{2}\right\}$, or $\left(-\infty, -\frac{3}{2}\right) \cup \left(\frac{17}{2}, \infty\right)$;
65. $\left\{x \mid x \leq -\frac{5}{4} \text{ or } x \geq \frac{23}{4}\right\}$, or $\left(-\infty, -\frac{5}{4}\right] \cup \left[\frac{23}{4}, \infty\right)$;
67. $\{y \mid -9 < y < 15\}$, or $(-9, 15)$;
69. $\left\{x \mid -\frac{7}{2} \leq x \leq \frac{1}{2}\right\}$, or $\left[-\frac{7}{2}, \frac{1}{2}\right]$;
71. $\left\{y \mid y < -\frac{4}{3} \text{ or } y > 4\right\}$, or $\left(-\infty, -\frac{4}{3}\right) \cup (4, \infty)$;
73. $\left\{x \mid x \leq -\frac{5}{4} \text{ or } x \geq \frac{23}{4}\right\}$, or $\left(-\infty, -\frac{5}{4}\right] \cup \left[\frac{23}{4}, \infty\right)$;
75. $\left\{x \mid -\frac{9}{2} < x < 6\right\}$, or $\left(-\frac{9}{2}, 6\right)$;
77. $\left\{x \mid x \leq -\frac{25}{6} \text{ or } x \geq \frac{23}{6}\right\}$, or $\left(-\infty, -\frac{25}{6}\right] \cup \left[\frac{23}{6}, \infty\right)$;
79. $\{x \mid -5 < x < 19\}$, or $(-5, 19)$;
81. $\left\{x \mid x \leq -\frac{2}{15} \text{ or } x \geq \frac{14}{15}\right\}$, or $\left(-\infty, -\frac{2}{15}\right] \cup \left[\frac{14}{15}, \infty\right)$;
83. -39 **85.** $\left\{-\frac{11}{2}, \frac{3}{4}\right\}$ **87.** $\left\{-\frac{3}{2}\right\}$ **89.** $\left\{0, \frac{24}{23}\right\}$ **91.** $\{x \mid x \text{ is a real number}\}$ **93.** $\{x \mid x \text{ is a real number}\}$ **95.** $|x| < 3$
97. $|x| \geq 6$ **99.** $|x + 2| < 3$ **101.** $|x + 3| > 5$

Review Exercises: Chapter 2, pp. 80–82

1. [2.1] Yes **2.** [2.1] -0.6 **3.** [2.1] 11 **4.** [2.1] $\frac{16}{7}$
5. [2.1] No solution **6.** [2.1] $\left\{\frac{6}{7}\right\}$ **7.** [2.2] 0.25 hr
8. [2.2] 50 ft, 75 ft **9.** [2.2] 4120 **10.** [2.2] 63 lb, 105 lb, 275 lb **11.** [2.2] 21, 47 **12.** [2.3] $m = PS$ **13.** [2.3] $x = \frac{c}{m - r}$ **14.** [2.3] 2 cm **15.** [2.3] $6.\overline{6}\%$
16. [2.4] -6, no; 0, no; 2, yes; 11, yes
17. [2.4] $\{t \mid t > -5\}$ or $(-5, \infty)$
18. [2.4] $\{y \mid y \geq -3\}$ or $[-3, \infty)$
19. [2.4] $\{x \mid x < 3\}$ or $(-\infty, 3)$ **20.** [2.4] $\{x \mid x \geq 4.5\}$ or $[4.5, \infty)$ **21.** [2.4] \$10,000 **22.** [2.4] 7 **23.** [2.5] $\{9, 18\}$
24. [2.5] $\{4, 8, 12, 14, 16, 18\}$
25. [2.5] **26.** [2.5]

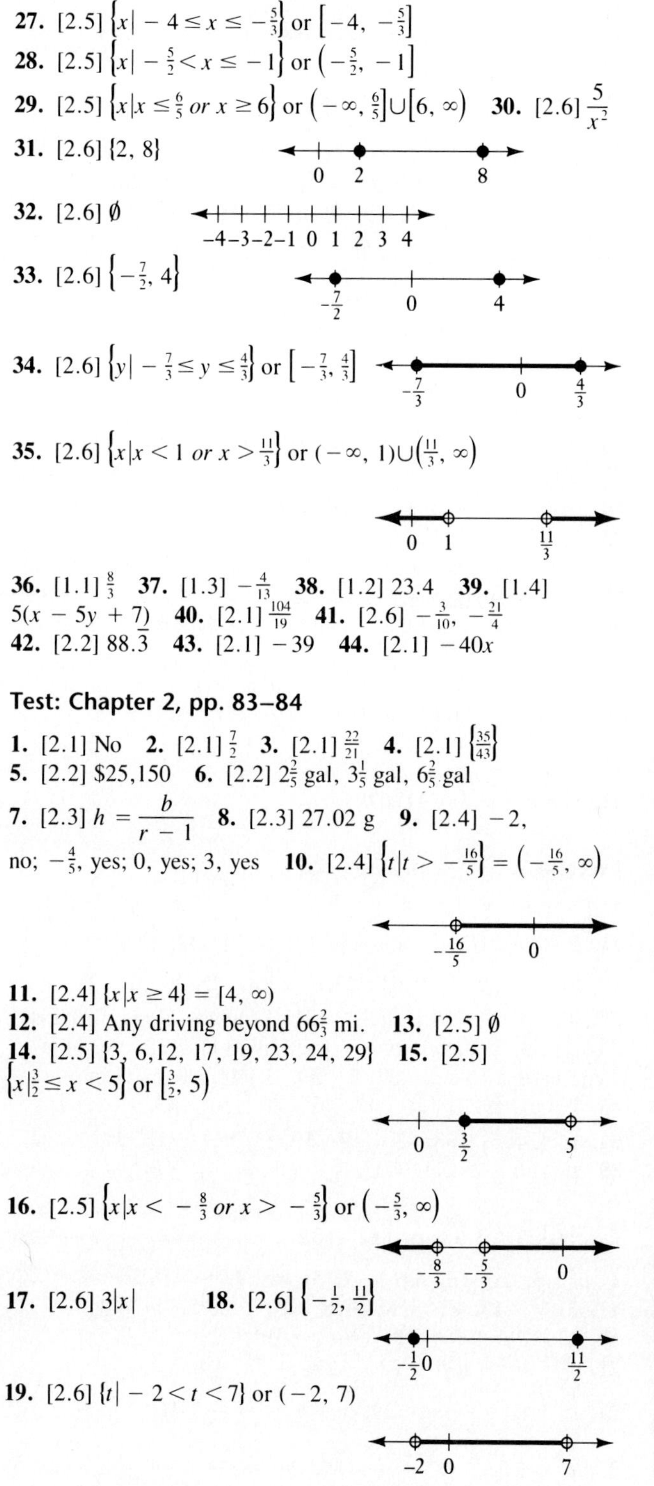

27. [2.5] $\{x \mid -4 \le x \le -\frac{5}{3}\}$ or $[-4, -\frac{5}{3}]$
28. [2.5] $\{x \mid -\frac{5}{2} < x \le -1\}$ or $(-\frac{5}{2}, -1]$
29. [2.5] $\{x \mid x \le \frac{6}{5}$ *or* $x \ge 6\}$ or $(-\infty, \frac{6}{5}] \cup [6, \infty)$ **30.** [2.6] $\frac{5}{x^2}$
31. [2.6] $\{2, 8\}$
32. [2.6] $\emptyset$
33. [2.6] $\{-\frac{7}{2}, 4\}$
34. [2.6] $\{y \mid -\frac{7}{3} \le y \le \frac{4}{3}\}$ or $[-\frac{7}{3}, \frac{4}{3}]$
35. [2.6] $\{x \mid x < 1$ *or* $x > \frac{11}{3}\}$ or $(-\infty, 1) \cup (\frac{11}{3}, \infty)$

36. [1.1] $\frac{8}{3}$ **37.** [1.3] $-\frac{4}{13}$ **38.** [1.2] 23.4 **39.** [1.4] $5(x - 5y + 7)$ **40.** [2.1] $\frac{104}{19}$ **41.** [2.6] $-\frac{3}{10}, -\frac{21}{4}$
42. [2.2] $88.\overline{3}$ **43.** [2.1] -39 **44.** [2.1] $-40x$

Test: Chapter 2, pp. 83–84

1. [2.1] No **2.** [2.1] $\frac{7}{2}$ **3.** [2.1] $\frac{22}{21}$ **4.** [2.1] $\{\frac{35}{43}\}$
5. [2.2] \$25,150 **6.** [2.2] $2\frac{2}{5}$ gal, $3\frac{1}{5}$ gal, $6\frac{2}{5}$ gal
7. [2.3] $h = \dfrac{b}{r - 1}$ **8.** [2.3] 27.02 g **9.** [2.4] -2, no; $-\frac{4}{5}$, yes; 0, yes; 3, yes **10.** [2.4] $\{t \mid t > -\frac{16}{5}\} = (-\frac{16}{5}, \infty)$

11. [2.4] $\{x \mid x \ge 4\} = [4, \infty)$
12. [2.4] Any driving beyond $66\frac{2}{3}$ mi. **13.** [2.5] $\emptyset$
14. [2.5] {3, 6, 12, 17, 19, 23, 24, 29} **15.** [2.5] $\{x \mid \frac{3}{2} \le x < 5\}$ or $[\frac{3}{2}, 5)$

16. [2.5] $\{x \mid x < -\frac{8}{3}$ *or* $x > -\frac{5}{3}\}$ or $(-\frac{5}{3}, \infty)$

17. [2.6] $3|x|$ **18.** [2.6] $\{-\frac{1}{2}, \frac{11}{2}\}$

19. [2.6] $\{t \mid -2 < t < 7\}$ or $(-2, 7)$

20. [2.6] $\{x \mid x \le -\frac{1}{5}$ *or* $x \ge 1\}$ or $(-\infty, -\frac{1}{5}] \cup [1, \infty)$

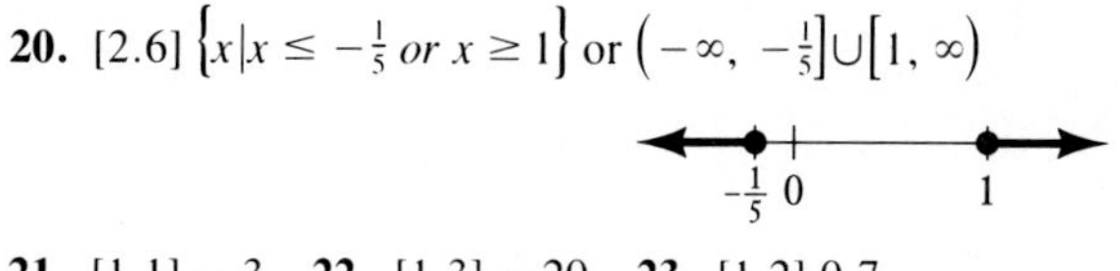

21. [1.1] -3 **22.** [1.3] -20 **23.** [1.2] 0.7
24. [1.4] $21x - 8$ **25.** [2.2], [2.3] 0.0000003%
26. [2.2], [2.3] The 17 in. costs less per square inch.
27. [2.2], [2.3] 729 cm³ **28.** [2.3] $z = y - \dfrac{x}{m}$

CHAPTER 3

Exercise Set 3.1, pp. 91–94

1. 4, 3, 2, 1, 0; 4 **3.** 3, 7, 6, 0; 7 **5.** 5, 6, 2, 1, 0; 6
7. $-4y^3 - 6y^2 + 7y + 23$; $-4y^3$; -4
9. $3x^7 + 5x^2 - x + 12$; $3x^7$; 3
11. $-a^7 + 8a^5 + 5a^3 - 19a^2 + a$; $-a^7$; -1
13. $12 + 4x - 5x^2 + 3x^4$ **15.** $3xy^3 + x^2y^2 - 9x^3y + 2x^4$
17. $-7ab + 4ax - 7ax^2 + 4x^6$ **19.** -18 **21.** 19
23. -12 **25.** 2 **27.** 4 **29.** 11 **31.** About 449
33. 1024 ft **35.** \$18,750 **37.** \$8375 **39.** 44.46 in²
41. $2x^2$ **43.** $3x + y$ **45.** $a + 6$ **47.** $-6a^2b - 2b^2$
49. $9x^2 + 2xy + 15y^2$ **51.** $-x^2y + 4y + 9xy^2$
53. $5x^2 + 2y^2 + 5$ **55.** $6a + b + c$
57. $-4a^2 - b^2 + 3c^2$ **59.** $-2x^2 + x - xy - 1$
61. $5x^2y - 4xy^2 + 5xy$ **63.** $9r^2 + 9r - 9$
65. $1.7x^2y - \frac{2}{15}xy + \frac{19}{12}xy^2$ **67.** $-(5x^3 - 7x^2 + 3x - 6)$, $-5x^3 + 7x^2 - 3x + 6$ **69.** $-(-12y^5 + 4ay^4 - 7by^2)$, $12y^5 - 4ay^4 + 7by^2$ **71.** $13x - 6$ **73.** $-4x^2 - 3x + 13$
75. $2a - 4b + 3c$ **77.** $-2x^2 + 6x$
79. $-4a^2 + 8ab - 5b^2$ **81.** $8a^2b + 16ab + 3ab^2$
83. $0.06y^4 + 0.032y^3 - 0.94y^2 + 0.93$ **85.** $x^4 - x^2 - 1$
87. $3y - 6$ **89.** $5x^2 - 8x$
91. $47x^{4a} + 40x^{3a} + 30x^{2a} + x^a + 4$ **93.** No; the coefficients of the *n*th degree terms could be additive inverses. For example, the sum of the third-degree polynomials $x^3 - 1$ and $-x^3 + 3x$ is $3x - 1$, a first-degree polynomial.

Exercise Set 3.2, pp. 100–101

1. $42x^2$ **3.** x^4 **5.** $-x^8$ **7.** $10y^3$ **9.** $-20x^3y$
11. $-10x^5y^6$ **13.** $6x - 2x^2$ **15.** $3a^2b + 3ab^2$
17. $15c^3d^2 - 25c^2d^3$ **19.** $6x^2 + x - 12$ **21.** $s^2 - 9t^2$
23. $x^2 - 2xy + y^2$ **25.** $2y^2 + 9xy - 56x^2$
27. $a^4 - 5a^2b^2 + 6b^4$ **29.** $x^3 - 64$ **31.** $x^3 + y^3$
33. $a^4 + 5a^3 - 2a^2 - 9a + 5$
35. $4a^3b^2 + 4a^3b - 10a^2b^2 - 2a^2b + 3ab^3 + 7ab^2 - 6b^3$
37. $x^2 - \frac{3}{4}x + \frac{1}{8}$ **39.** $3.25x^2 - 0.9xy - 28y^2$
41. $a^2 + 5a + 6$ **43.** $y^2 + y - 6$ **45.** $x^2 + 6x + 9$
47. $x^2 - 4xy + 4y^2$ **49.** $b^2 - \frac{5}{6}b + \frac{1}{6}$

51. $2x^2 + 13x + 18$ **53.** $400a^2 - 6.4ab + 0.0256b^2$
55. $4x^2 - 4xy - 3y^2$ **57.** $4a^2 + \frac{4}{3}a + \frac{1}{9}$ **59.** $c^2 - 4$
61. $4a^2 - 1$ **63.** $9m^2 - 4n^2$ **65.** $m^4 - m^2n^2$
67. $x^4 - 1$ **69.** $a^4 - 2a^2b^2 + b^4$ **71.** $a^2 + 2ab + b^2 - 1$
73. $4x^2 + 12xy + 9y^2 - 16$ **75.** $A = P + 2Pi + Pi^2$
77. $-\frac{15}{16}$ **79.** $(y + x)^3$ **81.** $-16s^8t + 24s^6t^5 - 8s^4t^8 + 8s^3t^{11}$ **83.** $16x^4 - 32x^3 + 16x^2$
85. $-a^4 - 2a^3b + 25a^2 + 2ab^3 - 25b^2 + b^4$
87. $r^8 - 2r^4s^4 + s^8$ **89.** $a^2 + 2ac - b^2 - 2bd + c^2 - d^2$
91. $10y^2 - 38xy + 24x^2 - 212y + 306x + 930$
93. $16x^4 + 4x^2y^2 + y^4$ **95.** $x^{4a} - y^{4b}$
97. $a^3 - b^3 + 3b^2 - 3b + 1$ **99.** $\frac{1}{81}x^{12} - \frac{8}{81}x^6y^4 + \frac{16}{81}y^8$

Exercise Set 3.3, pp. 109–110

1. $2a(2a + 1)$ **3.** $y(y - 5)$ **5.** $y^2(y + 9)$
7. $3x^2(2 - x^2)$ **9.** $4xy(x - 3y)$ **11.** $3(y^2 - y - 3)$
13. $2a(2b - 3c + 6d)$ **15.** $5(2a^4 + 3a^2 - 5a - 6)$
17. $-3(x - 4)$ **19.** $-6(y + 12)$
21. $-2(x^2 - 2x + 6)$ **23.** $(b - 2)(a + c)$
25. $(x - 2)(2x + 13)$ **27.** $2a^2(x - y)$
29. $(a + b)(c + d)$ **31.** $(b^2 + 2)(b - 1)$
33. $(a^2 + 2)(a - 3)$ **35.** $x^2(x^4 + x^3 - x + 1)$
37. $(2y^2 + 5)(y^2 + 3)$ **39.** $(x + 5)(x + 4)$
41. $(t - 5)(t - 3)$ **43.** $(x - 9)(x + 3)$
45. $2(y - 4)(y - 4)$ **47.** $(p + 9)(p - 6)$
49. $(x + 9)(x + 5)$ **51.** $(y + 9)(y - 7)$
53. $(t - 7)(t - 4)$ **55.** $(x + 5)(x - 2)$
57. $(x + 2)(x + 3)$ **59.** $(8 - x)(7 + x)$
61. $y(8 - y)(4 + y)$ **63.** $(x^2 + 16)(x^2 - 5)$
65. Not factorable using integers **67.** $(x + 9y)(x + 3y)$
69. $(x - 7)(x - 7)$ **71.** $(x^2 + 1)(x^2 + 49)$
73. $(x^3 - 7)(x^3 + 9)$ **75.** $(3x + 2)(x - 6)$
77. $x(3x - 5)(2x + 3)$ **79.** $(3a - 4)(a - 2)$
81. $(5y + 2)(7y + 4)$ **83.** $2(5t - 3)(t + 1)$
85. $4(2x + 1)(x - 4)$ **87.** $x(3x - 4)(4x - 5)$
89. $x^2(7x + 1)(2x - 3)$ **91.** $(3a - 4)(a + 1)$
93. $(3x + 1)(3x + 4)$ **95.** $(1 + 12z)(3 - z)$
97. $-(2t + 5)(2t - 3)$, or $(5 + 2t)(3 - 2t)$
99. $x(3x + 1)(x - 2)$ **101.** $(24x + 1)(x - 2)$
103. $(7x + 3)(3x + 4)$ **105.** $4(10x^4 + 4x^2 - 3)$
107. $(4a - 3b)(3a - 2b)$ **109.** $(2x - 3y)(x + 2y)$
111. $(2x - 7y)(3x - 4y)$ **113.** $(3x - 5y)(3x - 5y)$
115. $(3x^3 + 2)(2x^3 - 1)$ **117.** 2.5 hr
119. $2(2y^{2a} + 5)(y^{2a} + 3)$ **121.** $x^a(4x^b + 7x^{-b})$
123. $(x^a + 8)(x^a - 3)$ **125.** $a^2(p^a + 2)(p^a - 1)$
127. $(2x - 9)(3x - 22)$ **129.** $31, -31, 14, -14, 4, -4$

Exercise Set 3.4, pp. 118–119

1. 2^{28} **3.** a^{15} **5.** $9x^2$ **7.** $16m^4$ **9.** $25b^6$ **11.** $-27a^6b^9$
13. $\dfrac{x^{10}}{y^6}$ **15.** $\dfrac{27m^6}{n^{21}}$ **17.** $(x + 4)(x - 4)$
19. $(p + 7)(p - 7)$ **21.** $(pq + 5)(pq - 5)$
23. $6(x + y)(x - y)$ **25.** $4x(y^2 + z^2)(y + z)(y - z)$
27. $a(2a + 7)(2a - 7)$
29. $3(x^4 + y^4)(x^2 + y^2)(x + y)(x - y)$
31. $a^2(3a + 5b^2)(3a - 5b^2)$ **33.** $\left(\frac{1}{5} - x\right)\left(\frac{1}{5} + x\right)$
35. $(0.2x + 0.3y)(0.2x - 0.3y)$, or $0.01(2x + 3y)(2x - 3y)$
37. $(y - 3)^2$ **39.** $(x + 7)^2$ **41.** $(x + 1)^2$ **43.** $2(a + 2)^2$
45. $y(y - 9)^2$ **47.** $3(2a + 3)^2$ **49.** $2(x - 10)^2$
51. $(y^2 + 4)^2$ **53.** $(p - q)^2$ **55.** $(a + 2b)^2$
57. $(5a - 3b)^2$ **59.** $(x^2 + y^2)^2$
61. $(m - 7)(m + 2)(m - 2)$ **63.** $(a - 2)(a + b)(a - b)$
65. $(a + b + 10)(a + b - 10)$ **67.** $(a + b + 3)(a + b - 3)$
69. $(r - 1 - 2s)(r - 1 + 2s)$ **71.** $(3 - a - b)(3 + a + b)$
73. $(x + 2)(x^2 - 2x + 4)$ **75.** $(y - 4)(y^2 + 4y + 16)$
77. $(w + 1)(w^2 - w + 1)$ **79.** $(2a + 1)(4a^2 - 2a + 1)$
81. $(y - 2)(y^2 + 2y + 4)$ **83.** $(2 - 3b)(4 + 6b + 9b^2)$
85. $(4y + 1)(16y^2 - 4y + 1)$ **87.** $(2x + 3)(4x^2 - 6x + 9)$
89. $\left(a + \frac{1}{2}\right)\left(a^2 - \frac{1}{2}a + \frac{1}{4}\right)$ **91.** $2(y - 4)(y^2 + 4y + 16)$
93. $5(x - 2z)(x^2 + 2xz + 4z^2)$
95. $8(2x^2 - t^2)(4x^4 + 2x^2t^2 + t^4)$
97. $2y(y - 4)(y^2 + 4y + 16)$
99. $(z + 1)(z^2 - z + 1)(z - 1)(z^2 + z + 1)$ **101.** -3.1
103. $(3x^{2s} + 4y^t)(9x^{4s} - 12x^{2s}y^t + 16y^{2t})$
105. $(2x + y - r + 3s)(2x + y + r - 3s)$
107. $(c^2d^2 + a^8)(cd - a^4)(cd + a^4)$ **109.** $c(c^w - 1)^2$
111. $y(y^4 + 1)(y^2 + 1)(y + 1)(y - 1)$
113. $3(a + b - c - d)(a + b + c + d)$
115. $2m(m^2 + 3)$ **117.** 0

Exercise Set 3.5, pp. 124–126

1. $\{-7, 4\}$ **3.** $\{6\}$ **5.** $\{-5, -4\}$ **7.** $\{0, -8\}$ **9.** $\{-3, 3\}$
11. $\{-6, 6\}$ **13.** $\{-9, -5\}$ **15.** $\{4, 7\}$ **17.** $\left\{\frac{1}{2}, \frac{3}{4}\right\}$
19. $\{0, 6\}$ **21.** $\{-2, 2\}$ **23.** $\left\{-\frac{5}{7}, \frac{2}{3}\right\}$ **25.** $\{0, 3, -2\}$
27. $\{0, -8, 8\}$ **29.** $\frac{7}{2}$ or $-\frac{3}{2}$ **31.** Length is 12 cm; width is 7 cm. **33.** Length is 100 m; width is 75 m. **35.** Height is 7 cm; base is 16 cm. **37.** 2 **39.** 3 **41.** 40 m, 41 m
43. 7 sec **45.** $3x - 6$ **47.** 40°, 30°, 110° **49.** $\{-1, 11\}$
51. $\left\{-\frac{5}{3}, 4, 5\right\}$ **53.** $\{2, -2\}$ **55.** $\{3, -3, -2\}$ **57.** 4 cm
59. Length is 28 cm; width is 14 cm.

Exercise Set 3.6, pp. 131–132

1. a^6 **3.** $2x^3$ **5.** m^5n^4 **7.** $5x^6y^4$ **9.** $-7a^3b^{10}$
11. $3x^6y^5$ **13.** $6x^4 - 3x^2 + 8$ **15.** $-2a^2 + 4a - 3$
17. $y^3 - 2y^2 + 3y$ **19.** $-6x^5 + 3x^3 + 2x$
21. $1 - ab^2 - a^3b^4$ **23.** $-2pq + 3p - 4q$ **25.** $x + 7$
27. $a - 12 + \dfrac{32}{a + 4}$ **29.** $x - 6 + \dfrac{-7}{x - 5}$ **31.** $y - 5$
33. $y^2 - 2y - 1 + \dfrac{-8}{y - 2}$ **35.** $2x^2 - x + 1 + \dfrac{-5}{x + 2}$

37. $a^2 + 4a + 15 + \frac{72}{a-4}$ **39.** $4x^2 - 6x + 9$
41. $x^2 + 6$ **43.** $x^3 + x^2 - 1 + \frac{1}{x-1}$
45. $2y^2 + 2y - 1 + \frac{8}{5y-2}$ **47.** $2x^2 - x - 9 + \frac{3x+12}{x^2+2}$
49. $2x^3 + x^2 - 1 + \frac{x+1}{x^2+1}$ **51.** 2.7 **53.** 92
55. $x^2 + 2y$ **57.** $x^3 + x^2y + xy^2 + y^3$ **59.** $\frac{14}{3}$

Review Exercises: Chapter 3, pp. 134–135

1. [3.1] 6, 5, 3, 2, 0; 6 **2.** [3.1] 4, 5, 3, 2, 0; 5
3. [3.1] $-7b^4 + 2b^3 - 4b^2 - 21$; $-7b^4$; -7
4. [3.1] $3x^5 + x^4 - 2x^3 - 5x + 17$; $3x^5$; 3
5. [3.1] $-3x^2 + 2x^3 + 3x^6y - 7x^8y^3$ **6.** [3.1] 62, -108
7. [3.1] 24, 8 **8.** [3.1] (a)0, (b)64, (c)100, (d)64, (e)0
9. [3.1] $-x^2y - 2xy^2$ **10.** [3.1] $12ab^2 + ab + 4$
11. [3.1] $-x^3 + 2x^2 + 5x + 2$ **12.** [3.1] $x^3 + 6x^2 - x - 4$
13. [3.1] $13x^2y + 4xy - 8xy^2$ **14.** [3.1] $-2a + 6b + 7c$
15. [3.1] $6x^2 - 7xy + 3y^2$ **16.** [3.1] $\frac{13}{24}y^3 + \frac{5}{6}y^2 - 0.9y - \frac{1}{28}$
17. [3.2] $-18x^3y^4$ **18.** [3.2] $14a^4b^3 - 21a^2b^5$
19. [3.2] $8a^2b^2 + 2abc - 3c^2$ **20.** [3.2] $4x^2 - 25y^2$
21. [3.2] $4x^2 - 20xy + 25y^2$
22. [3.2] $20x^4 - 18x^3 - 47x^2 + 69x - 27$
23. [3.2] $x^4 + 8x^2y^3 + 16y^6$ **24.** [3.2] $x^2 - \frac{1}{2}x + \frac{1}{18}$
25. [3.2] $2y^3 - 11y^2 - y + 30$ **26.** [3.6] $\frac{4m^3n^4}{3} - \frac{2n^2}{3}$
27. [3.6] $y - 14 + \frac{-20}{y-6}$ **28.** [3.6] $6x^2 + 27 + \frac{5x+112}{x^2-4}$
29. [3.6] $x^3 + 2x^2 + x + 2$ **30.** [3.3] $x(6x + 5)$ **31.** [3.3] $3y^2(3y^2 - 1)$ **32.** [3.3] $3x(5x^3 - 6x^2 + 7x - 3)$ **33.** [3.3] $(a - 9)(a - 3)$ **34.** [3.3] $(3m + 2)(m + 4)$ **35.** [3.4] $(5x + 2)^2$ **36.** [3.4] $4(y + 2)(y - 2)$ **37.** [3.4] $(a + 9)(a - 9)$ **38.** [3.3] $(a + 2b)(x - y)$ **39.** [3.3] $(y + 2)(3y^2 - 5)$ **40.** [3.4] $(a^2 + 9)(a + 3)(a - 3)$
41. [3.3] $4(x^2 + 1)(x^2 + 5)$
42. [3.4] $(3x - 2)(9x^2 + 6x + 4)$
43. [3.4] $\left(\frac{1}{2}b + \frac{1}{5}c\right)\left(\frac{1}{4}b^2 - \frac{1}{10}bc + \frac{1}{25}c^2\right)$
44. [3.3] $2z^6(z^2 - 8)$
45. [3.4] $2y(3x^2 - 1)(9x^4 + 3x^2 + 1)$
46. [3.4] $(a - b + 2t)(a - b - 2t)$
47. [3.3] $(3t + p)(2t + 5p)$
48. [3.4] $(x + 2)(x + 3)(x - 3)$ **49.** [3.5] $\{10\}$ **50.** [3.5] $\left\{\frac{2}{3}, \frac{3}{2}\right\}$ **51.** [3.5] $\left\{\frac{5}{4}, \frac{1}{2}\right\}$ **52.** [3.5] 5 **53.** [3.5] 3, 5, 7 and $-7, -5, -3$ **54.** [3.5] 5 in. × 8 in. **55.** [1.2] -4.03 **56.** [1.4] $nm + 7$, $mn + 7$, or $7 + mn$
57. [2.2] a 7-ft piece and a 13-ft piece
58. [2.4] $\left\{x \middle| x > -\frac{11}{2}\right\}$ or $\left(-\frac{11}{2}, \infty\right)$ **59.** [3.4] $(2x - y)(4x^2 + 2xy + y^2)(2x + y)(4x^2 - 2xy + y^2)$
60. [3.4] $-2(3x^2 + 1)$

Test: Chapter 3, pp. 135–136

1. [3.1] 4, 3, 9, 5; 9 **2.** [3.1] $-4a^3 + a^2 + 8a - 2$; $-4a^3$; -4 **3.** [3.1] 461, -61 **4.** [3.1] $3xy + 3xy^2$ **5.** [3.1] $7m^3 + 2m^2n + 3mn^2 - 7n^3$ **6.** [3.1] $6a - 2b + 2c$
7. [3.1] $\frac{2}{21}y^2 + 1.3y + \frac{7}{3}$ **8.** [3.2] $64x^5y^3$ **9.** [3.2] $10x^2 + 14xyz - 12y^2z^2$ **10.** [3.2] $x^2 - 4y^2$
11. [3.2] $-3m^4 - 13m^3 + 5m^2 + 26m - 10$
12. [3.2] $-44x^2 + 57x + 35$ **13.** [3.6] $5y^3 - 3x^2$
14. [3.6] $x^2 + 2x + 2 - \frac{-1}{x+3}$ **15.** [3.3] $x(9x + 7)$
16. [3.4] $(3x - 5)^2$ **17.** [3.3] $(6m + 1)(2m + 3)$ **18.** [3.3] $(3c + d)(y - x)$ **19.** [3.4] $(b^2 + 4)(b + 2)(b - 2)$
20. [3.4] $(4x - 3)(16x^2 + 12x + 9)$
21. [3.4] $3x(1 + 2y)(1 - 2y + 4y^2)$
22. [3.4] $(x - 3)(x + 5)(x - 5)$
23. [3.5] $\left\{-\frac{1}{2}, 0, \frac{2}{3}\right\}$ **24.** [3.5] $\left\{\frac{3}{7}\right\}$
25. [3.5] 8 cm × 5 cm **26.** [1.2] 2.7
27. [1.4] $7x + 11$ **28.** [2.4] $\left\{x \middle| x \le \frac{12}{5}\right\}$ or $\left(-\infty, \frac{12}{5}\right]$
29. [2.2] 31 multiple-choice, 26 true–false, 13 fill-in
30. [3.2] **(a)** $x^5 + x + 1$; **(b)** [3.3] $(x^2 + x + 1)(x^3 - x^2 + 1)$
31. [3.2], [3.3] $(3x^n + 4)(2x^n - 5)$

CUMULATIVE REVIEW: CHAPTERS 1–3, pp. 136–137

1. [1.1] 23 **2.** [1.1] $\{x | x$ is an integer and $7 \le x \le 11\}$
3. [1.4] $2x + (5 + 7)$ **4.** [1.4] $9x - 9$ **5.** [1.4] $-5a + 24$ **6.** [1.3] 12 **7.** [1.3] 10 **8.** [1.3] 36
9. [1.2] 3.9 **10.** [1.2] $20m/45$ **11.** [2.1] equivalent
12. [2.1] -1.9 **13.** [2.1] $-\frac{16}{3}$ **14.** [2.4] $\{x | x \ge 12\}$ or $[12, \infty)$ **15.** [2.4] $\left\{x \middle| x < \frac{1}{6}\right\}$ or $\left(-\infty, \frac{1}{6}\right)$ **16.** [2.5] $\{x | -6 < x < 4\}$ or $(-6, 4)$ **17.** [2.5] $\left\{x \middle| \frac{3}{2} \le x < \frac{26}{3}\right\}$ or $\left[\frac{3}{2}, \frac{26}{3}\right)$
18. [2.5] $\{x | x < -12$ *or* $x > 11\}$ or $(-\infty, -12) \cup (11, \infty)$
19. [2.6] $\left\{-\frac{8}{3}, 8\right\}$ **20.** [2.6] $\{x | 3 < x < 11\}$ or $(3, 11)$
21. [2.6] $\left\{x \middle| x < -\frac{22}{3} \text{ or } x > 4\right\}$ or $\left(-\infty, -\frac{22}{3}\right) \cup (4, \infty)$
22. [3.5] $\{2, 7\}$ **23.** [3.5] $\{-5, 5\}$ **24.** [3.5] $\{-3\}$
25. [2.2] 8, 18 **26.** [2.2] 3, 5, 7 **27.** [3.5] 3, 4, 5
28. [3.5] 30 ft **29.** [2.3] $r = \frac{3m}{n} - p$
30. [2.3] 24 cm **31.** [3.1] 4, 2, 8, 1, 0; 8
32. [3.1] $-3x^8 - 3x^4 + x^3 + 4x^2 + 5x - 17$; $-3x^8$; -3
33. [3.1] 34 **34.** [3.1] $7x^3 - 7x^2 + 2x - 12$
35. [3.1] $-10a^2 + 3a - 8$ **36.** [3.2] $28x^3y^6$
37. [3.2] $3a^2 + 2ab - 8b^2$ **38.** [3.2] $9x^2 + 3x + \frac{1}{4}$
39. [3.6] $2x^{10}y^6$ **40.** [3.6] $18m^2 - 5m + 2$
41. [3.6] $3x^2 - 4x + 2 + \frac{3}{x-2}$ **42.** [3.3] $3a^2(3a^3 - 2)$
43. [3.3] $(x^2 - 2)(x + 1)$ **44.** [3.3] $(m - 12)(m + 2)$
45. [3.3] $2(3x^3 - 2)(2x^3 + 1)$ **46.** [3.4] $(6a - 7)(6a + 7)$
47. [3.4] $(m + 5)^2$ **48.** [3.4] $(x - 3 - y)(x - 3 + y)$

49. [3.4] $(2z - 1)(4z^2 + 2z + 1)$
50. [3.4] $3(x + 5)(x^2 - 5x + 25)$
51. [3.2] $-15x^{2a-2}y^{2b+3}$
52. [2.6] $\{x \mid -3 \le x \le -1 \text{ or } 9 \le x \le 11\}$ or $[-3, -1] \cup [9, 11]$

CHAPTER 4

Exercise Set 4.1, pp. 145–147

1. $\frac{2}{3}, \frac{23}{6}, \frac{163}{10}$ **3.** 0, -184, not meaningful **5.** $-\frac{9}{4}$, not meaningful, $-\frac{11}{9}$ **7.** 0, $\frac{9}{5}$, not meaningful **9.** $\frac{1}{15}$, not meaningful **11.** $-\frac{9}{4}, -\frac{4}{3}$
13. $\frac{7}{8}, \frac{1}{5}$ **15.** 0, not meaningful **17.** $\frac{3x(x+1)}{3x(x+3)}$
19. $\frac{(t-3)(t+3)}{(t+2)(t+3)}$ **21.** $\frac{(x^2-3)(x+6)}{(x-6)(x+6)}$
23. $\frac{(t^2-3)(t^2+3)}{(t^2-3)(t^2-4)}$ **25.** $\frac{3y}{5}$ **27.** $\frac{2}{t^4}$ **29.** $a - 3$
31. $\frac{2x-3}{4}$ **33.** $\frac{y-3}{y+3}$ **35.** $\frac{6}{5}$ **37.** $-\frac{6}{5}$ **39.** $\frac{t+4}{t-4}$
41. $\frac{x+8}{x-4}$ **43.** $\frac{4+t}{4-t}$ **45.** $\frac{3t^3x}{5}$ **47.** $\frac{3x^2}{25}$ **49.** $\frac{y+4}{2}$
51. $\frac{(x+4)(x-4)}{x(x+3)}$ **53.** $-\frac{2t^2}{(t+2)(t+3)}$
55. $\frac{(x+5)(2x+3)}{7x}$ **57.** $c - 2$ **59.** $\frac{a^2+ab+b^2}{3(a+2b)}$
61. $\frac{1}{2x+3y}$ **63.** $\frac{4a^4}{b^4}$ **65.** 3 **67.** $\frac{(y-3)(y+2)}{y}$
69. $\frac{2a+1}{a+2}$ **71.** $-x^2$ **73.** $\frac{(x+4)(x+2)}{3(x-5)}$
75. $\frac{y(y^2+3)}{(y+3)(y-2)}$ **77.** $\frac{x^2+4x+16}{(x+4)^2}$ **79.** $\frac{(2a+b)^2}{2(a+b)}$
81. $-\frac{47}{9}$ **83.** $w = \frac{g+s}{mr}$ **85.** $\frac{2s}{r+2s}$
87. $\frac{x-3}{(x+1)(x+3)}$ **89.** $\frac{m-t}{m+t+1}$
91. $\frac{x^2+xy+y^2+x+y}{x-y}$ **93.** $\frac{-2x}{x-1}$

Exercise Set 4.2, pp. 153–155

1. $\frac{4}{a}$ **3.** $-\frac{1}{a^2b}$ **5.** 2 **7.** $\frac{3y+5}{y-2}$ **9.** $\frac{1}{x-4}$
11. $\frac{1}{a+3}$ **13.** $a + b$ **15.** $\frac{11}{x}$ **17.** $\frac{1}{x+5}$ **19.** $\frac{1}{t^2+4}$
21. $\frac{1}{m^2+mn+n^2}$ **23.** $\frac{2y^2+22}{(y-5)(y+4)}$ **25.** $\frac{3x-1}{x+1}$
27. $\frac{x+y}{x-y}$ **29.** $\frac{3x-4}{(x-2)(x-1)}$ **31.** $\frac{8x+1}{(x+1)(x-1)}$
33. $\frac{2x-14}{15(x+5)}$ **35.** $\frac{-a^2+7ab-b^2}{(a-b)(a+b)}$ **37.** $\frac{x-5}{(x+5)(x+3)}$
39. $\frac{y}{(y-2)(y-3)}$ **41.** $\frac{3y-10}{(y+4)(y-5)}$
43. $\frac{3y^2-3y-29}{(y-3)(y+8)(y-4)}$ **45.** $\frac{2x^2-13x+7}{(x+3)(x-1)(x-3)}$
47. $\frac{10a-16}{(a^2-5a+4)(a^2-4)}$ **49.** $\frac{4t^2-2t-14}{(t+2)(t-2)}$ **51.** 0
53. $\frac{-3x^2-3x-4}{(x-1)(x+1)}$ **55.** $\frac{4t+26}{(t+2)(t+3)(t-4)(t+1)}$
57. $7bc + 5a$ **59.** $\left\{x \mid -1 < x < \frac{5}{3}\right\}$, or $\left(-1, \frac{5}{3}\right)$
61. $\frac{3}{x+8}$ **63.** $\frac{3t^2(t+3)}{-2t^2+13t-7}$
65. $8a^4, 8a^4b, 8a^4b^2, 8a^4b^3, 8a^4b^4, 8a^4b^5, 8a^4b^6, 8a^4b^7$
67. 420 years

Exercise Set 4.3, pp. 159–161

1. $\frac{1+4x}{1-3x}$ **3.** $\frac{x^2-1}{x^2+1}$ **5.** $\frac{3y+4x}{4y-3x}$ **7.** $\frac{x+y}{x}$
9. $\frac{a^2(b-3)}{b^2(a-1)}$ **11.** $\frac{1}{a-b}$ **13.** $-\frac{1}{x(x+h)}$ **15.** $\frac{y-3}{y+5}$
17. $\frac{4x-7}{7x-9}$ **19.** $\frac{a^2-3a-6}{a^2-2a-3}$ **21.** $\frac{x+2}{x+3}$ **23.** $\frac{1}{y+3}$
25. $\frac{a+1}{2a+5}$ **27.** $\frac{-1-3x}{8-2x}$ **29.** $-y$
31. $\frac{2(5a^2+4a+12)}{5(a^2+6a+18)}$ **33.** $\frac{3x^2(x+2y)}{y(x^2+3xy+3y^2)}$
35. $\frac{(2x+2)(x-2)}{(x+3)(2x-1)}$ **37.** $\frac{-y-1}{y+8}$ **39.** $\frac{2a+2)(a+1)}{(a+3)(2a-1)}$
41. $\frac{x^2+21x+8}{x^2+3x-34}$ **43.** $x = \frac{a}{b} - y$ **45.** Full price: \$8
47. $\frac{11x+8}{4x+3}$ **49.** $\frac{x}{x^3-1}$ **51.** $\frac{1}{(a^3-b^3)(a^2-ab+b^2)}$

Exercise Set 4.4, pp. 166–167

1. $\frac{51}{2}$ **3.** 144 **5.** -2 **7.** 5 **9.** No solution **11.** 2
13. $-5, -1$ **15.** $2, -\frac{3}{2}$ **17.** No solution **19.** 2 **21.** $\frac{17}{4}$
23. 11 **25.** $-\frac{10}{3}$ **27.** $\frac{3}{5}$ **29.** 5 **31.** 3, 2 **33.** -145
35. -3 **37.** $-6, 5$ **39.** No solution
41. $(3x - y)(3x + y)(9x^2 + y^2)$
43. 25 multiple-choice, 15 true–false, 10 fill-in
45. All real numbers except 1 and -1 are solutions.
47. 0.9465556
49. All real numbers except -2 are solutions.
51. Identity **53.** $\frac{1}{5}$

Exercise Set 4.5, pp. 173–176

1. $\frac{35}{12}$ **3.** $-3, -2$ **5.** 8 and 9, -9 and -8
7. $3\frac{3}{14}$ hours **9.** $8\frac{4}{7}$ hours **11.** $3\frac{9}{52}$ hours
13. 10 days for A; 40 days for B **15.** $1\frac{1}{5}$ hours

17. 6 hours for Jake; 12 hours for Skyler **19.** 5 hr
21. $22\frac{1}{2}$ hr for the new machine; 45 hr for the old machine
23. 7 mph **25.** $10\frac{3}{13}$ ft/sec
27. Rosanna: $3\frac{1}{3}$ mph; Simone: $5\frac{1}{3}$ mph
29. Train A: 46 mph; train B: 58 mph
31. Boat A: 30 km/h; boat B: 20 km/h **33.** 9 km/h
35. 2 km/h **37.** 20 mph **39.** 20 mph **41.** 6000
43. $-7, 11$ **45.** $21\frac{9}{11}$ minutes after 4:00 **47.** Speed of boat: 14 km/h; speed of stream: 10 km/h **49.** $3\frac{3}{4}$ km/h
51. $49\frac{1}{2}$ hours **53.** 48 km/h

Review Exercises: Chapter 4, pp. 177–178

1. [4.1] No, division by 0 is undefined.
2. [4.1] **(a)** $-\frac{15}{8}$; **(b)** 0; **(c)** 12 **3.** [4.2] $48x^3$
4. [4.2] $(x + 5)(x - 2)(x - 4)$ **5.** [4.2] $x^2 - 2x + 4$
6. [4.2] $\frac{1}{x - 4}$ **7.** [4.1] $\frac{b^2c^6d^2}{a^5}$ **8.** [4.2] $\frac{3np + 4m}{18m^2n^4p^2}$
9. [4.1] $\frac{y - 8}{2}$ **10.** [4.1] $\frac{(x - 2)(x + 5)}{x - 5}$
11. [4.1] $\frac{3a - 1}{a - 3}$ **12.** [4.1] $\frac{(x^2 + 4x + 16)(x - 6)}{(x + 4)(x + 2)}$
13. [4.2] $\frac{x - 3}{(x + 1)(x + 3)}$ **14.** [4.2] $\frac{x^2 + 11xy + y^2}{(x - y)(x + y)}$
15. [4.2] $\frac{2x^3 + 2x^2y + 2xy^2 - 2y^3}{(x - y)(x + y)}$
16. [4.2] $\frac{-y}{(y + 4)(y - 1)}$ **17.** [4.3] $\frac{3}{4}$
18. [4.3] $\frac{a^2b^2}{2(b^2 - ba + a^2)}$ **19.** [4.3] $\frac{(y + 11)(y + 5)}{(y - 5)(y + 2)}$
20. [4.3] $\frac{(14 - 3x)(x + 3)}{2x^2 + 16x + 6}$ **21.** [4.4] 2
22. [4.4] $\frac{28}{11}$ **23.** [4.4] 6
24. [4.4] No solution **25.** [4.4] 3 **26.** [4.5] $5\frac{1}{7}$ hr
27. [4.5] 24 mph **28.** [4.5] Motorcycle, 62 mph; car, 70 mph **29.** [1.3] -72 **30.** [2.3] $q = \frac{nx - t}{4}$
31. [2.5] $\{x \mid x < 8 \text{ or } x > 11\}$, or $(-\infty, 8) \cup (11, \infty)$
32. [2.6] $\left\{-\frac{2}{3}, 4\right\}$ **33.** [4.4] All real numbers except 0 and 13 **34.** [4.3], [4.4] 45 **35.** [4.5] Sara, 56; Stephen, 42

Test: Chapter 4, p. 179

1. [4.1] $\frac{4(y - 1)}{3}$ **2.** [4.1] $\frac{x^2 - 3x + 9}{x + 4}$
3. [4.2] $(x - 4)(x + 4)(x^2 + 4x + 16)$
4. [4.2] $(x - 3)(x + 11)(x - 9)$ **5.** [4.2] $\frac{25x + x^3}{x + 5}$
6. [4.2] $3(a - b)$ **7.** [4.2] $\frac{a^3 - a^2b + 4ab + ab^2 - b^3}{(a - b)(a + b)}$
8. [4.2] $\frac{-2(2x^2 + 5x + 20)}{(x - 4)(x + 4)(x^2 + 4x + 16)}$
9. [4.2] $\frac{y - 4}{(y + 3)(y - 2)}$ **10.** [4.3] $\frac{5y - 3x}{2y + 3x}$
11. [4.3] $\frac{(x - 9)(x - 6)}{(x + 6)(x - 3)}$ **12.** [4.3] $\frac{4x^2 - 14x + 2}{3x^2 + 7x - 11}$
13. [4.4] 1 **14.** [4.4] $-\frac{21}{4}$ **15.** [4.5] 1.97 hr **16.** [4.5] 5 and 6; -6 and -5 **17.** [4.5] $3\frac{3}{11}$ mph **18.** [4.5] $7\frac{1}{2}$ days
19. [1.3] 26 **20.** [2.6] $\{x \mid -1 < x < 8\}$, or $(-1, 8)$;

21. [2.5] $\{-2, 1, 3, 5, 7\}$ **22.** [2.3] $x = \frac{z}{y - 1}$ **23.** [4.4] All real numbers except 0 and 15 **24.** [4.2] $1 - t^6$

CHAPTER 5

Exercise Set 5.1, pp. 187–190

1. $4, -4$ **3.** $12, -12$ **5.** $20, -20$ **7.** $7, -7$ **9.** $-\frac{7}{6}$
11. 14 **13.** $-\frac{4}{9}$ **15.** 0.3 **17.** -0.07 **19.** $p^2 + 4$
21. $\frac{x}{y + 4}$ **23.** $4|x|$ **25.** $7|c|$ **27.** $|a + 1|$
29. $|x - 2|$ **31.** $|2x + 7|$ **33.** $4x$ **35.** $6b$ **37.** $a + 1$
39. $2(x + 1)$, or $2x + 2$ **41.** $3t - 2$ **43.** 3 **45.** $-4x$
47. -6 **49.** $5y$ **51.** $0.7(x + 1)$ **53.** 5 **55.** -1
57. $-\frac{2}{3}$ **59.** $|x|$ **61.** $5|a|$ **63.** 6 **65.** $|a + b|$ **67.** y
69. $x - 2$ **71.** No, yes **73.** Yes, no **75.** Yes, no
77. Yes, yes **79.** $\sqrt{6}$ **81.** $\sqrt[3]{10}$ **83.** $\sqrt[4]{72}$ **85.** $\sqrt{30ab}$
87. $\sqrt[5]{18t^3}$ **89.** $\sqrt{x^2 - a^2}$ **91.** $\sqrt[3]{0.06x^2}$ **93.** $\sqrt[4]{x^3 - 1}$
95. $3\sqrt{3}$ **97.** $3\sqrt{5}$ **99.** $2\sqrt{2}$ **101.** $2\sqrt{6}$ **103.** $6x^2\sqrt{5}$
105. $2\sqrt[3]{100}$ **107.** $-2x^2\sqrt[3]{2}$ **109.** $3x^2\sqrt[3]{2x^2}$
111. $2x^2\sqrt[3]{10x^2}$ **113.** $2\sqrt[4]{2}$ **115.** $3\sqrt[4]{10}$ **117.** $2a^2\sqrt[4]{6}$
119. $3cd\sqrt[4]{2d^2}$ **121.** $(x + y)\sqrt[3]{x + y}$
123. $20(m + n)^2\sqrt[3]{(m + n)^2}$
125. $-ab^2c^3\sqrt[5]{abc^2}$ **127.** $3\sqrt{2}$ **129.** $6\sqrt{5}$
131. $4\sqrt{3}$ **133.** $3\sqrt[3]{2}$ **135.** $5bc^2\sqrt{2b}$
137. $-5x^2\sqrt[3]{6x}$ **139.** $2y^3\sqrt[3]{2}$ **141.** $(b + 3)^2$
143. $4a^3b\sqrt{6ab}$ **145.** $a(b + c)^2\sqrt[5]{a(b + c)}$
147. $5x(z + y)$ **149.** x^8 **151.** **(a)** 20 mph; **(b)** 37.4 mph; **(c)** 42.4 mph **153.** 10 **155.** $x \geq -\frac{3}{2}$

Exercise Set 5.2, pp. 193–194

1. $\frac{4}{5}$ **3.** $\frac{4}{3}$ **5.** $\frac{7}{y}$ **7.** $\frac{5y\sqrt{y}}{x^2}$ **9.** $\frac{2x\sqrt[3]{x^2}}{3y}$ **11.** $\frac{2a}{3}$

13. $\frac{ab^2}{c^2}\sqrt[4]{\frac{a}{c^2}}$ **15.** $\frac{2x}{y^2}\sqrt[5]{\frac{x}{y}}$ **17.** $\frac{xy}{z^2}\sqrt[6]{\frac{y^2}{z^3}}$ **19.** $\sqrt{7}$ **21.** 3
23. $y\sqrt{5y}$ **25.** $2\sqrt[3]{a^2b}$ **27.** $3\sqrt{xy}$ **29.** $2xy^2\sqrt[4]{y}$
31. $6a\sqrt{6a}$ **33.** $64b^3$ **35.** $54a^3b\sqrt{2b}$ **37.** $3c^2d\sqrt[3]{3c^2d}$
39. $x\sqrt[3]{25xy^2}$ **41.** $x\sqrt[4]{x^2y^3}$ **43.** $4a^2\sqrt[3]{a^2b^2}$ **45.** $4xy\sqrt[4]{y^2}$
47. 8 **49.** Length is 20; width is 5. **51.** a^3bxy^2
53. $2yz\sqrt{2z}$ **55.** Odd roots exist for all real numbers.

Exercise Set 5.3, pp. 198–200

1. $8\sqrt{3}$ **3.** $3\sqrt[3]{5}$ **5.** $13\sqrt[3]{y}$ **7.** $7\sqrt{2}$ **9.** $6\sqrt[3]{3}$
11. $21\sqrt{3}$ **13.** $38\sqrt{5}$ **15.** $122\sqrt{2}$ **17.** $9\sqrt[3]{2}$
19. $(1+6a)\sqrt{5a}$ **21.** $(2-x)\sqrt[3]{3x}$ **23.** $3\sqrt{2y-2}$
25. $(x+3)\sqrt{x-1}$ **27.** $15\sqrt[3]{4}$ **29.** 0
31. $(3x^2+x)\sqrt[4]{x-1}$ **33.** $2\sqrt{6}-18$ **35.** $\sqrt{6}-\sqrt{10}$
37. $2\sqrt{15}-6\sqrt{3}$ **39.** -6 **41.** $3a\sqrt[3]{2}$
43. $x^2+\sqrt[4]{3x^3}$ **45.** 18 **47.** -3 **49.** -19 **51.** $a-b$
53. $1+\sqrt{5}$ **55.** -6 **57.** $a+\sqrt{3a}+\sqrt{2a}+\sqrt{6}$
59. $7+4\sqrt{3}$ **61.** $a^2+2a\sqrt{b}+b$
63. $4x^2-4x\sqrt{y}+y$ **65.** $m+2\sqrt{mn}+n$ **67.** $\frac{\sqrt{30}}{5}$
69. $\frac{2\sqrt[3]{6}}{3}$ **71.** $\frac{y\sqrt[3]{180x^2y}}{6x^2}$ **73.** $\frac{3\sqrt{5y}}{10xy}$ **75.** $\frac{2}{\sqrt{6}}$
77. $\frac{7}{\sqrt[3]{98}}$ **79.** $\frac{ab}{3\sqrt{ab}}$ **81.** $\frac{5(8+\sqrt{6})}{58}$
83. $\frac{x-2\sqrt{xy}+y}{x-y}$ **85.** $\frac{6x-7\sqrt{xy}-3y}{4x-9y}$
87. $\frac{-11}{4(\sqrt{3}-5)}$ **89.** $\frac{7}{43\sqrt{2}+66}$ **91.** $\frac{3-4x}{3-3\sqrt{3x}+2x}$
93. $4x^2-18x-10$ **95.** $\frac{5x+2}{(x-2)(x+2)}$
97. $-28\sqrt{3}-22\sqrt{5}$ **99.** $ac(3ac+2)\sqrt{ab}-2ab^2\sqrt[3]{a}$
101. $\frac{ab+(a-b)\sqrt{a+b}-a-b}{a+b-b^2}$ **103.** $\frac{b^2+\sqrt{b}}{b^2+b+1}$
105. $\frac{(2+x^2)\sqrt{1+x^2}}{1+x^2}$ **107.** 0.0188 m

Exercise Set 5.4, pp. 207–209

1. 1 **3.** 1 **5.** 1 **7.** 5 **9.** 1 **11.** $\frac{1}{6^3}$ **13.** $\frac{3}{2}$
15. $\frac{1}{-11}$ **17.** $\frac{1}{(5x)^3}$ **19.** $\frac{x^2}{y^3}$ **21.** $\frac{y^5}{x^2}$ **23.** x^3y^2
25. $\frac{1}{x^2y^5}$ **27.** 3^{-4} **29.** $(-16)^{-2}$ **31.** $\frac{1}{6^{-4}}$ **33.** $\frac{6}{x^{-2}}$
35. $(5y)^{-3}$ **37.** $\frac{y^{-4}}{3}$ **39.** $8^{-4}; \frac{1}{8^4}$ **41.** $b^{-3}; \frac{1}{b^3}$ **43.** a^3
45. $6x^{-8}; \frac{6}{x^8}$ **47.** $-14x^{-11}; \frac{-14}{x^{11}}$ **49.** 6^{12}
51. $(-2)^{-2}x^{-6}y^8; \frac{y^8}{4x^6}$ **53.** a^5 **55.** $\frac{-4x^9y^{-2}}{3}; \frac{-4x^9}{3y^2}$
57. $\sqrt[4]{x}$ **59.** 2 **61.** $\sqrt[3]{xyz}$ **63.** $\sqrt[5]{a^2b^2}$ **65.** 8
67. 243 **69.** $27\sqrt[4]{x^3}$ **71.** $125x^6$ **73.** $20^{1/3}$ **75.** $17^{1/2}$
77. $x^{3/2}$ **79.** $m^{2/5}$ **81.** $(cd^3)^{1/4}$ **83.** $(3mn)^{3/2}$ **85.** $5^{7/8}$
87. $7^{1/4}$ **89.** $a^{23/12}$ **91.** $x^{2/7}$ **93.** $m^{1/6}n^{1/8}$ **95.** $a^4b^{3/2}$
97. $\sqrt[3]{a^2}$ **99.** $2y^2$ **101.** $\sqrt[3]{2x}$ **103.** $2c^2d^3$ **105.** $\frac{m^2n^4}{2}$
107. $\sqrt[4]{r^2s}$ **109.** $\sqrt[6]{392}$ **111.** $\sqrt[6]{4x^5}$
113. $\sqrt[6]{x^5-4x^4+4x^3}$ **115.** $\sqrt[6]{a+b}$ **117.** $\sqrt[10]{x^9y^7}$
119. $abc^2\sqrt[6]{a^5bc^2}$ **121.** $-3, 3$ **123.** $6x^6+7x^3-3$
125. $\sqrt[10]{x^5y^3}$ **127.** $(a+b)\sqrt[12]{(a+b)^{11}}$
129. $\sqrt[3]{9}+\sqrt[3]{6}+\sqrt[3]{4}$ **131.** $x^5y^7z\sqrt[p]{yz^3}$

Exercise Set 5.5, pp. 215–216

1. 2 **3.** 12 **5.** 168 **7.** No solution **9.** 3 **11.** 19
13. 397 **15.** No solution **17.** 3 **19.** -6 **21.** 5
23. 3 **25.** 9 **27.** 7 **29.** -1 **31.** 6, 2 **33.** $\pm\sqrt{10}$
35. $-2+\sqrt[3]{5}$ **37.** $5\pm\sqrt{6}$ **39.** $7\pm\sqrt{7}$
41. $-3\pm\sqrt[4]{12}$ **43.** $\pm\sqrt{5}$ **45.** $\pm\sqrt{3}, \pm3$
47. $\pm1, \pm\frac{\sqrt{5}}{3}$ **49.** $-\frac{1}{2}, \frac{1}{3}$ **51.** $-1, 2$ **53.** $-27, 8$
55. 16 **57.** 32, -243 **59.** 1 **61.** 2 **63.** About 208 miles **65.** $-\frac{8}{9}$ **67.** $-8, 8$ **69.** 1, 4 **71.** 19

Exercise Set 5.6, pp. 221–222

1. $\sqrt{15}i$ **3.** $4i$ **5.** $-6i$ **7.** $-2\sqrt{3}i$ **9.** $5\sqrt{10}i$
11. $8+i$ **13.** $9-5i$ **15.** $7+4i$ **17.** $-2-3i$
19. $-1+i$ **21.** $11+6i$ **23.** $-12-6i$ **25.** $7+6i$
27. $6-5i$ **29.** $4-6i$ **31.** -30 **33.** $-\sqrt{30}$
35. $-5\sqrt{6}$ **37.** $-12\sqrt{2}$ **39.** -40 **41.** 35
43. $10+15i$ **45.** $-12-21i$ **47.** $1+5i$ **49.** $18+14i$
51. $38+9i$ **53.** $2-46i$ **55.** $5-37i$ **57.** $-11-16i$
59. $13-47i$ **61.** $5-12i$ **63.** $-5+12i$
65. $-5-12i$ **67.** $-i$ **69.** 1 **71.** i **73.** -1 **75.** 6
77. 14 **79.** $-1-i$ **81.** $-4i$ **83.** 58 **85.** 89 **87.** 25
89. $\frac{3}{2}+\frac{1}{2}i$ **91.** $\frac{3}{29}+\frac{7}{29}i$ **93.** $-\frac{7}{6}i$ **95.** $-\frac{3}{7}-\frac{8}{7}i$
97. $\frac{8}{5}+\frac{1}{5}i$ **99.** $-i$ **101.** $\frac{19}{13}-\frac{9}{13}i$ **103.** 7
105. $3x^3+5x^2+x-1$ **107.** $\frac{250}{41}+\frac{200}{41}i$ **109.** $\frac{3\sqrt{2}}{2}i$
111. 0 **113.** $-1-\sqrt{5}i$ **115.** $-\frac{2}{3}i$

Review Exercises: Chapter 5, pp. 225–226

1. [5.1] $-\frac{2}{3}$ **2.** [5.1] 0.07 **3.** [5.1] $9|a|$ **4.** [5.1] $|c+8|$
5. [5.1] $|x-3|$ **6.** [5.1] $|2x+1|$ **7.** [5.1] -2 **8.** [5.1] $-\frac{1}{3}$ **9.** [5.1] $|x|$ **10.** [5.1] 3 **11.** [5.1] -2: yes; 6: no

12. [5.1] $3xy\sqrt{2y}$ **13.** [5.1] $3a\sqrt[3]{a^2b^2}$ **14.** [5.2] $y\sqrt[3]{6}$
15. [5.2] $\dfrac{5\sqrt[3]{x}}{2}$ **16.** [5.2] $8xy^2$ **17.** [5.2] $2a\sqrt[3]{2ab^2}$
18. [5.3] $30\sqrt[3]{5}$ **19.** [5.3] $15\sqrt{2}$ **20.** [5.3] $9 - \sqrt[3]{4}$
21. [5.3] $-43 - 2\sqrt{10}$ **22.** [5.3] $8 - 2\sqrt{7}$ **23.** [5.3] $\dfrac{10a\sqrt{3} - 10\sqrt{3ab}}{a - b}$ **24.** [5.3] $\dfrac{30a}{a\sqrt{3} + \sqrt{3ab}}$ **25.** [5.4] 1
26. [5.4] $\dfrac{-4x^5}{3}$ **27.** [5.4] $\dfrac{x^6}{4}$ **28.** [5.4] $\dfrac{-2y^2}{x^3}$ **29.** [5.4] $\dfrac{x^{-6}}{5}$ **30.** [5.4] $(8x^6y^2)^{4/5}$ or $8^{4/5}x^{24/5}y^{8/5}$ **31.** [5.4] $\sqrt[4]{(5a)^3}$
32. [5.4] $\sqrt[12]{x^4y^3}$ **33.** [5.4] $\sqrt[12]{x^3(x-3)^4}$ **34.** [5.5] 4
35. [5.5] 13 **36.** [5.5] $3 \pm \sqrt{19}$ **37.** [5.5] $-\frac{1}{3}, -\frac{1}{2}$
38. [5.6] $-2\sqrt{2}i$ **39.** [5.6] $-2 - 9i$ **40.** [5.6] $1 + i$
41. [5.6] 29 **42.** [5.6] i **43.** [5.6] $9 - 12i$
44. [5.6] $\frac{2}{5} + \frac{3}{5}i$ **45.** [5.6] $\frac{9}{5} - \frac{12}{5}i$ **46.** [3.2] $9m^2 - 4n^2$
47. [4.4] $\frac{23}{7}$ **48.** [3.5] $-\frac{9}{2}, 3$ **49.** [4.1] $\dfrac{x(x+2)}{(x+y)(x-3)}$
50. [5.5] 3 **51.** [5.6] $-\frac{2}{5} + \frac{9}{10}i$

Test: Chapter 5, pp. 226–227

1. [5.1] $\frac{10}{7}$ **2.** [5.1] $6|y|$ **3.** [5.1] $|x + 5|$ **4.** [5.1] -2
5. [5.1] 4 **6.** [5.1] 0: no; 3: yes **7.** [5.1] $2x^3$ **8.** [5.2] $x\sqrt{5x}$ **9.** [5.2] $4a\sqrt[3]{4ab^2}$ **10.** [5.3] $38\sqrt{2}$ **11.** [5.3] -20
12. [5.3] $-\dfrac{13 + 8\sqrt{2}}{41}$ **13.** [5.4] $(5xy^2)^{5/2}$ **14.** [5.4] $\dfrac{-16}{x^2}$
15. [5.4] $\sqrt[4]{2y(x-3)^3}$ **16.** [5.5] 7 **17.** [5.5] $8, -512$
18. [5.6] $3\sqrt{2}i$ **19.** [5.6] $7 + 5i$ **20.** [5.6] $-2i$
21. [5.6] $-\frac{77}{50} + \frac{7}{25}i$ **22.** [5.6] $-1 + 5i$ **23.** [3.5] $-\frac{1}{3}, \frac{5}{2}$
24. [4.1] $\dfrac{x - 3}{x - 4}$ **25.** [4.4] No solution
26. [3.2] $x^4 + 6x^2 + 9$ **27.** [5.5] 3 **28.** [5.6] $-2 + \frac{15}{4}i$

CHAPTER 6

Exercise Set 6.1, pp. 236–238

1.

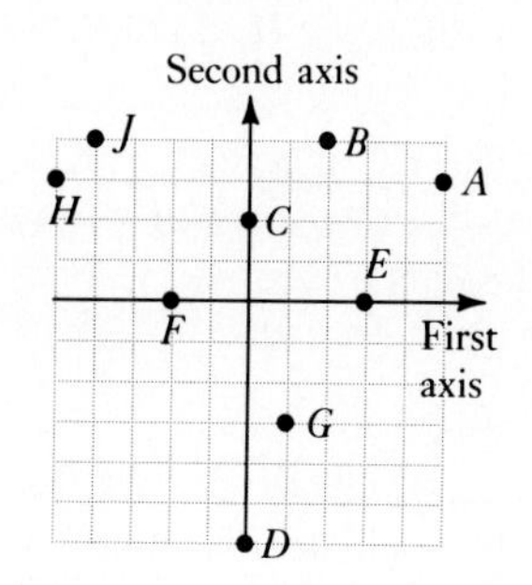

3.

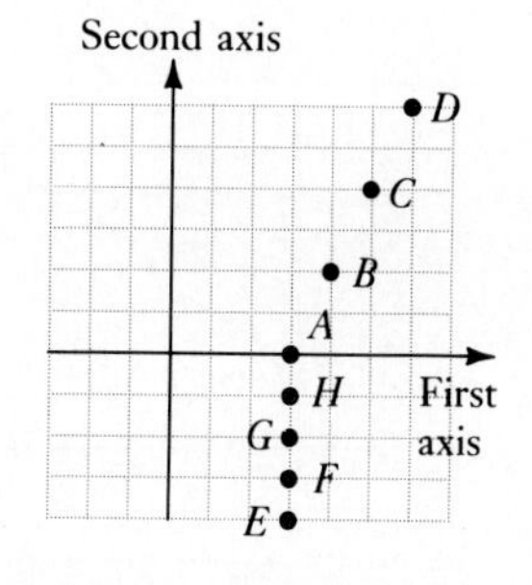

5. Triangle; area = 21 sq units

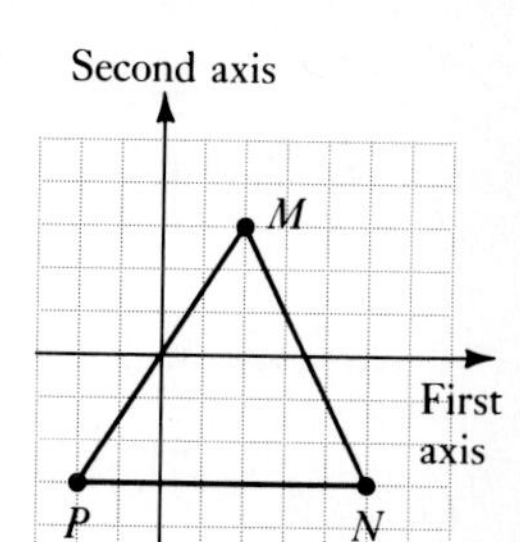

7. III **9.** II **11.** I **13.** IV **15.** Yes **17.** No **19.** Yes
21. Yes **23.** Yes **25.** No **27.** Yes **29.** No

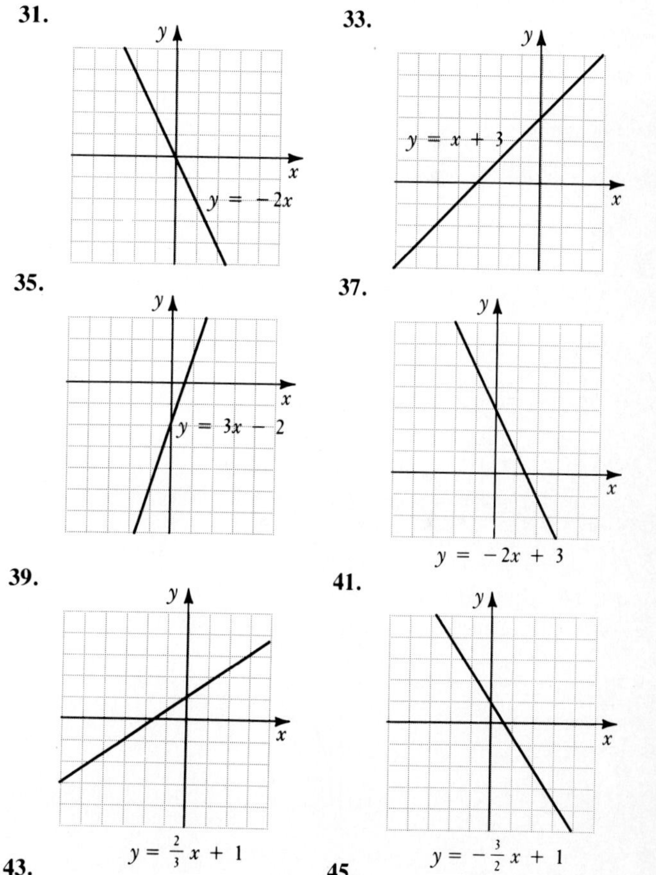

43.

45.

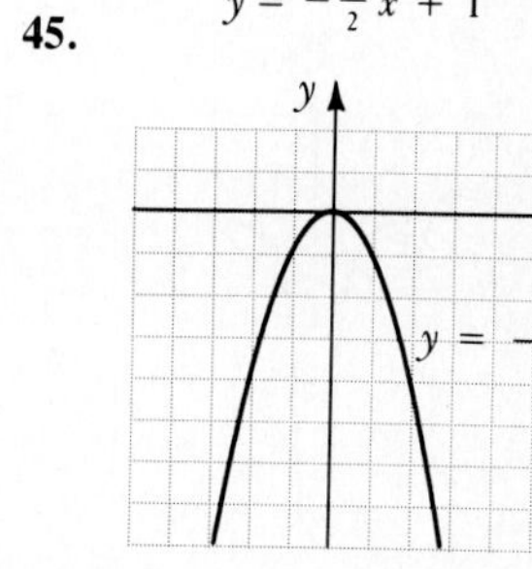

47.

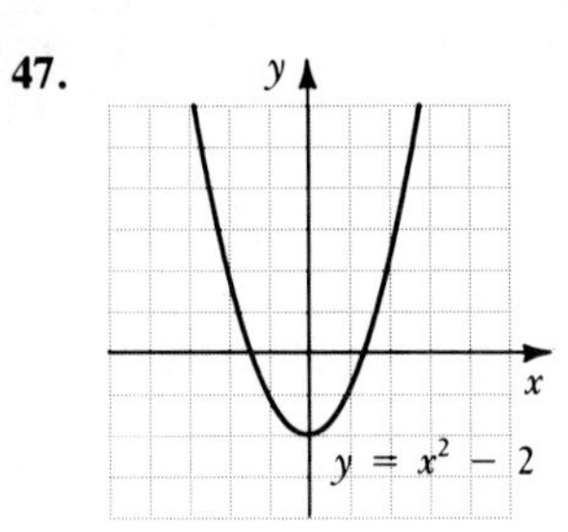

49. **53.**

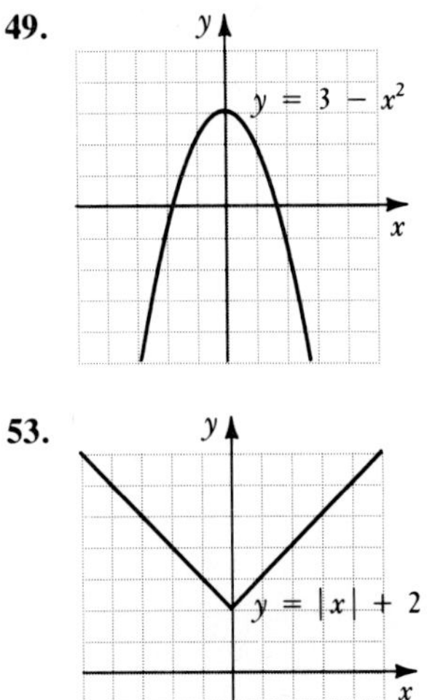

51.

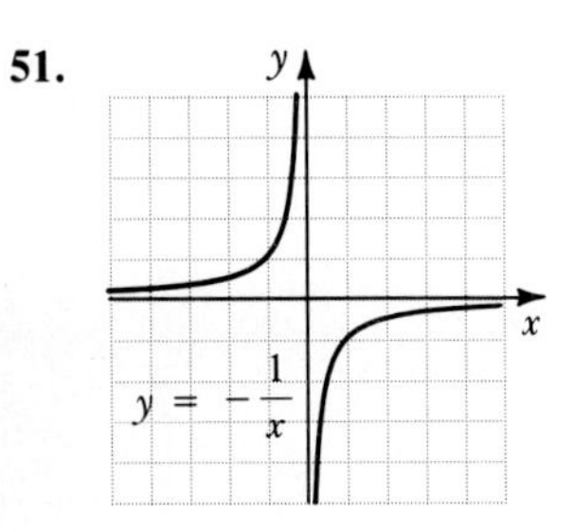

55. 26 ft **57.** -1.4

59.

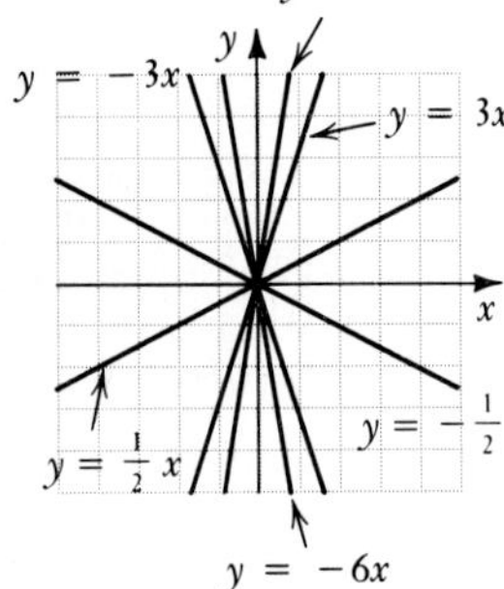

; the steepness and direction of the slant

61. (a); (d) **63.**

$y = x^3 - 6x^2 + 12x - 8$

65.

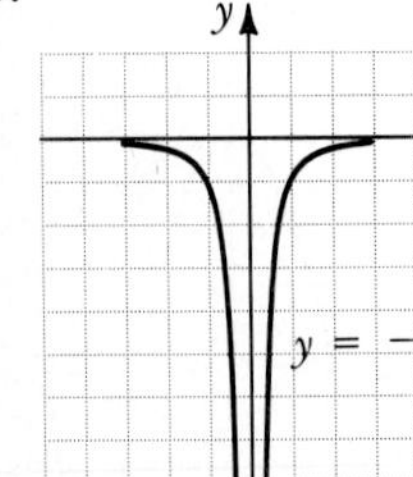

67.

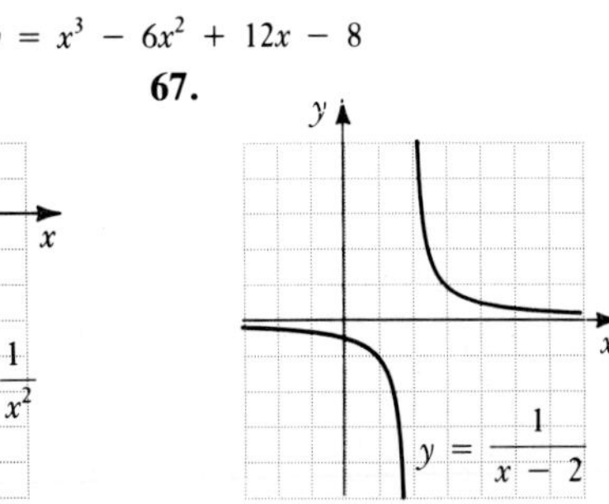

69. If the sign is negative, the graph is moved down. If the sign is positive, the graph is moved up. The three lines are parallel.

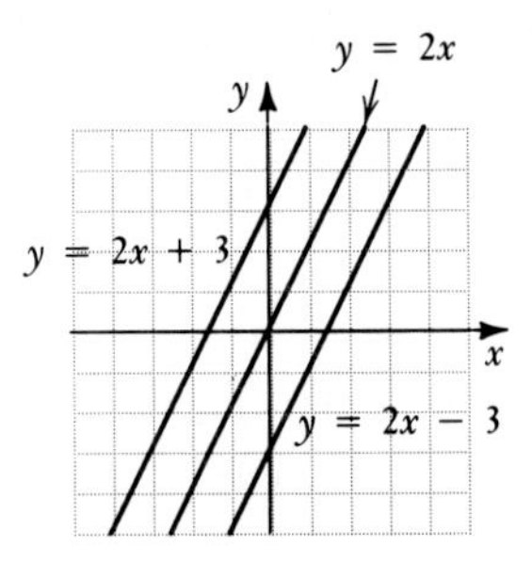

71.

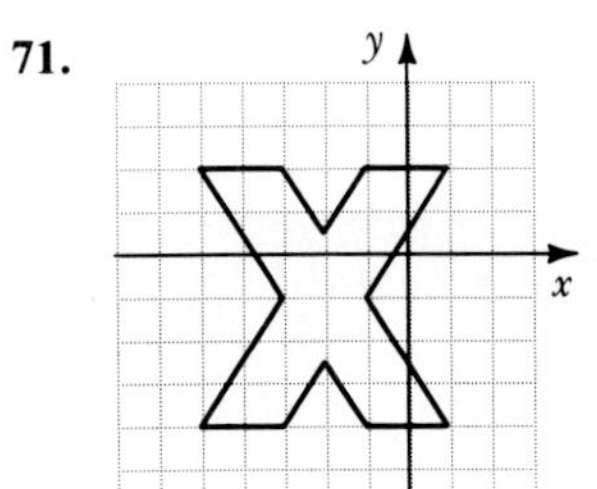

73.

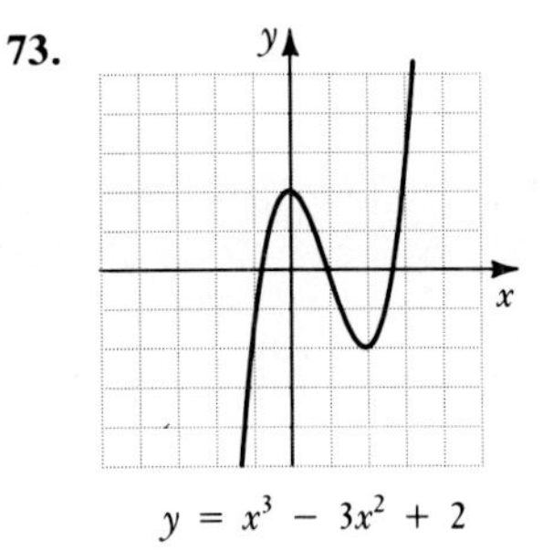

$y = x^3 - 3x^2 + 2$

75.

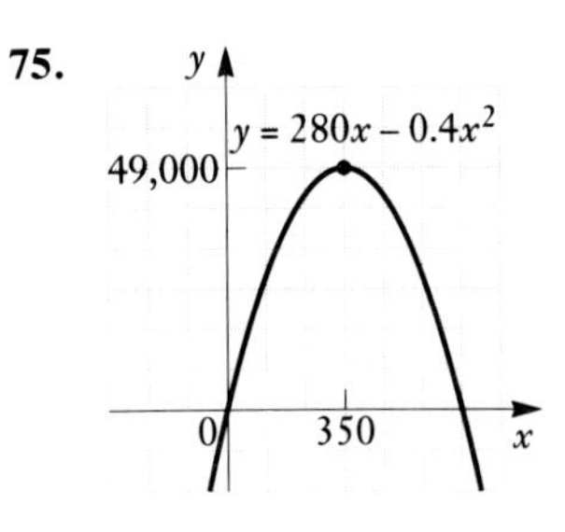

Exercise Set 6.2, pp. 244–250

1. No **3.** Yes **5.** Yes **7.** No **9.** Yes **11.** Function **13.** Function **15.** A relation, but not a function **17.** **(a)** 1; **(b)** -3; **(c)** -6; **(d)** 9; **(e)** $a + 3$ **19.** **(a)** 4; **(b)** 9; **(c)** 49; **(d)** $5t^2 + 4$; **(e)** $20a^2 + 4$ **21.** **(a)** 15; **(b)** 32; **(c)** 20; **(d)** 4; **(e)** $27r^2 + 6r - 1$ **23.** **(a)** $\frac{3}{5}$; **(b)** $\frac{1}{3}$; **(c)** $\frac{4}{7}$; **(d)** 0; **(e)** $\frac{x - 1}{2x - 1}$ **25.** -2 **27.** -2156 **29.** $4\sqrt{3}$ cm^2 **31.** 36π in$^2 \approx 113.04$ in^2 **33.** 14°F **35.** 181.48 cm **37.** \$23.55 **39.** ; 7 drinks **41.** \$1700

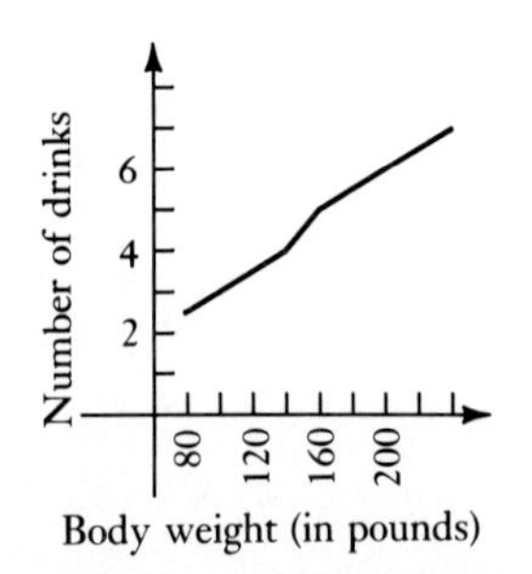

43. 75 **45.**

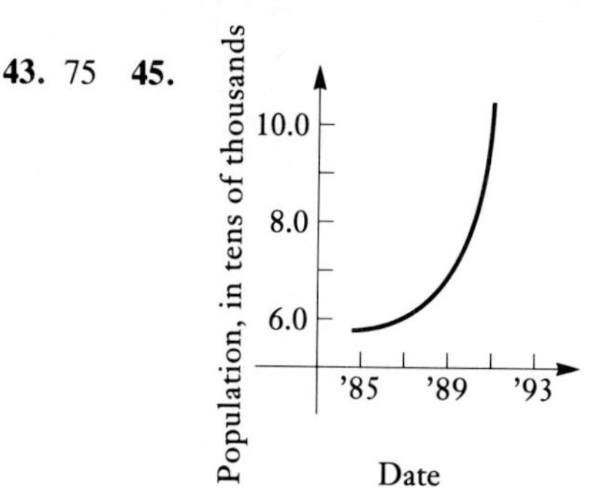

47. 64,258 **49.** Yes **51.** Yes **53.** No **55.** Yes **57.** No **59.** 18, 20, 22 **61.** $l = \dfrac{S - 2wh}{2h + 2w}$ **63. (a)** 3.1497708; **(b)** 55.7314683; **(c)** 3178.20675; **(d)** 1166.70323 **65.** Yes **67.** $g(x) = \frac{15}{4}x - \frac{13}{4}$ **69.** Yes; each activity has exactly one amount of calories associated with it. **71.** At 2 min, 40 sec and at 5 min, 40 sec **73.** 1 every 3 min, or $\frac{1}{3}$ per min

Exercise Set 6.3, pp. 258–259

1. Slope = 4; y-intercept = 5 **3.** Slope = −2; y-intercept = −6 **5.** Slope = $-\frac{3}{8}$; y-intercept = −0.2 **7.** Slope = 0.5; y-intercept = −9 **9.** Slope = 0; y-intercept = 7 **11.** Slope = 0; y-intercept = 3.7 **13.** $y = \frac{2}{3}x - 7$ **15.** $y = -4x + 2$ **17.** $y = -\frac{7}{9}x + 3$ **19.** $y = 5x + \frac{1}{2}$ **21.** $y = 0.7x + 3.8$

23. Slope = $\frac{5}{2}$; y-intercept = 1

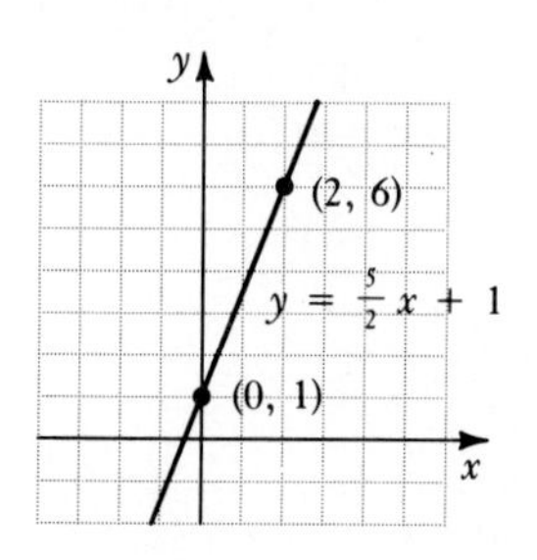

25. Slope = $-\frac{5}{2}$, y-intercept = 4

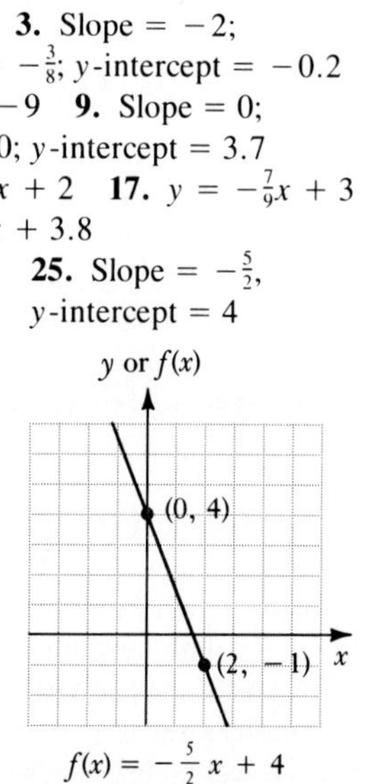

27. Slope = 2; y-intercept = −5

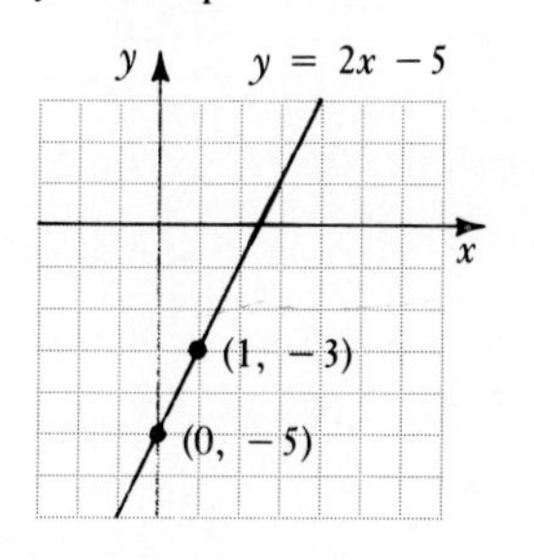

29. Slope = $\frac{1}{3}$; y-intercept = 6

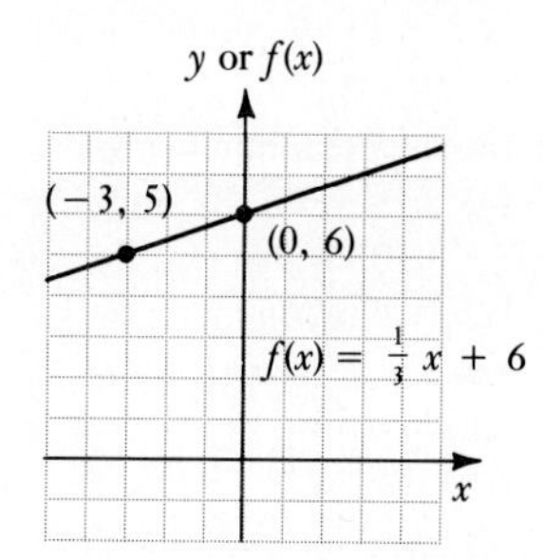

31. Slope = −0.25; y-intercept = 2

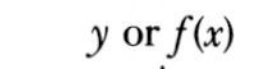

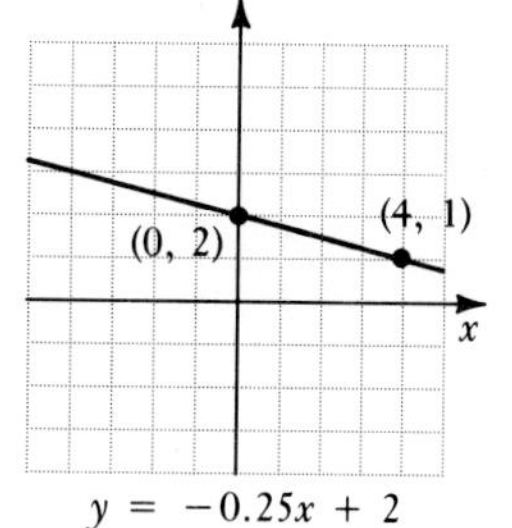

$y = -0.25x + 2$

33. Slope = $\frac{4}{5}$; y-intercept = −2

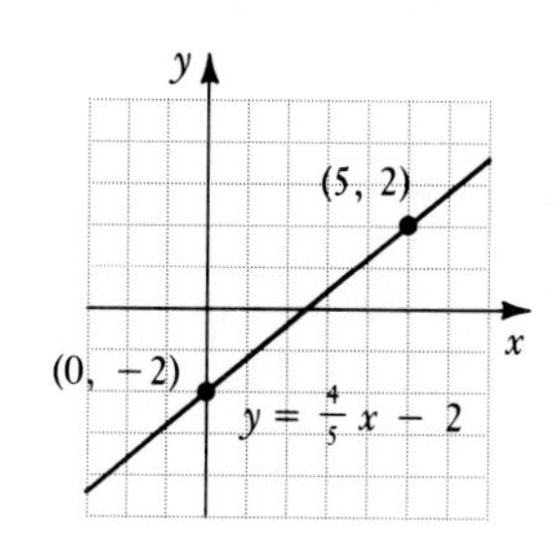

35. Slope = $\frac{5}{4}$; y-intercept = −2

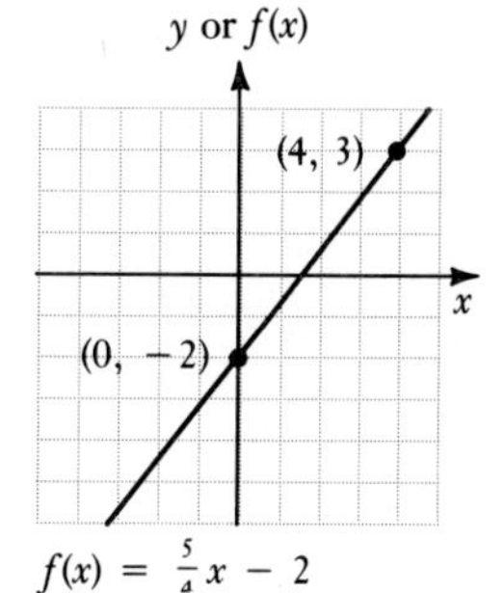

$f(x) = \frac{5}{4}x - 2$

37. Slope = 0; y-intercept = 4

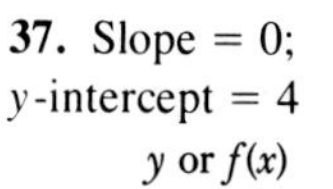

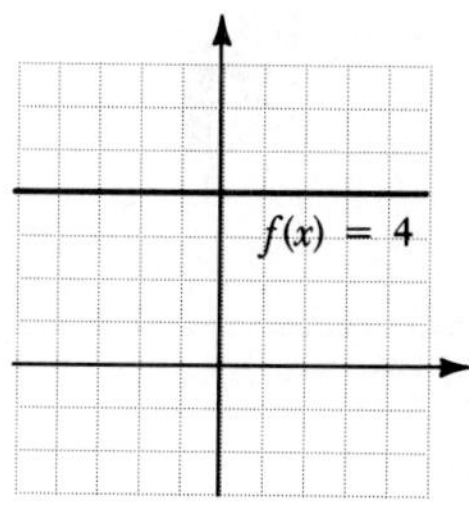

39. (a)

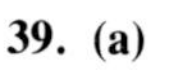

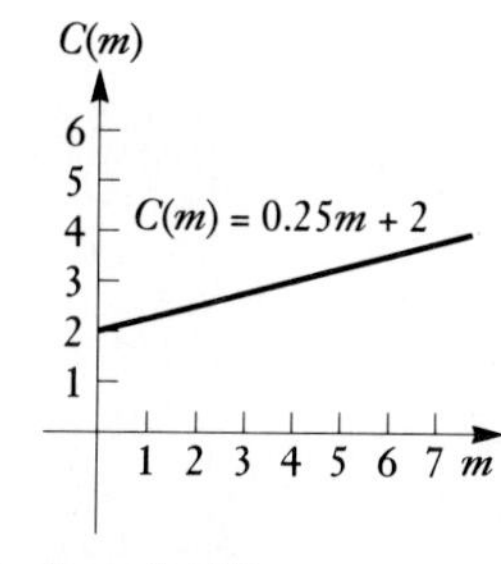

(b) about \$3.50

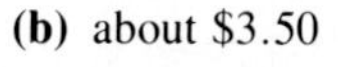

41. (a)

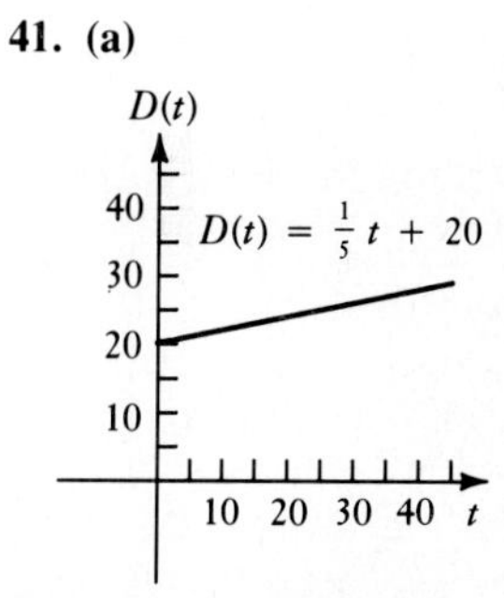

(b) about 27 quadrillion joules

43. (a)

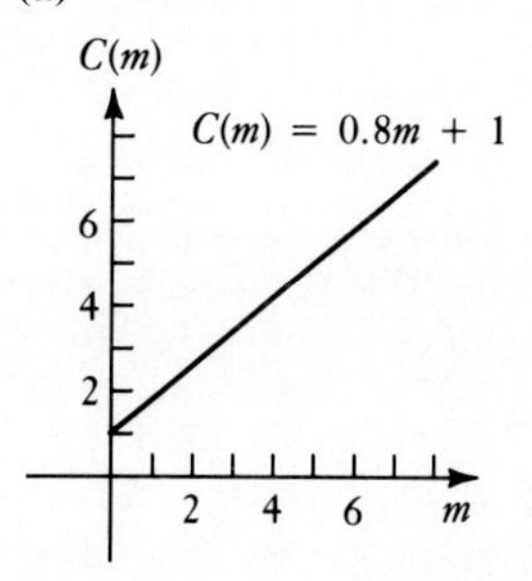

(b) about \$4.25; **(c)** about 8 min

45.

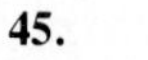

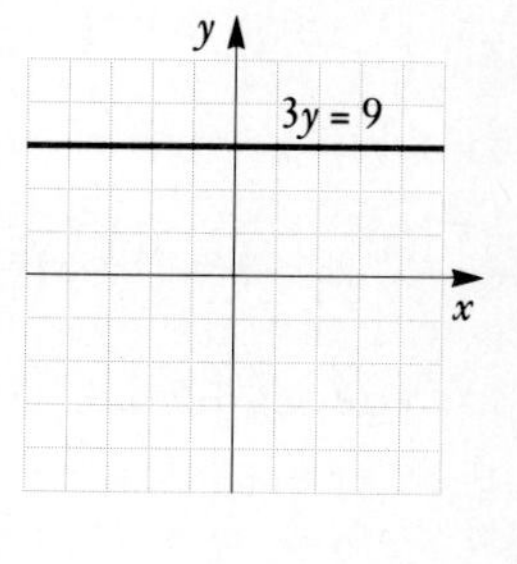

47.

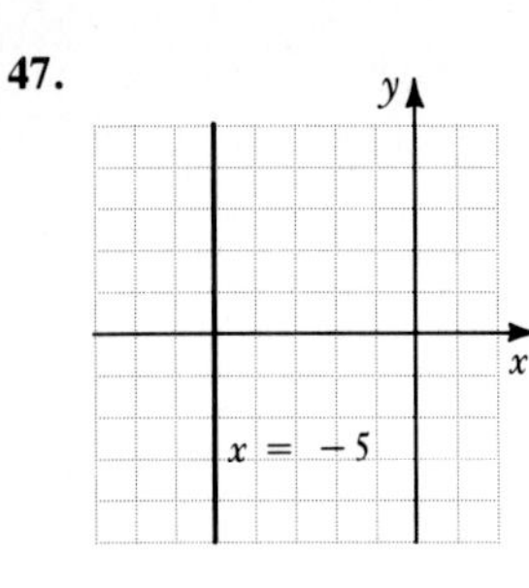

49.

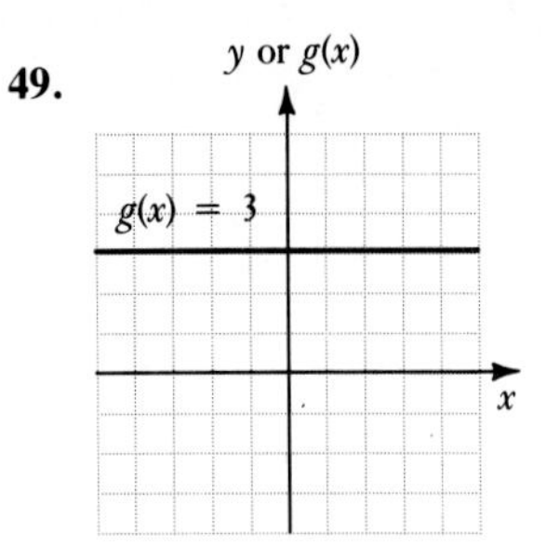

51.

$x = -\frac{8}{3}$

53. Linear; slope $= -\frac{3}{5}$

55. Linear **57.** Linear; slope $= -\frac{1}{2}$ **59.** Not linear **61.** Not linear **63.** Not linear **65.** $\sqrt{14x} - \sqrt{10}$ **67.** $45x + 54$ **69.** Slope $= -\dfrac{5}{a}$; y-intercept $= \dfrac{b}{a}$

71. Slope $= -\dfrac{a}{b}$; y-intercept $= \dfrac{c}{b}$ **73.** Linear **75.** Linear **77.** Linear **79.** $f(x) = \frac{47}{20}x + 3.1$

Exercise Set 6.4, pp. 265–266

For Exercises 1–15, each graph is a line through the intercepts given below.

1. $(0, -2), (2, 0)$ **3.** $(0, -1), \left(\frac{1}{3}, 0\right)$ **5.** $(0, -5), (4, 0)$ **7.** $(0, -5), (-1, 0)$ **9.** $(0, -3), (5, 0)$ **11.** $(0, 3), (-5, 0)$ **13.** $(0, -7), (2.8, 0)$ **15.** $\left(0, \frac{7}{2}\right), \left(\frac{7}{5}, 0\right)$ **17.** 2 **19.** -2 **21.** 5 **23.** $-\frac{1}{3}$ **25.** Slope is undefined. **27.** 0 **29.** -1 **31.** 10 km/h **33.** 0.6 ton per hour **35.** 300 ft per min **37.** \$0.4 million per year **39.** 3 **41.** Slope is undefined. **43.** 0 **45.** 0 **47.** Slope is undefined. **49.** $\frac{1}{2}$ **51.** Slope is undefined. **53.** 0 **55.** $7(x + 5)(x - 5)$

57. $F = \dfrac{f(c - v_s)}{c - v_0}$ **59.** Answers may vary: $(50, -2)$, $(25, -1)$, $(-25, 1)$, $(-50, 2)$ **61.** $\left(-\dfrac{b}{m}, 0\right)$

63. **(a)** $-\dfrac{5c}{4b}$; **(b)** slope is undefined; **(c)** $\dfrac{a + d}{f}$

Exercise Set 6.5, pp. 271–274

1. $y - 2 = 4(x - 3)$ **3.** $y - 7 = -2(x - 4)$ **5.** $y + 4 = 3(x + 2)$ **7.** $y = -2(x - 8)$ **9.** $y + 7 = 0$ **11.** $y + 1 = \frac{3}{4}(x - 5)$ **13.** $y = 5x - 13$

15. $y = -\frac{2}{3}x - \frac{13}{3}$ **17.** $y = -0.6x - 5.8$ **19.** $f(x) = \frac{1}{2}x + \frac{7}{2}$ **21.** $f(x) = x$ **23.** $f(x) = \frac{5}{2}x + 5$ **25.** $f(x) = \frac{1}{4}x + \frac{17}{4}$ **27.** $f(x) = \frac{2}{5}x$ **29.** $f(x) = 3x + 5$ **31.** **(a)** $R = -0.075t + 46.8$; **(b)** 42.3 seconds, 41.6 seconds; **(c)** 2021 **33.** **(a)** $A = -2.5x + 16.5$, where A represents millions of pounds of coffee sold and x is the price, in dollars; **(b)** 11.5 million lb **35.** **(a)** $A = 2x - 3$, where A is in millions of pounds and x is in dollars; **(b)** 1 million lb **37.** **(a)** $P = \frac{3}{100}d + 1$, where P is pressure in atmospheres and d is depth in feet; **(b)** 21.7 atm **39.** 3263 **41.** Yes **43.** No **45.** Yes **47.** $y = -\frac{1}{2}x + \frac{17}{2}$ **49.** $y = \frac{5}{7}x - \frac{17}{7}$ **51.** $y = \frac{1}{3}x + 4$ **53.** $y = -\frac{3}{2}x - \frac{13}{2}$ **55.** Yes **57.** No **59.** $y = \frac{1}{2}x + 4$ **61.** $y = \frac{4}{3}x - 6$ **63.** $y = \frac{5}{2}x + 9$

65. $y = -\frac{5}{3}x - \frac{41}{3}$ **67.** $\mathbb{R}$ **69.** $\dfrac{2x + 9}{(x + 3)(x - 3)}$

71. 100.03916 cm, 99.96796 cm **73.** **(a)** Plan A: $E = 600 + 0.04x$; plan B:

$$E = \begin{cases} 700, & \text{if } x \le 10{,}000, \\ 700 + 0.06(x - 10{,}000), & \text{if } x > 10{,}000; \end{cases}$$

(b) $x > 25{,}000$ or $x < 2500$ **75.** **(a)** $f(x) = \frac{1}{3}x + \frac{10}{3}$; **(b)** $f(3) = \frac{13}{3}$; **(c)** $a = 290$ **77.** $k = -\frac{40}{9}$

Exercise Set 6.6, pp. 280–282

1. 1 **3.** 7 **5.** -29 **7.** -41 **9.** -30 **11.** 110 **13.** $\frac{1}{2}$ **15.** $\frac{10}{11}$ **17.** 13 **19.** $x^2 - x + 1$ **21.** 5 **23.** 2 **25.** 42 **27.** $-\frac{3}{4}$ **29.** $\frac{1}{6}$ **31.** $\{x|x$ is a real number$\}$
33. $\{x|x$ is a real number and $x \neq 2\}$
35. $\{x|x$ is a real number and $x \neq 0\}$
37. $\{x|x$ is a real number and $x \neq 1\}$
39. $\{x|x$ is a real number and $x \neq 2$ and $x \neq 4\}$
41. $\{x|x$ is a real number and $x \neq -2$ and $x \neq 4\}$
43. $\{x|x$ is a real number and $x \neq 3\}$
45. $\{x|x$ is a real number and $x \neq 4\}$
47. $\{x|x$ is a real number and $x \neq 0$ and $x \neq 4\}$
49. $\{x|x$ is a real number and $x \neq 4$ and $x \neq 5\}$
51. $\{x|x$ is a real number and $x \neq -2.5$ and $x \neq -1\}$
53. $\{x|x$ is a real number and $x \neq 0$ and $x \neq 4\}$
55. $\{x|x$ is a real number and $x \neq 1$ and $x \neq 4\}$
57. $\{x|x$ is a real number and $x \neq 2$, $x \neq 3$, and $x \neq 4\}$
59. $\left\{x|x\right.$ is a real number and $x \neq -2$, $x \neq \frac{5}{3}$, and $\left.x \neq -1\right\}$
61. $\left\{x|x\right.$ is a real number and $x \neq \frac{5}{2}$, $x \neq 0$, and $\left.x \neq \frac{1}{3}\right\}$
63. $\frac{7}{5}$ **65.** $3(2 - x)(4 + 2x + x^2)$
67. $\{x|x$ is a real number and $x \neq 2$, $x \neq 3$, and $x \neq 1\}$
69. For $f + g$, $f - g$, and $f \cdot g$, domain $= \{-2, -1, 0, 1\}$; for f/g, domain $= \{-2, 0, 1\}$.
71. $\{x|x$ is a real number and $x > 5\}$
73. For $f + g$, $f - g$, and $f \cdot g$, domain $= \{x|-3 \le x \le 5\}$; for f/g, domain $= \{x|-3 \le x \le 5$ and $x \neq -1$ and $x \neq 3\}$.
75. Answers vary. Graph of g must have $g(1) = 0$ or $g(1)$ undefined and $g(x) \neq 0$ for $-2 \le x \le 3$.

Exercise Set 6.7, pp. 288–292

1. $k = 8; y = 8x$ **3.** $k = 3.6; y = 3.6x$ **5.** $k = 5; y = 5x$ **7.** $k = \frac{15}{4}; y = \frac{15}{4}x$ **9.** $k = 1.6; y = 1.6x$
11. 6 amperes **13.** 125,000 **15.** 532,500 tons
17. 40 lb **19.** 50 volts **21.** $k = 60; y = \dfrac{60}{x}$ **23.** $k = 12; y = \dfrac{12}{x}$ **25.** $k = 36; y = \dfrac{36}{x}$ **27.** $k = 9; y = \dfrac{9}{x}$
29. $\frac{2}{9}$ ampere **31.** 160 cm^3 **33.** $6\frac{2}{3}$ hours **35.** $y = 15x^2$
37. $y = \dfrac{0.0015}{x^2}$ **39.** $y = xz$ **41.** $y = 0.3xz^2$
43. $y = \dfrac{15x}{7z^2}$ **45.** $y = \dfrac{4wx^2}{z}$ **47.** $y = \dfrac{xz}{5wp}$ **49.** 36 mph
51. 6.25 km **53.** 94.03 lb **55.** 97 **57.** 2 mm
59. 8.17 mph **61.** $\dfrac{7\sqrt[3]{25x}}{5x}$ **63.** $(1 + 4x)(1 - 4x)$
65. y is tripled. **67.** y is multiplied by $\dfrac{1}{n^2}$. **69.** $\dfrac{\pi}{4}$
71. y is multiplied by 8. **73.** $\dfrac{6\sqrt{2}}{5} \approx 1.697$ m
75. Better to go from 5 to 10 employees. Expanding from 5 to 10 employees results in half as many complaints. Expanding from 20 to 25 results in four-fifths as many complaints.

Review Exercises: Chapter 6, pp. 293–294

1. [6.1] Yes **2.** [6.1] No **3.** [6.1] No **4.** [6.1] Yes
5. [6.3]

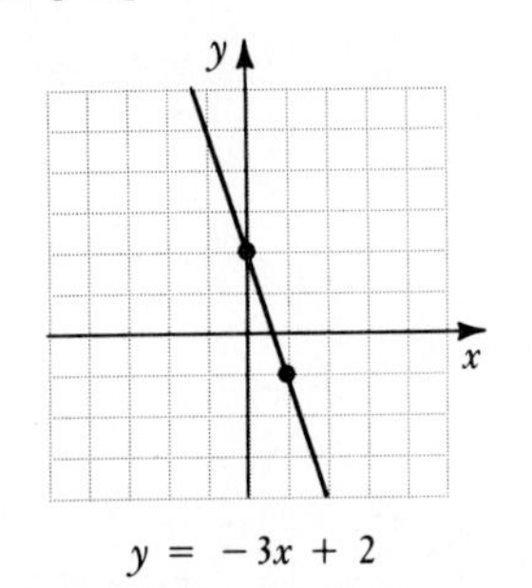

$y = -3x + 2$

6. [6.1]

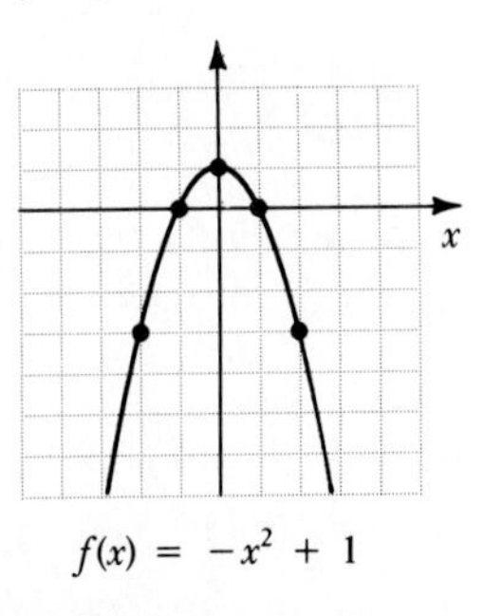

$f(x) = -x^2 + 1$

7. [6.3]

$8x + 32 = 0$

8. [6.3] Slope is -4; y-intercept is -9.

9. [6.3] Slope is $\frac{1}{3}$; y-intercept is $-\frac{7}{6}$. **10.** [6.3] Yes
11. [6.3] Yes **12.** [6.3] No **13.** [6.3] No

14. [6.4]—

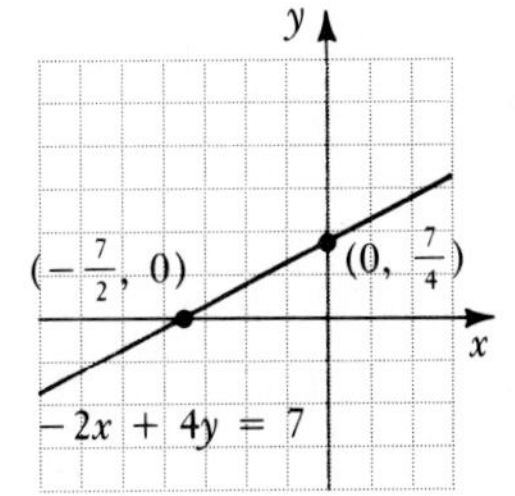

15. [6.4] $\frac{4}{7}$ **16.** [6.4] Slope is undefined.

17. [6.5] $y - 4 = -2(x + 3)$, or $y = -2x - 2$
18. [6.5] $y - 5 = \frac{4}{3}(x - 2), y + 3 = \frac{4}{3}(x + 4)$, or $y = \frac{4}{3}x + \frac{7}{3}$
19. [6.5] Perpendicular **20.** [6.5] Parallel
21. [6.5] Parallel **22.** [6.5] $y = \frac{3}{5}x - \frac{31}{5}$
23. [6.5] $y = -\frac{5}{3}x - \frac{5}{3}$ **24.** [6.2] -5 **25.** [6.2] -8
26. [6.6] 57 **27.** [6.6] -10 **28.** [6.6] $-\frac{7}{4}$
29. [6.2] $2a - 3$ **30.** [6.6] $\{x \mid x \text{ is a real number}\}$
31. [6.6] $\left\{x \mid x \text{ is a real number and } x \neq \frac{5}{2}\right\}$
32. [6.7] $y = \dfrac{5xz}{6w}$ **33.** [6.7] 500 watts **34.** [5.3] 1
35. [3.4] $25a^6b^2$ **36.** [2.1] -26 **37.** [4.2] $\dfrac{3(x - 4)}{x - 2}$
38. [6.4] -9 **39.** [6.5] $-\frac{9}{2}$

Test: Chapter 6, pp. 294–295

1. [6.1] Yes **2.** [6.1] No **3.** [6.1] No **4.** [6.1] No
5. [6.3] **6.** [6.1]

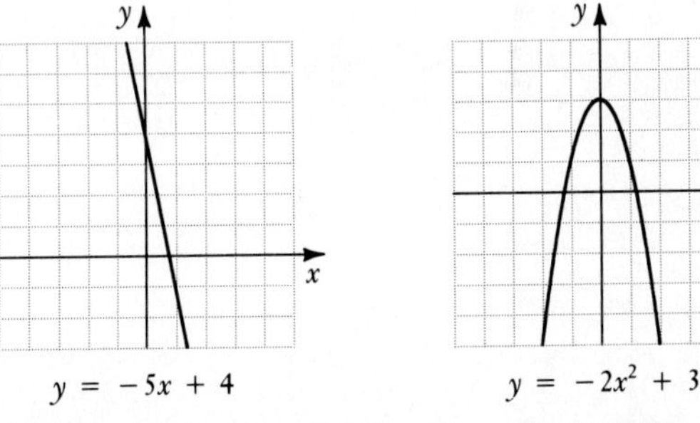

$y = -5x + 4$ $\qquad$ $y = -2x^2 + 3$

7. [6.3] Slope is 3; y-intercept is -5. **8.** [6.3] Slope is $\frac{4}{3}$; y-intercept is -3.
9. [6.3]

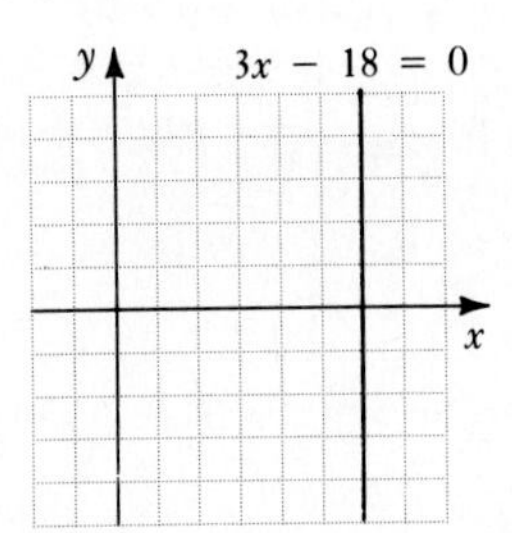

10. [6.3] **(a)** Linear; **(b)** nonlinear; **(c)** linear

11. [6.4]

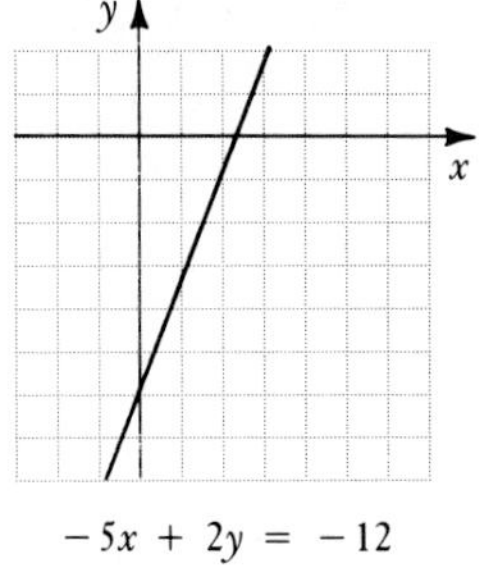

$-5x + 2y = -12$

12. [6.4] $\frac{5}{8}$ **13.** [6.4] 0

14. [6.4] **(a)** Zero slope; **(b)** undefined slope
15. [6.5] $y + 4 = 4(x + 2)$, or $y = 4x + 4$
16. [6.5] $y + 1 = -(x - 3)$, $y + 2 = -(x - 4)$, or $y = -x + 2$ **17.** [6.5] Parallel **18.** [6.5] Perpendicular
19. [6.5] $y = \frac{2}{5}x + \frac{16}{5}$ **20.** [6.5] $y = -\frac{5}{2}x - \frac{11}{2}$
21. [6.2], [6.6] **(a)** -4; **(b)** 5; **(c)** -2; **(d)** -5
22. [6.5] **(a)** $C = 0.3m + 25$; **(b)** \$175
23. [6.7] 6.664 cm^2 **24.** [2.1] All real numbers
25. [4.2] $\dfrac{3y - 7}{y - 5}$ **26.** [5.3] $\dfrac{\sqrt[4]{2x^2}}{x}$ **27.** [3.4] $(y - 4)^2$
28. [6.2], [6.4] **(a)** 30 miles; **(b)** 15 mph
29. [6.5] $s = -\frac{3}{2}r + \frac{27}{2}$

48. [6.3] **49.** [6.4]

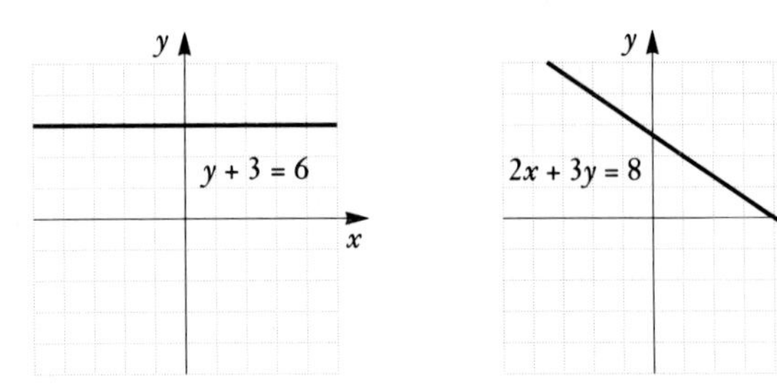

50. [6.3]

y
$y = -x + 4$
x

51. [6.7] 1.5 hr

52. [3.5] 7 cm, 10 cm **53.** [2.2] 3 hr **54.** [4.5] 6 mph
55. [5.6] $6 + 2i$ **56.** [6.6] $\{x \mid x$ is a real number and $x \neq 0\}$
57. [5.5], [5.6] $\pm\sqrt{5}$, $\pm\sqrt{2}i$ **58.** [4.4] $\{x \mid x$ is a real number and $x \neq 9$ and $x \neq -5\}$ **59.** [5.5] -6000
60. [3.5] 0, 1, -1, 5

CUMULATIVE REVIEW: CHAPTERS 1–6, pp. 296–297

1. [1.2] $-\frac{7}{10}$ **2.** [1.2] $\frac{1}{10}$ **3.** [1.3] $\frac{13}{4}$ **4.** [1.4] $yx + 3$, $3 + xy$, or $3 + yx$ **5.** [2.1] $\frac{61}{8}$ **6.** [2.1] $\frac{1}{6}$ **7.** [2.4] $\{x \mid x \geq -2\}$ or $[-2, \infty)$ **8.** [2.5] $\left\{x \mid -4 \leq x \leq -\frac{7}{2}\right\}$ or $\left[-4, -\frac{7}{2}\right]$ **9.** [2.5] $\left\{x \mid x < \frac{1}{2} \text{ or } x > \frac{5}{2}\right\}$ or $\left(-\infty, \frac{1}{2}\right) \cup \left(\frac{5}{2}, \infty\right)$
10. [2.6] $\left\{x \mid x \leq -\frac{33}{8} \text{ or } x \geq \frac{31}{8}\right\}$ or $\left(-\infty, -\frac{33}{8}\right] \cup \left[\frac{31}{8}, \infty\right)$
11. [2.6] 13, -23 **12.** [3.5] 0, 3 **13.** [5.5] $\pm 2\sqrt{5}$
14. [3.5] $-\frac{7}{2}, -\frac{2}{3}$ **15.** [4.4] $\frac{22}{7}$ **16.** [4.4] $-\frac{1}{4}$
17. [5.5] 12 **18.** [2.3] $a_2 = An - a_1 - a_3$ **19.** [3.1] -6
20. [3.4] $\dfrac{16x^8}{y^{12}}$ **21.** [3.1] $11x^2 + 3x - 4$ **22.** [3.1] $2y^2 + 5y$ **23.** [3.2] $6x^3 + x^2y - 4xy^2 + y^3$ **24.** [3.2] $6x^2 + 3x - 63$ **25.** [3.6] $2x^2 - 3xy + y^2$ **26.** [3.6] $x^2 + 4 + \dfrac{6}{x^2 - 1}$ **27.** [4.1] $\dfrac{(x + 2)(x - 2)}{(x - 3)(x - 1)}$ **28.** [4.1] $\dfrac{2(x + 2)}{(x - 1)^2}$ **29.** [4.2] $x + y$ **30.** [4.2] $\dfrac{x^2 + 1}{x(x - 1)}$ **31.** [3.4] $(2x - 3)^2$ **32.** [3.4] $y(3y + 2)(9y^2 - 6y + 4)$ **33.** [3.4] $3y^2(3y + 2)(3y - 2)$ **34.** [3.4] $(x + 1)(x - 1)(x + 2)$
35. [3.3] Not factorable **36.** [3.3] $(y^2 + 48)(y^2 + 1)$
37. [4.3] $\dfrac{5x - 1}{3x + 9}$ **38.** [5.1] $|6 - x|$ **39.** [5.1] 1
40. [5.2] $2xy^2\sqrt[3]{2xy^2}$ **41.** [5.1] $2\sqrt[3]{10}$ **42.** [5.3] $3 + 2\sqrt{x} - x$ **43.** [5.3] $21\sqrt{2}$ **44.** [5.4] $27x\sqrt{x}$
45. [5.4] $\dfrac{x^{-2}}{5}$ **46.** [6.5] $y = -x + 3$ **47.** [6.5] $y = \frac{1}{2}x$

CHAPTER 7

Exercise Set 7.1, pp. 305–306

1. $\pm\sqrt{5}$ **3.** 0 **5.** $\pm\frac{1}{4}$ **7.** ± 3 **9.** $\pm\dfrac{\sqrt{6}}{2}$
11. $\pm\dfrac{\sqrt{15}}{3}$ **13.** $\pm 10i$ **15.** $\pm i\sqrt{5}$ **17.** $\pm\frac{2}{5}i$
19. $\pm i\sqrt{7}$ **21.** $\pm\frac{3}{2}$ **23.** 0, 5 **25.** 0, -2 **27.** 0, $\frac{2}{3}$
29. 0, $-\frac{9}{14}$ **31.** 0, $\frac{11}{9}$ **33.** $x^2 + 10x + 25$, $(x + 5)^2$
35. $x^2 - 8x + 16$, $(x - 4)^2$ **37.** $x^2 - 24x + 144$, $(x - 12)^2$ **39.** $x^2 + 9x + \frac{81}{4}$, $\left(x + \frac{9}{2}\right)^2$ **41.** $x^2 - 7x + \frac{49}{4}$, $\left(x - \frac{7}{2}\right)^2$ **43.** $x^2 + \frac{2}{3}x + \frac{1}{9}$, $\left(x + \frac{1}{3}\right)^2$ **45.** $x^2 - \frac{5}{6}x + \frac{25}{144}$, $\left(x - \frac{5}{12}\right)^2$ **47.** $x^2 + \frac{9}{5}x + \frac{81}{100}$, $\left(x + \frac{9}{10}\right)^2$ **49.** $-7, -1$
51. $5 \pm \sqrt{47}$ **53.** $-5, -1$ **55.** 3, 7 **57.** $-3, -2$
59. $-2 \pm \sqrt{3}$ **61.** $5 \pm \sqrt{2}$ **63.** $-3 \pm 2i$ **65.** $-\frac{1}{2}, 3$
67. $-\frac{3}{2}, -\frac{1}{2}$ **69.** $-\frac{1}{2}, \frac{2}{3}$ **71.** $-1 \pm \dfrac{\sqrt{2}}{2}$
73. $\dfrac{5 \pm \sqrt{61}}{6}$

75.

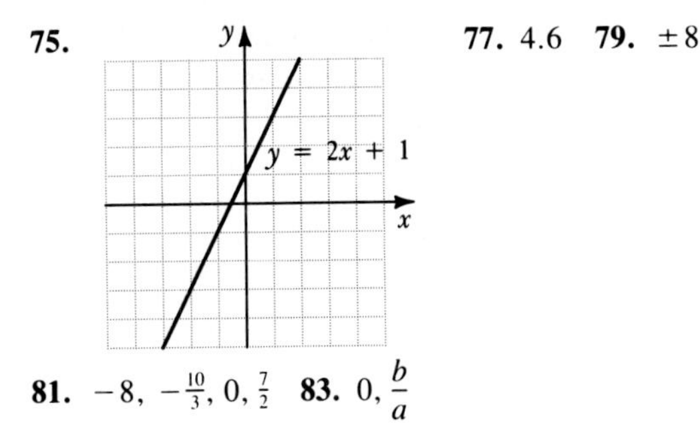

77. 4.6 **79.** ± 8

81. $-8, -\frac{10}{3}, 0, \frac{7}{2}$ **83.** $0, \dfrac{b}{a}$

Exercise Set 7.2, pp. 309–310

1. $-3 \pm \sqrt{5}$ **3.** $1, -5$ **5.** $3, -10$ **7.** $2, -\frac{1}{2}$ **9.** $-1, -\frac{5}{3}$ **11.** $\dfrac{1 \pm i\sqrt{3}}{2}$ **13.** $2 \pm 3i$ **15.** $\pm i\sqrt{5}$ **17.** $\dfrac{-3 \pm \sqrt{41}}{2}$ **19.** $-1 \pm 2i$ **21.** $1 \pm 2i$ **23.** $\pm\dfrac{\sqrt{10}}{2}$ **25.** $0, -1$ **27.** $\dfrac{-1 \pm 2i}{5}$ **29.** $-2, \frac{3}{4}$ **31.** $\dfrac{1 \pm 3i}{2}$ **33.** $\frac{3}{2}, \frac{2}{3}$ **35.** $1.3, -5.3$ **37.** $5.2, 0.8$ **39.** $2.8, -1.3$ **41.** 30 lb of cereal, 20 lb of granola **43.** $0.5700731, -0.7973459$ **45.** $-0.6567764, 0.4567764$ **47.** $1.1753905, -0.4253905$ **49.** $\dfrac{-1 \pm \sqrt{1 + 4\sqrt{2}}}{2}$ **51.** $\dfrac{-\sqrt{5} \pm \sqrt{5 + 4\sqrt{3}}}{2}$ **53.** $\dfrac{-3 \pm \sqrt{9 - 4i}}{2}$ **55.** $\dfrac{-y \pm \sqrt{-47y^2 + 108}}{6}$ **57.** $\frac{1}{2}$

Exercise Set 7.3, pp. 312–313

1. One real **3.** Two nonreal **5.** Two real **7.** One real **9.** Two nonreal **11.** Two real **13.** Two real **15.** Two real **17.** Two real **19.** Two nonreal **21.** One real **23.** Two nonreal **25.** $x^2 + 2x - 99 = 0$ **27.** $x^2 - 14x + 49 = 0$ **29.** $x^2 + 8x + 15 = 0$ **31.** $3x^2 - 14x + 8 = 0$ **33.** $6x^2 - 5x + 1 = 0$ **35.** $25x^2 - 20x - 12 = 0$ **37.** $x^2 - 4\sqrt{2}x + 6 = 0$ **39.** $x^2 + 3\sqrt{5}x + 10 = 0$ **41.** $x^2 + 9 = 0$ **43.** $x^2 + 2x + 2 = 0$ **45.** $x^2 - 10x + 29 = 0$ **47.** $2x^2 - 2x + 5 = 0$ **49.** Six 30-second commercials; six 60-second commercials **51. (a)** $k = -\frac{3}{5}$; **(b)** $-\frac{1}{3}$ **53. (a)** $k = 9 + 9i$; **(b)** $3 + 3i$ **55.** $x^2 - \sqrt{3}x + 8 = 0$ **57.** The solutions of $ax^2 + bx + c = 0$ are given by

$$x = \frac{-b \pm \sqrt{b^2 - 4ac}}{2a}.$$

(a) $\dfrac{-b + \sqrt{b^2 - 4ac}}{2a} + \dfrac{-b - \sqrt{b^2 - 4ac}}{2a} = \dfrac{-2b}{2a} = \dfrac{-b}{a}$;

(b) $\dfrac{-b + \sqrt{b^2 - 4ac}}{2a} \cdot \dfrac{-b - \sqrt{b^2 - 4ac}}{2a} =$

$\dfrac{(-b)^2 - (\sqrt{b^2 - 4ac})^2}{4a^2} = \dfrac{b^2 - b^2 + 4ac}{4a^2} = \dfrac{4ac}{4a^2} = \dfrac{c}{a}.$

Exercise Set 7.4, pp. 320–321

1.

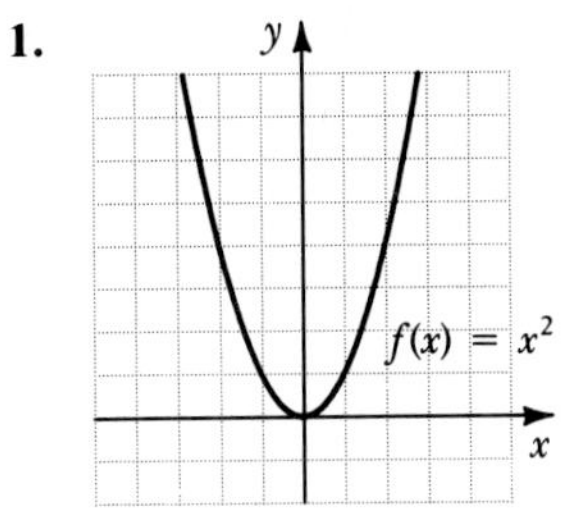

3.

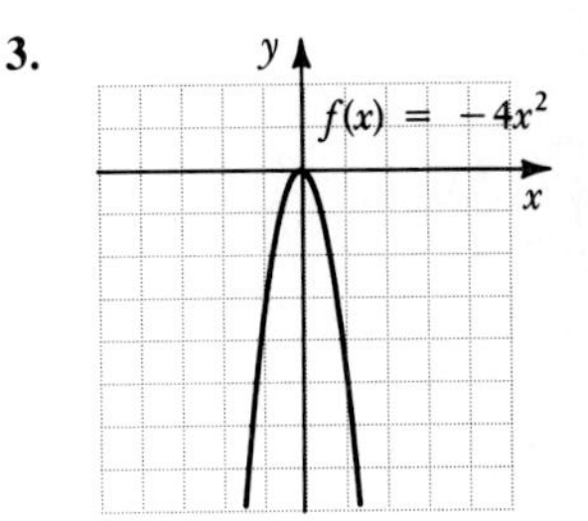

5.

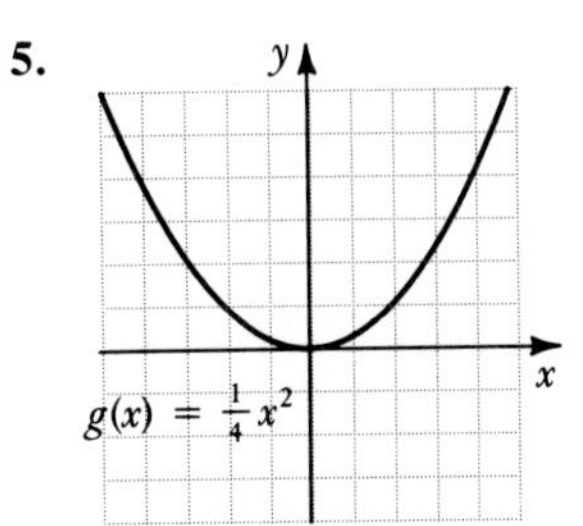

7

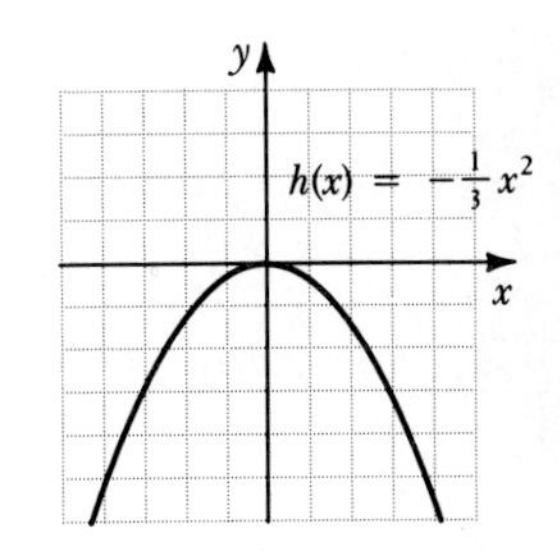

9.

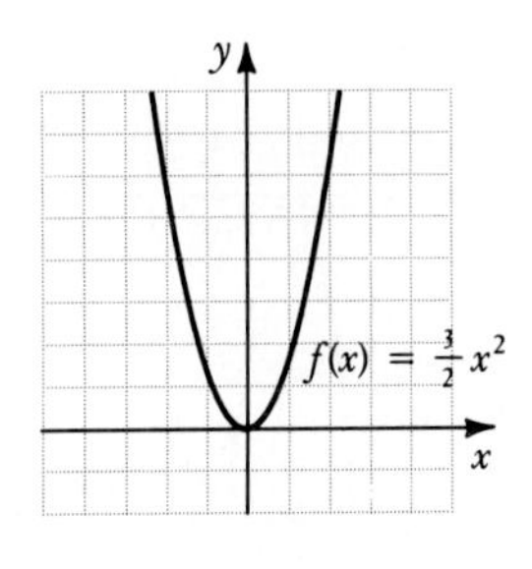

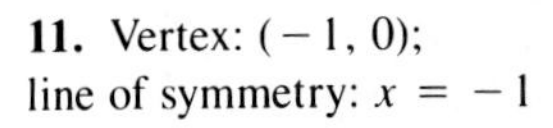
11. Vertex: $(-1, 0)$; line of symmetry: $x = -1$

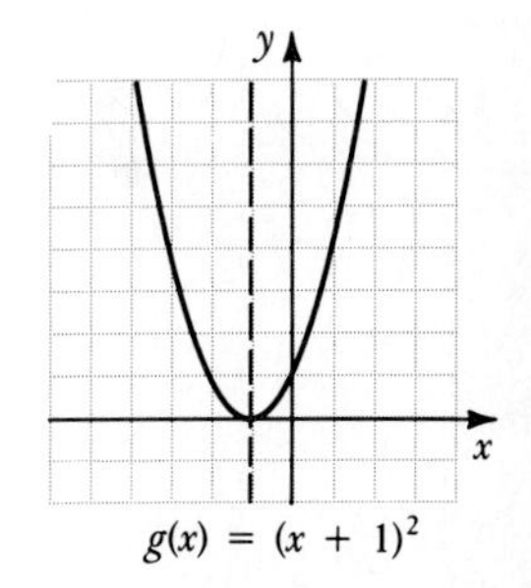

13. Vertex: $(4, 0)$; line of symmetry: $x = 4$

y

x

$f(x) = (x - 4)^2$

15. Vertex: $(3, 0)$; line of symmetry: $x = 3$

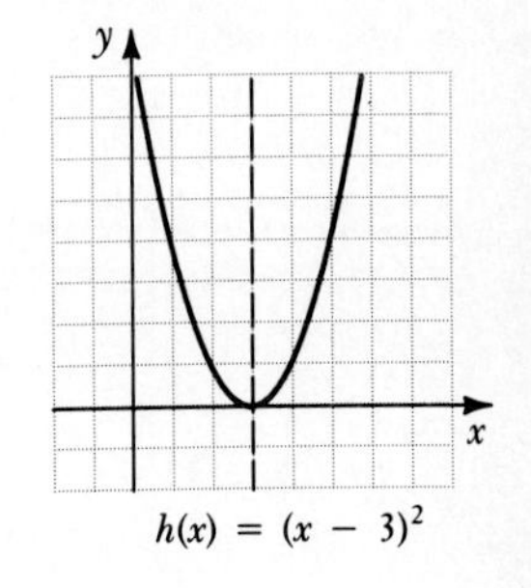

17. Vertex: $(-4, 0)$; line of symmetry: $x = -4$

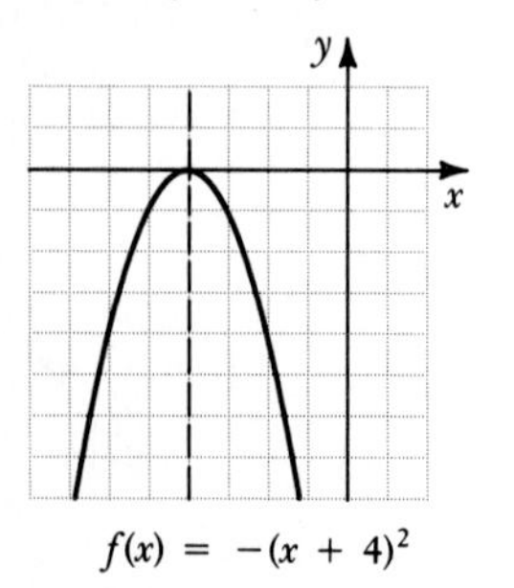

$f(x) = -(x + 4)^2$

19. Vertex: $(1, 0)$; line of symmetry: $x = 1$

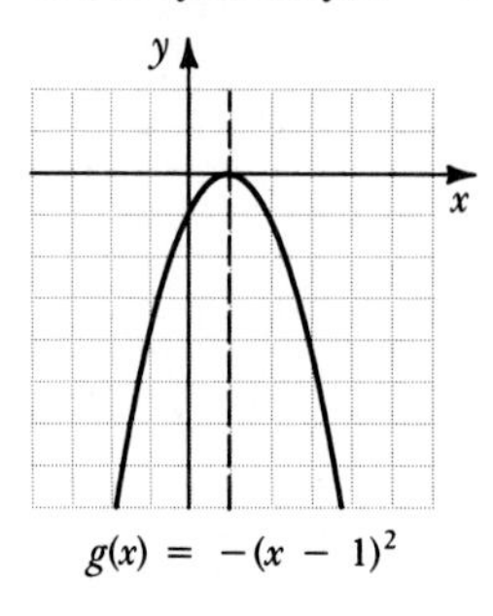

$g(x) = -(x - 1)^2$

21. Vertex: $(1, 0)$; line of symmetry: $x = 1$

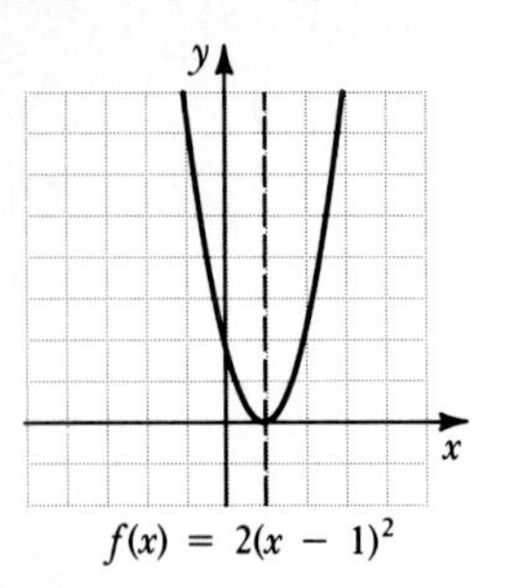

$f(x) = 2(x - 1)^2$

23. Vertex: $(3, 0)$; line of symmetry: $x = 3$

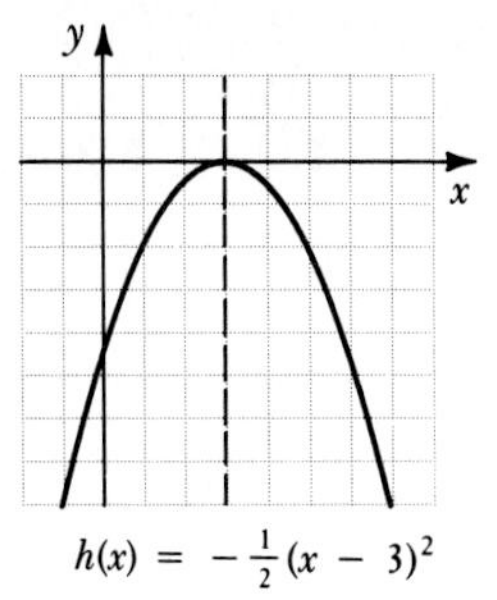

$h(x) = -\frac{1}{2}(x - 3)^2$

25. Vertex: $(-1, 0)$; line of symmetry: $x = -1$

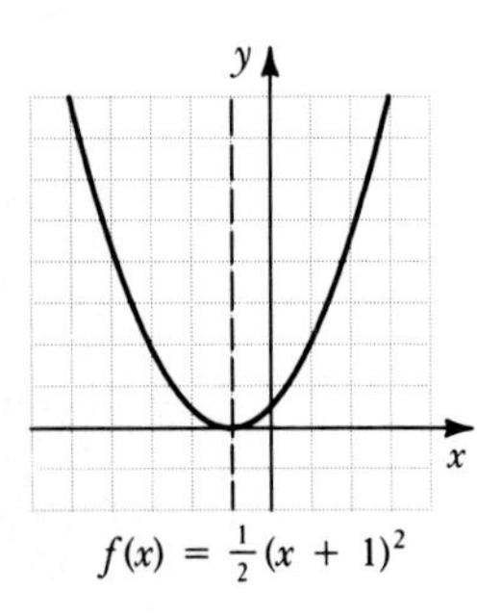

$f(x) = \frac{1}{2}(x + 1)^2$

27. Vertex: $(2, 0)$; line of symmetry: $x = 2$

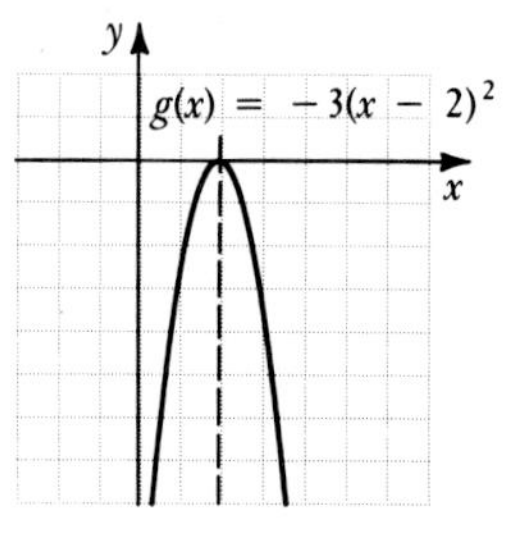

$g(x) = -3(x - 2)^2$

29. Vertex: $(-9, 0)$; line of symmetry: $x = -9$

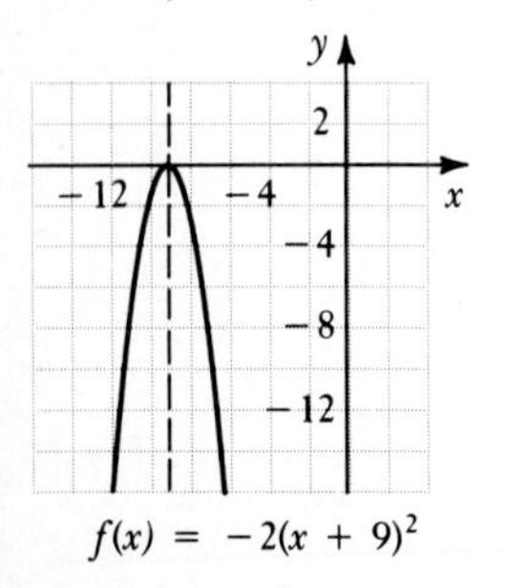

$f(x) = -2(x + 9)^2$

31. Vertex: $\left(\frac{1}{2}, 0\right)$; line of symmetry: $x = \frac{1}{2}$

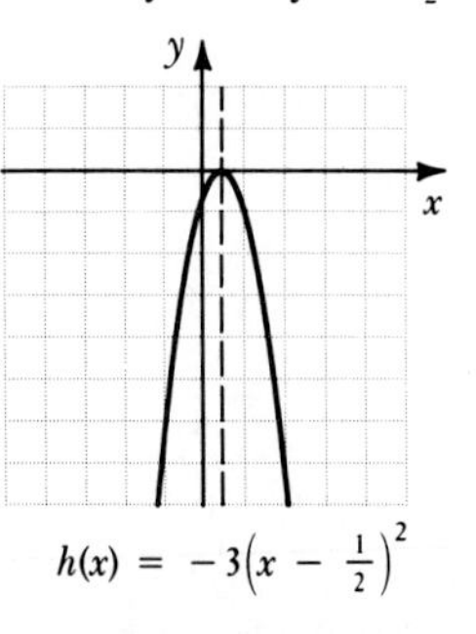

$h(x) = -3\left(x - \frac{1}{2}\right)^2$

33. Vertex: $(3, 1)$; line of symmetry: $x = 3$; minimum: 1

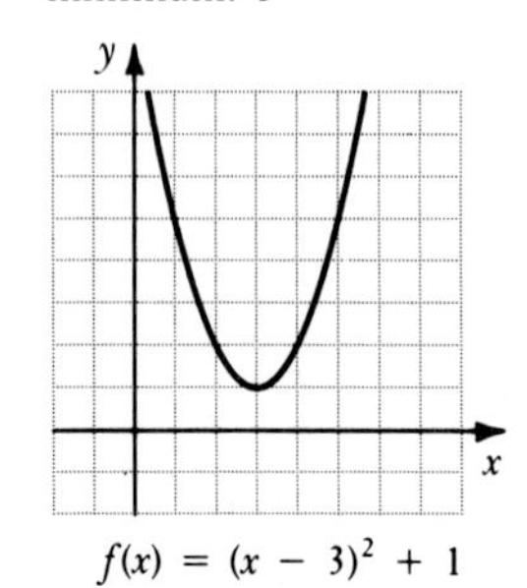

$f(x) = (x - 3)^2 + 1$

35. Vertex: $(-1, -2)$; line of symmetry: $x = -1$; minimum: -2

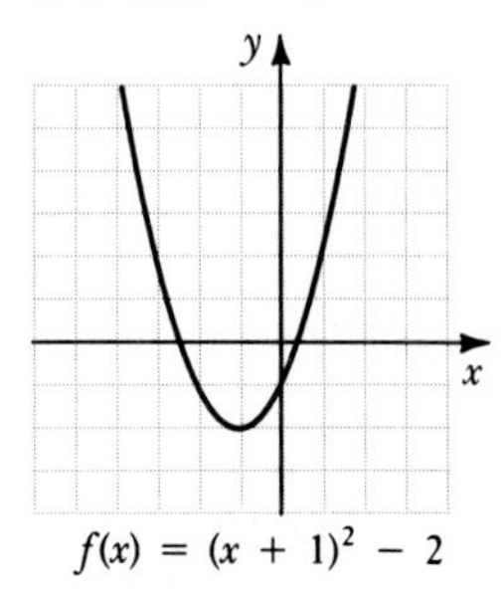

$f(x) = (x + 1)^2 - 2$

37. Vertex: $(-4, 1)$; line of symmetry: $x = -4$; minimum: 1

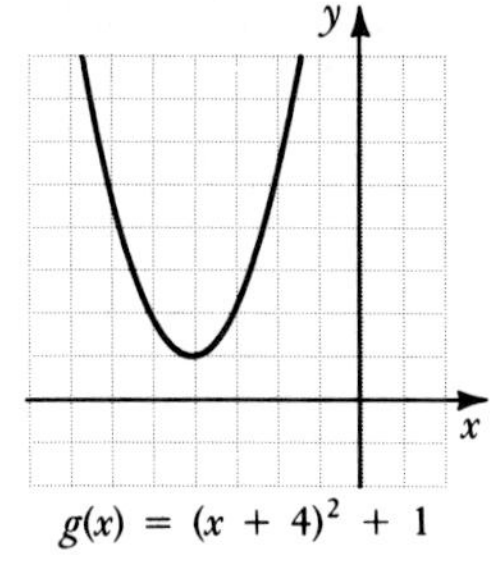

$g(x) = (x + 4)^2 + 1$

39. Vertex: $(5, 2)$; line of symmetry: $x = 5$; minimum: 2

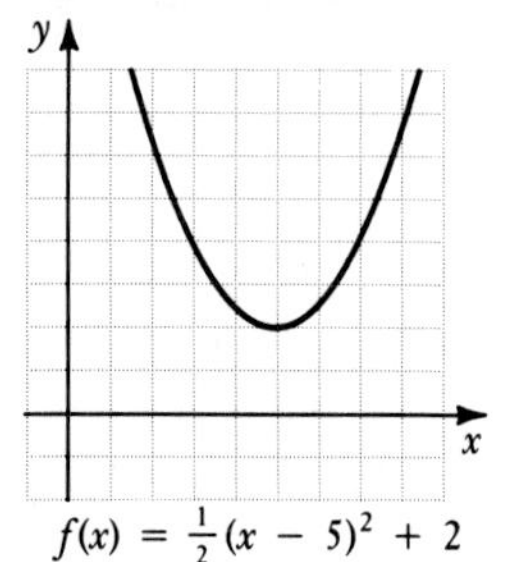

$f(x) = \frac{1}{2}(x - 5)^2 + 2$

41. Vertex: $(1, -3)$; line of symmetry: $x = 1$; maximum: -3

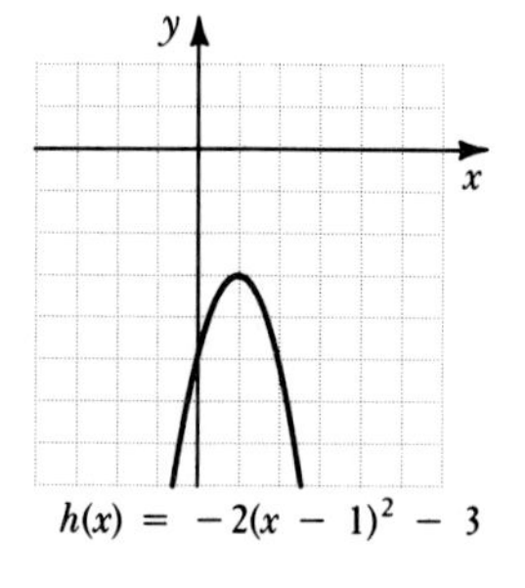

$h(x) = -2(x - 1)^2 - 3$

43. Vertex: $(-4, 1)$; line of symmetry: $x = -4$; maximum: 1

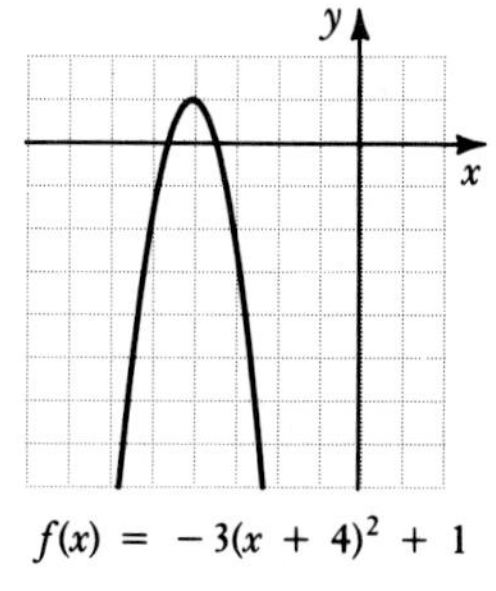

$f(x) = -3(x + 4)^2 + 1$

45. Vertex: $(1, 2)$; line of symmetry: $x = 1$; maximum: 2

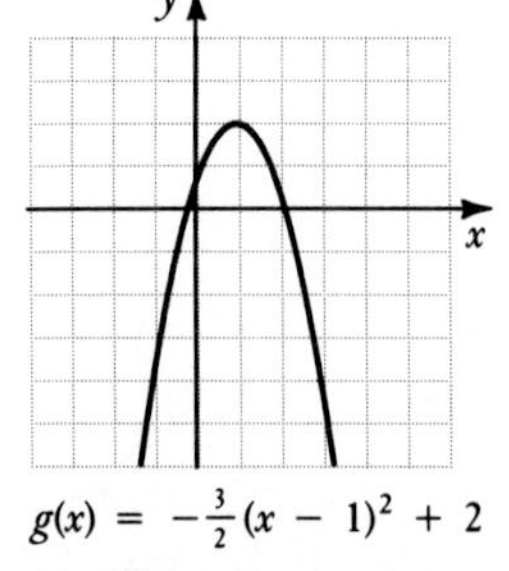

$g(x) = -\frac{3}{2}(x - 1)^2 + 2$

47. Vertex: (9, 5); line of symmetry: $x = 9$; minimum: 5 **49.** Vertex: $(-6, 11)$; line of symmetry: $x = -6$; maximum: 11 **51.** Vertex: $\left(-\frac{1}{4}, -13\right)$; line of symmetry: $x = -\frac{1}{4}$; minimum: -13 **53.** Vertex: $(10, -20)$; line of symmetry: $x = 10$; maximum: -20 **55.** Vertex: $(-4.58, 65\pi)$; line of symmetry: $x = -4.58$; minimum: 65π **57.** $7a^6b^6$
59. $f(x) = -2x^2 + 4$ **61.** $f(x) = 2(x - 6)^2$
63. $f(x) = -2(x - 3)^2 + 8$ **65.** $f(x) = 2(x + 3)^2 + 6$
67. $f(x) = 2(x - 2)^2 - 3$ **69.** $g(x) = -\frac{1}{2}(x - 1)^2 - 6$

71. **73.** **75.**

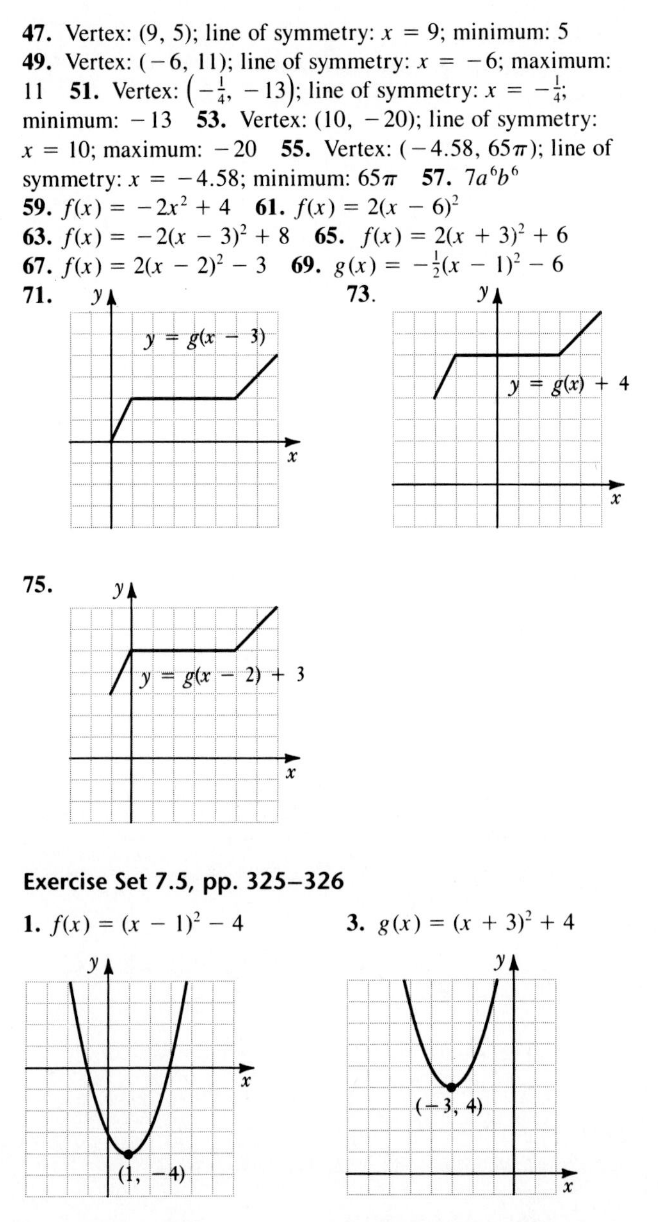

Exercise Set 7.5, pp. 325–326

1. $f(x) = (x - 1)^2 - 4$ **3.** $g(x) = (x + 3)^2 + 4$

5. $f(x) = (x + 2)^2 - 5$ **7.** $h(x) = 2(x + 4)^2 - 7$

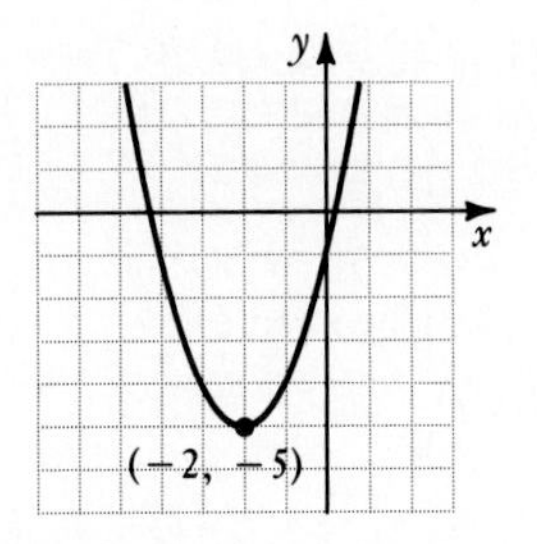

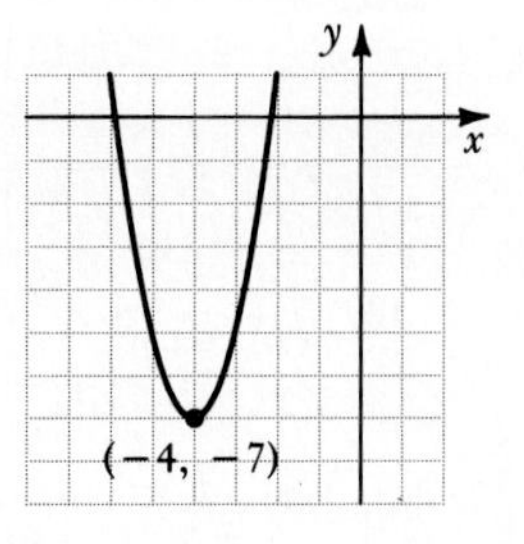

9. $f(x) = -(x - 2)^2 + 10$ **11.** $g(x) = \left(x + \frac{3}{2}\right)^2 - \frac{49}{4}$

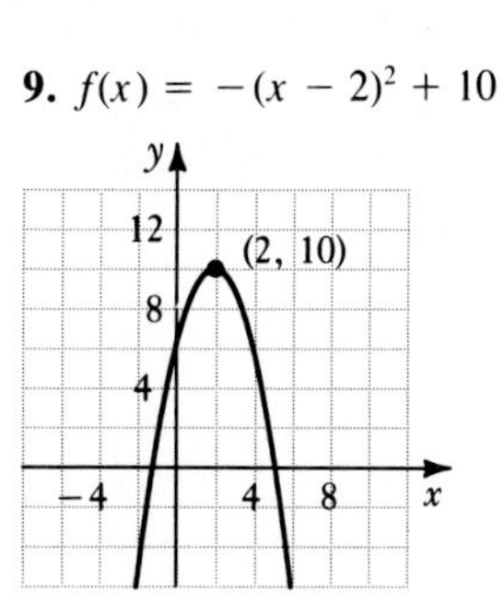

13. $f(x) = 3(x - 4)^2 + 2$ **15.** $h(x) = \left(x - \frac{9}{2}\right)^2 - \frac{81}{4}$

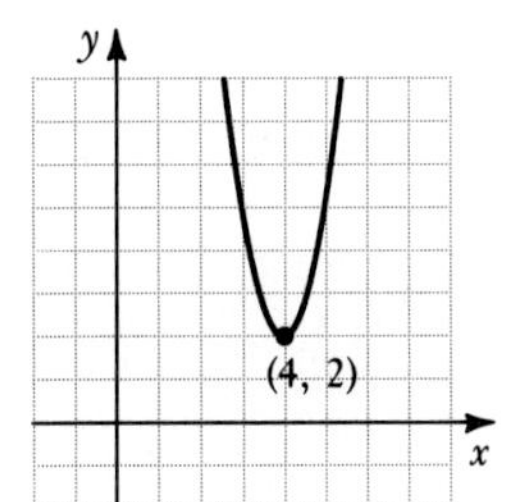

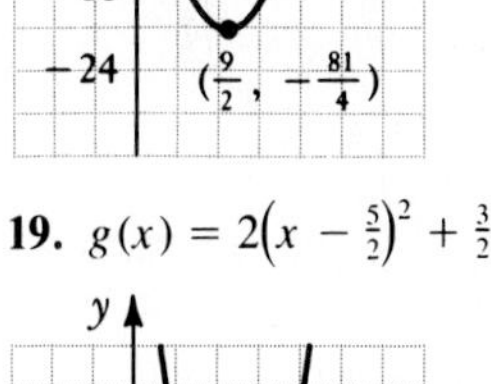

17. $f(x) = -2(x + 1)^2 - 4$ **19.** $g(x) = 2\left(x - \frac{5}{2}\right)^2 + \frac{3}{2}$

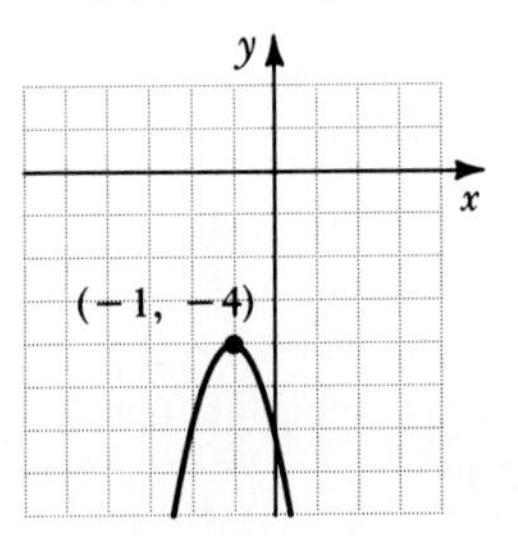

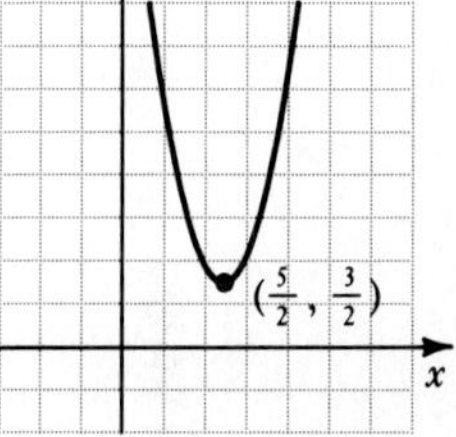

21. $f(x) = -3\left(x + \frac{1}{2}\right)^2 + \frac{7}{4}$ **23.** $h(x) = \frac{1}{2}(x + 4)^2 - \frac{5}{3}$

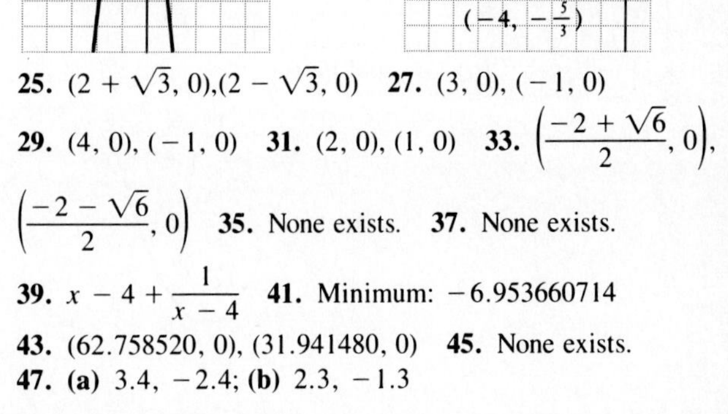

25. $(2 + \sqrt{3}, 0), (2 - \sqrt{3}, 0)$ **27.** $(3, 0), (-1, 0)$ **29.** $(4, 0), (-1, 0)$ **31.** $(2, 0), (1, 0)$ **33.** $\left(\frac{-2 + \sqrt{6}}{2}, 0\right)$, $\left(\frac{-2 - \sqrt{6}}{2}, 0\right)$ **35.** None exists. **37.** None exists. **39.** $x - 4 + \frac{1}{x - 4}$ **41.** Minimum: -6.953660714 **43.** $(62.758520, 0), (31.941480, 0)$ **45.** None exists. **47. (a)** $3.4, -2.4$; **(b)** $2.3, -1.3$

49. $f(x) = a\left[x - \left(-\frac{b}{2a}\right)\right]^2 + \frac{4ac - b^2}{4a}$

51.

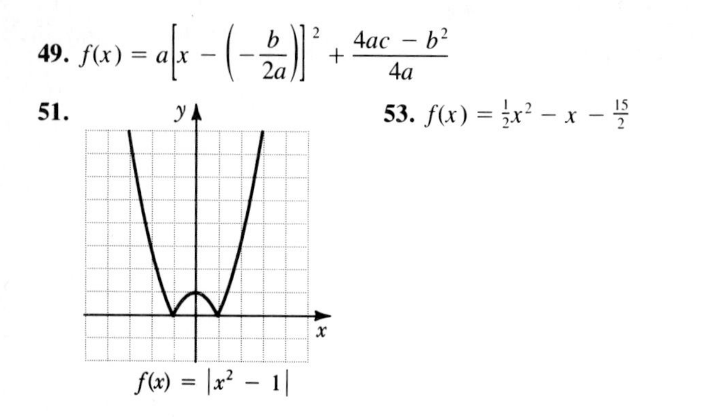

53. $f(x) = \frac{1}{2}x^2 - x - \frac{15}{2}$

Exercise Set 7.6, pp. 330–332

1. 19 ft by 19 ft; 361 ft^2 **3.** 64; 8 and 8 **5.** 121; 11 and 11 **7.** -4; 2 and -2 **9.** $-\frac{81}{4}$; $-\frac{9}{2}$ and $\frac{9}{2}$ **11.** $\frac{49}{4}$; $-\frac{7}{2}$ and $-\frac{7}{2}$ **13.** 200 ft^2; 10 ft by 20 ft **15.** 4 ft by 4 ft **17.** $s = \sqrt{\frac{A}{6}}$ **19.** $r = \sqrt{\frac{Gm_1m_2}{F}}$ **21.** $c = \sqrt{\frac{E}{m}}$ **23.** $b = \sqrt{c^2 - a^2}$ **25.** $k = \frac{3 + \sqrt{9 + 8N}}{2}$ **27.** $r = \frac{-\pi h + \sqrt{\pi^2h^2 + 2\pi A}}{2\pi}$ **29.** $n = \frac{1 + \sqrt{1 + 8N}}{2}$ **31.** $w = \frac{-2l + \sqrt{4l^2 + 2A}}{2}$ **33.** $g = \frac{4\pi^2 l}{T^2}$ **35.** $D = 4\sqrt{\frac{2LV}{g(P_1 - P_2)}}$ **37.** $v = \frac{c}{m}\sqrt{m^2 - m_0^2}$ **39.** $5xy^2\sqrt[4]{x}$ **41.** Base: 19 cm, Height: 19 cm; 180.5 cm^2 **43.** \$6 **45.** $P(x) = -x^2 + 192x - 5000$; a maximum profit of \$4216 occurs when 96 units are produced and sold. **47.** 78.4 ft **49.** $n = \pm\sqrt{\frac{r^2 \pm \sqrt{r^4 + 4m^4r^2p - 4mp}}{2m}}$

Review Exercises: Chapter 7, pp. 334–335

1. [7.1] $\pm\frac{\sqrt{14}}{2}$ **2.** [7.1] $0, -\frac{5}{14}$ **3.** [7.1] 3, 9 **4.** [7.2] $\frac{-3 \pm i\sqrt{7}}{8}$ **5.** [7.2] $\frac{7 \pm i\sqrt{3}}{2}$ **6.** [7.1] 3, 5 **7.** [7.2] $-0.3, -3.7$ **8.** [7.1] 36 **9.** [7.1] $\frac{9}{100}$ **10.** [7.1] 4, -2 **11.** [7.1] $3 \pm 2\sqrt{2}$ **12.** [7.3] Two real **13.** [7.3] Two nonreal **14.** [7.3] $25x^2 + 10x - 3 = 0$ **15.** [7.3] $x^2 + 8x + 16 = 0$ **16.** [7.4]

$f(x) = \frac{1}{2}(x - 1)^2$; $x = 1$; $(1, 0)$

17. [7.4]

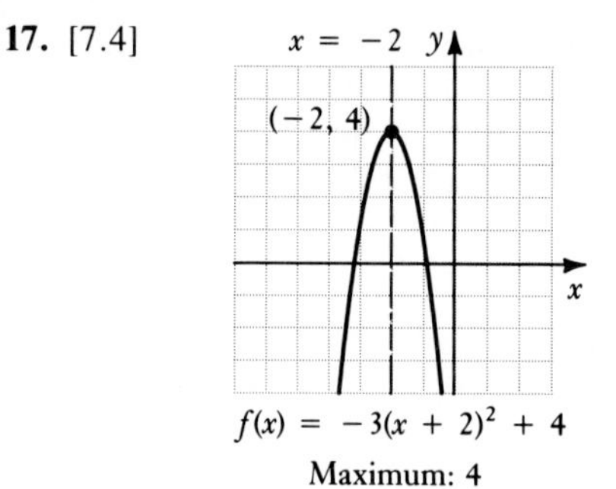

18. [7.5] **(a)** $f(x) = 2(x - 3)^2 + 5$; **(b)** vertex: (3, 5), line of symmetry: $x = 3$; **(c)**

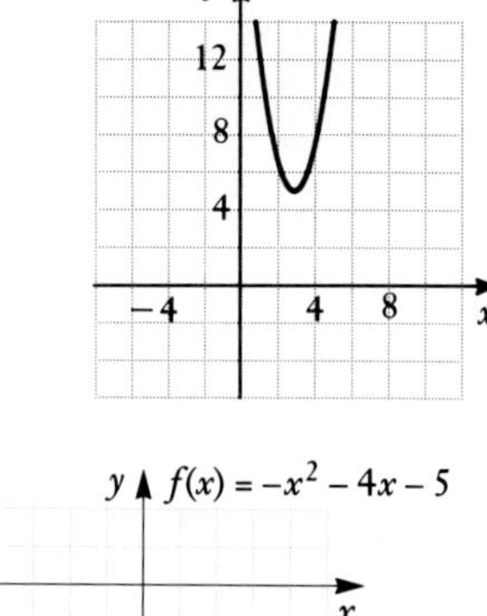

19. [7.5]

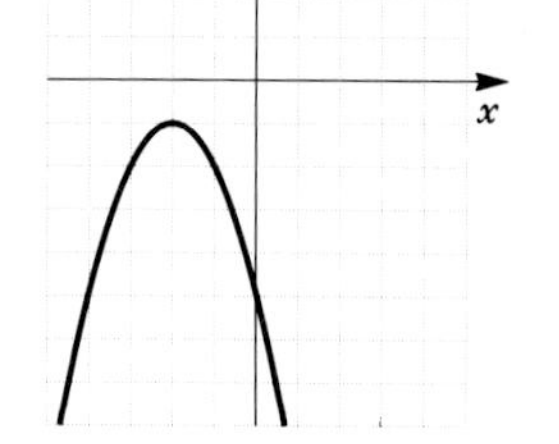

20. [7.5] (7, 0), (2, 0) **21.** [7.5] None exists. **22.** [7.6] $p = \frac{9\pi^2}{N^2}$ **23.** [7.6] $T = \frac{1 \pm \sqrt{1 + 24A}}{6}$ **24.** [7.6] -121; 11 and -11 **25.** [7.6] 20 ft by 20 ft; 400 ft^2 **26.** [3.6] $x^2 - x + 3 + \frac{-6}{x + 1}$ **27.** [5.1] $3st^5\sqrt[3]{s}$ **28.** [5.4] $3x^3y^4$ **29.** [6.5] $y = -\frac{3}{4}x - \frac{11}{4}$ **30.** [7.4] $f(x) = -2x^2 + 10$ **31.** [7.3] $h = 60$, $k = 60$ **32.** [7.6] 18 and 324

Test: Chapter 7, p. 335

1. [7.1] $\pm\frac{2\sqrt{3}}{3}$ **2.** [7.1] $0, \frac{7}{8}$ **3.** [7.1] 9, 2 **4.** [7.2] $\frac{-1 \pm i\sqrt{3}}{2}$ **5.** [7.2] $\frac{5 \pm \sqrt{37}}{6}$ **6.** [7.2] 0.4, -4.4 **7.** [7.1] $x^2 + 14x + 49$ **8.** [7.1] $x^2 - \frac{2}{7}x + \frac{1}{49}$ **9.** [7.1] -6, 3 **10.** [7.1] $-5 \pm \sqrt{10}$ **11.** [7.3] Two nonreal **12.** [7.3] $x^2 - 4\sqrt{3}x + 9 = 0$ **13.** [7.6] $r = \sqrt{\frac{3V}{\pi} - R^2}$ **14.** [7.5](3, 0) and $(-2, 0)$

15. [7.4]

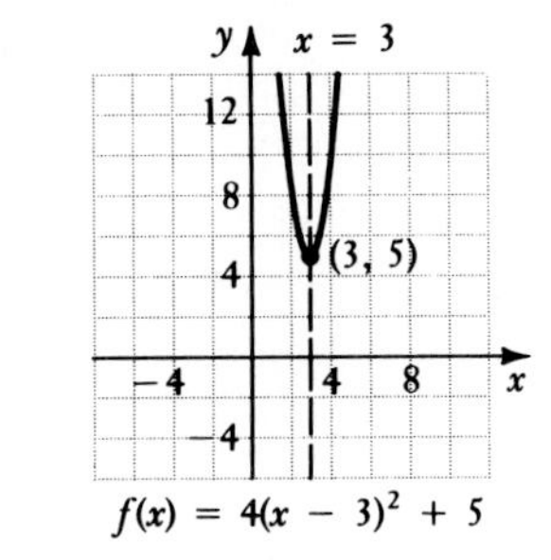

16. [7.5] **(a)** $f(x) = 2(x + 1)^2 - 8$; **(b)** vertex: $(-1, -8)$; line of symmetry: $x = -1$; **(c)**

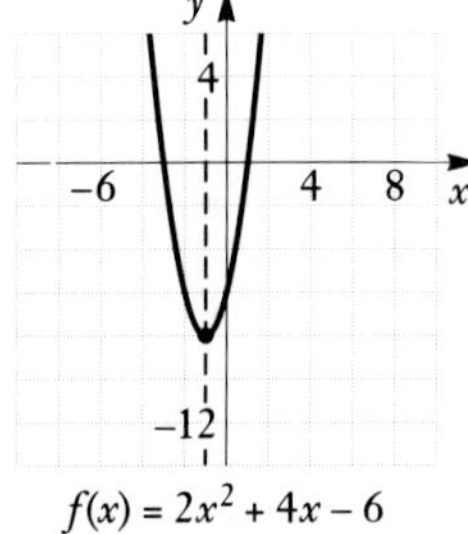

17. [7.6] -16 **18.** [7.6] 450 ft^2; 15 ft by 30 ft **19.** [3.6] $-8a^4$ **20.** [6.5] $y = -x + 3$ **21.** [5.1] $16|xy|$ **22.** [5.4] $\sqrt[4]{(x + y)^3}$ **23.** [7.3] $f(x) = 2x^2 - 4x - 6$ **24.** [7.2] $\dfrac{2 \pm \sqrt{4 + 3\sqrt{5}}}{3}$

CHAPTER 8

Exercise Set 8.1, p. 341

1. $x^2 - x + 1$, R -4 **3.** $a + 7$, R -47 **5.** $x^2 - 5x - 23$, R -43 **7.** $3x^2 - 2x + 2$, R -3 **9.** $y^2 + 2y + 1$, R 12 **11.** $3x^3 + 9x^2 + 2x + 6$ **13.** $x^2 + 3x + 9$ **15.** $y^4 + y^3 + y^2 + y + 1$ **17.** $3x^3 + 2x^2 - 2x - 3$, R 2 **19.** $3x^2 + 6x - 3$, R 2

21. **23.**

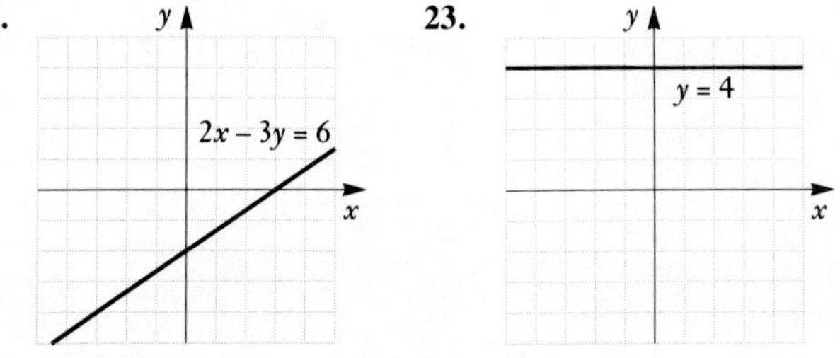

25. $3.41x^3 + 8.2181x^2 - 4.444379x - 10.7109533$, R -39.283397 **27.** **(a)** Remainder is 0; **(b)** If $f(x) = (x - 4) \cdot p(x)$ for some polynomial $p(x)$, then $f(4) = (4 - 4) \cdot p(4) = 0$; **(c)** $4^3 - 5(4)^2 + 5(4) - 4 = 0$

Exercise Set 8.2, pp. 344–346

1. Neither **3.** Even **5.** Neither **7.** Even **9.** Odd **11.** Neither **13.** Odd **15.** Even and odd **17.** Even **19.** y-axis **21.** x-axis **23.** Origin **25.** x-axis, y-axis, origin **27.** Even, y-axis **29.** Odd, origin **31.** Odd, origin **33.** Neither, none **35.** $-60i$ **37.** No

39. **41.**

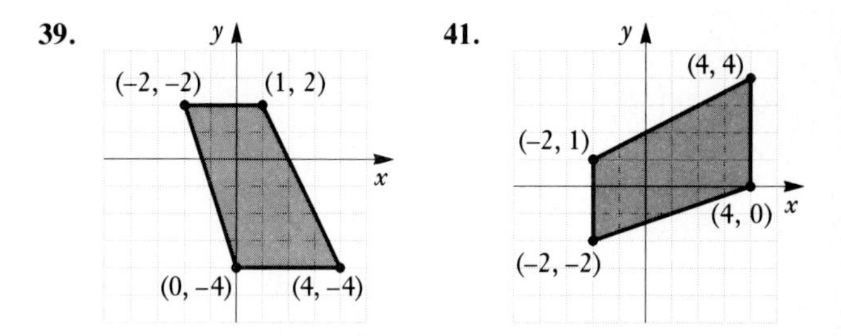

Exercise Set 8.3, pp. 350–351

1. $P(1) = 0$, $P(-2) = -60$, $P(3) = 0$ **3.** $P(10) = 73{,}120$; $P(-8) = -37{,}292$ **5.** $P(20) = 5{,}935{,}988$; $P(-3) = -772$ **7.** $P(2) = 0$, $P(-2) = 0$, $P(3) = 65$ **9.** -3: yes; 2: no **11.** -2: yes; 1: no; 5: yes **13.** -3: no; $\frac{1}{2}$: no **15.** $P(x) = (x - 1)(x + 2)(x + 3)$; $1, -2, -3$
17. $P(x) = (x - 2)(x - 5)(x + 1)$; $2, 5, -1$
19. $P(x) = (x - 2)(x - 3)(x + 4)$; $2, 3, -4$
21. $P(x) = (x - 2)(x + 5)(x + 3)$; $2, -5, -3$
23. $P(x) = (x + 4)(x + 3)(x - 1)(x + 1)$; $-4, -3, 1, -1$
25. $P(x) = (x - 1)(x + 2)(x - 3)(x + 4)$; $1, -2, 3, -4$
27. $P(x) = (x + 1)(x + 1)(x + 2)(x - 2)$; $-1, -2, 2$
29. $P(x) = (x - 1)(x - 2)(x - 3)(x + 5)$; $1, 2, 3, -5$
31. $a = \dfrac{b}{c} - b$ **33.** $\sqrt[4]{x^7}$ **35.** $-1, \frac{7}{6}$
37. $\frac{14}{3}$ **39.** $k = 0$

Exercise Set 8.4, pp. 357–358

1. $1, -1$ **3.** $\pm\left(1, \frac{1}{3}, \frac{1}{5}, \frac{1}{15}, 2, \frac{2}{3}, \frac{2}{5}, \frac{2}{15}\right)$ **5.** 3 **7.** 4 **9.** -1 **11.** -3 **13.** $-3, \sqrt{2}, -\sqrt{2}$; $(x + 3)(x - \sqrt{2})(x + \sqrt{2})$ **15.** $-2, 1$; $(x + 2)(x - 1)^2 = 0$ **17.** No rational roots **19.** $-\frac{1}{5}, 1, 2i, -2i$; $5(x + \frac{1}{5})(x - 1)(x - 2i)(x + 2i) = 0$
21. $-1, -2, 3 \pm \sqrt{13}$; $(x + 1)(x + 2)(x - 3 - \sqrt{13})(x - 3 + \sqrt{13})$
23. $2, 1 \pm \sqrt{3}$; $(x - 2)(x - 1 - \sqrt{3})(x - 1 + \sqrt{3}) = 0$
25. $-2, 1 \pm i\sqrt{3}$; $(x + 2)(x - 1 - i\sqrt{3})(x - 1 + i\sqrt{3})$
27. $\frac{1}{2}, \dfrac{1 \pm \sqrt{5}}{2}$; $\frac{1}{3}\left(x - \frac{1}{2}\right)\left(x - \dfrac{1 + \sqrt{5}}{2}\right)\left(x - \dfrac{1 - \sqrt{5}}{2}\right)$

29. None **31.** None **33.** None **35.** $-2, 1, 2$ **37.** $20 + 17i$ **39.** 3 or 1 **41.** 0 **43.** 2 or 0 **45.** 0 **47.** 3 or 1 **49.** 2 or 0 **51.** 3 cm, $\dfrac{7 - \sqrt{33}}{2}$ cm **53.** $\sqrt{5}$ is a root of $x^2 - 5 = 0$, which has no rational roots (since ± 1, ± 5 are not roots). Thus $\sqrt{5}$ must be irrational.

Exercise Set 8.5, pp. 364–365

1.

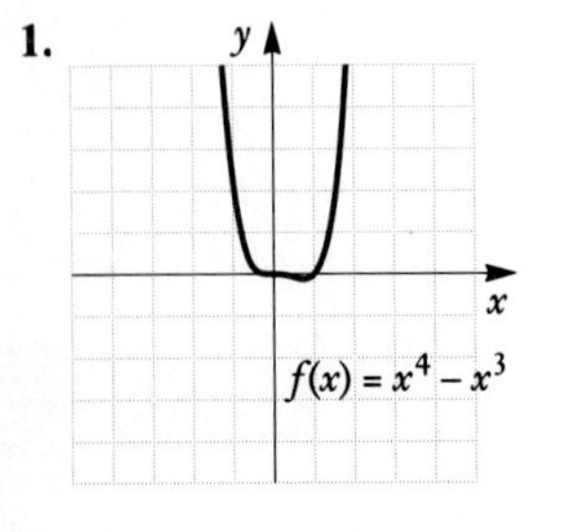

3. $f(x) = 9x^2 - x^4$

5.

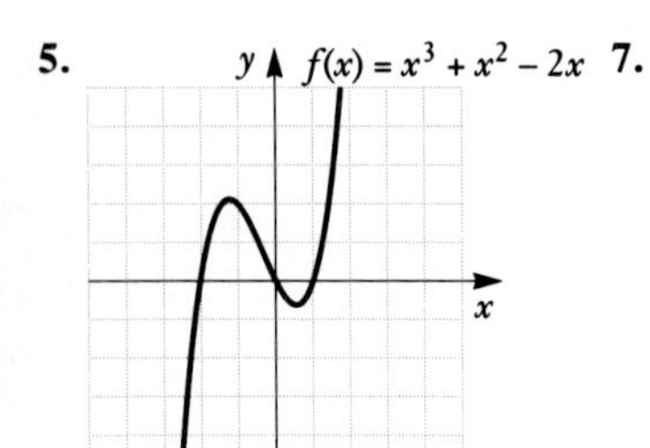

7.

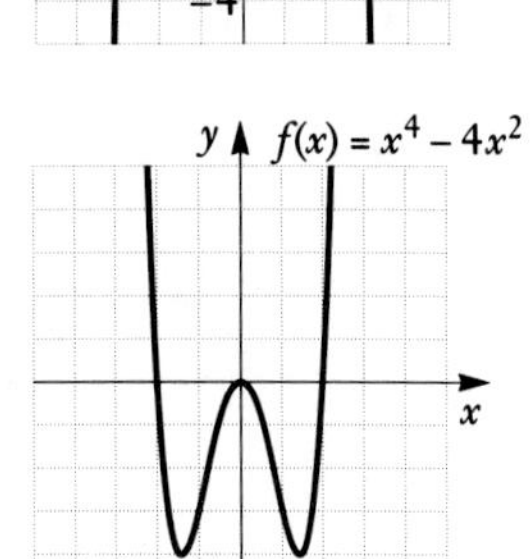

9.

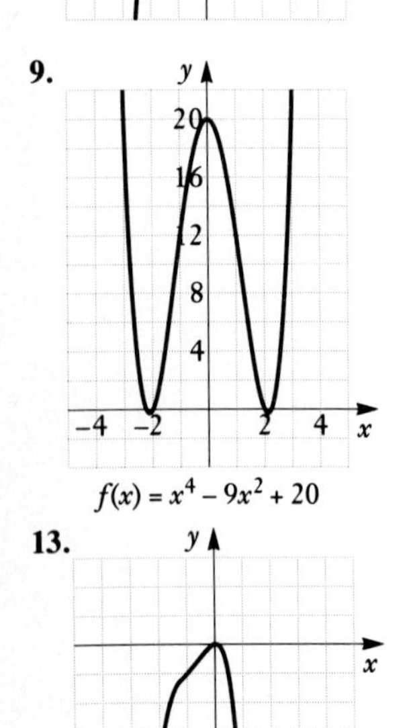

11.

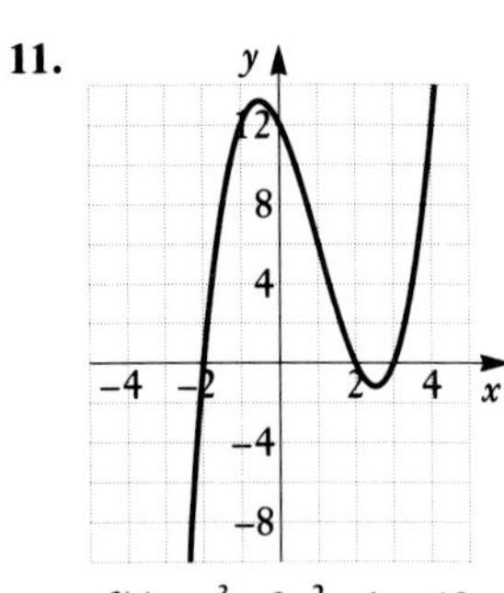

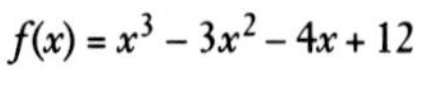

13.

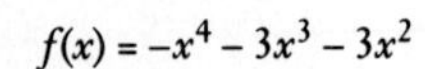

15.

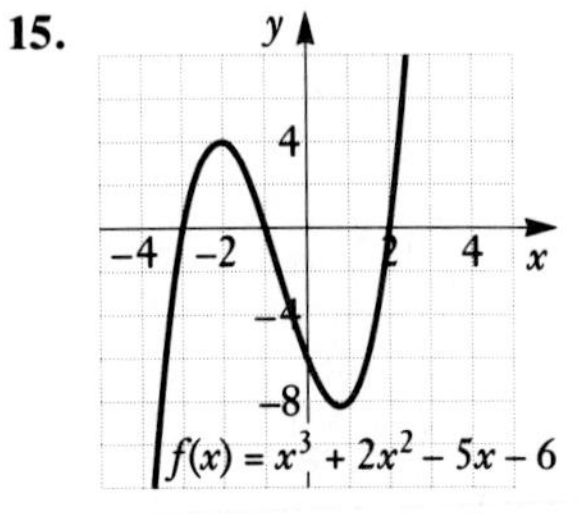

17. **19.**

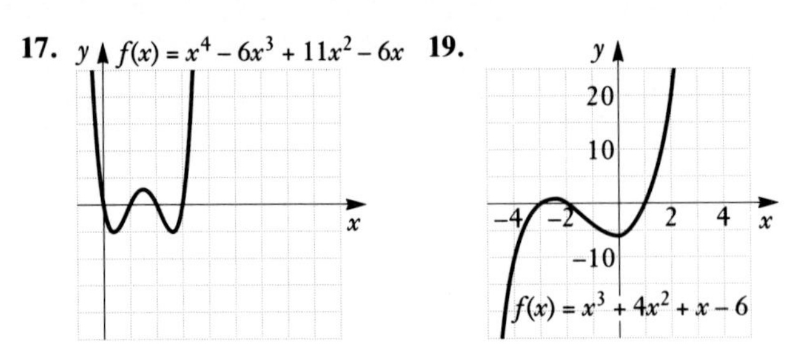

21.

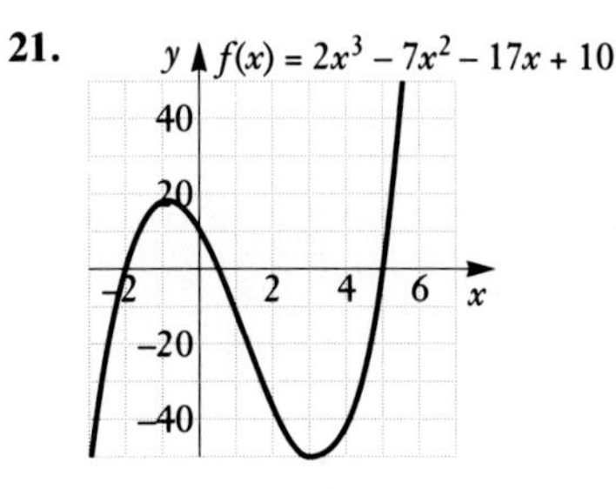

23.

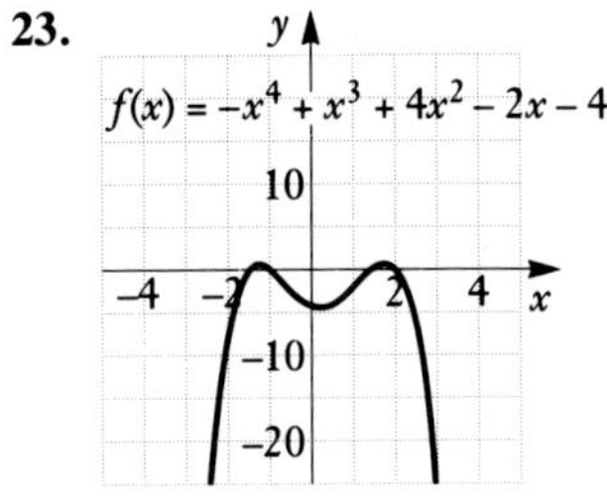

25.

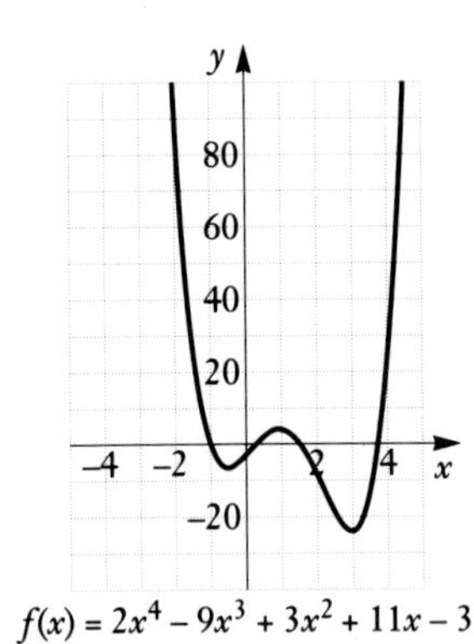

27. Two real **29.** $\dfrac{x^2 + 2x + 4}{(x + 2)(x^2 + 4)}$

31. $f(x) = x^3 - 9x^2 + 27x + 50$. Root: -1.3

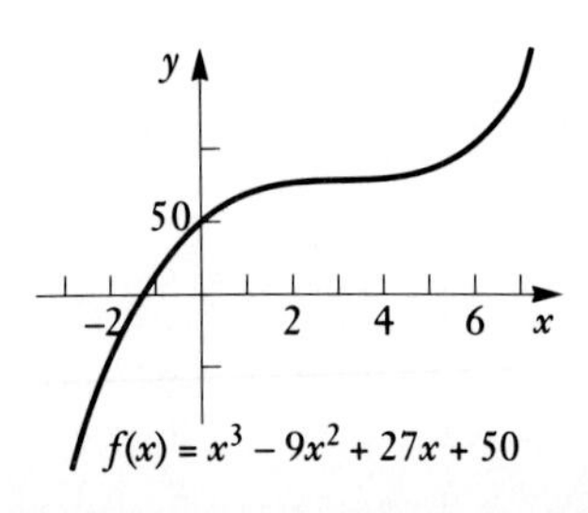

33. $f(x) = x^4 + 4x^3 - 36x^2 - 160x + 300$. Roots: 1.5, 5.7

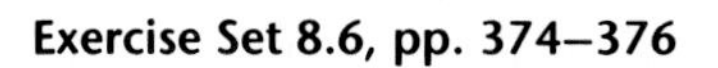

Exercise Set 8.6, pp. 374–376

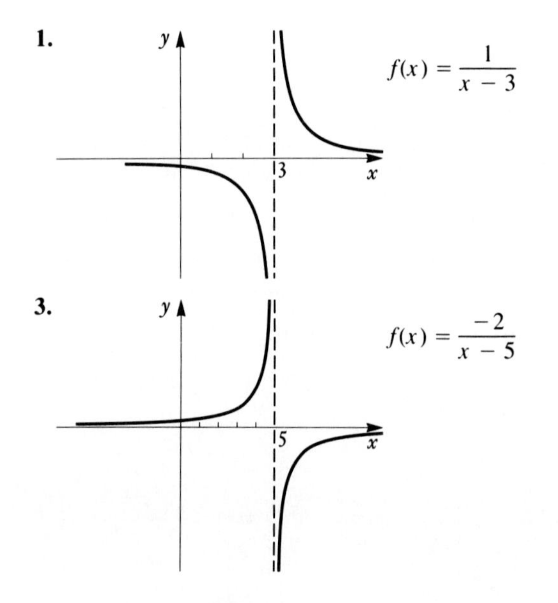

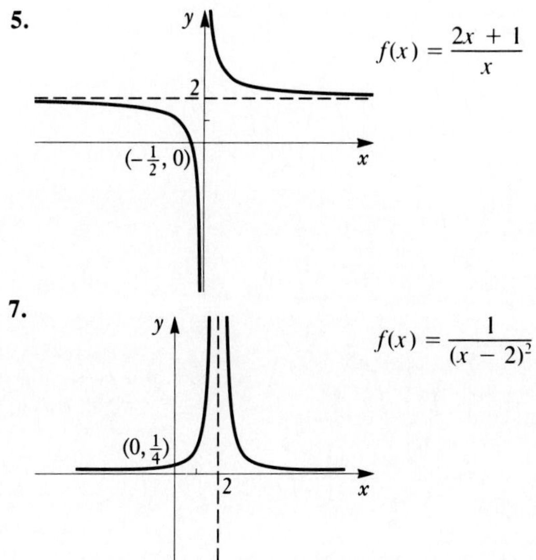

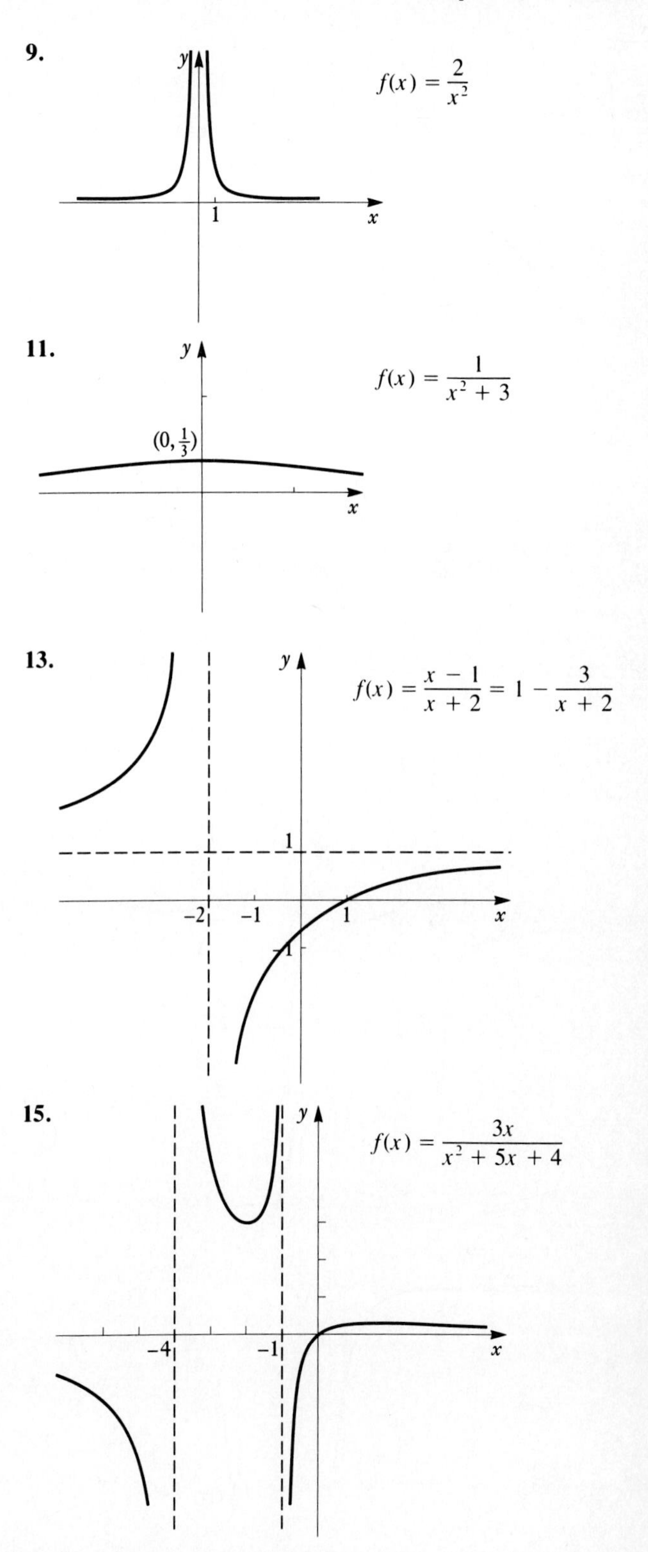

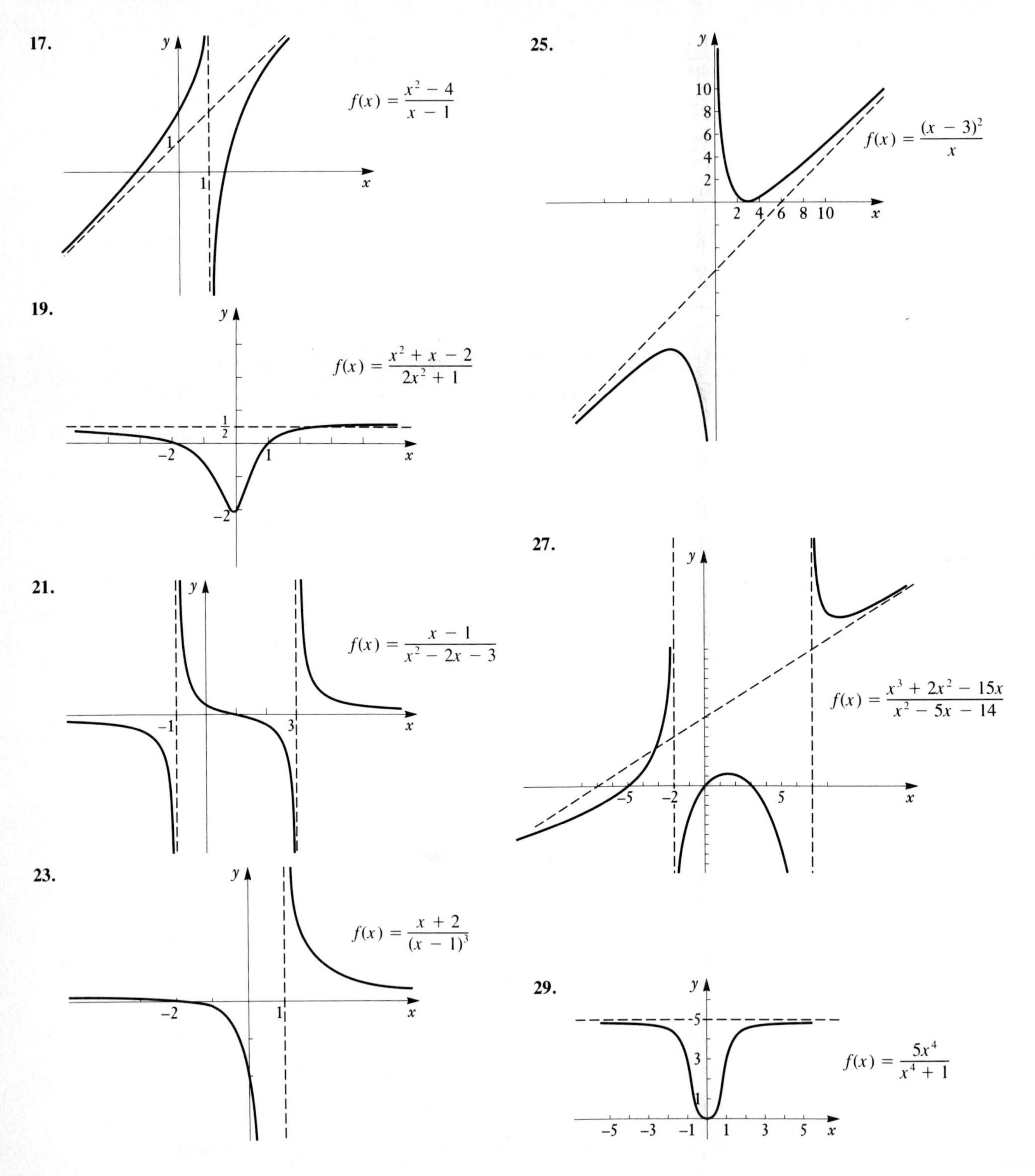
17.
$f(x) = \frac{x^2 - 4}{x - 1}$
y
x
1
1
19.
$f(x) = \frac{x^2 + x - 2}{2x^2 + 1}$
y
x
$\frac{1}{2}$
−2
1
−2
21.
$f(x) = \frac{x - 1}{x^2 - 2x - 3}$
y
x
−1
3
23.
$f(x) = \frac{x + 2}{(x - 1)^3}$
y
x
−2
1
25.
$f(x) = \frac{(x - 3)^2}{x}$
y
x
10
8
6
4
2
2
4
6
8
10
27.
$f(x) = \frac{x^3 + 2x^2 - 15x}{x^2 - 5x - 14}$
y
x
−5
−2
5
29.
$f(x) = \frac{5x^4}{x^4 + 1}$
y
x
5
3
1
−5
−3
−1
1
3
5

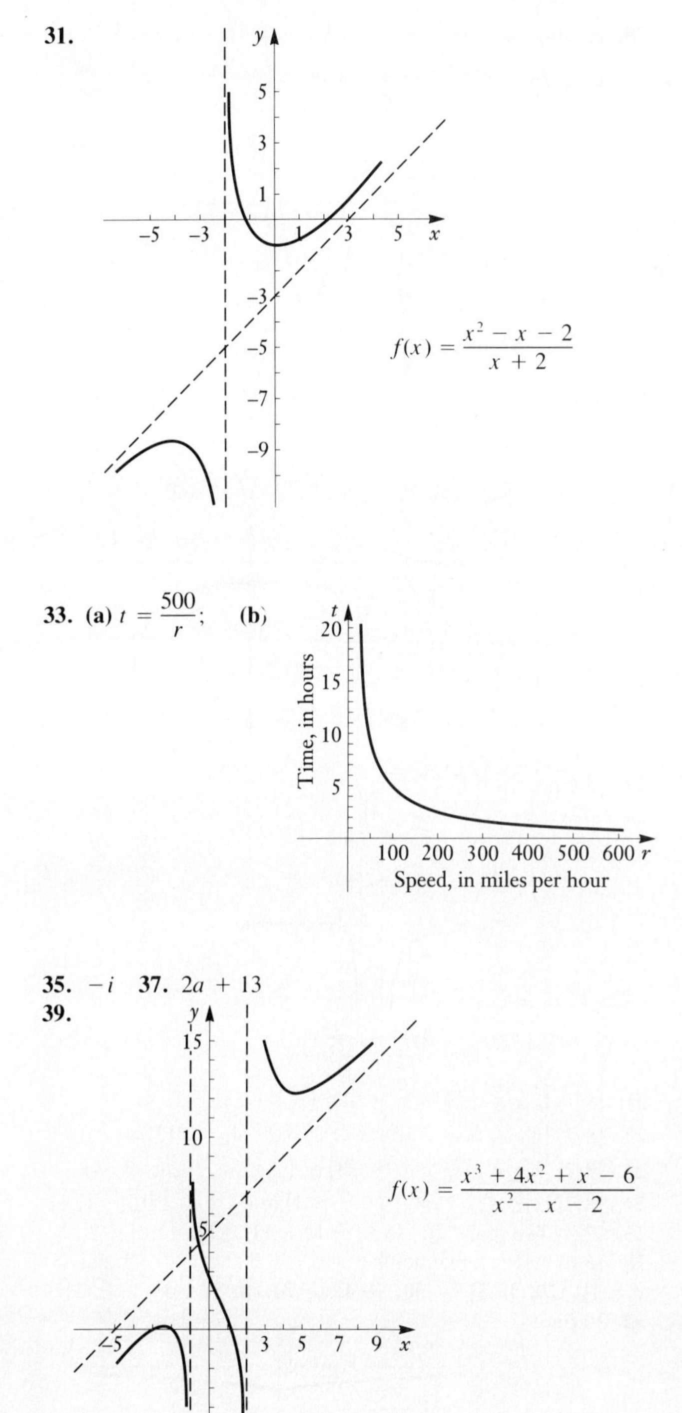

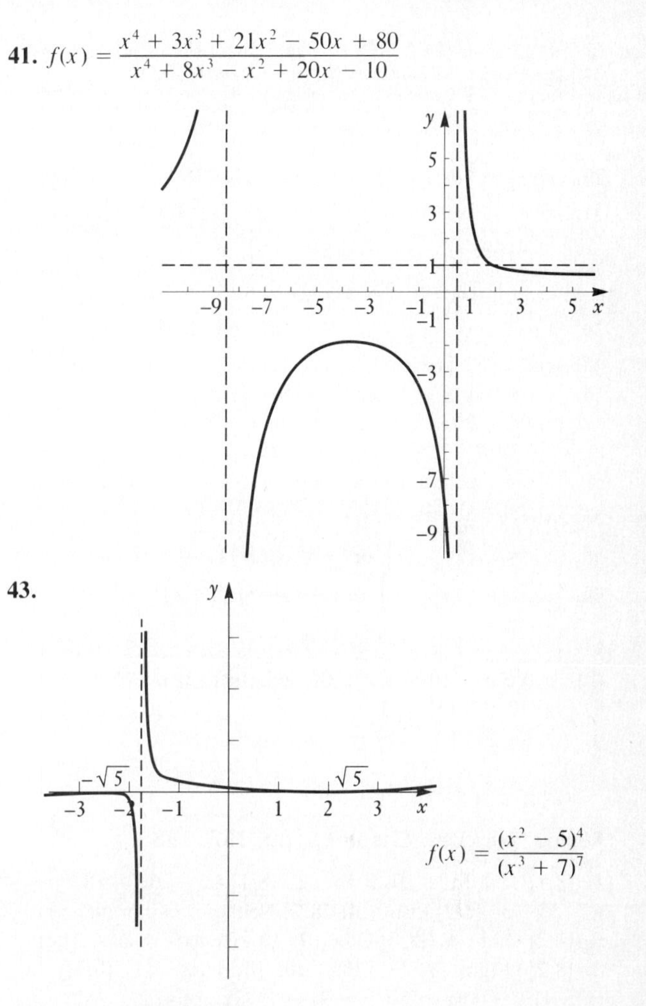

45. $f(x) = \dfrac{2x^2 - 9x + 4}{x^2 - 2x - 3}$

Exercise Set 8.7, pp. 383–384

1. $\{x \mid x < -5 \text{ or } x > 3\}$, or $(-\infty, -5) \cup (3, \infty)$
3. $\{x \mid -2 \le x \le 1\}$, or $[-2, 1]$ **5.** $\{x \mid -1 < x < 2\}$, or $(-1, 2)$ **7.** $\{x \mid x \le -1 \text{ or } x \ge 1\}$, or $(-\infty, -1] \cup [1, \infty)$
9. $\{x \mid x \le -3 \text{ or } x \ge 3\}$, or $(-\infty, -3] \cup [3, \infty)$
11. $\{x \mid x$ is a real number$\}$, or $(-\infty, \infty)$
13. $\{x \mid 2 < x < 4\}$, or $(2, 4)$ **15.** $\left\{x \mid -3 < x < \frac{5}{4}\right\}$, or $\left(-3, \frac{5}{4}\right)$
17. $\left\{x \mid x < \dfrac{-1 - \sqrt{41}}{4} \text{ or } x > \dfrac{-1 + \sqrt{41}}{4}\right\}$, or

$\left(-\infty, \dfrac{-1 - \sqrt{41}}{4}\right) \cup \left(\dfrac{-1 + \sqrt{41}}{4}, \infty\right)$

19. $\{x|x < -2 \text{ or } 0 < x < 2\}$, or $(-\infty, -2)\cup(0, 2)$
21. $\{x|-2 \le x \le 1 \text{ or } x \ge 5\}$, or $[-2, 1]\cup[5, \infty)$
23. $\{x|x < -3 \text{ or } -2 < x < 1\}$, or $(-\infty, -3)\cup(-2, 1)$
25. $\{x|x > 4\}$, or $(4, \infty)$ **27.** $\left\{x|0 < x < \frac{1}{3}\right\}$, or $\left(0, \frac{1}{3}\right)$
29. $\{x|x < -\frac{2}{3} \text{ or } x > 3\}$, or $\left(-\infty, -\frac{2}{3}\right)\cup(3, \infty)$
31. $\{x| -2 \le x < 0\}$, or $[-2, 0)$ **33.** $\left\{x|\frac{3}{2} < x \le 4\right\}$, or $\left(\frac{3}{2}, 4\right]$ **35.** $\left\{x|x \le -\frac{5}{2} \text{ or } x > -2\right\}$, or $\left(-\infty, -\frac{5}{2}\right]\cup(2, \infty)$ **37.** $\{x|x < 0\}$, or $(-\infty, 0)$
39. $\left\{x|x < -\frac{11}{7}\right\}$, or $\left(-\infty, -\frac{11}{7}\right)$ **41.** $\{x|x > 1\}$, or $(1, \infty)$
43. $\{x|0 < x < 2 \text{ or } x > 2\}$, or $(0, 2)\cup(2, \infty)$
45. $\{x|x < 0 \text{ or } x \ge 1\}$, or $(-\infty, 0)\cup[1, \infty)$
47. $\{x|x < -3 \text{ or } -2 < x < -1 \text{ or } x > 2\}$, or $(-\infty, -3)\cup(-2, -1)\cup(2, \infty)$ **49.** $\left\{-\frac{2}{3}, 4\right\}$
51. $\{x|x$ is a real number and $x \ne -5, x \ne -2$, and $x \ne 2\}$
53. $\left\{x|-\sqrt{2} \le x \le \sqrt{2}\right\}$, or $\left[-\sqrt{2}, \sqrt{2}\right]$
55. $\left\{x|x < \frac{5}{3} \text{ or } x > 11\right\}$, or $\left(-\infty, \frac{5}{3}\right)\cup(11, \infty)$ **57.** $\emptyset$
59. $\left\{x|x < -\frac{1}{4} \text{ or } x > \frac{1}{2}\right\}$, or $\left(-\infty, -\frac{1}{4}\right)\cup\left(\frac{1}{2}, \infty\right)$
61. $\left\{h|h > -2 + 2\sqrt{6} \text{ cm}\right\}$ **63.** **(a)** $\{x|10 < x < 200\}$; **(b)** $\{x|0 \le x < 10 \text{ or } x > 200\}$ **65.** $\{n|13 \le n \le 50\}$
67. **(a)** 10, 35; **(b)** $\{x|10 < x < 35\}$; **(c)** $\{x|x < 10 \text{ or } x > 35\}$

Review Exercises: Chapter 8, pp. 387–388

1. [8.1] $x^2 + 6x + 20$, R 54 **2.** [8.1] $4x^2 - 6x + 18$, R -59 **3.** [8.2] Even **4.** [8.2] Neither even nor odd
5. [8.2] Odd **6.** [8.2] Odd **7.** [8.2] x-axis, y-axis, origin
8. [8.2] Origin **9.** [8.3] 88 **10.** [8.3] Yes **11.** [8.3] $(x - 1)(x + 3)(x + 5)$; $1, -3, -5$ **12.** [8.4] 2
13. [8.4] -2 **14.** [8.4] $-3, 4, \pm 3i$
15. [8.4] $\pm\left(1, 2, 3, 4, 6, 12, \frac{1}{2}, \frac{3}{2}\right)$ **16.** [8.4] 2
17. [8.5]

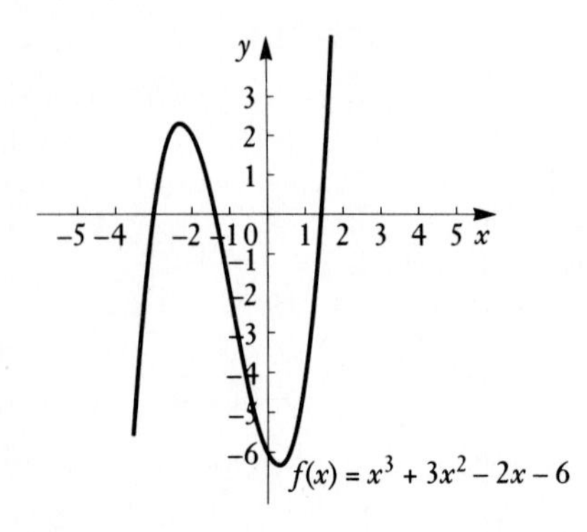

18. [8.5]

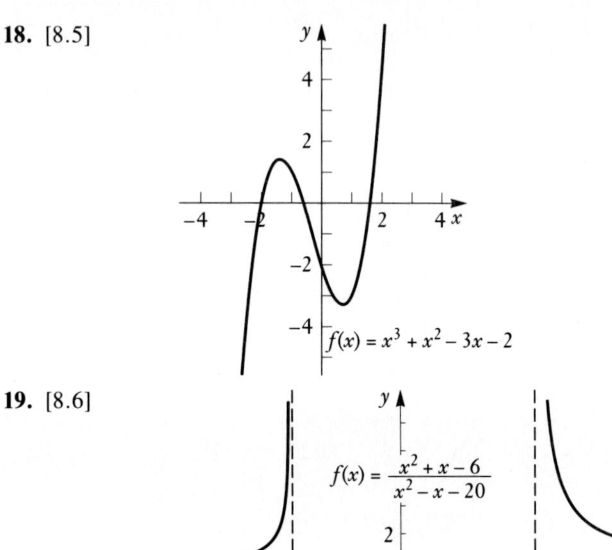

19. [8.6]

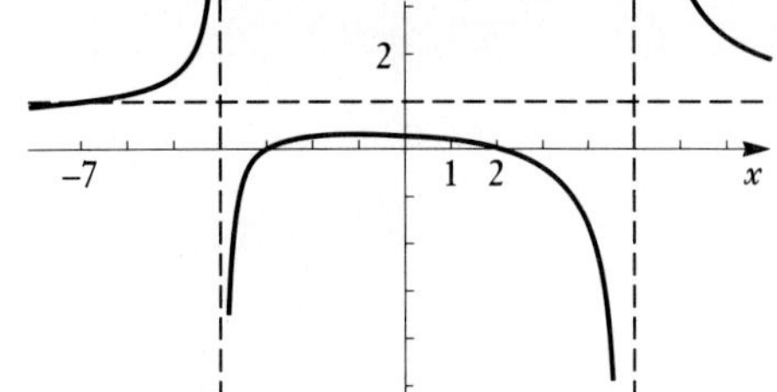

20. [8.6]

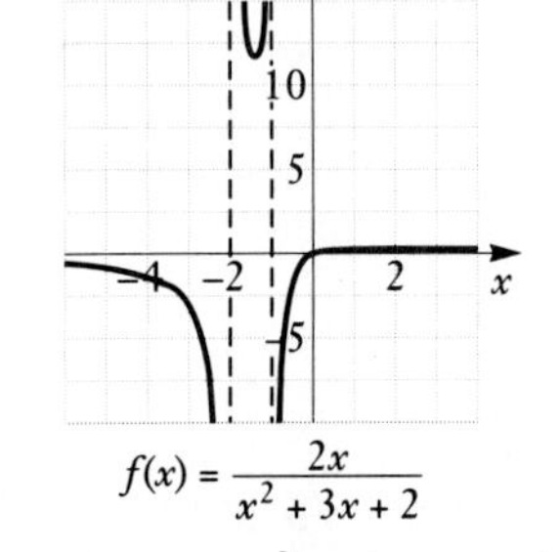

21. [8.7] $\left\{x|x < -\frac{1}{2} \text{ or } x > 2\right\}$ or $\left(-\infty, -\frac{1}{2}\right)\cup(2, \infty)$
22. [8.7] $\{x|-4 \le x \le 1 \text{ or } x \ge 2\}$ or $[-4, 1]\cup[2, \infty)$
23. [8.7] $\left\{x|x < -\frac{14}{3} \text{ or } x > -3\right\}$ or $\left(-\infty, -\frac{14}{3}\right)\cup(-3, \infty)$
24. [8.7] $\{x|x \le -2 \text{ or } -1 < x \le 1\}$ or $(-\infty, -2]\cup(-1, 1]$
25. [7.3] Two real **26.** [5.6] $-13 - 11i$ **27.** [6.2] -2
28. [6.6] $\{x|x$ is a real number and $x \ne 0$ and $x \ne -1$ and $x \ne 1\}$ **29.** [8.3] 7 **30.** [8.4] 4 **31.** [8.3] -4
32. [8.6]

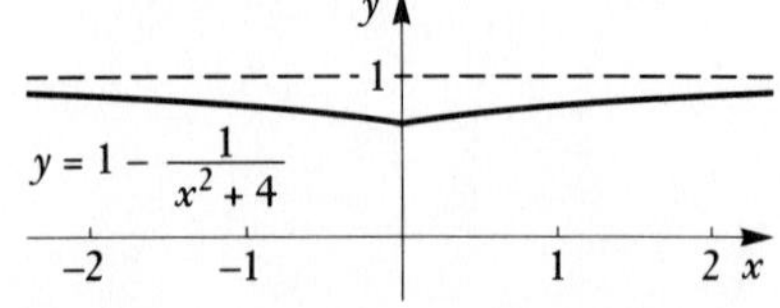

Test: Chapter 8, pp. 389–390

1. [8.1] $x^2 + 9x + 40$, R 153 **2.** [8.1] $2x^2 - 3x - 5$, R 2 **3.** [8.2] Neither; none **4.** [8.2] Odd; origin **5.** [8.2] Even; y-axis **6.** [8.3] 2315 **7.** [8.3] $(x - 2)(x + 1)(x - 3)$; $2, -1, 3$ **8.** [8.4] $\pm\left(\frac{1}{3}, \frac{2}{3}, 1, 2, 3, 6\right)$ **9.** [8.4] 3
10. [8.4] $-1, -\frac{3}{2}; -1 \pm \sqrt{5}$; $(x + 1)(2x + 3)(x + 1 - \sqrt{5})(x + 1 + \sqrt{5})$
11. [8.5]

$f(x) = x^4 - 5x^2$

12. [8.5]

$f(x) = x^4 - 3x^3 + 2x^2$

13. [8.6]

$f(x) = \dfrac{x - 2}{x^2 - 2x - 15}$

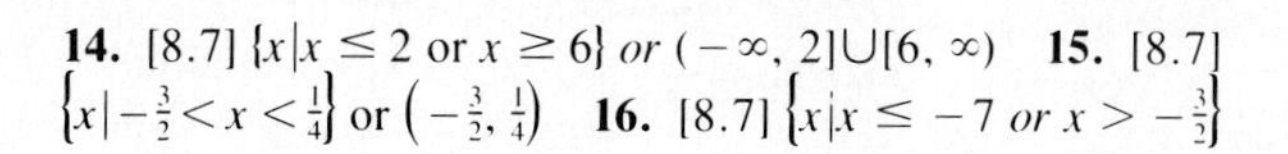
14. [8.7] $\{x \mid x \leq 2 \text{ or } x \geq 6\}$ *or* $(-\infty, 2] \cup [6, \infty)$ **15.** [8.7] $\left\{x \mid -\frac{3}{2} < x < \frac{1}{4}\right\}$ or $\left(-\frac{3}{2}, \frac{1}{4}\right)$ **16.** [8.7] $\left\{x \mid x \leq -7 \text{ or } x > -\frac{3}{2}\right\}$ or $(-\infty, -7] \cup \left(-\frac{3}{2}, \infty\right)$ **17.** [6.2] $\dfrac{x + 1}{x}$ **18.** [6.6] 41 **19.** [7.3] $3x^2 + 2x - 1$ **20.** [5.6] $-3 - 2i$
21. [2.6], [8.7] $\left\{x \mid x < -\frac{1}{8} \text{ or } x > \frac{1}{4}\right\}$ or $\left(-\infty, -\frac{1}{8}\right) \cup \left(\frac{1}{4}, \infty\right)$
22. [8.6]

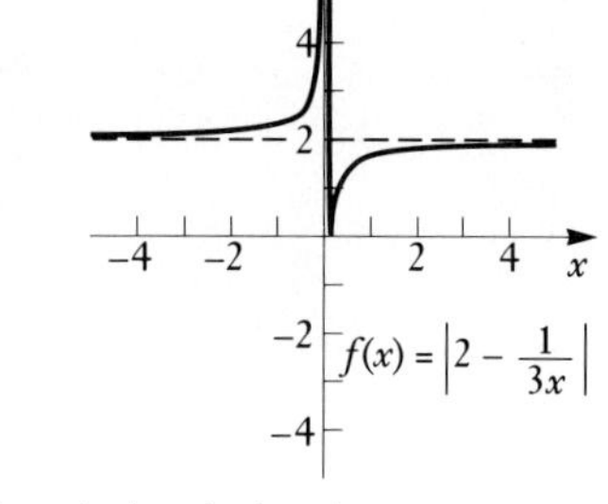

23. [8.4] $i, -i, 1 - i, 1 + i$

CHAPTER 9

Exercise Set 9.1, pp. 398–400

1.

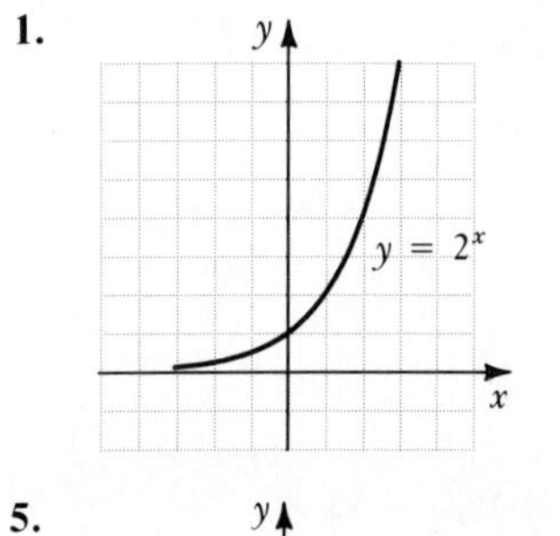

3.

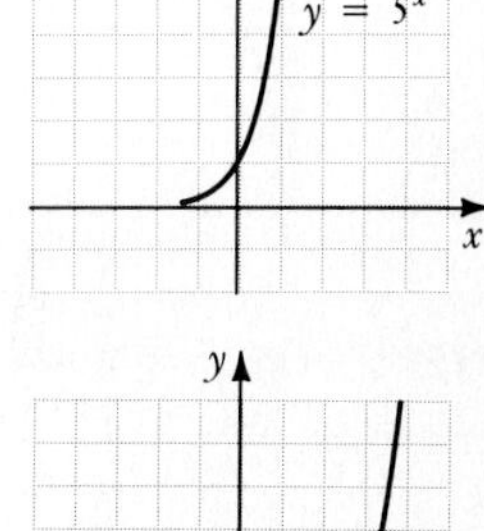

5.

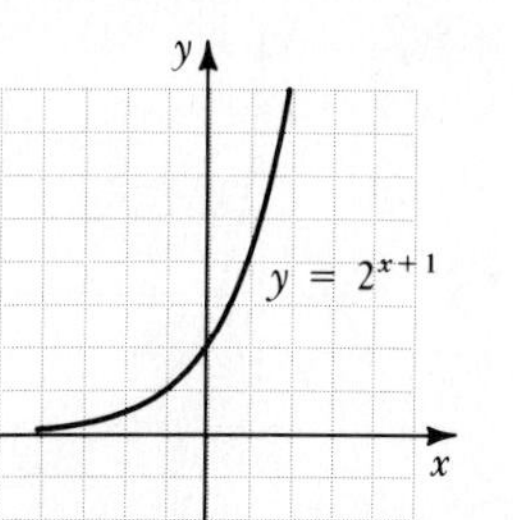

7.

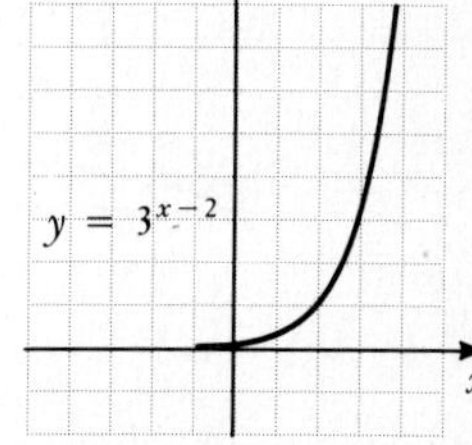

9.

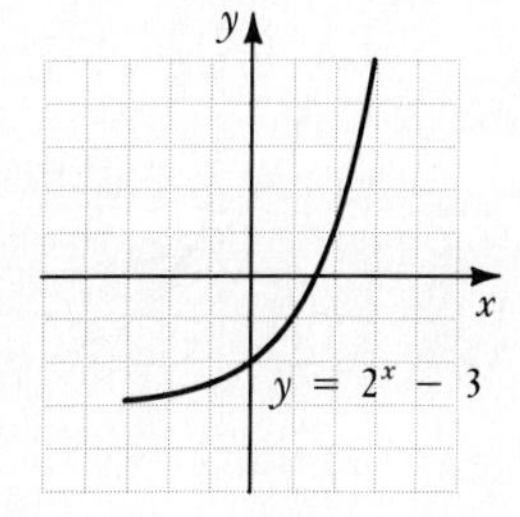

11.

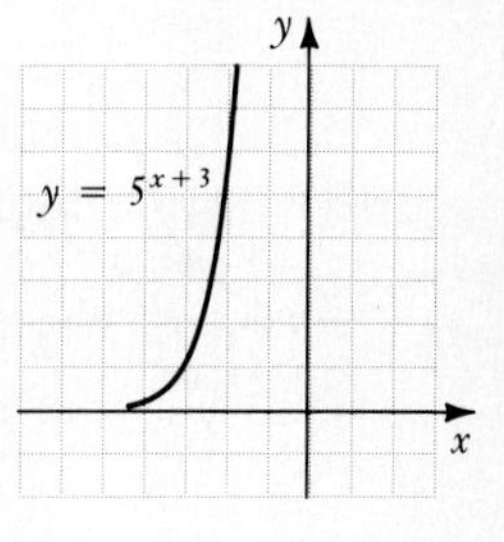

13.

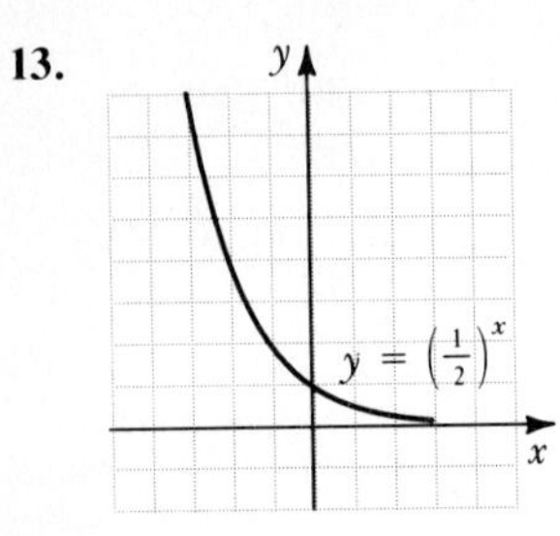

15.

$y = \left(\frac{1}{5}\right)^x$

17.

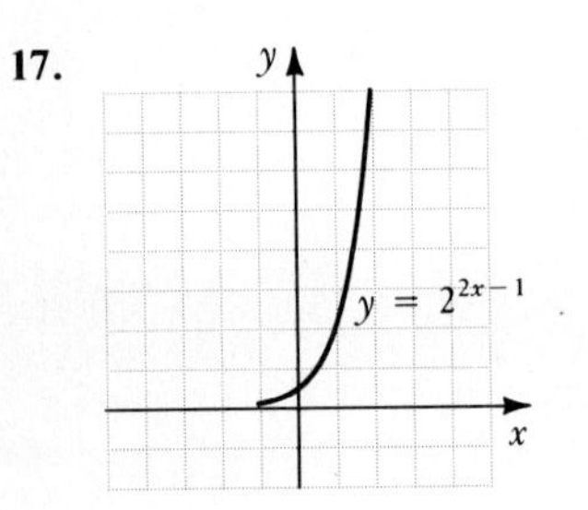

19.

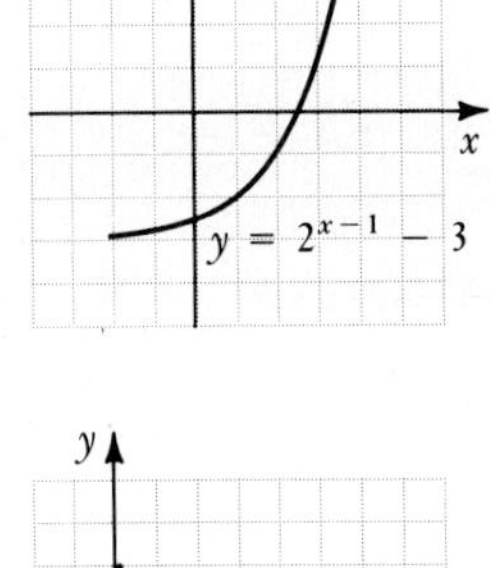

21.

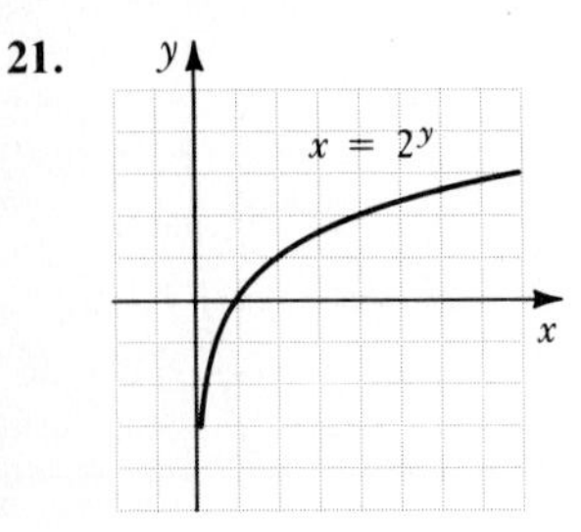

23.

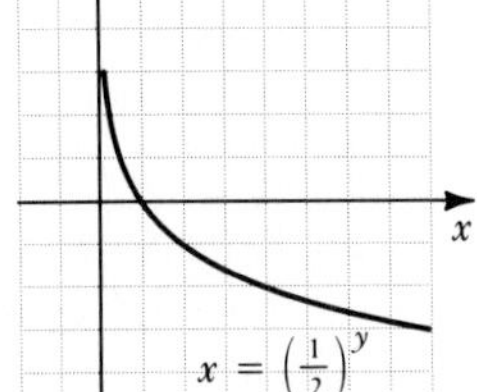

25.

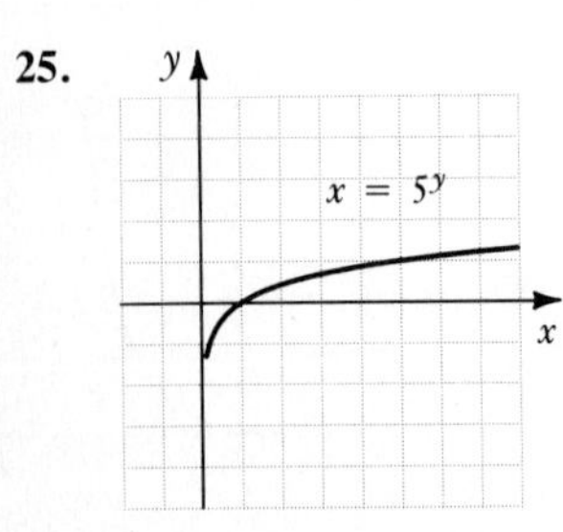

27.

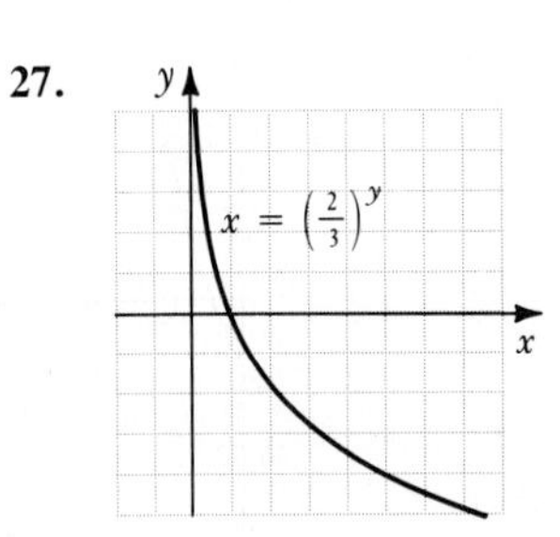

29.

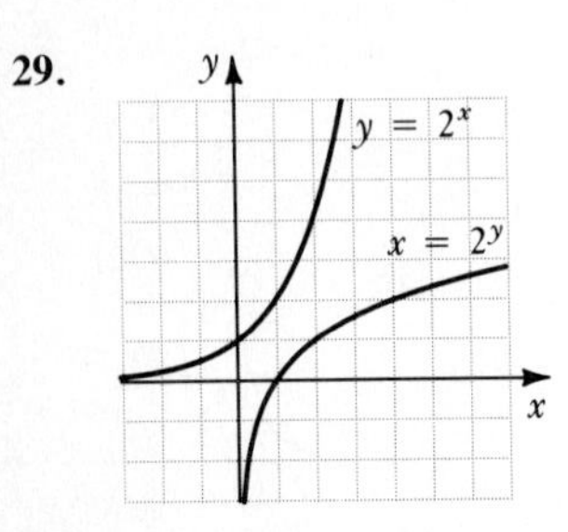

31.

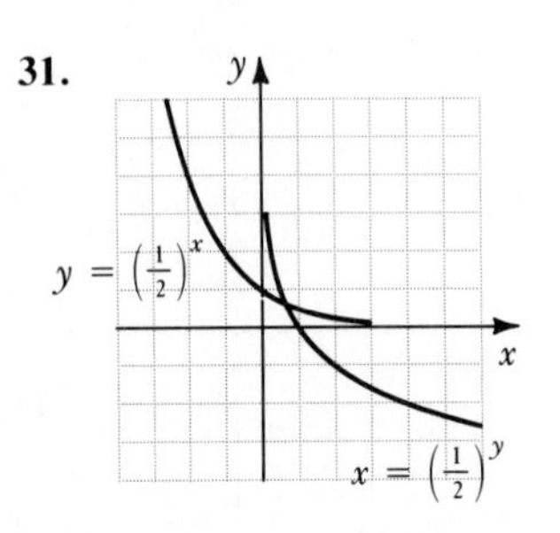

33. **(a)** $A(t) = \$50,000(1.09)^t$;
(b) \$50,000, \$70,579.08, \$99,628.13, \$118,368.18;
(c)

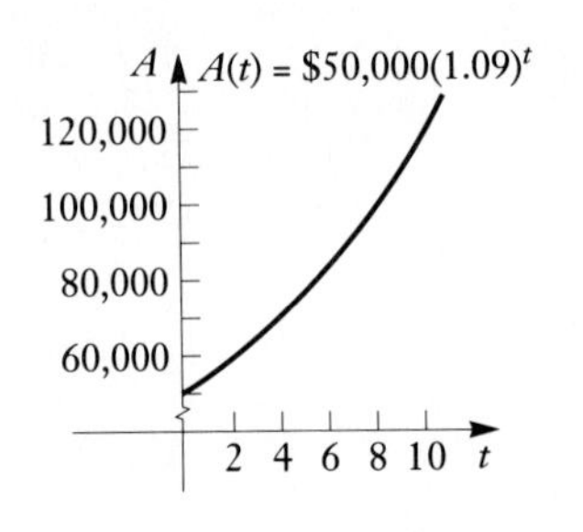

35. **(a)** 250,000, 62,500, 977, 0;
(b)

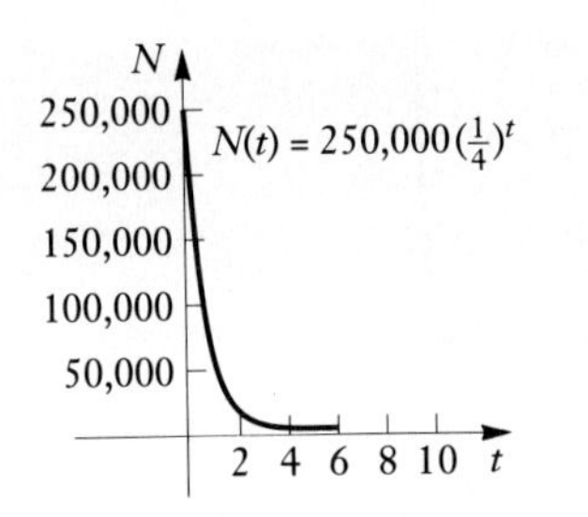

37. **(a)** 7.5 million, 18.4 million, 110.2 million, 661.4 million, 58,320 million, 5,142,752.7 million;
(b)

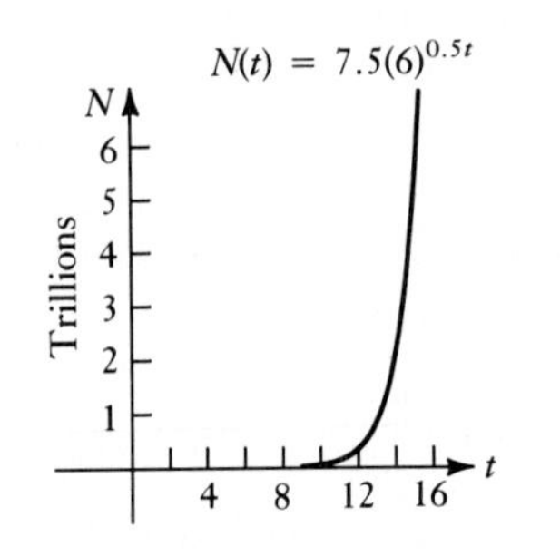

39. x^{-2}, or $\frac{1}{x^2}$ **41.** x^{-7}, or $\frac{1}{x^7}$

43. **(a)** 8; **(b)** 8.574188; **(c)** 8.815241; **(d)** 8.821353; **(e)** 8.824411; **(f)** 8.824962 **45.** $\pi^{2.4}$

47.

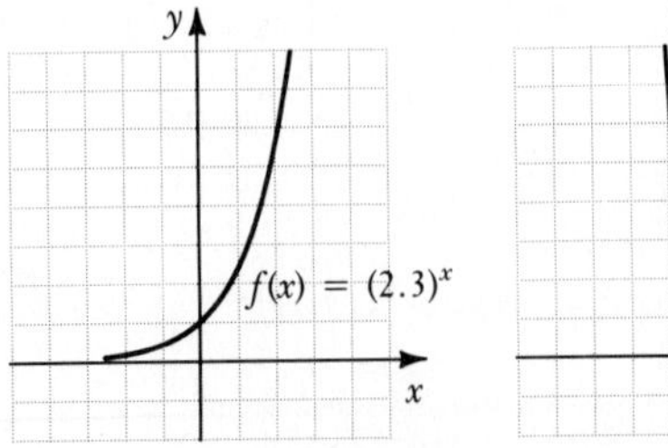

49.

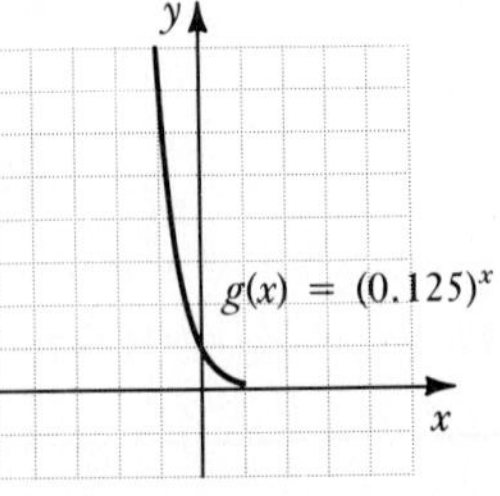

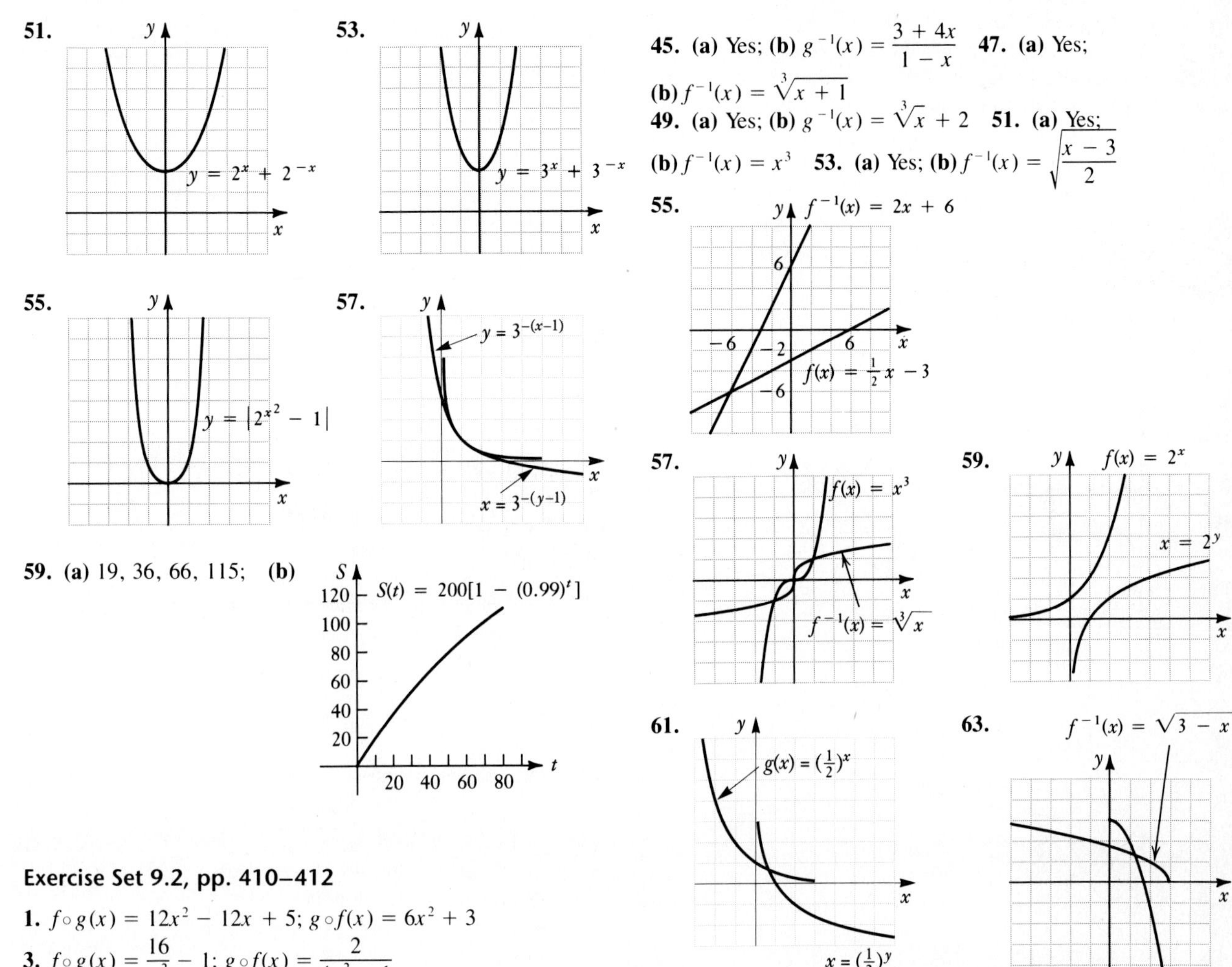

Exercise Set 9.2, pp. 410–412

1. $f \circ g(x) = 12x^2 - 12x + 5$; $g \circ f(x) = 6x^2 + 3$
3. $f \circ g(x) = \frac{16}{x^2} - 1$; $g \circ f(x) = \frac{2}{4x^2 - 1}$
5. $f \circ g(x) = x^4 - 2x^2 + 2$; $g \circ f(x) = x^4 + 2x^2$
7. $f(x) = x^2$; $g(x) = 5 - 3x$ **9.** $f(x) = x^5$; $g(x) = 3x^2 - 7$
11. $f(x) = \frac{1}{x}$; $g(x) = x - 1$
13. $f(x) = \frac{1}{\sqrt{x}}$; $g(x) = 7x + 2$
15. $f(x) = \frac{x + 1}{x - 1}$, $g(x) = x^3$ **17.** Yes **19.** No
21. Yes **23.** No **25.** No **27.** Yes **29.** **(a)** Yes; **(b)** $f^{-1}(x) = x - 2$ **31.** **(a)** Yes; **(b)** $f^{-1}(x) = 5 - x$
33. **(a)** Yes; **(b)** $g^{-1}(x) = x + 5$ **35.** **(a)** Yes; **(b)** $f^{-1}(x) = \frac{x}{3}$ **37.** **(a)** Yes; **(b)** $g^{-1}(x) = \frac{x - 2}{3}$
39. **(a)** Yes; **(b)** $h^{-1}(x) = \frac{4 - 3x}{x}$, or $\frac{4}{x} - 3$ **41.** **(a)** Yes; **(b)** $f^{-1}(x) = \frac{1}{x}$ **43.** **(a)** Yes; **(b)** $f^{-1}(x) = \frac{3x - 1}{2}$

45. **(a)** Yes; **(b)** $g^{-1}(x) = \frac{3 + 4x}{1 - x}$ **47.** **(a)** Yes; **(b)** $f^{-1}(x) = \sqrt[3]{x + 1}$
49. **(a)** Yes; **(b)** $g^{-1}(x) = \sqrt[3]{x} + 2$ **51.** **(a)** Yes; **(b)** $f^{-1}(x) = x^3$ **53.** **(a)** Yes; **(b)** $f^{-1}(x) = \sqrt{\frac{x - 3}{2}}$
55.
57.
59.
61.
63.

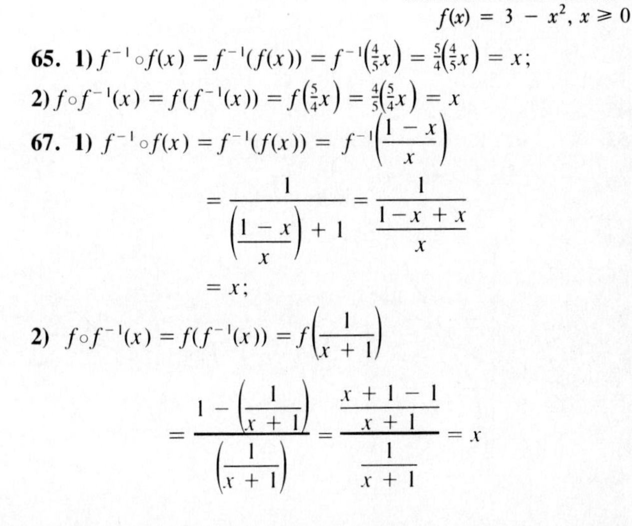

65. **1)** $f^{-1} \circ f(x) = f^{-1}(f(x)) = f^{-1}\left(\frac{4}{5}x\right) = \frac{5}{4}\left(\frac{4}{5}x\right) = x$;
2) $f \circ f^{-1}(x) = f(f^{-1}(x)) = f\left(\frac{5}{4}x\right) = \frac{4}{5}\left(\frac{5}{4}x\right) = x$
67. **1)** $f^{-1} \circ f(x) = f^{-1}(f(x)) = f^{-1}\left(\frac{1 - x}{x}\right)$

$$= \frac{1}{\left(\frac{1 - x}{x}\right) + 1} = \frac{1}{\frac{1 - x + x}{x}}$$

$$= x;$$

2) $f \circ f^{-1}(x) = f(f^{-1}(x)) = f\left(\frac{1}{x + 1}\right)$

$$= \frac{1 - \left(\frac{1}{x + 1}\right)}{\left(\frac{1}{x + 1}\right)} = \frac{\frac{x + 1 - 1}{x + 1}}{\frac{1}{x + 1}} = x$$

69. (a) 40, 42, 46, 50; (b) $f^{-1}(x) = x - 32$; (c) 8, 10, 14, 18 **71.** $y = 9x$ **73.** No; it fails the horizontal line test.

Exercise Set 9.3, pp. 419–420

1. **3.** **5.** **7.** **9.**

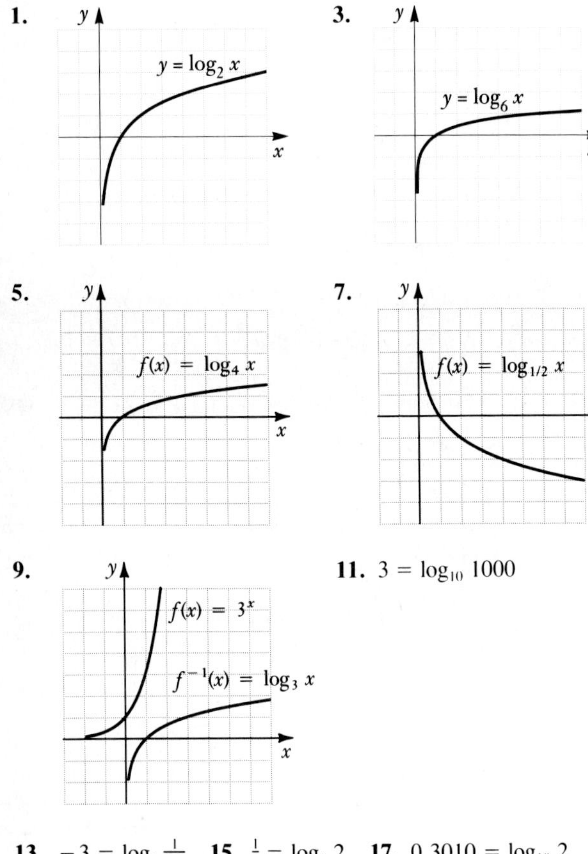

11. $3 = \log_{10} 1000$

13. $-3 = \log_5 \frac{1}{125}$ **15.** $\frac{1}{3} = \log_8 2$ **17.** $0.3010 = \log_{10} 2$
19. $2 = \log_e t$ **21.** $t = \log_Q x$ **23.** $2 = \log_e 7.3891$
25. $-2 = \log_e 0.1353$ **27.** $3^t = 8$ **29.** $5^2 = 25$
31. $10^{-1} = 0.1$ **33.** $10^{0.845} = 7$ **35.** $e^{2.9957} = 20$
37. $t^k = Q$ **39.** $e^{-1.3863} = 0.25$ **41.** $r^{-x} = T$ **43.** 9
45. 6 **47.** $\frac{1}{2}$ **49.** 2 **51.** 2 **53.** -1 **55.** 0 **57.** 4
59. -2 **61.** 0 **63.** 1 **65.** $\frac{2}{3}$ **67.** 3
69. t **71.** $\dfrac{x(3y-2)}{2y+x}$ **73.** $\dfrac{1}{4096}$ **75.** $\dfrac{1}{\sqrt[3]{t^2}}$
77. **79.**

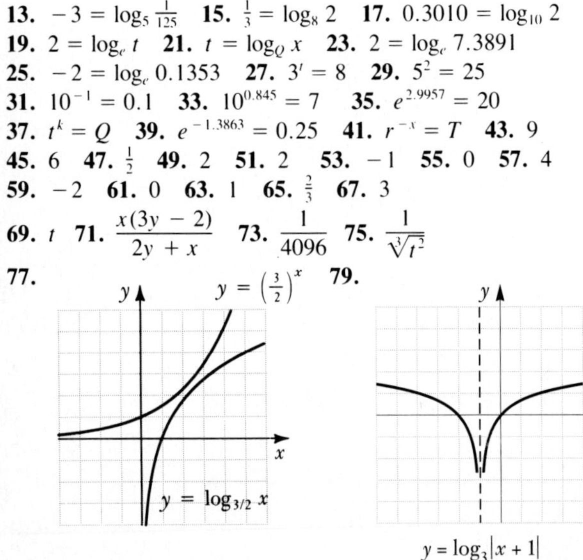

81. 25 **83.** $\dfrac{\sqrt{5}}{25}$ **85.** 6 **87.** 1296 **89.** 3 **91.** 0
93. -8

Exercise Set 9.4, pp. 424–426

1. $\log_2 32 + \log_2 8$ **3.** $\log_4 64 + \log_4 16$
5. $\log_c B + \log_c x$ **7.** $\log_a (6 \cdot 70)$ **9.** $\log_c (K \cdot y)$
11. $3 \log_a x$ **13.** $6 \log_c y$ **15.** $-3 \log_b C$
17. $\log_a 67 - \log_a 5$ **19.** $\log_b 3 - \log_b 4$
21. $\log_a \frac{15}{7}$ **23.** $2 \log_a x + 3 \log_a y + \log_a z$
25. $\log_b x + 2 \log_b y - 3 \log_b z$
27. $\frac{1}{3}(4 \log_c x - 3 \log_c y - 2 \log_c z)$
29. $\frac{1}{4}(8 \log_a x + 12 \log_a y - 3 - 5 \log_a z)$
31. $\log_a \dfrac{\sqrt[3]{x^2}\sqrt{y}}{y}$ **33.** $\log_a \dfrac{2x^4}{y^3}$ **35.** $\log_a \dfrac{\sqrt{a}}{x}$
37. 2.708 **39.** 0.51 **41.** -1.609 **43.** $\frac{3}{2}$ **45.** 2.609
47. 3.218 **49.** 9 **51.** m **53.** 4 **55.** -7 **57.** 18
59. $\pm\sqrt{6}, \pm\sqrt{2}$ **61.** $\log_a (x^6 - x^4y^2 + x^2y^4 - y^6)$
63. $\frac{1}{2} \log_a (1 - s) + \frac{1}{2} \log_a (1 + s)$ **65.** $\frac{10}{3}$ **67.** $\{x \mid x > 0\}$, or $(0, \infty)$ **69.** -2 **71.** False **73.** False **75.** True

77.
$$\frac{x + \sqrt{x^2 - 3}}{3} = \frac{x + \sqrt{x^2 - 3}}{3} \cdot \frac{x - \sqrt{x^2 - 3}}{x - \sqrt{x^2 - 3}} = \frac{x^2 - (x^2 - 3)}{3(x - \sqrt{x^2 - 3})} = \frac{1}{x - \sqrt{x^2 - 3}}.$$

Then
$$\log_a \left(\frac{x + \sqrt{x^2 - 3}}{3}\right) = \log_a \left(\frac{1}{x - \sqrt{x^2 - 3}}\right) = \log_a 1 - \log_a (x - \sqrt{x^2 - 3}) = 0 - \log_a (x - \sqrt{x^2 - 3}) = -\log_a (x - \sqrt{x^2 - 3}).$$

Exercise Set 9.5, pp. 431–432

1. 0.3010 **3.** 0.8021 **5.** 1.6532 **7.** 2.6405
9. 4.1271 **11.** -1.2840 **13.** 1000 **15.** 501.1872
17. 3.0001 **19.** 0.2841 **21.** 0.0011 **23.** 0.6931
25. 4.1271 **27.** 8.3814 **29.** -5.0832 **31.** 36.7890
33. 0.0023 **35.** 1.0057 **37.** 5.8346×10^{14} **39.** 7.6331
41. 2.5702 **43.** 3.3219 **45.** 0.6419 **47.** -2.3219
49. -2.3219 **51.** 3.5471
53. $f(x) = e^x$ **55.**

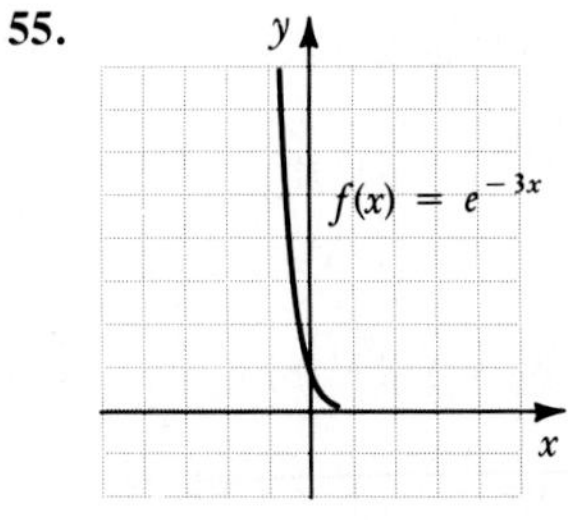

57. **61.** **65.**

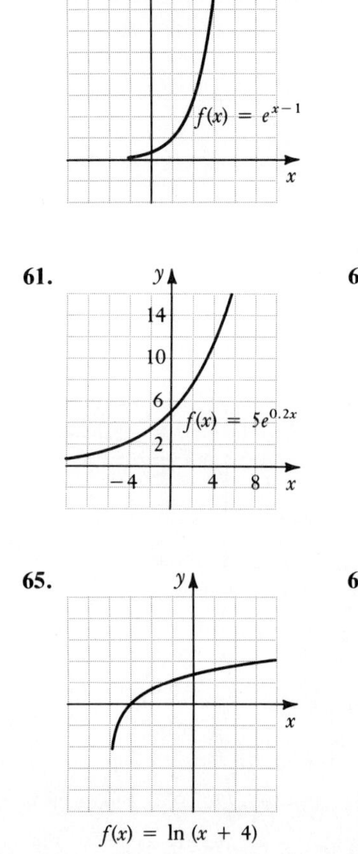

59. **63.**

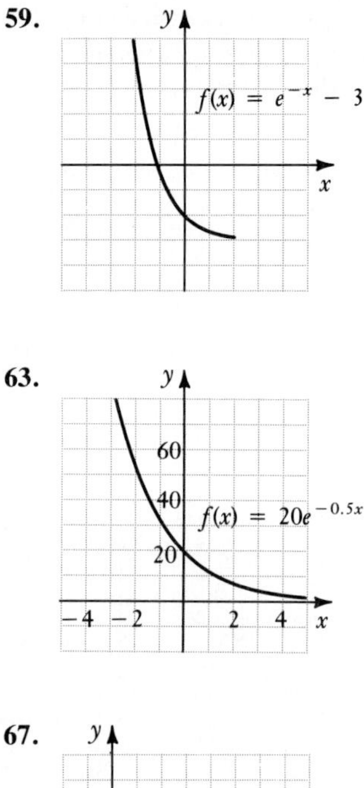

67.

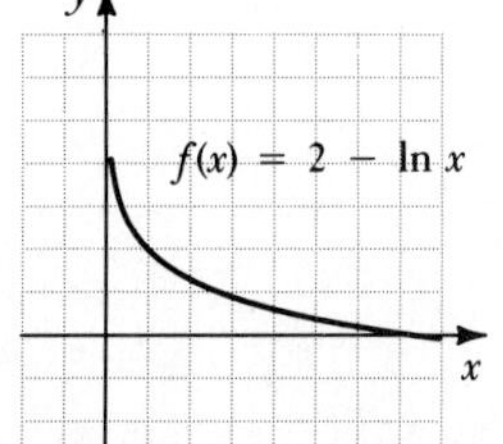

69.

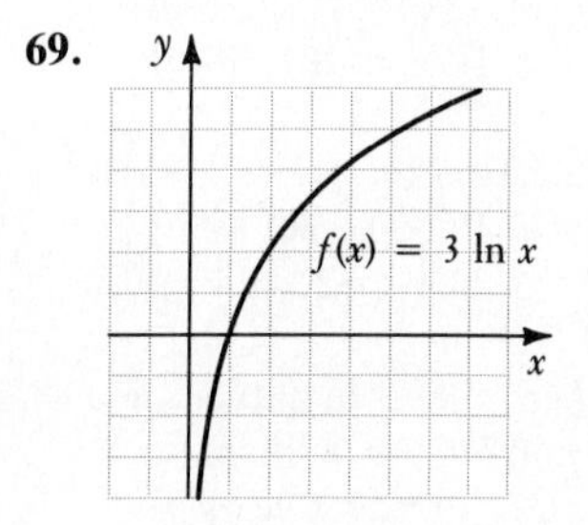

71.

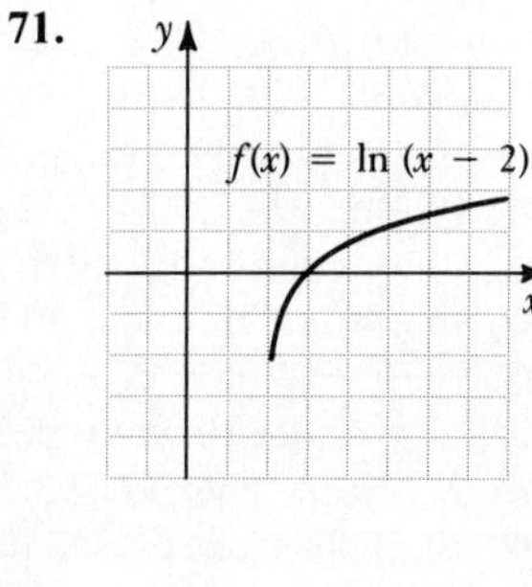

73.

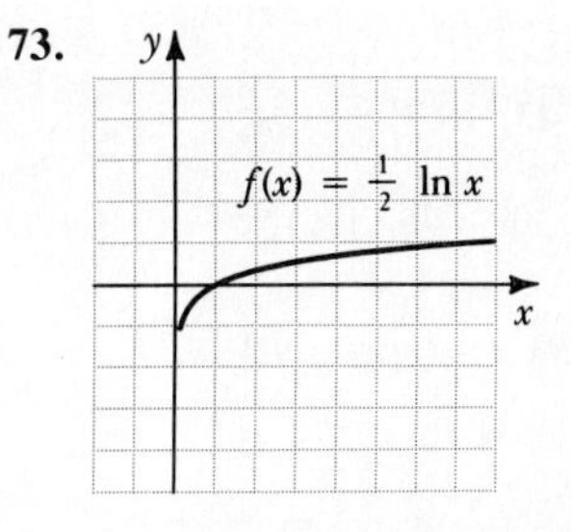

75.

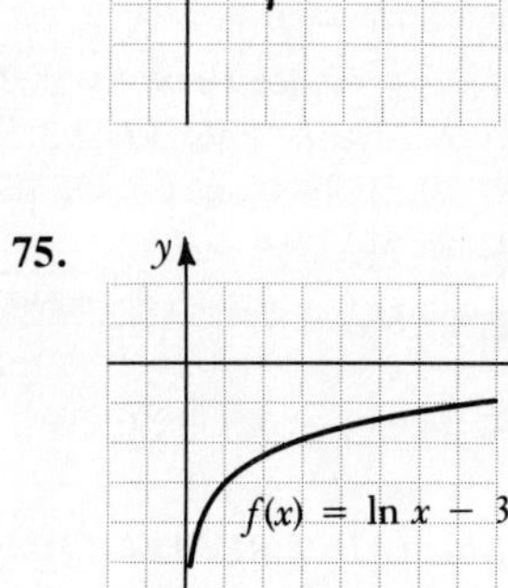

77.

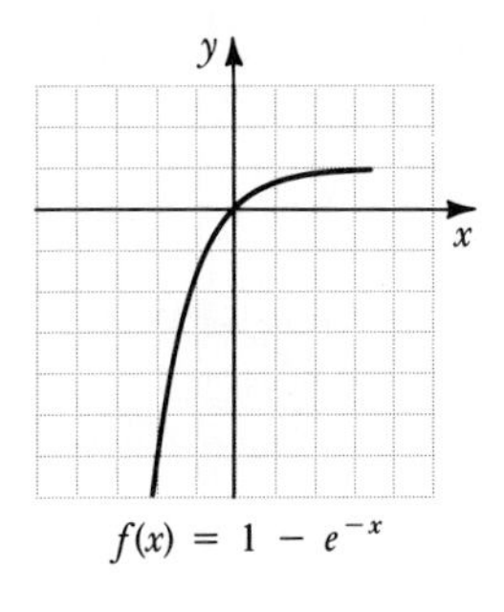

79. $\frac{5}{2}, -\frac{5}{2}$ **81.** 16, 256

83. $\ln M = \dfrac{\log M}{\log e}$ **85.** 52.5084 **87.** 4.9855

Exercise Set 9.6, pp. 440–445

1. 3 **3.** 4 **5.** $\frac{5}{2}$ **7.** $\frac{3}{5}$ **9.** 3.170 **11.** 3.322 **13.** $\frac{5}{2}$ **15.** −3, −1, **17.** 1.404 **19.** 4.605 **21.** 2.303 **23.** 140.671 **25.** 2.710 **27.** 3.607 **29.** 5.646 **31.** 27 **33.** $\frac{1}{8}$ **35.** 10 **37.** $\frac{1}{100}$ **39.** $e^2 \approx 7.389$ **41.** $\dfrac{1}{e} \approx 0.368$ **43.** 66 **45.** 10 **47.** $\frac{1}{3}$ **49.** 3 **51.** $\frac{2}{5}$ **53.** 5 **55.** **(a)** 37.7 years; **(b)** 11.9 years **57.** **(a)** 1 year; **(b)** 7.3 years **59.** **(a)** $P(t) = P_0e^{0.09t}$; **(b)** \$1094.17, \$1197.22; **(c)** 7.7 years **61.** 69.3 years **63.** **(a)** 5.5 years; **(b)** 0.8 year **65.** **(a)** 68%; **(b)** 54%, 40% **(c)**

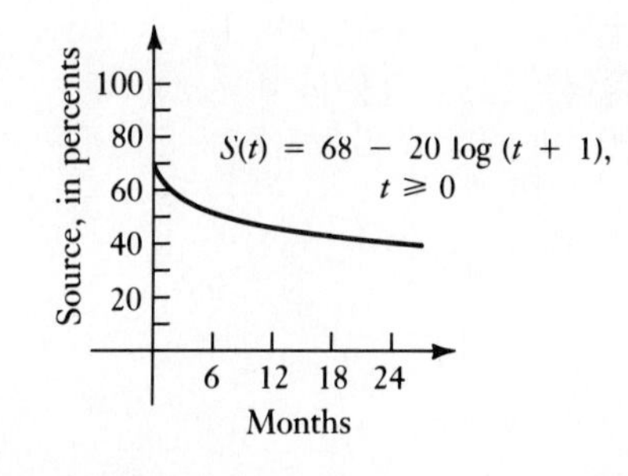

(d) 6.9 months **67.** 6.2 **69.** 7.8 **71.** 10^{-7} moles per liter **73.** 6.3×10^{-4} moles per liter **75.** 8.25 **77.** 1997 **79.** **(a)** 0.93%, $P(t) = 844{,}401e^{0.0093t}$; **(b)** 1,075,378 **81.** 1860 years **83.** 7.2 days **85.** 23% per minute **87.** **(a)** $k = 0.16$; $V(t) = 84e^{0.16t}$, where V is given in thousands of dollars; **(b)** \$250,400 thousand, or \$250,400,000; **(c)** 4.3 years; **(d)** 58.7 years **89.** $\dfrac{(x-5)(x-3)}{x^2 + 5x - 6}$ **91.** −16; 4 and −4 **93.** −4 **95.** 2 **97.** 1, 10^9 **99.** $\pm 2\sqrt{6}$ **101.** $\pm\sqrt{34}$ **103.** $10^{100{,}000}$ **105.** 1, 100 **107.** $\frac{1}{100{,}000}$, 100,000 **109.** −7, 3 **111.** **(a)** After studying about 69 hours; **(b)** about 138 hours

Review Exercises: Chapter 9, pp. 446–448

1. [9.1]

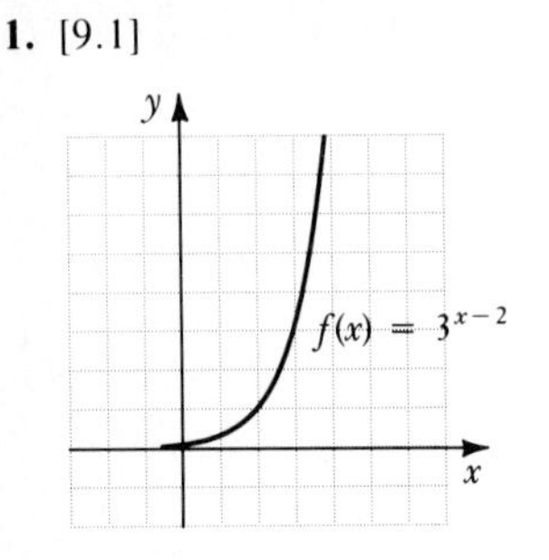

2. [9.1]

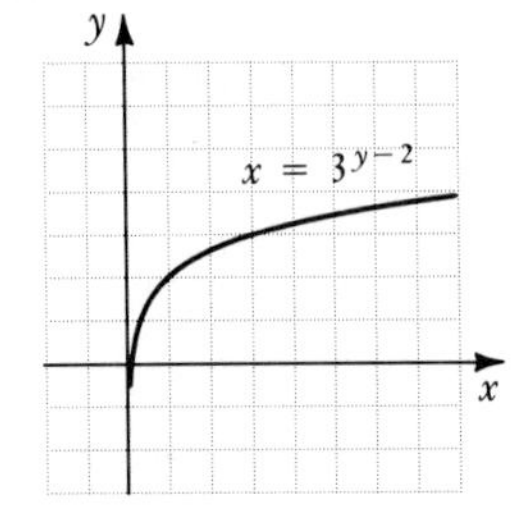

3. [9.3]

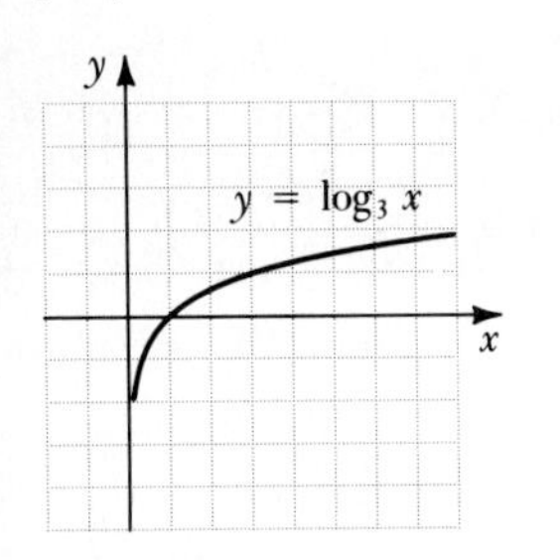

4. [9.3]

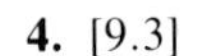

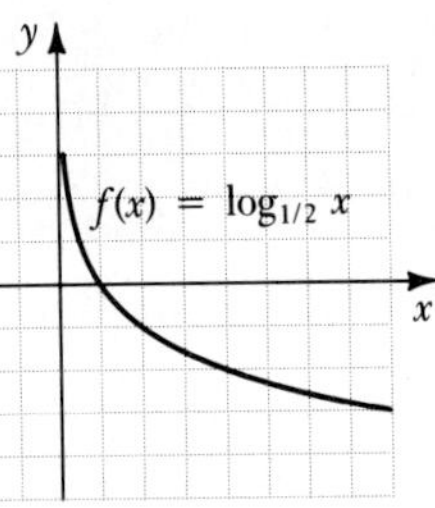

5. [9.2] $f \circ g(x) = 9x^2 - 30x + 25$, $g \circ f(x) = 3x^2 - 5$ **6.** [9.2] No **7.** [9.2] $f^{-1}(x) = x - 2$ **8.** [9.2] $g^{-1}(x) = \dfrac{7x + 3}{2}$ **9.** [9.2] $f^{-1}(x) = \dfrac{2 - 5x}{x}$, or $\dfrac{2}{x} - 5$ **10.** [9.2] $g^{-1}(x) = \dfrac{\sqrt[3]{x}}{2}$ **11.** [9.3] $4^x = 16$ **12.** [9.3] $10^{0.3010} = 2$ **13.** [9.3] $\left(\frac{1}{2}\right)^{-3} = 8$ **14.** [9.3] $16^{3/4} = 8$ **15.** [9.3] $\log_{10} 10{,}000 = 4$ **16.** [9.3] $\log_{25} 5 = \frac{1}{2}$ **17.** [9.3] $\log_7 \frac{1}{49} = -2$ **18.** [9.3] $\log_{2.718} 20.1 = 3$ **19.** [9.4] $4 \log_a x + 2 \log_a y + 3 \log_a z$ **20.** [9.4] $\log_a x + \log_a y - 2 \log_a z$ **21.** [9.4] $\frac{1}{4}(2 \log z - 3 \log x - \log y)$ **22.** [9.4] $2 \log_q x + \frac{1}{3} \log_q y - 4 \log_q z$ **23.** [9.4] $\log_a (8 \cdot 15)$, or $\log_a 120$ **24.** [9.4] $\log_a \frac{72}{12}$, or $\log_a 6$ **25.** [9.4] $\log \dfrac{a^{1/2}}{bc^2}$ **26.** [9.4] $\log_a \sqrt[3]{\dfrac{x}{y^2}}$ **27.** [9.3] 1 **28.** [9.3] 0 **29.** [9.4] 17 **30.** [9.4] −7 **31.** [9.4] 6.93 **32.** [9.4] −3.2698 **33.** [9.4] 8.7601 **34.** [9.4] 3.2698 **35.** [9.4] 2.54995 **36.** [9.4] −3.6602 **37.** [9.5] −2.2027 **38.** [9.5] 7.8621 **39.** [9.5] 29,798.88 **40.** [9.5] 0.0361 **41.** [9.5] 213.50 **42.** [9.5] −2.3065 **43.** [9.5] 5.5965 **44.** [9.5] 0.000000163 **45.** [9.5] 10.0821 **46.** [9.5] −2.6921 **47.** [9.5] 0.00002593 **48.** [9.5] 3.4934×10^{19} **49.** [9.5] 0.4307 **50.** [9.5] 1.7097 **51.** [9.3] $\frac{1}{9}$ **52.** [9.3] 2 **53.** [9.3] $\frac{1}{10{,}000}$ **54.** [9.6] $e^2 \approx 7.3891$ **55.** [9.6] $\frac{7}{2}$ **56.** [9.6] 1.5266 **57.** [9.3] 2 **58.** [9.6] 7 **59.** [9.6] 8 **60.** [9.6] 20 **61.** [9.6] $\sqrt{43}$ **62.** [9.6] **(a)** 62; **(b)** 46.8; **(c)** 35 months **63.** [9.6] **(a)** $k = 0.05$, $C(t) = \$4.65e^{0.05t}$; **(b)** \$34.36; **(c)** 1991; **(d)** 13.9 years **64.** [9.6] 4.3% **65.** [9.6] 8.25 years **66.** [9.6] 3463 years **67.** [9.6] 6.6 **68.** [9.6] 8.3 **69.** [7.6] $T = \dfrac{-b \pm \sqrt{b^2 + 4aQ}}{2a}$ **70.** [5.5] ± 1, ± 4 **71.** [4.3] $\dfrac{c - 2a}{2b + 3c}$ **72.** [6.7] $y = \dfrac{9}{x}$ **73.** [9.6] 3 **74.** [9.6] −2

Test: Chapter 9, pp. 448–449

1. [9.1] **2.** [9.3]

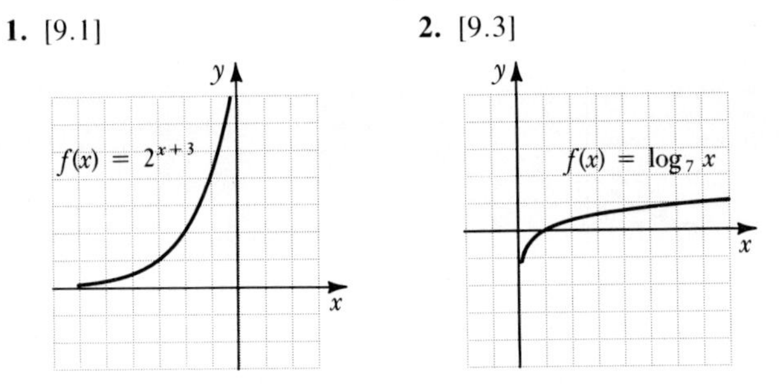

3. [9.2] $f \circ g(x) = 25x^2 - 15x + 2$, $g \circ f(x) = 5x^2 + 5x - 2$ **4.** [9.2] No **5.** [9.2] $f^{-1}(x) = \dfrac{x + 3}{4}$ **6.** [9.2] $g^{-1}(x) = 4x + 2$ **7.** [9.2] $f^{-1}(x) = \dfrac{1 + 2x}{x - 1}$ **8.** [9.3] $\log_4 x = -3$ **9.** [9.3] $\log_{256} 16 = \dfrac{1}{2}$ **10.** [9.3] $4^2 = 16$ **11.** [9.3] $7^m = 49$ **12.** [9.4] $3 \log a + \dfrac{1}{2} \log b - 2 \log c$ **13.** [9.4] $\log_a \dfrac{x^{1/3}z^2}{y^3}$ **14.** [9.4] 23 **15.** [9.3] 1 **16.** [9.3] 0 **17.** [9.4] −0.544 **18.** [9.4] 0.69 **19.** [9.4] 1.322 **20.** [9.5] −1.9101 **21.** [9.5] 445,040.98 **22.** [9.5] 0.000000054777 **23.** [9.5] 4.0913 **24.** [9.5] −4.3949 **25.** [9.5] 292.19 **26.** [9.5] 1.1881 **27.** [9.3] 5 **28.** [9.3] 2 **29.** [9.6] 10,000 **30.** [9.6] $\frac{1}{3}$ **31.** [9.6] 0.0937 **32.** [9.6] $e^{1/4} \approx 1.2840$ **33.** [9.6] 9 **34.** [9.6] **(a)** 2.45 ft/sec; **(b)** 984,262 **35.** [9.6] **(a)** $P(t) = 209e^{0.01t}$, where $P(t)$ is in millions and t is in years since 1959; **(b)** 302.6 million, 348 million; **(c)** 1977; **(d)** 69.3 years **36.** [9.6] 3.5% **37.** [9.6] 9.1 years **38.** [9.6] 4684 years **39.** [9.6] 4.7 **40.** [9.6] 7.0 **41.** [5.5] 1, 64 **42.** [7.6] $t = \dfrac{b \pm \sqrt{b^2 + 4aS}}{2a}$ **43.** [6.7] $y = \frac{2}{3}x$ **44.** [4.3] $\dfrac{1}{2x}$ **45.** [9.6] 316, − 309 **46.** [9.4] 2

CUMULATIVE REVIEW: CHAPTERS 1–9, pp. 449–451

1. [1.1] 48 **2.** [1.2] $2\frac{1}{4}$ **3.** [1.3] 16 **4.** [3.1] −5 **5.** [4.1] −3 **6.** [5.4] 1

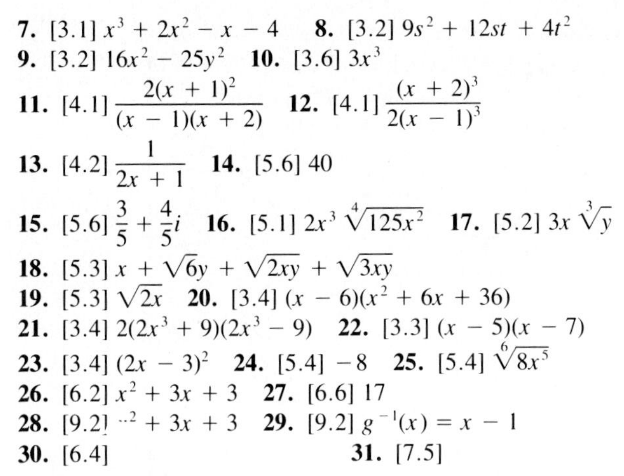

7. [3.1] $x^3 + 2x^2 - x - 4$ **8.** [3.2] $9s^2 + 12st + 4t^2$
9. [3.2] $16x^2 - 25y^2$ **10.** [3.6] $3x^3$
11. [4.1] $\dfrac{2(x + 1)^2}{(x - 1)(x + 2)}$ **12.** [4.1] $\dfrac{(x + 2)^3}{2(x - 1)^3}$
13. [4.2] $\dfrac{1}{2x + 1}$ **14.** [5.6] 40
15. [5.6] $\dfrac{3}{5} + \dfrac{4}{5}i$ **16.** [5.1] $2x^3 \sqrt[4]{125x^2}$ **17.** [5.2] $3x \sqrt[3]{y}$
18. [5.3] $x + \sqrt{6}y + \sqrt{2xy} + \sqrt{3xy}$
19. [5.3] $\sqrt{2x}$ **20.** [3.4] $(x - 6)(x^2 + 6x + 36)$
21. [3.4] $2(2x^3 + 9)(2x^3 - 9)$ **22.** [3.3] $(x - 5)(x - 7)$
23. [3.4] $(2x - 3)^2$ **24.** [5.4] -8 **25.** [5.4] $\sqrt[6]{8x^5}$
26. [6.2] $x^2 + 3x + 3$ **27.** [6.6] 17
28. [9.2] $x^2 + 3x + 3$ **29.** [9.2] $g^{-1}(x) = x - 1$
30. [6.4] **31.** [7.5]

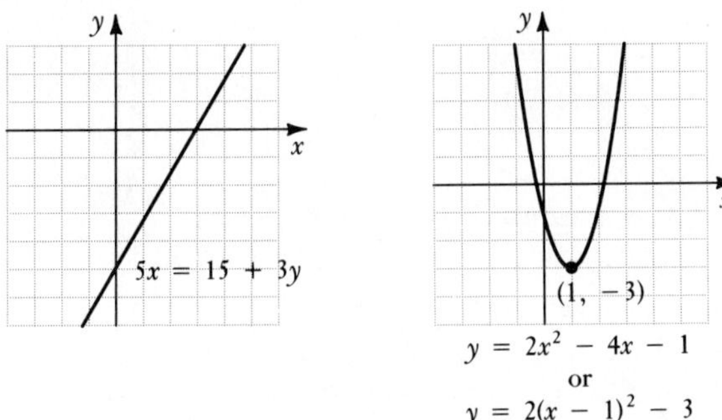

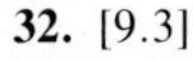
32. [9.3]

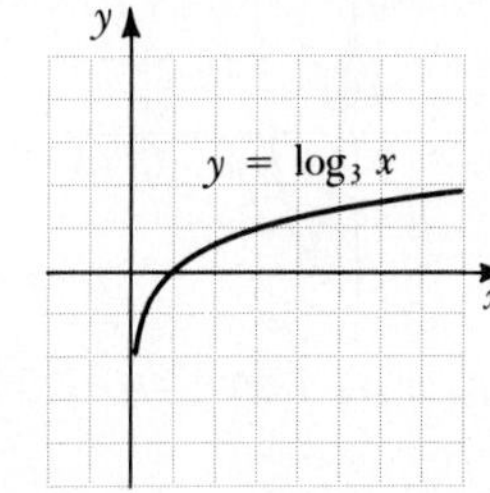

33. [9.1]

$y = 3^x$

34. [6.3] **35.** [8.5]

$5x = 25$

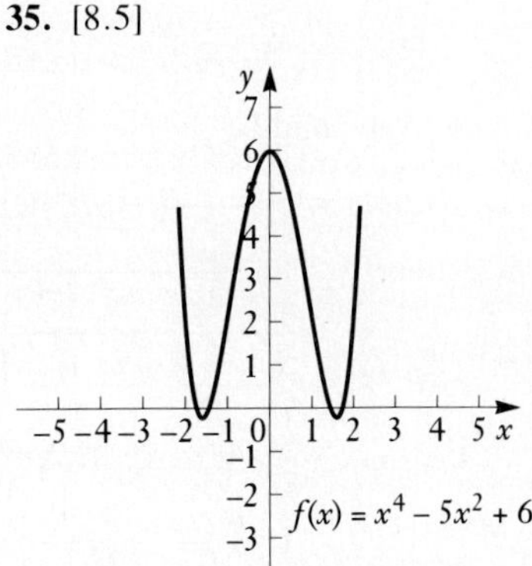

36. [7.4]

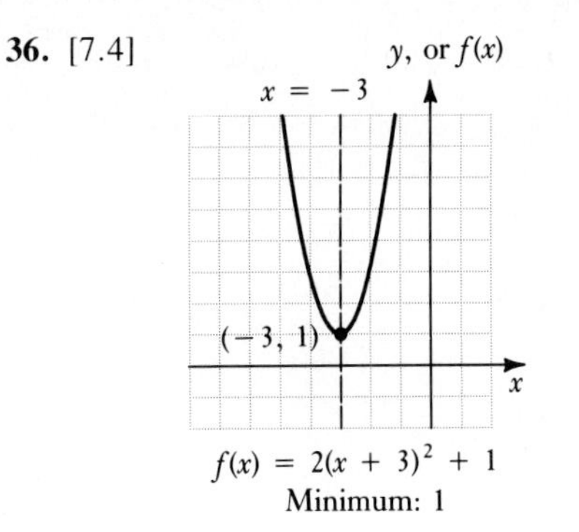

$f(x) = 2(x + 3)^2 + 1$
Minimum: 1

37. [6.4] $\frac{1}{2}$ **38.** [6.5] $y = -\frac{2}{3}x - 1$ **39.** [8.1] $2x^3 + 4x^2 + 7x + 15$, R 35 **40.** [8.2] Neither even nor odd
41. [8.3] $(x + 1)^2(x - 2)$; $-1, 2$ **42.** [8.4] $\pm\left(1, \frac{1}{7}\right)$
43. [9.3] $\log_{81} 3 = \frac{1}{4}$ **44.** [9.3] 0 **45.** [9.5] 4
46. [9.3] $\frac{1}{2}$ **47.** [9.5] 1.1314 **48.** [9.5] 15
49. [9.5] 1.5740 **50.** [9.4] $2 \log x + 3 \log z - \log y$
51. [9.4] $\log\left(\dfrac{x^3}{\sqrt{yz^2}}\right)$ **52.** [2.1] $\frac{11}{2}$ **53.** [3.5] $\frac{2}{5}, -5$
54. [5.5] 5 **55.** [5.5] 9, 25
56. [2.6] $\{x \mid x \leq -3 \text{ or } x \geq 6\}$ or $(-\infty, -3] \cup [6, \infty)$
57. [2.6] $\{x \mid -4 < x < -1\}$ or $(-4, -1)$
58. [3.5] $5, -2$ **59.** [7.2] $\dfrac{1 \pm i\sqrt{39}}{4}$ **60.** [4.4] $\frac{9}{2}$
61. [4.4] $\frac{5}{8}$ **62.** [7.1] $\pm 5i$ **63.** [7.2] $\dfrac{1 \pm \sqrt{7}}{2}$
64. [8.7] $\{x \mid x < -2 \text{ or } x > 5\}$ or $(-\infty, -2) \cup (5, \infty)$
65. [8.7] $\{x \mid -1 \leq x < 1\}$ or $[-1, 1)$ **66.** [9.3] 8
67. [9.3] 7 **68.** [9.6] $\frac{3}{2}$ **69.** [9.6] $\frac{80}{9}$ **70.** [9.6] 0.354
71. [2.2] 8 **72.** [4.5] $5\frac{5}{11}$ hr **73.** [4.5] $2\frac{7}{9}$ km/h
74. [7.6] -49; -7 and 7 **75.** [9.6] **(a)** 78; **(b)** 67.5
76. [9.6] **(a)** $P(t) = 430e^{0.01t}$, where P is in millions and t is in number of years after 1961; **(b)** 641 million, 702 million
77. [6.7] 18 **78.** [4.4] All real numbers except 1 and -2
79. [9.6] $\frac{1}{2}$, 5000 **80.** [2.6] $\{x \mid -3 \leq x \leq -1 \text{ or } 7 \leq x \leq 9\}$ or $[-3, -1] \cup [7, 9]$

CHAPTER 10

Exercise Set 10.1, pp. 458–460

1. 5 **3.** $3\sqrt{2} \approx 4.243$ **5.** 5 **7.** 8 **9.** 17.8
11. $\sqrt{49 + k^2}$ **13.** $\sqrt{13} \approx 3.606$ **15.** $8\sqrt{m^2 + n^2}$
17. $6\sqrt{2} \approx 8.485$ **19.** Yes **21.** No **23.** $\left(-\frac{1}{2}, -1\right)$
25. $\left(\frac{7}{2}, \frac{7}{2}\right)$ **27.** $(-1, -3)$ **29.** $(-0.25, -0.3)$
31. $\left(-\frac{1}{12}, \frac{1}{24}\right)$ **33.** $\left(\dfrac{\sqrt{2} + \sqrt{3}}{2}, \dfrac{3}{2}\right)$ **35.** (0, b)

37. (0, 0), 6

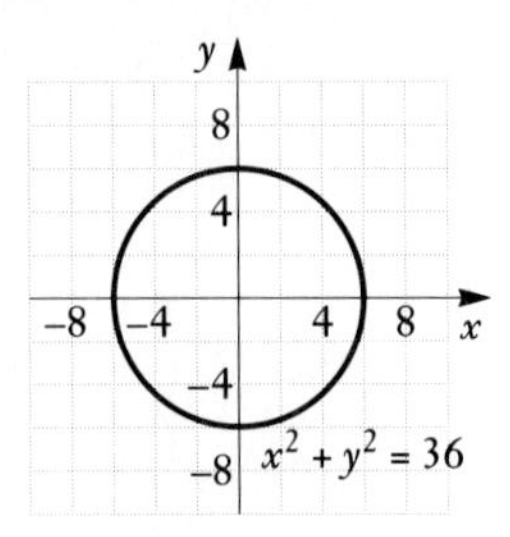

39. $(-1, -3)$, 2

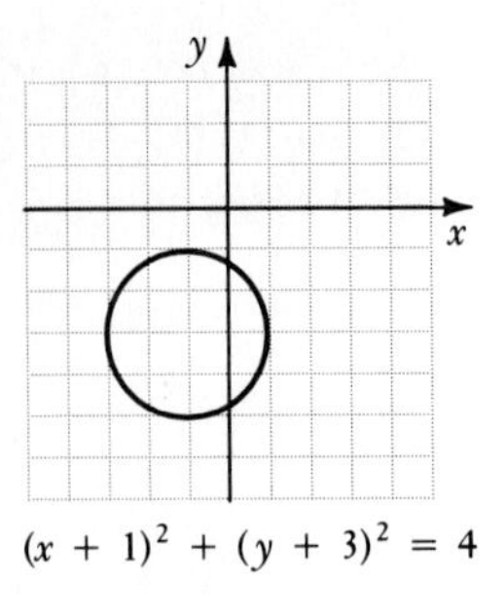

41. $(8, -3)$, $2\sqrt{10}$

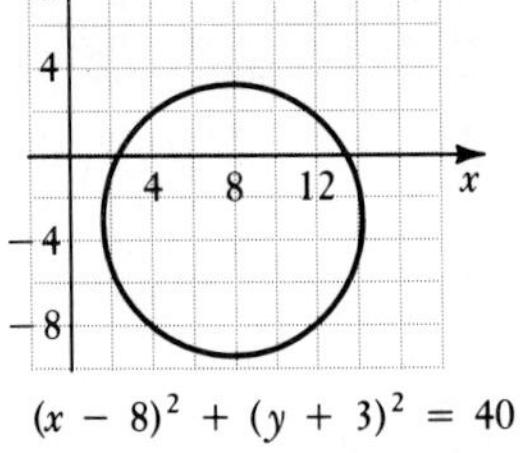

43. (0, 0), $\sqrt{2}$

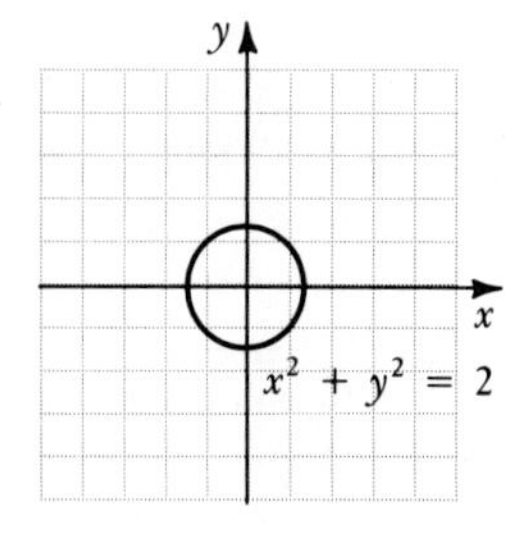

45. (5, 0), $\frac{1}{2}$

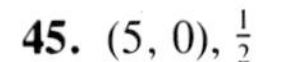

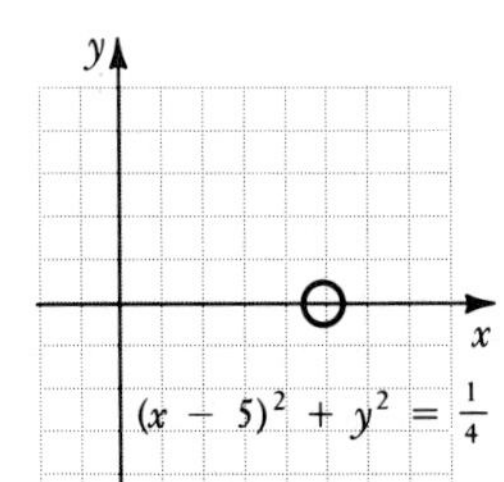

47. $(-4, 3)$, $2\sqrt{10}$

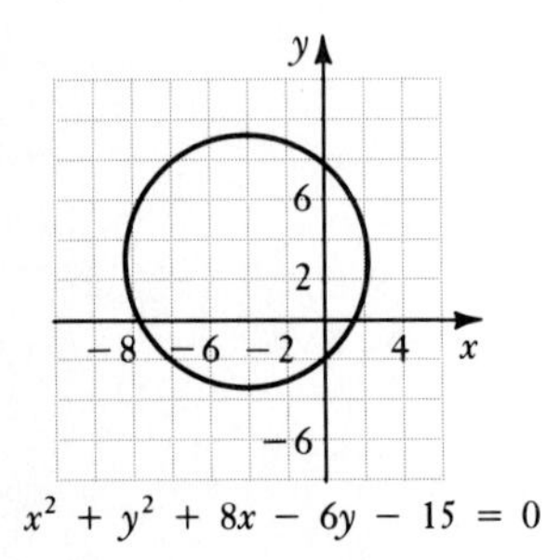

49. $(4, -1)$, 2

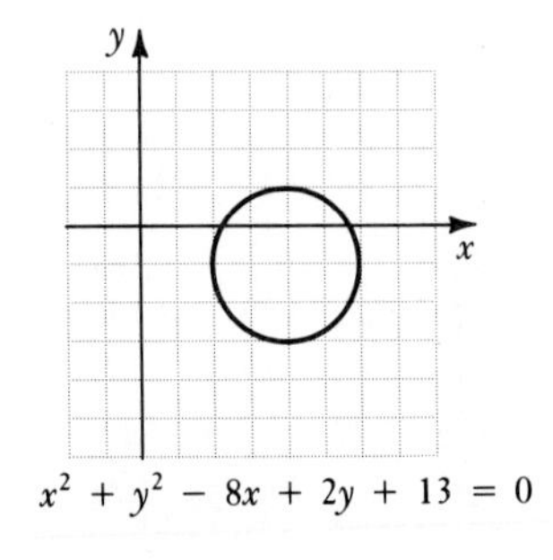

51. (2, 0), 2

$x^2 + y^2 - 4x = 0$

53. $(0, -5)$, 10

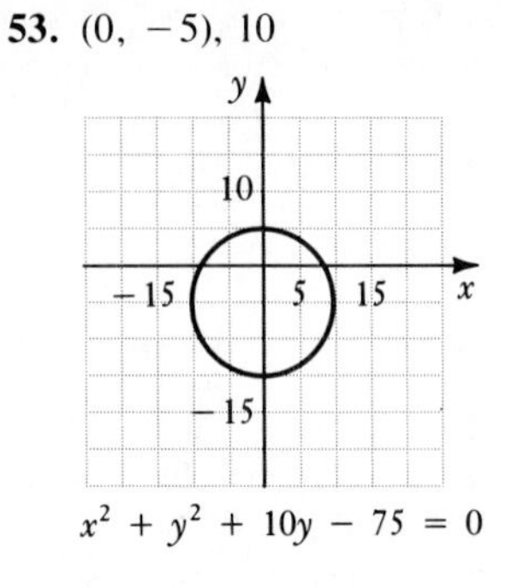

55. $\left(-\frac{7}{2}, \frac{3}{2}\right)$, $\frac{7\sqrt{2}}{2}$

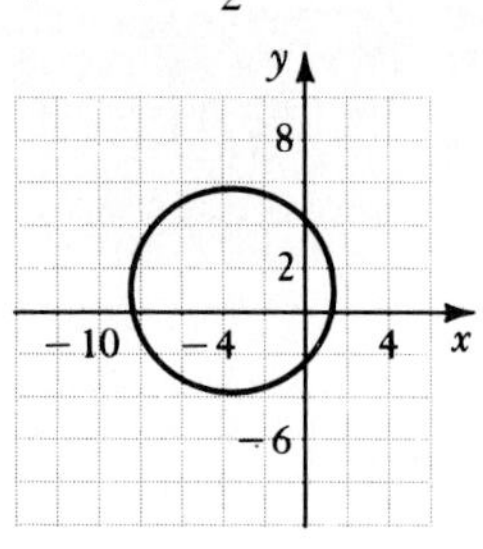

57. (0, 0), $\frac{1}{2}$

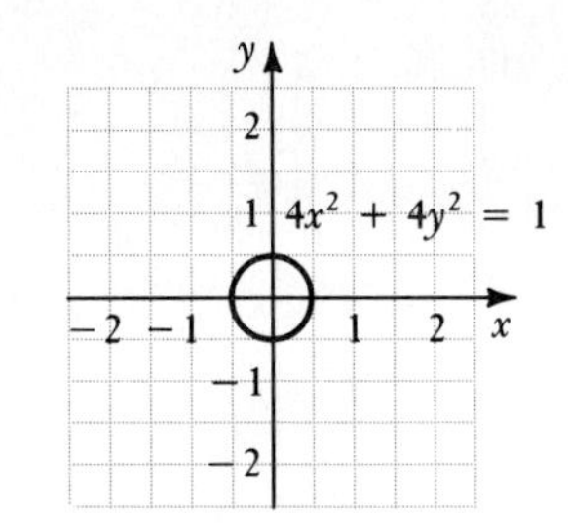

59. $x^2 + y^2 = 49$ **61.** $(x + 2)^2 + (y - 7)^2 = 5$
63. $(x + 4)^2 + (y - 3)^2 = 48$ **65.** $(x + 7)^2 + (y + 2)^2 = 50$
67. $(x + 8)^2 + (y - t)^2 = 1.69$
69. $x^2 + y^2 = 25$
71. $(x + 4)^2 + (y - 1)^2 = 20$
73. 4 in.
75. $f(x) = 2\left(x - \frac{5}{2}\right)^2 - \frac{11}{2}$

$f(x) = 2\left(x - \frac{5}{2}\right)^2 - \frac{11}{2}$

77. (0, 4) **79.** Yes **81.** Let $P_1 = (x_1, y_1)$, $P_2 = (x_2, y_2)$, and $M = \left(\frac{x_1 + x_2}{2}, \frac{y_1 + y_2}{2}\right)$. Let $d(AB)$ denote the distance from point A to point B.

i)
$$d(P_1M) = \sqrt{\left(\frac{x_1 + x_2}{2} - x_1\right)^2 + \left(\frac{y_1 + y_2}{2} - y_1\right)^2} = \frac{1}{2}\sqrt{(x_2 - x_1)^2 + (y_2 - y_1)^2};$$

$$d(P_2M) = \sqrt{\left(\frac{x_1 + x_2}{2} - x_2\right)^2 + \left(\frac{y_1 + y_2}{2} - y_2\right)^2} = \frac{1}{2}\sqrt{(x_1 - x_2)^2 + (y_1 - y_2)^2} = \frac{1}{2}\sqrt{(x_2 - x_1)^2 + (y_2 - y_1)^2} = d(P_1M).$$

ii) $d(P_1M) + d(P_2M) = \frac{1}{2}\sqrt{(x_2 - x_1)^2 + (y_2 - y_1)^2}$
$+ \frac{1}{2}\sqrt{(x_2 - x_1)^2 + (y_2 - y_1)^2}$
$= \sqrt{(x_2 - x_1)^2 + (y_2 - y_1)^2}$
$= d(P_1\,P_2).$

83. $(x + 7)^2 + (y + 4)^2 = 16$

85. $(x + 3)^2 + (y - 5)^2 = 16$

Exercise Set 10.2, p. 464

1.

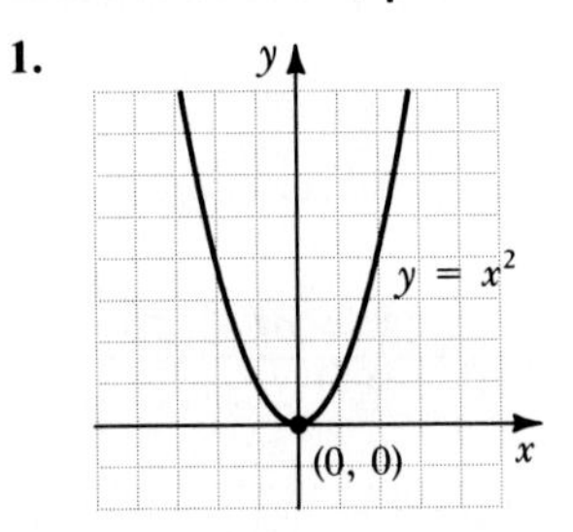

3.

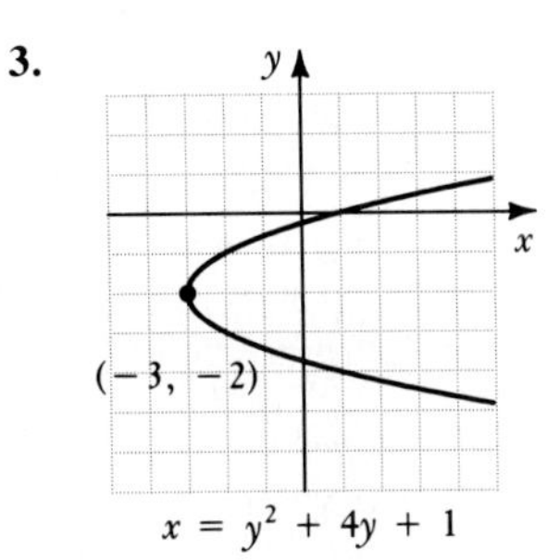

5.

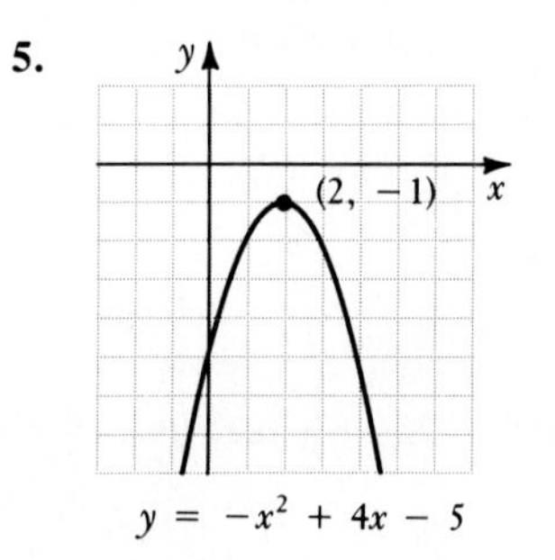

7.

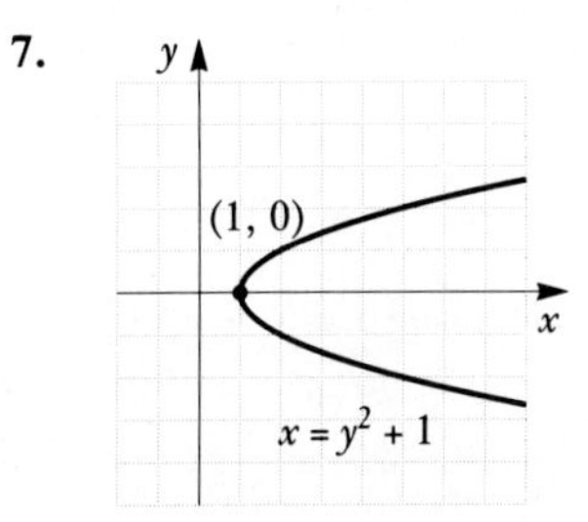

9.

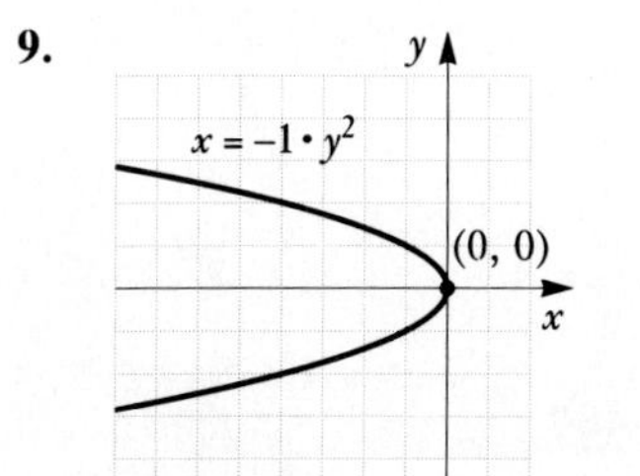

11. **15.**

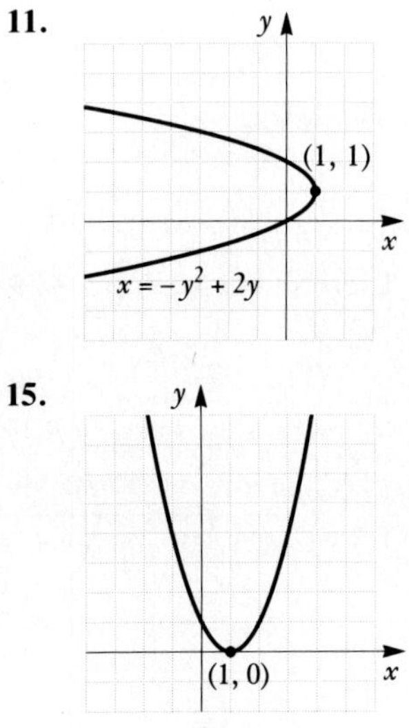

13.

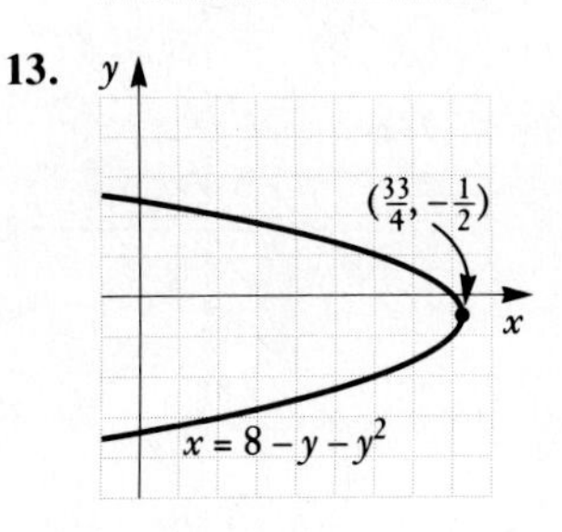

17. **19.**

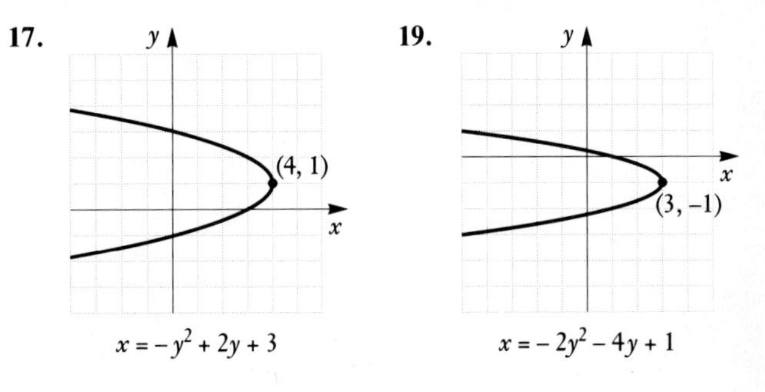

21. $\frac{2a^3\sqrt[3]{a^2}}{3b^2}$ **23.** $\pm i\sqrt{2}$ **25.** **(a)** 3.4, −2.4; **(b)** 2.3, −1.3

27. Reflect one graph across the line $y = x$ to obtain the other.

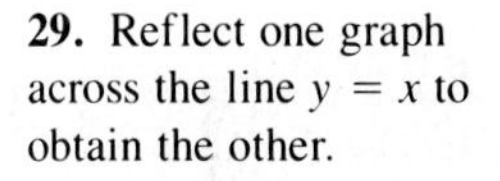

29. Reflect one graph across the line $y = x$ to obtain the other.

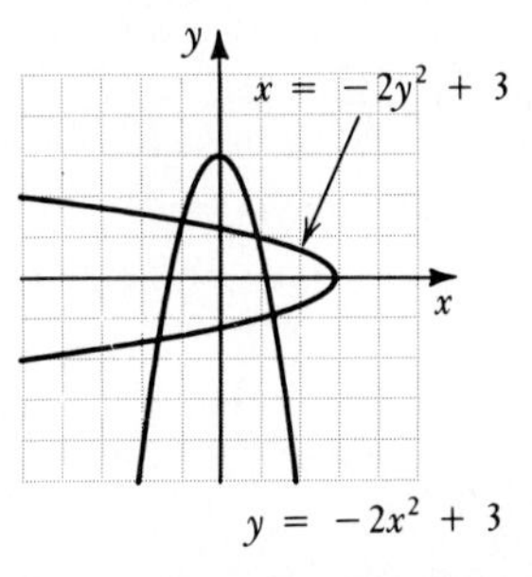

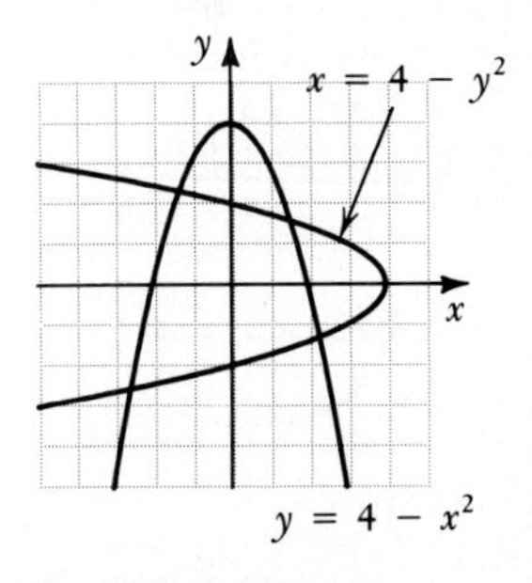

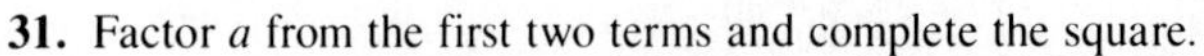

31. Factor a from the first two terms and complete the square.

$$y = a\left(x^2 + \frac{b}{a}x + \frac{b^2}{4a^2}\right) + c - \frac{b^2}{4a}$$
$$y = a\left(x + \frac{b}{2a}\right)^2 + c - \frac{b^2}{4a}.$$

Thus the x-coordinate of the vertex of the parabola is $-b/2a$.

Exercise Set 10.3, pp. 468–470

1.

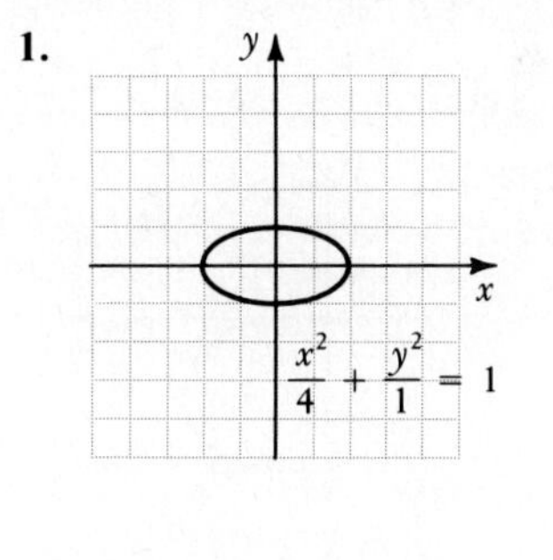

3.

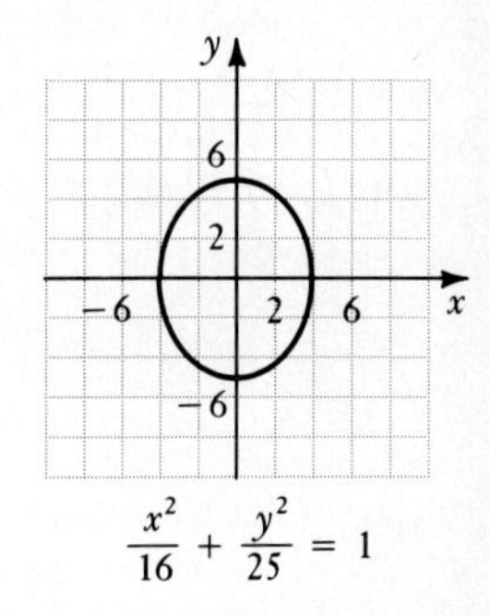

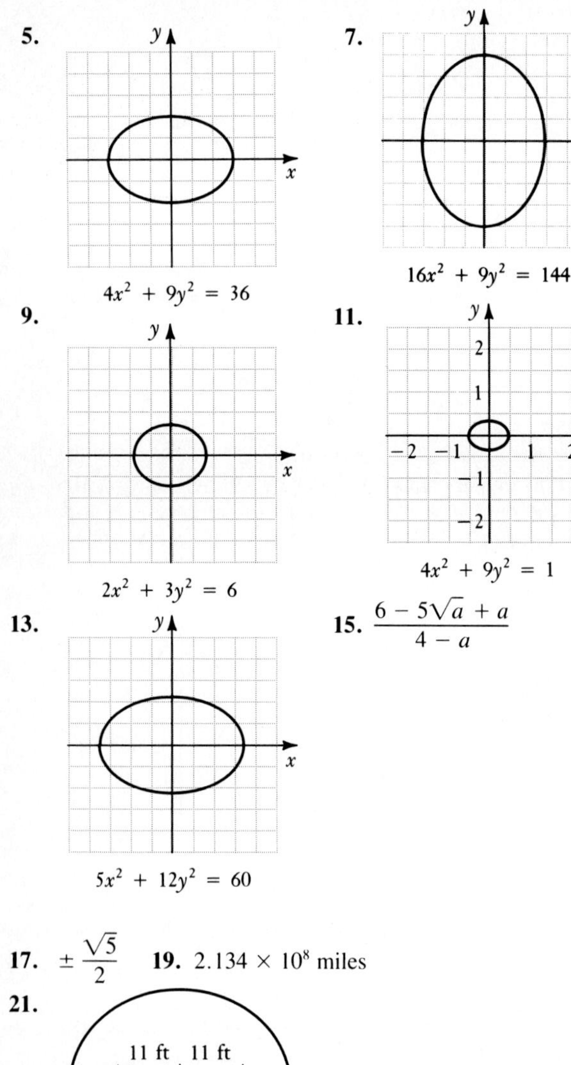

5. $4x^2 + 9y^2 = 36$

7. $16x^2 + 9y^2 = 144$

9. $2x^2 + 3y^2 = 6$

11. $4x^2 + 9y^2 = 1$

13. $5x^2 + 12y^2 = 60$

15. $\dfrac{6 - 5\sqrt{a} + a}{4 - a}$

17. $\pm \dfrac{\sqrt{5}}{2}$ **19.** 2.134×10^8 miles

21.

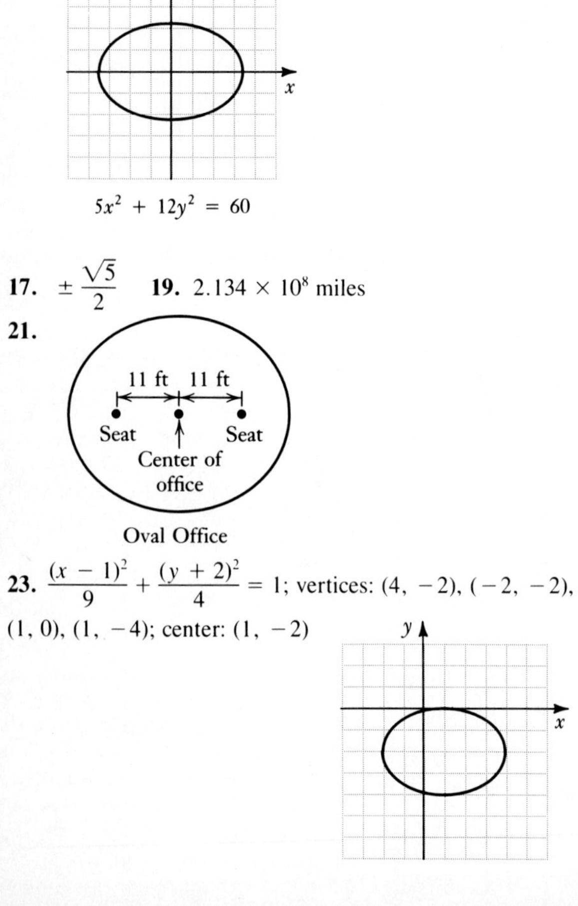

Oval Office

23. $\dfrac{(x-1)^2}{9} + \dfrac{(y+2)^2}{4} = 1$; vertices: $(4, -2)$, $(-2, -2)$, $(1, 0)$, $(1, -4)$; center: $(1, -2)$

25. $\dfrac{(x-1)^2}{25} + \dfrac{(y+1)^2}{4} = 1$; vertices: $(6, -1)$, $(-4, -1)$, $(1, 1)$, $(1, -3)$; center: $(1, -1)$

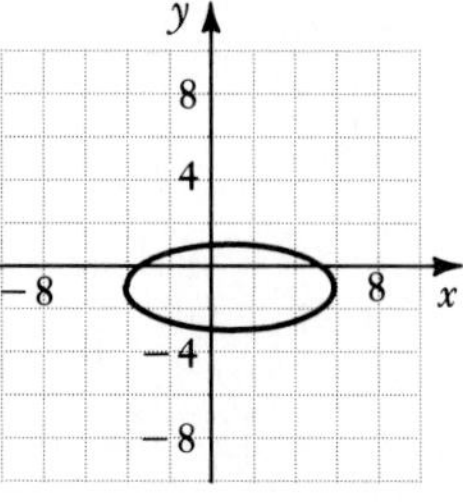

27. $\dfrac{x^2}{64} + \dfrac{y^2}{4} = 1$ **29.** $\dfrac{x^2}{4} + \dfrac{y^2}{16} = 1$ **31.** **(a)** Let $F_1 = (-c, 0)$ and $F_2 = (c, 0)$. Then the sum of the distances from the foci to P is $2a$. By the distance formula,

$$\sqrt{(x+c)^2 + y^2} + \sqrt{(x-c)^2 + y^2} = 2a, \quad \text{or}$$
$$\sqrt{(x+c)^2 + y^2} = 2a - \sqrt{(x-c)^2 + y^2}.$$

Squaring, we get

$$(x+c)^2 + y^2 = 4a^2 - 4a\sqrt{(x-c)^2 + y^2} + (x-c)^2 + y^2,$$
$$\text{or} \quad x^2 + 2cx + c^2 + y^2 = 4a^2 -$$
$$4a\sqrt{(x-c)^2 + y^2} + x^2 - 2cx + c^2 + y^2.$$

Thus

$$-4a^2 + 4cx = -4a\sqrt{(x-c)^2 + y^2}$$
$$a^2 - cx = a\sqrt{(x-c)^2 + y^2}.$$

Squaring again, we get

$$a^4 - 2a^2cx + c^2x^2 = a^2(x^2 - 2cx + c^2 + y^2)$$
$$a^4 - 2a^2cx + c^2x^2 = a^2x^2 - 2a^2cx + a^2c^2 + a^2y^2,$$

or

$$x^2(a^2 - c^2) + a^2y^2 = a^2(a^2 - c^2)$$
$$\frac{x^2}{a^2} + \frac{y^2}{a^2 - c^2} = 1.$$

(b) When P is at $(0, b)$, it follows that $b^2 = a^2 - c^2$. Substituting, we have

$$\frac{x^2}{a^2} + \frac{y^2}{b^2} = 1.$$

Exercise Set 10.4, pp. 479–480

1.

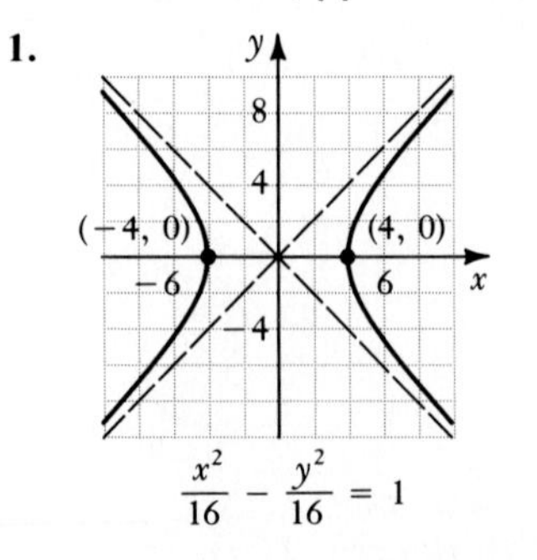

$\dfrac{x^2}{16} - \dfrac{y^2}{16} = 1$

3.

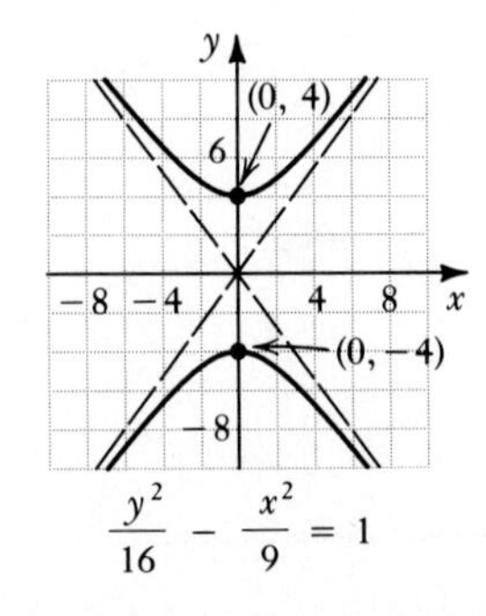

$\dfrac{y^2}{16} - \dfrac{x^2}{9} = 1$

5.

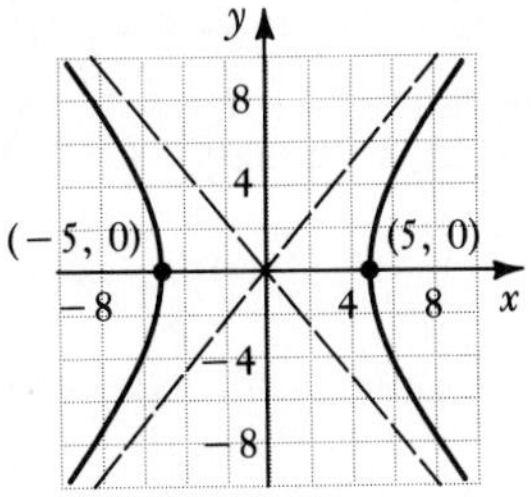

$$\frac{x^2}{25} - \frac{y^2}{36} = 1$$

7.

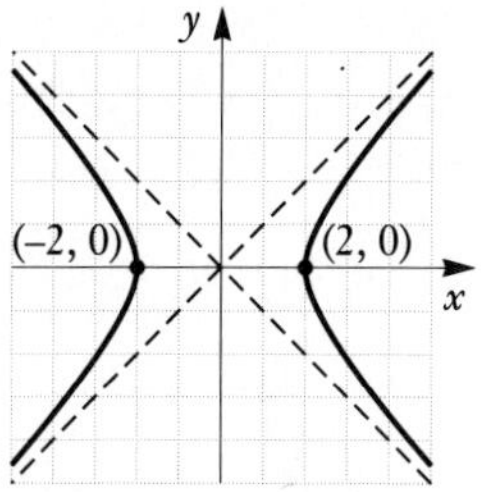

$$x^2 - y^2 = 4$$

9.

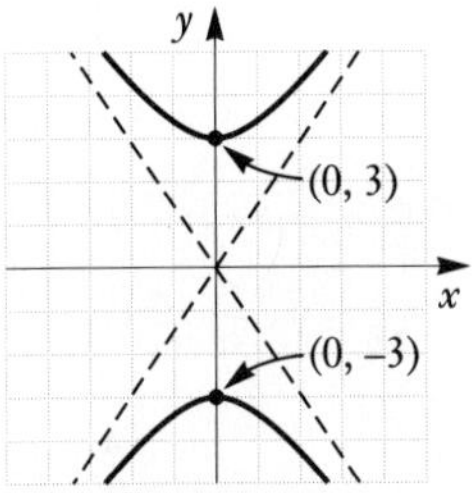

$$4y^2 - 9x^2 = 36$$

11.

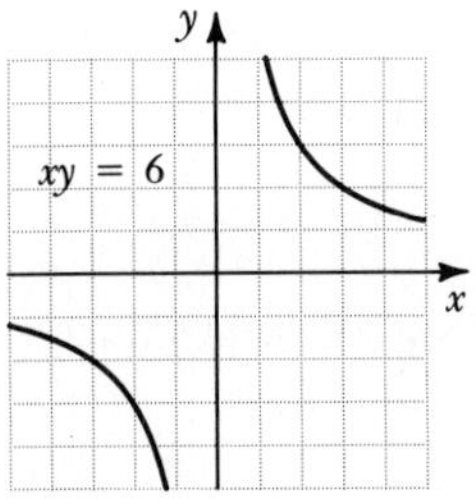

13.

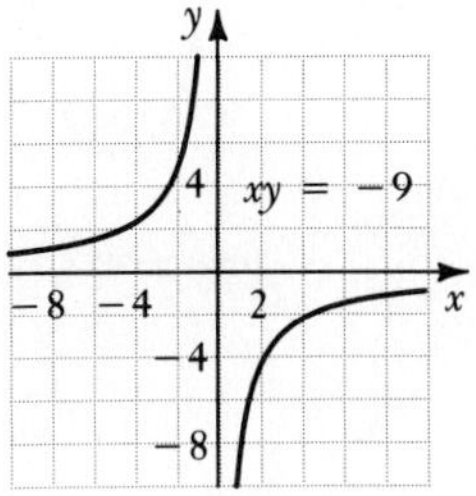

15.

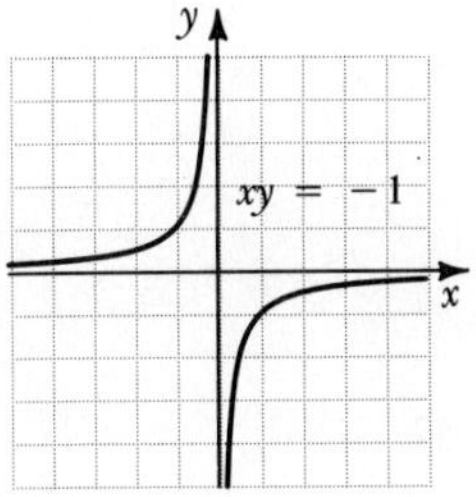

17.

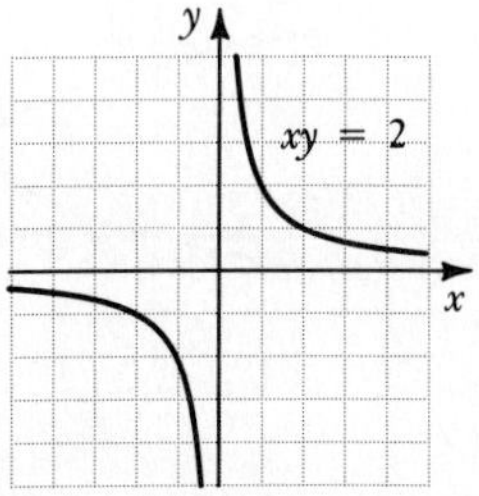

19. Circle **21.** Hyperbola **23.** Ellipse **25.** Circle **27.** Ellipse **29.** Hyperbola **31.** Parabola **33.** Hyperbola **35.** $5t^5$ **37.** $\frac{13 + 8\sqrt{6}}{10}$ **39.** $\frac{y^2}{64} - \frac{x^2}{4} = 1$

41. Center $(2, -1)$; vertices: $(-1, -1), (5, -1)$; asymptotes: $y + 1 = \frac{4}{3}(x - 2)$, $y + 1 = -\frac{4}{3}(x - 2)$;

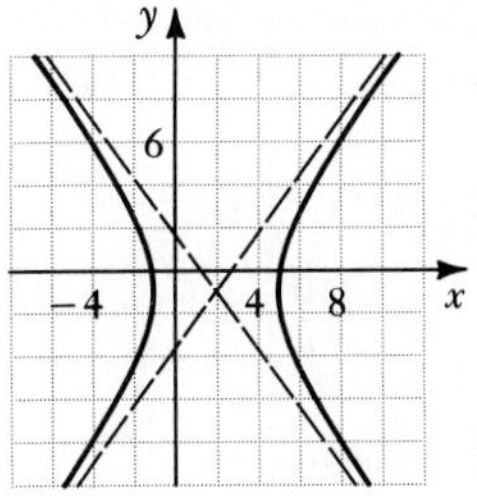

$$\frac{(x - 2)^2}{9} - \frac{(y + 1)^2}{16} = 1$$

43. $\frac{(y - 1)^2}{25} - \frac{(x + 2)^2}{4} = 1$; center: $(-2, 1)$; vertices: $(-2, 6)$, $(-2, -4)$; asymptotes: $y - 1 = \frac{5}{2}(x + 2)$, $y - 1 = -\frac{5}{2}(x + 2)$

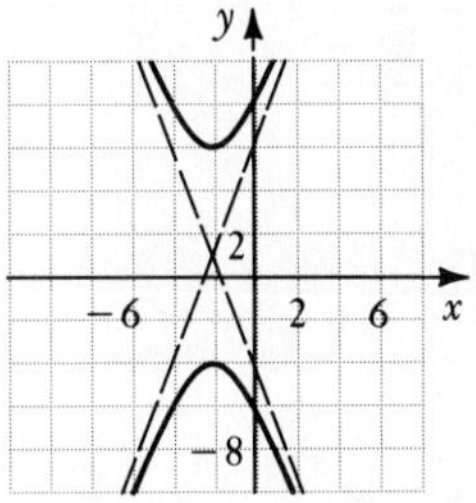

$$\frac{(y - 1)}{25} - \frac{(x + 2)^2}{4} = 1$$

Review Exercises: Chapter 10, pp. 481–482

1. [10.1] 4 **2.** [10.1] 5 **3.** [10.1] $\sqrt{130}$ **4.** [10.1] $\sqrt{9 + 4a^2}$ **5.** [10.1] No **6.** [10.1] Yes **7.** [10.1] (4, 6) **8.** [10.1] $\left(-3, \frac{5}{2}\right)$ **9.** [10.1] $\left(\frac{1}{2}, \frac{5}{2}\right)$ **10.** [10.1] $\left(\frac{1}{2}, 2a\right)$ **11.** [10.1] $(-2, 3)$, $\sqrt{2}$ **12.** [10.1] (5, 0), 7
13. [10.1] (3, 1), 3 **14.** [10.1] $(-4, 3)$, $\sqrt{35}$
15. [10.1] $(x + 4)^2 + (y - 3)^2 = 48$
16. [10.1] $(x - 7)^2 + (y + 2)^2 = 20$
17. [10.1] $(x - 3)^2 + (y - 4)^2 = 25$
18. [10.4], [10.1] Circle **19.** [10.4], [10.3] Ellipse

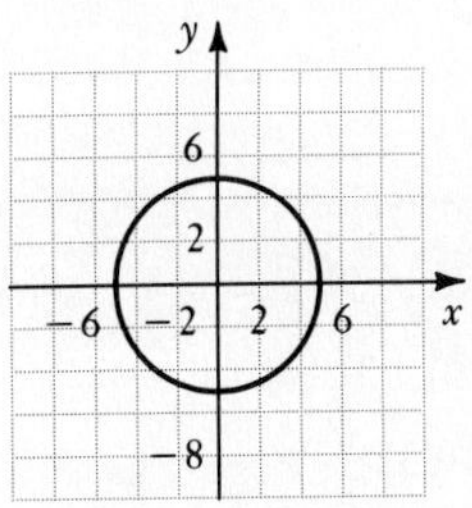

$$4x^2 + 4y^2 = 100$$

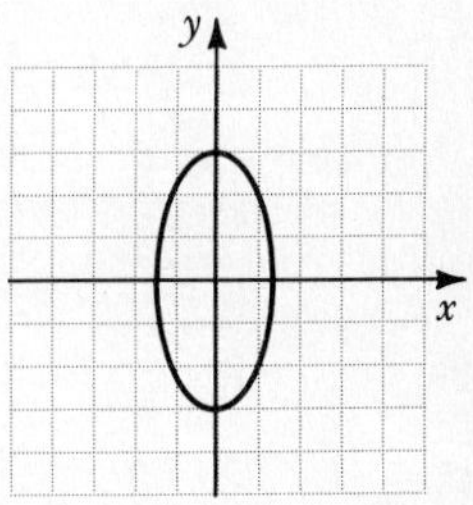

$$9x^2 + 2y^2 = 18$$

20. [10.4], [10.2] Parabola

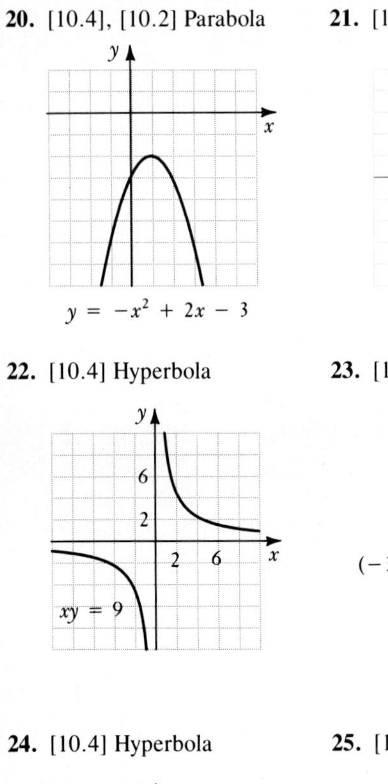

$y = -x^2 + 2x - 3$

21. [10.4] Hyperbola

$\dfrac{y^2}{9} - \dfrac{x^2}{4} = 1$

22. [10.4] Hyperbola

23. [10.4], [10.2] Parabola

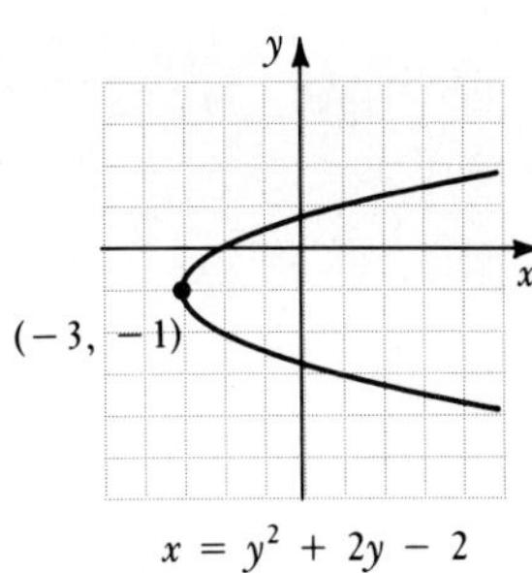

$x = y^2 + 2y - 2$

24. [10.4] Hyperbola

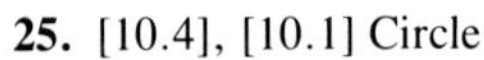

25. [10.4], [10.1] Circle

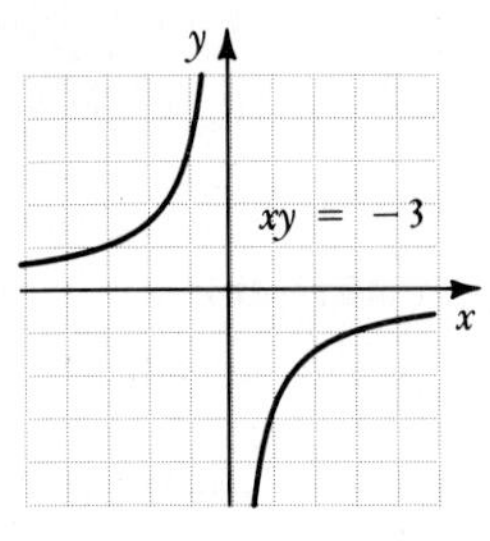

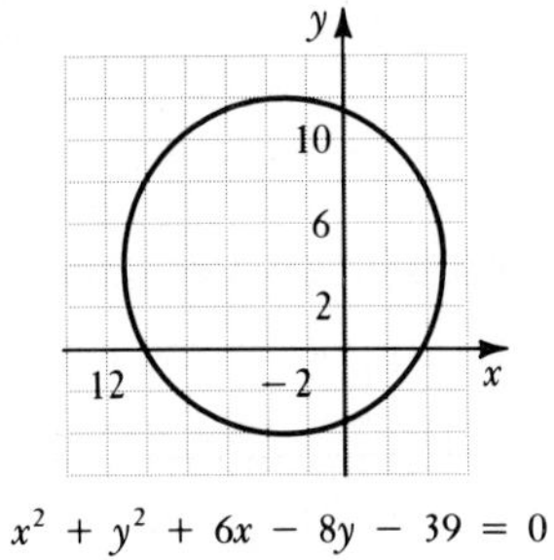

$x^2 + y^2 + 6x - 8y - 39 = 0$

26. [5.2] $3b^3\sqrt[3]{a^2}$ **27.** [7.2] $-1 \pm 2i$
28. [7.5]

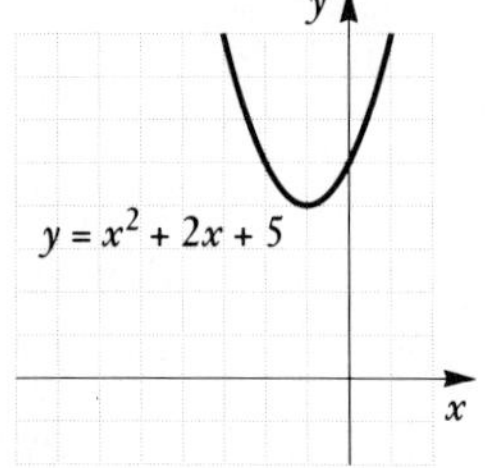

29. [4.5] 6 ft/sec **30.** [10.1] $\left(\frac{9}{4}, 0\right)$ **31.** [10.1] $(x - 2)^2 + (y - 4)^2 = 26$ **32.** [10.3] $\dfrac{x^2}{49} + \dfrac{y^2}{9} = 1$

Test: Chapter 10, pp. 482–483

1. [10.1] $9\sqrt{2}$ **2.** [10.1] $2\sqrt{9 + a^2}$ **3.** [10.1] $\left(-\frac{1}{2}, \frac{7}{2}\right)$
4. [10.1] (0, 0) **5.** [10.1] (−2, 3), 8 **6.** [10.1] (−2, 3), 3
7. [10.1] $(x - 2)^2 + (y + 5)^2 = 18$
8. [10.4], [10.2] Parabola **9.** [10.4], [10.1] Circle

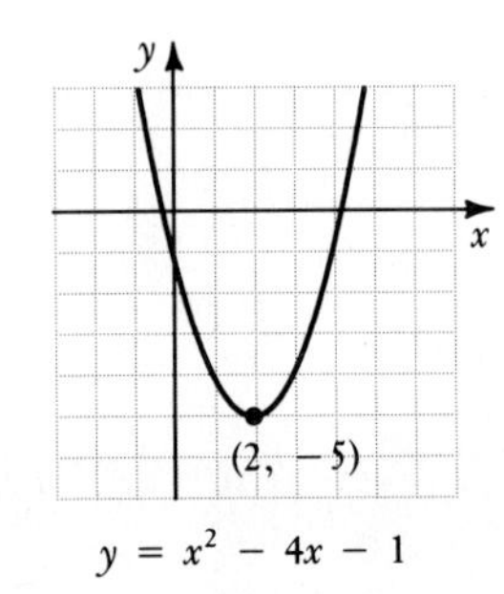

$y = x^2 - 4x - 1$

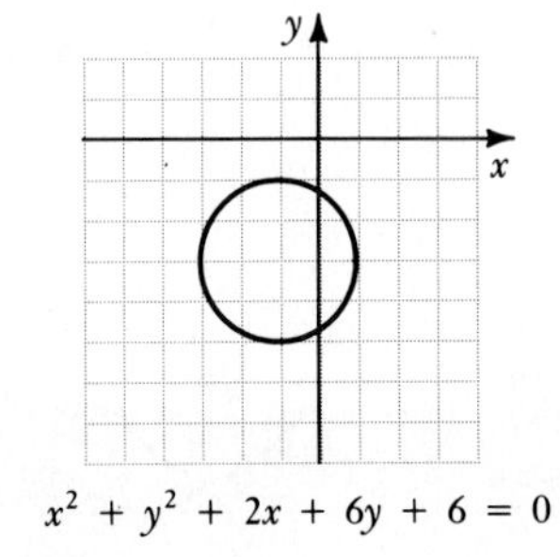

$x^2 + y^2 + 2x + 6y + 6 = 0$

10. [10.4] Hyperbola **11.** [10.4], [10.3] Ellipse

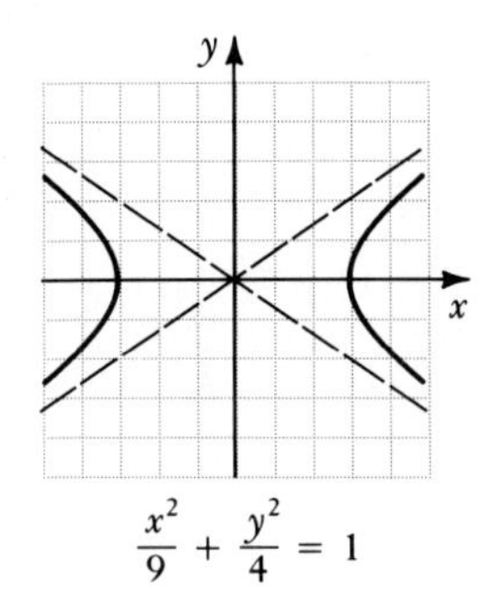

$\dfrac{x^2}{9} + \dfrac{y^2}{4} = 1$

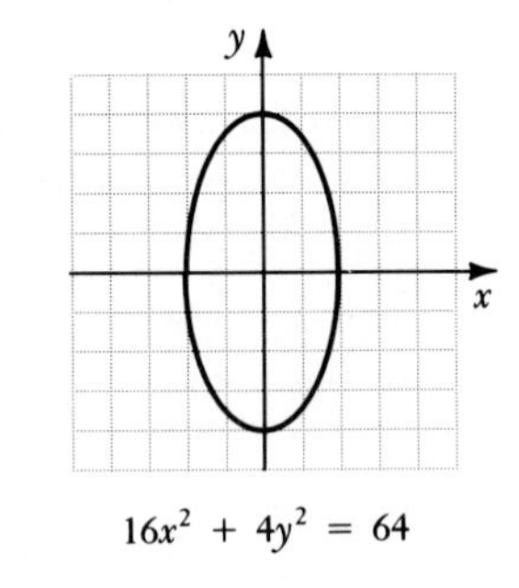

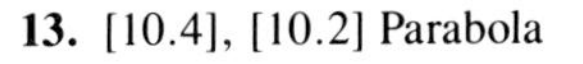

$16x^2 + 4y^2 = 64$

12. [10.4] Hyperbola **13.** [10.4], [10.2] Parabola

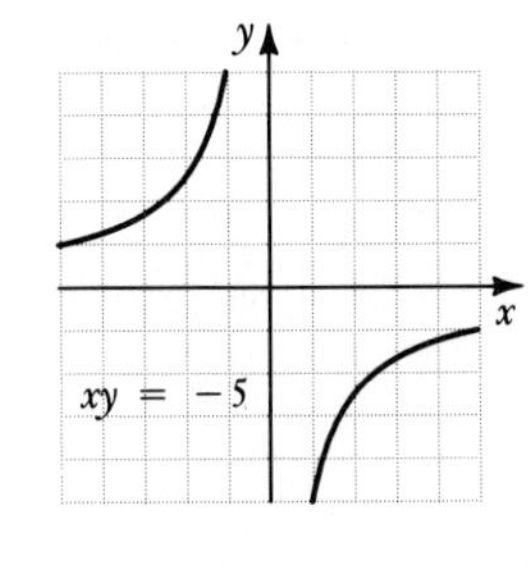

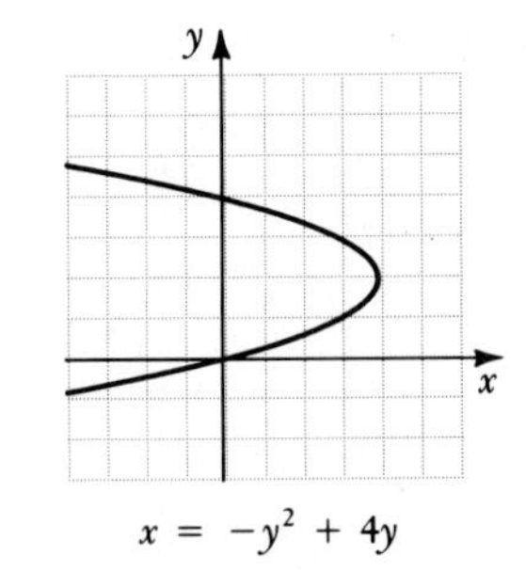

$x = -y^2 + 4y$

14. [7.2] $-1 \pm \sqrt{6}$ **15.** [5.2] $4\sqrt{x}$ **16.** [4.5] 6 mph
17. [7.5] $\left(\dfrac{1 + \sqrt{61}}{6}, 0\right), \left(\dfrac{1 - \sqrt{61}}{6}, 0\right)$ **18.** [10.3] $\dfrac{(x - 6)^2}{25} + \dfrac{(y - 3)^2}{9} = 1$ **19.** [10.1] $\left(0, -\frac{31}{4}\right)$

CHAPTER 11

Exercise Set 11.1, pp. 493–496

1. No **3.** $(-1, 3)$ **5.** No solution **7.** $(3, 1)$ **9.** $(3, 2)$ **11.** $(1, -5)$ **13.** $(-4, 3)$ **15.** $(-3, -15)$ **17.** $(2, -2)$ **19.** $(-2, 1)$ **21.** $\left(\frac{1}{2}, \frac{1}{2}\right)$ **23.** $\left(\frac{19}{8}, \frac{1}{8}\right)$ **25.** $(1, 2)$ **27.** $(3, 0)$ **29.** $(-1, 2)$ **31.** $\left(\frac{128}{31}, -\frac{17}{31}\right)$ **33.** $(6, 2)$ **35.** $\left(\frac{140}{13}, -\frac{50}{13}\right)$ **37.** $(4, 6)$ **39.** $-\frac{11}{2}, -\frac{9}{2}$ **41.** 20 km/h, 3 km/h **43.** 12.5 L of A, 7.5 L of B **45.** 3 hr **47.** 2 hr **49.** 12 white, 28 printed **51.** Paula is 32; Bob is 20. **53.** 76 m, 19 m **55.** 137 m, 55 m **57.** \$6800 at 9%, \$8200 at 10%
59.

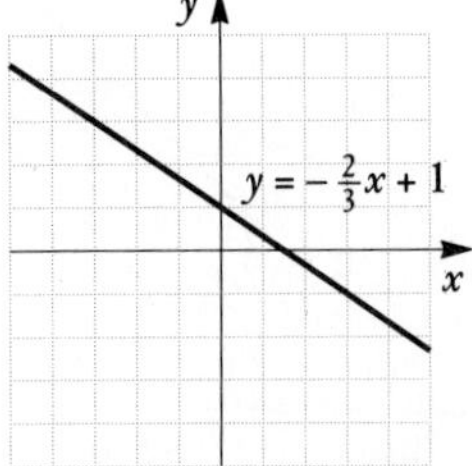

61. $x^2 - 4x + 10$, R -25 **63.** $(-12, 0)$ **65.** $\left(-\frac{1}{4}, -\frac{1}{2}\right)$ **67.** $\{(5, 3), (-5, 3), (5, -3), (-5, -3)\}$ **69.** 4 km **71.** 180 **73.** $4\frac{4}{7}$ L **75.** (370, \$10) **77.** (474, \$22) **79.** 40 mph

Exercise Set 11.2, pp. 502–505

1. Yes **3.** $(1, 2, 3)$ **5.** $(-1, 5, -2)$ **7.** $(3, 1, 2)$ **9.** $(-3, -4, 2)$ **11.** $(2, 4, 1)$ **13.** $(-3, 0, 4)$ **15.** $\left(\frac{1}{2}, \frac{2}{3}, -\frac{5}{6}\right)$ **17.** $(15, 33, 9)$ **19.** $\left(\frac{1}{4}, -\frac{1}{2}, -\frac{1}{4}\right)$ **21.** 8, 21, -3 **23.** A = 30°, B = 90°, C = 60° **25.** 20 on Monday, 35 on Tuesday, 32 on Wednesday **27.** A: 2200, B: 2500, C: 2700 **29.** A: 10, B: 12, C: 15 **31.** Steak: 2; baked potato: 1; broccoli: 2 **33.** $y = 2x^2 + 3x - 1$ **35.** **(a)** $E = -4t^2 + 40t + 2$; **(b)** \$98 **37.** **(a)** $f(x) = 104.5x^2 - 1501.5x + 6016$; **(b)** 1682, 769, 1451 **39.** $P(x) = (x - 1)(x + 1)^2$; 1, -1 **41.** 2 **43.** $\left(-\frac{1}{2}, -1, -\frac{1}{3}\right)$ **45.** A: 24 hr; B: 12 hr; C: $4\frac{4}{5}$ hr **47.** Men: 5; women: 1; children: 94

Exercise Set 11.3, pp. 512–513

1. $\left(\frac{y + 5}{3}, y\right)$ or $(x, 3x - 5)$; $(0, -5)$, $(1, -2)$, $(-1, -8)$, etc. **3.** $\emptyset$ **5.** $\left(\frac{5 - 2y}{3}, y\right)$ or $\left(x, \frac{5 - 3x}{2}\right)$; $(3, -2)$, $\left(\frac{5}{3}, 0\right)$, $\left(0, \frac{5}{2}\right)$, etc. **7.** $\left(\frac{6y - 3}{4}, y\right)$ or $\left(x, \frac{4x + 3}{6}\right)$; $\left(1, \frac{7}{6}\right)$, $\left(-\frac{3}{4}, 0\right)$, $\left(0, \frac{1}{2}\right)$, etc. **9.** $\emptyset$ **11.** $\left(\frac{10 + 11z}{9}, \frac{-11 + 5z}{9}, z\right)$; $\left(\frac{10}{9}, -\frac{11}{9}, 0\right)$, $\left(\frac{7}{3}, -\frac{2}{3}, 1\right)$, $\left(-\frac{1}{9}, -\frac{16}{9}, -1\right)$, etc. **13.** $\left(\frac{11}{9}z, \frac{5}{9}z, z\right)$; $\left(\frac{11}{9}, \frac{5}{9}, 1\right)$, $(0, 0, 0)$, $\left(-\frac{11}{9}, -\frac{5}{9}, -1\right)$, etc. **15.** $(4z - 5, -3z + 2, z)$; $(-1, -1, 1)$, $(-5, 2, 0)$, $(3, -4, 2)$, etc. **17.** $(0, 0, 0)$ **19.** Consistent: 1, 5, 7, 11, 13, 15, 17, the others are inconsistent; dependent: 1, 5, 7, 11, 13, 15, the others are independent **21.** $-\frac{3}{2}$, 4 **23.** 10.389 **25.** $\left(\frac{724y + 9160}{2013}, y\right)$ or $\left(x, \frac{2013x - 9160}{724}\right)$ **27.** **(a)** $\emptyset$; **(b)** inconsistent; **(c)** dependent **29.** $k = 2$ **31.** $(x, 18 - 2x, x)$, where $1 \le x \le 8$, x is an integer

Exercise Set 11.4, pp. 516–517

1. $\left(\frac{3}{2}, \frac{5}{2}\right)$ **3.** $(-1, 2, -2)$ **5.** $\left(\frac{1}{2}, \frac{3}{2}\right)$ **7.** $\left(\frac{3}{2}, -4, 3\right)$ **9.** $(r - 2, 3 - 2r, r)$ **11.** $(1, -3, -2, -1)$
13.

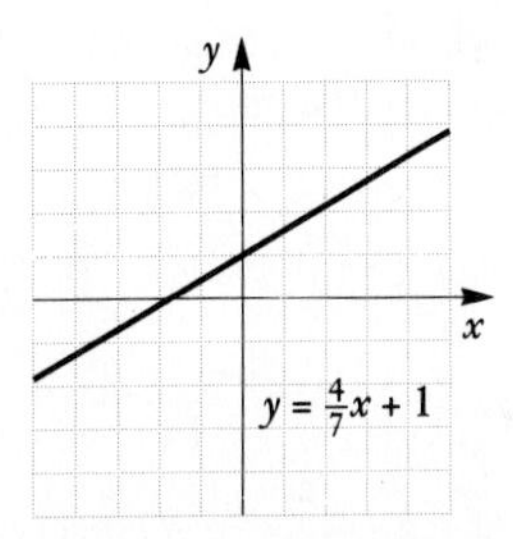

15. $-\frac{3}{5}$ **17.** 4 dimes, 30 nickels **19.** 10 nickels, 4 dimes, 8 quarters **21.** 5 lb of \$4.05; 10 lb of \$2.70 **23.** \$30,000 at $12\frac{1}{2}$%; \$40,000 at 13% **25.** $(1.0128, -4.8909)$ **27.** $(1.23, -2.11, 1.89)$

Exercise Set 11.5, pp. 521–522

1. -11 **3.** $x^3 - 4x$ **5.** -109 **7.** $-x^4 + x^2 - 5x$ **9.** $\left(-\frac{25}{2}, -\frac{11}{2}\right)$ **11.** $\left(\frac{4\pi - 5\sqrt{3}}{3 + \pi^2}, \frac{4\sqrt{3} + 5\pi}{-3 - \pi^2}\right)$ **13.** $\left(\frac{3}{2}, \frac{13}{14}, \frac{33}{14}\right)$ **15.** $\left(\frac{1}{2}, \frac{2}{3}, -\frac{5}{6}\right)$ **17.** 276 **19.** 10, -1 **21.** 2, -2 **23.** $\{x \mid x \le -\sqrt{3} \text{ or } x \ge \sqrt{3}\}$ **25.** -34 **27.** 4 **29.** $\begin{vmatrix} L & -W \\ 2 & 2 \end{vmatrix}$ **31.** $\begin{vmatrix} a & b \\ -b & a \end{vmatrix}$

Exercise Set 11.6, pp. 533–536

1. No **3.** No

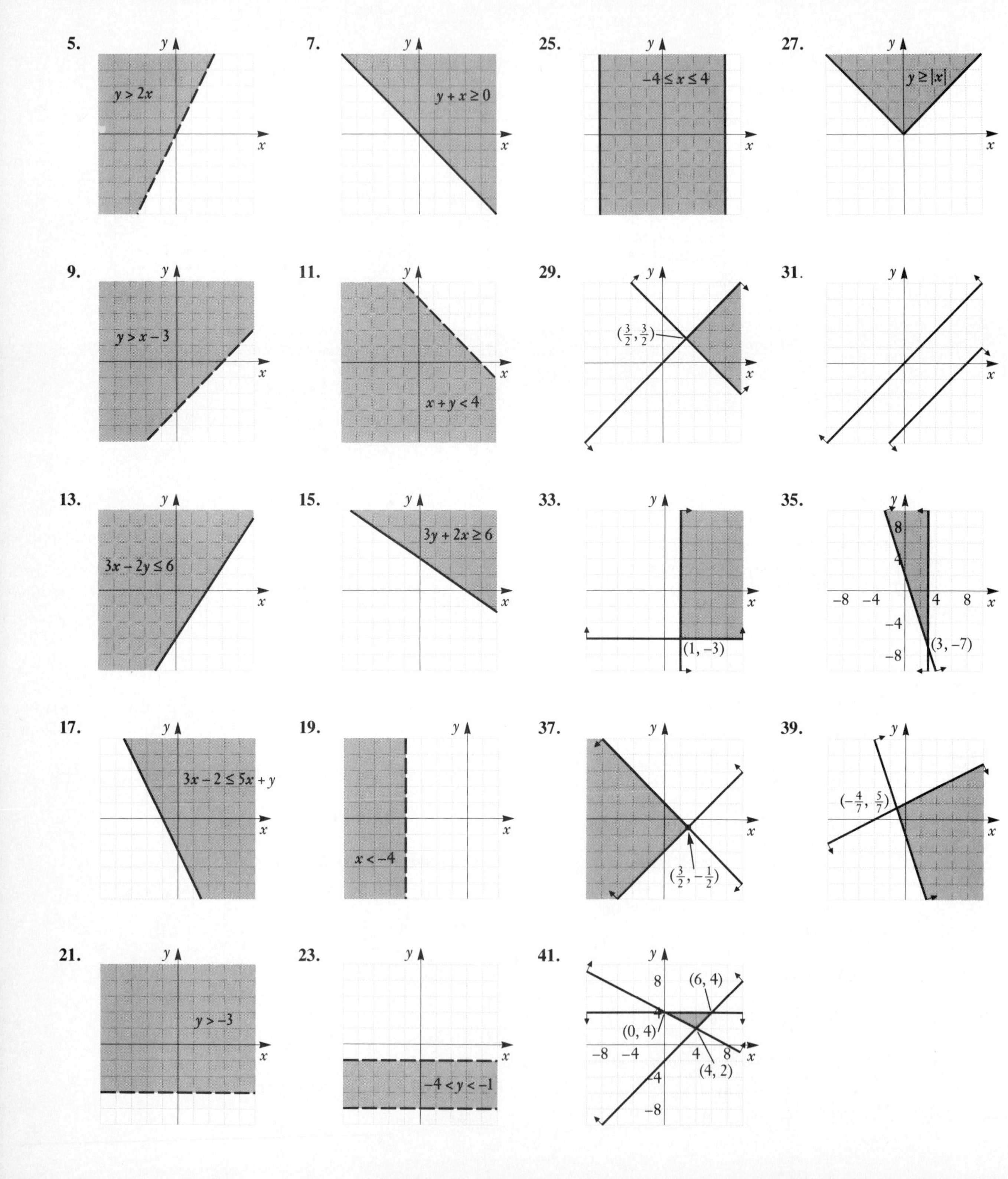

5.
y > 2x
7.
y + x ≥ 0
25.
−4 ≤ x ≤ 4
27.
y ≥ |x|
9.
y > x − 3
11.
x + y < 4
29.
(3/2, 3/2)
31.
13.
3x − 2y ≤ 6
15.
3y + 2x ≥ 6
33.
(1, −3)
35.
(3, −7)
17.
3x − 2 ≤ 5x + y
19.
x < −4
37.
(3/2, −1/2)
39.
(−4/7, 5/7)
21.
y > −3
23.
−4 < y < −1
41.
(6, 4)
(0, 4)
(4, 2)

43.

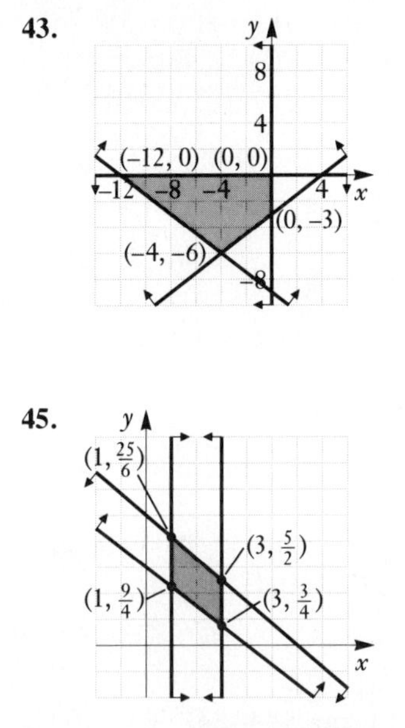

45.

47. The maximum value of P is 179 when $x = 7$ and $y = 0$. The minimum value of P is 48 when $x = 0$ and $y = 4$. **49.** The maximum value of F is 216 when $x = 0$ and $y = 6$. The minimum value of F is 0 when $x = 0$ and $y = 0$. **51.** Maximum income of \$18 when 100 of each type of biscuit is made. **53.** Maximum score of 425 when 5 questions of type A and 15 of type B are answered. **55.** Maximum profit of \$11,000 is achieved by producing 100 units of lumber and 300 units of plywood. **57.** Maximum income of \$3110 is achieved when \$22,000 is invested in corporate bonds and \$18,000 is invested in municipal bonds. **59.** Maximum profit of \$2520 when 125 batches of Smello and 187.5 batches of Roppo are made. **61.** $y = -\frac{2}{9}x + 4$ **63.** $\frac{89}{2}$

65.

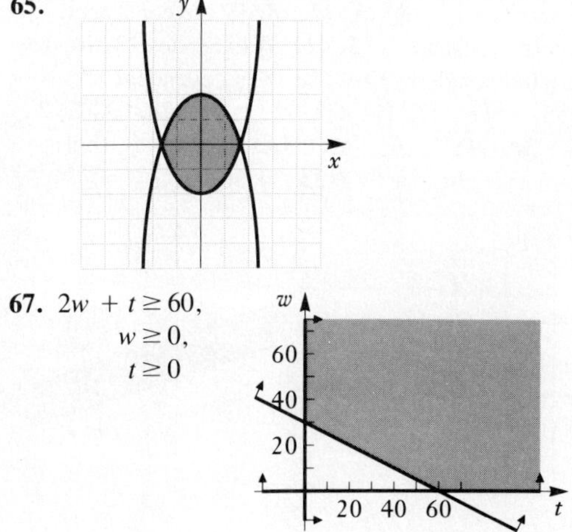

67. $2w + t \geq 60$,
$w \geq 0$,
$t \geq 0$

69. 25 chairs, 9 sofas

Exercise Set 11.7, pp. 541–542

1. $(-4, -3), (3, 4)$

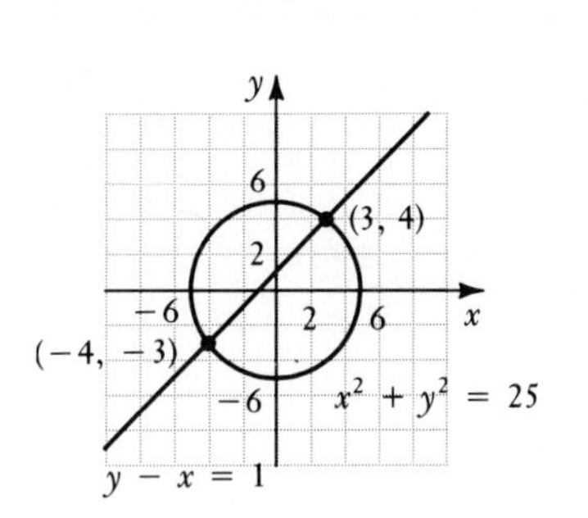

3. $(0, 2), (3, 0)$

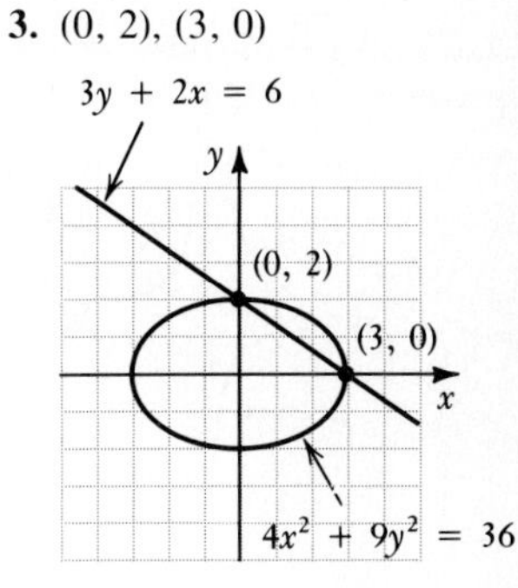

5. $(-2, 1)$

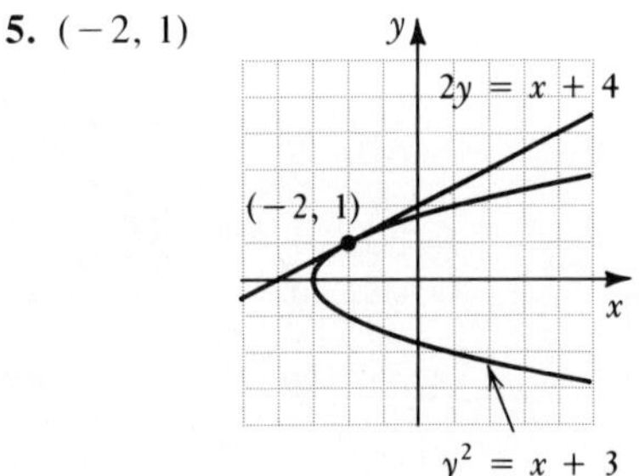

7.

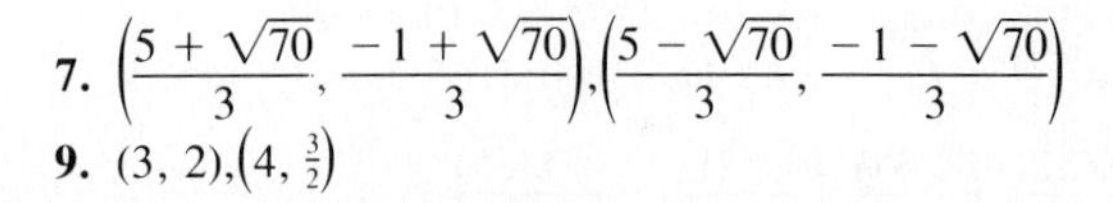

$\left(\frac{5+\sqrt{70}}{3}, \frac{-1+\sqrt{70}}{3}\right), \left(\frac{5-\sqrt{70}}{3}, \frac{-1-\sqrt{70}}{3}\right)$

9. $(3, 2), \left(4, \frac{3}{2}\right)$

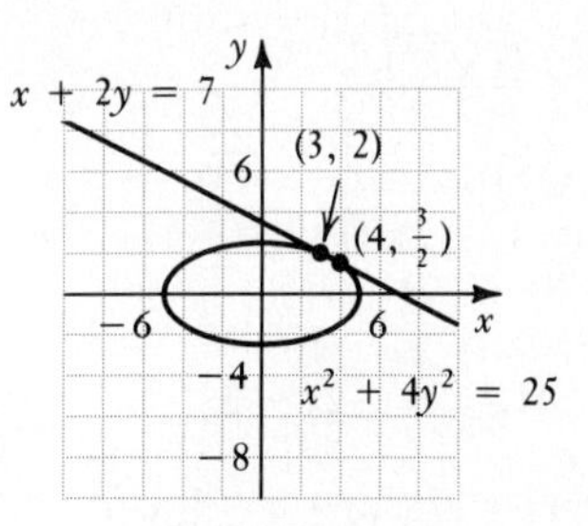

11. $\left(\frac{7}{3}, \frac{1}{3}\right), (1, -1)$

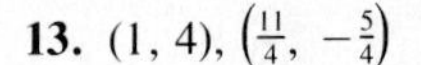

13. $(1, 4), \left(\frac{11}{4}, -\frac{5}{4}\right)$

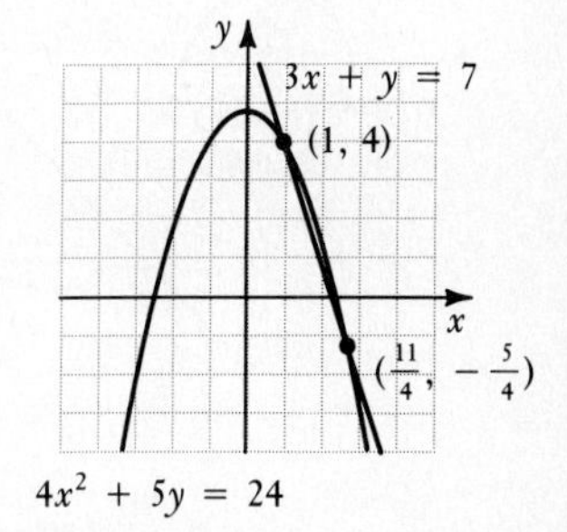

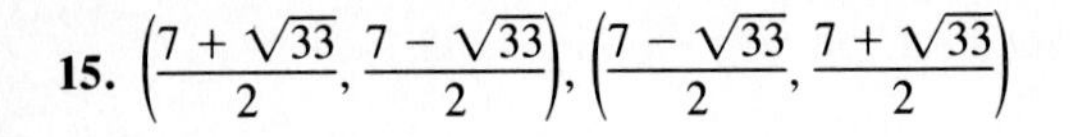
15. $\left(\frac{7+\sqrt{33}}{2}, \frac{7-\sqrt{33}}{2}\right), \left(\frac{7-\sqrt{33}}{2}, \frac{7+\sqrt{33}}{2}\right)$

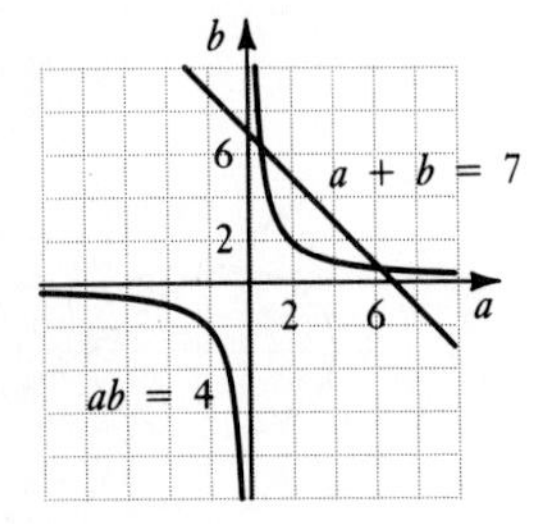

17. (3, −5), (−1, 3) **19.** (8, 5), (−5, −8)

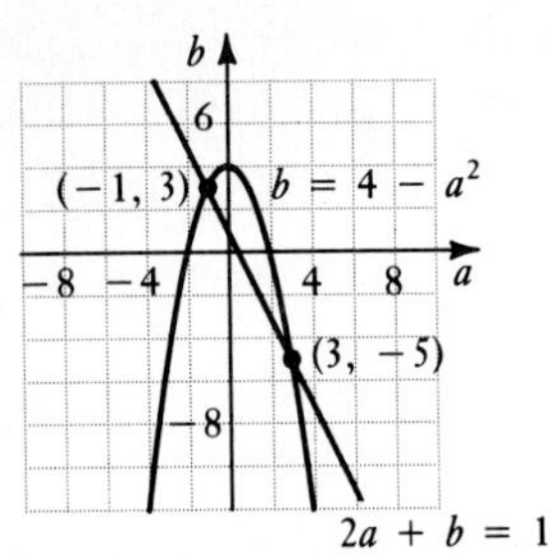

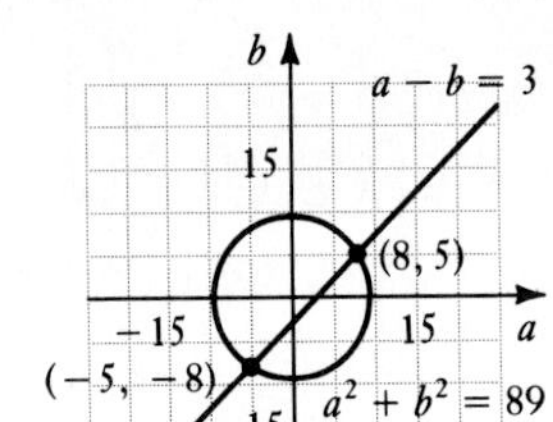

21. (4, −3), (4, 3), (−5, 0) **23.** (−3, 0), (3, 0)
25. (−4, −3), (−3, −4), (3, 4), (4, 3)

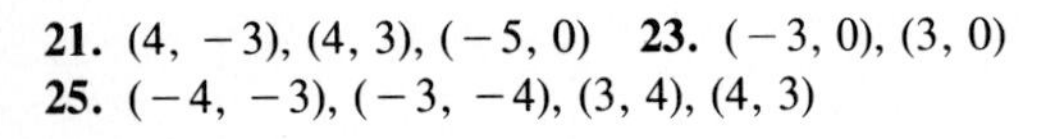
27. $\left(\frac{6\sqrt{21}}{7}, \frac{4i\sqrt{35}}{7}\right), \left(\frac{6\sqrt{21}}{7}, -\frac{4i\sqrt{35}}{7}\right),$
$\left(-\frac{6\sqrt{21}}{7}, \frac{4i\sqrt{35}}{7}\right), \left(-\frac{6\sqrt{21}}{7}, -\frac{4i\sqrt{35}}{7}\right)$
29. $(-\sqrt{2}, -\sqrt{14}), (-\sqrt{2}, \sqrt{14}), (\sqrt{2}, -\sqrt{14}),$ $(\sqrt{2}, \sqrt{14})$ **31.** (−2, −1), (−1, −2), (1, 2), (2, 1)
33. (−3, −2), (−2, −3), (2, 3), (3, 2)
35. (2, 5), (−2, −5) **37.** (3, 2), (−3, −2)
39. (−3, 4), (3, 4), (0, −5) **41.** $y = -2x - 3$
43. $(x - 1)(x + 2)(x - 3)$
45. $\left(4 + \frac{3}{2}i\sqrt{6}, -4 + \frac{3}{2}i\sqrt{6}\right), \left(4 - \frac{3}{2}i\sqrt{6}, -4 - \frac{3}{2}i\sqrt{6}\right);$

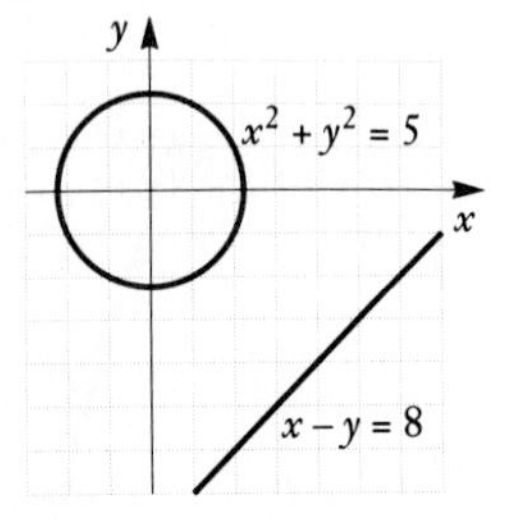

47. $\left(\frac{8i\sqrt{5}}{5}, \frac{3\sqrt{105}}{5}\right), \left(\frac{8i\sqrt{5}}{5}, -\frac{3\sqrt{105}}{5}\right), \left(-\frac{8i\sqrt{5}}{5}, \frac{3\sqrt{105}}{5}\right),$ $\left(-\frac{8i\sqrt{5}}{5}, -\frac{3\sqrt{105}}{5}\right)$

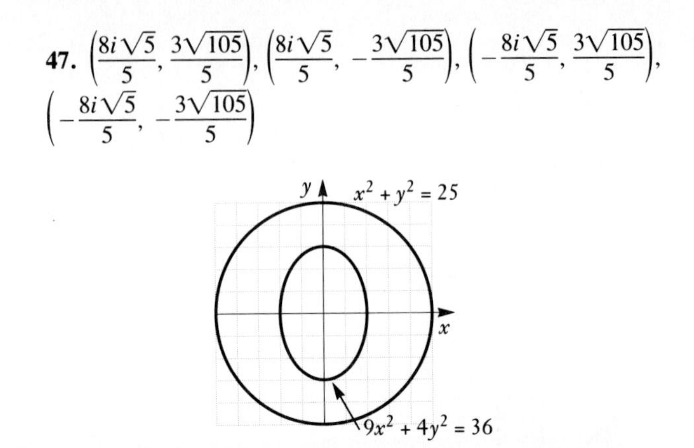

49. $\frac{1}{2}$ and $\frac{1}{4}$, $\frac{1}{2}$ and $-\frac{1}{4}$, $-\frac{1}{2}$ and $\frac{1}{4}$, $-\frac{1}{2}$ and $-\frac{1}{4}$ **51.** 10 in. by 7 in. by 5 in. **53.** $(a, -b)$ **55.** $\left(\frac{1}{3}, \frac{1}{2}\right), \left(\frac{1}{2}, \frac{1}{3}\right)$

Review Exercises: Chapter 11, pp. 544–547

1. [11.1] (3, 2) **2.** [11.1] (−2, 1) **3.** [11.1] $\left(\frac{2}{5}, -\frac{4}{5}\right)$
4. [11.3] ∅ **5.** [11.1] $\left(-\frac{11}{15}, -\frac{43}{30}\right)$ **6.** [11.1] $\left(\frac{37}{19}, \frac{53}{19}\right)$
7. [11.1] $\left(\frac{76}{17}, -\frac{2}{119}\right)$ **8.** [11.1] (2, 2) **9.** [11.1] Record album, \$7; poster, \$6 **10.** [11.1] 4 hours **11.** [11.1] 10 L of 30% alcohol, 30 L of 50% alcohol **12.** [11.1] \$1600 at 10%, \$3400 at 10.5% **13.** [11.2] (10, 4, −8) **14.** [11.2] $\left(-\frac{7}{3}, \frac{125}{27}, \frac{20}{27}\right)$ **15.** [11.2] (2, 0, 4) **16.** [11.3] ∅ **17.** [11.2] $\left(\frac{8}{9}, -\frac{2}{3}, \frac{10}{9}\right)$ **18.** [11.3] $\left(x, \frac{6-3x}{4}\right)$; $\left(1, \frac{3}{4}\right)$, (2, 0), $\left(0, \frac{3}{2}\right)$
19. [11.2] $\left(2, \frac{1}{3}, -\frac{2}{3}\right)$ **20.** [11.2] A = 90°, B = $67\frac{1}{2}°$, C = $22\frac{1}{2}°$ **21.** [11.2] 641 **22.** [11.2] \$20 bills: 5; \$5 bills: 15; \$1 bills: 19 **23.** [11.2] $f(x) = 2x^2 - x + 3$
24. [11.3] Exercises 13, 14, 15, 17, 18, 19 are consistent; Exercise 16 is inconsistent. **25.** [11.3] Exercises 13, 14, 15, 16, 17, 19 are independent; Exercise 18 is dependent.
26. [11.4] $\left(55, -\frac{89}{2}\right)$ **27.** [11.4] (−1, 1, 3) **28.** [11.5] 2
29. [11.5] 9 **30.** [11.5] (6, −2) **31.** [11.5] (−3, 0, 4)
32. [11.6] **33.** [11.6]

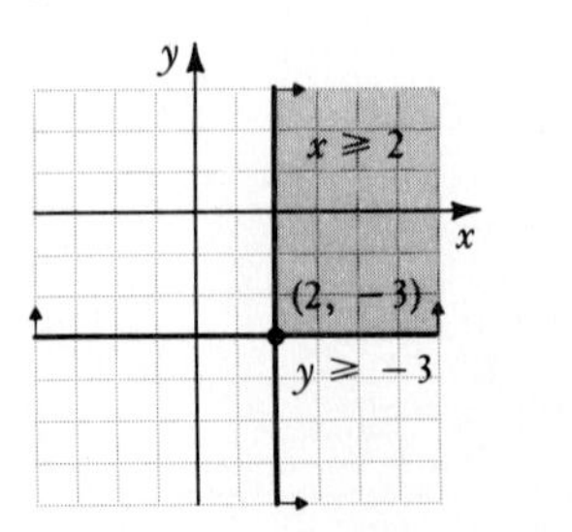

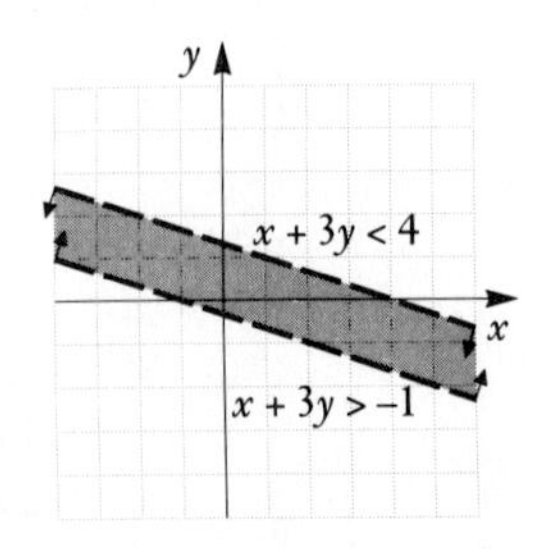

34. [11.6]

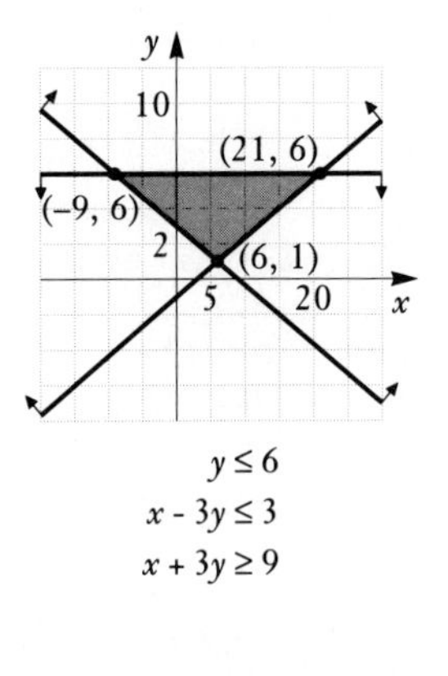

35. [11.6]

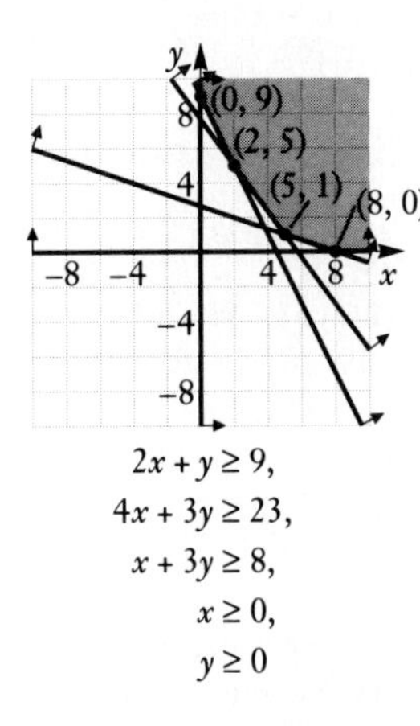

36. [11.6] Minimum = 52 at (2, 4); maximum = 92 at (2, 8)
37. [11.6] Type A: 0; type B: 10; maximum score = 120 points **38.** [11.7] (7, 4) **39.** [11.7] (2, 2), $\left(\frac{32}{9}, -\frac{10}{9}\right)$
40. [11.7] (0, −3), (2, 1)
41. [11.7] (4, 3), (4, −3), (−4, 3), (−4, −3)
42. [11.7] (2, 1), $(\sqrt{3}, 0)$, (−2, 1),$(-\sqrt{3}, 0)$
43. [11.7] (3, −3), $\left(-\frac{3}{5}, \frac{21}{5}\right)$
44. [11.7] (6, 8), (6, −8), (−6, 8), (−6, −8)
45. [11.7] (2, 2), (−2, −2), $(2\sqrt{2}, \sqrt{2})$, $(-2\sqrt{2}, -\sqrt{2})$
46. [6.3] $\frac{1}{3}$, (0, −2) **47.** [9.6] $-\frac{7}{2}$
48. [8.1] $5x^3 - 6x^2 + 12x - 21$, R 44 **49.** [8.3] −278,999
50. [9.6], [11.1] $\left(\frac{8}{3}, -\frac{2}{3}\right)$ **51.** [11.1] $\left(\frac{5}{18}, \frac{1}{7}\right)$ **52.** [11.7] $\left(\frac{-5 + i\sqrt{21}}{2}, \frac{5 + i\sqrt{21}}{2}\right), \left(\frac{-5 - i\sqrt{21}}{2}, \frac{5 - i\sqrt{21}}{2}\right)$
53. [11.7] 9

Test: Chapter 11, pp. 547–549

1. [11.1] $\left(3, -\frac{11}{3}\right)$ **2.** [11.3] $\emptyset$ **3.** [11.1] $\left(\frac{15}{7}, -\frac{18}{7}\right)$
4. [11.1] $\left(-\frac{3}{2}, -\frac{3}{2}\right)$ **5.** [11.2] $\left(2, -\frac{1}{2}, -1\right)$ **6.** [11.3] $(2 - 3y, y, 5y - 7)$; (2, 0, −7), (−1, 1, −2), (5, −1, −12)
7. [11.3] $\emptyset$ **8.** [11.2] (0, 1, 0) **9.** [11.3] Questions 2 and 7 are inconsistent; the rest are consistent. **10.** [11.3] Question 6 is dependent; the rest are independent. **11.** [11.1] 120 km/h
12. [11.1] A: $48\frac{8}{9}$ lb; B: $71\frac{1}{9}$ lb **13.** [11.2] $f(x) = -\frac{2}{3}x^2 - \frac{4}{3}x + 3$ **14.** [11.2] 74.5; 68.5; 82
15. [11.4] $\left(\frac{34}{107}, -\frac{104}{107}\right)$ **16.** [11.4] (3, 1, −2) **17.** [11.5] 2
18. [11.5] −2 **19.** [11.5] (4, −1) **20.** [11.5] (0, −2, 1)

21. [11.6] **22.** [11.6]

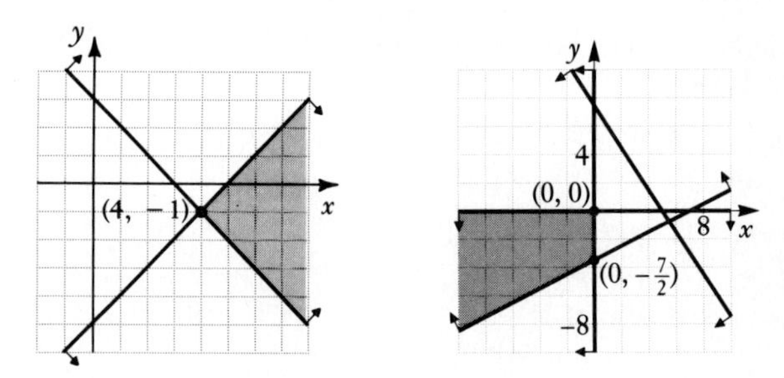

23. [11.6] Maximum is 10,040 at $x = 60$, $y = 100$; minimum is 1340 at $x = 10$, $y = 0$.
24. [11.6] Type A: 0; type B: 9; maximum score: 108
25. [11.7] (0, 3), (4, 0)
26. [11.7] (4, 0), (−4, 0)
27. [11.7] (8, 6), (0, 10)
28. [11.7] (3, 2), (2, 3), (−2, −3), (−3, −2)
29. [8.1] $x^4 + 2x^3 + x^2 + 2x + 2$, R 2
30. [8.3] 2: yes; −3: no **31.** [6.3]

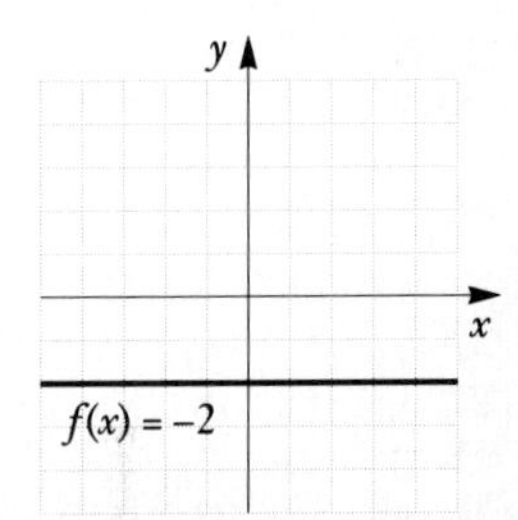

32. [9.6] 17 **33.** [10.1], [11.2] $(x + 2)^2 + (y - 2)^2 = 37$
34. [11.7] $\sqrt{5}$ m, $\sqrt{3}$ m **35.** [11.2], [11.7] $(1, \sqrt{2}, \sqrt{3})$, $(1, -\sqrt{2}, \sqrt{3})$, $(1, \sqrt{2}, -\sqrt{3})$, $(1, -\sqrt{2}, -\sqrt{3})$, $(-1, \sqrt{2}, \sqrt{3})$, $(-1, -\sqrt{2}, \sqrt{3})$, $(-1, \sqrt{2}, -\sqrt{3})$, $(-1, -\sqrt{2}, -\sqrt{3})$

CHAPTER 12

Exercise Set 12.1, pp. 556–558

1. 4, 7, 10, 13; 31; 46 **3.** $\frac{1}{2}, \frac{2}{3}, \frac{3}{4}, \frac{4}{5}; \frac{10}{11}; \frac{15}{16}$ **5.** −1, 0, 3, 8; 80; 195 **7.** 2, $2\frac{1}{2}$, $3\frac{1}{3}$, $4\frac{1}{4}$; $10\frac{1}{10}$; $15\frac{1}{15}$ **9.** −1, 4, −9, 16; 100; −225 **11.** −2, −1, 4, −7; −25; 40
13. $\frac{1}{2}, \frac{4}{7}, \frac{5}{8}, \frac{2}{3}; \frac{4}{5}; \frac{17}{20}$ **15.** 25 **17.** 225 **19.** −23.5
21. −33,880 **23.** $\frac{441}{400}$ **25.** 43 **27.** $\frac{1445}{1444}$ **29.** $2n - 1$

31. $(-1)^n 2(3)^{n-1}$ **33.** $\frac{n+1}{n+2}$ **35.** $3^{n/2}$ **37.** $-(3n-2)$
39. 28 **41.** 30 **43.** $\frac{1}{2}+\frac{1}{4}+\frac{1}{6}+\frac{1}{8}+\frac{1}{10}=\frac{137}{120}$
45. $2^0+2^1+2^2+2^3+2^4+2^5=63$
47. $\log 7+\log 8+\log 9+\log 10 \approx 3.7024$
49. $\frac{1}{2}+\frac{2}{3}+\frac{3}{4}+\frac{4}{5}+\frac{5}{6}+\frac{6}{7}+\frac{7}{8}+\frac{8}{9}=\frac{15{,}551}{2520}$
51. $(-1)^1+(-1)^2+(-1)^3+(-1)^4+(-1)^5=-1$
53. $(-1)^2 3^1+(-1)^3 3^2+(-1)^4 3^3+(-1)^5 3^4+(-1)^6 3^5+(-1)^7 3^6+(-1)^8 3^7+(-1)^9 3^8=-4920$ **55.** $\frac{2}{2}+\frac{2}{5}+\frac{2}{10}+\frac{2}{17}+\frac{2}{26}+\frac{2}{37}=\frac{75{,}581}{40{,}885}$ **57.** $3+2+3+6+11+18=43$
59. $\frac{1}{1\cdot 2}+\frac{1}{2\cdot 3}+\frac{1}{3\cdot 4}+\frac{1}{4\cdot 5}+\frac{1}{5\cdot 6}+\frac{1}{6\cdot 7}+\frac{1}{7\cdot 8}+\frac{1}{8\cdot 9}+\frac{1}{9\cdot 10}+\frac{1}{10\cdot 11}=\frac{10}{11}$ **61.** $\sum_{k=1}^{6}\frac{k}{k+1}$ **63.** $\sum_{k=1}^{6}(-2)^k$
65. $\sum_{k=2}^{n}(-1)^k k^2$ **67.** $\sum_{k=1}^{\infty}5k$ **69.** $\sum_{k=1}^{\infty}\frac{1}{k(k+1)}$ **71.** 1
73. 7 **75.** $\frac{3}{2}, \frac{3}{2}, \frac{9}{8}, \frac{3}{4}, \frac{15}{32}; \frac{171}{32}$ **77.** 0, 0.693, 1.792, 3.178, 4.787; 10.45 **79.** 2, 2.25, 2.370370, 2.441406, 2.488320, 2.521626 **81.** $\frac{n}{n+1}$ **83.** 0, 4, 20, 404, 163,220, 26,640,768,404 **85.** \$5200, \$3900, \$2925, \$2193.75, \$1645.31, \$1233.98, \$925.49, \$694.12, \$520.59, \$390.44

Exercise Set 12.2, pp. 565–567

1. $a_1=2, d=5$ **3.** $a_1=7, d=-4$ **5.** $a_1=\frac{3}{2}, d=\frac{3}{4}$
7. $a_1=\$2.12, d=\$0.12$ **9.** 46 **11.** -41
13. $-\$1628.16$ **15.** 27th **17.** 102nd **19.** 101 **21.** 5
23. 28 **25.** $a_1=8; d=-3$; 8, 5, 2, -1, -4 **27.** 670
29. 45,150 **31.** 2550 **33.** 735 **35.** 990 **37.** 3; 171
39. 1275 **41.** \$31,000 **43.** 6300 **45.** $a^k=P$
47. $x^2+y^2=81$ **49.** $\frac{n}{2}(1+n)$, or $\frac{n(n+1)}{2}$ **51.** 3, 5, 7
53. \$8760, \$7961.77, \$7163.54, \$6365.31, \$5567.08, \$4768.85, \$3970.62, \$3172.39, \$2374.16, \$1575.93
55. Let d = the common difference. Since p, m, and q form an arithmetic sequence, $m=p+d$ and $q=p+2d$. Then $\frac{p+q}{2}=\frac{p+(p+2d)}{2}=p+d=m$.

Exercise Set 12.3, pp. 575–577

1. 2 **3.** -1 **5.** $-\frac{1}{2}$ **7.** $\frac{1}{5}$ **9.** $\frac{1}{x}$ **11.** 1.1 **13.** 64
15. 162 **17.** 648 **19.** \$2331.64 **21.** $a_n=3^{n-1}$
23. $a_n=(-1)^{n-1}$ **25.** $a_n=\frac{1}{x^n}$ **27.** 762 **29.** $\frac{547}{18}$
31. $\frac{x^8-1}{x-1}$, or $(x+1)(x^2+1)(x^4+1)$ **33.** \$5134.51
35. 8 **37.** 125 **39.** $\frac{1000}{11}$ **41.** No **43.** $\frac{1}{3}$ **45.** \$25,000
47. $\frac{4}{9}$ **49.** $\frac{5}{9}$ **51.** $\frac{5}{33}$ **53.** $\frac{1}{256}$ ft **55.** 155,797
57. \$5236.19 **59.** 3100.35 ft **61.** About \$2,684,000
63. $2\log_a x+3\log_a y+5\log_a z$ **65.** $\left(-\frac{63}{29}, -\frac{114}{29}\right)$
67. $\frac{1-x^n}{1-x}$ **69.** 512 cm^2

Exercise Set 12.4, pp. 587–590

1. $4\cdot 3\cdot 2$, or 24 **3.** ${}_{10}P_7=10\cdot 9\cdot 8\cdot 7\cdot 6\cdot 5\cdot 4$, or 604,800 **5.** 120; 3125 **7.** 120 **9.** $9\cdot 9\cdot 8\cdot 7\cdot 6\cdot 5\cdot 4$, or 544,320 **11.** **(a)** 120; **(b)** 3840 **13.** $52\cdot 51\cdot 50\cdot 49=6{,}497{,}400$ **15.** $80\cdot 26\cdot 9999=20{,}797{,}920$ **17.** **(a)** 10^5, or 100,000; **(b)** 100,000 **19.** **(a)** 10^9, or 1,000,000,000; **(b)** yes **21.** 78 **23.** 78 **25.** 7 **27.** 10 **29.** 1 **31.** 12
33. $\frac{m!}{2!(m-2)!}$ **35.** $\binom{23}{4}=8855$ **37.** $\binom{10}{6}=210$
39. $\binom{8}{2}=28, \binom{8}{3}=56$ **41.** $\binom{10}{7}\cdot\binom{5}{3}=1200$
43. $\binom{58}{6}\cdot\binom{42}{4}$ **45.** $\binom{4}{3}\cdot\binom{48}{2}=4512$
47. **(a)** ${}_{33}P_3=32{,}736$; **(b)** $33^3=35{,}937$; **(c)** ${}_{33}C_3=5456$
49. $7\sqrt{2}$ **51.** 11 **53.** 9 **55.** $n-1$
57. $\binom{52}{5}=2{,}598{,}960$ **59.** $\binom{8}{3}=56$ **61.** $\binom{5}{2}\cdot\binom{8}{2}=280$

Exercise Set 12.5, pp. 595–596

1. $m^5+5m^4n+10m^3n^2+10m^2n^3+5mn^4+n^5$
3. $x^6-6x^5y+15x^4y^2-20x^3y^3+15x^2y^4-6xy^5+y^6$
5. $x^{10}-15x^8y+90x^6y^2-270x^4y^3+405x^2y^4-243y^5$
7. $729c^6-1458c^5d+1215c^4d^2-540c^3d^3+135c^2d^4-18cd^5+d^6$ **9.** $x^3-3x^2y+3xy^2-y^3$ **11.** $x^{-7}+7x^{-6}y+21x^{-5}y^2+35x^{-4}y^3+35x^{-3}y^4+21x^{-2}y^5+7x^{-1}y^6+y^7$ **13.** $a^9-18a^7+144a^5-672a^3+2016a-4032a^{-1}+5376a^{-3}-4608a^{-5}+2304a^{-7}-512a^{-9}$
15. $a^{10}+5a^8b^3+10a^6b^6+10a^4b^9+5a^2b^{12}+b^{15}$
17. $9-12\sqrt{3}t+18t^2-4\sqrt{3}t^3+t^4$ **19.** $x^{-8}+4x^{-4}+6+4x^4+x^8$ **21.** $15a^4b^2$ **23.** $-745, 472a^3$
25. $1120x^{12}y^2$ **27.** $-1{,}959{,}552u^5v^{10}$
29. **31.** Center: $(5, -1)$; radius: 2

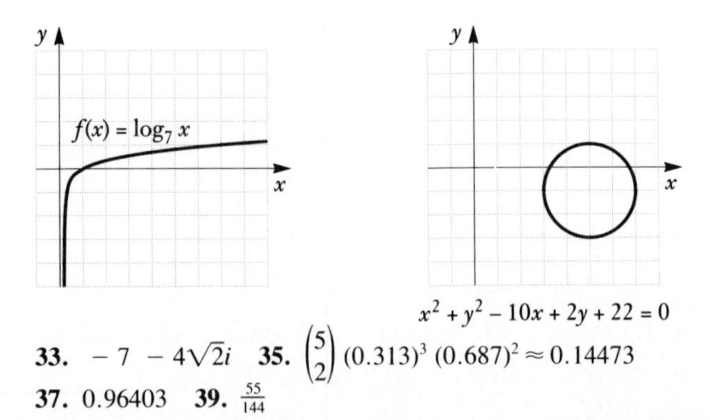

33. $-7-4\sqrt{2}i$ **35.** $\binom{5}{2}(0.313)^3(0.687)^2\approx 0.14473$
37. 0.96403 **39.** $\frac{55}{144}$

Exercise Set 12.6, pp. 602–604

1. 0.57, 0.43 **3.** 0.075, 0.134, 0.057, 0.071, 0.030 **5.** 0.633 **7.** 52 **9.** $\frac{1}{4}$ **11.** $\frac{1}{13}$ **13.** $\frac{1}{2}$ **15.** $\frac{2}{13}$ **17.** $\frac{5}{7}$ **19.** 0 **21.** $\frac{11}{4165}$ **23.** $\frac{28}{65}$ **25.** $\frac{1}{18}$ **27.** $\frac{1}{36}$ **29.** $\frac{30}{323}$ **31.** $\frac{9}{19}$ **33.** $\frac{1}{38}$ **35.** $\frac{1}{19}$ **37.** $c^7 = 15$ **39.** $(x + 5)^2 + (y - 3)^2 = 144$ **41.** 2,598,960 **43.** **(a)** 36; **(b)** 1.39×10^{-5} **45.** **(a)** $(13 \cdot {}_4C_3) \cdot (12 \cdot {}_4C_2) = 3744$; **(b)** $\frac{3744}{{}_{52}C_5} = \frac{3744}{2{,}598{,}960} \approx 0.00144$ **47.** **(a)** $13 \cdot \binom{4}{3} \cdot \binom{48}{2} - 3744 = 54{,}912$; **(b)** $\frac{54{,}912}{{}_{52}C_5} = \frac{54{,}912}{2{,}598{,}960} \approx 0.0211$ **49.** **(a)** $\binom{13}{2}\binom{4}{2}\binom{4}{2}\binom{44}{1} = 123{,}552$; **(b)** $\frac{123{,}552}{{}_{52}C_5} = \frac{123{,}552}{2{,}598{,}960} \approx 0.0475$

Review Exercises: Chapter 12, pp. 606–608

1. [12.2] $3\frac{3}{4}$ **2.** [12.2] $a + 4b$ **3.** [12.2] 531 **4.** [12.2] 465 **5.** [12.2] 11 **6.** [12.2] -4 **7.** [12.3] $n = 6$, $S_n = -126$ **8.** [12.3] $a_1 = 8$, $a_5 = \frac{1}{2}$ **9.** [12.3] No **10.** [12.3] Yes **11.** [12.3] $\frac{3}{8}$ **12.** [12.3] $\frac{211}{99}$ **13.** [12.2] 63 **14.** [12.3] About 50 ft **15.** [12.2] \$7.38, \$1365.10 **16.** [12.3] 7,680,000 **17.** [12.3] $\frac{64\sqrt{2}}{\sqrt{2} - 1}$ in., or $128 + 64\sqrt{2}$ in. **18.** [12.3] $23\frac{1}{3}$ cm **19.** [12.1] $\sum_{n=1}^{7} (n^2 - 1)$, or $\sum_{n=0}^{6} n(n + 2)$ **20.** [12.4] $6! = 720$ **21.** [12.4] $9 \cdot 8 \cdot 7 \cdot 6 = 3024$ **22.** [12.4] $\binom{15}{8} = 6435$ **23.** [12.4] $24 \cdot 23 \cdot 22 = 12{,}144$ **24.** [12.4] 36 **25.** [12.5] $\binom{18}{11} a^7x^{11}$ **26.** [12.5] $220a^9x^3$ **27.** [12.5] $m^7 + 7m^6n + 21m^5n^2 + 35m^4n^3 + 35m^3n^4 + 21m^2n^5 + 7mn^6 + n^7$ **28.** [12.5] $x^8 + 12x^6y + 54x^4y^2 + 108x^2y^3 + 81y^4$ **29.** [12.5] $-6624 + 16{,}280i$ **30.** [12.6] $\frac{86}{206} \approx 0.42$, $\frac{97}{206} \approx 0.47$, $\frac{23}{206} \approx 0.11$ **31.** [12.6] $\frac{1}{12}$, 0 **32.** [12.6] $\frac{1}{4}$ **33.** [12.6] $\frac{6}{5525}$ **34.** [9.3] -2 **35.** [9.4] 5 **36.** [10.1] $\left(2, \frac{1}{2}\right)$ **37.** [11.1] $\left(\frac{37}{25}, \frac{17}{25}\right)$ **38.** [12.2] $-2, 0, 2, 4$ **39.** [12.3] 0.27, 0.0027, 0.000027 **40.** [12.3] $\frac{1}{2}, -\frac{1}{6}, \frac{1}{18}$ **41.** [12.5] $\left(\log \frac{x}{y}\right)^{10}$ **42.** [12.4] 36 **43.** [12.4] 14

Test: Chapter 12, pp. 608–609

1. [12.2] 13 **2.** [12.2] 4 **3.** [12.2] 30th **4.** [12.3] 437.4 **5.** [12.3] $\frac{31}{64}$ **6.** [12.3] a **7.** [12.3] $\frac{125}{6}$ **8.** [12.2] \$2010 **9.** [12.3] 12,960 **10.** [12.3] $\frac{29}{225}$ **11.** [12.1] $\sum_{n=1}^{7} n^2 - n$, or $\sum_{n=1}^{7} n(n - 1)$ **12.** [12.4] $6! = 720$ **13.** [12.4] 35 **14.** [12.4] 103,740 **15.** [12.4] $20 \cdot 19 \cdot 18 = 6840$ **16.** [12.5] $243a^5 + 810a^4b^1 + 1080a^3b^2 + 720a^2b^3 + 240ab^4 + 32b^5$ **17.** [12.5] $672a^5b^2$ **18.** [12.5] $x^5 - 5\sqrt{2}x^4 + 20x^3 - 20\sqrt{2}x^2 + 20x - 4\sqrt{2}$ **19.** [12.6] $\frac{1}{6}$ **20.** [12.6] $\frac{2}{13}$ **21.** [12.6] $\frac{1}{14}$ **22.** [9.3] $4 = \log_5 625$ **23.** [9.4] $\log_b \frac{5}{7}$ **24.** [11.1] -4 and -11 **25.** [10.1] $6\sqrt{2}$ **26.** [12.2] 4, 7, 10, 13 **27.** [12.4] 54 **28.** [12.4] 18 **29.** [12.3] $-9, -\frac{9}{2} \pm \frac{9\sqrt{3}}{2}i, \frac{9}{2} \pm \frac{9\sqrt{3}}{2}i$

CUMULATIVE REVIEW: CHAPTERS 1–12, pp. 609–612

1. [5.4] $-45x^6y^{-4}$, or $\frac{-45x^6}{y^4}$ **2.** [1.2] 6.3 **3.** [1.4] $-3y + 17$ **4.** [1.3] 280 **5.** [1.3] $\frac{7}{6}$ **6.** [3.1] $3a^2 - 8ab - 15b^2$ **7.** [3.1] $13x^3 - 7x^2 - 6x + 6$ **8.** [3.2] $6a^2 + 7a - 5$ **9.** [3.2] $9a^4 - 30a^2y + 25y^2$ **10.** [4.2] $\frac{4}{x + 2}$ **11.** [4.1] $\frac{x - 4}{3(x + 2)}$ **12.** [4.1] $\frac{(x + y)(x^2 + xy + y^2)}{x^2 + y^2}$ **13.** [4.3] $x - a$ **14.** [3.4] $(2x - 3)^2$ **15.** [3.4] $(3a - 2)(9a^2 + 6a + 4)$ **16.** [3.3], [3.4] $(a + b)(a - b)(a + 3)$ **17.** [3.3] $3(y^2 + 3)(5y^2 - 4)$ **18.** [6.2] 20 **19.** [3.6], [8.1] $7x^3 + 9x^2 + 19x + 38 + \frac{72}{x - 2}$ **20.** [2.1] $\frac{2}{3}$ **21.** [3.5] 8, -6 **22.** [4.4] $-\frac{6}{5}$, 4 **23.** [4.4] No solution **24.** [11.1] $(1, -1)$ **25.** [11.2] $(2, -1, 1)$ **26.** [5.5] 9 **27.** [5.5] $\pm 5, \pm 2$ **28.** [11.7] $(\sqrt{5}, \sqrt{3}), (\sqrt{5}, -\sqrt{3}), (-\sqrt{5}, \sqrt{3}), (-\sqrt{5}, -\sqrt{3})$ **29.** [9.6] 1.2920 **30.** [9.6] 1005 **31.** [9.6] $\frac{1}{16}$ **32.** [9.6] $-\frac{1}{2}$ **33.** [2.6] $\{x \mid -2 \le x \le 3\}$, or $[-2, 3]$ **34.** [7.1] $\pm i\sqrt{2}$ **35.** [7.2] $-2 \pm \sqrt{7}$ **36.** [2.6] $\{y \mid y < -5$ or $y > 2\}$, or $(-\infty, -5) \cup (2, \infty)$ **37.** [8.7] $\left\{x \mid -1 < x < \frac{1}{2}\right\}$, or $\left(-1, \frac{1}{2}\right)$ **38.** [8.7] $\{x \mid x < -2$ or $x \ge 1\}$, or $(-\infty, -2) \cup [1, \infty)$ **39.** [11.6] 10 of type A; 4 of type B; maximum score 148 **40.** [2.2] \$3.34 **41.** [2.2] 65, 66, 67 **42.** [11.1] $11\frac{3}{7}$ **43.** [11.1] 24 L of A, 56 L of B **44.** [4.5] 350 mph **45.** [4.5] $8\frac{2}{5}$ min or 8 min, 24 sec **46.** [6.7] 20 **47.** [7.6] 1250 ft² **48.** [6.4]

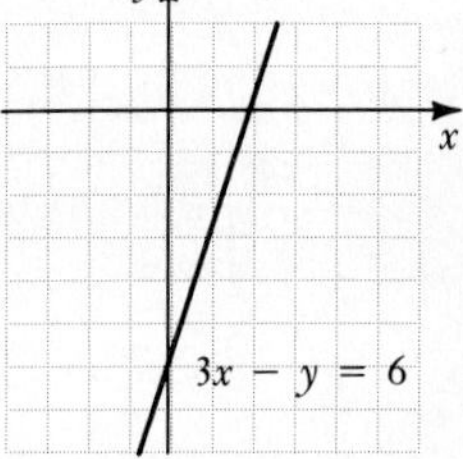

49. [10.1]

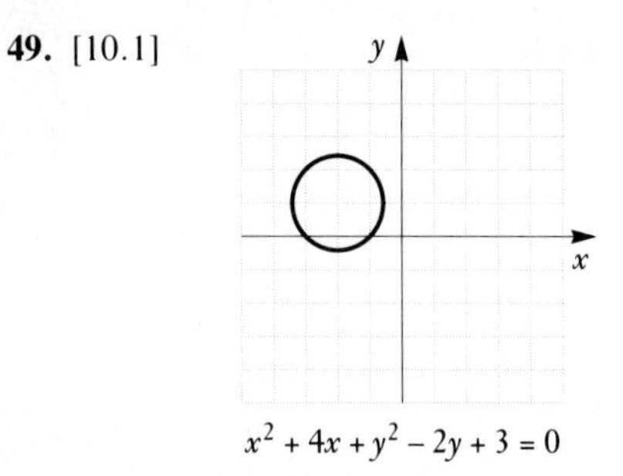

$x^2 + 4x + y^2 - 2y + 3 = 0$

54. [7.4]

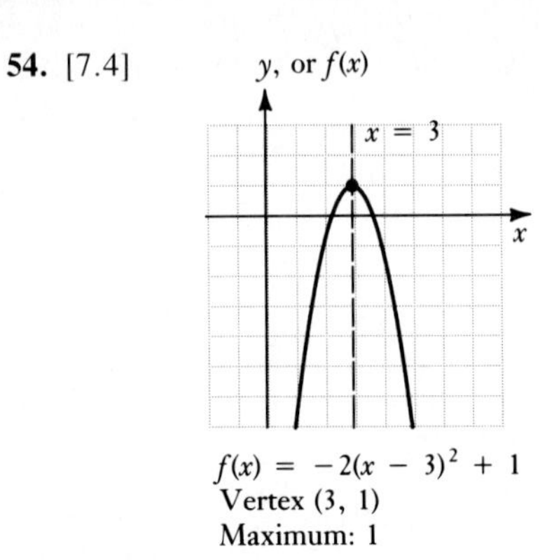

$f(x) = -2(x - 3)^2 + 1$
Vertex (3, 1)
Maximum: 1

50. [10.3]

51. [9.3]

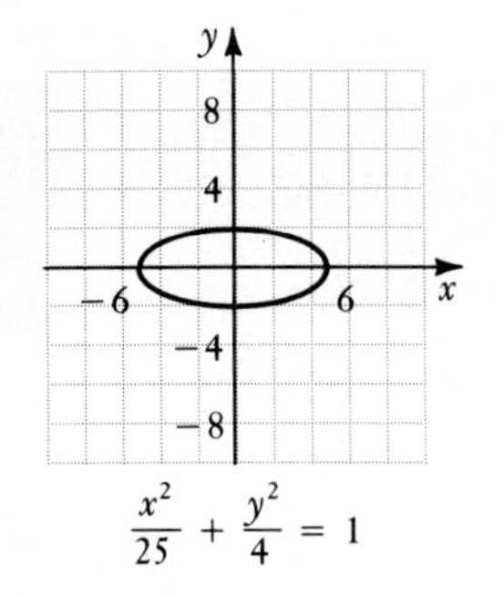

$\frac{x^2}{25} + \frac{y^2}{4} = 1$

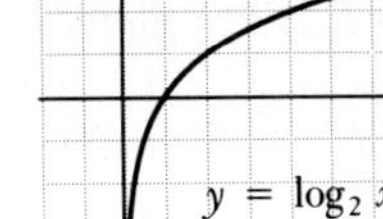
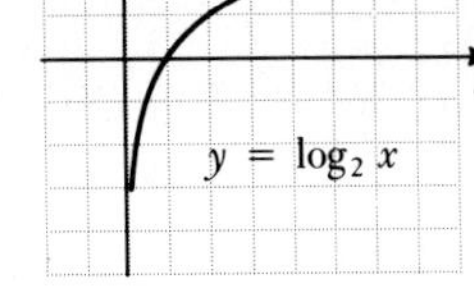

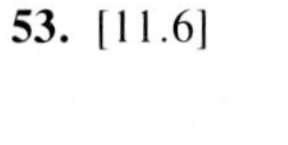

52. [8.6] $f(x) = \frac{x - 1}{x^2 - x - 2}$

53. [11.6]

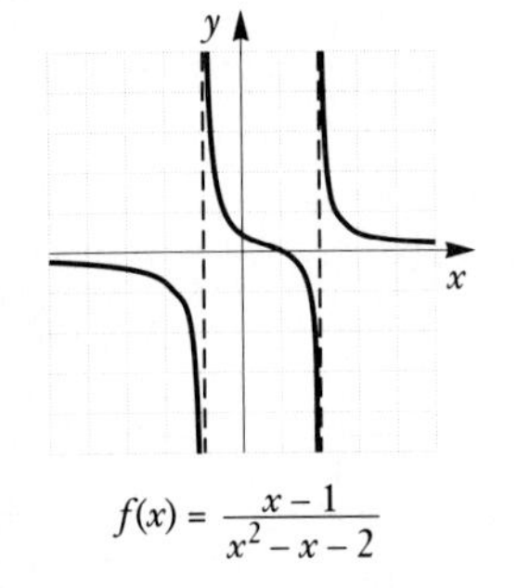

$f(x) = \frac{x - 1}{x^2 - x - 2}$

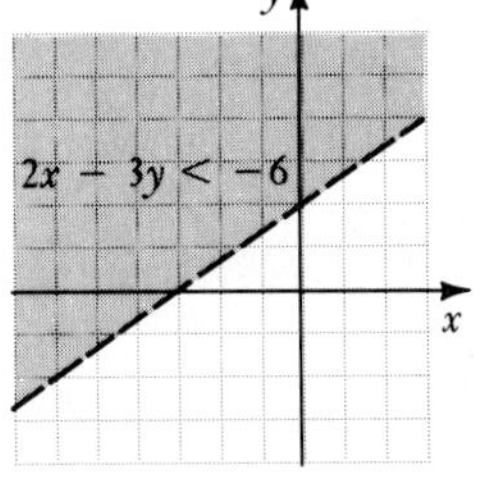

55. [2.3] $r = \frac{V - P}{-Pt}$ **56.** [6.5] $y = x + 4$ **57.** [6.5] $y = -\frac{1}{3}x + \frac{11}{3}$ **58.** [11.5] -2 **59.** [11.5] -2 **60.** [8.4] $\pm\left(1, 2, \frac{1}{2}\right)$ **61.** [8.4] $(x + 1)(x - 2)(2x + 1)$ **62.** [5.3] $16\sqrt{3}$ **63.** [5.1] $8x^2\sqrt{y}$ **64.** [5.4] $125x^2y^{3/4}$ **65.** [5.2] $\frac{\sqrt[3]{5xy}}{y}$ **66.** [5.3] $\frac{1 - 2\sqrt{x} + x}{1 - x}$ **67.** [5.4] $\sqrt[6]{x + 1}$ **68.** [5.6] $26 - 13i$ **69.** [7.3] $x^2 - 50 = 0$ **70.** [10.1] Center: $(2, -3)$; radius: 6 **71.** [9.4] $\log_a \frac{\sqrt[3]{x^2} \cdot z^5}{\sqrt{y}}$ **72.** [9.3] $a^5 = c$ **73.** [9.5] 3.7541 **74.** [9.5] 0.0000129 **75.** [9.5] 8.6442 **76.** [9.5] 0.0075 **77.** [9.6] $P(t) = 75{,}150e^{0.096t}$ **78.** [9.6] 828,392 **79.** [12.2] 81 **80.** [12.2] -121 **81.** [12.2] 875 **82.** [12.3] $16\left(\frac{1}{4}\right)^{n-1}$ **83.** [12.3] 40.96 **84.** [12.3] $74.88671874x$ **85.** [12.3] \$168.95 **86.** [12.4] 5040 **87.** [12.4] 210 **88.** [12.5] $2016x^4y^5$ **89.** [12.6] $\frac{1}{4}$ **90.** [4.4] All real numbers except 0 and -12 **91.** [9.6] 81 **92.** [6.7] y gets divided by 8 **93.** [5.6] $-\frac{7}{13} + \frac{2\sqrt{30}}{13}i$ **94.** [11.2] 84 yr

Index

L

M

N

S

For any real number $a \neq 0$ and any integer n, $a^{-n} = \frac{1}{a^n}$.

For any index n, $a^{1/n}$ means $\sqrt[n]{a}$. For $a < 0$, n must be odd.

For any natural numbers m and n ($n \neq 1$), and any real number a for which $\sqrt[n]{a}$ exists,

$$a^{m/n} \quad \text{means} \quad \left(\sqrt[n]{a}\right)^m \quad \text{or } \sqrt[n]{a^m}.$$

The Principle of Powers:

If an equation $a = b$ is true, then $a^n = b^n$ is true for any rational number n for which a^n and b^n exist.

The Principle of Positive and Negative Roots:

For any nonnegative number a, if $x^2 = a$, then $x = \pm\sqrt{a}$.

CHAPTER 6

The Vertical Line Test:

If it is possible for a vertical line to intersect a graph more than once, the graph is not that of a function.

$$\text{Slope} = m = \frac{\text{rise}}{\text{run}} = \frac{\text{change in } y}{\text{change in } x} = \frac{y_2 - y_1}{x_2 - x_1}$$

The slope–intercept equation of a line is $y = mx + b$.
The point–slope equation of a line is $y - y_1 = m(x - x_1)$.
The standard form of a linear equation is $Ax + By = C$.

Parallel lines: slopes equal, y-intercepts different.
Perpendicular lines: product of slopes $= -1$.

The Algebra of Functions

1. $(f + g)(x) = f(x) + g(x)$
2. $(f - g)(x) = f(x) - g(x)$
3. $(f \cdot g)(x) = f(x) \cdot g(x)$
4. $(f/g)(x) = f(x)/g(x)$, provided $g(x) \neq 0$.

CHAPTER 7

To solve a quadratic equation in x by completing the square:

1. Isolate the terms with variables on one side of the equation and arrange them in descending order.
2. Divide by the coefficient of x^2 on both sides, if that coefficient is not 1.
3. Complete the square by taking half of the coefficient of x and adding its square to both sides.
4. Factor one side. Find a common denominator on the other side and simplify.
5. Use the principle of positive and negative roots.
6. Solve for x by adding appropriately on both sides.

The Quadratic Formula: $x = \dfrac{-b \pm \sqrt{b^2 - 4ac}}{2a}$

CHAPTER 8

The Remainder Theorem:

For any polynomial $P(x)$, the remainder obtained by dividing $P(x)$ by $x - r$ is $P(r)$

The Factor Theorem:

For any polynomial $P(x)$, if $P(r) = 0$, then $x - r$ is a factor of $P(x)$; and if $x - r$ is a factor of $P(x)$, then $P(r) = 0$.

The Rational Roots Theorem:

Let $P(x) = a_n x^n + a_n - 1x^{n-1} + \cdots + a_1 x + a_0$, where all the coefficients are integers and n is a positive integer. Consider a rational number c/d that is written in reduced form. If c/d is a root of $P(x)$, then c is a factor of a_0 and d is a factor of a_n.

The Upper Bound Theorem:

If when a polynomial is divided by $x - a$, with $a > 0$, the remainder and coefficients of the quotient are either all nonnegative or all nonpositive, then a is an upper bound to the roots of the polynomial.

The Lower Bound Theorem:

If a polynomial is divided by $x - a$, where a is negative, and the remainder and all coefficients of the quotient alternate sign (with 0 considered a sign change), then a is a lower bound to the real roots of the polynomial. (No root is less than a.)

Any polynomial of degree n has n roots, where roots of multiplicity m are counted m times.

CHAPTER 9

Properties of Logarithms

$\log_a MN = \log_a M + \log_a N$, $\log_a \frac{M}{N} = \log_a M - \log_a N$, $\log_a M^p = p \cdot \log_a M$,

$\log_a 1 = 0$, $\log_a a = 1$, $\log_a a^k = k$,

$\log M = \log_{10} M$, $e \approx 2.7182818284\ldots$, $\ln M = \log_e M$,

$\log_b M = \dfrac{\log_a M}{\log_a b}$

To find a formula for the inverse of a function: If a function is one-to-one and has an inverse that is a function, a formula for its inverse can be found as follows:

1. Replace $f(x)$ by y.
2. Solve the equation for x.
3. Interchange x and y.
4. Replace y by $f^{-1}(x)$.

CHAPTER 10

(See also the summary of graphs at the end of Section 10.4)

Circle: Center at the origin   Center at (h, k)

Ellipse: Center at the origin

Foci on x-axis: $\dfrac{x^2}{a^2} + \dfrac{y^2}{b^2} = 1$, $a^2 > b^2$ Foci on y-axis: $\dfrac{x^2}{a^2} + \dfrac{y^2}{b^2} = 1$, $b^2 > a^2$